AF804064

High Dimensional Data Analysis

This book presents a rigorous and unified treatment of the analysis of high-dimensional data through the lens of distance-based methodology. It synthesizes recent advances in the field by framing them within a coherent paradigm: proximity-driven, non-parametric approaches for exploration and inference in complex multivariate spaces. Within this framework, interpoint distances serve as fundamental primitives, enabling the reduction of intricate, high-dimensional relationships to analytically tractable one-dimensional representations.

Designed for graduate students, advanced undergraduates, and researchers with training in matrix theory and mathematical statistics, the text assumes no prior exposure to high-dimensional techniques. Familiarity with classical multivariate analysis, while not required, will deepen appreciation of the material. The exposition balances theoretical development with practical insight, pairing formal proofs with illustrative examples and providing implementations in the R programming language to support hands-on engagement.

Key Features:

- Incorporates real-world data applications to ground theoretical concepts.
- More than 180 exercises, with solutions, available on the publisher's website.
- Provides accompanying R code for computational exploration.

Reza Modarres graduated from the American University with a B.S. in Mathematics and Computer Science. He later earned an M.S. in Computer Science and a Ph.D. in Statistics from the American University. His thesis focused on the analysis of correlation matrices. Upon graduation, Dr. Modarres worked as a post-doctoral fellow before joining George Washington University. Dr. Modarres is an elected member of the International Statistical Institute and has published more than 100 research articles on Statistics, Computer Science, and their interface. His research includes high-dimensional analysis, statistical computing, multivariate non-parametric analysis, and environmental statistics.

High Dimensional Data Analysis

An Interpoint Distance Approach

Reza Modarres

CRC Press
Taylor & Francis Group
Boca Raton London New York

CRC Press is an imprint of the
Taylor & Francis Group, an **informa** business
A CHAPMAN & HALL BOOK

Designed cover image: Shutterstock

First edition published 2027
by CRC Press
2385 NW Executive Center Drive, Suite 320, Boca Raton FL 33431

and by CRC Press
4 Park Square, Milton Park, Abingdon, Oxon, OX14 4RN

CRC Press is an imprint of Taylor & Francis Group, LLC

© 2027 Reza Modarres

Reasonable efforts have been made to publish reliable data and information, but the author and publisher cannot assume responsibility for the validity of all materials or the consequences of their use. The authors and publishers have attempted to trace the copyright holders of all material reproduced in this publication and apologize to copyright holders if permission to publish in this form has not been obtained. If any copyright material has not been acknowledged please write and let us know so we may rectify in any future reprint.

Except as permitted under U.S. Copyright Law, no part of this book may be reprinted, reproduced, transmitted, or utilized in any form by any electronic, mechanical, or other means, now known or hereafter invented, including photocopying, microfilming, and recording, or in any information storage or retrieval system, without written permission from the publishers.

For permission to photocopy or use material electronically from this work, access www.copyright.com or contact the Copyright Clearance Center, Inc. (CCC), 222 Rosewood Drive, Danvers, MA 01923, 978-750-8400. For works that are not available on CCC please contact mpkbookspermissions@tandf.co.uk

For Product Safety Concerns and Information please contact our EU representative GPSR@taylorandfrancis.com. Taylor & Francis Verlag GmbH, Kaufingerstraße 24, 80331 München, Germany.

Trademark notice: Product or corporate names may be trademarks or registered trademarks and are used only for identification and explanation without intent to infringe.

ISBN: 978-1-041-32194-1 (hbk)
ISBN: 978-1-041-33786-7 (pbk)
ISBN: 978-1-003-78958-1 (ebk)

DOI: 10.1201/9781003789581

Typeset in CMR10 font
by KnowledgeWorks Global Ltd.

Publisher's note: This book has been prepared from camera-ready copy provided by the author.

Access the Support Material: www.routledge.com/9781041321941

Dedication
To Bita, Arian, and Ideen

Contents

Preface

High-dimensional data, in which the number of variables d is comparable to or far exceeds the number of observations n, arise routinely in modern scientific and technological applications. Examples include genomics and proteomics, functional neuroimaging, finance, and network analysis. In such settings, classical multivariate methods—designed for regimes where $n \gg d$—often perform poorly or break down entirely. Phenomena such as singular covariance matrices, noise accumulation, and the concentration of measure fundamentally alter the behavior of standard procedures.

Over the past two decades, a large body of work has emerged to address these challenges. However, this literature is widely dispersed and often highly technical. The goal of this book is to present a coherent and unified treatment of high-dimensional inference through a single organizing principle: interpoint distances (IPDs) and dissimilarity-based methods.

Interpoint distances provide a natural and powerful way to analyze complex multivariate data. By reducing high-dimensional structure to functions of pairwise distances, one can construct procedures that are non-parametric, robust to model misspecification, and well-suited to high-dimensional regimes. This perspective not only simplifies analysis but also reveals deep connections among a wide range of methods, including hypothesis testing, classification, clustering, and dependence assessment.

The central theme of the book is that many seemingly distinct high-dimensional techniques can be understood, extended, and unified through distance-based representations. In particular, the concentration of distances in high dimensions, often viewed as a challenge, can instead be exploited to develop effective statistical procedures.

Chapters 2, 3, and 5 form the core theoretical foundation of the book. These chapters develop the properties of interpoint distances and introduce dissimilarity measures that underpin much of the subsequent methodology. The remaining chapters build on this foundation to address a variety of problems, including visualization, testing, classification, clustering, outlier detection, independence testing, and change-point analysis.

The chapters are designed to be largely self-contained, allowing for flexible use of the material. Depending on the audience and objectives, the book can support multiple course pathways. For example, a theory–oriented course may emphasize Chapters 2, 3, 5, and 11, while a more applied course may focus on Chapters 3, 4, 6, 10, and 13.

Each chapter includes examples, simulations, and exercises ranging from conceptual questions to more advanced problems. Several exercises involve computational implementation in the programming language R, encouraging hands-on engagement with the methods. Whenever possible, theoretical developments are accompanied by interpretation and discussion to highlight their practical implications.

The organization of the book is as follows:

- Chapter 1 introduces the challenges of high dimensional data analysis and provides essential background material.

- Chapter 2 develops the theory of interpoint distances in one-sample, two-sample, and mixture settings.

- Chapter 3 introduces dissimilarity indices and studies their properties in high dimensions.
- Chapter 4 presents visualization techniques and dimension reduction methods adapted to high-dimensional data.
- Chapter 5 develops non-parametric tests for equality of distributions based on distances and dissimilarities.
- Chapter 6 studies classification methods for discrete and high-dimensional data.
- Chapter 7 analyzes interpoint distances for multivariate power series distributions and mixtures.
- Chapter 8 discusses statistical depth functions and their applications.
- Chapter 9 revisits Hotelling's T^2 test in high-dimensional settings.
- Chapter 10 presents methods for outlier detection.
- Chapter 11 develops tests of independence based on distance-based ideas.
- Chapter 12 introduces change-point detection methods for high-dimensional data.
- Chapter 13 studies clustering methods and introduces new approaches based on dissimilarity.
- Chapter 14 analyzes the structure of distance matrices.
- Chapter 15 explores multivariate symmetry and associated tests.

This book is intended for graduate students, advanced undergraduates, and researchers in statistics, data science, bioinformatics, and related fields. A background in linear algebra and mathematical statistics is assumed. Prior exposure to classical multivariate analysis is helpful but not required.

It is our hope that this text serves both as an introduction to high-dimensional statistics from a distance-based perspective and as a foundation for further research in this rapidly evolving area.

Acknowledgment

I would like to express my sincere gratitude to my editor, Ms. Lara Spieker, for her guidance and support throughout the preparation of this manuscript. I am deeply thankful to my former students, Jade Freeman, Zhenyu Liu, Lingzhe Guo, Yu Song, Yong Wang, Haigang Liu, Mohammed Chowdhury, Mengta Yang, Chiraag Gohel, and many others whose curiosity, enthusiasm, and insights have been a constant source of inspiration for this book. I extend my special appreciation to Professors Hosam M. Mahmoud, Subrata Kundu, and Reza Drikvandi for their thoughtful comments and valuable suggestions on earlier drafts of this work. I am also grateful to Dr. Kiran Bhutani, Chairwoman of the Department of Mathematics and Statistics at The Catholic University of America, for providing a welcoming academic environment during my sabbatical, when portions of this book were written.

Reza Modarres

George Washington University
Washington, D. C.
March 2026

Author

Reza Modarres graduated from the American University with a B.S. in Mathematics and Computer Science. He later earned an M.S. in Computer Science and a Ph.D. in Statistics from the American University. His thesis focused on the analysis of correlation matrices. Upon graduation, Dr. Modarres worked as a post-doctoral fellow before joining George Washington University. Dr. Modarres is an elected member of the International Statistical Institute and has published more than 100 research articles on Statistics, Computer Science, and their interface. His research includes high-dimensional analysis, statistical computing, multivariate non-parametric analysis, and environmental statistics.

Glossary of Symbols and Abbreviations

Symbol	Description
IPD	interpoint distance
PDF	probability distribution function
PMF	probability mass function
CDF	cumulative distribution function
MGF	moment generating function
$tr(\mathbf{A})$	trace of matrix $\mathbf{A}$
$Q(\mathbf{X})$	quadratic form of random vector $\mathbf{X}$
$G_{\mathbf{X}}(\mathbf{z})$	probability generating function of random vector $\mathbf{X}$
$\mathbf{1}_d$	a d-dimensional vector of 1s
$\mathbb{E}(\mathbf{X})$	mean vector of random vector $\mathbf{X}$
$\mathbb{C}\text{ov}(\mathbf{X})$	covariance matrix of random vector $\mathbf{X}$
$\mathbb{C}\text{or}(\mathbf{X})$	correlation matrix of random vector $\mathbf{X}$
$A \oplus B$	direct sum of matrices
$A \otimes B$	Kronecker Product
$\mathbf{I}_d$	d-dimensional identity matrix
σ_{ij}	component of matrix $\mathbf{\Sigma}$ in i-th row and j-th column
$diag(a_1, \ldots, a_d)$	diagonal matrix with $a_1, \ldots, a_d$ on the main diagonal
$\mathbb{N}_d$	d-dimensional Normal Distribution
$\mathbb{Z}$	set of integers
$\|\|\mathbf{X}\|\|_s$	L_s norm of d-dimensional vector $\mathbf{X}$, $(\sum_{t=1}^{s} \|X_t\|^s)^{\frac{1}{s}}$
$\|\|\mathbf{X}\|\|$	L_2-norm or Euclidean norm of vector $\mathbf{X}$
$\|\|\mathbf{X}\|\|_F$	Frobenius norm of matrix $\mathbf{X}$, $\left(\sum_{i=1}^{d} \sum_{j=1}^{d} x_{ij}^2\right)^{\frac{1}{2}}$
$\mathbf{X}'$	the transpose of $\mathbf{X}$
$\|\mathbf{A}\|$	the determinant of matrix $\mathbf{A}$
$\mathbb{I}(\cdot)$	the indicator function
$\mathbb{R}^d$	d-dimensional real space
i.i.d.	independent and identically distributed
HDLSS	high dimensional, low sample size

1

Introduction

1.1 Abstract

This chapter provides a comprehensive introduction to the analysis of high-dimensional data, highlighting both the conceptual and methodological challenges that arise when the number of variables exceeds or is comparable to the number of observations. High-dimensional settings are increasingly common in modern applications, ranging from genomics and neuroimaging to finance and functional data analysis. The chapter emphasizes the implications of dimensionality for statistical inference, including data sparsity, concentration of distances, and the breakdown of classical geometric and probabilistic intuitions. A clear distinction is drawn between low-, high-, and ultra-high-dimensional regimes, characterized by the growth rate of the number of variables relative to the sample size, illustrating how these regimes influence the applicability of traditional statistical methods.

In addition to outlining these fundamental challenges, the chapter introduces a unifying framework based on interpoint distances to simplify complex multivariate relationships. This approach enables robust, non-parametric inference in high-dimensional spaces by exploiting the concentration of measure phenomenon. The chapter further discusses the impact of high dimensionality on algorithmic performance, including the diminishing contrast between nearest and farthest neighbors, as well as the curse of dimensionality, which necessitates careful consideration of sparsity, low-rank structure, and other regularization strategies.

The chapter surveys a suite of statistical distributions and methodological tools that underpin subsequent analyses. These include multivariate distributions, asymptotic approximations, variance estimation, and non-parametric inference techniques. By consolidating these foundational concepts, the chapter equips the reader with the theoretical and practical tools necessary for rigorous exploration, modeling, and interpretation of high-dimensional datasets, setting the stage for the distance-based methodologies developed in later chapters.

1.2 High-Dimensional Data

High-dimensional data, where the number of features d can be much larger than the number of independent observations n, are now commonplace. Classical multivariate analysis typically considers settings where $n \gg d$. However, modern advances in data acquisition, storage, and computational power have led to a paradigm shift in which high-dimensional datasets are ubiquitous. It is now common to encounter datasets consisting of at most a few hundred independent samples (subjects), each with tens or even hundreds of thousands of measured variables. This setting raises several fundamental questions:

- What happens to classical methods when d approaches n? What if $d \gg n$?

DOI: 10.1201/9781003789581-1

- Can classical procedures rely on the diagonal of the covariance matrix or on regularized estimators when the sample covariance matrix is singular?
- What is the concentration of measure phenomenon, and how can it be detected and exploited?
- How should distance or dissimilarity between observations be defined and interpreted?
- How can we compare two or more groups in terms of their distributions, means, or covariance structures?
- How can high-dimensional data be effectively visualized?
- How can we detect outliers, change points, or perform clustering and classification in high dimensions?
- How can multivariate observations be meaningfully ordered?

High-dimensional data pose unique challenges for analysis and interpretation due to several factors, including the concentration of Euclidean distances, data sparsity, computational complexity, and visualization limitations. Such data arise in diverse applications including microarrays, proteomics, neuroimaging, video data, functional data, and finance. In microarray studies, for example, the sample size n is often in the tens or hundreds, whereas the number of variables d can reach tens of thousands. In other applications, n may be in the thousands while d may be in the millions.

If $n \gg d$, i.e., $d/n \approx 0$, then there are many observations per parameter, and the problem is effectively low-dimensional. If $d/n \to c$ for some $c \in (0, 1)$, classical methods generally remain applicable. A problem becomes very high-dimensional when $d/n \to c$ with $1 < c < \infty$, implying fewer observations than parameters. More broadly, high-dimensional settings refer to cases where d diverges with n, for example $d = O(n^\alpha)$ for some $\alpha > 1$. Ultra-high-dimensional settings refer to cases where d grows at a non-polynomial rate, such as $d = O(\exp(\gamma n))$ for some $\gamma > 0$.

This book aims to bring coherence to these developments by organizing them around a unifying theme: distance-based, non-parametric methods for exploration and inference in multivariate spaces of very high dimension. Interpoint distances serve as fundamental building blocks in this framework, reducing complex multivariate relationships to tractable one-dimensional summaries. By focusing on distances and dissimilarities, we avoid restrictive distributional assumptions, accommodate singular covariance structures, and exploit the concentration of measure phenomenon.

1.3 Challenges of High-Dimensional Data

Although high dimensional data analysis remains inherently complex, the methodologies presented in this book, based on interpoint distances (IPDs), provide a versatile framework applicable to a wide range of problems, including classification, clustering, change-point detection, and outlier identification. These approaches address key challenges in high-dimensional hypothesis testing while offering computationally efficient and interpretable solutions.

In statistics, high-dimensional geometry studies the behavior of data in spaces where the number of variables is large, often comparable to or exceeding the number of observations.

As dimensionality increases, many familiar geometric and statistical intuitions from low-dimensional settings begin to fail, leading to phenomena that may appear surprising or even paradoxical. These effects have important implications for data analysis, machine learning, and statistical inference.

A central phenomenon is the concentration of measure. In high-dimensional spaces, the volume of a unit ball becomes increasingly concentrated near its surface. For example, in a 100-dimensional space, most of the volume of a unit sphere lies within a thin shell near its boundary. Consequently, randomly sampled points tend to be nearly equidistant from one another and from the origin. This reduces the effectiveness of traditional distance-based methods, such as k-nearest neighbors and clustering, which rely on meaningful distance contrasts. In particular, the ratio

$$\frac{d_{\text{nearest}}}{d_{\text{farthest}}} \to 1$$

as the dimension increases, making it difficult to distinguish between observations based on proximity.

Another counter-intuitive phenomenon is the curse of dimensionality, which describes how data become increasingly sparse as the number of dimensions grows. For instance, consider a unit cube in d-dimensional space. To capture even 1% of its volume near the origin, one must consider a region whose side length shrinks exponentially with d. As a result, exponentially more observations are required to adequately represent the space, making statistical estimation challenging without structural assumptions such as sparsity or low-rank representations.

Geometrically, angles between independent random vectors tend to concentrate around 90° in high dimensions. Let $\mathbf{X}, \mathbf{Y} \sim \mathbb{N}(0,1)^{100}$ be independent standard normal vectors. The angle θ between them, defined by

$$\cos(\theta) = \frac{\mathbf{X} \cdot \mathbf{Y}}{\|\mathbf{X}\|\,\|\mathbf{Y}\|},$$

is typically close to zero, implying near-orthogonality. This has important implications for methods such as principal component analysis and regression, which rely on projections and orthogonality. In high dimensions, such projections may lose discriminative power unless appropriate regularization is applied.

These phenomena highlight the importance of dimension reduction techniques, including principal component analysis (PCA), manifold learning, and random projection. Such methods aim to identify lower-dimensional structures that capture the essential features of the data, thereby mitigating the adverse effects of high dimensionality.

High-dimensional geometry plays a central role in modern statistics, data science, and machine learning. As the dimensionality increases, geometric quantities such as volume and surface area behave in unintuitive ways, often with most of the volume concentrated near the boundary. To illustrate the rapid growth in complexity, consider constructing a grid over $[0,1]^d$ using 10 evenly spaced points per dimension. When $d = 1$, there are 10 points; when $d = 2$, there are $10^2 = 100$ points; and when $d = 10$, there are 10^{10} points. This exponential growth quickly renders naive approaches computationally infeasible as d increases.

1.4 Geometry of High Dimensions

Zorich [371] and Chatterjee [57] illustrate the strangeness of high-dimensional space with several examples, a few of which are presented here.

Example 1.4.1 *Consider the volume of a t-dimensional hypersphere of radius r, given by*

$$V_t(r) = \int \cdots \int_{\sum_{i=1}^{t} x_i^2 \leq r^2} dx_1 \ldots dx_d = S_t \int_0^r q^{t-1} dq = S_t r^t / t \tag{1.1}$$

where $S_t = 2\pi^{t/2}/\Gamma(t/2)$ *is the surface area of* $(t-1)$*-dimensional hypersphere of unit radius. When* $t = 2$*, we have* $V_2(r) = S_2 r^2/2 = \pi r^2$ *and when* $t = 3$*, we have* $V_3(r) = S_3 r^3/3 = 4\pi r^3/3$ *since* $\Gamma(t/2) = (t/2-1)!$ *when t is even and* $\Gamma(t/2) = \frac{\sqrt{\pi}(t-2)!!}{2^{t/2-\frac{1}{2}}}$ *when t is odd, where* $t!! = t(t-2)\ldots 3.1$*. The surface area of the t-dimensional hypersphere is* $A_t(r) = dV_t(r)/dr = S_t r^{t-1} = V_t(r)t/r$*. Strangely enough, the hyper-surface area reaches a maximum and decreases toward* 0 *as t increases. The point of maximum surface area satisfies* $d(S_t)/d(t) = \pi^{t/2}(\ln \pi - \Psi(t/2))/\Gamma(t/2) = 0$ *where* $\Psi(x)$ *is the digamma function. Solving numerically for t, we obtain* $t = 7$ *so the seven-dimensional hypersphere has maximum surface area.*

Example 1.4.2 *Consider the hypersphere of radius r and volume* $V_d(r)$ *inscribed within a hypercube on* $(-r, r)^d$*. The ratio of the volume of the sphere to the volume of the cube is* $f(d) = \frac{1}{(2r)^d} \frac{2r^d \pi^{d/2}}{d\Gamma(d/2)}$.

d	1	2	3	4	5	6	7	∞
$f(d)$	1	0.785	0.524	0.308	0.164	0.80	0.037	0

As d increases the volume of the sphere becomes infinitesimal compared to that of the cube. Where did the volume go? As d becomes larger all the volume of the cube is in the spikes (star shaped).

Example 1.4.3 *Consider two points* $g = (r, r, \ldots, r)$ *and* $e = (r, 0, \ldots, 0)$ *in* $\mathbb{R}^d$*. Note that* $\frac{\|g\|^2}{\|e\|^2} = dr^2/r^2 = d \to \infty$*. Consider the angle* θ *between g and d,*

$$\cos(\theta) = \frac{g' e}{\|g\|^2 \|e\|^2} = r^2/\sqrt{dr^4} = 1/\sqrt{d} \to 0, \tag{1.2}$$

so that g and e become orthogonal as $d \to \infty$.

Example 1.4.4 *For another example of geometric concepts as d tends to infinity, consider the volume of the thin shell between hypersphere* S_1 *of radius* $r - \epsilon$ *inside a hypersphere* S_2 *of radius r. Note that*

$$Volume(S_1) = \frac{2\pi^{d/2}}{d\Gamma(d/2)}(r-\epsilon)^d, \quad Volume(S_2) = \frac{2\pi^{d/2}}{d\Gamma(d/2)} r^d. \tag{1.3}$$

The ratio of the volume of S_1 *to* S_2 *is* $(1-\epsilon/d)^d$*. No matter how small* ϵ *becomes, this ratio tends to zero as* $d \to \infty$.

High-dimensional datasets contain thousands, if not millions, of dimensions. Facial recognition algorithms depend on the use of high-dimensional data. Suppose there are n images, each has a resolution of r by c pixels. If we consider every pixel within a given image as a variable, each image is in $\mathbb{R}^{rtimesc}$. A high-resolution image may require at least a million pixels. For a large class of problems, the sample size n and the dimension d have the relationship $n < d$. These problems are not considered in classical multivariate analysis.

Specifying the relationship between n and d may change the analysis. High-dimensional data are usually classified as the situation where $d \to \infty$ as $n \to \infty$. High-dimensional low sample size data (HDLSS) refers to situations where $d \to \infty$ while n remains fixed. Applications are abundant in gene-expression and bioinformatics data. Recent research has been focused on a specific growth rate of d. Ultra-high-dimensional data are when d grows at a non-polynomial rate in n, namely $\log(d) = O(n^{\alpha})$ for $\alpha > 0$.

The term high-dimensional detectability refers to the fundamental question of whether two or more distributions can be statistically distinguished when the ambient dimension d is large relative to the sample size. In such regimes, differences between distributions may exist but remain statistically undetectable due to concentration phenomena and information-theoretic limits.

Let F_1 and F_2 be two distributions on $\mathbb{R}^d$. A difference between F_1 and F_2 is said to be detectable if there exists a test ϕ_n such that

$$\lim_{n\to\infty} \mathbb{P}_{F_1,F_2}(\phi_n = 1) = 1 \quad \text{and} \quad \lim_{n\to\infty} \mathbb{P}_{H_0}(\phi_n = 1) = \alpha,$$

for some fixed significance level $\alpha \in (0,1)$. In high-dimensional settings, detectability depends on how the magnitude of distributional differences scales with d and n. For example, consider a mean-shift model:

$$\mathbf{X} \sim \mathbb{N}(\mathbf{0}, \mathbf{I}_d), \qquad \mathbf{Y} \sim \mathbb{N}(\boldsymbol{\mu}, \mathbf{I}_d),$$

where $\boldsymbol{\mu} \in \mathbb{R}^d$. Classical results show that consistent detection is possible if and only if $\|\mu\|_2^2 \gtrsim \frac{d}{n}$, highlighting the curse of dimensionality.

1.5 Chapters Descriptions

Chapter 2 is dedicated to the exploration of IPDs, quadratic forms in normal variates, and the distributions of distances in one–sample, two–sample, and mixture contexts. Mastery of these concepts is essential for subsequent chapters. Interpoint distances (IPDs) constitute the foundational building blocks of many multivariate methods discussed herein. These concepts are essential for a better understanding of the chapters that follow.

Chapter 3 addresses a pivotal phenomenon in high-dimensional statistics: the concentration of distances. As d increases, pairwise distances tend to concentrate around a common value, undermining the discriminative power of traditional Euclidean-based techniques. This chapter introduces robust dissimilarity indices designed to exploit, rather than succumb to, this phenomenon. These indices have broad utility across tasks such as hypothesis testing, classification, clustering, outlier detection, and change point analysis.

In chapter 4, we delve into the visualization of high-dimensional data, a critical yet often underdeveloped aspect of statistical analysis. We present methods for reducing dimensionality (e.g., PCA, SVD), and discuss innovative graphical displays including biplots, Chernoff faces, Andrews plots, and parallel coordinate plots. A novel contribution includes the simultaneous plotting of empirical cumulative distribution functions (ECDFs) of IPDs, which effectively highlight differences in location, scale, and shape across groups.

Chapter 5 focuses on statistical methods for comparing high-dimensional groups. We explore several non-parametric procedures, including the multivariate runs test and the triangle test, both of which use IPD-based criteria. Permutation testing serves as the backbone for inference. We review existing metrics such as the energy statistic and introduce a refined version tailored to overcome the loss of power associated with Euclidean distance in high dimensions.

In chapter 6, the focus shifts to supervised classification in high dimensions. We propose a test-based classification scheme utilizing IPDs, specifically targeting discrete distributions like multivariate Bernoulli, multinomial, and Poisson. This chapter also explores depth-based classification and addresses the pitfalls of classical methods when the covariance matrix becomes singular in HDLSS regimes. The proposed methods offer robust, non-parametric alternatives for these challenging scenarios.

Chapter 7 investigates the behavior of IPDs within the framework of the Power Series Distributions (MPSDs). Furthermore, we derive the distributions of squared Euclidean distances for MPSDs, encompassing models like the multinomial, negative multinomial, multivariate Poisson, and multivariate logarithmic distributions. We also address the problem of testing homogeneity in multivariate finite mixture models, a fundamental issue in unsupervised learning. We propose a novel, IPD-based approach for detecting mixtures, applicable to both continuous and discrete data. The method is bolstered by theoretical distributional results and a bootstrap framework for models lacking closed-form solutions.

In chapter 8, we discuss the use of statistical depth functions as non-parametric tools for ranking and inference in multivariate settings. Depth functions quantify the centrality of observations, facilitating robust outlier detection, classification, and the construction of confidence regions. Applications span diverse fields including medical imaging, environmental science, and spectroscopy.

Chapter 9 reconsiders the Hotelling T^2 test under the constraints of high dimensionality and small sample sizes. By partitioning the variables into blocks and applying the union–intersection principle, we develop a more powerful testing framework. This strategy is extended to Wilks' method for outlier detection, resulting in improved performance in high-dimensional contexts.

Chapter 10 is dedicated to outlier detection methods. We examine both mode-based and IPD-based approaches, including distance, density, and depth-based techniques. Applications include fraud detection, epidemiology, finance, and geospatial analysis. Particular attention is given to the robustness and scalability of methods in high dimensions.

Chapter 11 discusses non-parametric tests of independence. Testing for independence is a cornerstone of statistical inference, with a wide range of applications across disciplines. High-dimensional data are pervasive in fields such as finance, robotics, network analysis, medical imaging, machine learning, and genomics. In such domains, independence assumptions are often central: for instance, in microarray studies, independence among gene expression levels is a foundational assumption, and in functional magnetic resonance imaging (fMRI), one may seek to test independence between signals from distinct brain regions. We develop and study a class of non-parametric procedures that exploit the geometry of high-dimensional data through IPDs. These IPD-based methods are particularly well-suited for high-dimensional inference and retain power even when classical techniques fail due to the curse of dimensionality.

Chapter 12 introduces a distribution-free, distance-based method for detecting change points in HDLSS data. The method accommodates both sparse and non-sparse distributional changes and is underpinned by novel dissimilarity measures. Algorithms for estimating single and multiple change points are presented, along with a permutation-based test for assessing their significance. The methodology is supported by theoretical guarantees and implemented in the R package `HDDchangepoint`.

Chapter 13 presents a comprehensive discussion of clustering methodologies tailored for HDLSS data. We propose a novel IPD–based outlier detection method that takes advantage of the phenomenon of distance concentration and the limiting behavior of dissimilarity indices to form clusters. An ordering algorithm is introduced that reorders each row of the dissimilarity matrix to estimate change points, which in turn define cluster boundaries. To evaluate clustering quality, we construct an agreement matrix using Rand indices, and identify optimal clusters by minimizing the row sum of this matrix. Theoretical analysis demonstrates that the proposed method achieves perfect clustering as the number of features diverges, while the sample size remains fixed. We present several illustrative examples and benchmark the new method against four established clustering approaches, including high-dimensional k-means, minimum spanning trees, and hierarchical scan techniques, using the Lymphoma dataset as a case study.

Chapter 14 is devoted to the analysis of distance matrices, which serve as fundamental representations in domains where direct data access is impractical, or data are inherently relational, such as social networks or transportation systems. In large–scale settings, distance or dissimilarity matrices provide structured representations of pairwise relationships. We explore the largely uncharted territory connecting the eigen–structure of distance matrices to the presence of outliers. Statistical tests are introduced to assess whether a distance matrix exhibits a constant structure. We study the eigenvalues of two–sample distance matrices in the context of testing for distributional equality, and further examine dissimilarity matrices and their spectral properties under HDLSS conditions. We investigate how mean shifts and scale changes influence the eigenvalues, and develop asymptotic results for IPD–based summaries. We also propose a visualization framework for detecting outliers and change points via the row sums of distance matrices and dissimilarity plots, providing intuitive insights into deviations from the null hypothesis.

In a lucid discussion on the construction of symmetric multivariate distributions, Fang et al. [96] write "the multivariate aspect characterizes the model of the dependent outcome of random experiments while symmetry goes, of course, to the heart of the theory and philosophy of the laws of nature and mathematics." Even though there is a unique way of defining symmetry in one dimension, there exists more than one definition in the multidimensional case. Chapter 15 considers various notations of multivariate symmetry and offers testing methods under diagonal symmetry. Multivariate symmetry plays a foundational role in many areas of statistics, such as distributional modeling, shape analysis, classification, hypothesis testing, and robust estimation.

1.6 Statistical Distributions

In this section we present a review of a number of statistical distributions that appear in various chapters.

1.6.1 Multinomial Distribution

Suppose we perform n independent trials, where each trial results in one of d categories with fixed probabilities $\mathbf{p} = (p_1, p_2, \ldots, p_d)$ such that $\sum_{i=1}^{d} p_i = 1$. Let X_i be the number of outcomes in category i. The vector $\mathbf{X} = (X_1, X_2, \ldots, X_d)$ has a Multinomial distribution denoted by $\mathbf{X} \sim \text{Multinomial}(n, \mathbf{p})$. A Multinomial distribution has probability mass function (PMF)

$$\mathbb{P}(X_1 = x_1, \ldots, X_d = x_d) = \frac{n!}{x_1! x_2! \ldots x_d!} \prod_{i=1}^{d} p_i^{x_i}, \quad \text{where } \sum_{i=1}^{d} x_i = n \tag{1.4}$$

The mean and covariance are given by

$$\mathbb{E}[X_i] = np_i, \tag{1.5}$$

$$\mathbb{C}ov(X_i, X_j) = \begin{cases} np_i(1-p_i), & \text{if } i = j; \\ -np_ip_j, & \text{if } i \neq j. \end{cases} \tag{1.6}$$

1.6.2 Hypergeometric Distribution

Let a population consist of N items, of which K are labeled successes, and $N - K$ are labeled failures. A sample of size n is drawn without replacement. Let X be the number of successes in the sample. Then X has a Hypergeometric distribution, denoted by $X \sim$ Hypergeometric(N, K, n). Its PMF is given by

$$\mathbb{P}(X = k) = \frac{\binom{K}{k}\binom{N-K}{n-k}}{\binom{N}{n}}, \quad \text{for } \max(0, n-(N-K)) \leq k \leq \min(n, K). \tag{1.7}$$

The mean and covariance are given by

$$\mathbb{E}[X] = n \cdot \frac{K}{N}, \tag{1.8}$$

$$\mathbb{V}ar((X) = n \cdot \frac{K}{N} \cdot \left(1 - \frac{K}{N}\right) \cdot \frac{N-n}{N-1}. \tag{1.9}$$

1.6.3 Chi-Squared Distribution

Let $Z_1, Z_2, \ldots, Z_d$ be independent standard normal random variables, i.e., $Z_i \sim \mathbb{N}(0, 1)$ for $i = 1, \ldots, d$. The random variable $X = \sum_{i=1}^{d} Z_i^2$ follows a Chi-Squared distribution with d degrees of freedom, denoted by $X \sim \chi^2(d)$. Its PDF is given by

$$f_X(x) = \frac{1}{2^{d/2}\Gamma(d/2)} x^{d/2-1} e^{-x/2}, \quad x > 0 \tag{1.10}$$

where $\Gamma(\cdot)$ is the Gamma function. The mean and variance are given by

$$\mathbb{E}[X] = d, \tag{1.11}$$

$$\mathbb{V}ar((X) = 2d. \tag{1.12}$$

Its MGF is given by

$$M_X(t) = (1-2t)^{-d/2}, \quad \text{for } t < \frac{1}{2}. \tag{1.13}$$

1.6.4 Non-Central Chi-Squared Distribution

Let $Z_1, Z_2, \ldots, Z_d$ be independent standard normal random variables, and let $\mu_1, \mu_2, \ldots, \mu_d$ be fixed constants. Define the random variable $X = \sum_{i=1}^{d}(Z_i + \mu_i)^2$. Then, X follows a Non–Central Chi-Squared distribution with d degrees of freedom and non–centrality parameter $\lambda = \sum_{i=1}^{d} \mu_i^2$, denoted by $X \sim \chi^2(d, \lambda)$. Its PDF is given by

$$f_X(x) = \frac{1}{2} e^{-(x+\lambda)/2} \left(\frac{x}{\lambda}\right)^{(d-2)/4} I_{(d-2)/2}(\sqrt{\lambda x}), \quad x > 0, \tag{1.14}$$

where $I_\nu(\cdot)$ is the modified Bessel function of the first kind of order ν. It has mean and variance

$$\mathbb{E}[X] = d + \lambda, \tag{1.15}$$
$$\mathbb{V}ar((X) = 2(d + 2\lambda). \tag{1.16}$$

Its MGF is given by

$$M_X(t) = \frac{\exp\left(\frac{\lambda t}{1-2t}\right)}{(1-2t)^{d/2}}, \quad \text{for } t < \frac{1}{2}. \tag{1.17}$$

1.6.5 Multivariate Normal Distribution

A d-dimensional random vector $\mathbf{X} \in \mathbb{R}^d$ is said to follow a multivariate normal distribution with mean vector $\boldsymbol{\mu} \in \mathbb{R}^d$ and covariance matrix $\boldsymbol{\Sigma} \in \mathbb{R}^{d\times d}$ (symmetric and positive definite), denoted by $\mathbf{X} \sim \mathbb{N}_d(\boldsymbol{\mu}, \boldsymbol{\Sigma})$. Its PDF is given by

$$f_{\mathbf{X}}(\mathbf{X}) = \frac{1}{(2\pi)^{d/2}|\boldsymbol{\Sigma}|^{1/2}} \exp\left(-\frac{1}{2}(\mathbf{X}-\boldsymbol{\mu})'\boldsymbol{\Sigma}^{-1}(\mathbf{X}-\boldsymbol{\mu})\right), \quad \mathbf{X} \in \mathbb{R}^d. \tag{1.18}$$

The MGF $M_{\mathbf{X}}(\mathbf{t}) = \mathbb{E}[e^{\mathbf{t}'\mathbf{X}}]$ of a multivariate normal distribution is given by

$$M_{\mathbf{X}}(\mathbf{t}) = \exp\left(\mathbf{t}'\boldsymbol{\mu} + \frac{1}{2}\mathbf{t}'\boldsymbol{\Sigma}\mathbf{t}\right), \quad \mathbf{t} \in \mathbb{R}^d. \tag{1.19}$$

Let $\mathbf{X} \sim \mathbb{N}_d(\boldsymbol{\mu}, \boldsymbol{\Sigma})$. Then the standardized random vector $\mathbf{Z} = \boldsymbol{\Sigma}^{-1/2}(\mathbf{X} - \boldsymbol{\mu})$ follows a standard multivariate normal distribution: $\mathbf{Z} \sim \mathbb{N}_d(\mathbf{0}, \mathbf{I})$. Here, $\boldsymbol{\Sigma}^{-1/2}$ denotes a symmetric matrix such that $\boldsymbol{\Sigma}^{-1/2}\boldsymbol{\Sigma}\boldsymbol{\Sigma}^{-1/2} = \mathbf{I}$, which can be obtained via the spectral decomposition.

1.6.6 Wishart Distribution

Let $\mathbf{X}_1, \mathbf{X}_2, \ldots, \mathbf{X}_n$ be independent random vectors from a multivariate normal distribution $\mathbb{N}_d(\mathbf{0}, \boldsymbol{\Sigma})$. Define the matrix $\mathbf{W} = \sum_{i=1}^n \mathbf{X}_i\mathbf{X}_i'$. Then, $\mathbf{W}$ has a Wishart distribution with n degrees of freedom and scale matrix $\boldsymbol{\Sigma}$, denoted by $\mathbf{W} \sim \mathbb{W}_d(n, \boldsymbol{\Sigma})$. This distribution is supported on the space of $d \times d$ symmetric positive semi-definite matrices. If $\mathbf{W} \sim \mathbb{W}_d(n, \boldsymbol{\Sigma})$ with $n \geq d$, then its PDF is given by

$$f(\mathbf{W}) = \frac{1}{2^{nd/2}|\boldsymbol{\Sigma}|^{n/2}\Gamma_d\left(\frac{n}{2}\right)}|\mathbf{W}|^{(n-d-1)/2} \exp\left(-\frac{1}{2}tr(\boldsymbol{\Sigma}^{-1}\mathbf{W})\right), \tag{1.20}$$

where $\mathbf{W}$ is a $d \times d$ positive definite and $\Gamma_d(\cdot)$ is the multivariate gamma function

$$\Gamma_d(a) = \pi^{d(d-1)/4} \prod_{j=1}^{d} \Gamma\left(a - \frac{j-1}{2}\right), \quad a > \frac{d-1}{2}. \tag{1.21}$$

The MGF of the Wishart distribution is given by

$$M_{\mathbf{W}}(\mathbf{T}) = |\mathbf{I} - 2\mathbf{T}\boldsymbol{\Sigma}|^{-n/2}, \quad \text{for } \mathbf{T} \text{ such that } \mathbf{I} - 2\mathbf{T}\boldsymbol{\Sigma} \text{ is positive definite.} \tag{1.22}$$

If $\mathbf{W} \sim \mathbb{W}_d(n, \boldsymbol{\Sigma})$, then the transformation $\mathbf{V} = \boldsymbol{\Sigma}^{-1/2}\mathbf{W}\boldsymbol{\Sigma}^{-1/2}$ gives $\mathbf{V} \sim \mathbb{W}_d(n, \mathbf{I})$, which is the standard Wishart distribution.

1.6.7 Matrix Normal Distribution

One can construct an MN distribution by vectorization of multivariate normal distribution. Let $vec(\cdot)$ be an operator that maps a matrix to a vector by stacking the columns of the matrix on top of one another. An $d \times m$ random matrix $\mathbf{X}$ follows a $\mathbb{MN}_{d\times m}(\mathbf{M}, \mathbf{\Sigma}, \mathbf{\Psi})$, if $vec(\mathbf{X}')$ follows a multivariate normal distribution with mean $vec(\mathbf{M}')$ and covariance $\mathbf{\Sigma} \otimes \mathbf{\Psi}$. The $d \times m$ matrix $\mathbf{M}$ is the expected value of $\mathbf{X}$, the $d \times d$ matrix $\mathbf{\Sigma}$ and the $m \times m$ matrix $\mathbf{\Psi}$ are covariance matrices of the rows and columns of $\mathbb{MN}$, respectively. So the matrix normal is simply a multivariate normal with structured covariance.

Gupta and Nagar [129] define the PDF of $\mathbf{X} \sim \mathbb{MN}_{d\times m}(\mathbf{M}, \mathbf{\Sigma}, \mathbf{\Psi})$ by

$$f(\mathbf{X}) = \frac{1}{(2\pi)^{md/2}|\mathbf{\Sigma}|^{m/2}|\mathbf{\Psi}|^{d/2}} e^{tr(-\frac{1}{2}\mathbf{\Sigma}^{-1}(\mathbf{X}-\mathbf{M})\mathbf{\Psi}^{-1}(\mathbf{X}-\mathbf{M})')}.$$

Suppose x_{ij} is the component of $\mathbf{X}$ on the ith row and jth column. One can show that $var(x_{ij}) = \sigma_{ii}\psi_{jj}$, where σ_{ii} is the ith diagonal component of $\mathbf{\Sigma}$ and ψ_{jj} is the jth diagonal component of $\mathbf{\Psi}$. Consider the components x_{ij} and x_{hk}, their covariances are $cov(x_{ij}, x_{hk}) = \sigma_{ih}\psi_{jk}$ and their correlations are $\rho(x_{ij}, x_{hk}) = \frac{\sigma_{ih}\psi_{jk}}{\sqrt{\sigma_{ii}\psi_{jj}\sigma_{hh}\psi_{kk}}}$. The correlation between two variables is the same on any unit, and the correlation between two units is the same on any variables, i.e. $\rho(x_{ij}, x_{ik}) = \frac{\psi_{jk}}{\sqrt{\psi_{jj}\psi_{kk}}}$ does not depend on i and $\rho(x_{ij}, x_{hj}) = \frac{\sigma_{ih}}{\sqrt{\sigma_{ii}\sigma_{hh}}}$ does not depend on j. Hence, $\mathbf{\Sigma}$ and $\mathbf{\Psi}$ contain variable and unit dependence parameters, respectively.

The covariance matrix of the vectorized form has a Kronecker separable covariance structure. This implies that for any two entries x_{ij} and x_{hk}, $\text{Cov}(x_{ij}, x_{hk}) = \sigma_{ih}\psi_{jk}$. Thus, All dependence between rows (units) is governed by $\mathbf{\Sigma}$. All dependence between columns (variables) is governed by $\mathbf{\Psi}$. The interaction between row and column dependence is multiplicative. The separable structure dramatically reduces the number of free covariance parameters from $\frac{dm(dm+1)}{2}$ (for a general dm-variate normal covariance matrix) to $\frac{d(d+1)}{2} + \frac{m(m+1)}{2}$.

1.6.8 Multivariate Bernoulli Distribution

Let $\mathbf{X} = (X_1, X_2, \ldots, X_d)'$ be a random vector taking values in $\{0,1\}^d$. The joint distribution of $\mathbf{X}$ is called a multivariate Bernoulli (MB) if all 2^d possible binary outcomes can occur with non–negative probabilities summing to one. Unlike the multivariate normal, it is not characterized by its mean and covariance. Each univariate marginal distribution X_i is Bernoulli with success probability $p_i = \mathbb{E}[X_i]$. The second moment matrix $\mathbb{E}[\mathbf{X}\mathbf{X}']$ has entries equal to $\mathbb{P}(X_i = 1, X_j = 1)$ (Modarres [227]). The covariance matrix $\mathbf{\Sigma}$ has entries $\mathbf{\Sigma}_{ij} = \mathbb{E}[X_iX_j] - p_ip_j$, which are determined by the bivariate mass function. The vector of central moments captures all 2-way or higher-level dependencies that might exist between the components of $\mathbf{X}$. The marginal distribution of any set of the components of $\mathbf{X}_i$ is MB with means, variances, and covariances obtained from the corresponding elements of $\mathbf{P}$ and $\mathbf{\Sigma}$. Since the sum of the 2^d probabilities must equal one, we need $2^d - 1$ parameters to fully describe a MB distribution. A d-variate MB distribution can be represented as a multinomial distribution with 2^d cells and $2^d - 1$ parameters.

1.6.9 Dirichlet Distribution

The Dirichlet distribution is a multivariate generalization of the Beta distribution, defined on the $(d-1)$-dimensional probability simplex

$$\Delta^{d-1} = \left\{ \mathbf{X} = (x_1, \ldots, x_d)' \in \mathbb{R}^d : x_i > 0, \sum_{i=1}^{d} x_i = 1 \right\}.$$

A random vector $\mathbf{X} = (X_1, \ldots, X_d)' \sim \text{Dirichlet}(\boldsymbol{\alpha})$, where $\boldsymbol{\alpha} = (\alpha_1, \ldots, \alpha_d)'$ with $\alpha_i > 0$, has the joint density

$$f(\mathbf{X}) = \frac{\Gamma\left(\sum_{i=1}^d \alpha_i\right)}{\prod_{i=1}^d \Gamma(\alpha_i)} \prod_{i=1}^d x_i^{\alpha_i - 1}, \quad \text{for } \mathbf{X} \in \Delta^{d-1}.$$

The distribution is parameterized by concentration parameters α_i, and its shape depends on both the relative and total magnitude of these parameters. The expectation and covariance of $\mathbf{X}$ are given by

$$\mathbb{E}[X_i] = \frac{\alpha_i}{\alpha_0}, \quad \text{where } \alpha_0 = \sum_{i=1}^d \alpha_i,$$

$$\text{Cov}(X_i, X_j) = \begin{cases} \dfrac{\alpha_i(\alpha_0 - \alpha_i)}{\alpha_0^2(\alpha_0 + 1)}, & \text{if } i = j; \\ -\dfrac{\alpha_i \alpha_j}{\alpha_0^2(\alpha_0 + 1)}, & \text{if } i \neq j. \end{cases}$$

The covariance matrix is singular with rank $d-1$, since $\sum X_i = 1$. The Dirichlet distribution is used in Bayesian statistics (as a conjugate prior for multinomial models), in compositional data analysis, and in modeling proportions. When all $\alpha_i = \alpha$, the distribution is symmetric and exhibits permutation symmetry about the center $(1/d, \ldots, 1/d)$.

1.7 Statistical Tools

In this section we review a number of statistical tools, which will be used throughout the book.

1.7.1 Jensen's Inequality

Jensen's inequality states that if ϕ is a convex function and X is a random variable such that $\mathbb{E}[X]$ exists, then,

$$\phi(\mathbb{E}[X]) \leq \mathbb{E}[\phi(X)]. \tag{1.23}$$

If ϕ is concave, the inequality is reversed,

$$\phi(\mathbb{E}[X]) \geq \mathbb{E}[\phi(X)]. \tag{1.24}$$

1.7.2 Multivariate Delta Method

Let $\mathbf{X}_n \in \mathbb{R}^d$ be a sequence of estimators such that

$$\sqrt{n}(\mathbf{X}_n - \boldsymbol{\theta}) \xrightarrow{d} \mathbb{N}(\mathbf{0}, \boldsymbol{\Sigma}), \tag{1.25}$$

and let $g : \mathbb{R}^d \to \mathbb{R}^m$ be a differentiable function at $\boldsymbol{\theta}$. Then,

$$\sqrt{n}(g(\mathbf{X}_n) - g(\boldsymbol{\theta})) \xrightarrow{d} \mathbb{N}(0, Dg(\boldsymbol{\theta})\boldsymbol{\Sigma}Dg(\boldsymbol{\theta})'), \tag{1.26}$$

where $Dg(\boldsymbol{\theta})$ is the Jacobian matrix of g at $\boldsymbol{\theta}$.

Example 1.7.1 *Let $\bar{X}_n$ and $\bar{Y}_n$ be sample means of two independent samples with true means $\mu_X = 2$ and $\mu_Y = 4$. Suppose*

$$\sqrt{n}\left(\begin{bmatrix}\bar{X}_n\\ \bar{Y}_n\end{bmatrix} - \begin{bmatrix}2\\4\end{bmatrix}\right) \xrightarrow{d} \mathbb{N}\left(\begin{bmatrix}0\\0\end{bmatrix}, \begin{bmatrix}1 & 0\\0 & 1\end{bmatrix}\right).$$

Let $g(x, y) = \frac{x}{y}$. Then,

$$g(\mu_X, \mu_Y) = \frac{2}{4} = 0.5, \quad Dg(x, y) = \left[\frac{1}{y}, -\frac{x}{y^2}\right].$$

Evaluating at $(2, 4)$,

$$Dg(2, 4) = \left[\frac{1}{4}, -\frac{2}{16}\right] = \left[\frac{1}{4}, -\frac{1}{8}\right].$$

Then the asymptotic variance is

$$Var = \left[\frac{1}{4}, -\frac{1}{8}\right]\begin{bmatrix}1 & 0\\0 & 1\end{bmatrix}\begin{bmatrix}\frac{1}{4}\\ -\frac{1}{8}\end{bmatrix} = \frac{1}{16} + \frac{1}{64} = \frac{5}{64}.$$

Therefore, we have

$$\sqrt{n}\left(\frac{\bar{X}_n}{\bar{Y}_n} - \frac{1}{2}\right) \xrightarrow{d} \mathbb{N}\left(0, \frac{5}{64}\right).$$

1.7.3 Kronecker Product and Direct Sum

The Kronecker product of two square matrices $\mathbf{A}, \mathbf{B} \in \mathbb{R}^{n\times n}$ is defined as

$$A \otimes \mathbf{B} = \begin{pmatrix} a_{11}\mathbf{B} & a_{12}\mathbf{B} & \dots & a_{1n}\mathbf{B} \\ a_{21}\mathbf{B} & a_{22}\mathbf{B} & \dots & a_{2n}\mathbf{B} \\ \vdots & \vdots & \dots & \vdots \\ a_{n1}\mathbf{B} & a_{n2}\mathbf{B} & \dots & a_{nn}\mathbf{B} \end{pmatrix}.$$

Example 1.7.2 *If*

$$\mathbf{A} = \begin{pmatrix} a_{11} & a_{12} \\ a_{21} & a_{22} \end{pmatrix}, \quad \mathbf{B} = \begin{pmatrix} b_{11} & b_{12} \\ b_{21} & b_{22} \end{pmatrix}.$$

then,

$$\mathbf{A} \otimes \mathbf{B} = \begin{pmatrix} a_{11}B & a_{12}\mathbf{B} \\ a_{21}B & a_{22}\mathbf{B} \end{pmatrix} = \begin{pmatrix} a_{11}b_{11} & a_{11}b_{12} & a_{12}b_{11} & a_{12}b_{12} \\ a_{11}b_{21} & a_{11}b_{22} & a_{12}b_{21} & a_{12}b_{22} \\ a_{21}b_{11} & a_{21}b_{12} & a_{22}b_{11} & a_{22}b_{12} \\ a_{21}b_{21} & a_{21}b_{22} & a_{22}b_{21} & a_{22}b_{22} \end{pmatrix}.$$

The operator $\oplus$ denotes the direct sum of matrices. For two square matrices, we define

$$\mathbf{A} \oplus \mathbf{B} = \begin{bmatrix} \mathbf{A} & 0 \\ 0 & \mathbf{B} \end{bmatrix}.$$

1.7.4 Big-O and Small-o Notations

Definition: Big-O. Let $f(n)$ and $g(n)$ be functions from the natural numbers to the positive real numbers. We say that $f(n) = O(g(n))$ if there exist constants $c > 0$ and n_0 such that $0 \le f(n) \le c \cdot g(n)$ for all $n \ge n_0$.

Big-O notation gives an upper bound on the growth rate of a function(Serfling [303]). It describes the worst-case asymptotic behavior.

Example 1.7.3 *If $f(n) = 3n^2 + 5n + 7$, then $f(n) = O(n^2)$, because for large n, the n^2 term dominates the lower-order terms. If $f(n) = 10n$, then $f(n) = O(n^2)$ because $10n \leq n^2$ for sufficiently large n.*

Definition: Small-o. Let $f(n)$ and $g(n)$ be functions from the natural numbers to the positive real numbers. We say that $f(n) = o(g(n))$ if for every constant $c > 0$, there exists an n_0 such that $0 \leq f(n) < c \cdot g(n) \quad \text{for all } n \geq n_0$.

Equivalently, $\lim_{n\to\infty} \frac{f(n)}{g(n)} = 0$. Small-o notation describes a function that grows strictly slower than another function.

Example 1.7.4 *If $f(n) = n$, then $f(n) = o(n^2)$ because $\lim_{n\to\infty} \frac{n}{n^2} = 0$. If $f(n) = \log n$, then $f(n) = o(n)$ because logarithmic growth is slower than linear growth.*

The main differences are

- *$f(n) = O(g(n))$ means $f(n)$ grows no faster than $g(n)$.*
- *$f(n) = o(g(n))$ means $f(n)$ grows strictly slower than $g(n)$.*

1.7.5 Big-O and Small-o Notations in Probability

Definition: $O_{\mathbb{P}}$. Let $\{X_n\}_{n\geq 1}$ be a sequence of random variables and let $\{a_n\}_{n\geq 1}$ be a sequence of positive real numbers. We say that $X_n = O_{\mathbb{P}}(a_n)$ if for every $\varepsilon > 0$ there exist constants $M > 0$ and N such that $\mathbb{P}\left(|X_n| > Ma_n\right) < \varepsilon \quad \text{for all } n \geq N$.

Example 1.7.5 *If $X_n \sim \mathbb{N}(0, 1)$ for all n, then $X_n = O_{\mathbb{P}}(1)$ because the standard normal distribution has bounded tails independent of n. Let $X_n = \sqrt{n}(\bar{Y}_n - \mu)$, where $Y_1, Y_2, \ldots$ are i.i.d. with mean μ and finite variance σ^2. By the Central Limit Theorem, $X_n \xrightarrow{d} \mathbb{N}(0, \sigma^2)$, so $X_n = O_{\mathbb{P}}(1)$. Hence, X_n/a_n is bounded in probability.*

Definition: $o_{\mathbb{P}}$. We say that $X_n = o_{\mathbb{P}}(a_n)$ if for every $\varepsilon > 0$, $\mathbb{P}\left(|X_n| > \varepsilon a_n\right) \to 0 \quad \text{as } n \to \infty$. Hence, X_n/a_n converges to 0 in probability.

Example 1.7.6 *Let $X_n = \bar{Y}_n - \mu$, where $Y_1, Y_2, \ldots$ are i.i.d. with finite variance. By the Law of Large Numbers, $X_n \xrightarrow{p} 0$, so $X_n = o_{\mathbb{P}}(1)$. If $Z_n = \sqrt{n}(\bar{Y}_n - \mu)$ as above, then $\frac{Z_n}{\sqrt{n}} = \bar{Y}_n - \mu = o_{\mathbb{P}}(1)$, which implies $Z_n = o_{\mathbb{P}}(\sqrt{n})$.*

If $X_n = o_{\mathbb{P}}(a_n)$, then automatically $X_n = O_{\mathbb{P}}(a_n)$, but the converse is not necessarily true.

1.7.6 Univariate Central Limit Theorem

Theorem 1.7.1 (Lindeberg–Lévy Central Limit Theorem) *Let $\{X_i\}_{i=1}^{\infty}$ be a sequence of independent and identically distributed random variables with $\mathbb{E}[X_i] = \mu$, and $\mathbb{V}ar(X_i) = \sigma^2 < \infty$. Define the normalized sum*

$$Z_n = \frac{1}{\sqrt{n}\sigma} \sum_{i=1}^{n} (X_i - \mu).$$

Then, $Z_n \xrightarrow{d} \mathbb{N}(0, 1)$, that is, Z_n converges in distribution to the standard normal distribution.

1.7.7 Lindeberg Condition (Triangular Arrays)

Let $\{X_{n,k} : 1 \le k \le n\}$ be a triangular array of independent random variables with

$$\mathbb{E}[X_{n,k}] = 0, \qquad s_n^2 = \sum_{k=1}^{n} \mathbb{V}ar(X_{n,k}).$$

Definition 1.7.1 (Lindeberg Condition) *The Lindeberg condition holds if for every $\varepsilon > 0$,*

$$\frac{1}{s_n^2} \sum_{k=1}^{n} \mathbb{E}\Big[X_{n,k}^2 \, \mathbf{1}\big(|X_{n,k}| > \varepsilon s_n\big)\Big] \longrightarrow 0 \quad \text{as } n \to \infty.$$

Theorem 1.7.2 (Lindeberg–Feller CLT) *If the Lindeberg condition holds and $s_n^2 \to \infty$, then*

$$\frac{1}{s_n} \sum_{k=1}^{n} X_{n,k} \xrightarrow{d} \mathbb{N}(0,1).$$

Remark 1 *For i.i.d. random variables with finite variance, the Lindeberg condition is automatically satisfied. Hence the Lindeberg–Lévy CLT is a special case of the Lindeberg–Feller theorem.*

1.7.8 Multivariate Central Limit Theorem

Let $\{\boldsymbol{X}_i\}_{i=1}^{\infty}$ be i.i.d. random vectors in $\mathbb{R}^d$ with $\mathbb{E}[\boldsymbol{X}_i] = \boldsymbol{\mu}$, and $\mathrm{Cov}(\boldsymbol{X}_i) = \boldsymbol{\Sigma}$, where $\boldsymbol{\Sigma}$ is a finite $d \times d$ covariance matrix.

Theorem 1.7.3 (Multivariate Central Limit Theorem) *Define*

$$\boldsymbol{Z}_n = \frac{1}{\sqrt{n}} \sum_{i=1}^{n} (\boldsymbol{X}_i - \boldsymbol{\mu}).$$

Then, $\boldsymbol{Z}_n \xrightarrow{d} \mathbb{N}_d(\mathbf{0}, \boldsymbol{\Sigma})$, where $\mathbb{N}_d(\mathbf{0}, \boldsymbol{\Sigma})$ denotes the d-dimensional multivariate normal distribution with mean $\mathbf{0}$ and covariance matrix $\boldsymbol{\Sigma}$.

1.7.9 Multivariate Lindeberg Condition

Let $\{\boldsymbol{X}_{n,k} : 1 \le k \le n\}$ be a triangular array of independent random vectors in $\mathbb{R}^d$ with $\mathbb{E}[\boldsymbol{X}_{n,k}] = \mathbf{0}$, and $S_n = \sum_{k=1}^{n} \mathrm{Cov}(\boldsymbol{X}_{n,k})$.

Definition 1.7.2 (Multivariate Lindeberg Condition) *The multivariate Lindeberg condition holds if for every $\varepsilon > 0$,*

$$\frac{1}{\mathrm{tr}(S_n)} \sum_{k=1}^{n} \mathbb{E}\Big[\|\boldsymbol{X}_{n,k}\|^2 \, \mathbf{1}\big(\|\boldsymbol{X}_{n,k}\| > \varepsilon\sqrt{\mathrm{tr}(S_n)}\big)\Big] \longrightarrow 0 \quad \text{as } n \to \infty,$$

where $\|\cdot\|$ denotes the Euclidean norm.

Theorem 1.7.4 (Multivariate Lindeberg–Feller CLT) *If the multivariate Lindeberg condition holds and $S_n \to \mathbf{\Sigma}$ (a finite covariance matrix), then*

$$\sum_{k=1}^{n} \boldsymbol{X}_{n,k} \xrightarrow{d} \mathbb{N}_d(\mathbf{0}, \mathbf{\Sigma}).$$

Example 1.7.7 (Univariate i.i.d. Example) *Let $X_1, \ldots, X_n$ be i.i.d. Bernoulli(p) random variables: $\mathbb{P}(X_i = 1) = p$ and $\mathbb{P}(X_i = 0) = 1 - p$. Then, $\mu = p$, and $sigma^2 = p(1-p)$. The normalized sum*

$$\frac{\sum_{i=1}^{n}(X_i - p)}{\sqrt{np(1-p)}}$$

converges in distribution to $\mathbb{N}(0, 1)$ as $n \to \infty$.

Example 1.7.8 (Non-i.i.d. Triangular Array) *Let $X_{n,k}$ be independent with $\mathbb{P}(X_{n,k} = \pm 1) = \frac{1}{2}$, and $\mathbb{V}ar(X_{n,k}) = 1$. Then, $s_n^2 = n$. Since the variables are uniformly bounded, $|X_{n,k}| \leq 1$, the Lindeberg condition holds automatically. Hence*

$$\frac{1}{\sqrt{n}} \sum_{k=1}^{n} X_{n,k} \xrightarrow{d} \mathbb{N}(0, 1).$$

Example 1.7.9 *Let $\boldsymbol{X}_i = (X_{i1}, X_{i2})'$ and $\mathbf{\Sigma} = \begin{pmatrix} 1 & \rho \\ \rho & 1 \end{pmatrix}$, where $\mathbb{E}[\boldsymbol{X}_i] = \mathbf{0}$ and $\mathrm{Cov}(\boldsymbol{X}_i) = \mathbf{\Sigma}$. Then,*

$$\frac{1}{\sqrt{n}} \sum_{i=1}^{n} \boldsymbol{X}_i \xrightarrow{d} \mathbb{N}_2\left(\mathbf{0}, \mathbf{\Sigma}\right).$$

The following points are noteworthy.

- The Lindeberg–Lévy CLT applies to i.i.d. random variables with finite variance.
- The Lindeberg condition generalizes the CLT to non-identically distributed triangular arrays.
- The multivariate CLT extends convergence to $\mathbb{R}^d$.
- The multivariate Lindeberg condition prevents a single component from dominating asymptotically.

1.7.10 U-Statistics

Let $X_1, \ldots, X_n$ be i.i.d. random variables with common distribution P. Let $h : \mathbb{X}^m \to \mathbb{R}$ be a symmetric kernel of order $m \geq 1$. Denote

$$\theta = \mathbb{E}[h(X_1, \ldots, X_m)].$$

A U-statistic with kernel h is

$$U_n = \binom{n}{m}^{-1} \sum_{1 \leq i_1 < \cdots < i_m \leq n} h(X_{i_1}, \ldots, X_{i_m}),$$

which is an unbiased estimator of θ (Hoeffding [150] and Serfling [303]). Examples include

- Sample variance,

$$U_n = \frac{1}{\binom{n}{2}} \sum_{i<j} \frac{(X_i - X_j)^2}{2}, \quad \mathbb{E}[U_n] = \mathbb{V}ar(X).$$

- Gini mean difference,

$$U_n = \frac{1}{\binom{n}{2}} \sum_{i<j} |X_i - X_j|, \quad \theta = \mathbb{E}|X_1 - X_2|.$$

1.7.11 V-Statistics

A V-statistic with kernel h is

$$V_n = \frac{1}{n^m} \sum_{i_1=1}^{n} \cdots \sum_{i_m=1}^{n} h(X_{i_1}, \ldots, X_{i_m}),$$

which allows repeated indices and is generally biased but consistent. Examples include

- Second moment estimator,

$$V_n = \frac{1}{n^2} \sum_{i,j=1}^{n} X_i X_j = \left(\frac{1}{n} \sum_{i=1}^{n} X_i \right)^2.$$

- Mean absolute difference (V-form),

$$V_n = \frac{1}{n^2} \sum_{i,j=1}^{n} |X_i - X_j|.$$

1.7.12 Hájek Projection

Let T_n be a symmetric statistic. Its Hájek projection

$$\widehat{T}_n = \mathbb{E}[T_n] + \sum_{i=1}^{n} \Big(\mathbb{E}[T_n \mid X_i] - \mathbb{E}[T_n] \Big),$$

is the best L_2 (square-integrable function) approximation of T_n by a sum of independent terms. Examples include

- U-statistic,

$$\widehat{U}_n = \theta + \frac{m}{n} \sum_{i=1}^{n} \Big(\mathbb{E}[h(X_i, X_2, \ldots, X_m) \mid X_i] - \theta \Big).$$

- Sample variance,

$$\sum_{i=1}^{n} ((X_i - \mu)^2 - \mathbb{V}ar(X)),$$

which governs asymptotic normality.

1.7.13 Asymptotic Variance Formulas

Define the first-order projection

$$h_1(x) = \mathbb{E}[h(x, X_2, \ldots, X_m)] - \theta.$$

Non-degenerate kernel ($h_1 \not\equiv 0$) is

$$\sqrt{n}\,(U_n - \theta) \xrightarrow{d} \mathbb{N}\Big(0,\, m^2 \mathbb{V}ar(h_1(X_1))\Big),$$

$$\text{AVar}(U_n) = \frac{m^2}{n} \mathbb{V}ar(h_1(X_1)).$$

The same formula holds for V_n under non-degeneracy.

Degenerate Kernels: h is degenerate of order 1 if

$$h_1(x) \equiv 0 \quad \Leftrightarrow \quad \mathbb{E}[h(x, X_2, \ldots, X_m)] = \theta, \quad \forall x.$$

Asymptotics under degeneracy: If h is degenerate,

$$n(U_n - \theta) \xrightarrow{d} \sum_{k=1}^{\infty} \lambda_k (Z_k^2 - 1),$$

where $\{Z_k\}$ are i.i.d. standard normals and $\{\lambda_k\}$ depend on the kernel.

Examples of degenerate kernels are

- Centered covariance kernel:

$$h(x, y) = (x - \mu)(y - \mu), \quad \mu = \mathbb{E}[X].$$

  Then $h_1(x) = 0$, so h is degenerate.

- Hoeffding independence kernel is

$$h\big((x_1, y_1), (x_2, y_2)\big) = \mathbf{1}\{x_1 < x_2, y_1 < y_2\} - \mathbf{1}\{x_1 < x_2\}\mathbf{1}\{y_1 < y_2\},$$

2

Interpoint Distances

2.1 Abstract

This chapter provides a comprehensive treatment of Euclidean IPDs and their central role in multivariate and high-dimensional statistical analysis. IPDs serve as fundamental building blocks for a wide range of methodologies, including distributional testing, classification, clustering, statistical depth, spatial statistics, and change-point detection. Unlike traditional multivariate techniques that rely on covariance estimation and suffer from the curse of dimensionality, IPD-based methods reduce complex multivariate structure to scalar quantities, offering robustness, computational efficiency, and theoretical tractability in high-dimensional settings. The chapter reviews key applications of IPDs in testing distributional homogeneity, model discrimination, graph-based inference, and pattern recognition. It develops the mathematical foundations of IPDs through norms, quadratic forms, and moment calculations, and derives exact and asymptotic distributions of IPDs under multivariate normal and mixture models in one-sample and two-sample settings. Properties such as pairwise independence, asymptotic normality, and implications for inference are discussed, along with supporting tools including Jensen's inequality and the multivariate delta method.

2.2 Introduction

Euclidean IPDs play a pivotal role across a wide spectrum of scientific disciplines. Serving as fundamental constructs in multivariate analysis, IPDs underpin numerous statistical methodologies including distributional comparisons, clustering algorithms, classification procedures, correspondence analysis, and multidimensional scaling. The utility of IPDs extends into spatial statistics and pattern recognition, where they are integral to point process analysis (Ripley, [280], [279]), minimum spanning trees, and spatial clustering of disease incidence [41]. Additionally, IPDs form the basis of numerous depth-based techniques and tests for distributional homogeneity. For example, Lukens [205] explores IPD densities derived from mixtures of multivariate normal distributions to examine structural features and outlier behavior in high-dimensional data. Similarly, Marozzi [213] develops and applies multivariate hypothesis tests based on IPDs to analyze magnetic resonance imaging data, illustrating their practical applicability in biomedical contexts. Modarres and Song [239] discuss IPDs, their applications, properties, and visualization.

Classical multivariate statistical methodologies are based on the assumption that observations arise from multivariate normal (or more broadly, elliptical) distributions. However, many of these methods such as nearest neighbor techniques or the Hotelling's T^2 [155] test tend to deteriorate in performance as dimensionality increases, particularly when the number of variables exceeds the number of observations. In such high-dimensional settings,

DOI: 10.1201/9781003789581-2

inferential procedures that rely on the invertibility of the covariance matrix become infeasible, and density estimation approaches require prohibitively large sample sizes due to the curse of dimensionality. IPD-based methods, by contrast, offer a robust alternative. Because IPDs are scalar quantities, their dimensionality remains fixed regardless of the underlying space in which the data resides. This dimensional reduction facilitates both theoretical tractability and computational feasibility, making IPD-based techniques especially attractive for high dimensional data analysis.

A primary application of IPDs is in the context of testing the homogeneity of multivariate distributions. In a significant theoretical contribution, Maa et al. [206] demonstrated that two multivariate distributions are identical if and only if the within-sample and between-sample IPDs follow the same univariate distribution. This result holds under relatively mild regularity conditions and applies to both continuous and discrete settings. A number of researchers, including Friedman and Rafsky [109], Henze and Penrose [147], Hall and Tajvidi [134], Liu and Modarres [201], and Jurečková and Kalina [165], have used IPDs to devise powerful non-parametric tests for equality of distributions. Beyond hypothesis testing, IPDs are also instrumental in classification problems. For instance, Silverman and Brown [311] employed IPDs to discriminate between stationary point processes, exploiting the fact that the expected values and distributions of IPDs are sensitive to underlying model differences. This property makes IPDs a natural candidate for both model discrimination and pattern recognition tasks.

Detection of change points in data are another application of IPDs. Suppose we observe a sequence of data vectors $\{\mathbf{X}_i\}_{i=1}^n$, indexed by some ordering, such as time or location. For detection of a single change point at τ, one tests $H_0: \ \mathbf{X}_i \sim F_0, \quad i = 1, \ldots, n$ against the alternative, $H_a: \ \text{exists } 1 \leq \tau < n, \quad \text{such that } \mathbf{X}_i \sim F_1, \quad \text{if } i > \tau, \text{ and } F_0, \quad \text{otherwise}$, where F_0 and F_1 are different probability distribution functions. Chen and Zhang [60] propose a test to detect change points based on the Minimum spanning tree (MST), which is constructed using Euclidean IPDs as weights. Other graphs-based methods, such as nearest neighborhood graph and minimum distance pairing graph, can also be used to define a test statistic in a similar manner. IPDs are also important in the detection of hot-spots on disease and crime clusters where the underlying distributions are often discrete (Patil and Modarres[263]. Atkinson et al. [17] use IPDs to compare an observed sample with a high-dimensional simulated sample. Another application of IPD concerns statistical depth functions. Data depth provides a systematic and non-parametric approach for defining quantitative and graphical multivariate distributional characteristics and developing inferential methods based on center-outward ranking of observations in $\mathbb{R}^d$. A depth function measures the closeness or centrality of a point $\mathbf{X} \in \mathbb{R}^d$ with respect to the center of a distribution function $F(\cdot)$. If a rank order is defined with respect to $F(\cdot)$, then observations which are further from the center of $F(\cdot)$ have low ranks whereas observations close to the center receive higher ranks. While there are many notions of data depth in the literature, a great majority of them are functions of IPDs (Modarres [227]. Interpoint distances are the building blocks of many clustering strategies. The family of agglomerative clustering methods contains some of the most frequently used clustering methods. We will describe a new clustering method for high-dimensional data. We discuss analysis of distance matrices, which contain groups of IPDs. Various notations of multivariate symmetry are also intimately related to IPDs.

We will define vector and matrix norms next. We discuss non-singular quadratic forms and their moments to obtain the distribution of IPDs in the one and two-sample cases. We define the moments of IPDs and obtain their distribution under a multivariate normal mixture model. We discuss the asymptotic normality of IPDs as $d \to \infty$. We also discuss the pairwise independence of IPDs.

2.3 L_s Norms

Suppose $\mathbf{X} = (X_1, \dots, X_d)'$ is an $\mathbb{R}^d$-valued random vector. The L_s norm of $\mathbf{X}$ is defined by

$$||\mathbf{X}||_s = \Big(\sum_{t=1}^{d} |X_t|^s\Big)^{\frac{1}{s}}. \tag{2.1}$$

L_s norm is a vector norm for specific s if it satisfies some certain properties.

Definition 2.3.1 (Vector Norm) *A function $||\cdot|| : \mathbb{R}^d \to \mathbb{R}$ is called a vector norm if it satisfies the following properties.*
1. $||\boldsymbol{X}|| \geq 0$ for any vector $\boldsymbol{X} \in \mathbb{R}^d$, and $||\boldsymbol{X}|| = 0$ if $\boldsymbol{X} = \boldsymbol{0}$.
2. $||\alpha \boldsymbol{X}|| = |\alpha| \cdot ||\boldsymbol{X}||$ for any vector $\boldsymbol{X} \in \mathbb{R}^d$ and any scalar $\alpha \in \mathbb{R}$.
3. Triangle inequality: $||\boldsymbol{X} + \boldsymbol{Y}|| \leq ||\boldsymbol{X}|| + ||\boldsymbol{Y}||$ for any vectors $\boldsymbol{X}, \boldsymbol{Y} \in \mathbb{R}^d$.

The following s-norms are of particular interest.

- $s = 1$: the L_1 norm or Manhattan norm, $||\mathbf{X}||_1 = \sum_{t=1}^{d} |x_t|$.
- $s = 2$: the L_2 norm or Euclidean norm, $||\mathbf{X}||_2 = \Big(\sum_{t=1}^{d} x_t^2\Big)^{\frac{1}{2}}$.
- $s = \infty$: the L_∞ norm, $||\mathbf{X}||_\infty = \max_{1 \leq t \leq d} |x_t|$. The set of vectors, the infinity norm of which is a given constant c, forms the surface of a hypercube with edge length $2c$.

It can be proved that for $s \geq 1$, $||\cdot||_s$ satisfies the triangle inequality (Minkowski inequality). However, for $0 < s < 1$, the triangle inequality does not hold. For $0 < r < s$, it follows from Cauchy-Schwarz inequality that $||\mathbf{X}||_s \leq ||\mathbf{X}||_r \leq n^{(\frac{1}{r} - \frac{1}{s})} \cdot ||\mathbf{X}||_s$. This is often called the Lyapunov inequality. Suppose $\mathbf{b}$ and $\mathbf{d}$ be two $d \times 1$ vectors. The Cauchy-Schwarz inequality states that

$$(\mathbf{d}'\mathbf{d})^2 \leq (\mathbf{b}'\mathbf{b})(\mathbf{dd}) \tag{2.2}$$

2.4 Frobenius Norm

The Frobenius norm for a $d \times m$ matrix $\mathbf{A} = (a_{ij})$ is defined as

$$||\mathbf{A}||_F = \Big(\sum_{i=1}^{d} \sum_{j=1}^{m} a_{ij}^2\Big)^{1/2}. \tag{2.3}$$

The squared Frobenius norm is a monotonically increasing function of the eigenvalues of a square matrix and invariant to orthogonal transformations. We use $\|\boldsymbol{u}\|$ to denote the Euclidean norm for a vector $\boldsymbol{u} = (u_1, \dots, u_d)$, i.e. $\|\boldsymbol{u}\| = (\sum_{i=1}^{d} u_i^2)^{1/2}$. Thus, the Frobenius norm is the Euclidean norm on the space of matrices $\mathbb{R}^{dm}$.

2.5 Notation for Interpoint Distances

Let $\mathbb{X} = \{\mathbf{X}_1, \ldots, \mathbf{X}_m\}$ be d-dimensional random vectors drawn from distribution F_1. For $i \neq j$, define the (Euclidean) IPD and squared IPD as

$$d^2_{(x)ij} = d^2(\mathbf{X}_i, \mathbf{X}_j) = \|\mathbf{X}_i - \mathbf{X}_j\|^2, \tag{2.4}$$

and

$$d_{(x)ij} = d(\mathbf{X}_i, \mathbf{X}_j) = \|\mathbf{X}_i - \mathbf{X}_j\|. \tag{2.5}$$

There are $\binom{m}{2}$ distinct within-sample pairs. The corresponding averages are

$$\bar{d}^2_x = \binom{m}{2}^{-1} \sum_{i=1}^{m} \sum_{j=i+1}^{m} d^2(\mathbf{X}_i, \mathbf{X}_j), \tag{2.6}$$

$$\bar{d}_x = \binom{m}{2}^{-1} \sum_{i=1}^{m} \sum_{j=i+1}^{m} d(\mathbf{X}_i, \mathbf{X}_j). \tag{2.7}$$

Similarly, let $\mathbb{Y} = \{\mathbf{Y}_1, \ldots, \mathbf{Y}_n\}$ be d-dimensional random vectors drawn from distribution F_2. For $i \neq j$, define

$$d^2_{(y)ij} = d^2(\mathbf{Y}_i, \mathbf{Y}_j) = \|\mathbf{Y}_i - \mathbf{Y}_j\|^2, \tag{2.8}$$

$$d_{(y)ij} = d(\mathbf{Y}_i, \mathbf{Y}_j) = \|\mathbf{Y}_i - \mathbf{Y}_j\|. \tag{2.9}$$

With $\binom{n}{2}$ distinct pairs, the averages are

$$\bar{d}^2_y = \binom{n}{2}^{-1} \sum_{i=1}^{n} \sum_{j=i+1}^{n} d^2(\mathbf{Y}_i, \mathbf{Y}_j), \tag{2.10}$$

$$\bar{d}_y = \binom{n}{2}^{-1} \sum_{i=1}^{n} \sum_{j=i+1}^{n} d(\mathbf{Y}_i, \mathbf{Y}_j). \tag{2.11}$$

For cross-sample pairs $(\mathbf{X}_i, \mathbf{Y}_j)$, define the squared IPD and IPD as

$$d^2_{(xy)ij} = d^2(\mathbf{X}_i, \mathbf{Y}_j) = \|\mathbf{X}_i - \mathbf{Y}_j\|^2, \tag{2.12}$$

$$d(\mathbf{X}_i, \mathbf{Y}_j) = \|\mathbf{X}_i - \mathbf{Y}_j\|. \tag{2.13}$$

The corresponding cross-sample averages are

$$\bar{d}^2_{xy} = \frac{1}{mn} \sum_{i=1}^{m} \sum_{j=1}^{n} d^2(\mathbf{X}_i, \mathbf{Y}_j), \tag{2.14}$$

$$\bar{d}_{xy} = \frac{1}{mn} \sum_{i=1}^{m} \sum_{j=1}^{n} d(\mathbf{X}_i, \mathbf{Y}_j). \tag{2.15}$$

2.6 Mean and Variance of IPDs

The following identities are useful in the proofs of the theorems.

$$\mathbb{P}\left(d_{(x)}^2 = u\right) = \sum_s \mathbb{P}\left(d_{(x)}^2 = u \mid S = s\right) \cdot \mathbb{P}(S = s), \tag{2.16}$$

$$\mathbb{E}\left(d_{(x)}^2\right) = \sum_s \mathbb{E}\left(d_{(x)}^2 \mid S = s\right) \cdot \mathbb{P}(S = s), \tag{2.17}$$

$$Var\left(d_{(x)}^2\right) = \sum_s \left[\left(\mathbb{E}\left(d_{(x)}^2 \mid S = s\right) + Var\left(d_{(x)}^2 \mid S = s\right)\right) \cdot \mathbb{P}(S = s)\right] - \left[E\left(d_{(x)}^2\right)\right]^2, \tag{2.18}$$

where S is the categorical random variable representing different cases in mixture models.

2.7 Quadratic Forms

Mathai and Provost [220] define quadratic forms and find their moment generating function. We draw on their work to find the distribution of squared IPDs in one-sample and two-sample cases under different multivariate distributions.

Definition 2.7.1 *Suppose $\boldsymbol{X} = (X_1, \ldots, X_d)'$ is a d-dimensional random vector with mean vector $\boldsymbol{\mu} = \mathbb{E}(\boldsymbol{X}) = (\mu_1, \ldots, \mu_d)'$ and covariance matrix $\boldsymbol{\Sigma} = \mathbb{C}\text{ov}(\boldsymbol{X}) = \mathbb{E}\left[(\boldsymbol{X} - \boldsymbol{\mu})(\boldsymbol{X} - \boldsymbol{\mu})'\right]$. The quadratic form in the random variables $X_1, \ldots, X_d$ associated with an $d \times d$ symmetric matrix $A = (a_{ij})$ is defined as*

$$Q(\boldsymbol{X}) = Q(X_1, \ldots, X_d) = \boldsymbol{X}'\boldsymbol{A}\boldsymbol{X} = \sum_{i=1}^{d}\sum_{j=1}^{d} a_{ij} X_i X_j. \tag{2.19}$$

Random vectors will be denoted by bold capital letters. When $\boldsymbol{X}$ is centered at $\boldsymbol{\mu}$, the quadratic form $Q(\boldsymbol{X}) = (\boldsymbol{X} - \boldsymbol{\mu})' \boldsymbol{A} (\boldsymbol{X} - \boldsymbol{\mu})$ is called the central quadratic form.

Let $\mathbf{X}$ be a d-dimensional random vector with mean $\boldsymbol{\mu}$ and positive definite covariance matrix $\boldsymbol{\Sigma}$. Let $\mathbf{Z} = \boldsymbol{\Sigma}^{-\frac{1}{2}}(\mathbf{X} - \boldsymbol{\mu})$. It follows that $\mathbb{E}(\mathbf{Z}) = \mathbf{0}$ and $\mathbb{C}\text{ov}(\mathbf{Z}) = \mathbf{I}_\text{d}$. Let $\mathbf{P}$ be a $d \times d$ orthogonal matrix which diagonalizes $\boldsymbol{\Sigma}^{\frac{1}{2}}\boldsymbol{A}\boldsymbol{\Sigma}^{\frac{1}{2}}$, where $\mathbf{P}\mathbf{P}' = \mathbf{I}_d$,

$$\boldsymbol{P}'\boldsymbol{\Sigma}^{\frac{1}{2}}\boldsymbol{A}\boldsymbol{\Sigma}^{\frac{1}{2}}\boldsymbol{P} = diag(\lambda_1, \ldots, \lambda_d), \tag{2.20}$$

where $\lambda_1, \ldots, \lambda_d$ are the eigenvalues of $\boldsymbol{\Sigma}^{\frac{1}{2}}\boldsymbol{A}\boldsymbol{\Sigma}^{\frac{1}{2}}$ and $diag(\lambda_1, \ldots, \lambda_d)$ is the diagonal matrix formed by them. Let $\mathbf{U} = \boldsymbol{P}'\boldsymbol{Z}$, it follows that $\mathbf{Z} = \mathbf{P}\mathbf{U}$, $\mathbb{E}(\mathbf{U}) = \mathbf{0}$, and $\mathbb{C}\text{ov}(\mathbf{U}) = \mathbf{I}_\text{d}$. Therefore, the non–singular quadratic form can be written as

$$\begin{aligned} Q(\mathbf{X}) &= \left(\mathbf{Z} + \boldsymbol{\Sigma}^{-\frac{1}{2}}\boldsymbol{\mu}\right)' \boldsymbol{\Sigma}^{\frac{1}{2}}\boldsymbol{A}\boldsymbol{\Sigma}^{\frac{1}{2}} \left(\mathbf{Z} + \boldsymbol{\Sigma}^{-\frac{1}{2}}\boldsymbol{\mu}\right) \\ &= (\mathbf{U} + \mathbf{b})' \boldsymbol{P}'\boldsymbol{\Sigma}^{\frac{1}{2}}\boldsymbol{A}\boldsymbol{\Sigma}^{\frac{1}{2}}\boldsymbol{P}(\mathbf{U} + \mathbf{b}) \\ &= (\mathbf{U} + \mathbf{b})' diag(\lambda_1, \ldots, \lambda_d)(\mathbf{U} + \mathbf{b}), \end{aligned} \tag{2.21}$$

where $\mathbf{U} = (U_1, \ldots, U_d)'$ and $\mathbf{b} = \boldsymbol{P}'\boldsymbol{\Sigma}^{-\frac{1}{2}}\boldsymbol{\mu} = (b_1, \ldots, b_d)'$. Hence, we find the following representation,

$$Q(\mathbf{X}) = \boldsymbol{X}'\boldsymbol{A}\boldsymbol{X} = \begin{cases} \sum_{t=1}^{d} \lambda_t (U_t + b_t)^2, & \text{if } \boldsymbol{\mu} \neq \mathbf{0}; \\ \sum_{t=1}^{d} \lambda_t U_t^2, & \text{if } \boldsymbol{\mu} = \mathbf{0}. \end{cases} \tag{2.22}$$

Note that $U_t = \frac{X_t - \mu_t}{\sqrt{\sigma_{tt}}}$ is the standardized version of X_t, for $t = 1, \ldots, d$. For example, if $\mathbf{X} \sim \mathbb{N}_d(\boldsymbol{\mu}, \boldsymbol{\Sigma})$ and $\boldsymbol{\Sigma}$ is positive definite, then $Q(\mathbf{X})$ is a linear combination of independent central chi-square variables when $\boldsymbol{\mu} = \mathbf{0}$ and of non–central chi-square variables when $\boldsymbol{\mu} \neq \mathbf{0}$. The moments of this form can then be written as

$$\mathbb{E}\left[Q(\mathbf{X})^r\right] = \sum_{r \in \mathbf{R}^+} \frac{r! \lambda_1^{r_1} \ldots \lambda_d^{r_d}}{r_1! \ldots r_d!} \mathbb{E}\left((U_1 + b_1)^{2r_1} \ldots (U_d + b_d)^{2r_d}\right), \tag{2.23}$$

where

$$\mathbf{R}^+ = \{r \mid r = r_1 + \ldots + r_d,\ r_j \in \{0, 1, 2, \ldots\},\ j = 1, \ldots, d\}.$$

It is important to note that the above representations for $Q(\mathbf{X})$ or its moments do not assume a distribution for $\mathbf{X}$. In fact, the representations are applicable to both continuous and discrete distributions. Moreover, if $\mathbf{X}$ has a multivariate normal distribution, then an exact form is found for the r-th moment of $Q(\mathbf{X})$. We evaluate the moments explicitly for specific values of r in Exercise 2.1.

2.8 Quadratic Forms: Mean, Variance

Quadratic forms are expressions of the type $Q = \boldsymbol{X}'\boldsymbol{A}\boldsymbol{X}$ where $\boldsymbol{X} \in \mathbb{R}^d$ is a random vector and $\boldsymbol{A} \in \mathbb{R}^{d \times d}$ is symmetric. We derive the mean and variance of Q for multiple multivariate distributions. Let $\boldsymbol{X}$ have mean $\boldsymbol{\mu} = \mathbb{E}[\boldsymbol{X}]$ and covariance $\boldsymbol{\Sigma} = \mathbb{V}ar[\boldsymbol{X}]$. The mean of a quadratic form is

$$\mathbb{E}[\boldsymbol{X}'\boldsymbol{A}\boldsymbol{X}] = \operatorname{tr}(\boldsymbol{A}\boldsymbol{\Sigma}) + \boldsymbol{\mu}'\boldsymbol{A}\boldsymbol{\mu}$$

The variance of a quadratic form is given by

$$\mathbb{V}ar(\boldsymbol{X}'\boldsymbol{A}\boldsymbol{X}) = 2\operatorname{tr}((\boldsymbol{A}\boldsymbol{\Sigma})^2) + 4\boldsymbol{\mu}'\boldsymbol{A}\boldsymbol{\Sigma}\boldsymbol{A}\boldsymbol{\mu} + \operatorname{Kurt}(\boldsymbol{X}, \boldsymbol{A}),$$

where the term $\operatorname{Kurt}(\boldsymbol{X}, \boldsymbol{A})$ accounts for the fourth-order cumulant contribution of the quadratic form,

$$\operatorname{Kurt}(\boldsymbol{X}, \boldsymbol{A}) = \mathbb{E}\left[((\boldsymbol{X} - \boldsymbol{\mu})'\boldsymbol{A}(\boldsymbol{X} - \boldsymbol{\mu}))^2\right] - (\operatorname{tr}(\boldsymbol{A}\boldsymbol{\Sigma}))^2 - 2\operatorname{tr}((\boldsymbol{A}\boldsymbol{\Sigma})^2)$$

It is zero for multivariate normal vectors and non-zero for distributions with non-normal fourth moments.

For the multivariate normal distribution

$$\mathbb{E}[\boldsymbol{X}'\boldsymbol{A}\boldsymbol{X}] = \operatorname{tr}(\boldsymbol{A}\boldsymbol{\Sigma}) + \boldsymbol{\mu}'\boldsymbol{A}\boldsymbol{\mu}, \quad \text{with Kurt} = 0,$$

$$\mathbb{V}ar(\boldsymbol{X}'\boldsymbol{A}\boldsymbol{X}) = 2\operatorname{tr}((\boldsymbol{A}\boldsymbol{\Sigma})^2) + 4\boldsymbol{\mu}'\boldsymbol{A}\boldsymbol{\Sigma}\boldsymbol{A}\boldsymbol{\mu}$$

The vector $\boldsymbol{X}$ is elliptically distributed if $\boldsymbol{X} \stackrel{d}{=} \boldsymbol{\mu} + R\boldsymbol{\Sigma}^{1/2}\boldsymbol{U}$, where $R \geq 0$ is a scalar random variable (radial component), $\boldsymbol{U} \in \mathbb{R}^d$ is uniform on the unit sphere S^{d-1}, and R and $\boldsymbol{U}$ are independent.

The mean is given by

$$\mathbb{E}[\boldsymbol{X}'\boldsymbol{A}\boldsymbol{X}] = \boldsymbol{\mu}'\boldsymbol{A}\boldsymbol{\mu} + \frac{\mathbb{E}[R^2]}{d}\text{tr}(\boldsymbol{A}\boldsymbol{\Sigma}),$$

and the variance is given by

$$\mathbb{V}ar(\boldsymbol{X}'\boldsymbol{A}\boldsymbol{X}) = \frac{\mathbb{E}[R^4]}{d(d+2)}\left(2\,\text{tr}((\boldsymbol{A}\boldsymbol{\Sigma})^2) + (\text{tr}(\boldsymbol{A}\boldsymbol{\Sigma}))^2\right) + 4\frac{\mathbb{E}[R^2]}{d}\boldsymbol{\mu}'\boldsymbol{A}\boldsymbol{\Sigma}\boldsymbol{A}\boldsymbol{\mu} - (\mathbb{E}[\boldsymbol{X}'\boldsymbol{A}\boldsymbol{X}])^2$$

2.9 Distribution of IPD: Normal Distribution

Song [320] considers IPD distributions and their applications. We discuss the distribution of IPD between vectors that are drawn from a multivariate normal distribution. The distribution of IPD within one sample and across two samples is determined. We derive the distribution of the norm of multivariate normal and distance from fixed vector in $\mathbb{R}^d$. We also determine the distribution of IPD between vectors that are drawn from a mixture of k multivariate distributions.

Let $\mathbf{X}$ be a d-variate multivariate normal vector with mean $\boldsymbol{\mu}$ and covariance $\boldsymbol{\Sigma}$, $\mathbf{X} \sim \mathbb{N}_d(\boldsymbol{\mu}, \boldsymbol{\Sigma})$, then the PDF of $\mathbf{X}$ is given in section 1.6.5.

2.9.1 Distribution of One Sample Normal IPD

Lemma 2.9.1 *If $\boldsymbol{Y} \sim \mathbb{N}_2(\boldsymbol{0}, \boldsymbol{\Sigma})$, where $\boldsymbol{\Sigma}$ has unit variances and correlation coefficient ρ, then* $\mathbb{C}\text{ov}\left[\text{Y}_1^2, \text{Y}_2^2\right] = 2\rho^2$.

Proof: Since $Y_1, Y_2 \sim \mathbb{N}(0,1)$, we have $\mathbb{E}\left(Y_1^2\right) = \mathbb{E}\left(Y_2^2\right) = 1$. Since $Y_1 \mid Y_2 \sim \mathbb{N}(\rho Y_2, 1-\rho^2)$, then $\mathbb{E}\left(Y_1^2 \mid Y_2\right) = 1 - \rho^2 + \rho^2 Y_2^2$. Therefore,

$$\begin{aligned}\mathbb{C}\text{ov}\left[\text{Y}_1^2, \text{Y}_2^2\right] &= \mathbb{E}\left(Y_1^2 Y_2^2\right) - \mathbb{E}\left(Y_1^2\right)\mathbb{E}\left(Y_2^2\right) = \mathbb{E}\left(Y_2^2(1-\rho^2+\rho^2 Y_2^2)\right) - 1 \\ &= \mathbb{E}\left(Y_2^2\right) - \rho^2\mathbb{E}\left(Y_2^2\right) + \rho^2\mathbb{E}\left(Y_2^4\right) - 1 \\ &= 1 + 2\rho^2 - 1 = 2\rho^2.\end{aligned} \tag{2.24}$$

□

The following theorem exposes the distribution of IPDs in samples from a multivariate normal distribution.

Theorem 2.9.1 (One sample multivariate normal IPD) *Suppose $\{\mathbf{X}_i\}_{i=1}^m$ is a sample of d-dimensional random vectors drawn from $\mathcal{N}_d(\boldsymbol{\mu}, \boldsymbol{\Sigma})$, where $\boldsymbol{\Sigma} = (\sigma_{rs})$. Then the squared interpoint distance*

$$d^2(\mathbf{X}_i, \mathbf{X}_j) = (\mathbf{X}_i - \mathbf{X}_j)'(\mathbf{X}_i - \mathbf{X}_j)$$

is a linear combination of independent $\chi^2(1)$ random variables. Moreover,

$$\mathbb{E}[d^2(\mathbf{X}_i, \mathbf{X}_j)] = 2\,\text{tr}(\boldsymbol{\Sigma}), \qquad \mathbb{V}ar[d^2(\mathbf{X}_i, \mathbf{X}_j)] = 8\,\text{tr}(\boldsymbol{\Sigma}^2),$$

and for distinct i, j, l,

$$\mathbb{C}ov\left(d^2(\mathbf{X}_i, \mathbf{X}_j), d^2(\mathbf{X}_i, \mathbf{X}_l)\right) = 2\,\text{tr}(\boldsymbol{\Sigma}^2), \qquad \mathbb{C}or\left(d^2(\mathbf{X}_i, \mathbf{X}_j), d^2(\mathbf{X}_i, \mathbf{X}_l)\right) = \frac{1}{4}.$$

It follows from (2.19) with $\mathbf{A} = \mathbf{I}$ and $\mathbf{X} = \mathbf{X}_i - \mathbf{X}_j$ that $\mathbb{E}\left(\bar{d}_x^2\right) = 2tr\left(\boldsymbol{\Sigma}\right)$. To compute $\bar{d}_x^2$, one needs to find $\frac{m(m-1)}{2}$ IPDs so the computational complexity is $O\left(m^2\right)$. There is another approach to compute $\bar{d}_x^2$ with a lower-computational complexity. See Exercise 2.3.

Mathai and Provost [220] derive the following theorem about the moment generating function of a quadratic form of normal random variables.

Theorem 2.9.2 (Moments of Normal Quadratic Form) *Suppose* $\boldsymbol{A}$ *is a symmetric* $d \times d$ *matrix and* $\boldsymbol{X} \sim \mathbb{N}_d(\boldsymbol{\mu}, \boldsymbol{\Sigma})$, $\boldsymbol{\Sigma}$ *is positive definite. Let*

$$diag(\lambda_1, \ldots, \lambda_d) = \boldsymbol{P}'\boldsymbol{\Sigma}^{\frac{1}{2}}\boldsymbol{A}\boldsymbol{\Sigma}^{\frac{1}{2}}\boldsymbol{P},$$

$$\boldsymbol{P}'\boldsymbol{P} = \boldsymbol{I}_d, \quad \boldsymbol{b} = (b_1, \ldots, b_d)' = \boldsymbol{P}'\boldsymbol{\Sigma}^{-\frac{1}{2}}\boldsymbol{\mu},$$

and $\theta_j = 1 - 2t\lambda_j$. *The moment generating function of* $Q = \boldsymbol{X}'\boldsymbol{A}\boldsymbol{X}$ *is*

$$\begin{aligned} M_Q(t) &= \exp\Big(-\frac{1}{2}\sum_{j=1}^{d} b_j^2\Big) \exp\left(\frac{1}{2}\sum_{j=1}^{d} b_j^2\theta_j^{-1}\right)\prod_{j=1}^{d}\theta_j^{-\frac{1}{2}} \\ &= \begin{cases} \exp\left(t\sum_{j=1}^{d} b_j^2\lambda_j\theta_j^{-1}\right)\prod_{j=1}^{d}\theta_j^{\frac{1}{2}}, & \text{for } \boldsymbol{\mu} \neq \boldsymbol{0}; \\ \prod_{j=1}^{d}\theta_j^{\frac{1}{2}}, & \text{for } \boldsymbol{\mu} = \boldsymbol{0}, \end{cases} \end{aligned} \tag{2.25}$$

The r*-th moment of* $Q(\boldsymbol{X})$ *is given by*

$$\mathbb{E}\left[Q^r(\boldsymbol{X})\right] = \left[\sum_{r_1=0}^{r-1}\binom{r-1}{r_1} g^{(r-1-r_1)} \sum_{r_2=0}^{r_1-1}\binom{r_1-1}{r_2} g^{(r_1-1-r_2)} \cdots\right], \tag{2.26}$$

where

$$g^{(k)} = 2^k k! \left[tr(\boldsymbol{A}\boldsymbol{\Sigma})^{k+1} + (k+1)\boldsymbol{\mu}'(\boldsymbol{A}\boldsymbol{\Sigma})^k\boldsymbol{A}\boldsymbol{\mu}\right], \quad k = 0, 1, 2, \ldots. \tag{2.27}$$

2.9.2 Distribution of Two-Sample Normal IPD

Theorem 2.9.3 *Suppose* $\{\boldsymbol{X}\}_{i=1}^m$ *and* $\{\boldsymbol{Y}\}_{j=1}^n$ *are independent samples of d-dimensional random vectors that are drawn from* $\mathbb{N}_d\left(\boldsymbol{\mu}_x, \boldsymbol{\Sigma}_x\right)$ *and* $\mathbb{N}_d\left(\boldsymbol{\mu}_y, \boldsymbol{\Sigma}_y\right)$ *respectively. The distribution of IPD between any* $\boldsymbol{X}_i$ *and* $\boldsymbol{Y}_j$ *is a linear combination of independent non-central chi-square random variables with 1 degree of freedom. The expected value and variance of the IPDs are*

$$\begin{aligned} \mathbb{E}\left[d^2_{(xy)ij}\right] &= tr\left(\boldsymbol{\Sigma}_x + \boldsymbol{\Sigma}_y\right) + \left(\boldsymbol{\mu}_x - \boldsymbol{\mu}_y\right)'\left(\boldsymbol{\mu}_x - \boldsymbol{\mu}_y\right), \\ \mathbb{V}\mathrm{ar}\left[\mathrm{d}^2_{(\mathrm{xy})\mathrm{ij}}\right] &= 2tr\left(\boldsymbol{\Sigma}_x + \boldsymbol{\Sigma}_y\right)^2 + 4\left(\boldsymbol{\mu}_x - \boldsymbol{\mu}_y\right)'\left(\boldsymbol{\Sigma}_x + \boldsymbol{\Sigma}_y\right)\left(\boldsymbol{\mu}_x - \boldsymbol{\mu}_y\right). \end{aligned} \tag{2.28}$$

Proof: Since $\mathbb{E}\left(\mathbf{X}_i - \mathbf{Y}_j\right) = \boldsymbol{\mu}_x - \boldsymbol{\mu}_y$ and $\mathbb{C}\mathrm{ov}\left(\mathbf{X}_\mathrm{i} - \mathbf{Y}_\mathrm{j}\right) = \Sigma_\mathrm{x} + \Sigma_\mathrm{y}$, it follows that,

$$d^2_{(xy)ij} = \sum_{t=1}^{d}\lambda_t(U_t + b_t)^2, \tag{2.29}$$

where $\lambda_t's$ are eigenvalues of $\boldsymbol{\Sigma}_x + \boldsymbol{\Sigma}_y$,

$$\mathbf{U} = \mathbf{P}'\left(\boldsymbol{\Sigma}_x + \boldsymbol{\Sigma}_y\right)^{-\frac{1}{2}}\left(\mathbf{X}_i - \mathbf{Y}_j - \boldsymbol{\mu}_x - \boldsymbol{\mu}_y\right) \text{ and } \mathbf{b} = \mathbf{P}'\left(\boldsymbol{\Sigma}_x + \boldsymbol{\Sigma}_y\right)^{-\frac{1}{2}}\left(\boldsymbol{\mu}_x - \boldsymbol{\mu}_y\right).$$

Note that $\mathbf{P}$ is an orthogonal matrix such that

$$\mathbf{P}'\left(\boldsymbol{\Sigma}_x + \boldsymbol{\Sigma}_y\right)\mathbf{P} = diag(\lambda_1, \ldots, \lambda_d).$$

Therefore, for $t = 1, \ldots, d$, $U_t \sim \mathbb{N}(0, 1)$ and $\left(U_t + b_t\right)^2 \sim$ non-central $\chi^2(1)$. One can find the mean and variance of the IPD using (2.26). See Exercise 2.1. □

2.9.3 Normal Norm Distribution

Theorem 2.9.4 (Norm of multivariate normal) *Suppose $\boldsymbol{X}$ is a d-dimensional random vector that is drawn from $\mathbb{N}_d(\boldsymbol{\mu}_x, \boldsymbol{\Sigma}_x)$. The distribution of the Euclidean squared norm is given by*

$$||\boldsymbol{X}||^2 = \sum_{t=1}^{d} \lambda_t (U_t + b_t)^2, \tag{2.30}$$

where λ_t's are eigenvalues of $\boldsymbol{\Sigma}_x$, $\boldsymbol{U} = \boldsymbol{P}'\boldsymbol{\Sigma}_x^{\frac{1}{2}}(\boldsymbol{X} - \boldsymbol{\mu}_x)$ and $\boldsymbol{b} = \boldsymbol{P}'\boldsymbol{\Sigma}_x^{\frac{1}{2}}\boldsymbol{\mu}_x$ and $\boldsymbol{P}$ is an orthogonal matrix such that $\boldsymbol{P}\boldsymbol{\Sigma}_x\boldsymbol{P}' = diag(\lambda_1, \ldots, \lambda_d)$. The mean and variance of the squared norm are

$$\mathbb{E}\left(||\boldsymbol{X}||^2\right) = tr\left(\boldsymbol{\Sigma}_x\right) + \boldsymbol{\mu}_x'\boldsymbol{\mu}_x, \quad \mathbb{V}\text{ar}\left(||\boldsymbol{X}||^2\right) = 2\text{tr}\left(\Sigma_{\text{x}}^2\right) + 4\boldsymbol{\mu}_{\text{x}}'\Sigma_{\text{x}}\boldsymbol{\mu}_{\text{x}}. \tag{2.31}$$

Note that the squared Euclidean norm is a linear combination of independent non-central chi-squared variables with 1 degree of freedom and non-centrality b_t.

Proof: Because $\mathbb{E}(\mathbf{X}) \neq \mathbf{0}$, we use Theorem 2.9.1 to obtain the expression in (2.30). The mean and variance follows (2.26). □

2.9.4 Distribution of Distance from a Fixed Vector

Theorem 2.9.5 (Distance from fixed point to multivariate normal vector) *Suppose $\boldsymbol{X}$ is a d-dimensional random vector that is drawn from $\mathbb{N}_d(\boldsymbol{\mu}_x, \boldsymbol{\Sigma}_x)$. The squared Euclidean distance from a fixed vector $\boldsymbol{q}$ is given by,*

$$||\boldsymbol{X} - \boldsymbol{q}||^2 = \sum_{t=1}^{d} \lambda_t (U_t + b_t)^2, \tag{2.32}$$

where λ_t's are eigenvalues of $\boldsymbol{\Sigma}_x$,

$$\boldsymbol{U} = \boldsymbol{P}'\boldsymbol{\Sigma}_x^{\frac{1}{2}}(\boldsymbol{X} - \boldsymbol{\mu}_x), \text{ and } \quad \boldsymbol{b} = \boldsymbol{P}'\boldsymbol{\Sigma}_x^{\frac{1}{2}}(\boldsymbol{\mu}_x - \boldsymbol{q}).$$

Here, $\boldsymbol{P}$ is an orthogonal matrix such that $\boldsymbol{P}\boldsymbol{\Sigma}_x\boldsymbol{P}' = diag(\lambda_1, \ldots, \lambda_d)$. The mean and variance of the squared distance are

$$\begin{aligned} \mathbb{E}\left(||\boldsymbol{X} - \boldsymbol{q}||^2\right) &= tr\left(\boldsymbol{\Sigma}_x\right) + (\boldsymbol{\mu}_x - \boldsymbol{q})'(\boldsymbol{\mu}_x - \boldsymbol{q}), \\ \mathbb{V}\text{ar}\left(||\boldsymbol{X} - \boldsymbol{q}||^2\right) &= 2tr\left(\boldsymbol{\Sigma}_x^2\right) + 4(\boldsymbol{\mu}_x - \boldsymbol{q})'\boldsymbol{\Sigma}_x(\boldsymbol{\mu}_x - \boldsymbol{q}). \end{aligned} \tag{2.33}$$

Similarly, the squared Euclidean distance from a fixed vector can also be expressed as a linear combination of independent non-central chi-square random variables with 1 degree of freedom. The non-centrality values are given by the components of $\mathbf{b}$.

Proof: One applies the quadratic form to $\mathbf{X} - \mathbf{q}$ in Theorem 2.9.4. □

2.9.5 Distribution of Normal IPD in Mixtures

Theorem 2.9.6 (Distribution of IPDs from Multivariate Normal Mixtures) *Suppose $\boldsymbol{X}$ and $\boldsymbol{X}^*$ are d-dimensional vectors that are drawn independently from a mixture of multivariate normal distributions $f(\boldsymbol{x}) = \sum_{i=1}^{k} \alpha_i \, \mathbb{N}_d(\boldsymbol{\mu}_i, \boldsymbol{\Sigma}_i)$, where $0 \leq \alpha_i \leq 1$*

and $\sum_{i=1}^{k}\alpha_i = 1$, the IPD $d^2_{(x)}$ can be expressed as a linear combination of central and non–central chi-square random variables,

$$d^2_{(x)} = \sum_{i=1}^{k}\alpha_i^2\sum_{t=1}^{d}\lambda_{it}U_{it}^2 + 2\sum_{i=1}^{k-1}\sum_{j=i+1}^{k}\alpha_i\alpha_j\sum_{t=1}^{d}\xi_{ijt}\left(W_{ijt}+b_{ijt}\right)^2, \tag{2.34}$$

where $\lambda'_{it}s$ are eigenvalues of $2\boldsymbol{\Sigma}_i$, and $\xi'_{ijt}s$ are eigenvalues of $\boldsymbol{\Sigma}_i+\boldsymbol{\Sigma}_j$. Note that the vectors

$$\boldsymbol{U}_i = \boldsymbol{P}'\left(2\boldsymbol{\Sigma}_i\right)^{-\frac{1}{2}}\left(\boldsymbol{X}-\boldsymbol{X}^*\right),$$

$$\boldsymbol{W}_{ij} = \boldsymbol{R}'\left(\boldsymbol{\Sigma}_i+\boldsymbol{\Sigma}_j\right)^{-\frac{1}{2}}\left(\boldsymbol{X}-\boldsymbol{X}^*-\boldsymbol{\mu}_i+\boldsymbol{\mu}_j\right),$$

and $\boldsymbol{b}_{ij} = \boldsymbol{R}'\left(\boldsymbol{\Sigma}_i+\boldsymbol{\Sigma}_j\right)^{-\frac{1}{2}}\left(\boldsymbol{\mu}_i-\boldsymbol{\mu}_j\right)$, where $\boldsymbol{P}$ and $\boldsymbol{R}$ are orthogonal matrices such that

$$2\boldsymbol{P}\boldsymbol{\Sigma_i}\boldsymbol{P}' = (\lambda_{i1},\ldots,\lambda_{id})' \quad \text{and} \quad \boldsymbol{R}\left(\boldsymbol{\Sigma_i}+\boldsymbol{\Sigma_j}\right)\boldsymbol{R}' = (\xi_{ij1},\ldots,\xi_{ijd})'.$$

The random variables U_{it}^2 are i.i.d. central and $\left(W_{ijt}+b_{ijt}\right)^2$ are i.i.d. non–central chi-square random variables with one degree of freedom for $t = 1,\ldots,d$.

We also obtain

$$\mathbb{E}\left[d^2_{(x)}\right] = 2\sum_{i=1}^{k}\alpha_i^2\cdot tr\left(\boldsymbol{\Sigma}_i\right) + 2\sum_{i=1}^{k-1}\sum_{j=1+1}^{k}\alpha_i\alpha_j\left[tr\left(\boldsymbol{\Sigma}_i+\boldsymbol{\Sigma}_j\right)+\left(\boldsymbol{\mu}_i-\boldsymbol{\mu}_j\right)'\left(\boldsymbol{\mu}_i-\boldsymbol{\mu}_j\right)\right], \tag{2.35}$$

$$\mathbb{V}\mathrm{ar}\left[\mathrm{d}^2_{(\mathrm{x})}\right] = 2\sum_{\mathrm{i}=1}^{\mathrm{k}}\alpha_{\mathrm{i}}^2\left[\mathrm{tr}\left(\Sigma_{\mathrm{i}}\right)+4\mathrm{tr}\left(\Sigma_{\mathrm{i}}^2\right)\right] + 2\sum_{\mathrm{i}=1}^{\mathrm{k}-1}\sum_{\mathrm{j}=\mathrm{i}+1}^{\mathrm{k}}\alpha_{\mathrm{i}}\alpha_{\mathrm{j}}\left[\mathrm{tr}\left(\Sigma_{\mathrm{i}}+\Sigma_{\mathrm{j}}\right)\right. \tag{2.36}$$

$$+\left(\boldsymbol{\mu}_i-\boldsymbol{\mu}_j\right)'\left(\boldsymbol{\mu}_i-\boldsymbol{\mu}_j\right)+2tr\left(\boldsymbol{\Sigma}_i+\boldsymbol{\Sigma}_j\right)^2$$
$$\left.+4\left(\boldsymbol{\mu}_i-\boldsymbol{\mu}_j\right)'\left(\boldsymbol{\Sigma}_i+\boldsymbol{\Sigma}_j\right)\left(\boldsymbol{\mu}_i-\boldsymbol{\mu}_j\right)\right] - \mathbb{E}^2\left[d^2_{(x)}\right]. \tag{2.37}$$

Example 2.9.1 *Suppose $\boldsymbol{X}$ and $\boldsymbol{X}^*$ are d-dimensional random vectors that are drawn from $\alpha\ \mathbb{N}_d(\boldsymbol{\mu}_1,\ \boldsymbol{\Sigma}_1)+(1-\alpha)\ \mathbb{N}_d(\boldsymbol{\mu}_2,\ \boldsymbol{\Sigma}_2)$. The IPD between $\boldsymbol{X}$ and $\boldsymbol{X}^*$ can be written as*

$$d^2_{(x)} = \alpha^2\sum_{t=1}^{d}\lambda_{1t}U_{1t}^2 + (1-\alpha)^2\sum_{t=1}^{d}\lambda_{2t}U_{2t}^2 + 2\alpha(1-\alpha)\sum_{t=1}^{d}\xi_t(W_t+b_t)^2, \tag{2.38}$$

where $\lambda'_{1t}s$ are the eigenvalues of $2\boldsymbol{\Sigma}_1$, $\lambda'_{2t}s$ are the eigenvalues of $2\boldsymbol{\Sigma}_2$ and $\xi'_t s$ are eigenvalues of $\boldsymbol{\Sigma}_1+\boldsymbol{\Sigma}_2$. The random vectors are

$$\boldsymbol{U}_1 = 2\boldsymbol{P}'\left(\boldsymbol{\Sigma}_1\right)^{-\frac{1}{2}}\left(\boldsymbol{X}-\boldsymbol{X}^*\right), \quad \boldsymbol{U}_2 = 2\boldsymbol{P}'\left(\boldsymbol{\Sigma}_2\right)^{-\frac{1}{2}}\left(\boldsymbol{X}-\boldsymbol{X}^*\right),$$

$$\boldsymbol{W} = \boldsymbol{R}'\left(\boldsymbol{\Sigma}_1+\boldsymbol{\Sigma}_2\right)^{-\frac{1}{2}}\left(\boldsymbol{X}-\boldsymbol{X}^*-\boldsymbol{\mu}_1+\boldsymbol{\mu}_2\right),$$

and $\boldsymbol{b} = \boldsymbol{R}'\left(\boldsymbol{\Sigma}_1+\boldsymbol{\Sigma}_2\right)^{-\frac{1}{2}}\left(\boldsymbol{\mu}_1-\boldsymbol{\mu}_2\right)$. When $\boldsymbol{\mu}_1 = \boldsymbol{\mu}_2$ with $\boldsymbol{\Sigma}_1 = \boldsymbol{I}_d$ and $\boldsymbol{\Sigma}_2 = 2\boldsymbol{I}_d$ we obtain

$$\begin{aligned} d^2_{(x)} &= \alpha^2\sum_{t=1}^{d}\lambda_{1t}U_{1t}^2 + (1-\alpha)^2\sum_{t=1}^{d}\lambda_{2t}U_{2t}^2 + 2\alpha(1-\alpha)\sum_{t=1}^{d}\xi_t W_t^2 \\ &= 2\alpha^2\chi^2(d) + 4(1-\alpha)^2\chi^2(d) + 6\alpha(1-\alpha)\chi^2(d) \\ &= \Gamma\Big(\frac{d}{2},\ 4\alpha^2\Big) + \Gamma\Big(\frac{d}{2},\ 8(1-\alpha)^2\Big) + \Gamma\Big(\frac{d}{2},\ 12\alpha(1-\alpha)\Big), \end{aligned} \tag{2.39}$$

which is a linear combination of independent Gamma variables. Note that when $\alpha = 1/2$, the cross-component IPDs cause the most variability in the distribution of IPDs.

Example 2.9.2 *Consider the squared norm of* $\mathbf{X}$, $Q = \|\mathbf{X}\|^2$ *where* $\mathbf{X}$ *is distributed as* $\mathbb{N}_d(\mathbf{0}, \boldsymbol{\Sigma})$ *and* $\boldsymbol{\Sigma}$ *has unit variances and constant correlation* ρ. *We will obtain an expression for*

$$\theta = \mathbb{V}ar_{\boldsymbol{\Sigma}}(Q)/\mathbb{V}ar_{\mathbf{I}}(Q)$$

where $\mathbb{V}ar_{\mathbf{I}}(Q)$ *is the variance of* Q *when* $\boldsymbol{\Sigma} = \mathbf{I}_d$. *Using* θ *we will show that the squared norm of vectors with independent components is more concentrated (less variable) than that of vectors with dependent components We have* $Q = \mathbf{X}'\mathbf{X} = \sum_{i=1}^d \lambda_i \chi^2(1)$. *Hence,*

$$\begin{aligned}\mathbb{E}(Q) &= \sum_{i=1}^d \lambda_i(1) = \sum_{i=1}^d \lambda_i = tr(\boldsymbol{\Sigma}),\\ \mathbb{V}ar(Q) &= \sum_{i=1}^d \lambda_i^2(2) = 2\sum_{i=1}^d \lambda_i^2 = 2tr(\boldsymbol{\Sigma}^2).\end{aligned} \tag{2.40}$$

When $\boldsymbol{\Sigma} = \mathbf{I}_d$; *i.e. under the independence of the components, we obtain* $\mathbb{E}(Q) = d$ *and* $\mathbb{V}ar(Q) = 2d$. *When* $\boldsymbol{\Sigma}$ *has 1 on the main diagonal and constant correlation* ρ *on the non–diagonal entries, the eigenvalues of* $\boldsymbol{\Sigma}$ *are*

$$\lambda_1 = 1 + (d-1)\rho, \qquad \lambda_2 = \ldots = \lambda_d = 1 - \rho.$$

Hence,

$$\mathbb{V}ar(Q) = 2\sum_{i=1}^d \lambda_i^2 = 2d(1 + (d-1)\rho^2)$$

Let $\theta = \mathbb{V}ar_{\boldsymbol{\Sigma}}(Q)/\mathbb{V}ar_{\mathbf{I}}(Q) = 1 + (d-1)\rho^2 > 0$. *For* $\boldsymbol{\Sigma}$ *to be positive definite we must have* $-\frac{1}{d-1} < \rho < 1$.

Since $\theta \geq 1$ *for all admissible* ρ, *the variance of* Q *is minimal when* $\rho = 0$. *Thus, the squared norm of vectors with independent components is more concentrated (less variable) than that of vectors with dependent components.*

Notice that the variance depends on $\sum_{i=1}^d \lambda_i^2$. *The eigenvalues satisfy* $\sum_{i=1}^d \lambda_i = d$. *Among all real numbers with fixed sum, the sum of squares is minimized when they are equal. Thus,* $\sum_{i=1}^d \lambda_i^2 \geq d$, *with equality if and only if* $\lambda_1 = \cdots = \lambda_d = 1$. *This occurs precisely when* $\boldsymbol{\Sigma} = \mathbf{I}_d$, *that is, when the components are independent. Therefore, independence minimizes* $\mathbb{V}ar(Q)$.

When $\boldsymbol{\Sigma} = \mathbf{I}_d$, *the distribution is rotationally symmetric. The probability cloud is spherical and variance is evenly distributed across all directions. The squared norm follows a* χ_d^2 *distribution. Random fluctuations are averaged across d independent directions. When there is elliptical stretching and* $\rho > 0$, *one eigenvalue increases while the remaining* $d-1$ *eigenvalues decrease. The normal cloud becomes elongated along one dominant direction. Most randomness is concentrated in that direction. Consequently,* Q *fluctuates more strongly.*

In the extreme case $\rho \to 1$, $\lambda_1 \to d$, $\lambda_2, \ldots, \lambda_d \to 0$, *so* $\mathbf{X}$ *collapses onto a one-dimensional subspace and* $Q \approx dZ^2$. *Thus,*$\mathbb{V}ar(Q) \to 2d^2$, *which is d times larger than in the independent case.*

When $\rho \to -\frac{1}{d-1}$, *one eigenvalue approaches zero. The distribution becomes flat in one direction. Again, unequal eigenvalues increase* $\sum \lambda_i^2$, *so variance increases. Under independence,* $Q \sim \chi_d^2$ *and* $\mathbb{V}ar(Q) = 2d$. *Relative fluctuations scale as*

$$\frac{\sqrt{\mathbb{V}ar(Q)}}{\mathbb{E}(Q)} = \frac{\sqrt{2d}}{d} = O(d^{-1/2}),$$

so Q concentrates around d as $d \to \infty$. Under strong dependence, effective dimension decreases. If variance is concentrated in one direction, relative fluctuations remain $O(1)$ and concentration disappears.

For fixed trace, $\mathbb{V}ar(Q) = 2\sum_{i=1}^{d} \lambda_i^2$ is minimized when the eigenvalues are equal. Thus, independence yields maximal concentration (minimal variability) of the squared norm.

2.10 Distribution of IPD: Non-Normal Distributions

Let $\mathbf{X}_1, \mathbf{X}_2 \in \mathbb{R}^d$ be independent random vectors. We are interested in finding the distribution of IPD $D(\mathbf{X}_1, \mathbf{X}_2) = \|\mathbf{X}_1 - \mathbf{X}_2\|$. We study non-Normal distributions for which closed-form expressions of $D(\mathbf{X}_1, \mathbf{X}_2)$ exist. We also consider independent components from diagonally symmetric distributions (IPDS).

2.10.1 IPDS for Symmetric Location–Scale Families

Let $\mathbf{X}_1, \mathbf{X}_2$ and $\mathbf{Y}_1, \mathbf{Y}_2$ be independent d-dimensional random vectors with independent components. Assume that the components of $\mathbf{X}$ and $\mathbf{Y}$ belong to the same symmetric location–scale family. That is, there exists a symmetric random variable Z with density f_0 satisfying $f_0(Z) = f_0(-Z)$ such that

$$X^{(k)} = \mu_1 + \sigma_1 Z^{(k)}, \qquad Y^{(k)} = \mu_2 + \sigma_2 W^{(k)},$$

where $Z^{(k)}, W^{(k)}$ are i.i.d. copies of Z and all coordinates are mutually independent. We determine the distributions of the IPDs $D^2(\mathbf{X}_1, \mathbf{X}_2)$, $D^2(\mathbf{Y}_1, \mathbf{Y}_2)$, and $D^2(\mathbf{X}_1, \mathbf{Y}_1)$.

For the same sample distances and for one coordinate, $X_1^{(k)} - X_2^{(k)} = \sigma_1(Z_1^{(k)} - Z_2^{(k)})$. Let $U_k = Z_1^{(k)} - Z_2^{(k)}$. Since Z is symmetric about zero, U_k is also symmetric. The variables $U_1, \ldots, U_d$ are independent and identically distributed. Hence, $\mathbf{X}_1 - \mathbf{X}_2 = \sigma_1(U_1, \ldots, U_d)$. The squared distance is

$$D^2(\mathbf{X}_1, \mathbf{X}_2) = \sigma_1^2 \sum_{k=1}^{d} U_k^2.$$

Thus the distribution of $D^2(\mathbf{X}_1, \mathbf{X}_2)$ is the d-fold convolution of the distribution of $\sigma_1^2 U_1^2$. Similarly,

$$D^2(\mathbf{Y}_1, \mathbf{Y}_2) = \sigma_2^2 \sum_{k=1}^{d} V_k^2,$$

where $V_k = W_1^{(k)} - W_2^{(k)}$ are i.i.d. symmetric. No further simplification is available in general.

For cross sample distances and for one coordinate,

$$X^{(k)} - Y^{(k)} = (\mu_1 - \mu_2) + \sigma_1 Z^{(k)} - \sigma_2 W^{(k)}.$$

Define $T_k = (\mu_1 - \mu_2) + \sigma_1 Z^{(k)} - \sigma_2 W^{(k)}$. The variables $T_1, \ldots, T_d$ are independent. They are symmetric only when $\mu_1 = \mu_2$. The squared cross-distance is

$$D^2(\mathbf{X}, \mathbf{Y}) = \sum_{k=1}^{d} T_k^2.$$

Thus $D(\mathbf{X}, \mathbf{Y})$ is a sum of independent squared variables. In general there is no closed-form density.

We know the following about the moment structure. Assume $\mathbb{E}[Z^2] < \infty$. Then $\mathbb{E}[U_k^2] = 2\,\mathbb{V}ar(Z)$. Hence,

$$\mathbb{E}[D^2(\mathbf{X}_1, \mathbf{X}_2)] = 2d\sigma_1^2 \mathbb{V}ar(Z),$$
$$\mathbb{E}[D^2(\mathbf{Y}_1, \mathbf{Y}_2)] = 2d\sigma_2^2 \mathbb{V}ar(Z).$$

For cross distances,

$$\mathbb{E}[D^2(\mathbf{X}, \mathbf{Y})] = d(\mu_1 - \mu_2)^2 + d(\sigma_1^2 + \sigma_2^2)\mathbb{V}ar(Z).$$

Higher moments exist only if the base distribution has sufficiently high finite moments.

For the special case of the normal distribution note that if $Z \sim \mathbb{N}(0, 1)$, then $U_k \sim \mathbb{N}(0, 2)$. Thus, $\mathbf{X}_1 - \mathbf{X}_2 \sim \mathbb{N}(0, 2\sigma_1^2 \mathbf{I}_d)$, and

$$D^2(\mathbf{X}_1, \mathbf{X}_2) \sim 2\sigma_1^2 \chi_d^2.$$

Similarly, $D^2(\mathbf{Y}_1, \mathbf{Y}_2) \sim 2\sigma_2^2 \chi_d^2$. For cross distances,

$$\mathbf{X} - \mathbf{Y} \sim \mathbb{N}(\mu_1 - \mu_2, (\sigma_1^2 + \sigma_2^2)\mathbf{I}_d),$$

so $D^2(\mathbf{X}, \mathbf{Y}) \sim (\sigma_1^2 + \sigma_2^2)\chi_d^2(\delta)$, where $\delta = \frac{d(\mu_1 - \mu_2)^2}{\sigma_1^2 + \sigma_2^2}$. This is the only case in the symmetric location–scale class that yields an exact chi-square representation.

Independence of coordinates does not imply spherical symmetry. Except in the normal case, the joint distribution of $\mathbf{X}_1 - \mathbf{X}_2$ is not rotationally invariant. For heavy-tailed families such as Student-t, the squared distance is a sum of independent heavy-tailed variables. Moments exist only when the base distribution has sufficient degrees of freedom. In general, the distribution of the Euclidean distance is determined by a d-fold convolution of squared independent symmetric differences, without a closed-form elementary density.

For symmetric location–scale families with independent coordinates,

$$D^2(\mathbf{X}_1, \mathbf{X}_2) = \sigma_1^2 \sum_{k=1}^{d} U_k^2, \qquad D^2(\mathbf{Y}_1, \mathbf{Y}_2) = \sigma_2^2 \sum_{k=1}^{d} V_k^2,$$

$$D(\mathbf{X}, \mathbf{Y})^2 = \sum_{k=1}^{d} \left((\mu_1 - \mu_2) + \sigma_1 Z_k - \sigma_2 W_k\right)^2.$$

Closed-form distributions arise only in the normal case. Otherwise, the distances are sums of independent squared symmetric differences.

2.10.2 Uniform Distribution on the Sphere S^{d-1}

A vector $\mathbf{X} \in \mathbb{R}^d$ is uniform on S^{d-1} if it is rotationally invariant and $\|\mathbf{X}\| = 1$ almost surely. Equivalently, all directions are equally likely on the unit sphere. To find the distribution of $D(\mathbf{X}_1, \mathbf{X}_2)$ we express the squared distance in terms of the angle between two points on the sphere.

$$D^2 = \|\mathbf{X}_1 - \mathbf{X}_2\|^2 = 2 - 2\cos\Theta, \quad \Theta = \text{angle between } \mathbf{X}_1, \mathbf{X}_2.$$

Lemma 2.10.1 *Distribution of the Angle Between Two Random Points on the Sphere*
Let $\mathbf{X}_1, \mathbf{X}_2$ be independent random vectors uniformly distributed on the unit sphere

$$S^{d-1} = \{\mathbf{x} \in \mathbb{R}^d : \|\mathbf{x}\| = 1\}.$$

Let Θ denote the angle between them, $\cos\Theta = \mathbf{X}_1 \cdot \mathbf{X}_2$. Then, $\Theta \in [0, \pi]$ has density

$$f_\Theta(\theta) = C_d(\sin\theta)^{d-2}, \quad 0 \le \theta \le \pi, \tag{2.41}$$

where $C_d = \frac{\Gamma(d/2)}{\sqrt{\pi}\,\Gamma((d-1)/2)}$.

Hence, the distribution of the angle Θ is determined by spherical symmetry and given in (2.41). We define distance as a function of Θ and compute the Jacobian for the change of variable. Using the half-angle identity $1 - \cos\Theta = 2\sin^2\left(\frac{\Theta}{2}\right)$ we can write

$$D = 2\sin(\Theta/2) \implies \Theta = 2\arcsin(D/2), \quad \frac{d\Theta}{dD} = \frac{2}{\sqrt{4 - D^2}}.$$

We apply the change-of-variable formula to obtain

$$f_D(d) = f_\Theta(\Theta(d))\frac{d\Theta}{dD} = 2C_d\, d^{d-2}(1 - D^2/4)^{(d-3)/2}, \quad 0 < d < 2.$$

2.10.3 Uniform Distribution in the Ball B_d

A vector $\mathbf{X} \in \mathbb{R}^d$ is uniform in the unit ball $B_d = \{\mathbf{x} : \|\mathbf{x}\| \le 1\}$ if all points in the ball are equally likely. The marginal radial distribution is

$$R = \|\mathbf{X}\| \sim f_R(r) = d\, r^{d-1}, \quad 0 \le r \le 1.$$

To find the distribution of $D(\mathbf{X}_1, \mathbf{X}_2)$ we separate distance computation into radial components R_1, R_2 and angular component Θ, $R_1 = \|\mathbf{X}_1\|$, $R_2 = \|\mathbf{X}_2\|$. Given the lengths R_1, R_2, we use the law of cosines for the distance to find

$$D^2 = R_1^2 + R_2^2 - 2R_1R_2\cos\Theta, \quad f_{\Theta|R_1,R_2}(\theta) = C_d(\sin\theta)^{d-2}.$$

We now express $\cos\Theta$ in terms of D and compute the Jacobian.

$$\cos\Theta = \frac{R_1^2 + R_2^2 - D^2}{2R_1R_2}, \quad \left|\frac{d\Theta}{dD}\right| = \frac{D}{\sqrt{4R_1^2R_2^2 - (R_1^2 + R_2^2 - D^2)^2}}.$$

We compute the conditional density using the Jacobian.

$$f_{D|R_1,R_2}(d|r_1, r_2) = C_d\frac{D(\sin\Theta)^{d-3}}{r_1r_2}.$$

We integrate the conditional density over independent radial components.

$$f_D(d) = \int_0^1\int_0^1 f_{D|R_1,R_2}(d|r_1, r_2)f_R(r_1)f_R(r_2)dr_1dr_2.$$

2.10.4 Spherically Symmetric Laplace Distribution

A vector $\mathbf{X} \in \mathbb{R}^d$ is spherically symmetric Laplace if its density depends only on $\|\mathbf{X}\|$:

$$f_X(\mathbf{x}) = c_d\exp(-\|\mathbf{x}\|), \quad \mathbf{x} \in \mathbb{R}^d,$$

with normalization constant c_d. Radial decomposition: $\mathbf{X} = R\mathbf{U}$, with $\mathbf{U}$ uniform on S^{d-1} and

$$f_R(r) = \frac{2}{\Gamma(d)}r^{d-1}e^{-r}, \quad r > 0.$$

Here, R is the random radius (magnitude of $\mathbf{X}$), and r is a possible value of R.

Note that $\mathbf{X}_1 - \mathbf{X}_2$ is radial; distance is magnitude of difference vector. Let $\mathbf{X}_1, \mathbf{X}_2$ be independent spherically symmetric Laplace vectors: $\mathbf{X}_i = R_i\mathbf{U}_i$, where $\mathbf{U}_i$ is uniform on S^{d-1} and

$$R_i \sim \text{Gamma}(d, 1), \quad f_R(r) = \frac{1}{\Gamma(d)} r^{d-1} e^{-r}, \quad r > 0.$$

Because the Laplace distribution is spherically symmetric and stable under convolution, the difference $\mathbf{Z} = \mathbf{X}_1 - \mathbf{X}_2$ is again spherically symmetric Laplace. To see the radial law, compute the characteristic function. For a d—dimensional Laplace vector,

$$\varphi_{\mathbf{X}}(t) = \frac{1}{(1 + \|t\|^2)^{d/2}}.$$

Independence gives

$$\varphi_{\mathbf{Z}}(t) = \varphi_{\mathbf{X}_1}(t)\varphi_{\mathbf{X}_2}(-t) = \frac{1}{(1 + \|t\|^2)^d}.$$

But this is the characteristic function of a spherical Laplace distribution with scale parameter 2, whose radial density is Gamma$(d, 2)$,

$$f_D(r) = \frac{1}{2^d\Gamma(d)} r^{d-1} e^{-r/2}, \quad r > 0.$$

Thus the distance $D = \|\mathbf{X}_1 - \mathbf{X}_2\|$ has a Gamma$(d, 2)$ distribution.

2.10.5 Elliptically Symmetric Distributions

Definition:

A random vector $\mathbf{X} \in \mathbb{R}^d$ is elliptically symmetric if it can be written as $\mathbf{X} = \boldsymbol{\mu} + R\mathbf{A}\mathbf{U}$, where

- $\boldsymbol{\mu} \in \mathbb{R}^d$ is a location vector,
- $\mathbf{A} \in \mathbb{R}^{d\times d}$ is a deterministic matrix such that $\boldsymbol{\Sigma} = \mathbf{A}\mathbf{A}'$ is positive definite (scale/shape matrix),
- $R \geq 0$ is a scalar radial random variable (determines the distance from the center),
- $\mathbf{U} \in S^{d-1}$ is uniform on the unit sphere and independent of R (determines direction).

Let $\mathbf{X}_1, \mathbf{X}_2$ be independent copies of $\mathbf{X}$. To find the distribution of $D(\mathbf{X}_1, \mathbf{X}_2)$ write

$$\mathbf{X}_1 - \mathbf{X}_2 = (\boldsymbol{\mu} + R_1\mathbf{A}\mathbf{U}_1) - (\boldsymbol{\mu} + R_2\mathbf{A}\mathbf{U}_2) = \mathbf{A}(R_1\mathbf{U}_1 - R_2\mathbf{U}_2),$$

where R_1, R_2 and $\mathbf{U}_1, \mathbf{U}_2$ are independent. The location $\boldsymbol{\mu}$ cancels when computing differences, leaving a linear transform of a spherically symmetric difference. The squared Euclidean distance is

$$D^2 = \|\mathbf{X}_1 - \mathbf{X}_2\|^2 = \|A(R_1\mathbf{U}_1 - R_2\mathbf{U}_2)\|^2 = (R_1\mathbf{U}_1 - R_2\mathbf{U}_2)'\boldsymbol{\Sigma}(R_1\mathbf{U}_1 - R_2\mathbf{U}_2),$$

where $\boldsymbol{\Sigma} = \mathbf{A}\mathbf{A}'$. Special cases where closed forms exist include

- Spherical case ($\boldsymbol{\Sigma} = \mathbf{I}$): recovers the spherical symmetry derivations. Distance depends only on the angle between $\mathbf{U}_1$ and $\mathbf{U}_2$ and the radial variables R_1, R_2.

- Diagonal $\mathbf{\Sigma} = \text{diag}(\sigma_1^2, \ldots, \sigma_d^2)$,

$$D^2 = \sum_{i=1}^{d} \sigma_i^2 (R_1 \mathbf{U}_1^{(i)} - R_2 \mathbf{U}_2^{(i)})^2.$$

  Distance is a weighted sum of independent spherically symmetric components, analogous to the IPDS case with different scales.

- General $\mathbf{\Sigma}$ (non-diagonal): Apply eigen decomposition $\mathbf{\Sigma} = Q\Lambda Q'$. Then

$$\begin{aligned} D^2 &= (R_1\mathbf{U}_1 - R_2\mathbf{U}_2)' Q\Lambda Q' (R_1\mathbf{U}_1 - R_2\mathbf{U}_2) \\ &= (Q'(R_1\mathbf{U}_1 - R_2\mathbf{U}_2))' \Lambda (Q'(R_1\mathbf{U}_1 - R_2\mathbf{U}_2)). \end{aligned}$$

  After rotation, this reduces to a diagonal case in the eigen basis of $\mathbf{\Sigma}$.

2.10.6 Multivariate Cauchy Distribution

A d-dimensional Cauchy with location $\boldsymbol{\mu}$ and scale $\mathbf{\Sigma}$ has density

$$f_{\mathbf{X}}(\mathbf{x}) = \frac{\Gamma((d+1)/2)}{\pi^{d/2}\Gamma(1/2)\,|\mathbf{\Sigma}|^{1/2}} \left(1 + (\mathbf{x} - \boldsymbol{\mu})'\mathbf{\Sigma}^{-1}(\mathbf{x} - \boldsymbol{\mu})\right)^{-(d+1)/2}.$$

The multivariate Cauchy distribution is elliptically symmetric. To find the distribution of $D(\mathbf{X}_1, \mathbf{X}_2)$ note that the difference of independent Cauchy vectors with common scale $\mathbf{\Sigma}$ is still elliptically symmetric. If $\mathbf{X}_1, \mathbf{X}_2 \overset{\text{i.i.d.}}{\sim} \text{Cauchy}_d(\mathbf{0}, \mathbf{\Sigma})$, then by stability of the Cauchy distribution, $\mathbf{X}_1 - \mathbf{X}_2 \sim \text{Cauchy}_d(\mathbf{0}, 2\mathbf{\Sigma})$. Hence, the density of $\mathbf{Y} = \mathbf{X}_1 - \mathbf{X}_2$ is

$$f_{\mathbf{Y}}(\mathbf{y}) = \frac{\Gamma((d+1)/2)}{\pi^{d/2}\Gamma(1/2)\,|2\mathbf{\Sigma}|^{1/2}} \left(1 + \mathbf{y}'(2\mathbf{\Sigma})^{-1}\mathbf{y}\right)^{-(d+1)/2}.$$

Let $\mathbf{\Sigma}^{1/2}$ denote the symmetric square root of $\mathbf{\Sigma}$. Write

$$\mathbf{Y} = \sqrt{2}\,\mathbf{\Sigma}^{1/2}\mathbf{Z}, \qquad \mathbf{Z} \sim \text{Cauchy}_d(\mathbf{0}, \mathbf{I}_d).$$

The Euclidean distance is

$$R = \|\mathbf{X}_1 - \mathbf{X}_2\| = \|\mathbf{Y}\| = \sqrt{2}\left(\mathbf{Z}'\mathbf{\Sigma}\mathbf{Z}\right)^{1/2}.$$

Thus the distance is no longer purely radial unless $\mathbf{\Sigma} = \sigma^2\mathbf{I}_d$ (spherical case).

Spherical case. If $\mathbf{\Sigma} = \sigma^2\mathbf{I}_d$, then $\mathbf{X}_1 - \mathbf{X}_2 \sim \text{Cauchy}_d(\mathbf{0}, 2\sigma^2\mathbf{I}_d)$, and

$$R = \|\mathbf{X}_1 - \mathbf{X}_2\| = \sqrt{2}\sigma\|\mathbf{Z}\|.$$

Using radial decomposition, the density of $\|\mathbf{Z}\|$ is

$$f_{\|\mathbf{Z}\|}(r) = \frac{2\Gamma((d+1)/2)}{\sqrt{\pi}\Gamma(d/2)} \frac{r^{d-1}}{(1+r^2)^{(d+1)/2}}, \qquad r > 0.$$

Therefore, by change of variables,

$$f_R(r) = \frac{2\Gamma((d+1)/2)}{\sqrt{\pi}\Gamma(d/2)} \frac{r^{d-1}}{(2\sigma^2)^{d/2}} \left(1 + \frac{r^2}{2\sigma^2}\right)^{-(d+1)/2}, \qquad r > 0.$$

General elliptical case. For general positive definite $\mathbf{\Sigma}$, $R^2 = 2\,\mathbf{Z}'\mathbf{\Sigma}\mathbf{Z}$. Let $\mathbf{\Sigma} = \mathbf{Q}\mathbf{\Lambda}\mathbf{Q}'$ be the spectral decomposition with eigenvalues $\lambda_1, \ldots, \lambda_d$. Since the Cauchy distribution is rotationally invariant, $\mathbf{Q}'\mathbf{Z} \stackrel{d}{=} \mathbf{Z}$, and hence $R^2 = 2\sum_{i=1}^{d} \lambda_i Z_i^2$. Thus, the squared distance is a weighted sum of dependent heavy—tailed components.

Elliptical symmetry generalizes spherical symmetry by introducing scaling. Distance distributions are generally not available in closed form except in spherical or diagonal cases. When $\mathbf{\Sigma}$ is diagonal, distance reduces to a weighted sum of independent squared radial differences. The technique of radial decomposition plus linear transformation is crucial for deriving distances in elliptical distributions.

Example 2.10.1 *Distribution of Pairwise Distances for Exponential Samples. Let X_1, X_2 be i.i.d. from* $\text{Exp}(\theta_1)$ *and Y_1, Y_2 be i.i.d. from* $\text{Exp}(\theta_2)$*, where*

$$f_X(x) = \theta_1 e^{-\theta_1 x}, \qquad f_Y(y) = \theta_2 e^{-\theta_2 y}, \qquad x, y \geq 0.$$

We want to find the exact distribution of $D(X_1, X_2), D(Y_1, Y_2)$ and $D(X_1, Y_1)$.

Let $X_1, X_2 \stackrel{i.i.d.}{\sim}$ $\text{Exp}(\theta_1)$*. Define $Z = X_1 - X_2$. The difference of two independent* $\text{Exp}(\theta_1)$ *random variables has the Laplace distribution with density*

$$f_Z(z) = \frac{\theta_1}{2} e^{-\theta_1 |z|}, \qquad z \in \mathbb{R}.$$

Now define $W = |Z| = |X_1 - X_2|$. For $w \geq 0$,

$$f_W(w) = f_Z(w) + f_Z(-w) = \frac{\theta_1}{2} e^{-\theta_1 w} + \frac{\theta_1}{2} e^{-\theta_1 w} = \theta_1 e^{-\theta_1 w}.$$

Therefore, $D(X_1, X_2) \sim$ $\text{Exp}(\theta_1)$*.*

By the same argument, if $Y_1, Y_2 \stackrel{i.i.d.}{\sim}$ $\text{Exp}(\theta_2)$*, then $D(Y_1, Y_2) \sim$* $\text{Exp}(\theta_2)$*. Let $X \sim$* $\text{Exp}(\theta_1)$ *and $Y \sim$* $\text{Exp}(\theta_2)$ *independently. Define $Z = X - Y$. The density of Z is obtained by convolution:*

$$f_Z(z) = \int_{-\infty}^{\infty} f_X(z + y) f_Y(y)\, dy.$$

Evaluating separately for $z \geq 0$ and $z < 0$ gives

$$f_Z(z) = \begin{cases} \dfrac{\theta_1 \theta_2}{\theta_1 + \theta_2} e^{-\theta_1 z}, & z \geq 0, \\[2ex] \dfrac{\theta_1 \theta_2}{\theta_1 + \theta_2} e^{\theta_2 z}, & z < 0. \end{cases}$$

Now define $W = |Z| = |X - Y|$. For $w \geq 0$,

$$f_W(w) = f_Z(w) + f_Z(-w) = \frac{\theta_1 \theta_2}{\theta_1 + \theta_2} \left(e^{-\theta_1 w} + e^{-\theta_2 w}\right).$$

Thus the density of $D(X_1, Y_2)$ is

$$f_{D(X,Y)}(w) = \frac{\theta_1 \theta_2}{\theta_1 + \theta_2} \left(e^{-\theta_1 w} + e^{-\theta_2 w}\right), \qquad w \geq 0.$$

This is a mixture of two exponential densities. In the special case $\theta_1 = \theta_2 = \theta$, it reduces to $f(w) = \theta e^{-\theta w}$, so $D(X_1, Y_1) \sim$ $\text{Exp}(\theta)$ *when the rates are equal.*

2.11 Pairwise Independence of IPDs

The following result helps in obtaining a random sample from the m possible sample IPDs. Shurygin [309] shows that IPDs are asymptotically pairwise independent. If the size m of sample $\{\mathbf{X}\}_{i=1}^{m}$ gets large enough, the probability of sampling two dependent IPDs is approximately zero. Given that there are m observations in sample $\mathbf{X}$. There are $t = \binom{m}{2} = \frac{m(m-1)}{2}$ IPDs in total for this sample. One can consider 2 of them at a time, so there are

$$u = \binom{t}{2} = \frac{(m+1)(m-2)(m-1)m}{8}$$

possible pairs. Among them, there are

$$w = \frac{2t(m-2)}{2} = \frac{m(m-1)(m-2)}{2}$$

dependent pairs.

Let the number of dependent pairs divide the total number of pairs and let m tend to infinity,

$$\lim_{m\to\infty} \frac{w}{u} = \lim_{m\to\infty} \frac{4m(m-1)(m-2)}{m(m-1)(m-2)(m+1)} = \lim_{m\to\infty} \frac{4}{m+1} = 0.$$

Hence, pairwise distances are asymptotically independent.

One may need to obtain a sample of independent IPDs from m observations. Suppose m is even, one can easily obtain $\frac{1}{2}m$ independent pairs of IPDs. The total number of ways to obtain $\frac{1}{2}m$ independent pairs is $\frac{m!}{2^{m/2}\left(\frac{m}{2}\right)!}$. To obtain $m/2$ independent pairs of IPD, the first 2 observations can be selected from the m observations. The next 2 observations are selected from the remaining $m-2$ observations. The process continues until there are no remaining observations. It is not difficult to obtain a group of mutually independent IPDs. For example, if $m = 6$, the mutually independent group is $\{d_{(x)12}, d_{(x)34}, d_{(x)56}\}$. The total number of ways to obtain this kind of group is

$$\frac{\binom{m}{2}\cdot\binom{m-2}{2}\cdots\binom{4}{2}\cdot\binom{2}{2}}{\left(\frac{m}{2}\right)!} = \frac{m!}{2^{m/2}\left(\frac{m}{2}\right)!}.$$

Two IPDs are dependent if they have an index in common. Otherwise, they are independent.

2.12 Independent of k distances in a Distance Matrix

Let $\mathbf{X}_1, \ldots, \mathbf{X}_m$ be i.i.d. observation vectors. Define the pairwise distances $d_{ij} = d(\mathbf{X}_i, \mathbf{X}_j)$, $1 \le i < j \le m$. There are $t = \binom{m}{2} = \frac{m(m-1)}{2}$ distinct distances. Suppose we select k distances uniformly at random without replacement from these t distances. What is the probability that they are independent?

A collection of k distances is mutually independent if and only if the $2k$ indices appearing in them are all distinct. Equivalently, the selected distances correspond to k disjoint edges in the complete graph K_m on m vertices. A necessary condition for independence is $2k \le m$. If $2k > m$ then independence is impossible.

2.12.1 Graph-Theoretic Interpretation

Selecting k edges from K_m can also be viewed as generating a random graph with m vertices and exactly k edges. Within this random graph model, the event defining $\mathbb{P}(m, k)$ corresponds to the graph having maximum degree at most one. Such graphs consist of disjoint edges together with isolated vertices. In other words, they are unions of matchings and isolated points. Thus, the independence probability describes how likely it is that a random sparse graph contains no vertex collisions.

This viewpoint connects the problem to classical questions in random graph theory, where researchers study the probability that a random graph avoids certain structures such as cycles or triangles. Here, the forbidden structure is simply two edges sharing a vertex.

Construct the complete graph K_m with vertices $1, \ldots, m$, where each edge corresponds to a distance d_{ij} in the distance matrix. Selecting k distances uniformly at random is equivalent to selecting k edges in K_m. Independence occurs precisely when no two edges share a vertex; i.e., the k edges form a matching.

Each distance corresponds to two indices (vertices). Independence requires these $2k$ indices to be distinct. Hence, selecting k independent distances is equivalent to selecting a k-edge matching uniformly at random:

$$\mathbb{P}(m, k) = \mathbb{P}(\text{random } k\text{-edge subgraph of } K_m \text{ forms a matching}).$$

Lemma 2.12.1 *The number of matchings of size k in K_m is*

$$N_{m,k} = \frac{m!}{(m-2k)!2^k k!}.$$

Proof: The number of ways to choose $2k$ vertices to participate is $\binom{m}{2k} = \frac{m!}{(2k)!(m-2k)!}$ ways. The number of ways to partition these $2k$ vertices into k unordered pairs is $\frac{(2k)!}{2^k k!}$ ways. Upon multiplication, we have

$$N_{m,k} = \binom{m}{2k}\frac{(2k)!}{2^k k!} = \frac{m!}{(m-2k)!2^k k!}.$$

□ The total number of k-edge subsets of K_m is $\binom{\binom{m}{2}}{k}$. Therefore the probability that a random k-edge subset forms a matching is

$$\mathbb{P}(m, k) = \frac{N_{m,k}}{\binom{\binom{m}{2}}{k}} = \frac{\frac{m!}{(m-2k)!2^k k!}}{\binom{\binom{m}{2}}{k}}. \tag{2.42}$$

For $m = 6$, $k = 2$, $N_{6,2} = 45$. Total 2-subsets: 105, giving $\mathbb{P}(6, 2) = 0.4286$.

2.12.2 Sequential Sampling Interpretation

Lemma 2.12.2 *Suppose that the distances are selected sequentially. The probability that a random k-edge subset forms a matching is*

$$\mathbb{P}(m, k) = \prod_{j=0}^{k-1} \frac{(m-2j)(m-2j-1)}{m(m-1)-2j}.$$

Proof: Suppose we choose the distances sequentially. After j independent distances are selected, $2j$ indices are used. The next distance must avoid these indices. The number of valid pairs is $\binom{m-2j}{2}$. The total remaining distances in the pool is $\binom{m}{2} - j$. Therefore, the conditional probability is

$$\frac{\binom{m-2j}{2}}{\binom{m}{2} - j}.$$

The combined probability is

$$\mathbb{P}(m,k) = \prod_{j=0}^{k-1} \frac{\binom{m-2j}{2}}{\binom{m}{2} - j} = \prod_{j=0}^{k-1} \frac{(m-2j)(m-2j-1)}{m(m-1)-2j}. \tag{2.43}$$

□ To see $\mathbb{P}(m,k)$ given in (2.42) and (2.43) provides the same result, consider $m = 6$, $k = 2$. Sequential probabilities produce $j = 0 : 1$, $j = 1$, give $\frac{\binom{4}{2}}{14} = 0.4286$.

2.12.3 Connection to the Occupancy Problem

In the occupancy problem (Kolchin [170]), balls are randomly placed into bins, and one studies the distribution of the number of balls in each bin. This model is widely used in probability theory, computer science, and combinatorics.

The matching probability $\mathbb{P}(m,k)$ can be interpreted in this framework. Each selected edge (i,j) contributes two "occupancies" to bins labeled i and j. Thus, selecting k edges corresponds to placing $2k$ tokens into m bins. Independence requires that each bin contain at most one token. Therefore, $\mathbb{P}(m,k)$ equals the probability that the occupancy vector contains only zeros and ones. Specifically,

$$\mathbb{P}(m,k) = \mathbb{P}(N_i \leq 1 \text{ for all } i)$$

where N_i denotes the number of tokens in bin i. This interpretation reveals a direct connection between matching probabilities and classical occupancy distributions.

2.12.4 Connection to the Coupon Collector Problem

The coupon collector problem studies the number of samples required to observe all categories in a finite set. In its classical form, coupons are drawn sequentially from a set of m possible types.

The connection with $\mathbb{P}(m,k)$ arises when we consider the early stage of the collection process. Suppose $2k$ coupons are drawn. The event that all coupons are distinct corresponds to observing $2k$ different types. This probability is exactly the same combinatorial quantity that appears in the independence problem. In both settings the relevant event is that a sequence of selections produces no repeated labels. Consequently,

$$\mathbb{P}(m,k) = \mathbb{P}(\text{first } 2k \text{ coupon draws contain no duplicates}).$$

This interpretation links the matching probability to the early growth phase of the coupon collector process.

2.12.5 Expected Number of Independent Edges

Let Y be the number of independent edges selected sequentially from K_m,

$$Y = \sum_{i=1}^{k} \mathbf{1}_i.$$

Here, $\mathbf{1}_i$ is an indicator variable. It equals 1 if the i-th edge in the sequential selection does not share any vertex with previously selected edges, and 0 otherwise. The random variable Y counts how many of the first k selected edges are independent.

Lemma 2.12.3 *By linearity of expectation we find*

$$\mathbb{E}[Y] = \sum_{i=1}^{k} \mathbb{E}\left[\mathbf{1}_i\right] = \sum_{j=0}^{k-1} \frac{(m-2j)(m-2j-1)}{m(m-1)-2j}.$$

Proof: Consider the sequential process below.

- For the first edge ($i = 1$), all $m(m-1)/2$ edges are available, so the probability of independence is 1.
- For the j-th edge ($j \geq 2$), $2(j-1)$ vertices are already used. The number of edges avoiding these vertices is

$$\binom{m-2(j-1)}{2} = \frac{(m-2j+2)(m-2j+1)}{2}.$$

  The total remaining edges in the pool: $\binom{m}{2} - (j-1) = \frac{m(m-1)-2(j-1)}{2}$.

Hence, the expected number of independent edges is

$$\mathbb{E}[Y] = \sum_{j=0}^{k-1} \frac{(m-2j)(m-2j-1)}{m(m-1)-2j}.$$

□

2.12.6 Distribution

For small k relative to m, covariances are weak, and Y can be approximated as

$$Y \approx \text{Binomial}\Big(k, p_{\text{indep}}\Big), \quad p_{\text{indep}} = \frac{\mathbb{E}[Y]}{k}.$$

Lemma 2.12.4 *The exact variance of Y is*

$$\mathbb{V}ar(Y) = \sum_{i=1}^{k} p_i(1-p_i) + 2 \sum_{1 \leq i < j \leq k} \Big(\mathbb{E}[\mathbf{1}_i \mathbf{1}_j] - p_i p_j\Big),$$

where $\mathbb{E}[\mathbf{1}_i \mathbf{1}_j]$ is defined in (2.44).

Proof: Variance of Y is

$$\mathbb{V}ar(Y) = \sum_{i=1}^{k} \mathbb{V}ar(\mathbf{1}_i) + 2 \sum_{1 \leq i < j \leq k} \text{Cov}(\mathbf{1}_i, \mathbf{1}_j).$$

The variance of each indicator is

$$\mathbb{V}ar(\mathbf{1}_i) = p_i(1-p_i), \quad p_i = \mathbb{E}[\mathbf{1}_i] = \frac{(m-2(i-1))(m-2(i-1)-1)}{m(m-1)-2(i-1)}.$$

For $i < j$, define the covariance,

$$\text{Cov}(\mathbf{1}_i, \mathbf{1}_j) = \mathbb{E}[\mathbf{1}_i \mathbf{1}_j] - \mathbb{E}[\mathbf{1}_i]\mathbb{E}[\mathbf{1}_j].$$

The joint probability $\mathbb{E}[\mathbf{1}_i \mathbf{1}_j]$ is the probability that both edges i and j are independent of all previous selections:

$$\mathbb{E}[\mathbf{1}_i \mathbf{1}_j] = \frac{\binom{m-2(i-1)}{2} \cdot \binom{m-2(j-1)}{2}}{\left(\binom{m}{2} - (i-1)\right)\left(\binom{m}{2} - (j-1)\right)}, \tag{2.44}$$

where the numerator counts edges avoiding the previously used vertices for i and j, and the denominator is the total remaining edges at each step. Thus, the exact variance is

$$\mathbb{V}ar(Y) = \sum_{i=1}^{k} p_i(1-p_i) + 2 \sum_{1 \le i < j \le k} \Big(\mathbb{E}[\mathbf{1}_i \mathbf{1}_j] - p_i p_j\Big).$$

□

For $k \ll m$, covariances are small, and $\mathbb{V}ar(Y) \approx \sum_{i=1}^{k} p_i(1-p_i)$, recovering the simplified variance formula used in many asymptotic and numerical calculations.

An algorithm to efficiently select a subset of k independent distance from m vertices follows.

1. Randomly permute the m vertices.
2. Partition the vertices into consecutive pairs: $(1,2), (3,4), \ldots$. This ensures that each pair is disjoint and thus forms an independent edge. We line up the vertices in a random order and treat adjacent pairs as edges.
3. Select the first k pairs for a uniform k-edge matching.

Alternatively, one may select edges randomly and reject any edge that shares a vertex with previously chosen edges.

2.12.7 Asymptotic Analysis

Lemma 2.12.5 *Suppose k is fixed while $m \to \infty$. Each term in the sequential product in* (2.43) *is*

$$\frac{(m-2j)(m-2j-1)}{m(m-1)-2j} = 1 - \frac{4j}{m} + O\left(\frac{j^2}{m^2}\right), \quad j = 0, \ldots, k-1.$$

Proof: Expand numerator and denominator:

$$(m-2j)(m-2j-1) = m^2 - 4jm + O(j^2),$$

$$m(m-1) - 2j = m^2 - m - 2j = m^2\left(1 - \frac{1}{m} - \frac{2j}{m^2}\right) = m^2 - 2jm + O(m),$$

so the ratio is

$$1 - \frac{4j}{m} + O\left(\frac{j^2}{m^2}\right).$$

Taking logarithms and summing:

$$\log \mathbb{P}(m,k) = \sum_{j=0}^{k-1} \log\left(1 - \frac{4j}{m} + O(j^2/m^2)\right) \approx -\sum_{j=0}^{k-1} \frac{4j}{m} = -\frac{2k(k-1)}{m}.$$

Hence, as $m \to \infty$ for fixed k:

$$\mathbb{P}(m, k) \approx \exp\Big(- \frac{2k(k-1)}{m}\Big) \to 1.$$

For intermediate scaling suppose $k \sim c\sqrt{m}$, then

$$\log \mathbb{P}(m, k) \approx -2c^2 \implies \mathbb{P}(m, k) \approx e^{-2c^2}.$$

□ When k is large proportional to m independence probability decays quickly. Hence, we have three regimes.
Small k: If $k \ll \sqrt{m}$, $\mathbb{P}(m, k) \approx 1$.
Intermediate regime: If $k \sim c\sqrt{m}$, then $\mathbb{P}(m, k) \approx e^{-2c^2}$.
Large k: If k grows proportionally with m, the exponential approximation breaks down, but the exact product shows $\mathbb{P}(m, k) \to 0$ rapidly as $m \to \infty$. Hence, when m is large pairwise distances are asymptotically independent. If fact, for fixed $k < m/2$ as $m \to \infty$, k-tuples of distances are asymptotically independent. The largest possible number of mutually independent distances is $k_{\max} = \lfloor \frac{m}{2} \rfloor$. Near this limit almost every subset of distances shares indices, so the probability of independence is extremely small.

Figure 2.1 is a plot of $\mathbb{P}(m, k)$ as a function of m for several values of k. Top curve is for $k = 1$ while the bottom curve is for $k = 5$.

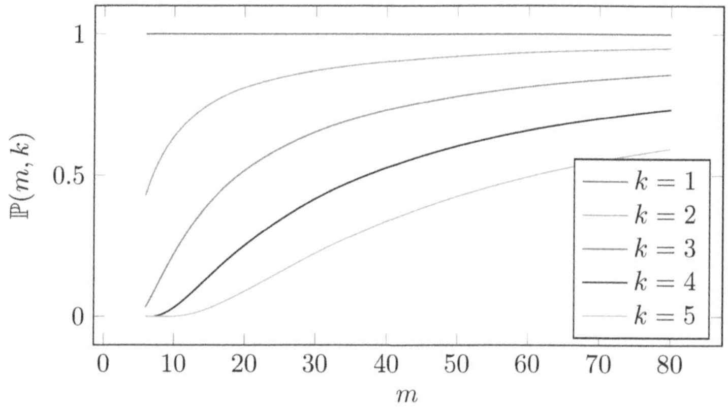

FIGURE 2.1: Plot of $\mathbb{P}(m, k) = \prod_{j=0}^{k-1} \frac{(m-2j)(m-2j-1)}{m(m-1)-2j}$ as a function of m for $k = 1$ (top) through $k = 5$ (bottom).

2.12.8 U-Statistics Implications

Many statistics constructed from independent observations involve pairwise quantities such as distances, similarities, or kernel evaluations. Examples include energy statistics, kernel methods, graph-based clustering, and distance-based hypothesis tests. A common structure is the U-statistic (Serfling, [303])

$$U_m = \frac{1}{\binom{m}{2}} \sum_{i<j} h(X_i, X_j)$$

where $X_1, \ldots, X_m$ are independent observations and h is a symmetric function.

Although the observations are independent, the individual terms $h(X_i, X_j)$ are not mutually independent. Dependence arises whenever two terms share an index. For example, $h(X_i, X_j)$, and $h(X_i, X_k)$ share the variable X_i and are therefore dependent. However, $h(X_i, X_j)$ and $h(X_k, X_\ell)$ are independent whenever the index sets are disjoint.

The probability $\mathbb{P}(m, k)$ computed earlier measures how likely it is that a randomly chosen collection of k kernel terms in a U-statistic is mutually independent. For large m and moderate k, most subsets of terms behave nearly independently.

Hoeffding Decomposition

Let the kernel have the decomposition $h(x, y) = \theta + \psi(x) + \psi(y) + \phi(x, y)$, where $\mathbb{E}[\psi(X)] = 0$, and $\mathbb{E}[\phi(X, Y)|X] = 0$. Then, it follows from Hoeffding [150] that

$$U_m = \theta + \frac{2}{m} \sum_{i=1}^{m} \psi(X_i) + R_m,$$

where

$$R_m = \frac{1}{\binom{m}{2}} \sum_{i<j} \phi(X_i, X_j).$$

The term R_m is called the degenerate remainder. It is of interest co compute the variance of R_m. Let $\sigma_\phi^2 = \mathbb{V}ar(\phi(X_1, X_2))$. Then, using degeneracy of ϕ, we know $\mathbb{E}[\phi(X_i, X_j)] = 0$. Therefore,

$$\mathbb{E}[R_m] = \frac{1}{\binom{m}{2}} \sum_{i<j} \mathbb{E}[\phi(X_i, X_j)] = 0,$$

$$\mathbb{V}ar(R_m) = \frac{1}{\binom{m}{2}^2} \sum_{i<j} \sum_{k<\ell} \mathbb{C}ov(\phi_{ij}, \phi_{k\ell}).$$

The covariance depends on whether indices overlap. For independent pairs, if $\{i, j\} \cap \{k, \ell\} = \emptyset$ then $\mathbb{C}ov(\phi_{ij}, \phi_{k\ell}) = 0$. The proportion of such independent terms is approximately $\mathbb{P}(m, 2)$. For overlapping pairs, if pairs share one index, for example $(i, j) = (1, 2)$, and $(k, \ell) = (1, 3)$, then the covariance is non-zero. If $(i, j) = (k, \ell)$, then $\mathbb{C}ov(\phi_{ij}, \phi_{ij}) = \sigma_\phi^2$. There are $\binom{m}{2}$ such terms. These overlapping configurations dominate the variance of R_m.

When m is large, R_m is a sum of many weakly dependent terms. Central limit theorems for U-statistics imply

$$\sqrt{m}\,(U_m - \theta) \;\Rightarrow\; N(0, \sigma^2),$$

where $\sigma^2 = 4\,\mathrm{Var}(\psi(X_1))$. Because the linear part $\frac{2}{m} \sum \psi(X_i)$ dominates, the remainder R_m contributes only a smaller variance term of order $O(m^{-1})$. Thus the distribution of U_m is approximately normal.

Another perspective is to consider the number of overlapping pairs. Define

$$Z = \text{number of index overlaps among selected kernel terms.}$$

For large m the probability of overlap is small, and the number of overlaps behaves approximately like a Poisson random variable, $Z \approx \text{Poisson}(\lambda)$. The parameter λ depends on the expected number of collisions, $\lambda \approx k(k-1)\frac{4}{m}$.

To obtain λ suppose we select k kernel terms from the $\binom{m}{2}$ possible pairs. Fix a pair (i, j). A second pair overlaps if it contains i or j. The number of such pairs is $2(m-2)$. Thus,

$$\mathbb{P}(\text{overlap}) \approx \frac{2(m-2)}{\binom{m}{2}} \approx \frac{4}{m}.$$

Among k selected pairs there are $\binom{k}{2}$ comparisons. Therefore the expected number of overlaps is

$$\lambda \approx \binom{k}{2}\frac{4}{m} \approx k(k-1)\frac{4}{m}.$$

Hence, $Z \sim \text{Poisson}(\lambda)$, where $\lambda = k(k-1)\frac{4}{m}$. Under this approximation the contribution of dependent terms to R_m can be analyzed using Poisson limit theory.

2.13 Asymptotic Distribution of IPDs

Li [192] investigates the asymptotic distribution of IPDs as $d \to \infty$. Suppose $\mathbf{X}_1, \ldots, \mathbf{X}_m \sim F$ with mean $\boldsymbol{\mu}_x$ and covariance matrix $\boldsymbol{\Sigma}_x$. Let the random variable $d^2_{(x)}$ denote any squared IPD from the $\mathbb{X}$ sample. In particular, she shows that, as $d \to \infty$ and for fixed m,

$$(d^2_{(x)} - \mathbb{E}(d^2_{(x)}))/(\mathbb{V}\text{ar}(\mathrm{d}^2_{(\mathrm{x})}))^{1/2} \xrightarrow{\mathrm{d}} \mathbb{N}(0,1), \tag{2.45}$$

where $\mathbb{E}(d^2_{(x)}) = 2tr\left(\boldsymbol{\Sigma}_x\right)$ and $\mathbb{V}\text{ar}\left(\mathrm{d}^2_{(\mathrm{x})}\right)$ depend on F. Let $t = m(m-1)/2$ and

$$\mathbf{Q} = (d^2_{(x)12}, d^2_{(x)13}, \ldots, d^2_{(x)(m-1)m})' \tag{2.46}$$

denote the m-dimensional vector of all IPDs. It follows that $\mathbb{E}(\mathbf{Q}) = 2tr\left(\boldsymbol{\Sigma}_x\right)\mathbf{1}_t$ where $\mathbf{1}_t$ is a t-dimensional vector of 1s. When F is the multivariate normal distribution, it follows that covariance the $\boldsymbol{\Sigma}_Q$ is a $t \times t$ matrix with the main diagonal $8tr\left(\boldsymbol{\Sigma}^2_x\right)$. The off-diagonal elements of $\boldsymbol{\Sigma}_Q$ are $2tr\left(\boldsymbol{\Sigma}^2_x\right)$ if the two IPDs have an index in common or zero, otherwise. Hence,

$$\boldsymbol{\Sigma}_Q^{-1/2}\left(\mathbf{Q} - \mathbb{E}(\mathbf{Q})\right) \xrightarrow{d} \mathbb{N}(0,1), \tag{2.47}$$

as $d \to \infty$ for fixed m. Li [192] shows that the asymptotic distribution of IPDs in the high-dimension, low-sample- size setting converges in distribution to multivariate normal under some regularity conditions.

Example 2.13.1 *Let $\mathbf{X}_1, \ldots, \mathbf{X}_m \in \mathbb{R}^d$ be i.i.d. from distribution F_1, and $\mathbf{Y}_1, \ldots, \mathbf{Y}_n \in \mathbb{R}^d$ be i.i.d. from distribution F_2, where F_1 and F_2 may differ in their mean and variance structures. Define the squared Euclidean distances*

$$d^2_{ij}(x) = \|\mathbf{X}_i - \mathbf{X}_j\|^2, \qquad d^2_{ij}(y) = \|\mathbf{Y}_i - \mathbf{Y}_j\|^2, \qquad d^2_{ij}(xy) = \|\mathbf{X}_i - \mathbf{Y}_j\|^2.$$

Assume that the vector components are independent with bounded fourth moments, and let the dimension $d \to \infty$ while the sample sizes m and n remain fixed. Under mild moment and independence assumptions, a central limit theorem yields

$$\frac{1}{d}d^2_{(xy)ij} \xrightarrow{d} \mathbb{N}\left(\mu_{xy}, \frac{\sigma^2_{xy}}{d}\right).$$

Analogous results hold for $\frac{1}{d}d^2_{ij}(x)$ and $\frac{1}{d}d^2_{ij}(y)$. If $\mathbf{X} \sim \mathbb{N}(\boldsymbol{\mu}_1, \sigma^2\mathbf{I}_d)$ and $\mathbf{Y} \sim \mathbb{N}(\boldsymbol{\mu}_2, \sigma^2\mathbf{I}_d)$, independently, then

$$\mu_{xy} = 2\sigma^2 + \frac{1}{d}\|\boldsymbol{\mu}_1 - \boldsymbol{\mu}_2\|^2, \qquad \sigma^2_{xy} = 8\sigma^4.$$

The same asymptotic variance constant applies to within-sample squared distances $d^2_{ij}(x)$ and $d^2_{ij}(y)$ under the same normal assumptions. An asymptotic normality result for $d_{ij}(x) = \|\mathbf{X}_i - \mathbf{X}_j\|$ can be obtained via the delta method, but with a different limiting variance.

Example 2.13.2 *Identical Normal Populations: Let $\mathbf{X}_i, \mathbf{Y}_j \sim \mathbb{N}(0, \mathbf{I}_d)$ for all i, j. Then,*

$$\frac{1}{d}\|\mathbf{X}_i - \mathbf{Y}_j\|^2 = \frac{1}{d}\sum_{k=1}^{d}(\mathbf{X}_{ik} - \mathbf{Y}_{jk})^2 \xrightarrow{d} \mathbb{N}(2, \frac{4}{d}).$$

Even though each pairwise distance is random, as $d \to \infty$, they concentrate around 2 with vanishing variance.

Example 2.13.3 *Shift in Mean: Suppose $\mathbf{X}_i \sim \mathbb{N}(0, \mathbf{I}_d)$ and $\mathbf{Y}_j \sim \mathbb{N}(\mu \cdot \mathbf{1}_d, \mathbf{I}_d)$ with $\mu \neq 0$.*

$$\mathbb{E}\|\mathbf{X}_i - \mathbf{Y}_j\|^2 = \mathbb{E}\|\mathbf{X}_i\|^2 + \|\mathbf{Y}_j\|^2 - 2\mathbb{E}[\mathbf{X}_i'\mathbf{Y}_j] = 2d + d\mu^2,$$
$$\frac{1}{d}\|\mathbf{X}_i - \mathbf{Y}_j\|^2 \xrightarrow{d} \mathbb{N}(2 + \mu^2, \frac{4}{d}).$$

The additional mean shift μ increases the distance location but not the variance order.

Example 2.13.4 *Heteroscedastic Variances: Let $\mathbf{X}_i \sim \mathbb{N}(0, \mathbf{I}_d)$ and $\mathbf{Y}_j \sim \mathbb{N}(0, \sigma^2\mathbf{I}_d)$ with $\sigma^2 \neq 1$. Then,*

$$\mathbb{E}\|\mathbf{X}_i - \mathbf{Y}_j\|^2 = d(1 + \sigma^2),$$
$$\mathbb{V}ar(\|\mathbf{X}_i - \mathbf{Y}_j\|^2) = 2d(1 + \sigma^2)^2,$$
$$\Rightarrow \frac{1}{d}\|\mathbf{X}_i - \mathbf{Y}_j\|^2 \xrightarrow{d} \mathbb{N}(1 + \sigma^2, \frac{2(1 + \sigma^2)^2}{d}).$$

Differences in scale manifest in the limiting mean and variance.

Notice that the above results do not depend on m and n. Li's [192] theoretical results enable the use of normal approximation to develop asymptotic tests for equality of distributions using IPDs, in HDLSS settings. Her theory supports construction of analytic two-sample tests without the need for resampling.

Example 2.13.5 *Let $d_{(x)} = \|\mathbf{X}_1 - \mathbf{X}_2\|$, $d_{(y)} = \|\mathbf{Y}_1 - \mathbf{Y}_1\|$ and $d_{(xy)} = \|\mathbf{X}_1 - \mathbf{Y}_2\|$, respectively. Let $F_1 \sim \mathbb{N}(\boldsymbol{\mu}_x, \sigma_x^2\mathbf{I}_d)$ and $F_2 \sim \mathbb{N}(\boldsymbol{\mu}_y, \sigma_y^2\mathbf{I}_d)$, respectively. Let $\boldsymbol{\Delta} = \boldsymbol{\mu}_x - \boldsymbol{\mu}_y$ and $v^2 = \boldsymbol{\Delta}'\boldsymbol{\Delta}/d$. It is not difficult to show that*

$$\frac{1}{\sqrt{d}}\left(d_{(x)}, D_{(y)}, d_{(xy)}\right) \to \left(\sqrt{2}\sigma_x, \sqrt{2}\sigma_y, \sqrt{\sigma_x^2 + \sigma_y^2 + v^2}\right),$$

in probability as $d \to \infty$ and they have a joint trivariate normal distribution.

Example 2.13.6 *We examine normality of interpoint Euclidean distances (IPDs) for i.i.d. samples $\mathbf{X}_1, \ldots, \mathbf{X}_m \in \mathbb{R}^d$, with $d = 2^i$ for $i = 2, \ldots, 10$. Coordinates are i.i.d. from $\mathbb{N}(0, 1)$, $U(0, 1)$, and $\chi^2(1)$ distributions. A random subsample of 200 ordered IPDs is tested for normality with Shapiro–Wilk test. We repeat the experiment 30 times and report the average p-value per (d, m) and distribution. Tables 2.1 and 2.2 show that for all values of d and m normality is achieved. Table 2.3 shows d must be at least in the hundreds for normality to take effect.*

TABLE 2.1: Average Shapiro–Wilk p-values (approx.), distribution: $\mathbb{N}(0, 1)$.

Dimension d	$m = 30$	$m = 50$	$m = 75$
4	0.1185	0.2231	0.2061
8	0.2392	0.0690	0.3110
16	0.1000	0.3561	0.4272
32	0.1312	0.4171	0.2870
64	0.1056	0.1924	0.5687
128	0.3411	0.2462	0.4844
256	0.3656	0.4050	0.4542
512	0.2918	0.4696	0.4909
1024	0.3001	0.6254	0.5370

TABLE 2.2: Average Shapiro–Wilk p-values (approx.), distribution: U(0,1).

Dimension d	$m = 30$	$m = 50$	$m = 75$
4	0.0739	0.3517	0.5020
8	0.4763	0.2239	0.3848
16	0.3027	0.5077	0.4606
32	0.2661	0.4953	0.6940
64	0.3389	0.3382	0.6119
128	0.1278	0.2730	0.3376
256	0.2227	0.3820	0.4737
512	0.3201	0.3061	0.3487
1024	0.1101	0.4087	0.3290

TABLE 2.3: Average Shapiro–Wilk p-values (approx.), distribution: $\chi^2(1)$.

Dimension d	$m = 30$	$m = 50$	$m = 75$
4	0.0001	0.0000	0.0000
8	0.0001	0.0002	0.0006
16	0.0662	0.0096	0.0320
32	0.0734	0.0225	0.0876
64	0.0068	0.1407	0.1119
128	0.4339	0.1722	0.0873
256	0.2460	0.2245	0.2264
512	0.1260	0.1482	0.2349
1024	0.1064	0.1220	0.5503

2.14 Summary

This chapter develops a rigorous and unified theory of Euclidean IPDs and establishes their foundational role in modern multivariate and high-dimensional statistics. By reducing multivariate observations to scalar distances, IPDs provide a powerful alternative to covariance-based methods, which are often unstable or infeasible in high-dimensional regimes. The chapter systematically introduces the mathematical framework underpinning IPDs, beginning with L_s norms and the Frobenius norm, and formalizes notation for pairwise distances in random samples. Through the use of quadratic forms and moment calculations, explicit expressions for the mean, variance, and higher-order properties of IPDs are derived, laying the groundwork for precise distributional analysis.

A central contribution of the chapter is the detailed characterization of the distributional behavior of IPDs under multivariate normal models. Exact and asymptotic distributions are obtained for one-sample and two-sample IPDs, norms of normal vectors, distances from fixed vectors, and IPDs arising from mixture distributions. These results illuminate the probabilistic structure of distances in high dimensions and clarify conditions under which pairwise independence and asymptotic normality emerge. The chapter highlights how tools such as Jensen's inequality and the multivariate delta method facilitate tractable asymptotic approximations, even as dimensionality grows.

Beyond theory, the chapter underscores the broad applicability of IPDs across statistical inference and data analysis. IPD-based methodologies naturally support distributional testing, clustering, classification, graph-based inference, and change-point detection, while offering robustness and computational efficiency. By connecting exact distribution theory with asymptotic results, the chapter provides a principled foundation for distance-based methods used in contemporary statistical practice, particularly in settings where traditional multivariate techniques are hindered by the curse of dimensionality.

2.15 Proofs

Proof of Theorem 2.9.1:

Since $\mathbf{X}_i - \mathbf{X}_j \sim \mathcal{N}_d(\mathbf{0}, 2\mathbf{\Sigma})$, we may use the spectral decomposition

$$2\mathbf{\Sigma} = \mathbf{P}\mathbf{\Lambda}\mathbf{P}',$$

where $\mathbf{P}$ is orthogonal and $\mathbf{\Lambda} = \text{diag}(\lambda_1, \ldots, \lambda_d)$.

Define

$$\mathbf{U} = \mathbf{\Lambda}^{-1/2}\mathbf{P}'(\mathbf{X}_i - \mathbf{X}_j).$$

Then $\mathbf{U} \sim \mathcal{N}_d(\mathbf{0}, \mathbf{I}_d)$ and

$$d^2(\mathbf{X}_i, \mathbf{X}_j) = (\mathbf{X}_i - \mathbf{X}_j)'(\mathbf{X}_i - \mathbf{X}_j) = \mathbf{U}'\mathbf{\Lambda}\mathbf{U} = \sum_{t=1}^{d} \lambda_t U_t^2,$$

where $U_t \overset{i.i.d.}{\sim} \mathcal{N}(0, 1)$.

Since $\mathbb{E}[U_t^2] = 1$ and $\mathbb{V}ar(U_t^2) = 2$, we obtain

$$\mathbb{E}[d^2(\mathbf{X}_i, \mathbf{X}_j)] = \sum_{t=1}^{d} \lambda_t = \text{tr}(2\mathbf{\Sigma}) = 2\,\text{tr}(\mathbf{\Sigma}),$$

and

$$\mathbb{V}ar[d^2(\mathbf{X}_i, \mathbf{X}_j)] = \sum_{t=1}^{d} \lambda_t^2 \mathbb{V}ar(U_t^2) = 2\sum_{t=1}^{d} \lambda_t^2 = 2\,\text{tr}((2\mathbf{\Sigma})^2) = 8\,\text{tr}(\mathbf{\Sigma}^2).$$

To compute the covariance, write

$$d^2(\mathbf{X}_i,\mathbf{X}_j)=\sum_{t=1}^{d}(X_{it}-X_{jt})^2,\quad d^2(\mathbf{X}_i,\mathbf{X}_l)=\sum_{s=1}^{d}(X_{is}-X_{ls})^2.$$

Thus

$$\mathbb{C}ov=\sum_{t=1}^{d}\sum_{s=1}^{d}\mathbb{C}ov\big((X_{it}-X_{jt})^2,(X_{is}-X_{ls})^2\big).$$

For jointly normal zero-mean variables Y_1, Y_2,

$$\mathbb{C}ov(Y_1^2,Y_2^2)=2\,\mathbb{C}ov(Y_1,Y_2)^2.$$

Here

$$\mathbb{C}ov(X_{it}-X_{jt},X_{is}-X_{ls})=\sigma_{ts}.$$

Therefore

$$\mathbb{C}ov\big((X_{it}-X_{jt})^2,(X_{is}-X_{ls})^2\big)=2\sigma_{ts}^2.$$

Summing over t, s,

$$\mathbb{C}ov=2\sum_{t=1}^{d}\sum_{s=1}^{d}\sigma_{ts}^2=2\,\mathrm{tr}(\mathbf{\Sigma}^2).$$

Finally,

$$\mathbb{C}or=\frac{2\,\mathrm{tr}(\mathbf{\Sigma}^2)}{\sqrt{8\,\mathrm{tr}(\mathbf{\Sigma}^2)}\sqrt{8\,\mathrm{tr}(\mathbf{\Sigma}^2)}}=\frac{1}{4}.$$

Proof of Theorem 2.9.6:

The probability that $\mathbf{X}$ and $\mathbf{X}'$ are drawn from the same distribution is $\sum_{i=1}^{k}\alpha_i^2$ and the probability that they are drawn from $\mathbb{N}_d(\boldsymbol{\mu}_i,\mathbf{\Sigma}_i)$ and $\mathbb{N}_d(\boldsymbol{\mu}_j,\mathbf{\Sigma}_j)$ is $\alpha_i\alpha_j$. Hence, we use results for quadratic form to find

$$d_{(x)}^2=\begin{cases}\sum_{t=1}^{d}\lambda_{it}U_{it}^2, & \text{with probability } \alpha_i^2;\\ \sum_{t=1}^{d}\xi_{ijt}\left(W_{ijt}+b_{ijt}\right)^2, & \text{with probability } 2\alpha_i\alpha_j,\end{cases}\tag{2.48}$$

where all the terms of $\mathbf{U}_i$, $\mathbf{W}_{ij}$, and $\mathbf{b}_{ij}$ are defined in the theorem. One gets the expression in (2.34) by applying identity (2.16) to the cases above. We apply identity (2.17) to find the expected value of the IPD,

$$\begin{aligned}E\left[d_{(x)}^2\right]&=\sum_{i=1}^{k}\alpha_i^2\mathbb{E}\left(d_{(x)}^2\mid \mathbf{X}\text{ and }\mathbf{X}^*\text{ are drawn from }\mathbb{N}_d(\boldsymbol{\mu}_i,\mathbf{\Sigma}_i)\right)\\&+\sum_{i=1}^{k-1}\sum_{j=i+1}^{k}2\alpha_i\alpha_j\mathbb{E}\left(d_{(x)}^2\mid \mathbf{X}\text{ and }\mathbf{X}^*\text{ are drawn from }\mathbb{N}_d(\boldsymbol{\mu}_i,\mathbf{\Sigma}_i)\text{ and }\mathbb{N}_d(\boldsymbol{\mu}_j,\mathbf{\Sigma}_j)\right)\\&=2\sum_{i=1}^{k}\alpha_i^2\cdot tr\left(\mathbf{\Sigma}_i\right)+2\sum_{i=1}^{k-1}\sum_{j=1+1}^{k}\alpha_i\alpha_j\left[tr\left(\mathbf{\Sigma}_i+\mathbf{\Sigma}_j\right)+\left(\boldsymbol{\mu}_i-\boldsymbol{\mu}_j\right)'\left(\boldsymbol{\mu}_i-\boldsymbol{\mu}_j\right)\right].\end{aligned}\tag{2.49}$$

To obtain the variance, we apply (2.18).

Proof of Lemma 2.10.1:

Because the distribution is rotationally invariant, we may fix one vector. Condition on $\mathbf{X}_1$. By spherical symmetry, we may assume

$$\mathbf{X}_1 = e_1 = (1, 0, \ldots, 0).$$

Thus, $\cos\Theta = \mathbf{X}_1 \cdot \mathbf{X}_2 = X_{2,1}$, the first coordinate of $\mathbf{X}_2$. Therefore the distribution of Θ is determined by the distribution of the first coordinate of a uniformly distributed point on S^{d-1}.

In spherical coordinates, the surface measure on S^{d-1} can be written as

$$d\sigma = (\sin\theta)^{d-2}\, d\theta\, d\omega_{d-2},$$

where θ is the polar angle with respect to e_1, and $d\omega_{d-2}$ is the surface measure on S^{d-2}.

Hence, the proportion of surface area corresponding to angles between θ and $\theta + d\theta$ is proportional to $(\sin\theta)^{d-2} d\theta$. Thus the density of Θ must be

$$f_\Theta(\theta) = C_d(\sin\theta)^{d-2},$$

for some normalizing constant C_d. We require $\int_0^\pi f_\Theta(\theta)\, d\theta = 1$. Hence,

$$C_d^{-1} = \int_0^\pi (\sin\theta)^{d-2}\, d\theta.$$

We use the standard integral identity

$$\int_0^\pi (\sin\theta)^k\, d\theta = \frac{\sqrt{\pi}\,\Gamma\left(\frac{k+1}{2}\right)}{\Gamma\left(\frac{k}{2}+1\right)}.$$

Here $k = d - 2$, so

$$\int_0^\pi (\sin\theta)^{d-2}\, d\theta = \frac{\sqrt{\pi}\,\Gamma\left(\frac{d-1}{2}\right)}{\Gamma\left(\frac{d}{2}\right)}.$$

Therefore,

$$C_d = \frac{\Gamma(d/2)}{\sqrt{\pi}\,\Gamma((d-1)/2)}.$$

Thus,

$$f_\Theta(\theta) = \frac{\Gamma(d/2)}{\sqrt{\pi}\,\Gamma((d-1)/2)}(\sin\theta)^{d-2}, \quad 0 \le \theta \le \pi.$$

2.16 Exercises

Exercise 2.1 *Use Theorem 2.9.2 to show that*

$$\mathbb{E}\left[Q(\boldsymbol{X})\right] = tr(\boldsymbol{A\Sigma}) + \boldsymbol{\mu}'\boldsymbol{A\mu}, \tag{2.50}$$

$$\mathbb{V}\text{ar}\left[\mathrm{Q}\left(\boldsymbol{X}\right)\right] = 2\mathrm{tr}(\mathrm{A}\Sigma)^2 + 4\boldsymbol{\mu}'\mathrm{A}\boldsymbol{\Sigma}\mathrm{A}\boldsymbol{\mu}. \tag{2.51}$$

Obtain expressions for $\mathbb{E}\left[Q^3(\boldsymbol{X})\right]$ *and* $\mathbb{E}\left[Q^4(\boldsymbol{X})\right]$. *Hence, compute the skewness and kurtosis of* $Q\left(\boldsymbol{X}\right)$.

Exercise 2.2 *The Frobenius norm of a matrix* $\mathbf{A} \in \mathbb{R}^{n\times n}$ *is defined as* $\|\mathbf{A}\|_F = \sqrt{tr(\mathbf{A}'\mathbf{A})}$.

a) *Show that* $\|\mathbf{A}\|_F = \left(\sum_i \sum_j |\mathbf{A}_{ij}|^2\right)^{1/2}$.

b) *Show that* $\mathbf{U}$ *and* $\mathbf{V}$ *are orthogonal, then* $\|\mathbf{UA}\|_F = \|\mathbf{VA}\|_F = \|\mathbf{A}\|_F$.

c) *Show that* $\|\mathbf{A}\| = \sqrt{\lambda_1^2 + \ldots + \lambda_r^2}$, *where* $\lambda_1, \ldots, \lambda_r$ *are the eigenvalues of* $\mathbf{A}$.

d) *Show that* $\lambda_{max}(\mathbf{A}) \leq \|\mathbf{A}\|_F \leq \sqrt{r}\lambda_{max}(\mathbf{A})$. *In particular,* $\|\mathbf{A}x\| \leq \|\mathbf{A}\|_F\|x\|$ *for all* x.

Exercise 2.3 *Suppose* $\mathbf{X}_i$ *for* $i = 1, \ldots n$ *is a random sample from a d-dimensional distribution with mean* $\boldsymbol{\mu}$, *covariance* $\boldsymbol{\Sigma}$. *Show that the sample covariance* $\mathbf{S}$ *is given by*

$$\mathbf{S} = \frac{1}{n(n-1)} \sum_{1\leq i\leq j\leq n} (\mathbf{X}_i - \mathbf{X}_j)(\mathbf{X}_i - \mathbf{X}_j)'.$$

What do you obtain when $d = 1$?

Exercise 2.4 *Show that* $tr(\mathbf{S}) = \frac{1}{2}\bar{d}_x^2$. *Hence, the average sample IPD is twice the trace of the sample covariance matrix. The computational complexity of* $tr(\mathbf{S})$ *is* $O(nd)$. *This complexity is smaller than* $O\left(n^2\right)$ *unless* d *is larger than* n.

Exercise 2.5 *Orthant probabilities. Let* (V_1, V_2) *be bivariate normal vector with zero means and variances* σ_1^2, σ_2^2 *and covariance* $\rho\sigma_1\sigma_2$. *Show that*

$$\mathbb{P}(V_1 \leq 0, V_2 \leq 0) = \frac{1}{4} + \frac{1}{2\pi}\sin^{-1}\rho. \tag{2.52}$$

For the trivariate normal, show that

$$\mathbb{P}(V_1 \leq 0, V_2 \leq 0, V_3 \leq 0) = \frac{1}{8} + \frac{3}{4\pi}\sin^{-1}\rho. \tag{2.53}$$

Exercise 2.6 *Suppose* (X_1, Y_1) *and* (X_2, Y_2) *are independent and identically distributed bivariate random vectors from distribution* F. *Define Kendall's* $\tau = 2\mathbb{P}[(X_2 - X_1)(Y_2 - Y_1) > 0] - 1$. *Compute* τ *when* F *is* $\mathbb{N}_2(\mathbf{0}, \boldsymbol{\Sigma})$.

Exercise 2.7 *Suppose* $\mathbf{X} = (X_1, X_2, \ldots, X_d)'$ *has a* $\mathbb{N}_d(\mathbf{0}, \boldsymbol{\Sigma})$.

a) *Find the distribution of* $Y = \mathbf{C}'\mathbf{X}/\sqrt{\mathbf{C}'\boldsymbol{\Sigma}\mathbf{C}}$ *for* $\mathbf{C}$ *a non-zero constant vector in* $\mathbb{R}^d$.

b) *Let* $\mathbf{W} = \sum_{r=1}^t \mathbf{X}_r\mathbf{X}_r'$. *Find the distribution of* $Z = \frac{\mathbf{C}'\mathbf{W}\mathbf{C}}{\mathbf{C}'\boldsymbol{\Sigma}\mathbf{C}}$.

Exercise 2.8 *Suppose* $\mathbf{X} = (X_1, X_2)'$ *has a bivariate normal distribution with mean vector zero, unit variances and correlation* ρ. *What is the probability that both variables will be observed simultaneously at greater than their individual 95th percentile if the correlation is neglected and if the correlation is* $\rho = 0.9$?

Exercise 2.9 *Suppose* $\mathbf{X} = (X_1, X_2, X_3)$ *follows a trivariate normal distribution with mean vector* $\mathbf{0}$ *and covariance matrix*

$$\boldsymbol{\Sigma} = \begin{pmatrix} 1 & 1/2 & 0 \\ 1/2 & 1 & 1/2 \\ 0 & 1/2 & 1 \end{pmatrix}.$$

Find a) $\mathbb{P}(X_1 > \max(X_2, X_3))$ *and b)* $\mathbb{P}(X_1 < X_2 < X_3)$.

Exercise 2.10 *Suppose* $\mathbf{X}_i$ *are independent vectors from* $\mathbb{N}_d(\boldsymbol{\mu}_x, \sigma_x^2 \boldsymbol{I})$, *and* $\mathbf{Y}_i$ *are independent vectors from* $\mathbb{N}_d(\boldsymbol{\mu}_y, \sigma_y^2 \boldsymbol{I})$, *for* $i = 1, 2$ *and the two samples are independent. Use the squared IPDs within and across the two samples defined in* (2.4), (2.8), *and* (2.12) *to*

a) find the distribution of $d_{(x)}^2$ *and* $d_{(y)}^2$,

b) find the distribution of $d_{(xy)}^2$,

c) form $E^2 = 2\mathbb{E}\big(d_{(xy)}^2\big) - \mathbb{E}(d_{(x)}^2) - \mathbb{E}(d_{(y)}^2\big)$. *Under what conditions is* $E^2 \geq$*?*

Exercise 2.11 *Prove* $|\mathbf{C} + \mathbf{Y}\mathbf{Y}'| = (1 + \mathbf{Y}'\mathbf{C}^{-1}\mathbf{Y})|\mathbf{C}|$ *where* $\mathbf{Y} \in \mathbb{R}^d$ *and* $\mathbf{C}$ *is a non–singular* $d \times d$ *matrix.*

Exercise 2.12 *Find the eigenvalues of a* $d \times d$ *constant correlation matrix*

$$\boldsymbol{\Sigma} = (1 - \rho)\mathbf{I}_d + \rho \mathbf{1}\mathbf{1}'.$$

Exercise 2.13 *Suppose* $\mathbf{X}$ *is distributed as* $\mathbb{N}_d(\mathbf{0}, \sigma^2\mathbf{I})$ *and* $\mathbf{X}_1$ *and* $\mathbf{X}_2$ *be i.i.d. copies of* $\mathbf{X}$. *Find the distribution of*

a) the squared length of $\mathbf{X}$, $||\mathbf{X}||^2$.

b) the squared length of the IPD, $||\mathbf{X}_1 - \mathbf{X}_2||^2$.

c) the squared length of the interpoint sum, $||\mathbf{X}_1 + \mathbf{X}_2||^2$.

d) What can you say about the results for parts b and c?

Exercise 2.14 *Suppose* $\mathbf{R}_i \sim \mathbb{N}_d(\boldsymbol{\mu}, \boldsymbol{\Sigma}/n)$ *for* $i = 1, 2, 3$ *are independent. Define*

$$Q_{12} = \frac{n}{2}(\mathbf{R}_1 - \mathbf{R}_2)'\boldsymbol{\Sigma}^{-1}(\mathbf{R}_1 - \mathbf{R}_2),$$

$$Q_{23} = \frac{n}{2}(\mathbf{R}_2 - \mathbf{R}_3)'\boldsymbol{\Sigma}^{-1}(\mathbf{R}_2 - \mathbf{R}_3).$$

Find $\mathbb{C}orr(Q_{12}, Q_{23})$.

Exercise 2.15 *Suppose* $\mathbf{X}_1$ *and* $\mathbf{X}_2$ *are independent and identically distributed random vectors from a d dimensional contaminated normal distribution*

$$f(\mathbf{X}) = \alpha f_1(\mathbf{X}) + (1 - \alpha) f_2(\mathbf{X})$$

where $f_1(\mathbf{X}) \sim \mathbb{N}_d(\mathbf{0}, \mathbf{I})$, $f_2(\mathbf{X}) \sim \mathbb{N}_d(\mathbf{0}, \sigma^2\mathbf{I})$, *and* $0 < \alpha < 1$. *Find the distribution of squared IPDs* $d_{(x)}^2 = \|\mathbf{X}_1 - \mathbf{X}_2\|^2$, *its mean and variance.*

Exercise 2.16 *A random vector* $\mathbf{X}$ *has a d-variate normal mixture distribution when with probability p it is drawn from* $f_1(\mathbf{X}) \sim \mathbb{N}_m(\boldsymbol{\mu}_1, \boldsymbol{\Sigma}_1)$ *and with probability* $1 - p$ *from* $f_2(\mathbf{X}) \sim \mathbb{N}_m(\boldsymbol{\mu}_2, \boldsymbol{\Sigma}_2)$. *It has probability density function* $f(\mathbf{X}) = pf_1(\mathbf{X}) + (1 - p)f_2(\mathbf{X})$.

a) How many parameters are there in this distribution?

b) Find $\mathbb{E}(\mathbf{X})$ *and* $\mathbb{C}ov(\mathbf{X})$.

Exercise 2.17 *Suppose* $\mathbf{u} = (u_1, u_2)$ *and* $\mathbf{w} = (w_1, w_2)$ *are two standard bivariate standard uniform random vectors. The IPD is* $z = \sqrt{(u_1 - w_1)^2 + (u_2 - w_2)^2}$. *Find the distribution of* z, *its mean and variance.*

Exercise 2.18 *Find the covariance between two IPDs formed from a bivariate uniform distribution, which share one point.*

Exercise 2.19 *Let* $\mathbf{X}_i, \mathbf{X}_j, \mathbf{Y}_i, \mathbf{Y}_j \overset{iid}{\sim} \mathrm{N}(\mathbf{0}, \mathbf{I}_d)$ *be independent vectors in* $\mathbb{R}^d$. *Use the squared Euclidean distances to*

(a) show that as $d \to \infty$, $\frac{1}{d}(d_{ij}(x), d_{ij}(y), d_{ij}(xy)) \overset{d}{\to} \mathrm{N}(\boldsymbol{\mu}, \boldsymbol{\Sigma})$, *where* $\boldsymbol{\mu} \in \mathbb{R}^3$ *and* $\boldsymbol{\Sigma} \in \mathbb{R}^{3\times 3}$ *are the asymptotic mean vector and covariance matrix.*

(b) compute explicitly the asymptotic means, variances, and covariances of

$$\frac{1}{d}d_{ij}(x),\ \frac{1}{d}d_{ij}(y),\ \frac{1}{d}d_{ij}(xy).$$

Exercise 2.20 *Assume the components of* $\mathbf{X}_i, \mathbf{X}_j, \mathbf{Y}_i, \mathbf{Y}_j$ *are independent* χ_1^2 *distributed in* $\mathbb{R}^d$. *Define distances as Exercise 2.21.*

(a) Show joint asymptotic normality of the scaled distances.

(b) Compute the means, variances, and covariances explicitly.

Exercise 2.21 *Let* $\mathbf{B} \in \mathbb{R}^{m\times m}$ *and* $\mathbf{C} \in \mathbb{R}^{n\times n}$ *be square matrices with eigenvalues* $\lambda_1(\mathbf{B}), \lambda_2(\mathbf{B}), \ldots, \lambda_m(\mathbf{B})$ *and* $\mu_1(\mathbf{C}), \mu_2(\mathbf{C}), \ldots, \mu_n(\mathbf{C})$, *respectively.*
Define the matrix $A = \mathbf{B} \otimes \mathbf{C} \in \mathbb{R}^{mn\times mn}$ *as the Kronecker product of* $\mathbf{B}$ *and* $\mathbf{C}$.

a) Show that the eigenvalues of A *are exactly the pairwise products*

$$\lambda_i(\mathbf{B})\mu_j(\mathbf{C}), \quad i = 1, \ldots, m, \quad j = 1, \ldots, n.$$

b) Show how the multiplicities of eigenvalues of A *relate to those of* $\mathbf{B}$ *and* $\mathbf{C}$.

Exercise 2.22 *Let* $\mathbf{X}, \mathbf{Y} \sim \mathrm{N}(0, \mathbf{I}_d)$ *in* $\mathbb{R}^d$ *be independent. Show that for all* $d \geq 1$

$$\mathbb{E}(\|\mathbf{X} - \mathbf{Y}\| - \sqrt{2d}) \leq \frac{1}{\sqrt{2d}}.$$

Exercise 2.23 *Write an R function to replicate Tables 2.1, 2.2, and 2.3.*

Exercise 2.24 *Let* $\|x\|_s = \left(\sum_{i=1}^n |x_i|^s\right)^{1/s}$ *for* $x \in \mathbb{R}^n$, *where* $0 < s < 1$. *Prove that* $\|\cdot\|_s$ *does not satisfy the triangle inequality; that is, show that there exist vectors* $x, y \in \mathbb{R}^n$ *such that*

$$\|x + y\|_s > \|x\|_s + \|y\|_s.$$

Exercise 2.25 *Let* $\mathbf{X}_1 = (X_{11}, \ldots, X_{1d})$, *and* $\mathbf{X}_2 = (X_{21}, \ldots, X_{2d})$ *be independent random vectors in* $\mathbb{R}^d$ *whose coordinates are i.i.d.* $\mathrm{Exp}(\theta_1)$, *and let*

$$\mathbf{Y}_1 = (Y_{11}, \ldots, Y_{1d}), \quad \mathbf{Y}_2 = (Y_{21}, \ldots, Y_{2d})$$

be independent random vectors with i.i.d. $\mathrm{Exp}(\theta_2)$ *coordinates. Assume all coordinates are mutually independent. We consider the* L_1 *(Manhattan) distance* $D(x, y) = \sum_{j=1}^d |x_j - y_j|$. *Find the distribution of* $D(\mathbf{X}_1, \mathbf{X}_2)$, $D(\mathbf{Y}_1, \mathbf{Y}_2)$, *and* $D(\mathbf{X}_1, \mathbf{Y}_1)$.

3

High-Dimensional Dissimilarity Indices

3.1 Abstract

This chapter examines measures of dissimilarity for high-dimensional, low-sample-size data, with particular emphasis on their theoretical behavior and practical performance relative to the Euclidean distance. Dissimilarity indices are fundamental to multivariate analysis, underpinning clustering, classification, ordination, and visualization methods across diverse applications such as machine learning, genomics, and computer vision. While Euclidean and Mahalanobis are widely used for continuous data, they encounter serious limitations in HDLSS settings due to scale effects, singular covariance structures, and the phenomenon of distance concentration. The chapter reviews a broad class of dissimilarity indices, including Minkowski-type norms and other proximity measures tailored to different data types, and discusses their suitability for high-dimensional inference. Key theoretical properties, including concentration behavior and the distribution of row-wise summaries of dissimilarity matrices, are developed. The implications for clustering, nearest-neighbor methods, and multidimensional scaling are critically assessed. Applications to outlier detection, change-point analysis, and financial data illustrate how appropriate dissimilarity choices can alleviate high-dimensional pathologies and improve inferential performance.

3.2 Introduction

The purpose of this chapter is to discuss measures of dissimilarity for high-dimensional observations, in low-sample-size (HDLSS) setting, and compare their behavior with the Euclidean distance. A dissimilarity or proximity measure for analysis of high-dimensional observations is essential in many applications such as machine learning, data mining, classification, databases, natural language processing, DNA microarray, and computer vision. These applications involve HDLSS datasets that need analysis, comparison, and visualization (Modarres [229]).

In multivariate analysis, dissimilarity indices quantify the extent of difference between two observations, forming the basis for many exploratory and inferential techniques. Unlike distance measures that require strict mathematical properties like the triangle inequality, dissimilarity indices offer flexibility in handling various data types, including binary and continuous variables. These indices transform complex multivariate datasets into structured representations such as dissimilarity matrices, which serve as inputs for methods like clustering, classification, multidimensional scaling, and principal coordinate analysis.

A common choice for continuous data are the Euclidean distance, which computes the root of the sum of squared differences between variables. However, when variables differ

DOI: 10.1201/9781003789581-3

in scale or exhibit strong correlations, the Mahalanobis distance ensures that variables with large variances do not disproportionately influence the dissimilarity measure. However, the inverse of the sample covariance matrix does not exist when $d \gg n$, making the Mahalanobis distance unsuitable in HDLSS settings. For mixed or non–continuous data, alternative indices are used. The Gower's dissimilarity coefficient (Gower [122]) is particularly versatile. Dissimilarity indices directly affect the geometric configuration produced by ordination techniques like multidimensional scaling. The affinity of the resulting visualization to the original high-dimensional structure depends on the appropriateness of the dissimilarity measure used.

In hierarchical and partition-based clustering, dissimilarity indices determine how objects are grouped. Clustering outcomes can vary dramatically with different indices, even when applied to the same dataset. Thus, sensitivity analysis involving multiple dissimilarity measures is often recommended. Overall, dissimilarity indices play a foundational role in multivariate statistics, offering a bridge between raw, complex data and visual or structural representations.

Different similarity or dissimilarity measures have been proposed for particular data types. The most commonly used proximity measure for continuous measurements is the Minkowski metric or L_s-norm in (2.3). Different dissimilarity indices such as the Euclidean (L_1 and L_2 norms), edit, cosine, Hamming, and Jaccard distances are used in applications. However, these dissimilarity indices are inappropriate for HDLSS settings. Marozzi [215] discusses multivariate tests based on IPDs and applies them to magnetic resonance images. He shows that L_s norm in (2.3), where $1 \leq s \leq \infty$ can enhance the power of multivariate tests based on IPDs. For comparing means and variability of multivariate data, Marozzi et al. [216] proposed powerful tests based on IPDs.

Several multivariate techniques such as clustering, classification, and multidimensional scaling rely on the computation of the dissimilarity matrix, which is an $N \times N$ matrix containing all the pairwise dissimilarities among the N sample points. A dissimilarity matrix is central for obtaining k-means clustering, k-nearest neighbor classifier or support vector machine. In agglomerative clustering, a dissimilarity matrix is defined in terms of the Euclidean distance. A linkage function such as Ward, single, complete, and average linkage (Ward [350]) extends the dissimilarity between pairs of observations to clusters.

While some traditional clustering algorithms such as k-means and nearest neighbor suffer from the curse of dimensionality (Sarkar and Ghosh [299]), Borysov et al. [44]) show that hierarchical clustering by Euclidean distance and Yata and Aoshima [365] show that clustering by principal component analysis have consistency properties in HDLSS settings. The distance concentration phenomenon (Angiulli [11]) describes the tendency for distances to become almost indistinguishable in high-dimensional spaces. The coefficient of variation of the IPDs, which measures the concentration of the IPD distribution, under appropriate moment assumptions, tends to zero as d tends to infinity. In a sense, relative difference of the distances of the closest and farthest observations of an independently selected data point tends to zero as $d \to \infty$. Hence, the concept of the nearest neighbor of an observation becomes meaningless. Sarkar and Ghosh [299] show that violation of the neighborhood structure is responsible for the poor performance of the nearest neighbor methods in high dimensions based on the Euclidean distance.

3.3 Pseudo-Metric, Semi-Metric, and Dissimilarity Functions

Measures of difference between objects arise throughout mathematics, statistics, and machine learning. These measures vary in the strength of the mathematical structure they impose. This section presents metrics, pseudo-metrics, semi-metrics, and dissimilarity functions, along with formal implications between these classes.

Let $\mathcal{X}$ be a non-empty set. A distance-like function is a mapping $d : \mathcal{X} \times \mathcal{X} \to [0, \infty)$.

A function d is a metric on $\mathcal{X}$ if for all $x, y, z \in \mathcal{X}$,

1. $d(x, y) \geq 0$ (non-negativity)
2. $d(x, y) = 0 \iff x = y$ (distance identity)
3. $d(x, y) = d(y, x)$ (symmetry)
4. $d(x, z) \leq d(x, y) + d(y, z)$ (triangle inequality)

Example include, Euclidean distance $d(x, y) = (\sum_{i=1}^{n}(x_i - y_i)^2)^{\frac{1}{2}}$ and Manhattan (L_1) distance $d(x, y) = \sum_{i=1}^{n} |x_i - y_i|$

A function d is a pseudo-metric on $\mathcal{X}$ if it satisfies all metric axioms except distance identity. A function d is a semi-metric if it satisfies non-negativity, symmetry, and distance identity, but may violate the triangle inequality. A dissimilarity function is any mapping $\delta : \mathcal{X} \times \mathcal{X} \to [0, \infty)$ to quantify how unlike two objects are, with minimal structural requirements. Typically, dissimilarities satisfy non-negativity and $\delta(x, x) = 0$, but distance identity, and the triangle inequality are not required. Example include cosine dissimilarity $\delta(x, y) = 1 - \frac{x'y}{\|x\|\|y\|}$ and Kullback–Leibler divergence $D_{\mathrm{KL}}(P\|Q) = \sum_i P(i) \log \frac{P(i)}{Q(i)}$.

The different classes of distance-like functions are related by strict logical implications. Every metric is a pseudo-metric. A metric satisfies non-negativity, symmetry, and the triangle inequality by definition. Since a pseudo-metric requires exactly these properties (and allows but does not require distance identity), every metric is a pseudo-metric. Every pseudo-metric is a dissimilarity function. A pseudo-metric is a mapping into $[0, \infty)$ and satisfies $d(x, x) = 0$ and non-negativity. Since dissimilarity functions impose no further axiomatic requirements, every pseudo-metric qualifies as a dissimilarity. The converse implications do not hold in general.

A dissimilarity may violate the triangle inequality and therefore need not be a pseudo-metric. Likewise, a pseudo-metric may assign zero distance to distinct points, so it need not be a metric. These relationships can be summarized as,

$$\text{Metric} \Rightarrow \text{Pseudo-Metric} \Rightarrow \text{Dissimilarity},$$

with each implication being strict.

TABLE 3.1: Comparison of distance-like and dissimilarity functions

Property	**Metric**	**Pseudo-Metric**	**Semi-Metric**	**Dissimilarity**
Non-negativity	Yes	Yes	Yes	Yes
$d(x, x) = 0$	Yes	Yes	Yes	Yes
$d(x, y) = 0 \Rightarrow x = y$	Yes	No	Yes	Usually
Symmetry	Yes	Yes	Yes	Yes
Triangle inequality	Yes	Yes	No	No

Metrics provide the strongest structure, pseudo-metrics relax point separation, semi-metrics relax the triangle inequality, and dissimilarities are maximally flexible. The strict implications between these classes clarify which mathematical tools are available in each case. Throughout this book, we use Euclidean distances and dissimilarity functions.

3.4 Distance Concentration

The distance concentration phenomenon (Francois et al. [105]) describes the effect of high dimension d on IPDS. Let $\eta = \frac{\sqrt{\mathbb{V}ar(\|\mathbf{X}_1-\mathbf{X}_2\|)}}{\mathbb{E}(\|\mathbf{X}_1-\mathbf{X}_2\|)}$ be the coefficient of variation of the IPDs. Under appropriate moment assumptions, one can show that η tends to zero as d tends to infinity. Using Chebyshev's inequality, Biau and Mason [31] show that

$$\text{for all } \epsilon > 0, \quad \mathbb{P}\{|\frac{\|\mathbf{X}_1 - \mathbf{X}_2\|}{\mathbb{E}\|\mathbf{X}_1 - \mathbf{X}_2\|} - 1| \geq \epsilon\} \to 0 \quad as \quad d \to \infty. \tag{3.1}$$

Hence, $\|\mathbf{X}_1-\mathbf{X}_2\|$ approaches $\mathbb{E}\|\mathbf{X}_1-\mathbf{X}_2\|$ as $d \to \infty$ and the most information about $\|\mathbf{X}_1-\mathbf{X}_2\|$ is provided by $\mathbb{E}\|\mathbf{X}_1 - \mathbf{X}_2\|$. How does one determine whether distance concentration occur in the datasets? Durrant and Kabán [82] provide necessary conditions for the distance concentration phenomena to occur. Let

$$\text{DMin}_n(d) = \min_{1\leq j\leq n} \|\mathbf{X}_j\| \quad \text{and} \quad \text{DMax}_n(d) = \max_{1\leq j\leq n} \|\mathbf{X}_j\|. \tag{3.2}$$

Suppose d is large enough so that $\mathbb{E}(\mathbf{X}) \in [\text{DMin}_n(d), \text{DMax}_n(d)]$ holds. If for any $\epsilon > 0$,

$$\lim_{d\to\infty} P\{\text{DMax}_n(d) \leq (1+\epsilon)\text{DMin}_n(d)\} = 1,$$

then $\lim_{d\to\infty} \text{RV}_d = 0$, where the relative variance is defined as

$$\text{R}V_d = \frac{\mathbb{V}ar[\sum_{i=1}^d x_i^2]}{[\mathbb{E}(\sum_{i=1}^d x_i)^2]^2} = \frac{\sum_{i=1}^d \sum_{j=1}^d \mathbb{C}ov(x_i^2, x_j^2)}{\sum_{i=1}^d \sum_{j=1}^d \mathbb{E}(x_i^2)\ \mathbb{E}(x_j^2)}. \tag{3.3}$$

Consider a random vector $\mathbf{X} = (x_1, \ldots, x_d)$. Define $S_d = \sum_{i=1}^d x_i^2$. The squared Euclidean norm is $\|X\|^2 = S_d$. The relative variance statistic can be written as the squared coefficient of variation

$$RV_d = \frac{\text{Var}(S_d)}{(\mathbb{E}S_d)^2}. \tag{3.4}$$

The condition that the largest distance is no more than ϵ away from the minimum distance is equivalent to the condition that $\lim_{d\to\infty} \text{RV}_d = 0$. This provides a basis for the distance concentration phenomena to occur. If $\mathbb{V}ar[\sum_{i=1}^d x_i^2]$ grows faster than $\mathbb{E}(\sum_{i=1}^d x_i^2)$ as $d \to \infty$, then the relative variance does not vanish and the distance concentration does not occur. In general, the observations must be highly correlated for distance concentration not to occur. Durrant and Kabán [82] conclude that i.i.d. (or close to i.i.d.) samples precipitate distance concentration.

Example 3.4.1 *Let $\mathbf{X} \in \mathbb{R}^d$ be a random vector whose entries are i.i.d. standard normal variables. Then $\|\mathbf{X}\|^2 = \sum_{i=1}^d x_i^2 \sim \chi_d^2$. As $d \to \infty$, the law of large numbers implies that $\|\mathbf{X}\|^2/d \to 1$, and the relative variance tends to 0, i.e., $RV_d \to 0$. Hence, distances concentrate. Let $x_i \sim \text{N}(0,1)$, i.i.d. for $i = 1, \ldots, d$. It follows that*

$$\mathbb{E}[x_i^2] = 1, \quad \mathbb{V}ar((x_i^2) = 2 \quad (since\ x_i^2 \sim \chi_1^2), \quad \mathbb{C}ov(x_i^2, x_j^2) = 0 \quad for\ i \neq j.$$

We obtain

$$Numerator = \sum_{i=1}^d \mathbb{V}ar((x_i^2) = d \cdot 2 = 2d,$$

$$Denominator = \left(\sum_{i=1}^d \mathbb{E}[x_i^2]\right)^2 = (d \cdot 1)^2 = d^2,$$

$$\Rightarrow RV_d = \frac{2d}{d^2} = \frac{2}{d} \to 0 \quad as\ d \to \infty.$$

Example 3.4.2 *Let* $\mathbf{X}$ *have i.i.d. components from a uniform distribution on* $[-1, 1]$. *Then* $\mathbb{E}(x_i^2) = 1/3$ *and* $\mathbb{V}ar(x_i^2)$ *is finite. As* $d \to \infty$, *the law of large numbers again implies the sum* $\sum x_i^2$ *becomes tightly concentrated around its mean, and the relative variance tends to 0. Let* $x_i \sim Unif([-1,1])$, *i.i.d. for* $i = 1, \ldots, d$. *It follows that*

$$\mathbb{E}[x_i^2] = \int_{-1}^{1} x^2 \cdot \frac{1}{2}\, dx = \frac{1}{3}, \qquad \mathbb{E}[x_i^4] = \int_{-1}^{1} x^4 \cdot \frac{1}{2}\, dx = \frac{1}{5},$$

$$\mathbb{V}ar((x_i^2) = \mathbb{E}[x_i^4] - (\mathbb{E}[x_i^2])^2 = \frac{4}{45}, \qquad \mathbb{C}ov(x_i^2, x_j^2) = 0 \quad for\ i \neq j.$$

We obtain

$$Numerator = d \cdot \frac{4}{45}, \qquad Denominator = \left(\sum_{i=1}^{d} \mathbb{E}[x_i^2]\right)^2 = \left(d \cdot \frac{1}{3}\right)^2 = \frac{d^2}{9},$$

$$\Rightarrow RV_d = \frac{\frac{4}{45}d}{\frac{d^2}{9}} = \frac{4 \cdot 9}{45 \cdot d} = \frac{36}{45d} = \frac{4}{5d} \to 0, \quad as\ d \to \infty.$$

Example 3.4.3 *Relative Variance for Correlated Normal Distance. Consider two independent vectors* $\mathbf{X}, \mathbf{Y} \sim \mathbb{N}_d(\mathbf{0}, \boldsymbol{\Sigma})$, *where each component has unit variance and all pairs of distinct components have correlation* $\rho \in [0, 1)$. *We are interested in the squared Euclidean distance* $D^2 = \|\mathbf{X} - \mathbf{Y}\|^2$. *Define* $\mathbf{Z} = \mathbf{X} - \mathbf{Y}$. *Since* $\mathbf{X}$ *and* $\mathbf{Y}$ *are independent,* $\mathbf{Z} \sim \mathbb{N}_d(0, 2\boldsymbol{\Sigma})$. *Then,* $D^2 = \mathbf{Z}'\mathbf{Z}$ *is a quadratic form in a normal vector. For a normal vector* $\mathbf{Z} \sim \mathbb{N}(0, \boldsymbol{\Sigma}_Z)$,

$$\mathbb{E}[\mathbf{Z}'\mathbf{Z}] = \mathrm{tr}(\boldsymbol{\Sigma}_Z), \qquad \mathbb{V}ar((\mathbf{Z}'\mathbf{Z}) = 2\,\mathrm{tr}(\boldsymbol{\Sigma}_Z^2).$$

Here $\boldsymbol{\Sigma}_Z = 2\boldsymbol{\Sigma}$, *so* $\mathbb{E}[D^2] = \mathrm{tr}(2\boldsymbol{\Sigma}) = 2d$, *and* $\mathbb{V}ar((D^2) = 2\,\mathrm{tr}\big((2\boldsymbol{\Sigma})^2\big) = 8\,\mathrm{tr}(\boldsymbol{\Sigma}^2)$. *The correlation structure is* $\boldsymbol{\Sigma} = (1-\rho)\mathbf{I}_d + \rho\mathbf{J}_d$, *where* $\mathbf{I}_d$ *is the* $d \times d$ *identity and* $\mathbf{J}_d$ *the* $d \times d$ *matrix of ones. The eigenvalues are well known,*

$$\lambda_1 = 1 - \rho + \rho d \quad (multiplicity\ 1), \qquad \lambda_2 = 1 - \rho \quad (multiplicity\ d-1).$$

Therefore, $\mathrm{tr}(\boldsymbol{\Sigma}^2) = (1 - \rho + \rho d)^2 + (d-1)(1-\rho)^2$.

The relative variance of D^2 *is defined as* $RV_d = \frac{\mathbb{V}ar((D^2)}{(\mathbb{E}[D^2])^2}$. *Substituting,*

$$RV_d = \frac{8\,\mathrm{tr}(\boldsymbol{\Sigma}^2)}{(2d)^2} = \frac{2}{d^2}\Big[(1-\rho+\rho d)^2 + (d-1)(1-\rho)^2\Big].$$

As $d \to \infty$, $(1 - \rho + \rho d)^2 = \rho^2 d^2 + O(d)$, *so the leading contribution is* $RV_d \longrightarrow 2\rho^2$.

- *If* $\rho = 0$ *(independent components),* $RV_d \to 0$. *This means the squared distance* D^2 *concentrates around its mean, and high-dimensional geometry behaves regularly.*
- *If* $\rho > 0$, *then* $RV_d \to 2\rho^2 > 0$. *In this case, the distances do not concentrate. The correlation prevents the law of large numbers from averaging out variability.*

Example 3.4.4 *Let* $\mathbf{X} = (x_1, x_2, \ldots, x_d) \in \mathbb{R}^d$ *be a random vector with i.i.d. components having finite second and fourth moments. We will prove that* $\lim_{d\to\infty} RV_d = 0$.

Assume the x_i *are i.i.d. random variables with* $\mu_2 = \mathbb{E}[x_1^2] < \infty$ *and* $\mu_4 = \mathbb{E}[x_1^4] < \infty$. *It follows that* $\mathbb{V}ar(x_i^2) = \mu_4 - \mu_2^2 = \sigma^2$. *Since the* x_i *are independent, we have*

$$Cov(x_i^2, x_j^2) = \begin{cases} \sigma^2 & if\ i = j \\ 0 & if\ i \neq j \end{cases}$$

So the numerator becomes $\sum_{i=1}^{d}\sum_{j=1}^{d} Cov(x_i^2, x_j^2) = \sum_{i=1}^{d} \mathbb{V}ar(x_i^2) = d\sigma^2$

The denominator becomes $\sum_{i=1}^{d}\sum_{j=1}^{d}\mathbb{E}(x_i^2)\mathbb{E}(x_j^2) = \left(\sum_{i=1}^{d}\mu_2\right)^2 = (d\mu_2)^2 = d^2\mu_2^2$. *Thus,*

$$RV_d = \frac{d\sigma^2}{d^2\mu_2^2} = \frac{\sigma^2}{d\mu_2^2}.$$

Since $\mu_2 > 0$ *and* σ^2 *is finite, we conclude* $\lim_{d\to\infty} RV_d = 0$.

Example 3.4.5 *Let* $X \sim \mathbb{N}(\boldsymbol{\mu}, \boldsymbol{\Sigma})$. *Define* $Q = \mathbf{X}'\mathbf{X}$. *Then,*

$$\mathbb{E}(Q) = \text{tr}(\boldsymbol{\Sigma}) + \boldsymbol{\mu}'\boldsymbol{\mu}.$$

$$\mathbb{V}ar(Q) = 2\text{tr}(\boldsymbol{\Sigma}^2) + 4\boldsymbol{\mu}'\boldsymbol{\Sigma}\boldsymbol{\mu}.$$

Therefore,

$$RV_d = \frac{2\text{tr}(\boldsymbol{\Sigma}^2) + 4\boldsymbol{\mu}^\top\boldsymbol{\Sigma}\boldsymbol{\mu}}{(\text{tr}\boldsymbol{\Sigma} + \boldsymbol{\mu}'\boldsymbol{\mu})^2}.$$

If the eigenvalues of $\boldsymbol{\Sigma}$ *are* $\{\lambda_1, \ldots, \lambda_d\}$, *then*

$$RV_d = \frac{2\sum_i \lambda_i^2 + 4\mu'\boldsymbol{\Sigma}\mu}{(\sum_i \lambda_i + \mu'\mu)^2}.$$

Example 3.4.6 *To compute the relative variance for elliptical distributions suppose* $\mathbf{X} = \boldsymbol{\mu} + R\mathbf{A}\mathbf{U}$, $\mathbf{U}$ *uniform on the unit sphere,* $\mathbf{A}\mathbf{A}' = \boldsymbol{\Sigma}$. *We have* $\|\mathbf{X}\|^2 = R^2\mathbf{U}'\boldsymbol{\Sigma}\mathbf{U}$ *with expectation*

$$\mathbb{E}[\|\mathbf{X}\|^2] = \mathbb{E}[R^2]\,\text{tr}(\boldsymbol{\Sigma}) + \|\boldsymbol{\mu}\|^2$$

and variance

$$\text{Var}(\|\mathbf{X}\|^2) = \text{Var}(R^2\mathbf{U}'\boldsymbol{\Sigma}\mathbf{U}).$$

Under mild moment assumptions:

$$Var(\|X\|^2) = O(d), \quad (E\|X\|^2)^2 = O(d^2), \quad RV_d = O(1/d).$$

Therefore, the relative variance is

$$RV_d = \frac{\text{Var}(\|\mathbf{X}\|^2)}{(\mathbb{E}\|\mathbf{X}\|^2)^2} = O\left(\frac{1}{d}\right)$$

3.5 Data Sparsity

Analyzing high-dimensional data is made more challenging by several factors, including the concentration of Euclidean distances, data sparsity, computational complexity, and data visualization. Data sparsity refers to the presence of a large number of zero or missing values in a dataset. It is particularly prevalent in high-dimensional or user-generated data contexts where full coverage is infeasible. A sparse dataset contains mostly absent or null entries, making it difficult to extract meaningful patterns or relationships using standard machine learning techniques.

Data sparsity is defined as the proportion of zero or missing values in a data matrix. Given a matrix $\mathbf{A} \in \mathbb{R}^{m\times n}$, its sparsity is given by

$$\text{Sparsity}(\mathbf{A}) = \left(1 - \frac{\|\mathbf{A}\|_0}{mn}\right) \times 100\% \tag{3.5}$$

Here, $\|\mathbf{A}\|_0$ is the number of non-zero elements in $\mathbf{A}$. A higher sparsity percentage indicates that the matrix is mostly empty. Data sparsity is prevalent across various domains.

A recommender engine is an AI-powered tool that filters data to suggest personalized content or products to users. In recommender systems, user-item matrices are sparse due to limited interactions per user. Examples include Netflix, Amazon, Spotify, among others. In natural language processing, term-document matrices (e.g., bag-of-words) are sparse as each document contains a small subset of the vocabulary. In computer vision, sparse depth maps and point clouds in LiDAR (Light Detection and Ranging) data often include vast empty regions. LiDAR is a remote sensing method that uses laser light to measure distances between the sensor and objects in the environment. In healthcare systems, patients have sparse medical records; most possible tests or diagnoses are absent for each patient. In sensor networks, sensor readings are intermittently missing due to transmission loss or energy-saving strategies. In social networks, connections (edges) in a user graph are sparse in large-scale networks most users are not directly connected.

Sparsity can be detected using the sparsity ratio. Handling strategies include matrix factorization, which predicts missing entries via latent factor models. Auto-encoders, which learns dense representations from sparse inputs. Graph neural networks, which exploits structure in sparse graph data. Imputation techniques, which fills missing values using statistical or machine learning methods.

3.6 Dissimilarity Indices

Let $\mathbb{X} = \{\mathbf{x}_1, \ldots, \mathbf{x}_m\}$ and $\mathbb{Y} = \{\mathbf{y}_1, \ldots, \mathbf{y}_n\}$ be two sets of independent observations from d-dimensional continuous probability distributions F_1 and F_2 with mean vectors $\boldsymbol{\mu}_x$, $\boldsymbol{\mu}_y$, and covariance matrices $\boldsymbol{\Sigma}_x$, and $\boldsymbol{\Sigma}_y$, respectively. The IPDs in the $\mathbb{X}$, $\mathbb{Y}$, and the cross samples are given in (2.5), (2.9), and (2.12). Zhou [369] discusses the dissimilarity index $\Gamma(\mathbf{x}_i, \mathbf{x}_j) = \frac{1}{N-2}\sum_{k\neq i,j}^{N}(\|\mathbf{x}_i - \mathbf{x}_k\| - \|\mathbf{x}_j - \mathbf{x}_k\|)^2$. The index is designed for community detection based on the average IPD between the observations. While this index involves all of the observations and is sensitive to its neighborhood, it fails for HDLSS classification due to the distance concentration phenomenon as $d \to \infty$. Thus, this index is not appropriate for high-dimensional observations. Modarres [231] discusses high-dimensional measures of dissimilarity.

3.6.1 Dissimilarity Index ρ

Let $\mathbb{Z}_N = \mathbb{X} \cup \mathbb{Y}$ be the set of $N = m + n$ observations from both groups, where $\mathbf{z}_i = \mathbf{x}_i$ for $i = 1, \ldots, m$ and $\mathbf{z}_i = \mathbf{y}_{i-m}$ for $i = m+1, \ldots, N$. Pal et al. [261] and Sarkar and Ghosh [299] use the data cloud $\mathbb{Z}_N$ to measure the dissimilarity between two observations $\mathbf{x}$ and $\mathbf{y}$ in $\mathbb{Z}_N$ by

$$\rho(\mathbf{x}, \mathbf{y}) = \frac{1}{N-2}\frac{1}{\sqrt{d}}\sum_{\mathbf{z}\in\mathbb{Z}_N\setminus\{\mathbf{x},\mathbf{y}\}} \big|\|\mathbf{x} - \mathbf{z}\| - \|\mathbf{y} - \mathbf{z}\|\big|. \tag{3.6}$$

The dissimilarity index is based on the mean absolute difference of distances (MADD) and uses the scaling factor $\frac{1}{\sqrt{d}}$ since IPDs usually increases with the rate of $\sqrt{d}$ (Hall et al. [133]). Differences of norms remove the leading $O(\sqrt{d})$ term in Euclidean distances. MADD is derived from the lower bound of the triangle inequality. For d-dimensional vectors $\mathbf{x}$ and

$\mathbf{y}$, it is not difficult to show that

$$|\|\mathbf{x}\| - \|\mathbf{y}\|| \le \|\mathbf{x} - \mathbf{y}\| \le \|\mathbf{x}\| + \|\mathbf{y}\|. \tag{3.7}$$

Also called the reverse triangle inequality, the lower bound is obtained by writing $\mathbf{x} = (\mathbf{x}-\mathbf{y})+\mathbf{y}$, taking norm and applying the triangle inequality to produce $\|\mathbf{x}\| = \|\mathbf{x}-\mathbf{y}+\mathbf{y}\| \le \|\mathbf{x} - \mathbf{y}\| + \|\mathbf{y}\|$. For any $\mathbf{z} \in \mathbb{R}^d$ it follows that

$$|\|\mathbf{x} - \mathbf{z}\| - \|\mathbf{y} - \mathbf{z}\|| \le \|\mathbf{x} - \mathbf{y}\| \le \|\mathbf{x} - \mathbf{z}\| + \|\mathbf{y} - \mathbf{z}\|. \tag{3.8}$$

While the Euclidean distance is a metric the MADD index is not a metric since $\rho(\mathbf{x}, \mathbf{y}) = 0$ does not imply $\mathbf{x} = \mathbf{y}$. However, as noted by Pal et al. [261], if F_1 and F_2 are absolutely continuous, then for any $\mathbf{x} \neq \mathbf{y}$, $\rho(\mathbf{x}, \mathbf{y}) > 0$ holds with probability 1. The Euclidean distance only uses the information between the two points whereas the dissimilarity index ρ is affected by other $N-2$ observations. It is noteworthy that the index Γ in Zhou [369] is a transformation of the index ρ, using L_2 norm and the scaling factor $\frac{1}{d}$.

Let $D(\mathbb{X}, \mathbb{Y})$ denote the $N \times N$ distance matrix made up of the IPDs in $\mathbb{Z}_N$. Let $\mathbf{R}_{-i}$ and $\mathbf{R}_{-j}$, for $1 \le i < j \le N$ denote the i-th and j-th rows of $D(\mathbb{X}, \mathbb{Y})$ with the i-th and j-th distances removed from both rows. Hence, there are $N-2$ elements in each row. One can also write the MADD dissimilarity index ρ as

$$\rho(\mathbf{x}_i, \mathbf{y}_j) = \frac{1}{\sqrt{d}(N-2)} \sum_{t \neq i \neq j = 1}^{N-2} |\mathbf{R}_{-i}(t) - \mathbf{R}_{-j}(t)|.$$

Example 3.6.1 *To visualize of the dissimilarity measure $\rho(\mathbf{x}, \mathbf{y})$ let $\mathbb{Z}_N = \mathbb{X} \cup \mathbb{Y}$ be a small set of $N = 6$ points in $\mathbb{R}^2$,*

$$\mathbf{x}_1 = (1,1), \quad \mathbf{x}_2 = (2,1), \quad \mathbf{x}_3 = (1,2), \quad \mathbf{y}_1 = (4,4), \quad \mathbf{y}_2 = (5,4), \quad \mathbf{y}_3 = (4,5)$$

We want to compute the dissimilarity between $\mathbf{x} = \mathbf{x}_1$ and $\mathbf{y} = \mathbf{y}_1$. The dissimilarity $\rho(\mathbf{x}, \mathbf{y})$ measures how differently $\mathbf{x}$ and $\mathbf{y}$ are positioned relative to all other points in the set $\mathbb{Z}_N$. Each dashed line in Figure 3.1 represents a distance to another point $\mathbf{z}$, and the sum of absolute differences of distances defines ρ. The green and orange points are the reference points whose dissimilarity is being measured.

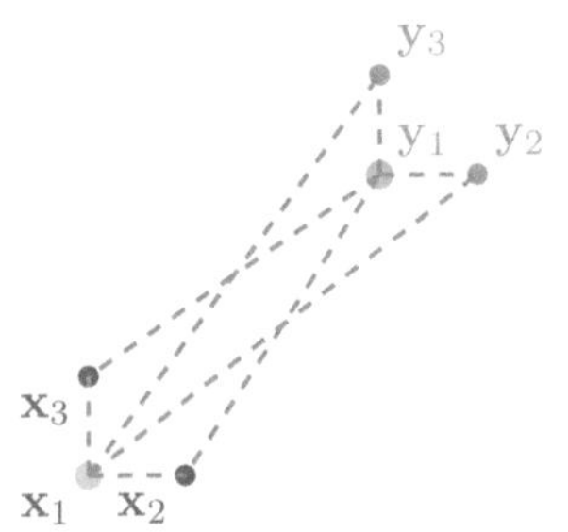

FIGURE 3.1: Illustration of the dissimilarity measure $\rho(\mathbf{x}, \mathbf{y})$. Blue points belong to group $\mathbb{X}$, red points to group $\mathbb{Y}$. Green ($\mathbf{X}_1$) and orange ($\mathbf{Y}_1$) points represent $\mathbf{x}$ and $\mathbf{y}$, respectively. Dashed lines connect $\mathbf{x}$ and $\mathbf{y}$ to the remaining points $\mathbf{z} \in \mathbb{Z}_N \setminus \{\mathbf{x}, \mathbf{y}\}$, visualizing the components $|\|\mathbf{x} - \mathbf{z}\| - \|\mathbf{y} - \mathbf{z}\||$ of $\rho(\mathbf{x}, \mathbf{y})$.

3.6.2 Dissimilarity Index δ

Modarres [231] defines the dissimilarity index δ. Consider the $N \times d$ data matrix of $\mathbb{Z}_N$. Let $\mathbf{x}_i = (x_{i1}, \ldots, x_{id})$, $\mathbf{y}_j = (y_{j1}, \ldots, y_{jd})$ and $\mathbf{z}_k = (z_{k1}, \ldots, z_{kd})$ be three different rows of the data matrix where $i \neq j \neq k = 1, \ldots, N$. Define

$$\begin{aligned} \bar{x}_i^d &= \frac{1}{d}\sum_{t=1}^{d} x_{it}, \quad [s_{(x)i}^d]^2 = \frac{1}{d}\sum_{t=1}^{d}(x_{it} - \bar{x}_i^d)^2, \quad \bar{\mu}_x^d = \mathbb{E}(\bar{x}_i^d), \quad [\sigma_x^d]^2 = \mathbb{E}([s_{(x)i}^d]^2), \\ \bar{y}_j^d &= \frac{1}{d}\sum_{t=1}^{d} y_{jt}, \quad [s_{(y)j}^d]^2 = \frac{1}{d}\sum_{t=1}^{d}(y_{jt} - \bar{y}_j^d)^2, \quad \bar{\mu}_y^d = \mathbb{E}(\bar{y}_j^d), \quad [\sigma_y^d]^2 = \mathbb{E}([s_{(y)j}^d]^2). \end{aligned} \tag{3.9}$$

Similarly, let $\bar{z}_k^d = \frac{1}{d}\sum_{t=1}^{d} z_{kt}$ and $[s_{(z)k}^d]^2 = \frac{1}{d}\sum_{t=1}^{d}(z_{kt} - \bar{z}_k^d)^2$. Note that the three rows $\mathbf{x}_i$, $\mathbf{y}_j$, and $\mathbf{z}_k$ are independent. In fact, any two rows that come from the same sample are identically distributed. Hence, $\bar{\mu}_x^d = \bar{\mu}_{x_i}^d = \bar{\mu}_{x_j}^d$ and $[\sigma_x^d]^2 = [\sigma_{x_i}^d]^2 = [\sigma_{x_j}^d]^2$. Similarly, $\bar{\mu}_y^d = \bar{\mu}_{y_i}^d = \bar{\mu}_{y_j}^d$ and $[\sigma_y^d]^2 = [\sigma_{y_i}^d]^2 = [\sigma_{y_j}^d]^2$.

Lemma 3.6.1 *Upper and lower bounds for the Euclidean distance between d-dimensional vectors $\mathbf{x}_i$ and $\mathbf{y}_j$ are given by*

$$\sqrt{d((\bar{x}_i^d - \bar{y}_j^d)^2 + (s_{(x)i}^d - s_{(y)j}^d)^2)} \le \|\mathbf{x}_i - \mathbf{y}_j\| \le \sqrt{d((\bar{x}_i^d - \bar{y}_j^d)^2 + (s_{(x)i}^d + s_{(y)j}^d)^2)}. \tag{3.10}$$

Define the following modified distances

$$\|\mathbf{x}_i - \mathbf{z}_k\|_D = \sqrt{(\bar{x}_i^d - \bar{z}_k^d)^2 + (s_{(x)i}^d - s_{(z)k}^d)^2}, \tag{3.11}$$

$$\|\mathbf{y}_i - \mathbf{z}_k\|_D = \sqrt{(\bar{y}_j^d - \bar{z}_k^d)^2 + (s_{(y)j}^d - s_{(z)k}^d)^2}. \tag{3.12}$$

We use the data cloud $\mathbb{Z}_N$ to define the dissimilarity measure

$$\delta(\mathbf{x}_i, \mathbf{y}_j) = \frac{1}{N-2}\sum_{k \neq i,j} \big|\|\mathbf{x}_i - \mathbf{z}_k\|_D - \|\mathbf{y}_j - \mathbf{z}_k\|_D\big|. \tag{3.13}$$

Since δ measures the mean absolute differences of modified distances $\|\mathbf{x}_i - \mathbf{z}_k\|_D$ and $\|\mathbf{y}_j - \mathbf{z}_k\|_D$ we refer to it as MADMD. While MADMD is similar to MADD, it uses the modified distances defined by the lower bound of (3.10) as compared to the lower bound of (3.8). Notice that dissimilarity ρ in (3.6) involves the term $\frac{1}{\sqrt{d}}$ while dissimilarity δ in (3.13) does not. The form of δ follows from the facts that differences of norms remove the leading $O(\sqrt{d})$ term in Euclidean distances and that the lower bound in (3.10) involves $\sqrt{d}$. Thus, they cancel each other. Hence, if we define $\|\mathbf{x}_i - \mathbf{z}_k\|_D = \frac{1}{\sqrt{d}}\|\mathbf{x}_i - \mathbf{z}_k\|$, then we obtain the dissimilarity index ρ.

The index $\delta(\mathbf{x}_i, \mathbf{y}_j)$ measures how differently two high-dimensional observations are positioned in the dataset, taking into account both their average levels (means) and variability (standard deviations) across dimensions. Figure 3.2 shows the modified distances $\|\cdot\|_D$ used to compute δ.

Example 3.6.2 *Consider the following* 2×2 *dataset,*

$$\mathbf{x}_1 = (1, 2), \quad \mathbf{x}_2 = (2, 1), \quad \mathbf{y}_1 = (4, 5), \quad \mathbf{y}_2 = (5, 4)$$

Compute means and standard deviations,

$$\bar{x}_1^2 = 1.5, \quad s_{(x)1}^2 = 0.5, \quad \bar{y}_1^2 = 4.5, \quad s_{(y)1}^2 = 0.5$$

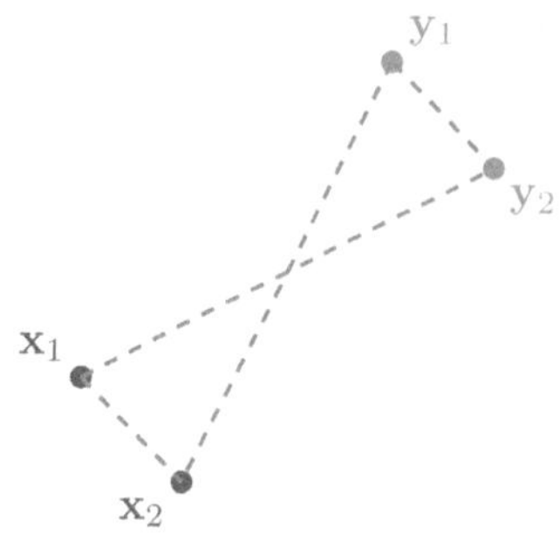

FIGURE 3.2: Illustration of the dissimilarity index $\delta(\mathbf{x}_1, \mathbf{y}_1)$. Blue points belong to group $\mathbb{X}$, red points to group $\mathbb{Y}$. Dashed lines represent the modified distances $\|\cdot\|_D$ for δ.

Take $\mathbf{z}_1 = \mathbf{x}_2$, $\mathbf{z}_2 = \mathbf{y}_2$ as the remaining points. Then the modified distances are,

$$\|\mathbf{x}_1 - \mathbf{z}_1\|_D = \sqrt{(1.5 - 1.5)^2 + (0.5 - 0.5)^2} = 0$$

$$\|\mathbf{y}_1 - \mathbf{z}_1\|_D = \sqrt{(4.5 - 1.5)^2 + (0.5 - 0.5)^2} = 3$$

$$\|\mathbf{x}_1 - \mathbf{z}_2\|_D = \sqrt{(1.5 - 4.5)^2 + (0.5 - 0.5)^2} = 3$$

$$\|\mathbf{y}_1 - \mathbf{z}_2\|_D = \sqrt{(4.5 - 4.5)^2 + (0.5 - 0.5)^2} = 0$$

The Euclidean distance is $\|\mathbf{x}_1 - \mathbf{y}_1\| = 3\sqrt{2}$ whereas the dissimilarity index is

$$\delta(\mathbf{x}_1, \mathbf{y}_1) = \frac{1}{2}(|0 - 3| + |3 - 0|) = \frac{6}{2} = 3.$$

To compute $\rho(\mathbf{x}_1, \mathbf{y}_1)$ note that

$$\|\mathbf{x}_1 - \mathbf{x}_2\| = \|(1, 2) - (2, 1)\| = \|(-1, 1)\| = \sqrt{(-1)^2 + 1^2} = \sqrt{2},$$

$$\|\mathbf{y}_1 - \mathbf{x}_2\| = \|(4, 5) - (2, 1)\| = \|(2, 4)\| = \sqrt{2^2 + 4^2} = \sqrt{20}.$$

Difference is $\left|\sqrt{2} - \sqrt{20}\right|$.

$$\|\mathbf{x}_1 - \mathbf{y}_2\| = \|(1, 2) - (5, 4)\| = \|(-4, -2)\| = \sqrt{(-4)^2 + (-2)^2} = \sqrt{20}.$$

$$\|\mathbf{y}_1 - \mathbf{y}_2\| = \|(4, 5) - (5, 4)\| = \|(-1, 1)\| = \sqrt{2}.$$

Difference is $\left|\sqrt{20} - \sqrt{2}\right|$. Both terms are equal, hence

$$\sum = \left|\sqrt{2} - \sqrt{20}\right| + \left|\sqrt{20} - \sqrt{2}\right| = 2(\sqrt{20} - \sqrt{2}).$$

$$\rho(\mathbf{x}_1, \mathbf{y}_1) = \frac{1}{N - 2}\frac{1}{\sqrt{d}} \cdot 2(\sqrt{20} - \sqrt{2}).$$

Since$N - 2 = 2$, and $\sqrt{d} = \sqrt{2}$,

$$\rho(\mathbf{x}_1, \mathbf{y}_1) = \frac{1}{2}\frac{1}{\sqrt{2}} \cdot 2(\sqrt{20} - \sqrt{2}) = \frac{1}{\sqrt{2}}(\sqrt{20} - \sqrt{2}).$$

We obtain $\rho(\mathbf{x}_1, \mathbf{y}_1) = \frac{2\sqrt{5}}{\sqrt{2}} - 1 = \sqrt{10} - 1$.

In terms of the modified distances, one can obtain a second form for δ as

$$\delta(\mathbf{x}_i, \mathbf{y}_j) = \frac{1}{N-2} \sum_{k \neq i,j} \max(\|\mathbf{x}_i - \mathbf{z}_k\|_D, \|\mathbf{y}_j - \mathbf{z}_k\|_D) - \min(\|\mathbf{x}_i - \mathbf{z}_k\|_D, \|\mathbf{y}_j - \mathbf{z}_k\|_D).$$

Similar to ρ, the MADMD index is not a metric since $\delta(\mathbf{x}, \mathbf{y}) = 0$ does not imply $\mathbf{x} = \mathbf{y}$. However, if F_1 and F_2 are absolutely continuous, then for any $\mathbf{x} \neq \mathbf{y}$, $\delta(\mathbf{x}, \mathbf{y}) > 0$ holds with probability 1. Using the triangle inequality, we obtain $\delta(\mathbf{x}, \mathbf{y}) \leq \frac{1}{\sqrt{d}}\|\mathbf{x}-\mathbf{y}\|$. Hence, proximity in terms of the Euclidean distance implies proximity in terms of δ, but the converse is not true. While the Euclidean distance only uses the information of the two observations the dissimilarity index δ is affected by the modified distances from $N-2$ other observations. The computational costs of δ and ρ are of order $O(dN^2)$.

To obtain a more efficient formulation for δ (and similarly ρ), define the ordered set of distances, $d_i^{(k)} = \|\boldsymbol{x}_i - \boldsymbol{z}_k\|_D$ for each $k \neq i, j$. The values $d_i^{(k)}$ can be arranged in increasing order to obtain $d_i^{(1)} \leq d_i^{(2)} \leq \ldots \leq d_i^{(N-2)}$. Similarly, we order $d_j^{(k)}$ to obtain $d_j^{(1)} \leq d_j^{(2)} \leq \cdots \leq d_j^{(N-2)}$. Since d_{ij} is the mean of absolute differences, we have

$$\delta(\mathbf{x}_i, \mathbf{y}_j) = \frac{1}{N-2} \sum_{k=1}^{N-2} |d_i^{(k)} - d_j^{(k)}|. \tag{3.14}$$

This reduces the number of operations from $O(dN^2)$ that uses brute-force pairwise comparisons to $O(dNlogN)$, which is required for sorting N values.

While the proposed dissimilarity measure is affected by modified distances from other $N-2$ observations they have the same weight $\frac{1}{N-2}$. One can give each observation a different weight. How to choose the weights and their effects on the dissimilarity measure need further research. The Euclidean distance is well-defined for two discrete vectors and the dissimilarity indices ρ and δ can be extended to the discrete case. However, if F_1 and F_2 are not absolutely continuous, then for any $\mathbf{x} \neq \mathbf{y}$, one can not ensure that $\rho(\mathbf{x}, \mathbf{y}) > 0$ or $\delta(\mathbf{x}, \mathbf{y}) > 0$ with probability 1. Hence, the dissimilarity indices may fail to detect differences in the two discrete distributions.

3.6.3 Properties of the Dissimilarity Indices

Lemma 3.6.2 *The dissimilarity index δ is a semi-metric on $\mathbb{Z}_N$ for $N \geq 3$.*

To obtain the asymptotic behavior of the dissimilarity index ρ and apply the law of large numbers Biswas and Ghosh [35] assume that

(A1) As $d \to \infty$, There exist constants $\sigma_{F_1}^2, \sigma_{F_2}^2 > 0$, such that

$$\frac{1}{d} tr\left(\boldsymbol{\Sigma}_x\right) \to \sigma_{F_1}^2 \text{ and } \frac{1}{d} tr\left(\boldsymbol{\Sigma}_y\right) \to \sigma_{F_2}^2,$$

respectively, and

$$\frac{1}{d}\|\boldsymbol{\mu}_x - \boldsymbol{\mu}_y\|^2 \to \nu^2.$$

(A2) The fourth moments of the components of $\mathbf{Z} = (Z_1, \ldots, Z_d)'$ are uniformly bounded.

(A3) Let vectors $\mathbf{X}_1, \mathbf{X}_2 \sim F_1$ and $\mathbf{Y}_1, \mathbf{Y}_2 \sim F_2$ be independent. For $\mathbf{W} = \mathbf{X}_1 - \mathbf{X}_2$, $\mathbf{Y}_1 - \mathbf{Y}_2$, and $\mathbf{X}_1 - \mathbf{Y}_2$,

$$C = \sum_{1 \leq s \neq t \leq d} \mathbb{C}or\left(W_r^2, W_s^2\right)$$

is of the order $o(d^2)$.

Assumptions (A1)–(A3) are needed in the HDLSS literature to establish convergence. Hall et al. [133], Jung and Marron [164], and Ahn et al. [6] have stated similar assumptions to study the asymptotic behavior of HDLSS techniques. When the variables are highly correlated, then C is of the order $O(d^2)$. Hence, (A2) places weak dependence among the variables. Assumption (A2) holds for stochastic processes with the ρ-mixing property and the process need not be stationary. When the variables are independent and identically distributed, (A1)–(A3) hold trivially.

Example 3.6.3 *Suppose* $\mathbf{X}_i \sim F_1 : \mathbb{N}(\boldsymbol{\mu}_x, \boldsymbol{\Sigma}_x)$ *is independent of* $\mathbf{Y}_i \sim F_2 : \mathbb{N}(\boldsymbol{\mu}_y, \boldsymbol{\Sigma}_y)$ *for* $i = 1, 2$. *One can show that*

$$\begin{aligned} \frac{1}{\sqrt{d}}\|\mathbf{X}_1 - \mathbf{X}_2\| &\to \frac{1}{\sqrt{d}}\,\mathbb{E}\|\mathbf{X}_1 - \mathbf{X}_2\| = \sqrt{2}\sigma_{F_1}, \\ \frac{1}{\sqrt{d}}\|\mathbf{Y}_1 - \mathbf{Y}_2\| &\to \sqrt{2}\sigma_{F_2}, \\ \frac{1}{\sqrt{d}}\|\mathbf{X}_1 - \mathbf{Y}_1\| &\to \sqrt{\sigma_{F_1}^2 + \sigma_{F_2}^2 + \nu^2} \end{aligned} \tag{3.15}$$

in probability as $d \to \infty$.

Sarkar and Ghosh [299] show that the limiting form of ρ as $d \to \infty$ is given by

$$\tilde{\rho}(F_1, F_2) = \frac{1}{N-2}[(m-1)|\sqrt{2}\sigma_{F_1} - \sqrt{\sigma_{F_1}^2 + \sigma_{F_2}^2 + \nu^2}| + (n-1)|2\sqrt{\sigma_{F_1}^2 + \sigma_{F_2}^2 + \nu^2} - \sqrt{2}\sigma_{F_2}|]. \tag{3.16}$$

In general, $\tilde{\rho}(F_1, F_2)$ depends on m and n. If $\sigma_{F_1} = \sigma_{F_2}$, one can show that $\tilde{\rho}(F_1, F_2) = |\sqrt{2}\sigma_{F_1} - \sqrt{2\sigma_{F_1}^2 + \nu^2}|$, which is independent of the sample sizes.

Example 3.6.4 *Let* $\mathbb{X} = \{\mathbf{X}_1, \dots, \mathbf{X}_m\} \subset \mathbb{R}^d$ *and* $\mathbb{Y} = \{\mathbf{Y}_1, \dots, \mathbf{Y}_n\} \subset \mathbb{R}^d$ *be two sets of independent observations from multivariate normal distributions,* $\mathbf{X}_i \sim \mathbb{N}_d(\boldsymbol{\mu}_1, \sigma_1^2\mathbf{I}_d)$ *and* $\mathbf{Y}_j \sim \mathbb{N}_d(\boldsymbol{\mu}_2, \sigma_2^2\mathbf{I}_d)$. *Let* $\mathbb{Z}_N = \mathbb{X} \cup \mathbb{Y}$ *be the combined data cloud, with* $N = m + n$. *Consider the dissimilarity between* $\mathbf{X} \in \mathbb{X}$ *and* $\mathbf{Y} \in \mathbb{Y}$ *by (3.6). Derive the expectation* $\mathbb{E}[\rho(\mathbf{X}, \mathbf{Y})]$ *under the above model.*

Define for fixed $\mathbf{z}$, $A_{\mathbf{z}} = |\|\mathbf{X} - \mathbf{z}\| - \|\mathbf{Y} - \mathbf{z}\||$. *Then,*

$$\mathbb{E}[\rho(\mathbf{X}, \mathbf{Y})] = \frac{1}{N-2} \cdot \frac{1}{\sqrt{d}} \sum_{\mathbf{z} \in \mathbb{Z}_N \setminus \{\mathbf{X}, \mathbf{Y}\}} \mathbb{E}[A_{\mathbf{z}}].$$

We distinguish two cases,

(i) $\mathbf{z} \in \mathbb{X}$, *Then* $\mathbf{z} \sim \mathbb{N}_d(\boldsymbol{\mu}_1, \sigma_1^2\mathbf{I}_d)$.

(ii) $\mathbf{z} \in \mathbb{Y}$, *Then* $\mathbf{z} \sim \mathbb{N}_d(\boldsymbol{\mu}_2, \sigma_2^2\mathbf{I}_d)$.

If $\mathbf{z} \sim \mathbb{N}_d(\boldsymbol{\mu}_1, \sigma_1^2\mathbf{I}_d)$, *then define the squared Euclidean norms with*

$$\begin{aligned} \|\mathbf{X} - \mathbf{z}\|^2 &\sim 2\sigma_1^2\chi_d^2(\delta_1^2), \quad \delta_1^2 = \frac{\|\boldsymbol{\mu}_1 - \boldsymbol{\mu}_1\|^2}{\sigma_1^2 + \sigma_1^2} = 0, \\ \|\mathbf{Y} - \mathbf{z}\|^2 &\sim 2\sigma_2^2\chi_d^2(\delta_2^2), \quad \delta_2^2 = \frac{\|\boldsymbol{\mu}_2 - \boldsymbol{\mu}_1\|^2}{\sigma_2^2 + \sigma_1^2}. \end{aligned}$$

It follows that,

$$\begin{aligned} \mathbb{E}[\|\mathbf{X} - \mathbf{z}\|] &\approx \sqrt{2\sigma_1^2 d}, \\ \mathbb{E}[\|\mathbf{Y} - \mathbf{z}\|] &\approx \sqrt{(\sigma_1^2 + \sigma_2^2)d + \|\boldsymbol{\mu}_2 - \boldsymbol{\mu}_1\|^2}. \end{aligned}$$

Thus, for the expected absolute difference we have

$$\mathbb{E}[A_{\mathbf{z}}] \approx \left| \sqrt{2\sigma_1^2 d} - \sqrt{(\sigma_1^2+\sigma_2^2)d + \|\boldsymbol{\mu}_2 - \boldsymbol{\mu}_1\|^2} \right|.$$

If $\mathbf{z} \in \mathbb{Y}z$ *in* Y*, then,*

$$\mathbb{E}[\|\mathbf{X} - \mathbf{z}\|] \approx \sqrt{(\sigma_1^2+\sigma_2^2)d + \|\boldsymbol{\mu}_2 - \boldsymbol{\mu}_1\|^2},$$
$$\mathbb{E}[\|\mathbf{Y} - \mathbf{z}\|] \approx \sqrt{2\sigma_2^2 d}.$$

Hence,

$$\mathbb{E}[A_{\mathbf{z}}] \approx \left| \sqrt{2\sigma_2^2 d} - \sqrt{(\sigma_1^2+\sigma_2^2)d + \|\boldsymbol{\mu}_2 - \boldsymbol{\mu}_1\|^2} \right|.$$

Simplifying, we obtain

$$\mathbb{E}[\rho(\mathbf{X},\mathbf{Y})] \approx \left[\frac{m-1}{N-2} \cdot \left| \sqrt{2}\sigma_1 - \sqrt{(\sigma_1^2+\sigma_2^2) + \|\boldsymbol{\mu}_2 - \boldsymbol{\mu}_1\|^2/d} \right| + \frac{n-1}{N-2} \cdot \left| \sqrt{2}\sigma_2 - \sqrt{(\sigma_1^2+\sigma_2^2) + \|\boldsymbol{\mu}_2 - \boldsymbol{\mu}_1\|^2/d} \right| \right].$$

Consider the $N \times d$ data matrix of $\mathbb{Z}_N$. Let $\mathbf{x}_i = (x_{i1}, \ldots, x_{id})$ and $\mathbf{y}_j = (y_{j1}, \ldots, y_{jd})$ be rows of the data matrix where $i \neq j = 1, \ldots, N$. To obtain the asymptotic behavior of the dissimilarity measure δ, we state assumptions (B1)–(B3).

(B1) Let $\boldsymbol{\mu}_{x_t}$ be the t-th component of $\boldsymbol{\mu}_x$. Let $\bar{\mu}_x$ be the mean of $\boldsymbol{\mu}_x$. There exist constants $\sigma_{F_1}^2, \sigma_{F_2}^2 > 0$ and $\nu_x, \nu_y \geq 0$, such that

$$\frac{1}{d} tr\left(\boldsymbol{\Sigma}_x\right) \to \sigma_{F_1}^2, \quad \frac{1}{d} tr\left(\boldsymbol{\Sigma}_y\right) \to \sigma_{F_2}^2,$$

$$\frac{1}{d}\sum_{t=1}^{d} (\boldsymbol{\mu}_{x_t} - \bar{\mu}_x)^2 \to \nu_x^2, \quad \frac{1}{d}\sum_{t=1}^{d} (\boldsymbol{\mu}_{y_t} - \bar{\mu}_y)^2 \to \nu_y^2,$$

respectively, as $d \to \infty$.

(B2) The fourth moments of the components of $\mathbf{X} = (X_1, \ldots, X_d)'$ and $\mathbf{Y} = (Y_1, \ldots, Y_d)'$ are uniformly bounded.

(B3) The following sums are of the order $o(d^2)$,

$$T_x = \mathbf{1}_d' \boldsymbol{\Sigma}_x \mathbf{1}_d = \sum_{i=1}^{d}\sum_{j=1}^{d} \sigma_{(x)ij}, \qquad T_y = \mathbf{1}_d' \boldsymbol{\Sigma}_y \mathbf{1}_d = \sum_{i=1}^{d}\sum_{j=1}^{d} \sigma_{(y)ij}.$$

Lemma 3.6.3 *Let* $\mathbf{x} = (x_1, \ldots, x_d) \sim F(\boldsymbol{\mu}_x, \boldsymbol{\Sigma}_x = (\sigma_{(x)ij})$. *Let* $\bar{x}^d = \frac{1}{d}\sum_{i=1}^{d} x_i$ *and* $[s_x^d]^2 = \frac{1}{d}\sum_{i=1}^{d}(x_i - \bar{x}^d)^2$. *One can show that* $\bar{x}^d = \frac{1}{d}\mathbf{1}'\mathbf{x}$ *has mean* $\bar{\mu}_x^d = \frac{1}{d}\mathbf{1}'\boldsymbol{\mu}_x$ *and variance* $\gamma_{\bar{x}}^2 = \frac{1}{d^2}T_x$ *where* $T_x = \sum_{r=1}^{d}\sum_{s=1}^{d}\sigma_{(x)rs}$. *Moreover,*

$$[\sigma_x^d]^2 = \mathbb{E}(s_x^2) \to \sigma_F^2 + \nu_x^2, \tag{3.17}$$

converges in probability as $d \to \infty$.

Applying Lemma 3.6.3 to the sample drawn from the Y process yields that $\bar{y}^d = \frac{1}{d}\mathbf{1}'\mathbf{y}$ has expectation $\bar{\mu}_y^d = \frac{1}{d}\mathbf{1}'\boldsymbol{\mu}_y$ and variance $\gamma_{\bar{y}}^2 = \frac{1}{d^2}T_y$, where $T_y = \sum_{r=1}^d \sum_{s=1}^d \sigma_{(y)rs}$. Furthermore, the quantity $[\sigma_y^d]^2 = \mathbb{E}(s_y^2)$ converges in probability to $\sigma_{F_2}^2 + \nu_y^2$ as $d \to \infty$.

Assumptions (B1)–(B3) are required to guarantee that both sample means and sample variances approach their corresponding population values. In the special case of independent and identically distributed variables, conditions (B1)–(B3) are automatically satisfied, and convergence in probability follows directly from the weak law of large numbers. In general, however, the variance of s_x^2 depends on mixed moments up to fourth order and admits a non-trivial form. For instance, in the i.i.d. setting, letting μ_4 denote the fourth central moment of the underlying distribution, one obtains $\mathbb{V}ar([s_x^d]^2) = \frac{\mu_4}{d} - \frac{d-3}{d(d-1)}\bar{\sigma}_x^4$ (O'Neill [259]). This expression illustrates the necessity of assumption (B2).

By contrast, in the presence of strong dependence across components, the quantity T_x grows on the order $O(d^2)$ not $o(d^2)$ as assumed in B3. Assumption (B3) therefore enforces a form of weak dependence sufficient to ensure that $\bar{x}^d \to \bar{\mu}_x^d$ and $\bar{y}^d \to \bar{\mu}_y^d$ as $d \to \infty$.

It is instructive to write the variance as the average squared differences of the variables,

$$[s_x^d]^2 = \frac{1}{d(d-1)} \sum_{r=1}^{d} \sum_{s=r+1}^{d} (x_r - x_s)^2.$$

For $r \neq s = 1, \dots d$, it follows that

$$\mathbb{E}(x_r - x_s)^2 = (\mu_{(x)r} - \mu_{(x)s})^2 + \sigma_{(x)r}^2 + \sigma_{(x)s}^2 - 2\rho_{(x)rs}\sigma_{(x)r}\sigma_{(x)s} = \bar{\delta}_0^2(x_r, x_s),$$

$$[\sigma_x^d]^2 = \mathbb{E}([s_x^d]^2) = \frac{1}{d(d-1)} \sum_{r=1}^{d} \sum_{s=r+1}^{d} \bar{\delta}^2(x_r, x_s) \to \sigma_F^2 + \nu_x^2, \tag{3.18}$$

as $d \to \infty$. As $\rho_{(x)rs}$ ranges in the $(-1, 1)$ interval, it follows that the upper and lower bounds for the expected Euclidean distance between x_r and x_s can be written

$$(\mu_{(x)r} - \mu_{(x)s})^2 + (\sigma_{(x)r} - \sigma_{(x)s})^2 \leq \mathbb{E}(x_r - x_s)^2 \leq (\mu_{(x)r} - \mu_{(x)s})^2 + (\sigma_{(x)r} + \sigma_{(x)s})^2. \tag{3.19}$$

This is, in fact, expected value of the squared univariate version of (3.10). The following lemma investigates the behavior of δ as d tends to infinity.

Lemma 3.6.4 *Suppose that $\mathbf{X}_1, \mathbf{X}_2 \sim F_1$, and $\mathbf{Y}_1, \mathbf{Y}_2 \sim F_2$ are independent. If F_1 and F_2 satisfy Assumptions (B1)–(B3), then as $d \to \infty$, $\delta(\mathbf{X}_1, \mathbf{X}_2)$, and $\delta(\mathbf{Y}_1, \mathbf{Y}_2)$ converge in probability to 0, and $\delta(\mathbf{X}_1, \mathbf{Y}_1)$ converges in probability to*

$$\tilde{\delta}(F_1, F_2) = |\sqrt{(\bar{\mu}_x^d - \bar{\mu}_y^d)^2 + (\bar{\sigma}_x^d - \bar{\sigma}_y^d)^2}| \geq 0 \tag{3.20}$$

and equals 0 if and only if $\bar{\mu}_x^d = \bar{\mu}_y^d$, and $\sigma_x^d = \sigma_y^d$.

It is left as an exercise to show that $\mathbb{E}(\delta(\mathbf{X}, \mathbf{Y})) = \tilde{\delta}$ where $\tilde{\delta}$ is defined in (3.20).

Example 3.6.5 *Let $\mathbf{x} = (x_1, \dots, x_d) \sim F_1 = \mathbb{N}_d(\mathbf{0}, \mathbf{I}_d)$ and $\mathbf{y} = (y_1, \dots, y_d) \sim F_2 = \mathbb{N}_d(\boldsymbol{\mu}_y, \mathbf{I}_d)$ be independent random vectors where $\boldsymbol{\mu}_y = (1, -1, 1, -1, \dots, 1, -1)$. If d is odd, then by Lemma 3.6.4 we have $\bar{\mu}_x^d = 0 \neq \bar{\mu}_y^d = \frac{1}{d}$. By Lemma 3.6.3 we have $\tilde{\delta}(F_1, F_2) = 1 \geq 0$ and the dissimilarity index is able to distinguish between F_1 and F_2. Suppose d is even so that $\bar{\mu}_x^d = \bar{\mu}_y^d = 0$. By Lemma 3.6.3, we have $[\sigma_x^d]^2 = \mathbb{E}(s_x^2) \to \sigma_{F_1}^2 + \nu_x^2 = 1 + 0 = 1$, and $[\sigma_y^d]^2 = \mathbb{E}(s_y^2) \to \sigma_{F_2}^2 + \nu_y^2 = 1 + 1 = 2$. It follows that $\bar{\mu}_x^d = \bar{\mu}_y^d$ while $\sigma_x^d \neq \sigma_y^d$ and the dissimilarity index distinguishes between the two distributions. Following Lemma 3.6.4, the limiting value of $\tilde{\delta}(F_1, F_2) = \sqrt{0 + (1 - \sqrt{2})^2} = 0.41$. It is interesting to determine the limiting value of $\tilde{\rho}(F_1, F_2)$. Since $\sigma_{F_1} = \sigma_{F_2} = 1$ and $\nu^2 = \frac{1}{d}\boldsymbol{\mu}_y'\boldsymbol{\mu}_y = 1$ we obtain $\tilde{\rho}(F_1, F_2) = |\sqrt{3} - \sqrt{2}| = 0.31$.*

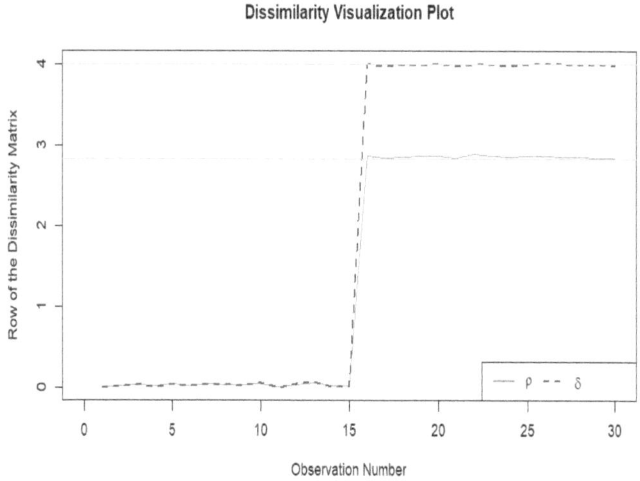

FIGURE 3.3: Row sum of the dissimilarity indices and the limiting values of δ and ρ.

The dissimilarity indices are unable to distinguish between the populations when the first two moments of the distributions are the same. If the one-dimensional marginals of the F_1 and F_2 are identical, then the dissimilarity indices will not be effective in separating the populations.

The dissimilarity measures ρ and δ can detect small differences in the first two-moments of the distributions. Suppose $\mathbf{x} = (x_1, \ldots, x_d)$ where x_k's are i.i.d. $t_{(f_x)}$, a t distribution with f_x degrees of freedom where $f_x > 2$, for $k = 1, \ldots d$. Suppose $\mathbf{y} = (y_1, \ldots, y_d)$ where y_k's are i.i.d. $t_{(f_y)}$, for $k = 1, \ldots d$. It is not difficult to show that $\sigma_{F_1} = \frac{f_x}{f_x - 2}$, $\sigma_{F_2} = \frac{f_x}{f_y - 2}$, and $\nu^2 = 0$. Hence, $\tilde{\rho}(F_1, F_2) > 0$ when $f_x \neq f_y$ as $d \to \infty$. Similarly, for δ index, we obtain $\bar{\mu}_x^d = \bar{\mu}_y^d = 0$, $\sigma_x^2 = \frac{f_x}{f_x - 2}$ and $\sigma_y^2 = \frac{f_y}{f_y - 2}$. Hence, $\tilde{\rho}(F_1, F_2) = |\frac{f_x}{f_x - 2} - \frac{f_y}{f_y - 2}| > 0$ when $f_x \neq f_y$ as $d \to \infty$.

Example 3.6.6 *We generated independent observations,* $\mathbf{X}_i \sim \mathbb{N}_d(-2\ \mathbf{1}, \mathbf{I})$ *for* $i = 1, \ldots 15$ *and* $\mathbf{Y}_i \sim \mathbb{N}_d(2\ \mathbf{1}, \mathbf{I})$ *for* $i = 16, \ldots 30$. *Figure 3.3 shows the row sum of the dissimilarity indices for* $n = 30$ *observations. For the given parameters and* $d = 1000$ *variables, Figure 3.3 also displays the limiting values of* $\tilde{\rho} = 2.8$ *and* $\tilde{\delta} = 4$.

Using Lemma 3.6.4, it is clear that δ does not depend on m and n, asymptotically. On the other hand, ρ depends on m and n, in general. Note that δ is quite robust to unbalanced samples such as $m \gg n$. We considered the following unbalanced case. A computer simulation with 500 replicates with m samples from $\mathbb{N}_d(\mathbf{0}, \mathbf{I}_d)$ and n samples from $\mathbb{N}_d(\mathbf{0}, 5\ \mathbf{I}_d)$ with $d = 100$ gives the values in Table 3.2. It is clear that δ does not depend on m and n,

TABLE 3.2: Estimated and asymptotic values of $\tilde{\rho}$ and $\tilde{\delta}$ for unbalanced samples with d=100.

m	n	$\tilde{\rho}$	$\hat{\tilde{\rho}}$	$\tilde{\delta}$	$\hat{\tilde{\delta}}$
50	15	0.781	0.778	1.271	1.275
15	50	0.961	0.964	1.271	1.273

asymptotically. On the other hand, ρ depends on m and n, in general. One can show δ is quite robust to unbalanced samples when $m \gg n$.

Example 3.6.7 *We compare the dissimilarity indices ρ and δ under $H_0 : F_1 = F_2$ and $H_a : F_1 \neq F_2$ by the following two sampling experiments. The dissimilarity matrices produce $N(N-1)/2$ dissimilarities.*

- *Figure 3.4 shows density plots of ρ and δ values with bandwidth of 0.05 when $m = n = 40$, $d = 1000$. The left panel is based on independent samples taken from $\mathbf{X}$, and $\mathbf{Y} \sim \mathbb{LN}(\mathbf{0}, \mathbf{I})$ under H_0. While ρ shows larger dissimilarities than δ neither one dominates the other.*

- *Considering shift alternatives, the right panel of Figure 3.4 shows density plots for samples from $\mathbf{X} \sim \mathbb{N}(\mathbf{0}, \mathbf{I})$, and $\mathbf{Y} \sim \mathbb{N}(\mathbf{1}, \mathbf{I})$. Both density estimates are bimodal. The first cluster of both dissimilarity measures are very close to each other while the second cluster of δ provides larger dissimilarities than the second cluster of ρ.*

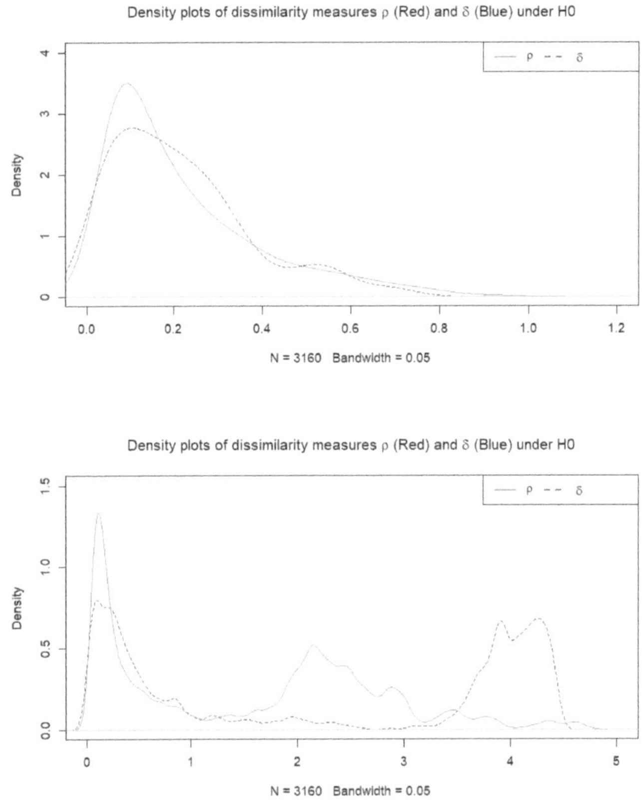

FIGURE 3.4: Density plots of ρ and δ. Left panel: $\mathbf{X}, \mathbf{Y} \sim \mathbb{LN}(\mathbf{0}, \mathbf{I})$. Right Panel: $\mathbf{X} \sim \mathbb{LN}(\mathbf{0}, \mathbf{I})$, $\mathbf{Y} \sim \mathbb{LN}(\mathbf{1}, \mathbf{I})$.

3.6.4 Finite-sample Behavior of $\tilde{\rho}$ and $\tilde{\delta}$

The limiting behavior of the dissimilarity indices ρ and δ as $d \to \infty$ in Lemmas 3.6.3 and 3.6.4. It is important to observe the behavior of $\tilde{\rho}$ and $\tilde{\delta}$ for fixed sample size as and finite d.

We generated $n = m = 20$ independent observations from the following distributions with i.i.d. components under H_0.

- $\mathbf{X}, \mathbf{Y} \sim \mathbb{N}(\mathbf{2}_d, \mathbf{I})$, $\quad \mathbf{X}, \mathbf{Y} \sim \beta(0.5, 0.5)$, and $\quad \mathbf{X}, \mathbf{Y} \sim \chi^2_{(2)}$.

We also generated $m = n = 20$ independent observations from the following distributions under the alternative H_a,

- $\mathbf{X} \sim \mathbb{N}(\mathbf{1}, \mathbf{I})$ and $\mathbf{Y} \sim \mathbb{N}(\mathbf{3}_d, \mathbf{I})$, $\quad \mathbf{X} \sim \beta(0.5, 0.5)$, and $\mathbf{Y} \sim \beta(0, 1) + 0.5$,
- $\mathbf{X} \sim \chi^2_{(2)}$ and $\mathbf{Y} \sim \chi^2_{(4)}$.

We repeat the experiment 1000 times and report the mean and standard deviations. Table 3.3 shows the values of $\tilde{\rho}$ and $\tilde{\delta}$ and their estimates $\hat{\tilde{\rho}}$ and $\hat{\tilde{\delta}}$. It is clear that the pug-in estimates $\hat{\tilde{\rho}}$ and $\hat{\tilde{\delta}}$ converge fast to $\tilde{\rho}$ and $\tilde{\delta}$, respectively, for large d and fixed sample sizes under different distributions.

TABLE 3.3: Estimated and asymptotic values of $\tilde{\rho}$ and $\tilde{\delta}$ for increasing d.

Dist.	d	$\tilde{\delta}$	$\hat{\tilde{\delta}}$	$SD(\hat{\tilde{\delta}})$	$\tilde{\rho}$	$\hat{\tilde{\rho}}$	$SD(\hat{\tilde{\rho}})$
H_0 : Normal	200	0	0.023	0.014	0	0.008	0.0067
	400	0	0.016	0.008	0	0.006	0.005
	800	0	0.011	0.006	0	0.004	0.003
H_0 : beta	200	0	0.007	0.004	0	0.001	0.001
	400	0	0.004	0.003	0	0.001	0.001
	800	0	0.003	0.001	0	0	0.000
$H_0 : \chi^2$	200	0	0.071	0.049	0	0.037	0.031
	400	0	0.045	0.026	0	0.025	0.018
	800	0	0.029	0.017	0	0.015	0.011
H_a : Normal	200	2	2	0.021	1.04	1.04	0.005
	400	2	2	0.017	1.04	1.04	0.003
	800	2	2	0.011	1.04	1.04	0.002
H_a : Beta	200	0.5	0.5	0.008	0.207	0.207	0.001
	400	0.5	0.5	0.005	0.207	0.207	0.001
	800	0.5	0.5	0.003	0.207	0.207	0.001
$H_a : \chi^2$	200	2.16	2.16	0.076	0.586	0.600	0.029
	400	2.16	2.16	0.045	0.586	0.595	0.019
	800	2.16	2.16	0.035	0.586	0.593	0.014

3.6.5 Finite-sample Guarantee

Definition 3.6.1 *A random variable X is called sub-Gaussian if there exists a constant $K > 0$ such that*

$$\mathbb{E}[\exp(t(X - \mathbb{E}X))] \leq \exp(Kt^2) \quad \textit{for all } t \in \mathbb{R}.$$

A sub-Gaussian variable has tails that decay at least as fast as a Gaussian distribution. We state the Bernstein inequality for sub-exponential variables. Let $W_1, \dots, W_d$ be independent centered variables with sub-exponential tails. Then

$$\mathbb{P}\left(\left|\frac{1}{d}\sum_{k=1}^{d} W_k\right| > \epsilon\right) \leq 2\exp(-cd\epsilon^2),$$

for some constant $c > 0$. Let

$$W_k = (X_k - Y_k)^2 - \mathbb{E}[(X_k - Y_k)^2].$$

Since X_k and Y_k are sub-Gaussian, their squares are sub-exponential. Applying the Bernstein inequality

$$\frac{1}{d}\sum_{k=1}^{d} W_k = \frac{1}{d}\|\mathbf{X} - \mathbf{Y}\|^2 - \mathbb{E}\frac{1}{d}\|\mathbf{X} - \mathbf{Y}\|^2$$

Thus,

$$\mathbb{P}\left(\left|\frac{1}{d}\|\mathbf{X} - \mathbf{Y}\|^2 - \mathbb{E}\frac{1}{d}\|\mathbf{X} - \mathbf{Y}\|^2\right| > \epsilon\right) \leq 2\exp(-cd\epsilon^2).$$

Hence, the squared distance concentrates sharply around its mean with exponentially small error probability. Concentration for Lipschitz functions of sub-Gaussian vectors follows.

Lemma 3.6.5 *Let* $\mathbf{X}$ *be a vector with independent sub-Gaussian coordinates and let* f *be L-Lipschitz. Then, it follows that*

$$\mathbb{P}(|f(\mathbf{X}) - \mathbb{E}f(\mathbf{X})| > \epsilon) \leq 2\exp\left(-\frac{c\epsilon^2}{L^2}\right).$$

Proof: Define

$$f(\mathbf{X}, \mathbf{Y}, \mathbf{Z}) = |\|\mathbf{X} - \mathbf{Z}\| - \|\mathbf{Y} - \mathbf{Z}\||.$$

By Lipschitz property, $|\|\mathbf{a}\| - \|\mathbf{b}\|| \leq \|\mathbf{a} - \mathbf{b}\|$. Thus f is Lipschitz with a constant proportional to 1. By applying the concentration inequality and after normalization by $\sqrt{d}$, we can say that $\frac{1}{\sqrt{d}}f$ concentrates with exponential tails. □

Hence, the dissimilarity indices are stable and do not fluctuate wildly. We present the result for ρ index, but the same applies to δ index. To obtain finite-sample separation guarantee we state the following. We define the individual dissimilarity contributions

$$D_i = |\|\mathbf{X}_i - \mathbf{Z}_i\| - \|\mathbf{Y}_i - \mathbf{Z}_i\||,$$

and define the overall dissimilarity index as the average $D = \rho(\mathbf{X}, \mathbf{Y}) = \frac{1}{\sqrt{d(N-2)}}\sum_{i=1}^{N-2} D_i$.

Theorem 3.6.1 *Assume* $\Delta = \frac{1}{\sqrt{d}}\|\boldsymbol{\mu}_1 - \boldsymbol{\mu}_2\| > 0$. *Then there exist constants* $c_1, c_2 > 0$ *such that*

1. *(Mean separation):* $\mathbb{E}[\rho(\mathbf{X}, \mathbf{Y})] - \mathbb{E}[\rho(\mathbf{X}, \mathbf{X}')] \geq c_1\Delta$.
2. *(Exponential error bound):* $\mathbb{P}(\rho(\mathbf{X}, \mathbf{Y}) \leq \rho(\mathbf{X}, \mathbf{X}')) \leq \exp(-c_2 d)$.

Proof: For fixed $\mathbf{Z}$, we have $\mathbb{E}\|\mathbf{X} - \mathbf{Z}\| - \mathbb{E}\|\mathbf{Y} - \mathbf{Z}\| \approx \|\boldsymbol{\mu}_1 - \boldsymbol{\mu}_2\|$. Averaging over $\mathbf{Z}$, $\mathbb{E}_{\mathbf{Z}}[\cdot]$ preserves this difference. Each term in ρ satisfies

$$\mathbb{P}(|D_i - \mathbb{E}D_i| > \epsilon) \leq \exp(-cd\epsilon^2)$$

By independence D also concentrates exponentially. Note that $\mathbb{P}(\rho(X, Y) \leq \rho(X, X'))$ is contained in

$$\mathbb{P}(|\rho(X, Y) - \mathbb{E}\rho(X, Y)| > \epsilon) + \mathbb{P}(|\rho(X, X') - \mathbb{E}\rho(X, X')| > \epsilon)$$

Each term is bounded by $\exp(-cd)$, so it is $\leq 2\exp(-cd)$. Both quantities are tightly concentrated, so the probability they cross is extremely small. □

What the finite-sample guarantee mean? The results guarantee that the method correctly ranks cross-sample dissimilarities above within-sample ones and the probability of error decays exponentially in dimension d. Even at finite d (not asymptotic), the method detects differences reliably and becomes extremely accurate as d increases. Even at finite d, the method detects differences reliably and becomes extremely accurate as d increases.

3.7 Mean and Variance of the j-th Row of a Data Matrix

Let $\mathbf{X} \in \mathbb{R}^{m \times d}$ be a data matrix whose rows

$$\mathbf{X}_j^d = (X_{j1}, \dots, X_{jd})' \in \mathbb{R}^d, \qquad j = 1, \dots, m,$$

are independent and identically distributed as F with mean vector $\boldsymbol{\mu} \in \mathbb{R}^d$ and covariance matrix $\boldsymbol{\Sigma} = (\sigma_{ij}) \in \mathbb{R}^{d \times d}$.

For the mean vector $\bar{X}_j^d = \frac{1}{d}\sum_{k=1}^{d} X_{jk}$. It follows that

$$\bar{X}_j^d = \frac{1}{d}\sum_{k=1}^{d} X_{jk}, \quad \mathbb{E}(\bar{X}_j^d) = \frac{1}{d}\sum_{k=1}^{d} \mathbb{E}(X_{jk}) = \frac{1}{d}\mathbf{1}'\boldsymbol{\mu}, \quad \mathbb{V}\mathrm{ar}(\bar{X}^d) = \frac{1}{d^2}\mathbf{1}'\boldsymbol{\Sigma}\mathbf{1}. \tag{3.21}$$

Hence the expected value of the j-th row mean is the average of the individual components of the mean vector. Define the centering matrix $\mathbf{P} = \mathbf{I}_d - \frac{1}{d}\mathbf{1}\mathbf{1}'$. The variance across the d entries of j-th row is

$$[s_j^d]^2 = \frac{1}{d}\sum_{k=1}^{d}\left(X_{jk} - \bar{X}_j^d\right)^2 = \frac{1}{d}\|\mathbf{X}_j^d - \bar{X}_j^d\mathbf{1}\|^2 = \frac{1}{d}\mathbf{X}_j'^d\mathbf{P}\mathbf{X}_j^d,$$

where $\mathbf{1} \in \mathbb{R}^d$ is the vector of ones. To compute $\mathbb{E}[[s_j^d]^2]$ a standard quadratic-form identity gives

$$\mathbb{E}[\mathbf{X}_j'^d\mathbf{P}\mathbf{X}_j^d] = \mathrm{tr}(\mathbf{P}\boldsymbol{\Sigma}) + \boldsymbol{\mu}'\mathbf{P}\boldsymbol{\mu}.$$

Hence,

$$\mathbb{E}[[s_j^d]^2] = \frac{1}{d}\left(\mathrm{tr}(\mathbf{P}\boldsymbol{\Sigma}) + \boldsymbol{\mu}'\mathbf{P}\boldsymbol{\mu}\right).$$

We compute the trace term and obtain $\mathrm{tr}(\mathbf{P}\boldsymbol{\Sigma}) = \mathrm{tr}(\boldsymbol{\Sigma}) - \frac{1}{d}\mathbf{1}'\boldsymbol{\Sigma}\mathbf{1}$. Therefore,

$$\mathbb{E}[[s_j^d]^2] = \frac{1}{d}\left(\mathrm{tr}(\boldsymbol{\Sigma}) - \frac{1}{d}\mathbf{1}'\boldsymbol{\Sigma}\mathbf{1} + \boldsymbol{\mu}'\mathbf{P}\boldsymbol{\mu}\right). \tag{3.22}$$

It is noteworthy that $\mathbb{E}[[s_j^d]^2]$ depends on both the mean vector and the covariance matrix. Any changes in the mean vector or covariance matrix is reflected in $\mathbb{E}[[s_j^d]^2]$. Next we look at some special cases.

Equal Means

If all components have the same mean ($\mu_1 = \cdots = \mu_d$), then $\mathbf{P}\boldsymbol{\mu} = 0$ and the formula simplifies to

$$\mathbb{E}[[s_j^d]^2] = \frac{1}{d}\left(\mathrm{tr}(\boldsymbol{\Sigma}) - \frac{1}{d}\mathbf{1}'\boldsymbol{\Sigma}\mathbf{1}\right).$$

Note that $\mathrm{tr}(\boldsymbol{\Sigma}) = \sum_{k=1}^{d}\sigma_{kk}$ (sum of marginal variances), and $\mathbf{1}'\boldsymbol{\Sigma}\mathbf{1} = \sum_{k=1}^{d}\sum_{\ell=1}^{d}\sigma_{k\ell}$ (total covariance). Thus, the expected row variance depends on the sum of marginal variances, the total covariance between coordinates, and the structure of the mean vector. Strong positive correlation reduces the row variance because of the subtraction term $\frac{1}{d^2}\mathbf{1}'\boldsymbol{\Sigma}\mathbf{1}$.

Zero Mean: $\mu_1 = \cdots = \mu_d = 0$

If $\boldsymbol{\mu} = 0$, then automatically $\boldsymbol{\mu}'\mathbf{P}\boldsymbol{\mu} = 0$. Thus,

$$\mathbb{E}[[s_j^d]^2] = \frac{1}{d}\left(\text{tr}(\boldsymbol{\Sigma}) - \frac{1}{d}\mathbf{1}'\boldsymbol{\Sigma}\mathbf{1}\right).$$

Equal Variances Only

Assume $\sigma_{kk} = \sigma^2$ for all k, but allow arbitrary covariances $\sigma_{k\ell}$ for $k \neq \ell$. Then, we get $\text{tr}(\boldsymbol{\Sigma}) = d\sigma^2$. Also,

$$\mathbf{1}'\boldsymbol{\Sigma}\mathbf{1} = \sum_{k=1}^{d} \sigma_{kk} + \sum_{k \neq \ell} \sigma_{k\ell} = d\sigma^2 + \sum_{k \neq \ell} \sigma_{k\ell}.$$

Substitute into the zero-mean formula and simplify to obtain

$$\mathbb{E}[[s_j^d]^2] = \sigma^2 - \frac{1}{d^2}\left(d\sigma^2 + \sum_{k \neq \ell} \sigma_{k\ell}\right).$$

Thus, even with equal marginal variances, correlations reduce the row variance.

Equal Correlations Only

Assume $\mathbb{V}ar(X_{jk}) = \sigma_k^2$, and for $k \neq \ell$, $\text{Corr}(X_{jk}, X_{j\ell}) = \rho$. Then,

$$\sigma_{k\ell} = \begin{cases} \sigma_k^2 & k = \ell, \\ \rho\sigma_k\sigma_\ell & k \neq \ell. \end{cases}$$

Compute $\mathbf{1}'\boldsymbol{\Sigma}\mathbf{1} = \sum_{k=1}^{d} \sigma_k^2 + \rho \sum_{k \neq \ell} \sigma_k\sigma_\ell$. Note that

$$\sum_{k \neq \ell} \sigma_k\sigma_\ell = \left(\sum_{k=1}^{d} \sigma_k\right)^2 - \sum_{k=1}^{d} \sigma_k^2.$$

Hence,

$$\mathbf{1}'\boldsymbol{\Sigma}\mathbf{1} = \sum_{k=1}^{d} \sigma_k^2 + \rho\left[\left(\sum_{k=1}^{d} \sigma_k\right)^2 - \sum_{k=1}^{d} \sigma_k^2\right].$$

Substitute into the zero-mean formula

$$\mathbb{E}[[s_j^d]^2] = \frac{1}{d}\left[\sum_{k=1}^{d} \sigma_k^2 - \frac{1}{d}\mathbf{1}'\boldsymbol{\Sigma}\mathbf{1}\right].$$

This explicitly shows how increasing ρ decreases the expected row variance.

3.8 Variance of the Row-Wise Sample Variance

Let $\mathbf{X}_j^d = (X_{j1}, \ldots, X_{jd})' \in \mathbb{R}^d$ denote the j-th row of a data matrix. Assume $\mathbb{E}(\mathbf{X}_j^d) = \boldsymbol{\mu}$, and $\mathbb{C}ov(\mathbf{X}_j^d) = \boldsymbol{\Sigma}$, and $\mathbb{E}\|\mathbf{X}_j^d\|^4 < \infty$. Define the row mean $\bar{X}_j^d = \frac{1}{d}\mathbf{1}'\mathbf{X}_j^d$, and the row variance

$$[s_j^d]^2 = \frac{1}{d}\sum_{k=1}^{d}(X_{jk} - \bar{X}_j^d)^2 = \frac{1}{d}\mathbf{X}_j^{'d}\mathbf{P}\mathbf{X}_j^d,$$

where $\mathbf{P} = \mathbf{I}_d - \frac{1}{d}\mathbf{1}\mathbf{1}'$ is the centering matrix. Note that $\mathbf{P}$ is idempotent, so that $\mathbf{P}' = \mathbf{P}$, and $\mathbf{P}^2 = \mathbf{P}$. Define

$$Q = \mathbf{X}_j^{'d}\mathbf{P}\mathbf{X}_j^d, \qquad [s_j^d]^2 = \frac{1}{d}Q.$$

Therefore,

$$\mathbb{V}ar([s_j^d]^2) = \frac{1}{d^2}\mathbb{V}ar(Q).$$

We will make the following centering decomposition,

$$\mathbf{X}_j^d = \boldsymbol{\mu} + \mathbf{Z}, \qquad \mathbb{E}(\mathbf{Z}) = \mathbf{0}, \qquad \mathbb{C}\mathbf{ov}(\mathbf{Z}) = \boldsymbol{\Sigma}.$$

It follows that

$$Q = \boldsymbol{\mu}'\mathbf{P}\boldsymbol{\mu} + 2\boldsymbol{\mu}'\mathbf{P}\mathbf{Z} + \mathbf{Z}'\mathbf{P}\mathbf{Z}.$$

Taking variance,

$$\mathbb{V}ar(Q) = 4\,\boldsymbol{\mu}'\mathbf{P}\boldsymbol{\Sigma}\mathbf{P}\boldsymbol{\mu} + \mathbb{V}ar(\mathbf{Z}'\mathbf{P}\mathbf{Z}) + 4\,\mathbb{C}ov(\boldsymbol{\mu}'\mathbf{P}\mathbf{Z}, \mathbf{Z}'\mathbf{P}\mathbf{Z}).$$

For any zero-mean vector with finite fourth moments,

$$\mathbb{V}ar(\mathbf{Z}'\mathbf{A}\mathbf{Z}) = 2\,\mathrm{tr}(\mathbf{A}\boldsymbol{\Sigma}\mathbf{A}\boldsymbol{\Sigma}) + \kappa(\mathbf{A}),$$

where

$$\kappa(\mathbf{A}) = \sum_{i,j,k,l}\mathbf{A}_{ij}\mathbf{A}_{kl}\mathrm{Cum}(Z_i, Z_j, Z_k, Z_l).$$

Hence,

$$\mathbb{V}ar([s_j^d]^2) = \frac{1}{d^2}\left[4\,\boldsymbol{\mu}'\mathbf{P}\boldsymbol{\Sigma}\mathbf{P}\boldsymbol{\mu} + 2\,\mathrm{tr}(\mathbf{P}\boldsymbol{\Sigma}\mathbf{P}\boldsymbol{\Sigma}) + \kappa(\mathbf{P}) + 4\,\mathbb{C}ov(\boldsymbol{\mu}'\mathbf{P}\mathbf{Z}, \mathbf{Z}'\mathbf{P}\mathbf{Z})\right].$$

If we assume $\mathbb{E}(Z_iZ_jZ_k) = 0$ for all i, j, k, then, $\mathbb{C}ov(\boldsymbol{\mu}'\mathbf{P}\mathbf{Z}, \mathbf{Z}'\mathbf{P}\mathbf{Z}) = 0$. Assume $Z_1, \ldots, Z_d$ are independent. Then

$$\mathrm{Cum}(Z_i, Z_j, Z_k, Z_l) = \begin{cases}\gamma_i, & i = j = k = l,\\ 0, & \text{otherwise.}\end{cases}$$

Thus,

$$\kappa(\mathbf{P}) = \sum_{i=1}^{d}P_{ii}^2\gamma_i = \left(1 - \frac{1}{d}\right)^2\sum_{i=1}^{d}\gamma_i.$$

Multivariate Normal Distribution

If we assume $\mathbf{X}_j^d \sim \mathcal{N}(\boldsymbol{\mu}, \boldsymbol{\Sigma})$, then it follows that

$$\kappa(\mathbf{P}) = 0, \qquad \mathbb{C}ov(\boldsymbol{\mu}'\mathbf{P}\mathbf{Z}, \mathbf{Z}'\mathbf{P}\mathbf{Z}) = 0.$$

Therefore, under the multivariate normal distribution,

$$\mathbb{V}ar([s_j^d]^2) = \frac{4}{d^2}\boldsymbol{\mu}'\mathbf{P}\boldsymbol{\Sigma}\mathbf{P}\boldsymbol{\mu} + \frac{2}{d^2}\mathrm{tr}(\mathbf{P}\boldsymbol{\Sigma}\mathbf{P}\boldsymbol{\Sigma}).$$

Under normality, the variance of the row-wise sample variance depends only on the second-order structure $(\boldsymbol{\mu}, \boldsymbol{\Sigma})$. It is interesting that the variance depends on all parameters of multivariate normal distribution, including the means, variances, and covariances. All higher-order moment contributions vanish exactly, not approximately. This makes the normal case analytically tractable and serves as a natural benchmark for assessing non-Gaussian effects.

Symmetric Elliptical Distributions

Assume $\mathbf{Z}$ is a symmetric elliptical with a radial kurtosis factor

$$\tau = \frac{\mathbb{E}(R^4)}{d(d+2)}.$$

It follows that

$$\mathbb{V}ar(\mathbf{Z}'\mathbf{P}\mathbf{Z}) = 2\tau\,\mathrm{tr}(\mathbf{P}\boldsymbol{\Sigma}\mathbf{P}\boldsymbol{\Sigma}) + (\tau - 1)\,\mathrm{tr}(\mathbf{P}\boldsymbol{\Sigma})^2.$$

Therefore,

$$\mathbb{V}ar([s_j^d]^2) = \frac{4}{d^2}\boldsymbol{\mu}'\mathbf{P}\boldsymbol{\Sigma}\mathbf{P}\boldsymbol{\mu} + \frac{1}{d^2}\left[2\tau\,\mathrm{tr}(\mathbf{P}\boldsymbol{\Sigma}\mathbf{P}\boldsymbol{\Sigma}) + (\tau - 1)\,\mathrm{tr}(\mathbf{P}\boldsymbol{\Sigma})^2\right]. \tag{3.23}$$

Under symmetric elliptical distributions, the variance of the row-wise sample variance also depends on the kurtosis of the distribution. The multivariate normal distribution corresponds to $\tau = 1$.

3.9 Exact Distribution of Row mean and Variance

Consider a single row of the matrix $\mathbf{X} = (X_1, \ldots, X_d) \in \mathbb{R}^d$.

Constant Correlation

Assume $\mathbf{X} \sim \mathbb{N}_d(\boldsymbol{\mu}, \boldsymbol{\Sigma})$, and $\boldsymbol{\Sigma} = (1-\rho)\mathbf{I}_d + \rho\,\mathbf{1}\mathbf{1}'$, where $-1/(d-1) < \rho < 1$, and define

$$\bar{X}^d = \frac{1}{d}\sum_{t=1}^{d} X_t, \qquad [s^d_{(X)}]^2 = \frac{1}{d}\sum_{t=1}^{d}(X_t - \bar{X}^d)^2.$$

Because $\mathbf{X}$ has mean $\boldsymbol{\mu}$, we have $\mathbb{E}(\bar{X}^d) = \frac{1}{d}\mathbf{1}'\boldsymbol{\mu}$. The variance is unaffected by the mean shift,

$$\mathbb{V}\mathrm{ar}(\bar{X}^d) = \frac{1}{d^2}\mathbf{1}'\boldsymbol{\Sigma}\mathbf{1} = \frac{1}{d}(1 + (d-1)\rho).$$

Thus,

$$\bar{X}^d \sim \mathbb{N}\left(\frac{1}{d}\mathbf{1}'\boldsymbol{\mu},\ \frac{1 + (d-1)\rho}{d}\right).$$

To find the distribution of row variance, we diagonalize $\boldsymbol{\Sigma}$,

$$\boldsymbol{\Sigma} = \mathbf{Q}\boldsymbol{\Lambda}\mathbf{Q}', \qquad \boldsymbol{\Lambda} = \mathrm{diag}(1-\rho, \ldots, 1-\rho, 1 + (d-1)\rho),$$

the last eigenvalue associated with $\mathbf{1}_d$. Write

$$\mathbf{X} = \boldsymbol{\mu} + \mathbf{Q}\boldsymbol{\Lambda}^{1/2}\mathbf{Z}, \qquad \mathbf{Z} \sim \mathbb{N}_d(\mathbf{0}, \mathbf{I}_d).$$

Define the projection onto the mean-free subspace $\mathbf{P} = \mathbf{I}_d - \frac{1}{d}\mathbf{1}\mathbf{1}'$. The quadratic form for the sample variance is

$$d\,[s^d_{(X)}]^2 = \mathbf{X}'\mathbf{P}\mathbf{X} = (\boldsymbol{\mu} + \mathbf{Q}\boldsymbol{\Lambda}^{1/2}\mathbf{Z})'\mathbf{P}(\boldsymbol{\mu} + \mathbf{Q}\boldsymbol{\Lambda}^{1/2}\mathbf{Z}).$$

Because $\mathbf{P}\boldsymbol{\mu}$ removes the projection onto $\mathbf{1}$, and the only eigenvector of $\boldsymbol{\Sigma}$ not in the residual subspace is exactly $\mathbf{1}$, we have $\mathbf{P}\boldsymbol{\mu}$ lying in the same $(d-1)$-dimensional subspace as the residuals. But $\mathbf{PQ}$ annihilates the last eigenvector and preserves the first $d-1$ eigenvectors.

$$\mathbf{P}\boldsymbol{\mu} \in \text{span}\{\text{eigenvectors with eigenvalue } 1-\rho\},$$

while the cross--term and mean--term vanish because $\mathbf{P}$ destroys the mean direction. Hence, the quadratic form is

$$d\,[s^d_{(X)}]^2 = (1-\rho)\sum_{i=1}^{d-1} Z_i^2.$$

Thus, $\frac{d}{1-\rho}[s^d_{(X)}]^2 \sim \chi^2_{d-1}$. The vector $\bar{X}^d$ depends only on the projection of $\mathbf{X}$ onto span$(\mathbf{1})$, while $(s^d_{(X)})^2$ depends only on the projection into the orthogonal complement. Normal orthogonal projections are independent. Thus we have independence, $\bar{X}^d \perp [s^d_{(X)}]^2$ and

$$\begin{aligned} &\bar{X}^d \sim \mathbb{N}\left(\frac{1}{d}\mathbf{1}'\boldsymbol{\mu},\ \frac{1+(d-1)\rho}{d}\right), \\ &\frac{d}{1-\rho}[s^d_{(X)}]^2 \sim \chi^2_{d-1}, \qquad \bar{X}^d \text{ independent of } [s^d_{(X)}]^2. \end{aligned} \tag{3.24}$$

Spatial Correlation

Consider a row $\mathbf{X} = (X_1, \ldots, X_d)$ with a spatially correlated Toeplitz covariance,

$$\mathbf{X} \sim \mathbb{N}_d(\boldsymbol{\mu}, \boldsymbol{\Sigma}), \qquad \boldsymbol{\Sigma} = [\sigma_{jj'}], \quad \sigma_{jj'} = \rho^{|j-j'|}, \quad |\rho| \le 1.$$

The row mean is $\bar{X}^d = \frac{1}{d}\sum_{j=1}^d X_j$.

To obtain the distribution of the row mean note that its expectation and variance are

$$\mathbb{E}(\bar{X}^d) = \frac{1}{d}\mathbf{1}'\boldsymbol{\mu}, \qquad \mathbb{V}\text{ar}(\bar{X}^d) = \frac{1}{d^2}\sum_{j,j'=1}^{d}\sigma_{jj'} = \frac{1}{d^2}\sum_{j,j'=1}^{d}\rho^{|j-j'|}.$$

Evaluating the Toeplitz sum gives

$$\mathbb{V}\text{ar}(\bar{X}^d) = \frac{1}{d^2}\left[d + 2\sum_{k=1}^{d-1}(d-k)\rho^k\right].$$

Hence,

$$\bar{X}^d \sim \mathbb{N}\left(\frac{1}{d}\mathbf{1}'\boldsymbol{\mu},\ \frac{1}{d^2}\left[d + 2\sum_{k=1}^{d-1}(d-k)\rho^k\right]\right).$$

To obtain the distribution of the row variance, the mean-centered residual vector is

$$\mathbf{r} = \mathbf{X} - \bar{X}^d\,\mathbf{1} = \mathbf{PX}, \qquad \mathbf{P} = \mathbf{I}_d - \frac{1}{d}\mathbf{1}\mathbf{1}'.$$

The row variance satisfies

$$d[s^d_{(X)}]^2 = \mathbf{r}'\mathbf{r} = \mathbf{X}'\mathbf{P}\mathbf{X}.$$

Write $\mathbf{X} = \boldsymbol{\mu} + \mathbf{Z}$, where $\mathbf{Z} \sim \mathbb{N}_d(\mathbf{0}, \boldsymbol{\Sigma})$. Then,

$$\mathbf{X}'\mathbf{P}\mathbf{X} = (\boldsymbol{\mu} + \mathbf{Z})'\mathbf{P}(\boldsymbol{\mu} + \mathbf{Z}) = \mathbf{Z}'\mathbf{P}\mathbf{Z} + 2\,\boldsymbol{\mu}'\mathbf{P}\mathbf{Z} + \boldsymbol{\mu}'\mathbf{P}\boldsymbol{\mu}.$$

Diagonalize the residual covariance,

$$\mathbf{P}\boldsymbol{\Sigma}\mathbf{P} = \mathbf{V}\,\mathrm{diag}(\lambda_1, \dots, \lambda_d)\,\mathbf{V}', \qquad \mathbf{V}'\mathbf{Z} = (Z_1, \dots, Z_d)',$$

where $Z_i \sim \mathbb{N}(0, 1)$ independent.

Projecting the mean,

$$\mathbf{P}\boldsymbol{\mu} = \mathbf{V}\boldsymbol{\delta}, \qquad \boldsymbol{\delta} = (\delta_1, \dots, \delta_d)'.$$

Thus, the quadratic form becomes the non-central weighted chi-square representation

$$d[s^d_{(X)}]^2 = \sum_{i=1}^{d} \lambda_i (Z_i + \delta_i)^2,$$

a linear combination of non-central χ^2_1 variables,

$$d[s^d_{(X)}]^2 \sim \sum_{i=1}^{d} \lambda_i\, \chi^2_1(\delta_i^2).$$

Because Toeplitz covariance couples the mean direction with residual directions, the row mean and row variance depend on overlapping linear combinations of $\mathbf{X}$. Therefore, $barX^d$ and $s^d_{(X)}]^2$ are not independent. For large d, one may approximate the distribution of $[s^d_{(X)}]^2$ using the eigenvalues of $\mathbf{P}\boldsymbol{\Sigma}\mathbf{P}$ and the projected mean $\mathbf{P}\boldsymbol{\mu}$.

3.10 Asymptotic Distributions of $\hat{A}^2_\delta$

We will see in forthcoming chapters that the distribution of the estimated asymptotic value of dissimilarity δ is of particular interest in test of hypothesis, clustering methods, and change-point problems. For example, a test statistic for testing the equality of two distribution functions $H_0 : F_1 = F_2$ can be constructed based on the square of the asymptotic form of $\tilde{\delta}(F_1, F_2)$ given in (3.20). In this section we study the asymptotic distribution

$$\hat{A}^2_\delta = (\bar{X}^d - \bar{Y}^d)^2 + (s^d_{\mathbf{X}} - s^d_{\mathbf{Y}})^2$$

where $\mathbf{X}_1, \dots, \mathbf{X}_m$ and $\mathbf{Y}_1, \dots, \mathbf{Y}_n$ are d-dimensional row vectors. Row-level summaries are

$$\bar{X}^d_i = \frac{1}{d}\mathbf{1}'_d\mathbf{X}_i, \quad (s^d_{\mathbf{X}_i})^2 = \frac{1}{d}\sum_{j=1}^{d}(X_{ij} - \bar{X}^d_i)^2,$$

and analogously for $\mathbf{Y}_i$. We analyze $\hat{A}^2_\delta$ when the sample sizes $m, n \to \infty$. For small m, n we recommend full or random permutation distribution. We study the case for the general mean vector $\boldsymbol{\mu}$ and covariance matrix $\boldsymbol{\Sigma}$, and for the spherical case, $\boldsymbol{\mu} = \mu\mathbf{1}_d$, $\boldsymbol{\Sigma} = \sigma^2 I_d$.

3.10.1 Moments of Row Means and Variances

Let $\mathbf{X}_i \in \mathbb{R}^d$ be i.i.d. with $\mathbb{E}[\mathbf{X}_i] = \boldsymbol{\mu}_1$, $\mathbb{C}ov(\mathbf{X}_i) = \boldsymbol{\Sigma}_1$, $\mathbb{E}\|\mathbf{X}_i\|^4 < \infty$. Similarly define $\mathbf{Y}_i$ with $\boldsymbol{\mu}_2$ and $\boldsymbol{\Sigma}_2$. Define row statistics

$$T^X_i = \begin{pmatrix} \bar{X}^d_i \\ s^d_{\mathbf{X}_i} \end{pmatrix}, \quad T^Y_i = \begin{pmatrix} \bar{Y}^d_i \\ s^d_{\mathbf{Y}_i} \end{pmatrix}.$$

The mean and variance of row statistics are given in (3.21), (3.22), and (3.23). If coordinates are i.i.d. with fourth moment μ_4,

$$\mathbb{V}ar((s^d_{\mathbf{X}_i})^2) = \frac{\mu_4 - \sigma^4}{d} + O(d^{-2}).$$

We know $(s^d_{\mathbf{X}_i})^2 = \frac{1}{d}\sum_{j=1}^d X_{ij}^2 - (\bar{X}_i^d)^2$ so to find variance of $(s^d_{\mathbf{X}_i})^2$ we write $(s^d_{\mathbf{X}_i})^2 = B_d - C_d$, where $B_d = \frac{1}{d}\sum X_{ij}^2$, and $C_d = (\bar{X}_i^d)^2$. Then, $[\mathbb{V}ar((s^d_{\mathbf{X}_i})^2) = \mathbb{V}ar(B_d) + \mathbb{V}ar(C_d) - 2\mathbb{C}ov(B_d, C_d)$. If coordinates are independent and identically distributed with

$$\mathbb{E}[X_{ij}] = \mu, \quad \mathbb{V}ar(X_{ij}) = \sigma^2, \quad \mathbb{E}[X_{ij}^4] = \mu_4,$$

then, $\mathbb{V}ar(B_d) = \frac{\mu_4 - \sigma^4}{d}$. Since $\mathbb{V}ar(\bar{X}_i^d) = \frac{\sigma^2}{d}$, we obtain $\mathbb{V}ar(C_d) = O(d^{-2})$. Also, $\mathbb{C}ov(B_d, C_d) = O(d^{-2})$. Therefore,

$$\mathbb{V}ar((s^d_{\mathbf{X}_i})^2) = \frac{\mu_4 - \sigma^4}{d} + O(d^{-2}).$$

3.10.2 Distribution of $\hat{A}^2_\delta$

The following theorem gives the asymptotic distribution of $\hat{A}^2_\delta$ for fixed d, as $m, n \to \infty$ under $H_0 : F_1 = F_2$.

Theorem 3.10.1 *Let* $\boldsymbol{T}_X = \frac{1}{m}\sum_{i=1}^m T_i^X$ *and* $\boldsymbol{T}_Y = \frac{1}{n}\sum_{i=1}^n T_i^Y$. *Hence,*

$$T_X = \begin{pmatrix} \bar{X}^d = \frac{1}{m}\sum_{i=1}^m \bar{X}_i^d \\ s^d_{\mathbf{X}} = \frac{1}{m}\sum_{i=1}^m s^d_{\mathbf{X}_i} \end{pmatrix},$$

and T_Y *is similarly defined. Since rows are i.i.d., by the multivariate CLT,*

$$\sqrt{m}(\boldsymbol{T}_X - \mathbb{E}T_i^X) \xrightarrow{d} \mathbb{N}(0, \boldsymbol{\Sigma}_X),$$

$$\sqrt{n}(\boldsymbol{T}_Y - \mathbb{E}T_i^Y) \xrightarrow{d} \mathbb{N}(0, \boldsymbol{\Sigma}_Y).$$

Define $\boldsymbol{\Sigma}_X = \mathbb{C}ov(T_i^X)$, *and* $\boldsymbol{\Sigma}_Y = \mathbb{C}ov(T_i^Y)$. *These are* 2×2 *matrices of the form*

$$\boldsymbol{\Sigma}_X = \begin{pmatrix} \mathbb{V}ar(\bar{X}_i^d) & \mathbb{C}ov(\bar{X}_i^d, s^d_{\mathbf{X}_i}) \\ \mathbb{C}ov(\bar{X}_i^d, s^d_{\mathbf{X}_i}) & \mathbb{V}ar(s^d_{\mathbf{X}_i}) \end{pmatrix}.$$

Then, $\mathbb{C}ov(\boldsymbol{T}_X) = \frac{1}{m}\boldsymbol{\Sigma}_X$ *and* $\mathbb{C}ov(\boldsymbol{T}_Y) = \frac{1}{n}\boldsymbol{\Sigma}_Y$. *Under* $H_0 : F_1 = F_2$, *we have* $\mathbb{E}T_i^X = \mathbb{E}T_i^Y$. *Since the two samples are independent,*

$$\boldsymbol{T}_X - \boldsymbol{T}_Y \xrightarrow{d} N\left(0, \frac{1}{m}\boldsymbol{\Sigma}_X + \frac{1}{n}\boldsymbol{\Sigma}_Y\right).$$

Define

$$\mathbf{Z} = \sqrt{\frac{mn}{m+n}}(\boldsymbol{T}_X - \boldsymbol{T}_Y).$$

If $F_1 = F_2$, *then* $\boldsymbol{\Sigma}_X = \boldsymbol{\Sigma}_Y = \boldsymbol{\Sigma}^*$. *It follows that*

$$\mathbb{C}ov(\mathbf{Z}) = \frac{mn}{m+n}\left(\frac{1}{m}\boldsymbol{\Sigma}_X + \frac{1}{n}\boldsymbol{\Sigma}_Y\right) = \frac{n}{m+n}\boldsymbol{\Sigma}_X + \frac{m}{m+n}\boldsymbol{\Sigma}_Y = \boldsymbol{\Sigma}^*.$$

Therefore, $\mathbf{Z} \xrightarrow{d} \mathbb{N}(0, \mathbf{\Sigma}^*)$. *Since* $\hat{A}_\delta^2 = \|\boldsymbol{T}_X - \boldsymbol{T}_Y\|^2$ *we can write*

$$\frac{mn}{m+n}\hat{A}_\delta^2 = \mathbf{Z}'\mathbf{Z}.$$

Since $\mathbf{Z} \sim \mathbb{N}(0, \mathbf{\Sigma}^*)$ *asymptotically, write the spectral decomposition* $\mathbf{\Sigma}^* = \mathbf{Q\Lambda Q}'$, *where*

$$\mathbf{\Lambda} = \begin{pmatrix} \lambda_1 & 0 \\ 0 & \lambda_2 \end{pmatrix}.$$

Define $\mathbf{U} = \mathbf{Q}'\mathbf{Z}$. *Then,* $\mathbf{U} \sim \mathbb{N}(0, \mathbf{\Lambda})$, *and* $\mathbf{Z}'\mathbf{Z} = \mathbf{U}'\mathbf{U} = \lambda_1 Z_1^2 + \lambda_2 Z_2^2$, *where* $Z_1, Z_2 \sim \mathbb{N}(0,1)$ *independent. Therefore,*

$$\frac{mn}{m+n}\hat{A}_\delta^2 \xrightarrow{d} \lambda_1\chi_1^2 + \lambda_2\chi_1^2.$$

The following theorem gives the asymptotic distribution of $\hat{A}_\delta^2$ for fixed m, n, as $d \to \infty$.

Theorem 3.10.2 *If coordinates* X_{ij} *are i.i.d. with* $\mathbb{E}[X_{ij}] = \mu$, $\mathbb{V}ar(X_{ij}) = \sigma^2$, *and* $\mathbb{E}[X_{ij}^4] < \infty$, *then*

$$\bar{X}_i^d \xrightarrow{a.s.} \mu, \quad s_{\mathbf{X}_i}^d \xrightarrow{a.s.} \sigma,$$

Therefore,

$$\hat{A}_\delta^2 \xrightarrow{a.s.} (\mu_1 - \mu_2)^2 + (\sigma_1 - \sigma_2)^2.$$

To show this we appeal to SLLN for $\bar{X}_i^d$ and for $(s_{\mathbf{X}_i}^d)^2 = \frac{1}{d}\sum(X_{ij} - \mu)^2 - (\bar{X}_i^d - \mu)^2$. We will next find the eigenvalues of $\mathbf{\Sigma}_X$.

Corollary 3.10.1 *Let* $T_i^X = (\bar{X}_i^d, s_{\mathbf{X}_i}^d)'$, $\mathbf{\Sigma}_X = \mathbb{C}ov(T_i^X)$. *Then the eigenvalues of* $\mathbf{\Sigma}_X$ *can be computed as*

$$\lambda_{1,2} = \frac{1}{2}\mathbb{V}ar(\bar{X}_i^d) + \mathbb{V}ar(s_{\mathbf{X}_i}^d) \pm \sqrt{(\mathbb{V}ar(\bar{X}_i^d) - \mathbb{V}ar(s_{\mathbf{X}_i}^d))^2 + 4\mathbb{C}ov(\bar{X}_i^d, S_{\mathbf{X_i}}^d)^2}.$$

This allows us to write the asymptotic distribution of $\hat{A}_\delta^2$ *as a weighted sum of independent chi-square variables,*

$$\frac{mn}{m+n}\hat{A}_\delta^2 \xrightarrow{d} \lambda_1 Z_1^2 + \lambda_2 Z_2^2, \quad Z_1, Z_2 \sim \mathbb{N}(0,1) \textit{ independent.}$$

Corollary 3.10.2 *For the spherical case,* $\boldsymbol{\mu} = \mu\mathbf{1}_d$, $\mathbf{\Sigma} = \sigma^2 I_d$ *we obtain*

$$\mathbb{E}[(s_{\mathbf{X}_i}^d)^2] = \sigma^2\left(1 - \frac{1}{d}\right), \quad \mathbb{V}ar((s_{\mathbf{X}_i}^d)^2) = \frac{\mu_4 - \sigma^4}{d}, \quad \mathbb{V}ar(s_{\mathbf{X}_i}^d) = \frac{\mu_4 - \sigma^4}{4d\sigma^2}.$$

Eigenvalues of $\mathbf{\Sigma}_X$ *are*

$$\lambda_1 = \frac{\sigma^2}{d}, \quad \lambda_2 = \frac{\mu_4 - \sigma^4}{4d\sigma^2}.$$

Corollary 3.10.3 *For fixed* d, $m, n \to \infty$,

$$\frac{mn}{m+n}\hat{A}_\delta^2 \xrightarrow{d} \lambda_1 Z_1^2 + \lambda_2 Z_2^2, \quad Z_i \sim \mathbb{N}(0,1) \textit{ independent.}$$

3.10.3 Saddlepoint Approximation for $\hat{A}_\delta$

We look for the asymptotic sampling distribution of $\hat{A}_\delta$ where

$$A_\delta = (\bar{\mu}_x^d - \bar{\mu}_y^d)^2 + (\bar{\sigma}_x^d - \bar{\sigma}_y^d)^2. \tag{3.25}$$

We estimate the population quantities by their sample analogues,

$$\bar{X}^d = \frac{1}{m}\sum_{i=1}^{m} \bar{X}_i^d, \qquad s_{\mathbf{X}}^d = \frac{1}{m}\sum_{i=1}^{m} s_{\mathbf{X}_i}^d.$$

Similarly, for the Y sample. The plug-in estimator of A_δ is then

$$\hat{A}_\delta^2 = (\bar{X}^d - \bar{Y}^d)^2 + (s_{\mathbf{X}}^d - s_{\mathbf{Y}}^d)^2,$$

Let $B_1 = \bar{X}^d - \bar{Y}^d$, and $B_2 = s_{\mathbf{X}}^d - s_{\mathbf{Y}}^d$. Then, $\hat{A}_\delta = B_1^2 + B_2^2$. Define

$$\Delta_1 = \mathbb{E}(\bar{X}^d - \bar{Y}^d) = \frac{1}{d}\mathbf{1}_d'(\boldsymbol{\mu}_1 - \boldsymbol{\mu}_2), \qquad \Delta_2 = \mathbb{E}(s_{\mathbf{X}}^d - s_{\mathbf{Y}}^d),$$
$$\tau_1^2 = \mathbb{V}ar(\bar{X}^d) + \mathbb{V}ar(\bar{Y}^d), \qquad \tau_2^2 = \mathbb{V}ar(s_{\mathbf{X}}^d) + \mathbb{V}ar(s_{\mathbf{Y}}^d).$$

If X_i and Y_j are normally distributed, then $B_1 \sim \mathbb{N}(\Delta_1, \tau_1^2)$ and $B_2 \sim \mathbb{N}(\Delta_2, \tau_2^2)$. It follows that $B_1^2 \sim \chi_1^2(\lambda_1)$ and $B_2^2 \sim \chi_1^2(\lambda_2)$, where $\lambda_1 = \Delta_1^2/\tau_1^2$ and $\lambda_2 = \Delta_2^2/\tau_2$. B_1 and B_2 are asymptotically independent for large d under spherical or elliptical distributions. Hence, $\hat{A}_\delta$ is the sum of two independent non-central chi-square variables. The moment generating function of $\hat{A}_\delta$ is

$$M_{\hat{A}_\delta}(t) = M_{B_1^2}(t) \cdot M_{B_2^2}(t),$$

where

$$M_{B_i^2}(t) = (1 - 2t\tau_i^2)^{-1/2} \exp\left(\frac{t\Delta_i^2}{1 - 2t\tau_i^2}\right), \quad i = 1, 2.$$

The exact MGF fully characterizes the distribution. We can invert the MGF numerically or use saddle–point methods for density approximations. This approach is suitable when d is large and components of X, Y are independent or weakly dependent.

Delta method provides a linear approximation for the distribution of $\hat{A}_\delta$ using CLT, and is valid primarily when $m, n \to \infty$. The MGF approach is exact or asymptotically exact for HDLSS setting, particularly under elliptical or normal distributions.

The MGF of $\hat{A}_\delta$ is

$$M_{\hat{A}_\delta}(t) = \prod_{i=1}^{2}(1 - 2t\tau_i^2)^{-1/2} \exp\left(\frac{t\Delta_i^2}{1 - 2t\tau_i^2}\right),$$

where $i = 1, 2$ corresponds to B_1^2 and B_2^2. The cumulant generating function is

$$K(t) = \log M_{\hat{A}_\delta}(t) = -\frac{1}{2}\sum_{i=1}^{2}\log(1 - 2t\tau_i^2) + \sum_{i=1}^{2}\frac{t\Delta_i^2}{1 - 2t\tau_i^2}.$$

The saddle–point equation is solve for $\hat{t}$ such that $K'(\hat{t}) = \hat{A}_\delta$, where

$$K'(t) = \sum_{i=1}^{2}\frac{\Delta_i^2 + \tau_i^2}{(1 - 2t\tau_i^2)^2} - \frac{\tau_i^2}{1 - 2t\tau_i^2}.$$

The density of $\hat{A}_\delta$ is approximated by

$$f_{\hat{A}_\delta}(x) \approx \frac{1}{\sqrt{2\pi K''(\hat{t})}} \exp\{K(\hat{t}) - \hat{t}x\},$$

where

$$K''(\hat{t}) = \sum_{i=1}^{2} \frac{2\tau_i^4 + 4\tau_i^2\Delta_i^2}{(1 - 2\hat{t}\tau_i^2)^3}.$$

This approach uses the exact MGF to generate a highly accurate approximation of the density. It does not rely on linearization (delta method) and is especially useful for finite m, n. The saddle–point automatically accounts for non-centrality and variance differences.

3.11 Asymptotic Distribution of $\hat{A}_\rho$

Consider two independent samples of d-dimensional vectors,

$$\mathbf{X}_1, \ldots, \mathbf{X}_m \sim F_1, \qquad \mathbf{Y}_1, \ldots, \mathbf{Y}_n \sim F_2,$$

with expectations and covariance matrices $\mathbb{E}[\mathbf{X}_i] = \boldsymbol{\mu}_x$, $\mathbb{E}[\mathbf{Y}_j] = \boldsymbol{\mu}_y$, $\mathbb{C}ov(\mathbf{X}_i) = \boldsymbol{\Sigma}_x$, and $\mathbb{C}ov(\mathbf{Y}_j) = \boldsymbol{\Sigma}_y$. Let $N = m + n$ denote the total sample size. Recall that

$$\sigma_{F_1}^2 = \frac{1}{d}\text{tr}(\boldsymbol{\Sigma}_x), \quad \sigma_{F_2}^2 = \frac{1}{d}\text{tr}(\boldsymbol{\Sigma}_y), \quad \nu^2 = \frac{1}{d}\|\boldsymbol{\mu}_x - \boldsymbol{\mu}_y\|^2.$$

Let $\Delta = \sigma_{F_1}^2 + \sigma_{F_2}^2 + \nu^2$. We look for the asymptotic sampling distribution of $\hat{A}_\rho$ where

$$A_\rho = \frac{1}{N-2}\Big[(m-1)\big|\sqrt{2}\sigma_{F_1} - \sqrt{\Delta}\big| + (n-1)\big|2\sqrt{\Delta} - \sqrt{2}\sigma_{F_2}\big|\Big]. \tag{3.26}$$

We estimate the population quantities by their sample analogues,

$$\begin{aligned} T_1 &= \hat{\sigma}_{F_1}^2 = \frac{1}{d}\text{tr}(\hat{\boldsymbol{\Sigma}}_x), \\ T_2 &= \hat{\sigma}_{F_2}^2 = \frac{1}{d}\text{tr}(\hat{\boldsymbol{\Sigma}}_y), \\ T_3 &= \hat{\nu}^2 = \frac{1}{d}(\bar{\mathbf{X}} - \bar{\mathbf{Y}})'(\bar{\mathbf{X}} - \bar{\mathbf{Y}}), \end{aligned} \tag{3.27}$$

where $\bar{\mathbf{X}}$ and $\bar{\mathbf{Y}}$ are the sample means of the respective groups. Let $T = T_1 + T_2 + T_3$. The plug-in estimator of A_ρ is then

$$\hat{A}_\rho = \frac{1}{N-2}\Big[(m-1)\big|\sqrt{2}\sqrt{T_1} - \sqrt{T}\big| + (n-1)\big|2\sqrt{T} - \sqrt{2}\sqrt{T_2}\big|\Big].$$

Under the multivariate Central Limit Theorem, as the dimension $d \to \infty$, the vector $\hat{\boldsymbol{\theta}} = (T_1, T_2, T_3)'$ satisfies $\sqrt{d}(\hat{\boldsymbol{\theta}} - \boldsymbol{\theta}) \to \mathbb{N}(0, \boldsymbol{\Sigma}_\theta)$, where $\boldsymbol{\theta} = (\sigma_{F_1}^2, \sigma_{F_2}^2, \nu^2)'$ and $\boldsymbol{\Sigma}_\theta$ is the asymptotic covariance matrix to be derived below.

Under multivariate normality, the sample covariance matrices have Wishart distributions,

$$\mathbf{W}_x = (m-1)\hat{\boldsymbol{\Sigma}}_x \sim \mathcal{W}_d(\boldsymbol{\Sigma}_x, m-1), \quad \mathbf{W}_y = (n-1)\hat{\boldsymbol{\Sigma}}_y \sim \mathcal{W}_d(\boldsymbol{\Sigma}_y, n-1).$$

Using the known properties of the Wishart distribution, we compute the variances

$$\mathbb{V}ar(T_1) = \frac{2}{d^2(m-1)}\text{tr}(\boldsymbol{\Sigma}_x^2), \quad \mathbb{V}ar(T_2) = \frac{2}{d^2(n-1)}\text{tr}(\boldsymbol{\Sigma}_y^2).$$

For T_3, consider $\mathbf{D} = \bar{\mathbf{X}} - \bar{\mathbf{Y}}$ with

$$\mathbb{C}ov(\mathbf{D}) = \boldsymbol{\Sigma}^* = \frac{\boldsymbol{\Sigma}_x}{m} + \frac{\boldsymbol{\Sigma}_y}{n}.$$

Then, using standard quadratic-form results for non-central normals,

$$\mathbb{V}ar(T_3) = \frac{2}{d^2}\text{tr}((\boldsymbol{\Sigma}^*)^2) + \frac{4}{d^2}\delta'\boldsymbol{\Sigma}^*\delta, \quad \delta = \boldsymbol{\mu}_x - \boldsymbol{\mu}_y.$$

Finally, the asymptotic covariance matrix is diagonal, $\boldsymbol{\Sigma}_\theta = \text{diag}(\mathbb{V}ar(T_1), \mathbb{V}ar(T_2), \mathbb{V}ar(T_3))$.

Delta Method for $\hat{A}_\rho$

To obtain the asymptotic distribution of the plug-in estimator $\hat{\hat{\rho}}$, we apply the multivariate Delta method. Let

$$g(T_1, T_2, T_3) = \frac{1}{N-2}\Big[(m-1)\big|\sqrt{2}\sqrt{T_1} - \sqrt{T}\big| + (n-1)\big|2\sqrt{T} - \sqrt{2}\sqrt{T_2}\big|\Big].$$

Introduce the sign functions to handle absolute values,

$$s_1 = \text{sgn}(\sqrt{2}\sqrt{T_1} - \sqrt{T}), \quad s_2 = \text{sgn}(2\sqrt{T} - \sqrt{2}\sqrt{T_2}).$$

Gradient Calculation

Derivative with respect to T_1, T^2, and T_3 are

$$\begin{aligned}
c_1 &= \frac{\partial g}{\partial T_1} = \frac{1}{N-2}\Big[(m-1)s_1\Big(\frac{\sqrt{2}}{2\sqrt{T_1}} - \frac{1}{2\sqrt{T}}\Big) + (n-1)s_2\frac{1}{\sqrt{T}}\Big],\\
c_2 &= \frac{\partial g}{\partial T_2} = \frac{1}{N-2}\Big[(m-1)s_1\Big(-\frac{1}{2\sqrt{T}}\Big) + (n-1)s_2\Big(\frac{1}{\sqrt{T}} - \frac{\sqrt{2}}{2\sqrt{T_2}}\Big)\Big],\\
c_3 &= \frac{\partial g}{\partial T_3} = \frac{1}{N-2}\Big[(m-1)s_1\Big(-\frac{1}{2\sqrt{T}}\Big) + (n-1)s_2\frac{1}{\sqrt{T}}\Big].
\end{aligned} \tag{3.28}$$

Using the Delta method, the asymptotic variance of $\hat{\hat{\rho}}$ is $\tau^2 = \nabla c(\boldsymbol{\theta})'\boldsymbol{\Sigma}_\theta \nabla c(\boldsymbol{\theta})$. Because $\boldsymbol{\Sigma}_\theta$ is diagonal, this simplifies to $\tau^2 = \sum_{i=1} c_i \mathbb{V}ar(T_i)$.

3.11.1 Moment Generating Functions

The moment generating function (MGF) of each component is derived using the Wishart and non-central chi-square properties. They are

$$M_{T_1}(t) = \det\left(\mathbf{I} - \frac{2t}{d(m-1)}\boldsymbol{\Sigma}_x\right)^{-(m-1)/2}, \quad M_{T_2}(t) = \det\left(\mathbf{I} - \frac{2t}{d(n-1)}\boldsymbol{\Sigma}_y\right)^{-(n-1)/2},$$

$$M_{T_3}(t) = \frac{\exp\left(\frac{t}{d}\delta'(\mathbf{I} - \frac{2t}{d}\boldsymbol{\Sigma}^*)^{-1}\delta\right)}{\sqrt{\det(\mathbf{I} - \frac{2t}{d}\boldsymbol{\Sigma}^*)}}.$$

3.11.2 Saddlepoint Approximation for $\hat{A}_\rho$

The density of $\hat{A}_\rho$ can be approximated using the saddle–point method. Let $K_{\hat{A}_\rho(t)} = \log \mathbb{E}[e^{t\hat{A}_\rho}]$ denote the cumulant generating function (CGF). Using the linear Delta-method approximation,

$$\hat{A}_\rho \approx c_1 T_1 + c_2 T_2 + c_3 T_3,$$

with $c_j = \partial g/\partial T_j$ evaluated at the mean. Let $g_i = \frac{M'_{T_i}(c_i t)}{M_{T_i}(c_i t)}$. The first derivative of the CGF is $K'_{\hat{A}_\rho}(t) = \sum_{i=1}^{3} g_i c_i$. Let

$$W_i = c_i^2 \left(\frac{M''_{T_i}(c_i t)}{M_{T_i}(c_i t)} - \left(\frac{M'_{T_i}(c_i t)}{M_{T_i}(c_i t)} \right)^2 \right).$$

The second derivative is $K''_{\hat{A}_\rho}(t) = W_1 + W_2 + W_3$.

Let t^* satisfy the saddle–point equation$K'_{\tilde{\rho}}(t^*) = a$, where a is the observed value of $\tilde{\rho}$. Then the density is approximated as

$$f_{\hat{A}_\rho}(a) \approx \frac{1}{\sqrt{2\pi K''_{\tilde{\rho}}(t^*)}} \exp\Big(K_{\hat{A}_\rho}(t^*) - t^* a \Big).$$

This formula captures the skewness and kurtosis of the distribution better than a normal approximation, especially in moderate sample sizes, by using the exact MGFs of the underlying T_j components.

3.12 Asymptotic Theory for Two-Sample Dissimilarity Statistics

We will introduce two new statistics that use dissimilarity indices ρ and δ to measure the average dissimilarity of $\mathbb{X}$ and $\mathbb{Y}$ samples. These averages are of the form (3.29), where h is replaced by ρ and δ, respectively.

Let $\mathbf{X}_1, \ldots, \mathbf{X}_m \sim F_1$ and $\mathbf{Y}_1, \ldots, \mathbf{Y}_n \sim F_2$, be independent random vectors in $\mathbb{R}^d$. Let $N = m + n$ and define $\lambda = \frac{m}{N}$, and $\lambda \to \lambda_0 \in (0, 1)$. Let $h : \mathbb{R}^d \times \mathbb{R}^d \to \mathbb{R}$ be measurable and define

$$U_{m,n} = \frac{1}{mn} \sum_{i=1}^{m} \sum_{j=1}^{n} h(\mathbf{X}_i, \mathbf{Y}_j). \qquad (3.29)$$

Let $\theta = \mathbb{E}[h(\mathbf{X}_1, \mathbf{Y}_1)]$. We assume that $\mathbb{E}[h(\mathbf{X}_1, \mathbf{Y}_1)^2] < \infty$ and $m, n \to \infty$ with $\lambda \to \lambda_0 \in (0, 1)$.

3.12.1 Hoeffding Decomposition

Using Hoeffding [150], define the first-order projections,

$$\psi_X(x) = \mathbb{E}[h(x, \mathbf{Y}_1)] - \theta, \qquad \psi_Y(y) = \mathbb{E}[h(\mathbf{X}_1, y)] - \theta.$$

Here the expectation in $\psi_X(x)$ is taken with respect to $\mathbf{Y}_1$ conditioning on $\mathbf{X}_1 = x$, that is,

$$\psi_X(x) = \mathbb{E}[h(x, \mathbf{Y}_1) \mid \mathbf{X}_1 = x] - \theta,$$

and similarly,

$$\psi_Y(y) = \mathbb{E}[h(\mathbf{X}_1, y) \mid \mathbf{Y}_1 = y] - \theta.$$

Define the second-order remainder

$$h_2(x, y) = h(x, y) - \theta - \psi_X(x) - \psi_Y(y).$$

Lemma 3.12.1 *We have* $\mathbb{E}[h_2(\mathbf{X}_1, \mathbf{Y}_1) \mid \mathbf{X}_1] = 0$ *and* $\mathbb{E}[h_2(\mathbf{X}_1, \mathbf{Y}_1) \mid \mathbf{Y}_1] = 0$.

By definition,

$$\mathbb{E}[h(\mathbf{X}_1, \mathbf{Y}_1) \mid \mathbf{X}_1] = \theta + \psi_X(\mathbf{X}_1).$$

Subtracting $\theta + \psi_X(\mathbf{X}_1)$ yields zero. The second identity is analogous, conditioning on $\mathbf{Y}_1$.

Proposition 3.12.1 (Exact Decomposition)

$$U_{m,n} - \theta = \frac{1}{m}\sum_{i=1}^{m} \psi_X(\mathbf{X}_i) + \frac{1}{n}\sum_{j=1}^{n} \psi_Y(\mathbf{Y}_j) + R_{m,n},$$

where

$$R_{m,n} = \frac{1}{mn}\sum_{i=1}^{m}\sum_{j=1}^{n} h_2(\mathbf{X}_i, \mathbf{Y}_j).$$

The proof follows upon substitute the identity $h = \theta + \psi_X + \psi_Y + h_2$ into the definition of $U_{m,n}$ and simplify.

To see the order of the degenerate term note that $\mathbb{V}ar(R_{m,n}) = O(\frac{1}{mn})$. Because h_2 is degenerate in both arguments, that is,

$$\mathbb{E}[h_2(\mathbf{X}_1, \mathbf{Y}_1) \mid \mathbf{X}_1] = 0 \quad \text{and} \quad \mathbb{E}[h_2(\mathbf{X}_1, \mathbf{Y}_1) \mid \mathbf{Y}_1] = 0,$$

cross-covariances vanish unless indices coincide in both coordinates. Hence

$$\mathbb{V}ar(R_{m,n}) = \frac{1}{m^2n^2}\sum_{i,j} \mathbb{V}ar(h_2(\mathbf{X}_i, \mathbf{Y}_j)) = O\left(\frac{1}{mn}\right).$$

It also follows that $\sqrt{N} R_{m,n} \xrightarrow{p} 0$.

3.12.2 Asymptotic Distribution

Define $\sigma_X^2 = \mathbb{V}ar(\psi_X(\mathbf{X}_1))$ and $\sigma_Y^2 = \mathbb{V}ar(\psi_Y(\mathbf{Y}_1))$.

Theorem 3.12.1 *Assuming that* $\mathbb{E}[h(\mathbf{X}_1, \mathbf{Y}_1)^2] < \infty$ *and* $m, n \to \infty$ *with* $\lambda \to \lambda_0 \in (0, 1)$, *we have*

$$\sqrt{N}(U_{m,n} - \theta) \xrightarrow{d} \mathbb{N}(0, \eta^2),$$

where

$$\eta^2 = \frac{\sigma_X^2}{\lambda} + \frac{\sigma_Y^2}{1-\lambda}. \tag{3.30}$$

Proof: By the decomposition,

$$\sqrt{N}(U_{m,n} - \theta) = \sqrt{N}\Big(\frac{1}{m}\sum_{i=1}^{m} \psi_X(\mathbf{X}_i) + \frac{1}{n}\sum_{j=1}^{n} \psi_Y(\mathbf{Y}_j)\Big) + o_p(1).$$

Rewrite

$$\sqrt{N}\frac{1}{m}\sum_{i=1}^{m}\psi_X(\mathbf{X}_i) = \frac{1}{\sqrt{m}}\sum_{i=1}^{m}\psi_X(\mathbf{X}_i)\cdot\sqrt{\frac{N}{m}}.$$

Since $m/N \to \lambda$, we have $N/m \to 1/\lambda$. By the classical CLT,

$$\frac{1}{\sqrt{m}}\sum_{i=1}^{m}\psi_X(\mathbf{X}_i) \xrightarrow{d} \mathrm{N}(0, \sigma_X^2).$$

Thus the first term converges to $\mathrm{N}\left(0, \frac{\sigma_X^2}{\lambda}\right)$. Similarly,

$$\sqrt{N}\frac{1}{n}\sum_{j=1}^{n}\psi_Y(\mathbf{Y}_j) \xrightarrow{d} \mathrm{N}\left(0, \frac{\sigma_Y^2}{1-\lambda}\right).$$

Independence of samples implies independence of limits, hence variances add. □

If $F_1 = F_2$ and $h(X, Y)$ is symmetric in distribution, then $\sigma_X^2 = \sigma_Y^2 = \tau_0^2$, and $\eta^2 = \frac{\tau_0^2}{\lambda(1-\lambda)}$. Under identical distributions, the laws of $\psi_X(X)$ and $\psi_Y(Y)$ coincide.

3.12.3 Estimation of the Asymptotic Variance

Define empirical projections,

$$\hat{\psi}_X(X_i) = \frac{1}{n}\sum_{j=1}^{n}h(X_i, Y_j) - U_{m,n}, \qquad \hat{\psi}_Y(Y_j) = \frac{1}{m}\sum_{i=1}^{m}h(X_i, Y_j) - U_{m,n},$$

where

$$U_{m,n} = \frac{1}{mn}\sum_{i=1}^{m}\sum_{j=1}^{n}h(X_i, Y_j).$$

Define the $m \times n$ matrix $H_{ij} = h(X_i, Y_j)$. Let

$$\bar{H}_{i\cdot} = \frac{1}{n}\sum_{j=1}^{n}H_{ij}, \qquad \bar{H}_{\cdot j} = \frac{1}{m}\sum_{i=1}^{m}H_{ij}, \qquad \bar{H}_{\cdot\cdot} = \frac{1}{mn}\sum_{i=1}^{m}\sum_{j=1}^{n}H_{ij}.$$

Then,

$$\hat{\psi}_X(X_i) = \bar{H}_{i\cdot} - \bar{H}_{\cdot\cdot}, \qquad \hat{\psi}_Y(Y_j) = \bar{H}_{\cdot j} - \bar{H}_{\cdot\cdot}.$$

Thus, $\hat{\psi}_X(X_i)$ is the row mean of the ith row minus the grand mean, and $\hat{\psi}_Y(Y_j)$ is the column mean of the jth column minus the grand mean.

Since row means and column means are generally different, $\hat{\psi}_X(X_i) \neq \hat{\psi}_Y(Y_j)$ in general. Recall the population projections

$$\psi_X(x) = \mathbb{E}[h(x, Y)] - \theta, \qquad \psi_Y(y) = \mathbb{E}[h(X, y)] - \theta.$$

The empirical versions replace expectations by sample averages:

$$\mathbb{E}[h(x, Y)] \longrightarrow \frac{1}{n}\sum_{j=1}^{n}h(x, Y_j), \qquad \mathbb{E}[h(X, y)] \longrightarrow \frac{1}{m}\sum_{i=1}^{m}h(X_i, y).$$

Thus, the two empirical projections estimate different population quantities. Both empirical projections are centered:

$$\sum_{i=1}^{m}\hat{\psi}_X(X_i) = 0, \qquad \sum_{j=1}^{n}\hat{\psi}_Y(Y_j) = 0.$$

To show this note that

$$\sum_{i=1}^{m} \hat{\psi}_X(X_i) = \sum_{i=1}^{m} \bar{H}_{i\cdot} - m\bar{H}_{\cdot\cdot}.$$

But,

$$\sum_{i=1}^{m} \bar{H}_{i\cdot} = \sum_{i=1}^{m} \frac{1}{n} \sum_{j=1}^{n} H_{ij} = \frac{1}{n} \sum_{i=1}^{m} \sum_{j=1}^{n} H_{ij} = m\bar{H}_{\cdot\cdot}.$$

Hence, the difference equals zero. The argument for $\hat{\psi}_Y$ is analogous.

3.12.4 Application to the ρ Statistic

Recall that

$$\rho(\mathbf{X}, \mathbf{Y}) = \frac{1}{N-2} \frac{1}{\sqrt{d}} \sum_{\mathbf{Z} \in \mathbb{Z}_N \setminus \{\mathbf{X}, \mathbf{Y}\}} \left| \|\mathbf{X} - \mathbf{Z}\| - \|\mathbf{Y} - \mathbf{Z}\| \right|.$$

Let

$$T_{m,n} = \frac{1}{mn} \sum_{i=1}^{m} \sum_{j=1}^{n} \rho(\mathbf{X}_i, \mathbf{Y}_j).$$

Theorem 3.12.2 *Assuming that* $\mathbb{E}[\rho(\mathbf{X}_1, \mathbf{Y}_1)^2] < \infty$ *and* $m, n \to \infty$ *with* $\lambda \to \lambda_0 \in (0, 1)$,

$$\sqrt{N}(T_{m,n} - \mathbb{E}[\rho(X, Y)]) \xrightarrow{d} \mathbb{N}(0, \eta^2),$$

with η^2 *given by* (3.30).

To estimate η^2 for $T_{m,n}$, treat $h(x, y) = \rho(x, y)$, and compute the empirical projections

$$\hat{\psi}_X^{\rho}(\mathbf{X}_i) = \frac{1}{n} \sum_{j=1}^{n} \rho(\mathbf{X}_i, \mathbf{Y}_j) - T_{m,n},$$

$$\hat{\psi}_Y^{\rho}(\mathbf{Y}_j) = \frac{1}{m} \sum_{i=1}^{m} \rho(\mathbf{X}_i, \mathbf{Y}_j) - T_{m,n}.$$

Then define

$$\hat{\sigma}_{X,\rho}^2 = \frac{1}{m-1} \sum_{i=1}^{m} (\hat{\psi}_X^{\rho}(\mathbf{X}_i))^2, \quad \hat{\sigma}_{Y,\rho}^2 = \frac{1}{n-1} \sum_{j=1}^{n} (\hat{\psi}_Y^{\rho}(\mathbf{Y}_j))^2,$$

and

$$\hat{\eta}_{\rho}^2 = \frac{\hat{\sigma}_{X,\rho}^2}{\lambda} + \frac{\hat{\sigma}_{Y,\rho}^2}{1-\lambda}.$$

Under $F_1 = F_2$, use the pooled version $\hat{\eta}_{\rho}^2 = \frac{\hat{\tau}_{0,\rho}^2}{\lambda(1-\lambda)}$, where $\hat{\tau}_{0,\rho}^2$ is the common sample variance of the empirical projections.

3.12.5 Application to the δ Statistic

Recall that using the data cloud $\mathbb{Z}_N$ we have

$$\delta(\mathbf{X}_i, \mathbf{Y}_j) = \frac{1}{N-2} \sum_{k \neq i,j} \left| \|\mathbf{X}_i - \mathbf{Z}_k\|_D - \|\mathbf{Y}_j - \mathbf{Z}_k\|_D \right|. \tag{3.31}$$

Let

$$\Delta_{m,n} = \frac{1}{mn} \sum_{i=1}^{m} \sum_{j=1}^{n} \delta(\mathbf{X}_i, \mathbf{Y}_j).$$

Theorem 3.12.3 *Assuming that* $\mathbb{E}[\delta(\mathbf{X}_1, \mathbf{Y}_1)^2] < \infty$ *and* $m, n \to \infty$ *with* $\lambda \to \lambda_0 \in (0, 1)$,

$$\sqrt{N}(\Delta_{m,n} - E[\delta(X, Y)]) \xrightarrow{d} \mathbb{N}(0, \eta^2).$$

where η^2 *is given in* (3.30).

To estimate η^2 for $\Delta_{m,n}$, set $h(x, y) = \delta(x, y)$, and define empirical projections

$$\hat{\psi}_X^\delta(\mathbf{X}_i) = \frac{1}{n}\sum_{j=1}^{n} \delta(\mathbf{X}_i, \mathbf{Y}_j) - \Delta_{m,n},$$

$$\hat{\psi}_Y^\delta(\mathbf{Y}_j) = \frac{1}{m}\sum_{i=1}^{m} \delta(\mathbf{X}_i, \mathbf{Y}_j) - \Delta_{m,n}.$$

Then compute

$$\hat{\sigma}_{X,\delta}^2 = \frac{1}{m-1}\sum_{i=1}^{m}(\hat{\psi}_X^\delta(\mathbf{X}_i))^2, \quad \hat{\sigma}_{Y,\delta}^2 = \frac{1}{n-1}\sum_{j=1}^{n}(\hat{\psi}_Y^\delta(\mathbf{Y}_j))^2,$$

and finally

$$\hat{\eta}_\delta^2 = \frac{\hat{\sigma}_{X,\delta}^2}{\lambda} + \frac{\hat{\sigma}_{Y,\delta}^2}{1-\lambda}.$$

Under $F_1 = F_2$, the symmetric reduction applies and we use $\hat{\eta}_\delta^2 = \frac{\hat{\tau}_{0,\delta}^2}{\lambda(1-\lambda)}$, where $\hat{\tau}_{0,\delta}^2$ is the common empirical variance of the projections.

3.13 Comparisons

Modarres [238] discusses tests of the equality of high-dimensional distributions. To test $H_0 : F_1 = F_2$ he proposed a $\hat{A}_\delta$ statistic based on the square of the asymptotic form of $\tilde{\delta}(F_1, F_2)$ given in (3.20). He also considered a test statistic $\hat{A}_\rho$ based on the asymptotic form of $\tilde{\rho}(F_1, F_2)$ given in (3.16). We will compare the statistics $T_{m,n}$ and $\Delta_{m,n}$ against $\hat{A}_\delta$ and $\hat{A}_\rho$ in the next section.

Several existing procedures, including the energy test (Székely and Rizzo [330]) and the BG statistic of Biswas and Ghosh [35], are effective in high-dimensional regimes where the dimension d far exceeds the sample size N. Nevertheless, these methods face serious computational challenges when N is moderate to large. A primary computational bottleneck for many IPD-based algorithms is the evaluation of all pairwise distances. For a sample of size N, the cost of computing IPDs scales as $O(dN^2)$. While this remains manageable when N is on the order of a few hundred even for large d, it becomes prohibitive when N reaches the thousands in conjunction with high dimensionality. The statistics $\hat{A}_\delta$ and $\hat{A}_\rho$ do not require computing IPDs.

In this section we compare test statistics $\hat{A}_\rho$, $\hat{A}_\delta$, $\Delta_{m,n}$, and $T_{m,n}$ in terms of observed significance level and power under several alternatives in two experiments. In the first experiment $d = 500$ is fixed and other parameters such as μ or σ^2 vary. In experiment 2 the parameters are fixed and $d = 2^i$ for $i = 2, \ldots, 10$. Each experiment is based on 1000 random random samples of sizes $m = 30$ and $n = 30$.

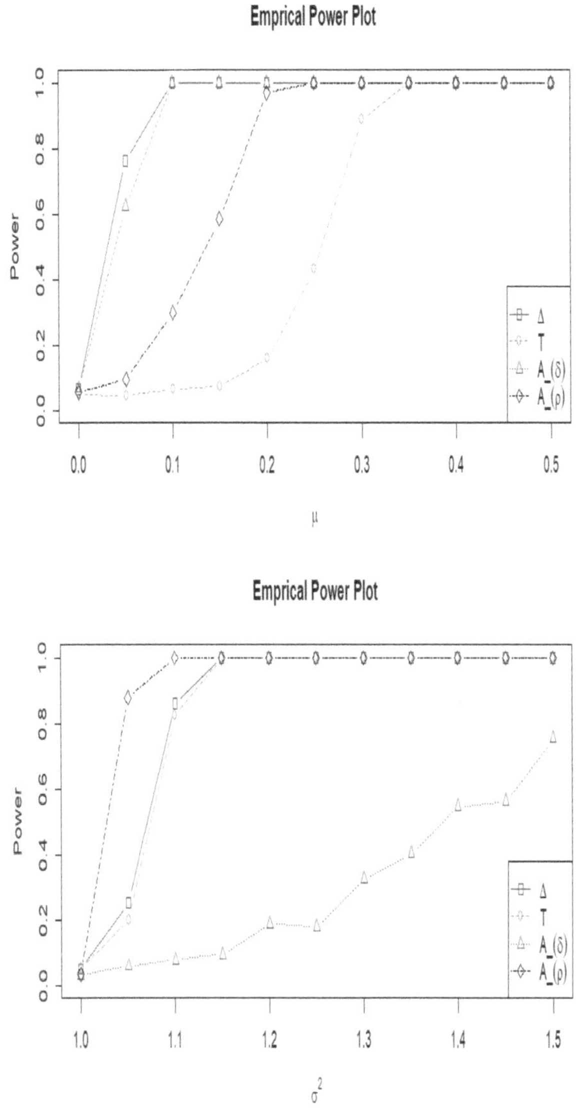

FIGURE 3.5: Empirical power of the statistics under normal shift (left panel) and scale (right panel) alternatives.

- $\mathbf{X} \sim \mathbb{N}(\mathbf{0}, \mathbf{I})$ and $\mathbf{Y} \sim \mathbb{N}(\mu \mathbf{1}_d, \mathbf{I})$ where $\mu \in \{0, 0.05, 0.1, \ldots, 0.5\}$.
- $\mathbf{X} \sim \mathbb{N}(\mathbf{0}, \mathbf{I})$ and $\mathbf{Y} \sim \mathbb{N}(\mathbf{0}_d, \sigma^2\ \mathbf{I})$ where $\sigma^2 \in \{1, 1.05, 1.1, \ldots, 1.5\}$.
- $\mathbf{X} \sim \chi_1^2$ and $\mathbf{Y} \sim \chi_1^2 + \mu$ where $\mu \in \{0, 0.05, 0.1, \ldots, 0.5\}$.
- $\mathbf{X} \sim t_{10}$ and $\mathbf{Y} \sim t_{10} + \mu$ where $\mu \in \{0, .05, 0.1, \ldots 0.5\}$.

Figures 3.5 and 3.6 show the empirical powers of the test statistics for the first experiment. The statistics Δ and A_δ performs the best for shift alternatives while A_ρ outperforms other statistics under scale alternatives. Figures 3.7 and 3.8 display the empirical power

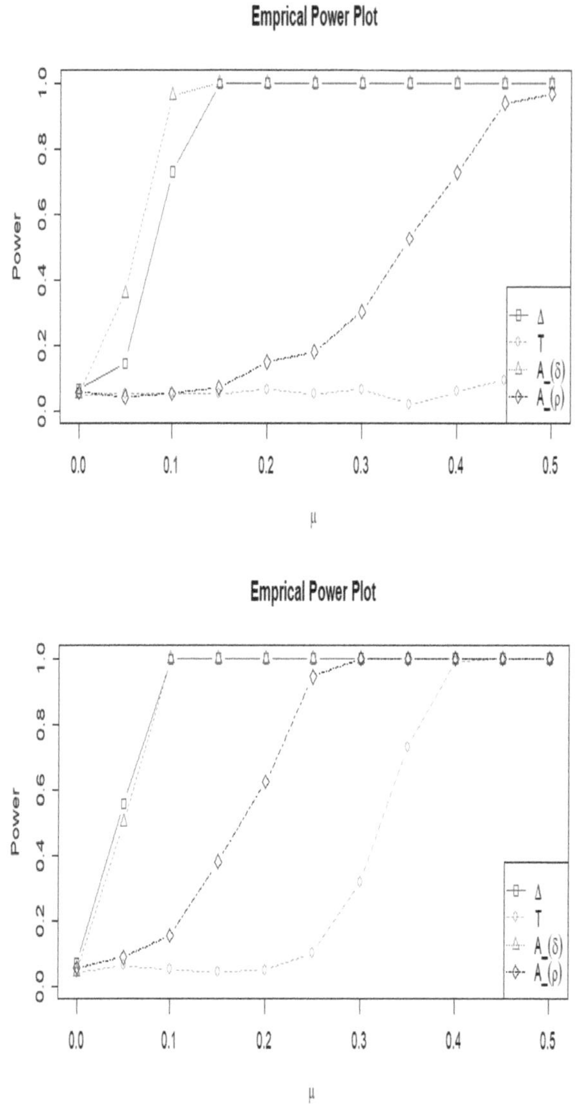

FIGURE 3.6: Empirical power of the statistics under $chi^2_{(1)}$ (left panel) and $t_{(10)}$ (right panel) shift alternatives.

for the second experiment where $d = 2^i$ for $i = 2, \ldots, 10$ for fixed parameter values. The same ordering holds in this experiment. The statistics Δ and A_δ perform the best for shift alternatives while $A_\rho$performs well under scale alternatives. If we rank all four statistics across both experiments, then Δ has an average rank of 1.5, followed by A_δ, A_ρ, and T with average ranks of 2, 2.4, and 3.5, respectively.

In experiment 2 the parameters μ or σ^2 are fixed and $d = 2^i$ for $i = 2, \ldots, 10$.

- $\mathbf{X} \sim \mathbb{N}_d(\mathbf{0}, \mathbf{I})$ and $\mathbf{Y} \sim \mathbb{N}(0.1\ \mathbf{1}_d, \mathbf{I})$.

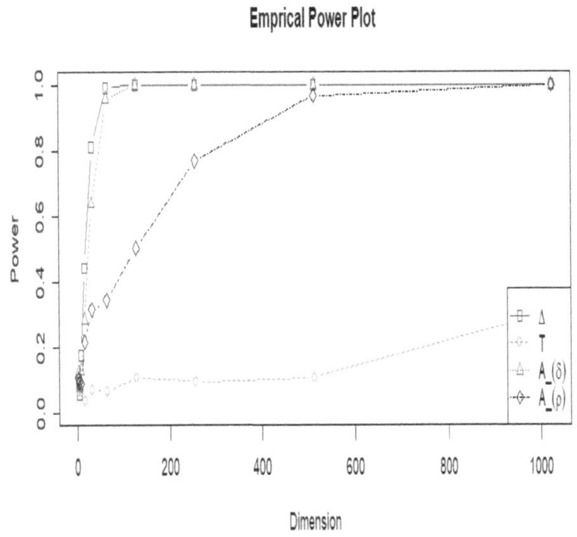

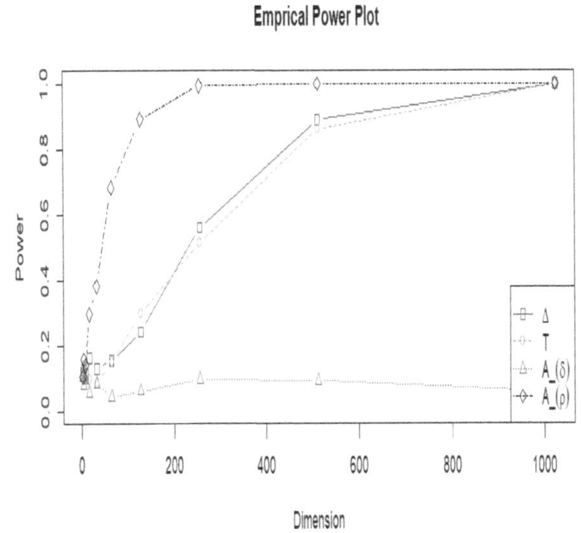

FIGURE 3.7: Empirical power of the statistics under normal shift (left panel) and scale change (right panel) with $d = 2^i$, $\mu = 0.1$, and $\sigma^2 = 1.1$.

- $\mathbf{X} \sim \mathbb{N}(\mathbf{0}, \mathbf{I})$ and $\mathbf{Y} \sim \mathbb{N}(\mathbf{0}, 1.1\ \mathbf{I})$.
- $\mathbf{X} \sim \chi_1^2$ and $\mathbf{X} \sim \chi_1^2 + 0.1$.
- $\mathbf{X} \sim t_{(10)})$ and $\mathbf{Y} \sim t_{(10)} + 0.1$.

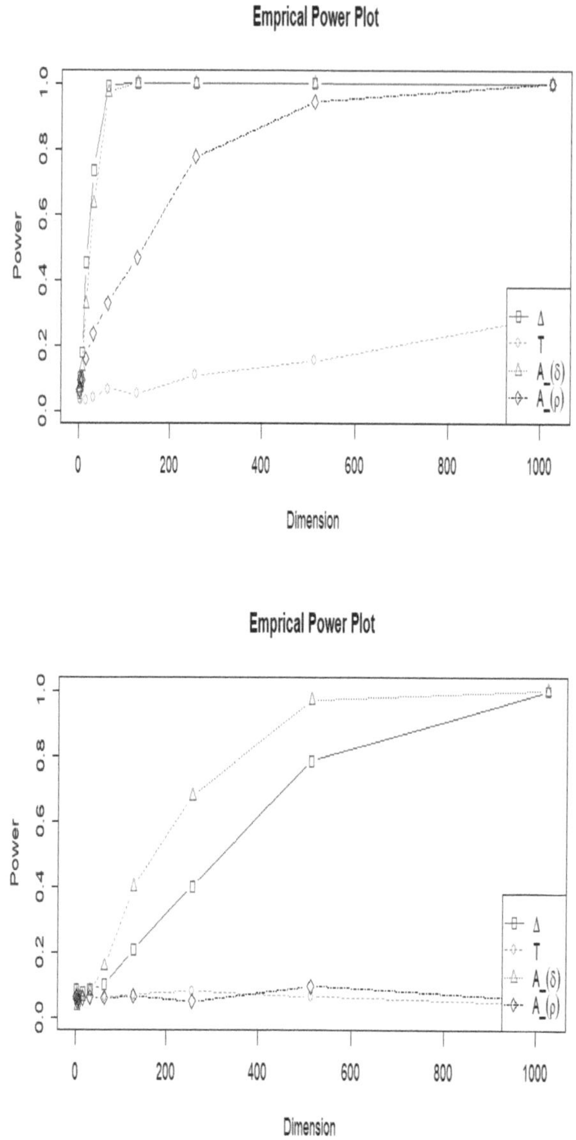

FIGURE 3.8: Empirical power of the statistics under $t_{(10)}$ (left panel) and $\chi^2_{(1)}$ (right panel) shift with $d = 2^i$ and $\mu = 0.1$.

3.14 Applications

3.14.1 Outlier and Change Point Detection

Outliers are observations that exhibit different characteristics than the rest of the observations in the dataset. Hadi et al. [131] present an overview of the methods of outlier detection. Singh and Xie [315] propose an outlier detection plot based on the bootstrap as a non-parametric graphical tool for detecting outliers. Let k be the number of outliers. We are interested in testing

$$H_0 : \mathbf{X}_i \sim F(\boldsymbol{\mu}_x, \boldsymbol{\Sigma}_x), \quad i = 1, \ldots, m, \tag{3.32}$$

against the alternative hypothesis

$$H_a : \mathbf{X}_i \sim F(\boldsymbol{\mu}_x, \boldsymbol{\Sigma}_x), \text{ for some } i, \text{ and } \mathbf{X}_j \sim F(\boldsymbol{\mu}_x + \mathbf{a}, \boldsymbol{\Sigma}_x), \text{ for all } j \neq i. \tag{3.33}$$

The slippage parameter $\mathbf{a}$, $\boldsymbol{\mu}_x$, and $\boldsymbol{\Sigma}_x$ matrix are unknown, so are the indices j of the outliers.

Change point analysis concerns the detection of changes in the distribution of the observations that are ordered by time or location. Chen and Gupta [61] summarize the parametric models for most single change point problems. Let $\mathbf{X}_1, \mathbf{X}_2, \ldots, \mathbf{X}_m$ be a sequence of independent random vectors with probability distributions $F_1, F_2, \ldots, F_m$, respectively. Consider the null and the alternative hypotheses

$$K_0 : F_1 = F_2 = \ldots = F_m, \tag{3.34}$$

$$K_a : F_1 = \ldots = F_{\tau_1} \neq F_{\tau_1+1} = \ldots = F_{\tau_2} \neq F_{\tau_2+1} = \ldots F_{\tau_k} \neq F_{\tau_k+1} = \ldots = F_m, \tag{3.35}$$

where $1 < \tau_1 < \ldots < \tau_k < m$ are the unknown change points that have to be estimated, and k is the unknown number of change points. Both the outlier and change point detection problems share a common goal of detecting the observations that create a disruption in the behavior of data under the null hypothesis. Neither problem has suffered from lack of attention, and elegant methods exist to test for outliers or change points.

Sarkar and Ghosh [299] demonstrate that multivariate runs and nearest-neighbor tests based on Euclidean distances are adversely affected by the distance concentration phenomenon, leading to a substantial loss of power as the dimension d increases, especially when compared with analogous procedures constructed using the dissimilarity measure ρ. Motivated by this observation, we investigate the effectiveness of the dissimilarity indices ρ and δ for detecting distributional change points through a dissimilarity visualization plot.

In high-dimensional settings, both ρ and δ yield relatively small dissimilarities among observations drawn from the same population, while producing markedly larger dissimilarities between observations originating from different populations. Let T_r denote the sum of the r-th row of the $\boldsymbol{\delta}$ matrix, and let S_r denote the corresponding sum of the r-th row of the $\boldsymbol{\rho}$ matrix, for $r = 1, \ldots, m$, defined as

$$T_r = \sum_{c=1}^{m} [\boldsymbol{\delta}]_{rc}, \quad S_r = \sum_{c=1}^{m} [\boldsymbol{\rho}]_{rc}. \tag{3.36}$$

Thus, T_r and S_r quantify the overall dissimilarity of the r-th observation relative to all remaining observations in the sample.

Assume that the dataset consists of m observations, including a single outlier or change point ($k = 1$) located at position t. By Lemma 3.6.4, as $d \to \infty$, it follows that T_r, for $r \neq t$ and $r = 1, \ldots, m$, converges to $\tilde{\delta}(F_1, F_2)$, whereas T_t converges to $(m-1)\tilde{\delta}(F_1, F_2)$. Consequently, the t-th observation attains the maximum total dissimilarity, rendering it a natural candidate for an outlier or change point. Plotting the values of T_r against the observation indices therefore provides a simple graphical tool for identifying such anomalies. An analogous argument applies to the $\boldsymbol{\rho}$ matrix, as $d \to \infty$, S_r, for $r \neq t$, converges to $\tilde{\rho}(F_1, F_2)$, while S_t converges to $(m-1)\tilde{\rho}(F_1, F_2)$. In fact, the larger the sample size m, the larger is the separation from the outlier.

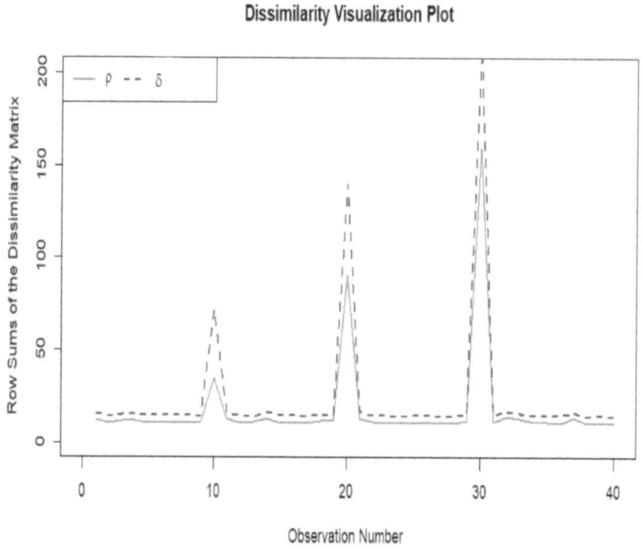

FIGURE 3.9: Row sums of the dissimilarity matrices ρ and δ from a sample of $m = 40$ observations distributed as χ_1^2 with $d = 500$. Three change points at positions 10, 20, and 30 show high dissimilarities whereas the remaining observations show low dissimilarities.

Example 3.14.1 *We evaluate the effectiveness of the dissimilarity indices ρ and δ for identifying outliers or change points through the following sampling experiment. A dataset of $m = 40$ observations and $d = 500$ variables was generated from a χ^2 distribution with 1 degree of freedom. Subsequently, $k = 3$ observations located at positions 10, 20, and 30 were perturbed by shifts of 2, 4, and 6 units, respectively. Figure 3.9 presents the corresponding dissimilarity visualization plot, in which the horizontal axis denotes the observation index and the vertical axis represents the row sums of the dissimilarity matrices based on ρ and δ. The three perturbed observations are clearly distinguishable, standing out as prominent candidates for outliers or change points.*

Let $\mathbf{Z}_i$, $i = 1, \ldots, N = m + n$, be independent and identically distributed random vectors with distributions F_i. We consider the null hypothesis K_0 against the alternative hypothesis K_a that there exists a single change point at position $m + 1$. Accordingly, the sample is partitioned into two groups, the first m observations, denoted by $\mathbf{X}_1, \ldots, \mathbf{X}_m$, and the remaining n observations, denoted by $\mathbf{Y}_1, \ldots, \mathbf{Y}_n$.

Define the $N \times N$ symmetric distance matrix

$$\mathbf{D}_\delta = \begin{bmatrix} \mathbf{R}_{m\times m} & \mathbf{H}_{m\times n} \\ \mathbf{H}'_{n\times m} & \mathbf{C}_{n\times n} \end{bmatrix}, \tag{3.37}$$

where $\mathbf{R}_{m\times m} = (r_1)$ and $\mathbf{C}_{n\times n} = (r_2)$ denote the within-group distance matrices for the $\mathbf{X}$ and $\mathbf{Y}$ samples, respectively, each having zeros along the main diagonal, and $\mathbf{H}_{m\times n} = (h)$ represents the between-group distance matrix. Specifically, the entries of $\mathbf{D}_\delta$ are given by $\delta(\mathbf{X}_i, \mathbf{X}_j)$ for $i \neq j = 1, \ldots, m$ in the matrix $\mathbf{R}$, by $\delta(\mathbf{Y}_i, \mathbf{Y}_j)$ for $i \neq j = m + 1, \ldots, N$ in the matrix $\mathbf{C}$, and by $\delta(\mathbf{X}_i, \mathbf{Y}_j)$ in the matrix $\mathbf{H}$ for $i = 1, \ldots, m$ and $j = m + 1, \ldots, N$.

Similarly, the distance matrix $\mathbf{D}_\rho$ admits the same block structure as in (3.37), with entries defined analogously by replacing $\delta(\cdot, \cdot)$ with $\rho(\cdot, \cdot)$. That is, the elements of $\mathbf{D}_\rho$ are

$\rho(\mathbf{X}_i, \mathbf{X}_j)$ in the matrix $\mathbf{R}$ for $i \neq j = 1, \ldots, m$, $\rho(\mathbf{Y}_i, \mathbf{Y}_j)$ in the matrix $\mathbf{C}$ for $i \neq j = m+1, \ldots, N$, and $\rho(\mathbf{X}_i, \mathbf{Y}_j)$ in the matrix $\mathbf{H}$ for $i = 1, \ldots, m$ and $j = m+1, \ldots, N$.

By performing an eigenvalue decomposition of the distance matrix (Mardia, Kent, and Bibby [211]), one can show that its eigenvalues are

$$\lambda_1 = \frac{1}{2N}[2mnh^2 - n(m-1)r_1^2 - m(n-1)r_2^2), \tag{3.38}$$

$$\lambda_2 = \ldots = \lambda_m = \frac{1}{2}r_1^2, \quad \lambda_{m+1} = \ldots = \lambda_{N-1} = \frac{1}{2}r_2^2 \text{ and } \lambda_N = 0.$$

Lemma 3.14.1 *As $d \to \infty$, the eigenvalues of $\mathbf{D}_\delta$ are $\lambda_1 = \frac{mn}{N}\tilde{\delta}^2$, and 0 with multiplicity $N-1$. The eigenvalues of $\mathbf{D}_\rho$, as $d \to \infty$ are $\lambda_1 = \frac{mn}{N}\tilde{\rho}^2$, and 0 with multiplicity $N-1$.*

Lemma 3.14.1 provides a principled mechanism for detecting a change point through the spectral properties of the dissimilarity matrix. Suppose that a change point occurs at position $m+1$ in the dissimilarity matrix $\mathbf{D}_\delta$. To locate this change point, we consider a sequence of candidate split locations $t = a, \ldots, N-a$, where the constant a is chosen to ensure that each segment contains a sufficient number of observations for meaningful comparison.

For a given t, the sample is partitioned into two groups consisting of the observations before and after t, yielding group sizes $m = t$ and $n = N-t$. The corresponding dissimilarity matrix $\mathbf{D}_\delta(t)$ is then constructed based on this partition, and its eigenvalues are computed. In the high-dimension, low-sample-size (HDLSS) regime, Lemma 3.14.1 implies that as $d \to \infty$, the presence of a change point is reflected in the leading eigenvalue of $\mathbf{D}_\delta(t)$. Specifically, the true change point is identified by the value of t for which the largest eigenvalue is maximized. At the true change point $t = m+1$, the largest eigenvalue satisfies $\lambda_1 = \frac{mn}{N}\tilde{\delta}^2$, thereby providing a clear criterion for detection. An analogous argument applies to the dissimilarity matrix $\mathbf{D}_\rho$, for which the change point is similarly characterized by the maximization of its leading eigenvalue.

3.14.2 Analysis of Christensen and Chowdary Gene Expression Datasets

The microarray dataset of Christensen et al. [67]. consists of molecular measurements collected from 217 non-pathologic (normal) human tissue samples obtained from 10 different anatomic sites. The study quantified methylation or gene expression levels at 1,413 loci (CpG sites or probes), resulting in a high-dimensional data matrix with 1,413 variables and 217 observations. Each variable corresponds to the molecular intensity measurement of a specific genomic locus. The primary grouping factor in the dataset is tissue type, with samples categorized according to their anatomical origin. There are 3 groups in the dataset named 'blood', 'placenta', and 'other'. We tested for the equality of the distributions of 'blood' and 'placenta' groups.

The data are publicly available through the NCBI Gene Expression Omnibus (GEO) repository and can be accessed within the R statistical computing environment using packages such as GEOquery and other Bioconductor tools. Due to its multiple tissue groups and moderate dimensionality, it serves as a valuable benchmark dataset in bioinformatics research and education.

We see the plot of the mean and standard deviations of the variables (VAR plot) and observations (OBS Plot) in the left and right panels of Figure 3.10. The VAR plot is cluttered with 2826 data points while the OBS plot shows the 217 observations and clearly separates the two groups.

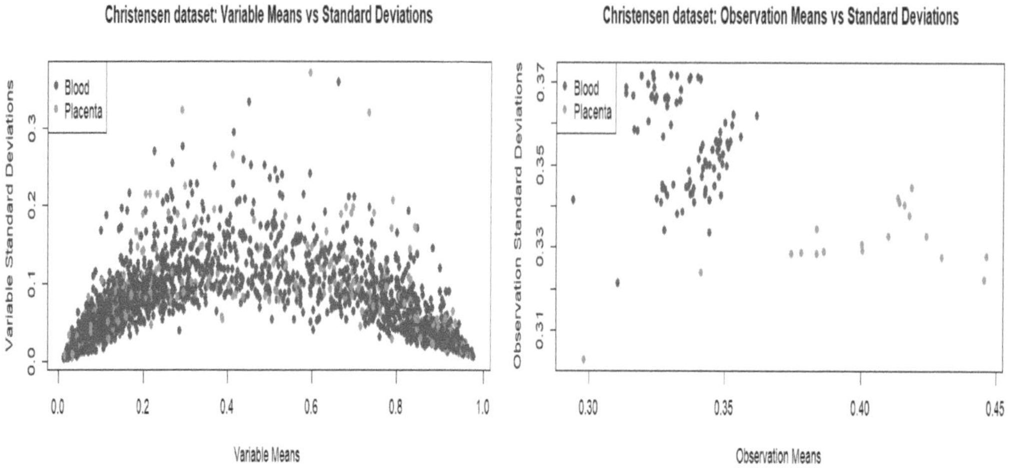

FIGURE 3.10: Christensen dataset: Plot of the means and standard deviations of variables (left panel) and observation (right panel).

The Chowdary microarray dataset (Chowdary et al., 2006) is a high-dimensional gene expression dataset generated using cDNA microarray technology to study tumor classification problems. The dataset contains 22,283 gene expression variables measured across 104 tissue samples. Each variable corresponds to the expression intensity of a specific gene probe. The samples are divided into two primary biological groups: 62 breast tumor samples and 42 colon samples. We tested for the equality of the distributions of the two groups.

This dataset is widely used in bioinformatics, machine learning, and statistical research for bench-marking classification and feature selection methods. In the R ecosystem, the dataset is available through packages such as `plsgenomics`, and it can also be accessed from the NCBI Gene Expression Omnibus (GEO) repository. We see the a plot of the mean and standard deviations of the variables (VAR plot) and observations (OBS Plot)in the left and right panels of Figure 3.11. The VAR plot is cluttered with 44566 data points while the OBS plot shows the 104 observations. The groups are not visually separated, but the OBS plot identifies a subgroup of 6 observations in the colon cancer group that are likely outliers.

Figure 3.12 is a plot of observation number against the row sums of the dissimilarity matrices δ and ρ for colon cancer group. The plot indicates that observations numbers 9-14 have much higher total dissimilarity compared to the other observations. Both indices indicate the same observations as potential outliers.

3.14.3 Analysis of S&P 500 Dataset

We apply the proposed dissimilarity indices to stock return data drawn from the U.S. equity market. The dataset is publicly available from Yahoo Finance and was retrieved using the `BatchGetSymbols` package in R. It consists of daily closing prices for $p = 500$ constituent stocks of the S&P 500 index, observed over $m = 125$ trading days spanning the period from January 1, 2020 to June 30, 2020. In the third week of March 2020, the S&P 500 index exhibited extreme volatility after declining by approximately 20% in early March, thereby entering bear market territory. This period included the steepest single-day

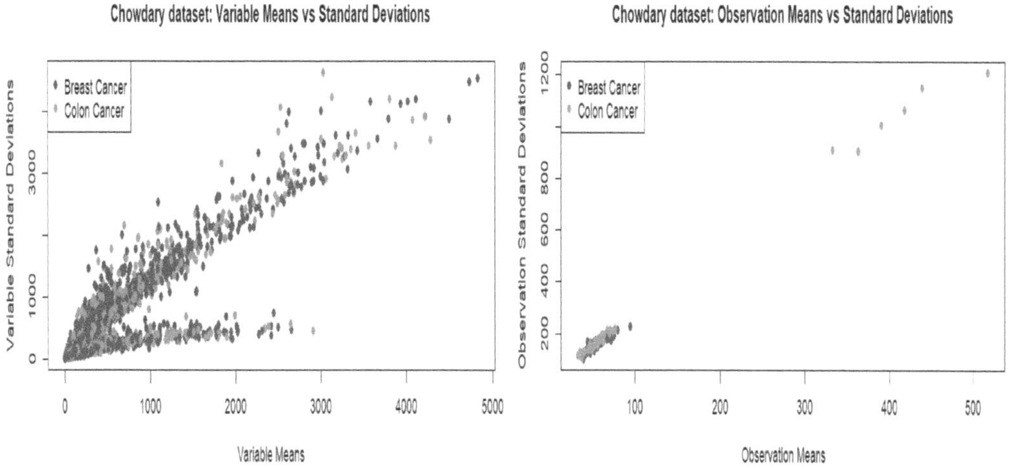

FIGURE 3.11: Chowdary dataset: Plot of the means and standard deviations of variables (left panel) and observation (right panel).

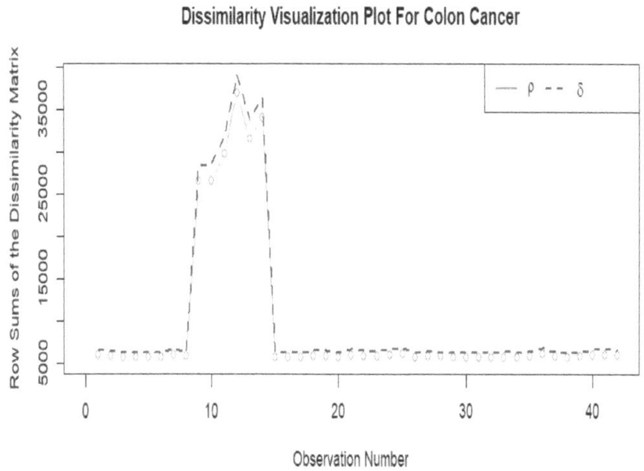

FIGURE 3.12: Cancer group dissimilarity plot. Observations 9-14 stand out as potential outliers.

percentage decline since 1987. Beginning in early April, the index started to recover. The sharp downturn in equity markets was driven by the global outbreak of the COVID-19 pandemic, as investors feared its rapid spread would severely undermine economic growth.

During the third week of March 2020, the S&P 500 experienced exceptionally high volatility, following a sharp decline of approximately 20% in early March that officially marked the onset of a bear market. This period included the steepest single-day percentage drop since 1987. Beginning in early April, the index showed signs of recovery. The pronounced market turbulence was largely driven by the global outbreak of the COVID-19 pandemic, as investors anticipated severe disruptions to economic activity and corporate earnings. Figure 3.13 displays the dissimilarity visualization plot for $S\&P500$. The row sums of the dissimilarity indices ρ and δ appear as the vertical axis and the trading days

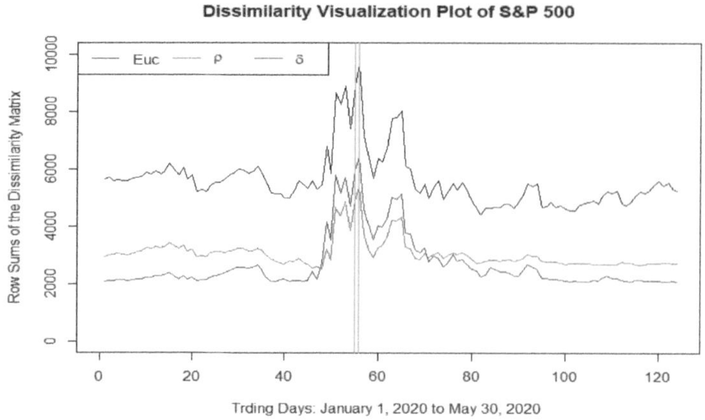

FIGURE 3.13: Row sums of the dissimilarity matrices for $S\&P$ 500 dataset: Edu (top), ρ (middle), δ (bottom).

in the first half of 2020 constitute the horizontal axis. We also included results based on the Euclidean distances divided by $\sqrt{d}$. It is interesting to note the similarity between the three indices and the fact that they all find the same change points. The plot also confirms that $\delta(\mathbf{x}, \mathbf{y}), \rho(\mathbf{x}, \mathbf{y}) \leq \frac{1}{\sqrt{d}}\|\mathbf{x} - \mathbf{y}\|$. We look for change points in the distribution of the closing costs of $d = 500$ stocks. The trading days Friday, March 20 and Monday, March 23 of 2020 show the largest total dissimilarity from all the other observations and are marked by vertical bars.

3.15 Summary

This chapter provides a systematic investigation of dissimilarity indices designed for HDLSS data, highlighting both the theoretical challenges and methodological opportunities that arise when dimension dominates sample size. It begins by formalizing the phenomena of distance concentration and data sparsity, which undermine the discriminative power of classical measures such as Euclidean and Mahalanobis distances in high dimensions. These effects are shown to distort neighborhood structure, weaken clustering and classification performance, and complicate visualization and ordination. Against this backdrop, the chapter motivates the need for alternative dissimilarity constructions that remain informative under HDLSS asymptotics.

A broad class of dissimilarity indices is introduced and analyzed, including two novel indices specifically designed to alleviate high-dimensional degeneracy. Their structural properties are examined in detail, with emphasis on invariance, scaling behavior, and sensitivity to dependence. Finite-sample and asymptotic properties of the proposed indices are established, including the distributional behavior of row-wise means and variances of dissimilarity matrices under both constant and spatial correlation models. These results provide theoretical insight into how dependence and dimensionality jointly shape the geometry of high-dimensional data clouds, and they offer principled guidance for comparing dissimilarity measures beyond heuristic criteria.

The chapter further demonstrates the practical relevance of these theoretical developments through applications to outlier detection, change-point analysis, and financial time series, including an analysis of S&P 500 data. These examples illustrate how carefully chosen dissimilarity indices can restore contrast, stabilize inference, and improve detection power in settings where traditional distance-based methods fail. By linking high-dimensional probability theory with applied data analysis, the chapter establishes a rigorous foundation for the use of dissimilarity-based methods in modern statistical learning and multivariate inference.

3.16 Proofs

Proof: Lemma 3.6.1 See 12.9 □

Proof: Lemma 3.6.2. See 12.17 □

Proof: Lemma 3.6.3:

Let $\mathbf{1}_d$ be a d-dimensional vector of 1s and $\bar{x} = \frac{1}{d}\mathbf{1}'\mathbf{x}$. It follows that

$$\bar{\mu}_x^d = \mathbb{E}(\bar{x}^d) = \frac{1}{d}\sum_{t=1}^{d}\mathbb{E}(x_t) = \frac{1}{d}\mathbf{1}'\boldsymbol{\mu}_x. \tag{3.39}$$

One can show that

$$\gamma_{\bar{x}}^2 = \mathbb{V}ar(\frac{1}{d}\mathbf{1}'\mathbf{x}) = \frac{1}{d^2}T_x \tag{3.40}$$

where $T_x = \mathbf{1}_d'\boldsymbol{\Sigma}_x\mathbf{1}_d = \sum_{i=1}^{d}\sum_{j=1}^{d}\sigma_{(x)ij}$. It is not difficult to show that

$$\begin{aligned}\mathbb{E}([s_x^d]^2) &= \frac{1}{d}\sum_{t=1}^{d}\mathbb{E}(x_t - \bar{x}^d)^2) = \frac{1}{d}\sum_{t=1}^{d}\mathbb{E}(x_t^2) - \frac{1}{d^2}\mathbb{E}(\sum_{t=1}^{d}x_t)^2 \\ &= \frac{1}{d}tr(\boldsymbol{\Sigma}_x) + \frac{1}{d}\boldsymbol{\mu}_x'\boldsymbol{\mu}_x - \frac{1}{d^2}tr(\boldsymbol{\Sigma}_x) - \frac{1}{d^2}\boldsymbol{\mu}_x'\boldsymbol{\mu}_x - \frac{2}{d^2}\sum_{t=1}^{d}\sum_{r=t+1}^{d}(\sigma_{rs} + \mu_r\mu_s) \\ &= [\sigma_x^d]^2 \to \sigma_{F_1}^2 + \nu_x^2,\end{aligned} \tag{3.41}$$

$$\frac{1}{d}\boldsymbol{\mu}_x'\boldsymbol{\mu}_x - \frac{1}{d^2}\boldsymbol{\mu}_x'\boldsymbol{\mu}_x - \frac{2}{d^2}\sum_{t=1}^{d}\sum_{r=t+1}^{d}\mu_t\mu_r = \frac{1}{d}\sum_{t=1}^{d}(\mu_{x_t} - \bar{\mu}_x)^2 \to \nu_x^2,$$

in probability as $d \to \infty$. Moreover, $\frac{1}{d}tr(\boldsymbol{\Sigma}_x) \to \sigma_{F_1}^2$, $\frac{1}{d^2}tr(\boldsymbol{\Sigma}_x) \to 0$, and

$$\frac{2}{d^2}\sum_{t=1}^{d}\sum_{r=t+1}^{d}\sigma_{tr} = \frac{1}{d^2}T_x \to 0$$

since T_x is of order $o(d^2)$ by Assumption (B3). □

Proof: Lemma 3.6.4:

Consider the modified distances (3.11) and (3.12). Since $\mathbf{X}_1, \mathbf{X}_2$ are i.i.d. F_1 and $\mathbf{Y}_1, \mathbf{Y}_2$ are i.i.d. F_2, by Assumptions (B1)–(B3) and Lemma 3.6.3, we have as $d \to \infty$,

$$\bar{x}_1 \to \bar{\mu}_x,\ \bar{y}_1 \to \bar{\mu}_y,\ s_x^2 \to \sigma_x^2 \text{ and } s_y^2 \to \sigma_y^2,$$

respectively. Hence, $\|\mathbf{x}_1 - \mathbf{z}\|_D - \|\mathbf{x}_2 - \mathbf{z}\|_D \to 0$, and, $\delta_0(\mathbf{X}_1, \mathbf{X}_2) = 0$.

Similarly, $\delta_0(\mathbf{Y}_1, \mathbf{Y}_2) = 0$ as $d \to \infty$. For the limiting value of $\delta_0(\mathbf{X}_1, \mathbf{Y}_1)$, if $\mathbf{z}$ comes from F_1 then (3.11) converges to 0 while (3.12) converges to $\tilde{\delta}(F_1, F_2)$. Similarly, if $\mathbf{z}$ comes from F_2 then (3.12) converges to 0 while (3.11) converges to $\tilde{\delta}(F_1, F_2)$. Hence,

$$\delta_0(\mathbf{X}_1, \mathbf{Y}_1) \to \frac{1}{N-2}((m-1)\tilde{\delta}(F_1, F_2) + (n-1)\tilde{\delta}(F_1, F_2)) = \tilde{\delta}(F_1, F_2).$$

□

3.17 Exercises

Exercise 3.1 *Suppose* $\mathbf{X} \sim F_1 = \mathbb{N}_d(\mathbf{0}, \sigma_1^2\mathbf{I})$, *and independently,* $\mathbf{Y} \sim F_2 = \mathbb{N}_d(\mu\mathbf{1}_d, \sigma_2^2\mathbf{I})$. *Let* $\mathbb{X} = \{\mathbf{X}_i\}_{i=1}^m$ *and* $\mathbb{Y} = \{\mathbf{Y}_j\}_{j=1}^n$ *be i.i.d. copies of* $\mathbf{X}$ *and* $\mathbf{Y}$*,respectively. Prove* (3.15). *Under what condition do all the observations in* $\mathbb{Y}$ *have their neighbors in* $\mathbb{X}$ *as* $d \to \infty$*?*

Exercise 3.2 *Suppose* $\mathbf{X} \sim F_1 = \mathbb{N}_d(\mathbf{0}, \sigma_1^2\mathbf{I})$, *and independently,* $\mathbf{Y} \sim F_2 = \mathbb{N}_d(\mu\mathbf{1}_d, \sigma_2^2\mathbf{I})$. *Find the limiting values of* ρ *in* (3.16) *and* δ *in* (3.20).

Exercise 3.3 *Prove Lemma 3.14.1.*

Exercise 3.4 *Suppose* $\mathbf{x} = (x_1, \ldots, x_d)$ *where* x_k*'s are i.i.d.* $\mathbb{N}(0, \frac{f}{f-2})$, $f > 2$, *for* $k = 1, \ldots d$. *Suppose* $\mathbf{y} = (y_1, \ldots, y_d)$ *where* y_k*'s are i.i.d.* $t_{(f)}$, *a t-distribution with* $f > 2$ *degrees of freedom, for* $k = 1, \ldots d$. *Show that* $\tilde{\rho} = \tilde{\delta} = 0$ *as* $d \to \infty$.

Exercise 3.5 *Verify the results of Example 3.6.5.*

Exercise 3.6 *Write algorithms to implement the dissimilarity indices* ρ *and* δ *as they appear in* (3.6) *and* (3.14).

Exercise 3.7 *Let* $\mathbf{X} \in \mathbb{R}^d$ *be a random vector with i.i.d. components from an exponential distribution with mean 1. Compute* $\mathbb{E}(\|\mathbf{X}\|^2)$ *and show that the relative variance* RV_d *tends to 0 as* $d \to \infty$.

Exercise 3.8 *Suppose* $\mathbf{X} \in \mathbb{R}^d$ *has dependent entries where* $x_i = Z + \epsilon_i$ *for* $Z \sim \mathbb{N}(0,1)$ *and* $\epsilon_i \sim \mathbb{N}(0,1)$ *independent. Discuss whether* RV_d *tends to 0 as* $d \to \infty$ *and whether distance concentration occurs.*

Exercise 3.9 *Show that the dissimilarity indices* $\rho(\mathbf{x}, \mathbf{y})$ *and* $\delta(\mathbf{x}, \mathbf{y})$ *are not metrics.*

Exercise 3.10 *Let* $\mathbb{X} = \{\mathbf{X}_1, \ldots, \mathbf{X}_m\} \subset \mathbb{R}^d$ *and* $\mathbb{Y} = \{\mathbf{Y}_1, \ldots, \mathbf{Y}_n\} \subset \mathbb{R}^d$ *be two sets of independent observations from multivariate normal distributions,* $\mathbf{X}_i \sim \mathbb{N}_d(\boldsymbol{\mu}_1, \sigma_1^2\mathbf{I}_d)$ *and* $\mathbf{Y}_j \sim \mathbb{N}_d(\boldsymbol{\mu}_2, \sigma_2^2\mathbf{I}_d)$. *Let* $\mathbb{Z}_N = \mathbb{X} \cup \mathbb{Y}$ *be the combined data cloud, with* $N = m+n$. *Consider the dissimilarity between* $\mathbf{X} \in \mathbb{X}$ *and* $\mathbf{Y} \in \mathbb{Y}$ *by* (3.13). *Derive the expectation* $\mathbb{E}[\delta(\mathbf{X}, \mathbf{Y})]$ *under the above model.*

Exercise 3.11 *Write a program to demonstrate the concentration phenomenon. Let* $n = 20$ *and* $d = \{50, 100, 500, 1000\}$. *Generate 100 d-dimensional normal samples with mean zero and identity covariance matrix and use them to obtain the IPDs (*L_2 *norm).*

a) Find and display the coefficient of variation of IPDs.

b) Find and display the ratio of the maximum IPD to the minimum IPD.

c) Apply shapiro.test of normality to IPDS and report the p-value.

Exercise 3.12 *Suppose* $\mathbf{X} = (X_1, \ldots, X_d)'$ *is a* $d \times 1$ *multivariate normal vector with mean zero, unit variances, and common correlation* ρ, *i.e.,*

$$\mathbb{E}[X_i] = 0, \quad \mathbb{V}ar(X_i) = 1, \quad Cov(X_i, X_j) = \rho \text{ for } i \neq j.$$

Let $\bar{X} = \frac{1}{d}\sum_{i=1}^{d} X_i$ *be the mean of the components. Determine whether the Weak Law of Large Numbers (WLLN) or the Strong Law of Large Numbers (SLLN) applies to* $\bar{X}$ *as* $d \to \infty$.

Exercise 3.13 *Suppose* $\boldsymbol{X}$ *is a mean-zero multivariate normal vector with covariance* $\boldsymbol{\Sigma}_{ij} = \sigma^2 \rho^{|i-j|}$. *Evaluate the relative variance of the squared Euclidean norm as* $d \to \infty$.

4

High-Dimensional Graphics

4.1 Abstract

This chapter presents a comprehensive treatment of graphical methods for visualizing high-dimensional data, addressing the challenges posed by the complexity and scale of modern datasets. As conventional low-dimensional plots become inadequate in high-dimensional (HD) settings, specialized visualization techniques are required to reveal the underlying structure, group separation, and multivariate dependence. The chapter surveys a broad range of HD graphics, beginning with graph-based and matrix-decomposition approaches such as minimum spanning trees, singular value decomposition, and principal component analysis, which provide principled dimension reduction and structure extraction. The use of biplots is discussed as an effective means of jointly visualizing observations and variables. Projection-based and iconic displays, including parallel coordinate plots, Andrews' plots, and the Dutta–Ghosh transformation, are introduced as flexible tools for exploring multivariate patterns. The chapter further emphasizes graphical methods for comparing distributions across two or multiple groups, incorporating techniques based on IPDs and cumulative distribution functions. Through illustrative applications to benchmark datasets, the chapter highlights how carefully chosen HD visualization methods facilitate exploratory analysis, diagnostic assessment, and inferential insight in complex multivariate settings.

4.2 Introduction

High-dimensional data visualization is a critical aspect of modern statistical analysis, especially as the amount of data and the number of measured variables continue to grow across fields such as bioinformatics, finance, and social sciences. Traditional two- or three-dimensional plots are insufficient for such complex datasets, leading to the development of specialized high-dimensional graphical techniques. These tools aim to reveal the structure, patterns, and relationships embedded in high-dimensional data, enabling researchers to interpret and analyze data more effectively.

One prominent category of HD visualization techniques includes iconic representations, such as Chernoff faces and parallel coordinate plots. Chernoff faces use human facial features to represent multivariate data, using our natural ability to recognize facial patterns. Parallel coordinate plots depict each variable on a parallel axis, allowing for the visualization of trends and clustering in high-dimensional spaces. Similarly, Andrews' plots transform multivariate data into functions, typically sinusoidal curves, offering another dimension-reducing perspective.

Matrix decomposition techniques such as Singular Value Decomposition (SVD) and Principal Component Analysis (PCA) are fundamental in HD data analysis. PCA, in particular,

DOI: 10.1201/9781003789581-4

reduces dimensionality by transforming original variables into a new set of uncorrelated variables that capture the maximum variance in the data. Biplots, which combine score plots and loading plots, are used in conjunction with PCA to visualize both observations and variables in a reduced dimension.

Other specialized transformations and visualization methods have been developed to enhance interpretability. The Dutta and Ghosh transformation [84] is one such method aimed at simplifying complex multivariate relationships. Visual analysis also involves assessing how data distributions change with respect to shift, scale, shape, and dependence. Recognizing these changes is crucial for hypothesis testing and model diagnostics in HD settings. Tools like Multidimensional Scaling (MDS) and Principal Coordinate Analysis help visualize the similarity or dissimilarity between data points by mapping them into a lower-dimensional space based on distance or dissimilarity matrices.

4.3 Minimum Spanning Tree

Graph-based methods such as Minimum Spanning Trees offer insights into data structure by connecting all data points in a graph with the shortest possible total edge length, without forming loops. MST algorithms, along with mean and variance analyses of MST structures, are useful in cluster detection and visualizing the topological relationships within data. MST plots thus provide an intuitive yet powerful approach for uncovering the underlying architecture of high-dimensional datasets. We briefly explain the two most utilized algorithms for construction of an MST. Prim's Algorithm (Prim [273]) constructs a Minimum Spanning Tree (MST) from a given vertex set $\mathbf{X}$ and edge set E. The procedure starts with an empty vertex set $\mathbf{X}^*$ and edge set E^*. An arbitrary vertex is added to $\mathbf{X}^*$ to begin, and the algorithm iteratively adds the shortest edge connecting $\mathbf{X}^*$ to $\mathbf{X} \setminus \mathbf{X}^*$, updating the sets accordingly. The process repeats until all vertices are included in $\mathbf{X}^*$.

Algorithm 4.3.1 Prim's Algorithm

Input: Vertex set $\mathbf{X}$, edge set E
Output: Minimum Spanning Tree (MST) with vertex set $\mathbf{X}^*$ and edge set E^*
Step 1: Initialization
$\mathbf{X}^* \leftarrow \{\text{arbitrary vertex from } \mathbf{X}\}$
$E^* \leftarrow \emptyset$
Step 2: Grow the MST
while $\mathbf{X}^* \neq \mathbf{X}$ **do**
    Find the shortest edge $(u, v) \in E$ such that $u \in \mathbf{X}^*$ and $v \in \mathbf{X} \setminus \mathbf{X}^*$
    $E^* \leftarrow E^* \cup \{(u, v)\}$
    $\mathbf{X}^* \leftarrow \mathbf{X}^* \cup \{v\}$
end while
Step 3: Termination
Return $\mathbf{X}^*$ and E^* as the MST

Kruskal's Algorithm (Kruskal [176]) constructs a Minimum Spanning Tree (MST) from a given vertex set $\mathbf{X}$ and edge set E. The algorithm iteratively adds the next shortest edge that does not form a cycle in the current tree, until the MST contains $n - 1$ edges.

Algorithm 4.3.2 Kruskal's Algorithm

Input: Vertex set $\mathbf{X}$, edge set E
Step 1: Initialization
Sort all edges in E in non-decreasing order of weight
$E^* \leftarrow \emptyset$
Step 2: Iteratively add edges
for each edge (u, v) in sorted order **do**
    if adding (u, v) to E^* does not create a cycle **then**
        $E^* \leftarrow E^* \cup \{(u, v)\}$
    end if
    if $|E^*| = |\mathbf{X}| - 1$ **then**
        break: MST is complete
    end if
end for
Step 3: Termination
Return E^* as the MST

The computational complexity of Prim's algorithm is $O(|E|log|\mathbf{X}|)$ using binary tree heap while Kruskal's is of $O(|E|log|E|)$ where $|\cdot|$ is the cardinality of a set. One major difference between the two algorithms is that Prim's maintains a connected tree in all stages while Kruskal's algorithm deals with a forest that eventually merges into a tree. Kruskal's algorithm performs better on average while Prim's algorithm has an advantage when dealing with complete graphs. A spanning tree provides a path between every two vertices of a graph and is a compact method of displaying and summarizing observations in $\mathbb{R}^d$. Depth functions defined on the MST allow for the use of colors for visualization purposes, identify those vertices in the outskirts of the sample or discriminate among several groups.

Example 4.3.1 *Figures 4.1 and 4.2 show an example of how Kruskal's and Prim's algorithm build an MST of nine vertices step by step. Table 4.1 shows the* $\binom{9}{2}$ *IPDs between the vertices considered.*

TABLE 4.1: The IPDs between vertices in Figures 4.1 and 4.2.

	$\mathbf{X}_1$	$\mathbf{X}_2$	$\mathbf{X}_3$	$\mathbf{X}_4$	$\mathbf{X}_5$	$\mathbf{X}_6$	$\mathbf{X}_7$	$\mathbf{X}_8$
$\mathbf{X}_2$	1.601							
$\mathbf{X}_3$	6.597	6.392						
$\mathbf{X}_4$	5.234	6.265	4.967					
$\mathbf{X}_5$	2.028	3.511	6.372	3.544				
$\mathbf{X}_6$	3.448	3.941	3.556	2.723	2.818			
$\mathbf{X}_7$	4.186	4.11	2.412	3.907	4.125	1.477		
$\mathbf{X}_8$	3.257	4.589	5.908	2.268	1.33	2.49	3.962	
$\mathbf{X}_9$	2.277	2.469	4.321	4.029	2.583	1.483	1.909	2.978

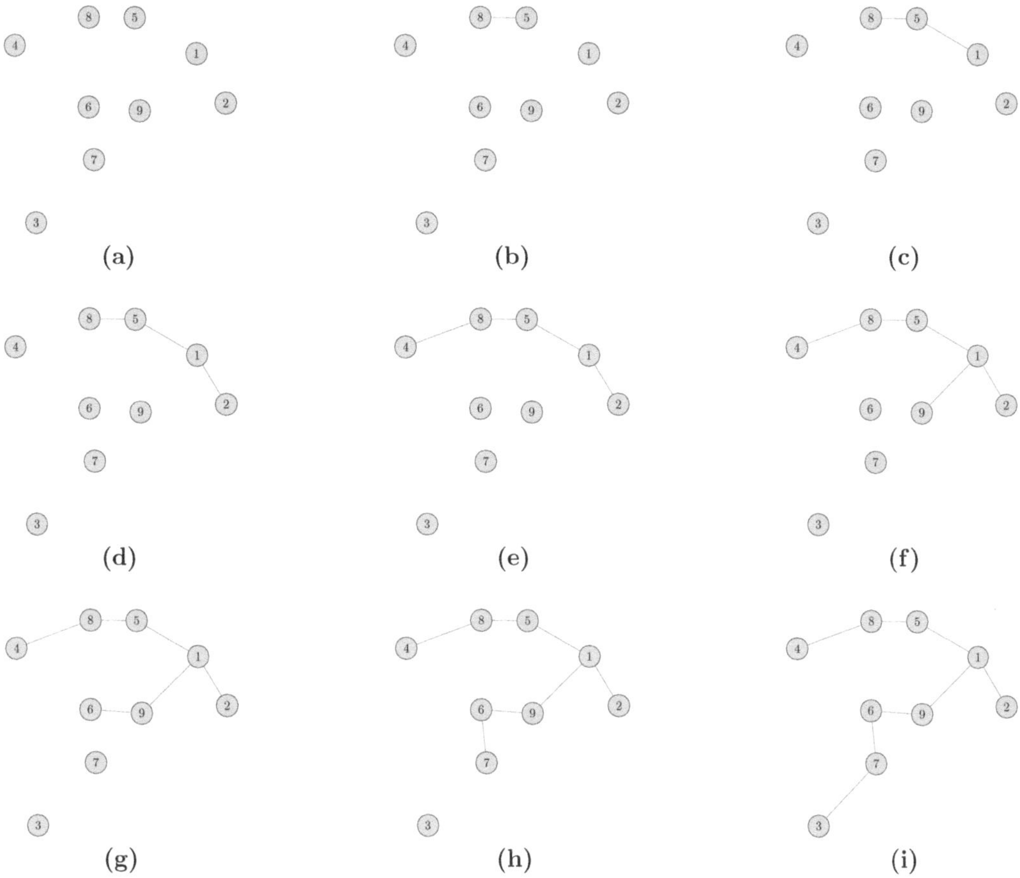

FIGURE 4.1: Construction of the MST using Prim's Algorithm.

4.4 Matrix Decomposition

Let $\mathbf{\Sigma}$ be a $d \times d$ square matrix and $\mathbf{I}$ be the $d \times d$ identity matrix. The scalars $\lambda_1, \ldots, \lambda_d$ satisfying the polynomial equation $|\mathbf{\Sigma} - \lambda \mathbf{I}| = 0$ are the eigenvalues of $\mathbf{\Sigma}$. If $\mathbf{e}_i$ is a $d \times 1$ non-zero vector such that $\mathbf{\Sigma}\mathbf{e}_i = \lambda \mathbf{e}_i$, then $\mathbf{e}_i$ is an eigenvector of $\mathbf{\Sigma}$ with associated eigenvalue λ_i. The spectral decomposition of

$$\mathbf{\Sigma} = \sum_{i=1}^{d} \lambda_i \mathbf{e}_i \mathbf{e}_i^{'} = \mathbf{E}\mathbf{\Lambda}\mathbf{E}^{'}. \tag{4.1}$$

Let $\mathbf{X}$ be $n \times d$ matrix of real numbers. There exist an $n \times n$ orthogonal matrix $\mathbf{U}$ and a $d \times d$ orthogonal matrix $\mathbf{V}$ such that

$$\mathbf{X} = \mathbf{U}\mathbf{\Lambda}\mathbf{V}^{'} = \sum_{i=1}^{r} \lambda_i \mathbf{u}_i \mathbf{v}_i^{'} \tag{4.2}$$

where the $n \times d$ matrix $\mathbf{\Lambda}$ is diagonal with (i, i) entry $\lambda_i \geq 0$ for $i = 1, \ldots, \min(n, d)$. The positive λ_i are called the singular values of $\mathbf{X}$ and $r = \text{rank}(X)$. Let $\mathbf{X}$ be $n \times d$ matrix

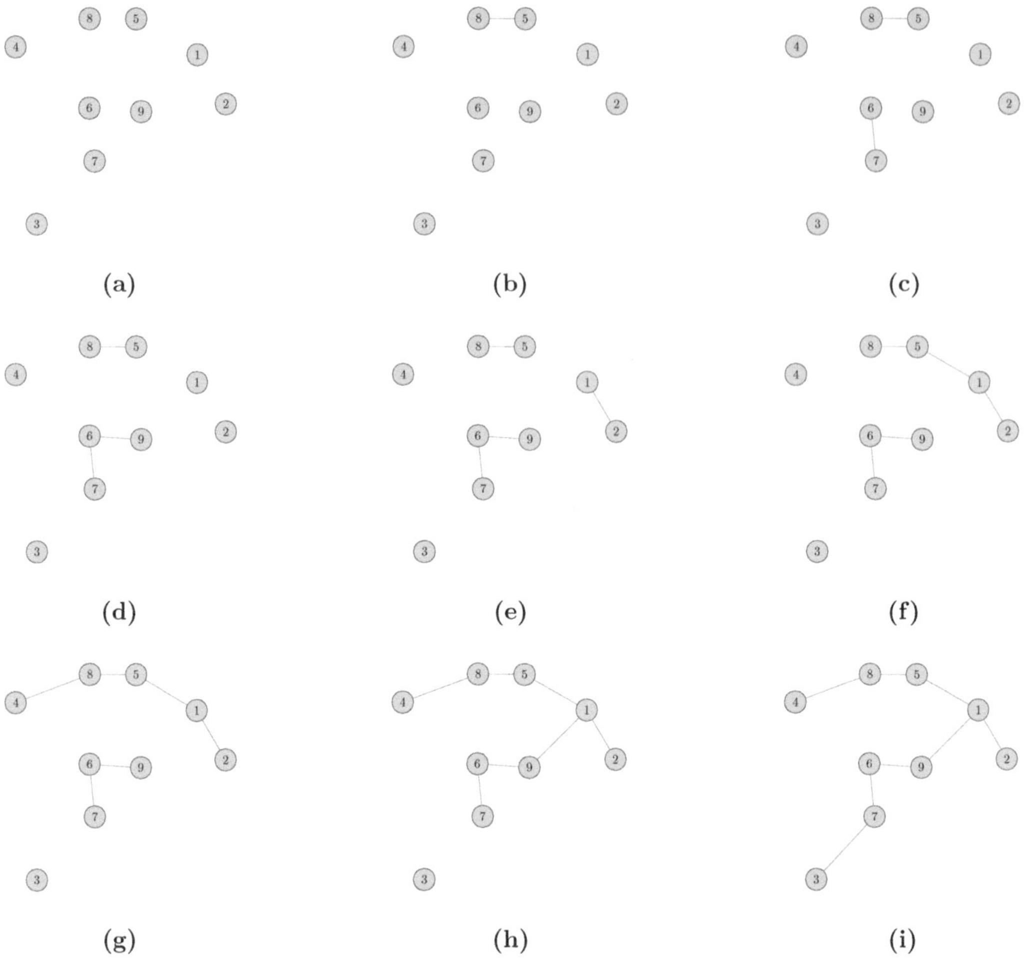

FIGURE 4.2: Construction of the MST using Kruskal's Algorithm.

of reals numbers with $n \geq d$ and the SVD $\mathbf{U}\mathbf{\Lambda}\mathbf{V}^{'}$. Let $s < d = \text{rank}(\mathbf{X})$. The rank s least squares approximation to $\mathbf{X}$ is $\mathbf{Y} = \sum_{i=1}^{s} \lambda_i \mathbf{u}_i \mathbf{v}_i^{'}$, which minimizes $tr[(\mathbf{X} - \mathbf{Y})(\mathbf{X} - \mathbf{Y})^{'}]$ over all $n \times d$ matrices $\mathbf{Y}$ having rank smaller or equal to s. The approximation error is $\sum_{i=s+1}^{d} \lambda_i^2$.

The spectral decomposition expresses $\mathbf{\Sigma}$ as stretching along orthogonal eigenvector directions. The SVD decomposes $\mathbf{X}$ into rotation $\rightarrow$ scaling $\rightarrow$ rotation. Spectral decomposition of $\mathbf{\Sigma}$ appears in Figure 4.3. SVD decomposition of $\mathbf{X} = \mathbf{U}\mathbf{\Lambda}\mathbf{V}'$ appears in Figure 4.4.

4.5 Principal Component Analysis

Multivariate Analysis often starts out with data involving a substantial number of correlated variables. We would like to reduce the data dimensionality while retaining most of the information. Principal Component Analysis (PCA) is a classical dimension-reduction tool that can be used to reduce a large set of variables to a small set that still contains most

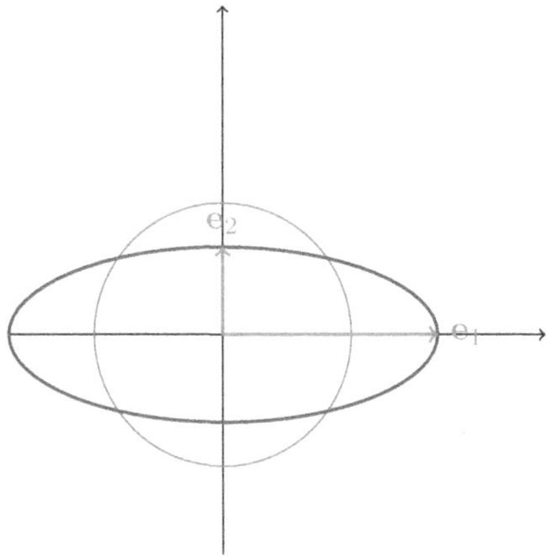

FIGURE 4.3: Spectral decomposition $\boldsymbol{\Sigma} = \mathbf{E}\Lambda\mathbf{E}'$. A unit circle is transformed into an ellipse. The eigenvectors $\mathbf{e}_1, \mathbf{e}_2$ define orthogonal directions, and eigenvalues determine stretching along those directions.

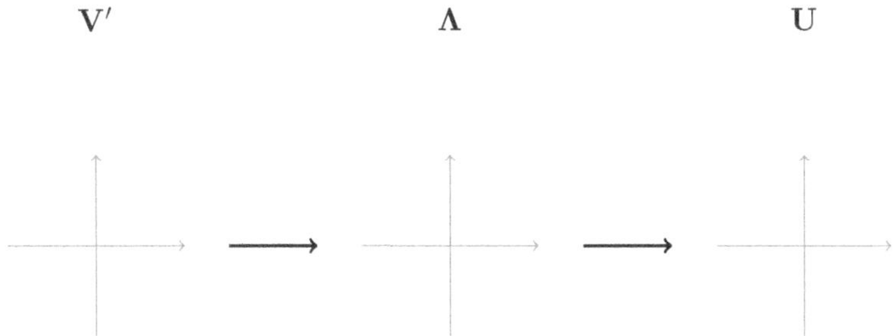

FIGURE 4.4: SVD decomposition $\mathbf{X} = \mathbf{U}\boldsymbol{\Lambda}\mathbf{V}'$. The matrix $\mathbf{V}'$ rotates the coordinate system, $\boldsymbol{\Lambda}$ stretches it along orthogonal axes (singular values), and $\mathbf{U}$ applies a final rotation.

of the information in the large set. Although d components are required to reproduce the total variability, often much of this variability can be accounted for by a small number k of the principal components that are orthogonal. If so, there is (almost) as much information in the k components as there is in the original d variables. The first principal component accounts for as much of the variability in the data as possible, and each succeeding component accounts for as much of the remaining variability as possible. Traditionally, principal component analysis is performed on a square symmetric matrix (Covariance or Correlation matrix). A correlation matrix is used if the variances of individual variates differ much, or if the units of measurement of the individual variates differ.

Let the random vector $\mathbf{X} = [X_1, \dots, X_d]$ have the covariance matrix $\boldsymbol{\Sigma}$ where $\boldsymbol{\Sigma}$ has the eigenvalue, eigenvector pairs $(\lambda_i, \mathbf{e}_i)$ for $\lambda_1 \geq \lambda_2 \geq \dots \geq \lambda_d \geq 0$. Suppose $Y_i = \mathbf{a}_i\mathbf{X}$ for $i = 1, \dots d$. It follows that

$$\mathbb{V}ar(Y_i) = \mathbf{a}_i'\boldsymbol{\Sigma}\mathbf{a}_i, \quad \mathbb{C}ov(Y_i, Y_k) = \mathbf{a}_i'\boldsymbol{\Sigma}\mathbf{a}_k. \tag{4.3}$$

The ith principal component is the linear combination $\mathbf{a}_i'\mathbf{X}$ that maximizes $\mathbb{V}ar(\mathbf{a}_i'\mathbf{X})$ subject to $\mathbf{a}_i'\mathbf{a}_i = 1$ for $i = 1, \dots, d$ and $\mathbb{C}ov(\mathbf{a}_i'\mathbf{X}, \mathbf{a}_k'\mathbf{X}) = 0$ for $k < i$. The ith principal component is given by $Y_i = \mathbf{e}_i'\mathbf{X}$. Using $\max_{\mathbf{x}\neq\mathbf{0}} \frac{\mathbf{x}'\boldsymbol{\Sigma}\mathbf{x}'}{\mathbf{x}}\mathbf{x} = \lambda_i$ at $\mathbf{x} = \mathbf{e}_i$ we obtain

$$\mathbb{V}ar(Y_i) = \mathbf{e}_i'\boldsymbol{\Sigma}\mathbf{e}_i = \lambda_i, \quad \mathbb{C}ov(Y_i, Y_k) = \mathbf{e}_i'\boldsymbol{\Sigma}\mathbf{e}_k = 0. \tag{4.4}$$

By spectral decomposition $\mathbf{\Sigma} = \mathbf{E\Lambda E'} = \sum_{i=1}^{d} \lambda_i \mathbf{e}_i \mathbf{e}_i^{'}$. Hence, $trace(\mathbf{\Sigma}) = \sum_{i=1}^{d} \lambda_i$. Proportion of the total variance due to the kth PC is $d_k = \frac{\lambda_k}{\sum_{i=1}^{d} \lambda_i}$ for $k = 1, 2 \ldots, d$. Sample principal components are, in general, not invariant with respect to changes in scale. We often use standardized PCs $\mathbf{z}_j = \mathbf{D}^{-1/2}(\mathbf{x}_j - \bar{\mathbf{x}})$ where $\mathbf{D}$ is a diagonal matrix of variances.

Example 4.5.1 *If $\mathbf{Z}_1, \ldots, \mathbf{Z}_n$ are trivariate vectors ($d = 3$) that are equally correlated with unit variance, or when PCA is conducted on standardized variables, the correlation matrix has unit variances and correlations ρ. It is not difficult to show that $\lambda_1 = 1 + (d-1)\rho$, $\lambda_2 = \lambda_3 = \ldots \lambda_d = 1 - \rho$ with eigenvectors*

$$\begin{aligned} \mathbf{e}_1^{'} &= (\frac{1}{\sqrt{d}}, \frac{1}{\sqrt{d}}, \ldots, \frac{1}{\sqrt{d}}), \\ \mathbf{e}_i^{'} &= (\frac{1}{\sqrt{(i-1)i}}, \ldots, \frac{1}{\sqrt{(i-1)i}}, \frac{-(i-1)}{\sqrt{(i-1)i}}, 0, \ldots, 0), \end{aligned} \tag{4.5}$$

for $i \geq 2$. When $d = 3$ we have

$$\begin{aligned} \mathbf{e}_1^{'} &= (1/\sqrt{3}, 1/\sqrt{3}, 1/\sqrt{3}), \\ \mathbf{e}_2^{'} &= (1/\sqrt{2}, -1/\sqrt{2}, 0), \\ \mathbf{e}_3^{'} &= (1/\sqrt{6}, 1/\sqrt{6}, -2/\sqrt{6}). \end{aligned} \tag{4.6}$$

Hence, the first principal component is $\mathbf{Y}_1 = \mathbf{e}_1^{'}\mathbf{Z} = \frac{1}{\sqrt{d}} \sum_{i=1}^{d} \mathbf{Z}_i$ is proportional to the sum of the standardized variables. The first principal component explains $\lambda_1/d = \frac{1+(d-1)\rho}{d} = \rho + \frac{1-\rho}{d}$ of the total population variance. For ρ close to 1 and large d, this proportion is close to ρ.

Example 4.5.2 *Suppose $\mathbf{X} \sim \mathbb{N}_d(\boldsymbol{\mu}, \mathbf{\Sigma})$. The density of $\mathbf{X}$ is constant on the $\boldsymbol{\mu}$ centered ellipsoids $(\mathbf{X} - \boldsymbol{\mu})^{'}\mathbf{\Sigma}^{-1}(\mathbf{X} - \boldsymbol{\mu}) = c^2$ which have axes $\pm c\sqrt{\lambda_i}\mathbf{e}_i$, for $i = 1, \ldots, d$, where $(\lambda_i, \mathbf{e}_i)$ are the eigenvalue and eigenvectors of $\mathbf{\Sigma}$. Using spectral decomposition, we can write $c^2 = \mathbf{X}^{'}\mathbf{\Sigma}^{-1}\mathbf{X} = \sum_{i=1}^{d} \frac{1}{\lambda_i}(\mathbf{e}_i^{'}\mathbf{x})^2$ where $\mathbf{Y}_i = \mathbf{e}_i^{'}\mathbf{X}$ are the principal components of $\mathbf{X}$. Note that $\mathbf{X}^{'}\mathbf{\Sigma}^{-1}\mathbf{X}$ has a $\chi^2(d)$ and χ^2 quantiles are used to obtain contour plots.*

Example 4.5.3 *We consider the 4×4 Toeplitz covariance matrix $\mathbf{\Sigma}$ with unit variances and correlations $\mathbb{C}ov(X_i, X_j) = \rho^{|i-j|}$. The principal components are given by the eigenvalue decomposition $\Sigma = Q\Lambda Q'$, where Λ contains the eigenvalues and the columns of Q are the orthonormal eigenvectors. For a 4×4 AR(1) covariance matrix, the eigenvalues are*

$$\lambda_k = \frac{1-\rho^2}{1 - 2\rho\cos\left(\frac{k\pi}{5}\right) + \rho^2}, \qquad k = 1, 2, 3, 4.$$

The eigenvalues are ordered $\lambda_1 \geq \lambda_2 \geq \lambda_3 \geq \lambda_4 > 0$. The principal component directions (eigenvectors) are

$$q_k' = \sqrt{\frac{2}{5}} \left(\sin\left(\tfrac{k\pi}{5}\right) \quad \sin\left(\tfrac{2k\pi}{5}\right) \quad \sin\left(\tfrac{3k\pi}{5}\right) \quad \sin\left(\tfrac{4k\pi}{5}\right)\right), \qquad k = 1, 2, 3, 4.$$

These vectors form an orthonormal basis $\mathbf{Q} = \left(q_1, q_2, q_3, q_4\right)$, and $Q'Q = I$. The first principal component corresponds to a smooth, slowly varying mode. Higher components represent increasingly oscillatory patterns. As $\rho \to 1$, most variance concentrates in the first component.

Figure 4.5 shows the level set of a centered bivariate normal distribution with covariance matrix $\mathbf{\Sigma}$. The ellipse represents the quadratic form $\mathbf{x}^\top \mathbf{\Sigma}^{-1} \mathbf{x} = c$. The Figure depicts the two principal components (major and minor axis of the ellipse), $\mathbb{E}(Y \mid X)$ and $\mathbb{E}(X \mid Y)$. While PC1 and PC2 point to the directions of the largest and smallest variability, respectively, $\mathbb{E}(Y \mid X)$ connects the vertical tangents of the level set, and $\mathbb{E}(X \mid Y)$ connects the horizontal tangents of the level set.

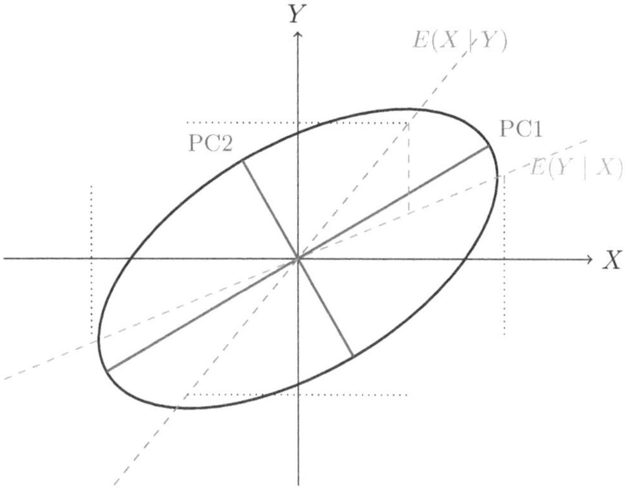

FIGURE 4.5: The solid blue lines (PC1 and PC2) are the eigenvectors of $\mathbf{\Sigma}$, corresponding to the largest and smallest eigenvalues. The dashed red and green lines denote the regression functions $\mathbb{E}(Y \mid X)$ and $\mathbb{E}(X \mid Y)$, whose slopes are $\mathbb{C}ov(X, Y)/\mathbb{V}ar(X)$ and $\mathbb{C}ov(X, Y)/\mathbb{V}ar(Y)$, respectively. The dotted horizontal and vertical lines indicate tangency points, illustrating geometrically that principal component directions and regression lines generally differ unless the covariance matrix is spherical or the correlation is ± 1.

4.6 Multidimensional Scaling

Multidimensional scaling (MDS) is a family of statistical and geometric techniques designed to recover a configuration of points in a low-dimensional Euclidean space from pairwise dissimilarity information. Unlike methods that operate directly on vector data, MDS assumes as primitive only a matrix of distances or dissimilarities, making it particularly well-suited to domains where explicit coordinates are unavailable, ill-defined, or conceptually secondary.

At its core, MDS addresses the inverse problem of geometry. Given IPDs, reconstruct an embedding whose induced Euclidean distances approximate the observed dissimilarities as closely as possible. This problem is deeply connected to classical geometry, spectral theory, and optimization, and has found enduring relevance across statistics, psychometrics, machine learning, bioinformatics, and the social sciences.

Let n objects be given, together with a symmetric dissimilarity matrix $\mathbf{\Delta} = (\delta_{ij})_{i,j=1}^{n}$, where $\delta_{ij} \geq 0$, $\delta_{ii} = 0$, and $\delta_{ij} = \delta_{ji}$. The goal of multidimensional scaling is to find points $\boldsymbol{x}_1, \ldots, \boldsymbol{x}_n \in \mathbb{R}^d$ for some target dimension $d \ll n$ such that the Euclidean distances $d_{ij} = \|\boldsymbol{x}_i - \boldsymbol{x}_j\|$ approximate the given dissimilarities δ_{ij} in a suitable sense.

Different variants of MDS arise depending on how the approximation error is defined and what assumptions are made about the relationship between δ_{ij} and d_{ij}. The two most prominent formulations are classical (metric) MDS and non-metric MDS.

4.6.1 Classical Multidimensional Scaling

Classical MDS assumes that the dissimilarities are Euclidean distances (possibly corrupted by noise). The method exploits the algebraic relationship between distances and inner products. Let $\mathbf{D}^{(2)}$ denote the matrix of squared dissimilarities, with entries δ_{ij}^2. Define the centering matrix $\mathbf{P} = \mathbf{I}_n - \frac{1}{n}\mathbf{1}\mathbf{1}'$, where $\mathbf{1}$ is the vector of ones. The double-centered matrix $\mathbf{B} = -\frac{1}{2}\mathbf{P}\mathbf{D}^{(2)}\mathbf{P}$ is called the Gram matrix. If the dissimilarities are exact Euclidean distances, then $\mathbf{B}$ is positive semi-definite and admits a factorization $\mathbf{B} = \mathbf{X}\mathbf{X}'$, where $\mathbf{X} \in \mathbb{R}^{n\times d}$ contains the coordinates of the embedded points. An eigenvalue decomposition $\mathbf{B} = \mathbf{Q}\boldsymbol{\Lambda}\mathbf{Q}'$ yields the configuration $\mathbf{X}_d = Q_d\boldsymbol{\Lambda}_d^{1/2}$, where $\boldsymbol{\Lambda}_d$ contains the largest d nonnegative eigenvalues. This embedding minimizes the Frobenius norm between $\mathbf{B}$ and its rank-d approximation.

4.6.2 Non-Metric Multidimensional Scaling

Non-metric multidimensional scaling relaxes the assumption that dissimilarities correspond directly to metric distances. Instead, it assumes only that the rank order of the dissimilarities is meaningful. Formally, non-metric MDS seeks an embedding for which the Euclidean distances d_{ij} are a monotonic transformation of the observed dissimilarities δ_{ij}. This formulation is particularly appropriate in settings where measurements are ordinal, subjective, or derived from comparative judgments rather than precise numerical assessments.

The solution to non-metric MDS is typically framed as an optimization problem involving a stress function, which quantifies the discrepancy between the distances in the embedding and the transformed dissimilarities. A commonly used criterion is Kruskal's stress function

$$\text{Stress}(\mathbf{X}) = \left(\frac{\sum_{i<j}\left(d_{ij} - \hat{\delta}_{ij}\right)^2}{\sum_{i<j} d_{ij}^2}\right)^{\frac{1}{2}},$$

where $\hat{\delta}_{ij}$ denotes a monotonic regression of δ_{ij} onto d_{ij}. The optimization is non-convex and is typically solved via iterative majorization or gradient-based procedures. Despite the absence of closed-form solutions, non-metric MDS is highly robust and excels in uncovering latent geometric structure when only qualitative similarity information is available.

Multidimensional scaling exhibits several important structural properties, including

- Invariance: The solution is unique only up to rigid transformations (translation, rotation, and reflection).
- Optimality: Classical MDS provides the best rank-d approximation to the centered inner-product matrix in the least-squares sense.
- Spectral nature: The embedding is governed by the leading eigenvalues and eigenvectors of B, linking MDS to principal component analysis.
- Dimensional interpretability: The magnitude of eigenvalues indicates the intrinsic dimensionality of the data.
- Stability: Small perturbations in distances lead to controlled perturbations in the embedding, provided eigenvalue gaps are well separated.

The abstraction of MDS as a distance-to-geometry mapping has made it broadly applicable in psychometrics for visualization of perceptual similarities among stimuli, and computational biology for Analysis of genetic or protein sequence dissimilarities. The method is also widely applicable in information retrieval for semantic visualization of documents, manifold learning as a linear precursor to non-linear methods, and exploratory data analysis for low-dimensional visualization of complex relational data.

Example 4.6.1 *Consider three objects with pairwise and squared distances*

$$\mathbf{D} = \begin{pmatrix} 0 & 1 & 2 \\ 1 & 0 & 1 \\ 2 & 1 & 0 \end{pmatrix}, \qquad \mathbf{D}^{(2)} = \begin{pmatrix} 0 & 1 & 4 \\ 1 & 0 & 1 \\ 4 & 1 & 0 \end{pmatrix}.$$

The centering matrix $\mathbf{P}$ *and* $\mathbf{B}$ *matrix are*

$$\mathbf{P} = \begin{pmatrix} \frac{2}{3} & -\frac{1}{3} & -\frac{1}{3} \\ -\frac{1}{3} & \frac{2}{3} & -\frac{1}{3} \\ -\frac{1}{3} & -\frac{1}{3} & \frac{2}{3} \end{pmatrix}, \qquad \mathbf{B} = -\frac{1}{2}\mathbf{P}\mathbf{D}^{(2)}\mathbf{P} = \begin{pmatrix} 1 & 0 & -1 \\ 0 & 0 & 0 \\ -1 & 0 & 1 \end{pmatrix}.$$

The eigenvalues of $\mathbf{B}$ *are* $\{2, 0, 0\}$, *with corresponding eigenvector*

$$\boldsymbol{q} = \frac{1}{\sqrt{2}}(1, 0, -1)'.$$

A one-dimensional embedding is given by $\mathbf{X}' = (1,\ 0\ -1)$, *which exactly reproduces the original distances.*

Multidimensional scaling provides a mathematically elegant and computationally efficient bridge between abstract dissimilarities and concrete geometric representations. Its spectral foundations, interpretability, and adaptability continue to make it a cornerstone technique in both classical statistics and modern data analysis.

4.7 Biplot

Biplot (Gabriel [112]) are graphical displays that represent both the observations and variables of a multivariate dataset in a reduced-dimensional space, typically constructed using principal component analysis (PCA). The biplot aims to provide a simultaneous visualization of the structure of the data and the relationships among variables.

Given a data matrix $\mathbf{X} \in \mathbb{R}^{nd}$, where n is the number of observations and d is the number of variables, the PCA of $\mathbf{X}$ results in the decomposition, $\mathbf{X} = \mathbf{UDV}'$, where $\mathbf{U} \in \mathbb{R}^{n \times r}$ is the matrix of left singular vectors (scores), $\mathbf{D} \in \mathbb{R}^{r \times r}$ is a diagonal matrix of singular values, $\mathbf{V} \in \mathbb{R}^{d \times r}$ is the matrix of right singular vectors (loadings), and $r = \min(n, d)$ is the rank of $\mathbf{X}$.

To construct a biplot for visualization, one usually selects the first two principal components, denoted by $\mathbf{U}_2\mathbf{D}_2$ for the scores and $\mathbf{V}_2$ for the loadings. The row scores are plotted as points, and the column vectors (variables) are plotted as arrows from the origin. The key idea of a biplot is to approximate the inner product, $x_{ij} \approx (\mathbf{U}_2\mathbf{D}_2)_{i,\cdot} \cdot (\mathbf{V}_2)_{j,\cdot}$, which allows interpretation of variable contributions and sample similarities.

There are different types of biplots based on the type of scaling used. GH Biplot uses scaled row and column markers to preserve the inner product. JK Biplot emphasizes preserving distances among observations, and HJ Biplot emphasizes the correlations among variables.

The angle between vectors in a biplot approximates the correlation between variables, while the projection of a point onto a variable vector estimates the value of that variable for the corresponding observation. Biplots are valuable tools for interpreting multivariate data by providing a joint visualization of observations and variables. Through PCA-based dimensionality reduction, biplots reveal underlying data structures, variable correlations, and groupings in an intuitive graphical format.

Example 4.7.1 *Given the covariance matrix*

$$\mathbf{C} = \begin{bmatrix} 1 & -0.5 & 0.9 & 0 \\ -0.5 & 1 & 0.707 & -0.9 \\ 0.9 & 0.707 & 1 & 0 \\ 0 & -0.9 & 0 & 1 \end{bmatrix},$$

we compute the first two principal components and project the variables to obtain the biplot. For a covariance biplot, variable coordinates are $\mathbf{L} = \mathbf{V}_2\mathbf{\Lambda}_2^{1/2}$.
PCA Variable Loadings are

Variable	*PC1*	*PC2*
X_1	0.123	1.005
X_2	1.002	−0.363
X_3	0.714	0.756
X_4	−0.786	0.381

PCA Biplot is

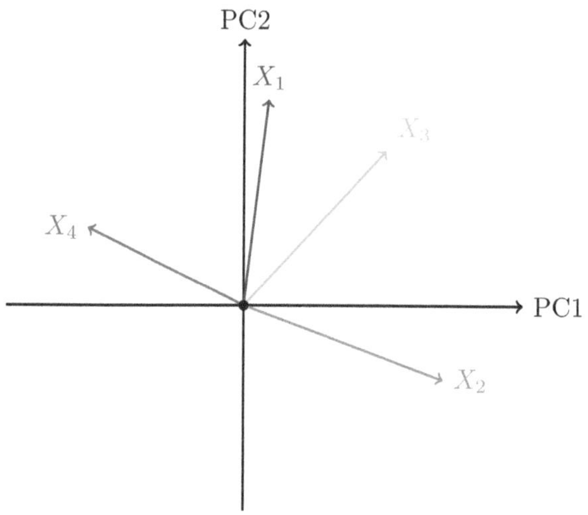

FIGURE 4.6: PCA Biplot of for Matrix **C**.

Interpretation of the biplot:

- X_1 *(blue) dominates the second principal component.*
- X_2 *(red) loads strongly on the first principal component.*
- X_3 *(green) contributes to both components.*
- X_4 *(purple) contrasts negatively with* X_2 *along PC1.*

4.8 Projection Method

In this section we consider groups of observations in $\mathbb{R}^d$ and present a simultaneous plot of the empirical cumulative distribution functions of the within and between IPDs to visualize and examine the equality of the underlying distribution functions of the observations. We provide several examples to illustrate how such plots can be utilized to envision and canvass the relationship between the two distributions under location, scale, dependence, and shape changes.

Chambers, Cleveland, Kleiner, and Tukey [55] famously observed that "there is no statistical tool that is as powerful as a well-chosen graph." Motivated by this insight, and building on the work of Modarres [229] on the graphical comparison of high-dimensional distributions, we propose a visualization framework tailored to high-dimensional datasets. Although graphical methods play a central role in facilitating the understanding of complex multivariate structures, effective visualization becomes increasingly challenging as the dimensionality of the data grows. High-dimensional data typically arise in settings where the dimension $d \to \infty$ as the sample size $n \to \infty$, while ultra-high-dimensional data refer to scenarios in which d increases at a non-polynomial rate relative to n, specifically when $\log(d) = O(n^\alpha)$ for some $\alpha > 0$. A wide variety of techniques have been proposed for the visualization of multivariate data; however, their effectiveness often deteriorates in such high- and ultra-high-dimensional regimes.

Two broad approaches to the visualization of multivariate observations are iconic and geometric data representations. Iconic representations employ visual symbols such as glyphs, stars, sunflowers, or Chernoff faces to encode multivariate information. Glyphs are compact graphical icons used to represent individual observations in $\mathbb{R}^d$, where variations in their shape, structure, or appearance convey differences and similarities among data points, thereby revealing salient features of the data. Although iconic representations can be effective for data with small to moderate dimensionality, their utility diminishes rapidly as d increases, rendering them impractical for datasets involving hundreds or thousands of variables.

4.8.1 Glyphs

A glyph is a mapping $G : \mathbb{R}^d \to \mathcal{A}$ that encodes a multivariate observation $\mathbf{o} = (x_1, \ldots, x_d)$ into visual attributes such as shape, color, or size. The main categories of glyphs are geometric, iconic, hybrid, and variable distance glyphs. A variable distance glyph encodes pairwise distances between variables, $G(\mathbf{o})_{ij} = D(x_i, x_j)$.

Example 4.8.1 *Suppose we have $n = 5$ observations and $d = 6$ variables in the dataset*

Obs	x_1	x_2	x_3	x_4	x_5	x_6
1	0.8	0.6	0.3	0.5	0.2	0.7
2	0.7	0.65	0.4	0.6	0.1	0.8
3	0.5	0.4	0.7	0.3	0.5	0.2
4	0.9	0.8	0.2	0.7	0.4	0.9
5	0.3	0.5	0.6	0.4	0.3	0.1

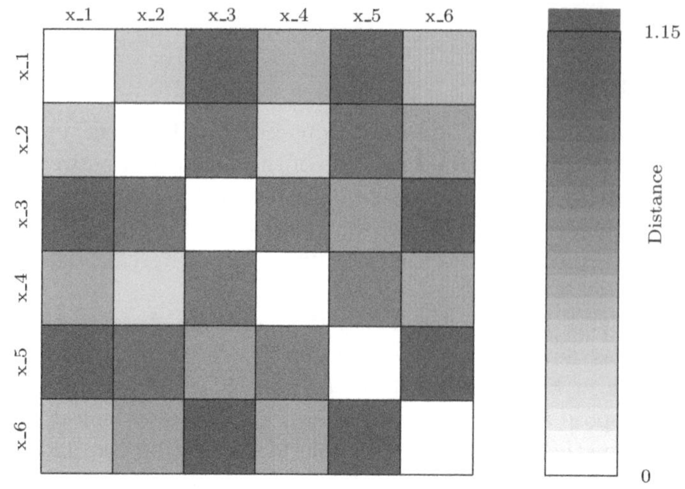

FIGURE 4.7: Variable distance glyph showing Euclidean distances between all 6 variables across 5 observations. Color intensity encodes distance magnitude, with a color bar indicating 0 to maximum distance.

The pairwise Euclidean Distance Matrix is

$$\begin{bmatrix} 0 & 0.320 & 0.985 & 0.436 & 0.985 & 0.387 \\ 0.320 & 0 & 0.782 & 0.206 & 0.820 & 0.492 \\ 0.985 & 0.782 & 0 & 0.735 & 0.520 & 1.145 \\ 0.436 & 0.206 & 0.735 & 0 & 0.693 & 0.469 \\ 0.985 & 0.820 & 0.520 & 0.693 & 0 & 1.058 \\ 0.387 & 0.492 & 1.145 & 0.469 & 1.058 & 0 \end{bmatrix}$$

The variable distance glyph (Heatmap) appears in Figure 4.7.

A star glyph encodes a multivariate observation $\mathbf{o} = (x_1, \ldots, x_d)$ as a star-shaped figure. Each variable x_i is represented as a ray (axis) radiating from a common center. The length along the ray is proportional to the variable value. Star glyphs belong to the class of geometric glyphs. They allow visual comparison of variable profiles across observations.

Star Glyph visualization for Observation 1 appears in Figure 4.8. The shape of the star glyph allows immediate recognition of dominant and weaker variables in Observation 1. Long rays x_1 (0.8) and x_6 (0.7) indicate the dominant variables. Medium-length rays x_2 (0.6) and x_4 (0.5) show moderate influence and short rays x_3 (0.3) and x_5 (0.2) represent weaker variables. The overall "shape" of the star acts as a visual fingerprint of the observation's profile.

4.8.2 Chernoff Faces

For more complex datasets, Chernoff faces provide an alternative iconic representation in which each d-dimensional observation is mapped to a stylized human face, with facial attributes corresponding to individual variables. Figure 4.9 represents a Chernoff face where each eye is drawn as a filled ellipse. Eye size may represent variability or dispersion, while eye separation can encode correlation strength. The nose is represented by a simple triangle. Its length or shape can encode scale-related features such as range or norm of the observation vector. The mouth is drawn using a Bezier curve. Its curvature represents location-related characteristics such as mean shift or skewness, an upward curve indicates positive deviation,

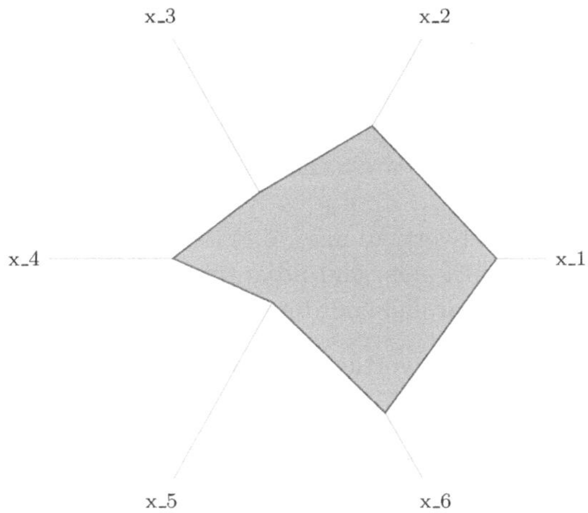

FIGURE 4.8: Star glyph representing Observation 1. Each ray corresponds to a variable; length is proportional to its value. Longer rays indicate higher variable values, showing which variables dominate the observation.

and downward curve indicates negative deviation. Eyebrow slope and position can encode higher-order distributional features such as kurtosis or tail behavior.

FIGURE 4.9: A Chernoff face representing a multivariate observation. Facial features encode variables such as dispersion, correlation, skewness, and scale.

Geometric approaches represent multivariate observations within a spatial framework, either through direct plots such as scatter plot matrices or parallel coordinates, or via projections onto a reduced-dimensional space, often two or three dimensions. Scatter plot matrices offer a systematic way to examine all pairs of variables simultaneously, making them effective for revealing relationships, identifying outlying observations, and summarizing the overall structure of the data. For high-dimensional datasets, it is common to apply dimensionality reduction methods, such as principal component analysis (PCA) or multidimensional scaling, before projecting the data for visualization.

4.8.3 Parallel Coordinate Plot

Parallel coordinate plots, as described by Inselberg [159] and Wegman [352], display each variable along a set of axes arranged in parallel, either vertically or horizontally. Each observation is represented by a line connecting its values across these axes, allowing patterns, unusual observations, or specific attribute combinations to be identified. Despite their usefulness, parallel coordinate plots are most effective for datasets of moderate dimensionality. As the number of variables grows, the plots can become overly complex and difficult to interpret. Moreover, the sequence in which the axes are arranged can influence how relationships among variables are perceived, potentially affecting the clarity of the visualization.

Parallel coordinate plots (PCPs) are a classical method for visualizing multivariate data. Given an observation $\mathbf{x}_i = (x_{i1}, x_{i2}, \ldots, x_{id}) \in \mathbb{R}^d$, the parallel coordinate plot represents $\mathbf{x}_i$ as a polyline crossing d parallel vertical axes, one for each variable. The jth coordinate x_{ij} is scaled and marked on the jth axis, and adjacent coordinates are connected by straight line segments. To construct a parallel coordinate plot

1. Place d equally spaced parallel vertical axes, each representing one variable.
2. Each variable X_j is linearly scaled so its values fit the axis range (often $[0, 1]$).
3. Each observation $\mathbf{x}_i$ becomes a polyline connecting the d axis positions $(x_{i1}, x_{i2}, \ldots, x_{id})$ in order.

Parallel coordinate plots possess several important properties.

1. They can represent arbitrarily high-dimensional data. Each additional variable adds one more parallel axis.
2. A set of points lying on a k-dimensional linear subspace in $\mathbb{R}^d$ maps to a $(d-k)$-dimensional structure in the PCP. Hence, they preserve linearity. In particular, colinear points in $\mathbb{R}^d$ map to lines intersecting at one point in the PCP.
3. Groups of observations that are similar across variables appear as groups of nearly overlapping polylines. Well-separated clusters in $\mathbb{R}^d$ remain visually distinct in the PCP.
4. Positive correlation between variables X_j and X_{j+1} appears as polylines with small slopes between axes j and $j+1$; negative correlation appears as crisscrossing patterns.
5. Anomalous observations manifest as polylines deviating strongly from the common groupings across one or more axes.
6. The interpretability of PCPs depends on the ordering of the axes. Different permutations of variable order can emphasize or hide patterns.
7. With large n, the display may suffer from clutter.

Parallel coordinate plots can handle high-dimensional data visually, highlight clusters, correlations, and outliers, and preserves some geometric properties of multivariate data. The limitations of this method include being sensitive to axis ordering and becoming cluttered with large d. Parallel coordinate plots provide a powerful framework for visualizing multivariate data. They reveal structure such as correlations, clusters, and outliers, but require careful axis scaling and ordering to be effective.

Example 4.8.2 *We consider ten observations* $\mathbf{x}_i \in \mathbb{R}^5$. *Each coordinate has been linearly scaled to the interval* $[0, 1]$.

i	X_1	X_2	X_3	X_4	X_5	*Class*	*i*	X_1	X_2	X_3	X_4	X_5	*Class*
1	*0.10*	*0.20*	*0.30*	*0.40*	*0.50*	C_1	*6*	*0.55*	*0.45*	*0.60*	*0.55*	*0.40*	C_2
2	*0.15*	*0.10*	*0.35*	*0.30*	*0.45*	C_1	*7*	*0.60*	*0.50*	*0.65*	*0.60*	*0.45*	C_2
3	*0.20*	*0.25*	*0.40*	*0.35*	*0.55*	C_1	*8*	*0.65*	*0.55*	*0.70*	*0.65*	*0.50*	C_2
4	*0.25*	*0.30*	*0.20*	*0.45*	*0.60*	C_1	*9*	*0.70*	*0.60*	*0.75*	*0.70*	*0.55*	C_2
5	*0.30*	*0.35*	*0.25*	*0.50*	*0.65*	C_1	*10*	*0.75*	*0.65*	*0.80*	*0.75*	*0.60*	C_2

The parallel coordinate plot for these points is given below. Each row of the data table corresponds to one polyline in the parallel coordinate plot. The dots indicate the coordinate values on each axis, while the connecting segments encode multivariate structure and class separation in $\mathbb{R}^5$.

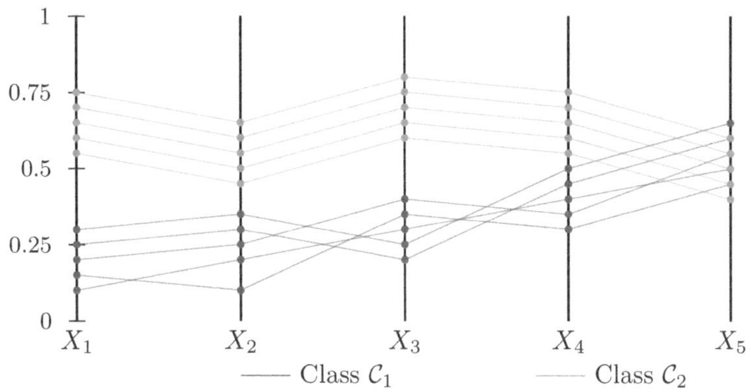

FIGURE 4.10: Parallel plot of 10 points in $\mathbb{R}^5$, Class $\mathcal{C}_1$ (blue, bottom), Class $\mathcal{C}_2$ (red, top).

4.8.4 Andrews' Plot

Andrews' plot utilizes the projection technique to map a d-dimensional data point to a plane. For each data point $\mathbf{X} = (x_1, \ldots, x_d)'$ one plots a finite Fourier series

$$h_{\mathbf{X}}(t) = x_1/\sqrt{2} + x_2 \sin(t) + x_3 \cos(t) + x_4 \sin(2t) + x_5 \cos(2t) + \ldots \tag{4.7}$$

where $-\pi < t < \pi$. Hence, the length of the projection of $\mathbf{X}_i$ onto the vector

$$(1/\sqrt{2}, \sin(t), \cos(t), \sin(2t), \cos(2t), \ldots) \tag{4.8}$$

is proportional to $h_{\mathbf{X}}(t)$. The squared L_2 distance between two curves $h_{\mathbf{X}_1}(t)$ and $h_{\mathbf{X}_2}(t)$ is $\pi\|\mathbf{X}_1 - \mathbf{X}_2\|$, which is proportional to the Euclidean IPD between the two vectors. Andrews' plot depends on the order in which the coordinates appear.

Example 4.8.3 *We obtain the Andrews' Plot for data in Example 4.8.2. The dataset consist of 10 points in* $\mathbb{R}^5$*:* $\mathbf{x}_1, \mathbf{x}_2, \ldots, \mathbf{x}_{10} \in \mathbb{R}^5$.

For each point $\mathbf{x}_i = (x_{i1}, x_{i2}, x_{i3}, x_{i4}, x_{i5})$*, define the Andrews' function*

$$f_{\mathbf{x}_i}(t) = \frac{x_{i1}}{\sqrt{2}} + x_{i2} \sin t + x_{i3} \cos t + x_{i4} \sin(2t) + x_{i5} \cos(2t), \quad t \in [-\pi, \pi].$$

The Andrews' plot is obtained by plotting $f_{\mathbf{x}_i}(t)$ *for all* $i = 1, \ldots, 10$*, optionally coloring curves by class.*

Most existing visualization approaches become increasingly cluttered and difficult to interpret as the dimensionality of the data grows. To date, the multivariate graphics literature lacks effective methods for displaying high-dimensional observations and facilitating comparisons across multiple groups. Traditional techniques are particularly limited when applied to datasets with hundreds or thousands of variables, providing little insight into the underlying structure.

To address this gap, we introduce a straightforward projection-based approach for visualizing the empirical cumulative distribution functions (ECDFs) of pairwise inter-point

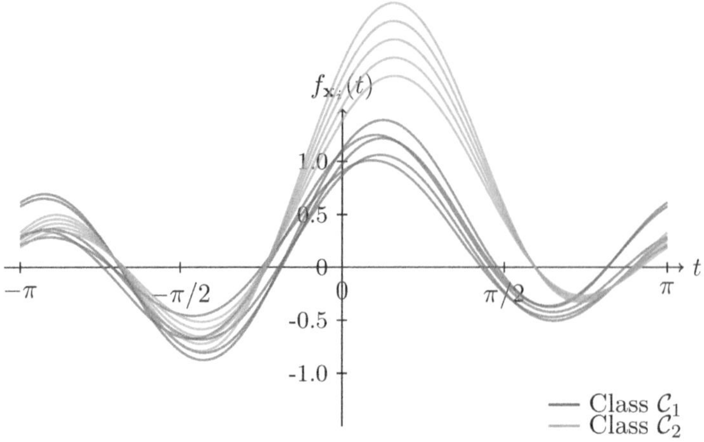

FIGURE 4.11: Andrews' plot of 10 points in $\mathbb{R}^5$, Class $\mathcal{C}_1$ (blue, bottom), Class $\mathcal{C}_2$ (red, top).

distances (IPDs) within and between high-dimensional observation vectors. Furthermore, we extend this method to enable simultaneous comparison of ECDFs across several groups, offering a clear and interpretable depiction of differences among high-dimensional datasets.

Suppose $\mathbb{X} = \{\mathbf{X}_i\}_{i=1}^m$ and $\mathbb{Y} = \{\mathbf{Y}_i\}_{i=1}^n$ represent i.i.d. observation vectors in $\mathbb{R}^d$ from distribution functions (DFs) F_1 and F_2, respectively. We are interested in graphics, transformations, and inference to examine the null hypothesis

$$H_0 : F_1 = F_2 \qquad \text{against} \qquad H_a : F_1 \neq F_2.$$

Both parametric and non–parametric tests have been proposed to investigate H_0. Some approaches reduce H_0 and assume F_1 and F_2 are equal except for their centers and/or scales. Hotelling T^2 [155] test assumes that F_1 and F_2 are multivariate normal and examines mean differences, assuming equal and unknown covariance matrices.

The Hotelling T^2 statistic [155] offers several advantages, including invariance under linear transformations, known exact and asymptotic distributions, and being the uniformly most powerful invariant test for evaluating the mean against two-sided alternatives. Its performance is strong when the number of variables d is small relative to the sample size. However, the Hotelling test exhibits two significant limitations. Bai and Saranadasa [19] demonstrated that the test's power diminishes as the ratio $d/(m+n)$ increases, even when $d/(m+n) \to \delta < 1$. This reduction arises because, when d and $m+n$ are of similar magnitude, the sample covariance matrix may fail to converge to the population covariance matrix, undermining its reliability. Since the Hotelling statistic is standardized by the sample covariance, this inconsistency directly impairs the test's effectiveness.

Additionally, the test cannot be computed when $d > m+n-2$, as the sample covariance matrix becomes singular and non-invertible. The Hotelling T^2 test is also sensitive to departures from normality, performing poorly under non-normal distributions (Everitt [91]). Thulin [336] proposed a computationally intensive approach that applies Hotelling tests to numerous lower-dimensional random subspaces. Furthermore, Fan [93] showed that even for a one-sample test of the mean under normality with known covariance $\sigma^2\mathbf{I}$, the powers of the likelihood ratio and Wald score tests decline toward the nominal level as the dimensionality d increases, reducing sensitivity to deviations from the null hypothesis.

4.8.5 Dutta-Ghosh Transformation

Dutta and Ghosh [83] propose an ingenuous transformation based on the IPDs to classify an observation $\mathbf{Z}$, whose class membership is unknown. Suppose that

$$\begin{aligned}
\mathbf{X}_i^* &= \left(\frac{\|\mathbf{X}_i - \mathbf{X}_1\|}{\sqrt{d}}, \ldots, \frac{\|\mathbf{X}_i - \mathbf{X}_m\|}{\sqrt{d}}, \frac{\|\mathbf{X}_i - \mathbf{Y}_1\|}{\sqrt{d}}, \ldots, \frac{\|\mathbf{X}_i - \mathbf{Y}_n\|}{\sqrt{d}}\right)', \\
\mathbf{Y}_j^* &= \left(\frac{\|\mathbf{Y}_j - \mathbf{X}_1\|}{\sqrt{d}}, \ldots, \frac{\|\mathbf{Y}_j - \mathbf{X}_m\|}{\sqrt{d}}, \frac{\|\mathbf{Y}_j - \mathbf{Y}_1\|}{\sqrt{d}}, \ldots, \frac{\|\mathbf{Y}_j - Y_n\|}{\sqrt{d}}\right)', \\
\mathbf{Z}^* &= \left(\frac{\|\mathbf{Z} - \mathbf{X}_1\|}{\sqrt{d}}, \ldots, \frac{\|\mathbf{Z} - \mathbf{X}_m\|}{\sqrt{d}}, \frac{\|\mathbf{Z} - \mathbf{Y}_1\|}{\sqrt{d}}, \ldots, \frac{\|\mathbf{Z} - \mathbf{Y}_n\|}{\sqrt{d}}\right)',
\end{aligned} \tag{4.9}$$

for $1 \le i \le m$ and $1 \le j \le n$. Note that the i-th component in $\mathbf{X}_i^*$ is 0, while the $(m+j)$-th component in $\mathbf{Y}_j^*$ is 0. Hence, the coordinates of $\mathbf{X}_i^*$ and $\mathbf{Y}_j^*$ contain a zero component, making them partially sparse. The transformation is invariant under rotation and translation of the original data. Dimensionality is reduced from d to $N = (m+n)$, which is favorable when $d \gg N$.

Example 4.8.4 *We consider a synthetic problem with $m = n = 5$ with $d = 100$ from, $\mathbf{X}_i \sim \mathbb{N}_d(\mathbf{0}, \mathbf{I}_d)$, $i = 1, \ldots, 5$, and $\mathbf{Y}_j \sim \mathbb{N}_d(\mathbf{1}, \mathbf{I}_d)$, $j = 1, \ldots, 5$. After applying the Dutta-Ghosh transformation to all data points, we can train a linear discriminant classifier on the transformed space. Figure 4.12 shows the transformed training and test data using the first two dimensions of the DG transformation. Test points are well-separated, indicating that the DG projection preserves class structure.*

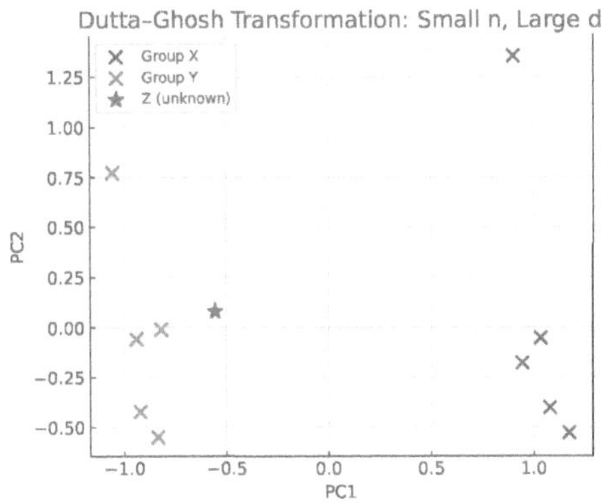

FIGURE 4.12: Transformed training and test data using the first two dimensions of Dutta-Ghosh transformation. The $\mathbf{Y}$ values are to the left of $\mathbf{Z}$, and the $\mathbf{X}$ values are to the right of $\mathbf{Z}$.

The Dutta-Ghosh transformation provides a powerful mechanism for high-dimensional classification when classical methods fail due to small sample sizes. By embedding data into a space of IPDs, the transformation facilitates the use of IPD-based or linear classifiers even in extreme dimensional regimes.

4.8.6 Multivariate Rosenblatt Transformation

Let $\mathbf{X} = (X_1, X_2, \ldots, X_d)$ be a continuous random vector with joint cumulative distribution function (CDF) $F_{\mathbf{X}}(x_1, \ldots, x_d)$ and marginal/conditional distributions

- $F_1(x_1) = \mathbb{P}(X_1 \le x_1)$,
- $F_2(x_2|x_1) = \mathbb{P}(X_2 \le x_2|X_1 = x_1)$,
- $F_d(x_d|x_1, \dots, x_{d-1}) = \mathbb{P}(X_d \le x_d|X_1 = x_1, \dots, X_{d-1} = x_{d-1})$.

The Rosenblatt transformation [285] defines a mapping $T : \mathbb{R}^d \to [0,1]^d$ and is given by

$$U_1 = F_1(X_1), \quad U_2 = F_2(X_2|X_1), \quad \dots, \quad U_d = F_d(X_d|X_1, \dots, X_{d-1}).$$

The vector $\mathbf{U} = (U_1, \dots, U_d)$ is uniformly distributed on the unit cube $[0,1]^d$. Yang and Modarres [362] offer multivariate tests of uniformity. Properties of the Rosenblatt transformation include

- $\mathbf{U} = T(\mathbf{X})$ is uniformly distributed on $[0,1]^d$.
- T is invertible if the joint distribution of $\mathbf{X}$ has a positive and smooth density.
- The transformation is sensitive to the ordering of variables.
- It is particularly useful for testing multivariate goodness-of-fit and simulation via inverse transform sampling.

Suppose $\mathbf{X}$ has a joint density $f_{\mathbf{X}}(x_1, \dots, x_d)$. Using the chain rule for densities, we have

$$f_{\mathbf{X}}(x_1, \dots, x_d) = f_1(x_1) f_2(x_2|x_1) \dots f_d(x_d|x_1, \dots, x_{d-1}).$$

We see that the Jacobian of the transformation is 1, and so $\mathbf{U} \sim \text{Uniform}[0,1]^d$.

Example 4.8.5 *Let $\mathbf{X} = (X_1, X_2)$ be bivariate normal with mean zero, unit variances, and correlation ρ. We have*

$$U_1 = \Phi(X_1), \qquad U_2 = \Phi\left(\frac{X_2 - \rho X_1}{\sqrt{1-\rho^2}}\right),$$

where Φ is the standard normal CDF. Then $\mathbf{U} = (U_1, U_2)$ is uniformly distributed on $[0,1]^2$.

Example 4.8.6 *Let (X_1, X_2) have a triangular density on the unit triangle $\{(x_1, x_2) : 0 < x_1 < 1,\ 0 < x_2 < x_1\}$. We have*

$$U_1 = F_1(X_1) = X_1^2, \qquad U_2 = F_2(X_2|X_1) = \frac{X_2}{X_1}.$$

Hence, (U_1, U_2) is uniformly distributed on $[0,1]^2$.

Applications of the Rosenblatt transformation in data analysis include multivariate goodness-of-fit testing, and copula-based dependence modeling, where it is a fundamental diagnostic tool. After fitting a copula to data, the transformation is applied to the observations. Systematic deviations from uniformity indicate model misspecification, making it particularly valuable for high-dimensional dependence analysis. Other applications include uncertainty quantification and sensitivity analysis, where it allows dependent inputs to be mapped to independent ones. The Rosenblatt transformation also underlies autoregressive density estimators.

4.9 Graphics for Comparing Two Groups

Let $\mathbf{X}_i, \mathbf{X}_j \in \mathbb{R}^d$ be i.i.d. from F_1, and $\mathbf{Y}_i, \mathbf{Y}_j \in \mathbb{R}^d$ be i.i.d. from F_2. Denote the within-sample distances by $d_{(x)}$ and $d_{(y)}$, and the between-sample distances by $d_{(xy)}$. In this section, we discuss the contribution of Maa, Pearl, and Bartoszyński [206] study of IPDs. We examine the IPDs for two groups and introduce a simultaneous ECDF plot of the within and between sample distances. Examples illustrate how these plots reveal differences in location, scale, dependence, and shape. We extend this approach to $k > 2$ groups, with examples comparing three distributions under various alternatives. New test statistics for equality of k distributions are proposed and compared with existing IPD-based methods. The final section summarizes the findings and offers concluding remarks.

4.9.1 Maa, Pearl, and Bartoszyński Equivalence

Let $\mathcal{K}_x$, $\mathcal{K}_y$, and $\mathcal{K}_{xy}$ denote the distribution functions of the IPDs $d_{(x)} = \|\mathbf{X}_1 - \mathbf{X}_2\|$, $d_{(y)} = \|\mathbf{Y}_1 - \mathbf{Y}_2\|$, and $d_{(xy)} = \|\mathbf{X}_1 - \mathbf{Y}_2\|$, respectively. Maa, Pearl, and Bartoszyński [206] made a significant contribution to the study of IPDs by showing that two distributions are identical if and only if the IPDs within and between the samples share the same univariate distribution. This equivalence holds for both discrete and continuous cases,

$$H_0' : \mathcal{K}_x = \mathcal{K}_y = \mathcal{K}_{xy} \iff H_0 : F_1 = F_2.$$

Consequently, rather than testing the original null hypothesis $H_0 : F_1 = F_2$, one can instead consider the equivalent hypothesis H_0'.

Example 4.9.1 *Let* $\mathbf{X} \sim \mathbb{N}(\mathbf{0}, \mathbf{I}_d)$ *and* $\mathbf{Y} \sim \mathbb{N}(\boldsymbol{\mu}, \mathbf{I}_d)$ *be independent random vectors in* $\mathbb{R}^d$. *For within-sample differences consider two independent copies* $\mathbf{X}_1, \mathbf{X}_2 \overset{iid}{\sim} \mathbb{N}(\mathbf{0}, \mathbf{I}_d)$. *It follows that* $\mathbf{D}_{XX} = \mathbf{X}_1 - \mathbf{X}_2 \sim \mathbb{N}(\mathbf{0}, 2\mathbf{I}_d)$. *Define* $\mathbf{Z} = \frac{\mathbf{D}_{XX}}{\sqrt{2}} \sim \mathbb{N}(\mathbf{0}, \mathbf{I}_d)$. *The squared norm satisfies* $\|\mathbf{Z}\|^2 = \sum_{i=1}^d Z_i^2 \sim \chi_d^2$, *the central chi-square distribution with d degrees of freedom. Therefore,* $\|\mathbf{D}_{XX}\| = \sqrt{2}\,\|\mathbf{Z}\| \sim \sqrt{2} \cdot \chi_d$, *where* χ_d *is the chi distribution with d degrees of freedom.*

For between-sample differences consider $\mathbf{X} \sim \mathbb{N}(\mathbf{0}, \mathbf{I}_d)$ *and* $\mathbf{Y} \sim \mathbb{N}(\boldsymbol{\mu}, \mathbf{I}_d)$ *independent. Hence,* $\mathbf{D}_{XY} = \mathbf{X} - \mathbf{Y} \sim \mathbb{N}(-\boldsymbol{\mu}, 2\mathbf{I}_d)$. *Define*

$$\mathbf{W} = \frac{\mathbf{D}_{XY}}{\sqrt{2}} \sim \mathbb{N}\left(-\frac{\boldsymbol{\mu}}{\sqrt{2}}, \mathbf{I}_d\right).$$

The squared norm $\|\mathbf{W}\|^2$ *follows a non-central chi-square distribution with degrees of freedom d and non-centrality parameter*

$$\lambda = \left\|\frac{\boldsymbol{\mu}}{\sqrt{2}}\right\|^2 = \frac{\|\boldsymbol{\mu}\|^2}{2}.$$

Hence, $\|\mathbf{D}_{XY}\|^2 = 2\|\mathbf{W}\|^2 \sim 2 \cdot \chi_d^2(\lambda)$, $\|\mathbf{D}_{XY}\| \sim \sqrt{2} \cdot \chi_d(\sqrt{\lambda})$, *a non-central chi distribution with non-centrality parameter* $\sqrt{\lambda} = \frac{\|\boldsymbol{\mu}\|}{\sqrt{2}}$. *When the non-centrality parameter* $\lambda = 0$, *so the between-sample distance distribution coincides with the within-sample distance distribution,* $\|\mathbf{D}_{XY}\| \sim \sqrt{2} \cdot \chi_d$. *This equality of distributions of IPDs within and between samples char-*

acterizes equality of distributions $F_1 = F_2$, consistent with Maa, Pearl, and Bartoszyński [206].

If the null hypothesis H_0 is rejected, this indicates that the distributions F_1 and F_2 are not identical; however, the specific source of the discrepancy, whether it arises from differences in location, scale, or other distributional characteristics, cannot be determined from the test alone. More generally, when considering $k \geq 2$ groups, one may wish to explore a potential ordering among the distributions under the alternative hypothesis. The approach we propose uses IPDs to both visualize differences between groups and to uncover their stochastic ordering, providing a graphical and interpretable means to compare multiple high-dimensional distributions.

Let μ_{xx} and μ_{yy} denote the expected values of the within-sample IPDs for the two groups, and let μ_{xy} represent the expected value of the between-sample IPDs. Consider the pair $(d_{(x)}, d_{(xy)})$, which follows a bivariate distribution, denoted by $\mathcal{K}_x$, with marginal distributions $\mathcal{K}_x$ and $\mathcal{K}_{xy}$. Similarly, the pair $(d_{(xy)}, d_{(y)})$ follows another bivariate distribution, $\mathcal{K}_y$, with marginals $\mathcal{K}_{xy}$ and $\mathcal{K}_y$.

Let $\boldsymbol{\mu}_{\mathcal{K}_x}$ and $\boldsymbol{\mu}_{\mathcal{K}_y}$ denote the mean vectors of $\mathcal{K}_x$ and $\mathcal{K}_y$, respectively. It follows that

$$\boldsymbol{\mu}_{\mathcal{K}_x} = \boldsymbol{\mu}_{\mathcal{K}_y} \iff \mu_{xx} = \mu_{xy} = \mu_{yy} \iff F_1 = F_2.$$

Thus, equality of the mean vectors of the bivariate IPD distributions is equivalent to equality of the underlying distributions. Consequently, rather than testing the original null hypothesis $H_0 : F_1 = F_2$, one can consider the equivalent formulation

$$H_0'' : \boldsymbol{\mu}_{\mathcal{K}_x} = \boldsymbol{\mu}_{\mathcal{K}_y} \quad \text{versus} \quad H_a'' : \boldsymbol{\mu}_{\mathcal{K}_x} \neq \boldsymbol{\mu}_{\mathcal{K}_y}.$$

Biswas and Ghosh [35] established that, provided μ_{xx}, μ_{yy}, and μ_{xy} exist, equality of these quantities $\mu_{xx} = \mu_{yy} = \mu_{xy}$ holds if and only if $F_1 = F_2$. In practice, we estimate μ_{xx}, μ_{yy}, and μ_{xy} using the corresponding sample averages $\bar{d}_{(x)}$, $\bar{d}_{(y)}$, and $\bar{d}_{(xy)}$, as defined in equations (2.7), (2.11), and (2.15), respectively.

Suppose that the observations in the samples $\mathbb{X}$ and $\mathbb{Y}$ are drawn from distributions with means $\boldsymbol{\mu}_x$ and $\boldsymbol{\mu}_y$, and covariances $\boldsymbol{\Sigma}_x$ and $\boldsymbol{\Sigma}_y$, respectively. Using quadratic forms, it can be shown that $D_x^2 = \sum_{r=1}^{d} \lambda_r U_r^2$, where U_r is the standardized version of the r-th component X_r of $\mathbf{X}_1$, for $r = 1, \ldots, d$, and $\lambda_1, \ldots, \lambda_d$ are the eigenvalues of $2\boldsymbol{\Sigma}_x$. Consequently, the expectation of D_x^2 is

$$\mathbb{E}(D_x^2) = \sum_{r=1}^{d} \lambda_r = 2\,\mathrm{tr}(\boldsymbol{\Sigma}_x),$$

which can be estimated using twice the total sample variance. Similarly, for the between-sample distances, one obtains $\mathbb{E}(D_{xy}^2) = \mathrm{tr}(\boldsymbol{\Sigma}_x + \boldsymbol{\Sigma}_y) + \|\boldsymbol{\mu}_x - \boldsymbol{\mu}_y\|^2$. Applying Jensen's inequality (Equation (1.7.1)) yields the bounds

$$\mathbb{E}(d_{(x)}) \leq \sqrt{2\,\mathrm{tr}(\boldsymbol{\Sigma}_x)}, \quad \mathbb{E}(d_{(y)}) \leq \sqrt{2\,\mathrm{tr}(\boldsymbol{\Sigma}_y)}, \mathbb{E}(d_{(xy)}) \leq \sqrt{\mathrm{tr}(\boldsymbol{\Sigma}_x + \boldsymbol{\Sigma}_y) + \|\boldsymbol{\mu}_x - \boldsymbol{\mu}_y\|^2}.$$

These results provide natural upper bounds on the expected within- and between-sample IPDs in terms of the underlying means and covariances.

Example 4.9.2 *Low-rank decomposition of squared IPD matrices: Let $\mathbb{X} = \{\mathbf{X}_1, \ldots, \mathbf{X}_m\}$ be a sample with $\mathbf{X}_i \in \mathbb{R}^d$. The squared IPD between $\mathbf{X}_i$ and $\mathbf{X}_j$ is*

$$d_{(x)ij}^2 = \|\mathbf{X}_i - \mathbf{X}_j\|^2 = \mathbf{X}_i'\mathbf{X}_i + \mathbf{X}_j'\mathbf{X}_j - 2\mathbf{X}_i'\mathbf{X}_j. \tag{4.10}$$

The within-sample distance matrix $D_{xx}(\mathbb{X}) \in \mathbb{R}^{m\times m}$ *is defined by*

$$D_{xx}(\mathbb{X}) = \begin{pmatrix} 0 & d^2_{(x)12} & d^2_{(x)13} & \cdots & d^2_{(x)1m} \\ d^2_{(x)21} & 0 & d^2_{(x)23} & \cdots & d^2_{(x)2m} \\ d^2_{(x)31} & d^2_{(x)32} & 0 & \cdots & d^2_{(x)3m} \\ \vdots & \vdots & \vdots & \ddots & \vdots \\ d^2_{(x)m1} & d^2_{(x)m2} & d^2_{(x)m3} & \cdots & 0 \end{pmatrix}. \tag{4.11}$$

Low-rank factorization: Define the matrices $\mathbf{U}_x \in \mathbb{R}^{m\times(d+2)}$ *and* $\mathbf{V}_x \in \mathbb{R}^{(d+2)\times m}$ *by*

$$\mathbf{U}_x = \begin{pmatrix} \mathbf{X}_1'\mathbf{X}_1 & 1 & -2\mathbf{X}_1' \\ \mathbf{X}_2'\mathbf{X}_2 & 1 & -2\mathbf{X}_2' \\ \vdots & \vdots & \vdots \\ \mathbf{X}_m'\mathbf{X}_m & 1 & -2\mathbf{X}_m' \end{pmatrix}, \tag{4.12}$$

and

$$\mathbf{V}_x = \begin{pmatrix} 1 & 1 & \cdots & 1 \\ \mathbf{X}_1'\mathbf{X}_1 & \mathbf{X}_2'\mathbf{X}_2 & \cdots & \mathbf{X}_m'\mathbf{X}_m \\ \mathbf{X}_1 & \mathbf{X}_2 & \cdots & \mathbf{X}_m \end{pmatrix}. \tag{4.13}$$

Note that the last block row of $\mathbf{V}_x$ *consists of* d *rows, with the* j*th column equal to the vector* $\mathbf{X}_j \in \mathbb{R}^d$*. Hence,* $D_{xx}(\mathbb{X}) = \mathbf{U}_x\mathbf{V}_x$*. Indeed, the* (i, j) *entry of* $\mathbf{U}_x\mathbf{V}_x$ *is*

$$(\mathbf{U}_x\mathbf{V}_x)_{ij} = \mathbf{X}_i'\mathbf{X}_i + \mathbf{X}_j'\mathbf{X}_j - 2\mathbf{X}_i'\mathbf{X}_j = \|\mathbf{X}_i - \mathbf{X}_j\|^2 = d^2_{(x)ij}. \tag{4.14}$$

Similarly, for a second sample $\mathbb{Y} = \{\mathbf{Y}_1, \ldots, \mathbf{Y}_n\}$ *with* $\mathbf{Y}_j \in \mathbb{R}^d$*, the within-sample distance matrix satisfies* $D_{yy}(\mathbb{Y}) = \mathbf{U}_y\mathbf{V}_y$*, where* $\mathbf{U}_y \in \mathbb{R}^{n\times(d+2)}$ *and* $\mathbf{V}_y \in \mathbb{R}^{(d+2)\times n}$ *are defined analogously.*

It is not difficult to show that

$$D(\mathbb{X}, \mathbb{Y}) = \begin{pmatrix} \mathbf{U}_x \\ \mathbf{U}_y \end{pmatrix} (\mathbf{V}_x \ \mathbf{V}_y) = \begin{pmatrix} D_{xx}(\mathbb{X}) = \mathbf{U}_x \ \mathbf{V}_x & d_{(xy)}(\mathbb{X}, \mathbb{Y}) = \mathbf{U}_x \ \mathbf{V}_y \\ D_{yx}(\mathbb{X}, \mathbb{Y}) = \mathbf{U}_y\mathbf{V}_x & D_{yy}(\mathbb{X}) = \mathbf{U}_y \ \mathbf{V}_y \end{pmatrix}. \tag{4.15}$$

Computational expense is a major obstacle when using IPDs with high-dimensional data with large sample size. The above low-rank decompositions enable efficient and parallel computation of distance matrices involving within- and between-sample IPDs.

4.9.2 IPD CDF Plot

Building on the work of Modarres [229] on the graphical comparison of high-dimensional distributions, we propose the following simultaneous IPD CDF plots. There are $m_x = \binom{m}{2}$ pairs of IPDs in $\mathbb{X}$, $m_y = \binom{n}{2}$ pairs of IPDs in $\mathbb{Y}$, and $m_{xy} = mn$ pairs of IPDS between samples. Let $R = \max D(\mathbb{X}, \mathbb{Y}) - \min D(\mathbb{X}, \mathbb{Y})$ denote the range of all IPDs that appear below the main diagonal of $D(\mathbb{X}, \mathbb{Y})$. We will divide the range into s evaluation points denoted by $\delta(t)$ for $t = 1, \ldots s$. Denote the cumulative distributions of $d_{(x)ij}$, $d_{(y)ij}$, and $d_{(xy)ij}$ evaluated at $\delta(t)$ by $H_x(t) = \mathbb{P}(d_{(x)ij} \leq \delta(t))$, $H_y(t) = \mathbb{P}(d_{(y)ij} \leq \delta(t))$ and $H_{xy}(t) = \mathbb{P}(d_{(xy)ij} \leq \delta(t))$, respectively. Let $\mathbb{I}(.)$ denote the indicator function and estimate the DFs by

$$\begin{aligned} \hat{H}_x(t) &= \frac{1}{m_x} \sum_{i=1}^{m-1} \sum_{j=i+1}^{m} \mathbb{I}(d_{(x)ij} \leq \delta(t)), \\ \hat{H}_y(t) &= \frac{1}{m_y} \sum_{i=1}^{n-1} \sum_{j=i+1}^{n} \mathbb{I}(d_{(y)ij} \leq \delta(t)), \\ \hat{H}_{xy}(t) &= \frac{1}{m_{xy}} \sum_{i=1}^{m} \sum_{i=1}^{n} \mathbb{I}(d_{(xy)ij} \leq \delta(t)). \end{aligned} \tag{4.16}$$

We obtain a simultaneous plot of

$$(\delta(t), \hat{H}_x(\delta(t))),\ (\delta(t), \hat{H}_y(\delta(t))),\ (\delta(t), \hat{H}_{xy}(\delta(t))),\ \text{for } t = 1, \ldots, s. \tag{4.17}$$

As $m, n \to \infty$, the empirical cumulative distribution functions converge in expectation to the true cumulative distribution functions. That is,

$$\begin{aligned} \mathbb{E}[\hat{H}_x(t)] &\longrightarrow H_x(t) = \mathbb{P}(d_{(x)ij} \leq \delta(t)), \\ \mathbb{E}[\hat{H}_y(t)] &\longrightarrow H_y(t) = \mathbb{P}(d_{(y)ij} \leq \delta(t)), \\ \mathbb{E}[\hat{H}_{xy}(t)] &\longrightarrow H_{xy}(t) = \mathbb{P}(d_{(xy)ij} \leq \delta(t)). \end{aligned}$$

Each empirical CDF (e.g., $\hat{H}_x(t)$) is a U-statistic of order 2, being an average over $\binom{m}{2}$ i.i.d. random variables of the form $\mathbb{I}(d_{(x)ij} \leq \delta(t))$. Each term is a Bernoulli random variable with mean equal to the true CDF value $H_x(t)$. Thus, by the law of large numbers for U-statistics, the empirical CDFs converge in mean (and almost surely) to their true counterparts as $m, n \to \infty$. As sample sizes grow, the empirical distribution functions $\hat{H}_x(t)$, $\hat{H}_y(t)$, and $\hat{H}_{xy}(t)$ provide increasingly accurate approximations of the true dissimilarity distributions. This convergence justifies using these empirical functions for hypothesis testing and distributional comparison.

Stochastic orderings of random vectors are of considerable interest for comparing multivariate distributions, with applications in statistical decision theory, clinical trials, experimental design, and portfolio analysis. Although there exists a rich literature on testing stochastic orderings for univariate distributions, the body of work addressing inference for non-parametric multivariate distributions remains relatively sparse. Giovagnoli and Wynn [119] introduced the multivariate dispersive order, also referred to as the weak D-ordering, for two random vectors $\mathbf{X}$ and $\mathbf{Y}$ in $\mathbb{R}^d$. Specifically, $\mathbf{X} \leq_D \mathbf{Y}$ is defined to hold if and only if $d_{(x)} \leq_{St} d_{(y)}$, where $\leq_{St}$ denotes the usual stochastic order between two random variables. It can be shown that $d_{(x)} \leq_{St} d_{(y)}$ if and only if $\mathcal{K}_x(t) \geq \mathcal{K}_y(t)$ for all $t \in \mathbb{R}$.

Giovagnoli and Wynn [119] show that $d_{(x)} \leq_{St} d_{(y)}$ if and only if the average probability content of balls of a given radius with center having the same distribution (independently), is more for $\mathbf{X}$ than for $\mathbf{Y}$. Under $K_0 : \mathbf{X} \leq_D \mathbf{Y}$, it follows that $d_{(x)} \leq_{St} d_{(y)}$ with probability 1. One can estimate $\mathbb{P}\left(d_{(x)} \leq d_{(y)}\right)$ and reject K_0 for small values of this probability. It is not difficult to show that the estimated probability is a U-statistic with a normal limiting distribution for large sample sizes. The hypotheses K_0 and K_0' are equivalent,

$$K_0 : \mathbf{X} \leq_D \mathbf{Y} \iff K_0^{'} : \mathcal{K}_x(t) \geq \mathcal{K}_y(t) \text{ for all } t \in \mathbb{R}. \tag{4.18}$$

It follows from $\mathbf{X} \leq_D \mathbf{Y}$, that $\mathbb{E}(r(d_{(x)})) \leq \mathbb{E}(r(d_{(y)}))$ for all non-decreasing functions r on $[0, \infty)$ and $tr(\mathbb{C}ov(\mathbf{X})) \leq tr(\mathbb{C}ov(\mathbf{Y}))$. Furthermore, the dispersion order $\leq_D$ is location and rotation free. If $d_{(x)} \leq_{St} d_{(y)}$, then $\mathbf{\Gamma X} + \mathbf{a} \leq_D \mathbf{\Lambda Y} + \mathbf{b}$ for all orthogonal matrices $\mathbf{\Gamma}$ and $\mathbf{\Lambda}$ and for all vectors $\mathbf{a}$ and $\mathbf{b}$.

Lemma 4.9.1 *When the* $\mathbb{X}$ *and* $\mathbb{Y}$ *groups differ in location* $\boldsymbol{\mu}$ *only; i.e.,* $F_1(\mathbf{X}) = F_2(\mathbf{Y}-\boldsymbol{\mu})$, *we have* $\mathcal{K}_x = \mathcal{K}_y \geq \mathcal{K}_{xy}$ *with equality holding if and only if* $\boldsymbol{\mu} = \mathbf{0}$.

If $\boldsymbol{\mu} = \mathbf{0}$, then the result follows since $F_1 = F_2$ is equivalent to $\mathcal{K}_x = \mathcal{K}_y = \mathcal{K}_{xy}$. The within-sample IPDs are invariant with respect to location shift so that $\mathcal{K}_x = \mathcal{K}_y$. The between-sample IPDs will become larger than within sample IPDs when $\boldsymbol{\mu} \neq \mathbf{0}$ since $\mathbf{X}-\mathbf{Y}$ is no longer centered at zero. Hence, $\mathcal{K}_x = \mathcal{K}_y \geq \mathcal{K}_{xy}$. As the following example shows, when ECDFs $\hat{H}_x(t)$ and $\hat{H}_y(t)$ are closer together than the ECDF $\hat{H}_{xy}(t)$; i.e., $\mathcal{K}_x = \mathcal{K}_y \geq \mathcal{K}_{xy}$, one can say that the two distributions F_1 and F_2 have a location shift difference.

Example 4.9.3 *To illustrate the utility of such simultaneous display of the ECDFs, we use* $d = 1000$ *and generate* $m = n = 50$ *observations from* $\mathbb{LN}(\mathbf{0}, \sigma_x^2\mathbf{I}_d)$ *and* $\mathbb{LN}(\boldsymbol{\mu}, \sigma_y^2\mathbf{I}_d)$, *respectively. Figure 4.13, left panel displays the simultaneous plot of 4.17 when* $H_0 : F_1 = F_2$ *is true, with* $\sigma_x^2 = \sigma_y^2 = 1$ *and* $\boldsymbol{\mu} = \mathbf{0}$. *As one expects, the ECDFs of the three IPDs concentrate in a narrow band since* $H_0' : \mathcal{K}_x = \mathcal{K}_x = \mathcal{K}_{xy}$ *is true. Equivalently,* $d_{(x)} \overset{(D)}{=} d_{(y)} \overset{(D)}{=} d_{(xy)}$ *where* $\overset{(D)}{=}$ *stands for equality in distribution. Figure 4.13, right panel displays the simultaneous plot of 4.17 under multivariate log-normal distributions with* $\sigma_x^2 = \sigma_y^2 = 1$ *and* $\boldsymbol{\mu} = \mathbf{11}$. *Here,* $H_0 : F_1 = F_2$ *is false, and we expect the ECDFs to differ. Moreover, the distribution* $\mathcal{K}_{xy}$ *of the between sample IPDs is stochastically larger than the within-sample IPDs* $\mathcal{K}_y$ *and smaller than* $\mathcal{K}_x$ *with probability 1 under a location shift.*

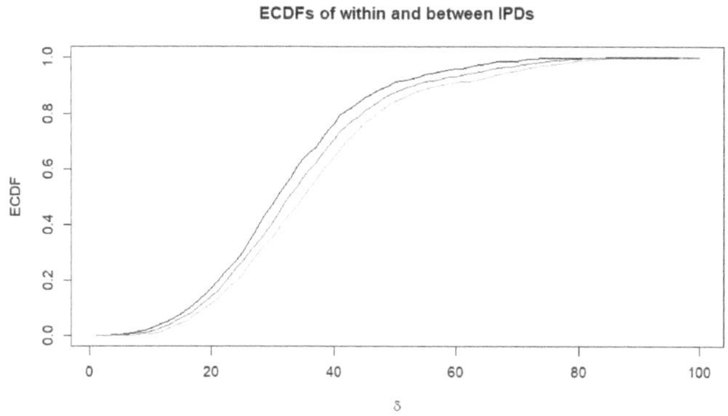

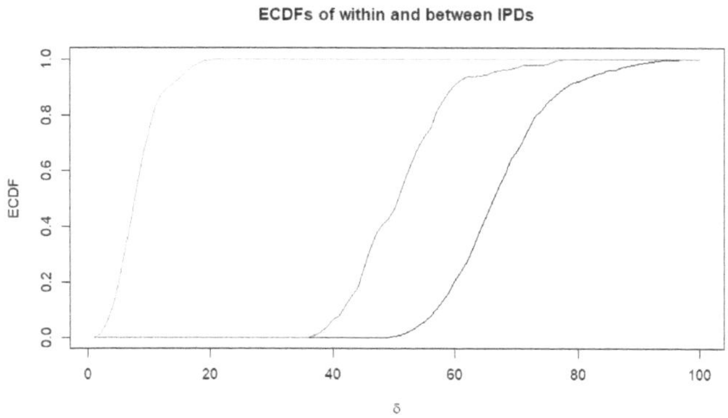

FIGURE 4.13: Left panel: Under $F_1 = F_2 \sim \mathbb{LN}(\mathbf{0}, \mathbf{I}_d)$: Empirical CDFs of IPDs (black, top), $\mathbb{X}$ sample IPDs (red, bottom), and between-sample IPDs (blue, middle). Right panel: Under $F_1 \sim \mathbb{LN}(\mathbf{0}, \mathbf{I}_d)$ and $F_2 \sim \mathbb{N}(1\ \mathbf{1}, \mathbf{I}_d)$: Empirical CDFs of IPDs (red, top), $\mathbb{Y}$ sample IPDs (black, bottom), and between-sample IPDs (blue, middle).

Lemma 4.9.2 *When the $\mathbb{X}$ and $\mathbb{Y}$ groups differ in scale only; i.e., $F_1(X) = F_2(\frac{1}{\sigma}Y)$, we have $\min(\mathcal{K}_x, \mathcal{K}_y) \le \mathcal{K}_{xy} \le \max(\mathcal{K}_x, \mathcal{K}_y)$ with equality holding if and only if $\sigma = 1$.*

If $\sigma = 1$, then the result follows since $F_1 = F_2$ is equivalent to $\mathcal{K}_x = \mathcal{K}_y = \mathcal{K}_{xy}$. Furthermore, $\mathcal{K}_y \le \mathcal{K}_{xy}$ and $\mathcal{K}_{xy} \le \mathcal{K}_x$ when $\sigma < 1$. Similarly, $\mathcal{K}_x \le \mathcal{K}_{xy}$ and $\mathcal{K}_{xy} \le \mathcal{K}_y$ when $\sigma > 1$. Equivalently, $d_{(x)} \le_{St} d_{(xy)} \le_{St} d_{(y)}$ if $\sigma > 1$ and $d_{(y)} \le_{St} d_{(xy)} \le_{St} d_{(x)}$ if $\sigma < 1$ with probability 1 under a scale change.

To illustrate the impact of shifts in location and/or scale, consider the squared within- and between-sample distances for observations drawn from $\mathbf{X} \sim F_1 = \mathbb{N}(\boldsymbol{\mu}_1, \sigma_1^2\mathbf{I}_d)$ and $\mathbf{Y} \sim F_2 = \mathbb{N}(\boldsymbol{\mu}_2, \sigma_2^2\mathbf{I}_d)$, with $\boldsymbol{\Delta} = \boldsymbol{\mu}_1 - \boldsymbol{\mu}_2$. It can be shown that

$$D_x^2 \sim 2\sigma_1^2\chi^2(d), \quad D_y^2 \sim 2\sigma_2^2\chi^2(d), \quad \text{and} \quad D_{xy}^2 \sim (\sigma_1^2 + \sigma_2^2)\chi_\eta'^2(d),$$

where $\chi_\eta'^2(d)$ denotes a non-central chi-square distribution with d degrees of freedom and non-centrality parameter $\eta = \frac{\boldsymbol{\Delta}'\boldsymbol{\Delta}}{\sigma_1^2+\sigma_2^2}$. The expected values are

$$\mathbb{E}(D_x^2) = 2\sigma_1^2 d, \quad \mathbb{E}(D_y^2) = 2\sigma_2^2 d, \quad \mathbb{E}(D_{xy}^2) = d(\sigma_1^2 + \sigma_2^2) + \boldsymbol{\Delta}'\boldsymbol{\Delta}.$$

In the special case where the variances are equal ($\sigma_1 = \sigma_2$) and $\boldsymbol{\Delta} \ne \mathbf{0}$, the stochastic ordering of the distances is $d_{(x)} = d_{(y)} \le_{St} d_{(xy)}$. Conversely, when the means coincide ($\boldsymbol{\Delta} = \mathbf{0}$) but the scale ratio is less than one, the ordering becomes $d_{(x)} \le_{St} d_{(xy)} \le_{St} d_{(y)}$. This demonstrates how differences in either mean vectors or variances between distributions affect the stochastic behavior of within- and between-sample IPDs.

Example 4.9.4 *To see the effects of dependence on the distributions on the IPDs assume $X \sim N_d(\mathbf{0}, \mathbf{R}_x)$ and $Y \sim N_d(\mathbf{0}, \mathbf{R}_y)$, where $\mathbf{R}_x$ and $\mathbf{R}_y$ are constant-correlation matrices with unit diagonal and off-diagonal correlations ρ_x and ρ_y, respectively.*

A $d \times d$ constant-correlation matrix with correlation ρ has eigenvalues

$$\lambda_1 = 1 + (d-1)\rho, \qquad \lambda_2 = \cdots = \lambda_d = 1 - \rho.$$

Hence, since quadratic forms in multivariate normal vectors decompose along eigen-directions, we obtain

$$d_{(x)}^2 = X^\top X \sim \sum_{i=1}^d \lambda_i \chi_{(1)}^2, \qquad \lambda_1 = 1 + (d-1)\rho_x, \quad \lambda_2 = \cdots = \lambda_d = 1 - \rho_x,$$

$$d_{(y)}^2 = Y^\top Y \sim \sum_{i=1}^d \theta_i \chi_{(1)}^2, \qquad \theta_1 = 1 + (d-1)\rho_y, \quad \theta_2 = \cdots = \theta_d = 1 - \rho_y.$$

If we define the average covariance matrix

$$\mathbf{R}_{xy} = \frac{\mathbf{R}_x + \mathbf{R}_y}{2},$$

then $\mathbf{R}_{xy}$ is also constant-correlation with correlation $\frac{\rho_x+\rho_y}{2}$ and eigenvalues

$$\gamma_1 = 1 + (d-1)\frac{\rho_x + \rho_y}{2}, \qquad \gamma_2 = \cdots = \gamma_d = 1 - \frac{\rho_x + \rho_y}{2}.$$

Therefore,

$$d_{(xy)}^2 \sim \sum_{i=1}^d \gamma_i \chi_{(1)}^2.$$

4.10 Asymptotic Ordering of IPDs

Let $\mathbf{X}_1, \mathbf{X}_2 \overset{\text{iid}}{\sim} F_1(\boldsymbol{\mu}_x, \boldsymbol{\Sigma}_x)$ and $\mathbf{Y}_1, \mathbf{Y}_2 \overset{\text{iid}}{\sim} F_2(\boldsymbol{\mu}_y, \boldsymbol{\Sigma}_y)$, with all vectors independent. Define the interpoint Euclidean distances scaled by $\sqrt{d}$ as

$$\tilde{d}_{xx} = \frac{1}{\sqrt{d}}\|\mathbf{X}_1 - \mathbf{X}_2\|, \qquad \tilde{d}_{yy} = \frac{1}{\sqrt{d}}\|\mathbf{Y}_1 - \mathbf{Y}_2\|, \qquad \tilde{d}_{xy} = \frac{1}{\sqrt{d}}\|\mathbf{X}_1 - \mathbf{Y}_1\|.$$

By Assumption 3.6.3 as $d \to \infty$, there exist constants $\sigma^2_{F_1}, \sigma^2_{F_2} > 0$, such that

$$\frac{1}{d}tr\left(\boldsymbol{\Sigma}_x\right) \to \sigma^2_{F_1} \text{ and } \frac{1}{d}tr\left(\boldsymbol{\Sigma}_y\right) \to \sigma^2_{F_2},$$

respectively, and $\frac{1}{d}\|\boldsymbol{\mu}_x - \boldsymbol{\mu}_y\|^2 \to \nu^2$. By the law of large numbers we obtain

$$\begin{aligned}
&\frac{1}{d}\|\mathbf{X}_1 - \mathbf{X}_2\|^2 \xrightarrow{p} 2\sigma^2_{F_1},\\
&\frac{1}{d}\|\mathbf{Y}_1 - \mathbf{Y}_2\|^2 \xrightarrow{p} 2\sigma^2_{F_2},\\
&\frac{1}{d}\|\mathbf{X}_1 - \mathbf{Y}_1\|^2 \xrightarrow{p} \nu^2 + \sigma^2_{F_1} + \sigma^2_{F_2}.
\end{aligned}$$

Taking square roots, we have

$$\tilde{d}_{xx} \xrightarrow{p} \delta_{xx} = \sqrt{2}\sigma_{F_1}, \quad \tilde{d}_{yy} \xrightarrow{p} \delta_{yy} = \sqrt{2}\sigma_{F_2}, \quad \tilde{d}_{xy} \xrightarrow{p} \delta_{xy} = \sqrt{\nu^2 + \sigma^2_{F_1} + \sigma^2_{F_2}}.$$

As $d \to \infty$, the distances concentrate around their respective limits. Therefore, the joint distribution of $(\tilde{d}_{yy}, \tilde{d}_{xx}, \tilde{d}_{xy})$ becomes degenerate. Each of the following probabilities converges to 1 or 0 depending on the strict ordering of the limits. In high dimensions, the distances become deterministic. Let us analyze each case.

1. $\mathbb{P}(\tilde{d}_{yy} < \tilde{d}_{xx} < \tilde{d}_{xy}) \to 1$ if and only if $\sqrt{2}\sigma_{F_2} < \sqrt{2}\sigma_{F_1} < \sqrt{\nu^2 + \sigma^2_{F_1} + \sigma^2_{F_2}}$.
2. $\mathbb{P}(\tilde{d}_{yy} < \tilde{d}_{xy} < \tilde{d}_{xx}) \to 1$ if and only if $\sqrt{2}\sigma_{F_2} < \sqrt{\nu^2 + \sigma^2_{F_1} + \sigma^2_{F_2}} < \sqrt{2}\sigma_{F_1}$.
3. $\mathbb{P}(\tilde{d}_{xx} < \tilde{d}_{yy} < \tilde{d}_{xy}) \to 1$ if and only if $\sqrt{2}\sigma_{F_1} < \sqrt{2}\sigma_{F_2} < \sqrt{\nu^2 + \sigma^2_{F_1} + \sigma^2_{F_2}}$.
4. $\mathbb{P}(\tilde{d}_{xx} < \tilde{d}_{xy} < \tilde{d}_{yy}) \to 1$ if and only if $\sqrt{2}\sigma_{F_1} < \sqrt{\nu^2 + \sigma^2_{F_1} + \sigma^2_{F_2}} < \sqrt{2}\sigma_{F_2}$.
5. $\mathbb{P}(\tilde{d}_{xy} < \tilde{d}_{xx} < \tilde{d}_{yy}) = 0$.
6. $\mathbb{P}(\tilde{d}_{xy} < \tilde{d}_{yy} < \tilde{d}_{xx}) = 0$.

We examine under what conditions each ordering of the deterministic distance limits can occur as $d \to \infty$, assuming $\sigma_{F_1}, \sigma_{F_2} > 0$, and $\nu^2 \geq 0$.

Case	Ordering	Condition
1	$\tilde{\delta}_{yy} < \tilde{\delta}_{xx} < \tilde{\delta}_{xy}$	$\sigma_{F_2} < \sigma_{F_1}$
2	$\tilde{\delta}_{yy} < \tilde{\delta}_{xy} < \tilde{\delta}_{xx}$	$\sigma^2_{F_1} > \sigma^2_{F_2} + \nu^2$
3	$\tilde{\delta}_{xx} < \tilde{\delta}_{yy} < \tilde{\delta}_{xy}$	$\sigma_{F_1} < \sigma_{F_2}$
4	$\tilde{\delta}_{xx} < \tilde{\delta}_{xy} < \tilde{\delta}_{yy}$	$\sigma^2_{F_2} > \sigma^2_{F_1} + \nu^2$
5	$\tilde{\delta}_{xy} < \tilde{\delta}_{xx} < \tilde{\delta}_{yy}$	Impossible
6	$\tilde{\delta}_{xy} < \tilde{\delta}_{yy} < \tilde{\delta}_{xx}$	Impossible

Maa, Pearl, and Bartoskzyński [206] proved that, under mild conditions, $F_1 = F_2$ if and only if the IPD within samples and between samples have the same univariate distribution. Hence, in cases 1 and 2 we have $F_2 < F_1$ and in cases 3 and 4 we have $F_1 < F_2$.

Example 4.10.1 *For any fixed k, the nearest neighbor (NN) test (Schilling [300]; Henze [147]; Sarkar and Ghosh [299]) uses the statistic*

$$T_{N,k} = \frac{1}{Nk}\Big[\sum_{t=1}^{m}\sum_{r=1}^{k}\mathbb{I}_r(\mathbf{X}_i) + \sum_{j=1}^{n}\sum_{r=1}^{k}\mathbb{I}_r(\mathbf{Y}_j)\Big] \tag{4.19}$$

where $\mathbf{X}_1,\ldots,\mathbf{X}_m \sim F_1$, $\mathbf{Y}_1,\ldots,\mathbf{Y}_n \sim F_2$, $\mathbb{I}_r(\mathbf{Z})$ *is an indicator that is 1 if* $\mathbf{Z}$ *and its r-th nearest neighbor come from the same distribution. We reject* $H_0 : F_1 = F_2$ *for large values of* $T_{N,k}$. *Consider* $\mathbf{X}_1, \mathbf{X}_2 \sim F_1$, *and* $\mathbf{Y}_1, \mathbf{Y}_2 \sim F_2$. *Under Assumptions A1-A3 when* $\sigma^2_{F_2} - \sigma^2_{F_1} > (\mu_1 - \mu_2)^2$, *then condition 4 is true as* $d \to \infty$. *When* $r < \min(m, n)$, *we have* $\mathbb{I}_r(\mathbf{X}_i) \to 1$ *for* $i = 1,\ldots,m$, *and* $\mathbb{I}_r(\mathbf{Y}_j) \to 0$ *in probability for* $j = 1,\ldots,n$ *as* $d \to \infty$. *Hence,* $T_{N,k} \to m/N$ *in probability. When* $m = n$ *it is the limiting value of* $T_{N,k}$ *under* H_0 *and far from the threshold value to reject* H_0. *Hence, NN method fails as* $d \to \infty$.

In high-dimensional settings, the IPDs between random vectors from different distributions concentrate around their means. Hence, relative orderings of distances converge in probability to deterministic orderings of their limiting values. The four probabilities considered converge to 1 if and only if the corresponding inequalities between the asymptotic distance limits hold. Conditions 5 and 6 are clearly impossible.

4.11 Geometric Structure of Distance Orderings

4.11.1 Elliptic Paraboloid Structure

Recall the deterministic limits

$$\delta_{xx} = \sqrt{2}\sigma_{F_1}, \qquad \delta_{yy} = \sqrt{2}\sigma_{F_2}, \qquad \delta_{xy} = \sqrt{\nu^2 + \sigma^2_{F_1} + \sigma^2_{F_2}}.$$

Eliminating $\sigma_{F_1}, \sigma_{F_2}$ gives

$$\sigma^2_{F_1} = \frac{\delta^2_{xx}}{2}, \qquad \sigma^2_{F_2} = \frac{\delta^2_{yy}}{2}.$$

Substituting into the expression for δ_{xy} yields

$$\delta^2_{xy} = \nu^2 + \frac{\delta^2_{xx} + \delta^2_{yy}}{2} \geq \min(\delta^2_{xx}, \delta^2_{yy}). \tag{4.20}$$

This surface is an elliptic paraboloid in $(\delta_{xx}, \delta_{yy}, \delta_{xy})$-–space. All admissible triples lie on this surface.

Let us check Case 5. Squaring the inequalities we obtain

$$\delta^2_{xy} < \delta^2_{xx} < \delta^2_{yy} \implies \delta^2_{xy} < \delta^2_{xx} \text{ and } \delta^2_{xy} < \delta^2_{yy}.$$

But from (4.20) we get $\delta^2_{xy} \geq \min(\delta^2_{xx}, \delta^2_{yy})$. This contradicts the requirement $\delta^2_{xy} < \delta^2_{xx}$ and $\delta^2_{xy} < \delta^2_{yy}$. Similarly, Case 6 fails for the same reason. Therefore, $\delta_{xy} < \min(\delta_{xx}, \delta_{yy})$ cannot hold, so Cases 5 and 6 are impossible.

4.11.2 Geometric Partition of the Surface

The orderings in the table arise from the intersection of the paraboloid with the planes

$$\delta_{xx} = \delta_{yy}, \qquad \delta_{xy} = \delta_{xx}, \qquad \delta_{xy} = \delta_{yy}.$$

These three planes intersect the paraboloid along curves. Together they partition the paraboloid into four connected regions, corresponding to Cases 1-–4:

- Case 1 (blue symbol, largest) corresponds to $\delta_{yy} < \delta_{xx} < \delta_{xy}$, where $\sigma_{F_1} > \sigma_{F_2}$ and δ_{xy} is largest.
- Case 2 (green symbol, medium) corresponds to $\delta_{yy} < \delta_{xy} < \delta_{xx}$, where δ_{xx} is largest.
- Case 3 (orange symbol, medium) corresponds to $\delta_{xx} < \delta_{yy} < \delta_{xy}$, where $\sigma_{F_2} > \sigma_{F_1}$ and δ_{xy} is largest.
- Case 4 (red symbol, medium) corresponds to $\delta_{xx} < \delta_{xy} < \delta_{yy}$, where δ_{yy} is largest.

Figure 4.14 illustrates the geometric structure of high-dimensional distance orderings. The admissible deterministic limits lie on a single elliptic paraboloid, and changes in ordering occur precisely when crossing one of the three separating curves. The color and size of the symbols explicitly indicate the corresponding region from the ordering table (Cases 1–4), with the largest blue symbol denoting Case 1.

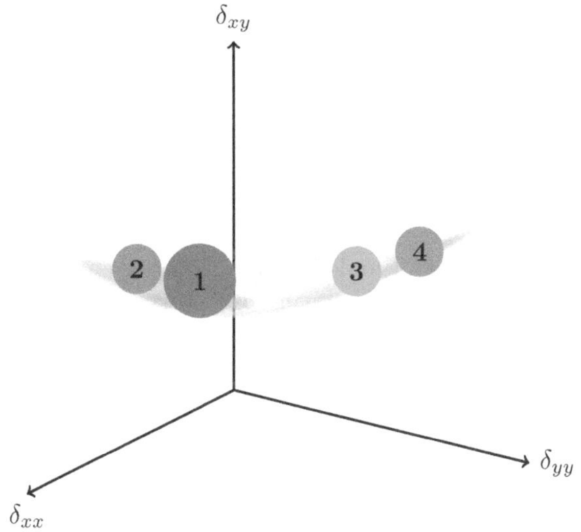

FIGURE 4.14: Elliptic paraboloid $\delta_{xy}^2 = \nu^2 + \frac{\delta_{xx}^2+\delta_{yy}^2}{2}$ (with $\nu = 0.4$). The gray surface represents all feasible deterministic distance limits. Thick black, blue, and red curves show the separating planes $\delta_{xx} = \delta_{yy}$, $\delta_{xy} = \delta_{xx}$, and $\delta_{xy} = \delta_{yy}$, respectively. Shaded regions correspond to the four feasible ordering cases: blue = Case 1, green = Case 2, orange = Case 3, red = Case 4. Large colored symbols with numbers indicate representative points for each case.

4.12 Comparing K Groups

Suppose that for each $i = 1, \ldots, k$, we observe n_i independent and identically distributed random vectors in $\mathbb{R}^d$ drawn from distribution functions F_i. The objective is to develop a visualization framework for comparing these distributions and for testing the null hypothesis

$$H_0 : F_1 = F_2 = \cdots = F_k$$

against the alternative hypothesis

$$H_a : F_i \neq F_j \quad \text{for at least one } i \neq j.$$

It can be shown that H_0 holds if and only if the within-sample and between-sample distributions of all IPDs are identical across groups. Specifically, there are k within-sample IPD distributions and $k(k-1)/2$ between-sample IPD distributions, yielding a total of $c_k = \binom{k}{2}$ distinct IPD distributions.

Let $\mathcal{K}_{F_i,F_i}$ denote the distribution of within-sample IPDs from group i, and let $\mathcal{K}_{F_i,F_j}$, for $i \neq j$ and $i, j = 1, 2, \ldots, k$, denote the distribution of between-sample IPDs from groups i and j. Instead of testing H_0 directly, we may consider the equivalent null hypothesis

$$H_0' : \mathcal{K}_{F_i,F_i} = \mathcal{K}_{F_j,F_j} = \mathcal{K}_{F_i,F_j} \quad \text{for all } i \neq j,\ i, j = 1, 2, \ldots, k.$$

Thus, the problem reduces to testing the equality of the on-diagonal and lower-diagonal IPD distributions in the collection $\mathbb{C}$, which contains all within- and between-sample IPD distributions $\mathcal{K}_{F_i,F_i}$ and $\mathcal{K}_{F_i,F_j}$.

$$\mathbb{C} = \begin{array}{c} \\ F_1 \\ F_2 \\ \\ F_k \end{array} \begin{array}{c} \begin{array}{ccccc} F_1 & F_2 & F_3 & \ldots & F_k \end{array} \\ \begin{pmatrix} \mathcal{K}_{F_1,F_1} & \mathcal{K}_{F_1,F_2} & \mathcal{K}_{F_1,F_3} & \ldots & \mathcal{K}_{F_1,F_k} \\ \mathcal{K}_{F_2,F_1} & \mathcal{K}_{F_2,F_2} & \mathcal{K}_{F_2,F_3} & \ldots & \mathcal{K}_{F_2,F_k} \\ \ldots & \ldots & \ldots & \ldots & \ldots \\ \mathcal{K}_{F_k,F_1} & \mathcal{K}_{F_k,F_2} & \mathcal{K}_{F_k,F_3} & \ldots & \mathcal{K}_{F_k,F_k} \end{pmatrix} \end{array}. \tag{4.21}$$

Suppose $\mathbb{X} = \{\mathbf{X}_i\}_{i=1}^m$, $\mathbb{Y} = \{\mathbf{Y}_i\}_{i=1}^{n_2}$ and $\mathbb{Z} = \{\mathbf{z}_i\}_{i=1}^{n_3}$ represent i.i.d. observation vectors in $\mathbb{R}^d$ from DFs F_x and F_y, and F_z, respectively. Let $R = \max D(\mathbb{X}, \mathbb{Y}, \mathbb{Z}) - \min D(\mathbb{X}, \mathbb{Y}, \mathbb{Z})$ denote the range of all positive IPDs and divide the range into s intervals $\delta(t)$ for $t = 1, \ldots s$. The ECDFs of $d_{(x)ij}$, $d_{(y)ij}$, $d_{(xy)ij}$ appear in (4.16). The ECDFs of $d_{(z)ij}$, $d_{(xz)ij}$, and $d_{(yz)ij}$ evaluated at $\delta(t)$ are

$$\begin{aligned} \hat{H}_z(t) &= \frac{1}{n_3} \sum_{i=1}^{n_3-1} \sum_{j=i+1}^{n_3} \mathbb{I}(d_{(z)ij} \leq \delta(t)), \\ \hat{H}_{xz}(t) &= \frac{1}{n_1 n_3} \sum_{i=1}^{n_1} \sum_{j=1}^{n_3} \mathbb{I}(d_{(xz)ij} \leq \delta(t)), \\ \hat{H}_{yz}(t) &= \frac{1}{n_2 n_3} \sum_{i=1}^{n_2} \sum_{j=1}^{n_3} \mathbb{I}(d_{(yz)ij} \leq \delta(t)). \end{aligned} \tag{4.22}$$

The next examples are on two well-known datasets with $k = 3$ groups and small d to determine the stochastic ordering of the IPDs of the groups.

4.12.1 Analysis of Iris Dataset

Example 4.12.1 *We examine the Iris dataset, which records four characteristics for three species of the Iris plant. The dataset includes measurements of sepal length, sepal width, petal length, and petal width for Iris Setosa ($\mathbb{X}$), Iris Versicolor ($\mathbb{Y}$), and Iris Virginica ($\mathbb{Z}$), with 50 observations for each species. To assess whether the mean vectors differ significantly among the three species, a MANOVA test was conducted. The test rejects the null hypothesis of equal mean vectors, providing strong evidence that the mean vectors are indeed different across the species.*

Next, to evaluate whether the covariance matrices differ significantly across species, Box's M-test was applied. The test rejects the hypothesis of homogeneity of covariance matrices, indicating that the covariance structures are not identical among the species. Bivariate plots of petal and sepal lengths and widths reveal that while Setosa is clearly distinct from the other two species, Versicolor and Virginica are not easily separable. Figures 4.15 and 4.16 display the parallel coordinate plot and Andrews' curves for the three Iris species, illustrating these differences.

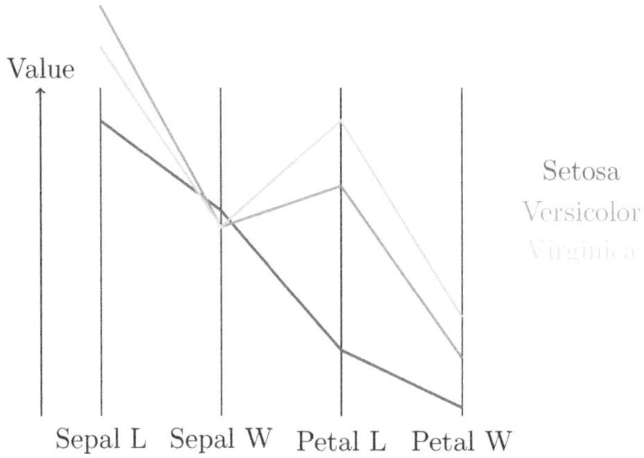

FIGURE 4.15: Parallel coordinates plot for the three Iris species: From top to bottom on the right, Virginica (top), Versicolor (middle), Setosa (bottom).

How do the groups differ in terms of their IPD structures? Figure 4.17 displays the ECDFs for the six comparison groups. The average IPDs, $\bar{d}_{(x)}$, $\bar{d}_z$, $\bar{d}_{(y)}$, $\bar{d}_{(xy)}$, $\bar{d}_{yz}$, *and* $\bar{d}_{xz}$, *are* 0.69, 0.71, 0.99, 3.3, 8.9, *and* 6.99, *respectively. The three species show little difference in their within-sample IPDs. The figure illustrates that the within-group distances* $d_{(x)}$, $d_{(y)}$, *and* d_z *are comparable, while the ordering of the between-group distances,* $d_{(xy)} \leq_{St} d_{yz} \leq_{St} d_{xz}$, *indicates that all pairs of Iris species are distinguishable. In particular, with* $\bar{d}_{(xy)} = 3.3$ *and* $\bar{d}_{yz} = 8.9$, *the ECDFs suggest that Versicolor exhibits an IPD structure that differs from both Setosa and Virginica. Notably, all between-sample IPDs exceed the within-sample IPDs, consistent with a location shift model, where between-group distances are generally larger than within-group distances.*

4.12.2 Analysis of Wine Dataset

Example 4.12.2 *We examine the wine dataset (Asuncion and Newman [16]), which contains 178 observations on the concentrations of 13 different chemicals in wines produced in the same region of Italy. The wines originate from three grape cultivars: Barolo, Grignolino,*

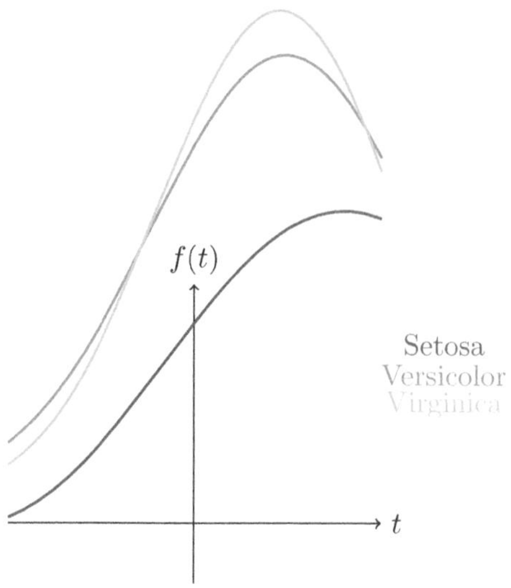

FIGURE 4.16: Andrews' curves for the three Iris species: Virginica (top), Versicolor (middle), Setosa (bottom).

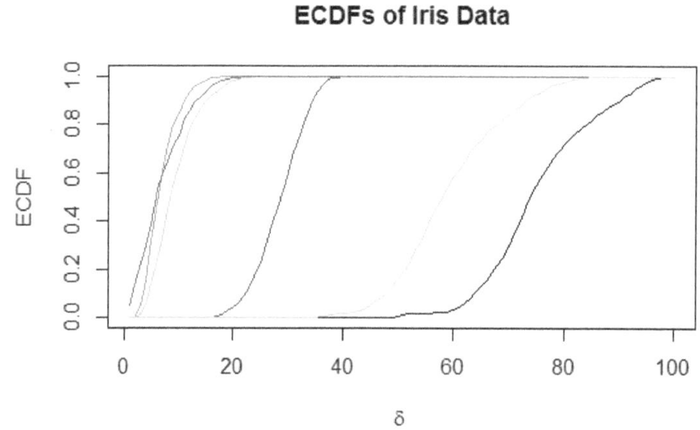

FIGURE 4.17: Empirical CDFs of sample IPDs for Iris data. From top to bottom on the left, the ECDFs are $\hat{H}_x$ (red), $\hat{H}_z$ (brown), $\hat{H}_y$ (green) grouped together followed by the ECDFs of $\hat{H}_{xy}$ (blue), $\hat{H}_{yz}$ (pink) and $\hat{H}_{xz}$ (black).

and Barbera, with 59, 71, and 48 observations for each type, respectively. Barolo is produced from Nebbiolo grapes, whereas the other two wine types are named after the grapes from which they are made. Univariate analysis reveals that the chemical measurements vary widely in scale and that the three wine types are well-separated. In particular, Barolo wines contain distinctly higher amounts of Proline compared to Barbera and Grignolino.

Let X, Y, and Z denote the three grape cultivars: Barolo, Grignolino, and Barbera, respectively. The parallel coordinate and Andrew' plots of the wine data appear in Figures 4.18 and 4.19, respectively. The corresponding mean IPDs are $\bar{d}_{(x)} = 253.25$, $\bar{d}_{(y)} = 176.69$, $\bar{d}_z = 133.86$, $\bar{d}_{(xy)} = 599.83$, $\bar{d}_{xz} = 488.99$, and $\bar{d}_{yz} = 186.07$. Figure 4.20 presents the six ECDF plots of the within- and between-sample IPDs. The axes in the plot include 5

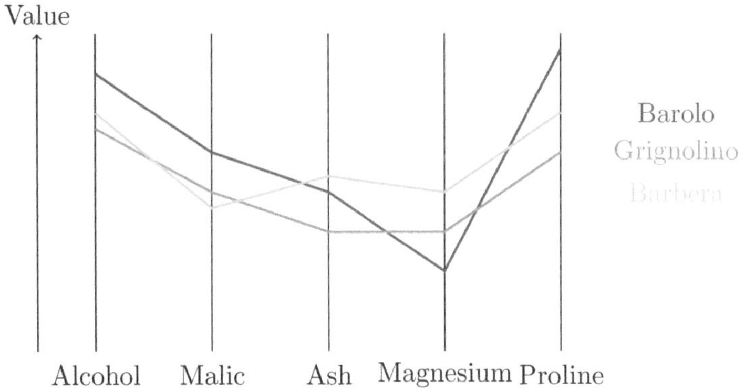

FIGURE 4.18: Parallel coordinates plot for representative chemical measurements in the three wine cultivars. From top to bottom on the right, Barolo (top), Barbera (middle), Grignolino (bottom).

features (Alcohol, Malic, Ash, Magnesium, Proline) instead of all 13. These plots clearly indicate that the null hypothesis $H_0 : F_x = F_y = F_z$ *is false, providing strong support for the alternative hypothesis* $H_a : F_z \leq F_y \leq F_x$.

Example 4.12.3 *Low-rank Decomposition for k groups: Let* $\{\mathbf{X}_i^{(t)}\}_{i=1}^{n_t} \subset \mathbb{R}^d$ *denote* n_t *observations from population* G_t, *for* $t = 1, \ldots, k$. *Let* $N = \sum_{t=1}^{k} n_t$, *and define the pooled sample* $\mathbb{X} = \bigcup_{t=1}^{k} \mathbb{X}^{(t)}$. *For any two observations* $\mathbf{X}_i, \mathbf{X}_j \in \mathbb{R}^d$, *the squared interpoint distance (IPD) is*

$$\|\mathbf{X}_i - \mathbf{X}_j\|^2 = \mathbf{X}_i'\mathbf{X}_i + \mathbf{X}_j'\mathbf{X}_j - 2\mathbf{X}_i'\mathbf{X}_j. \tag{4.23}$$

Define the matrices $\mathbf{U} \in \mathbb{R}^{N\times(d+2)}$ *and* $\mathbf{V} \in \mathbb{R}^{(d+2)\times N}$ *by*

$$\mathbf{U} = \begin{pmatrix} \mathbf{X}_1'\mathbf{X}_1 & 1 & -2\mathbf{X}_1' \\ \mathbf{X}_2'\mathbf{X}_2 & 1 & -2\mathbf{X}_2' \\ \vdots & \vdots & \vdots \\ \mathbf{X}_N'\mathbf{X}_N & 1 & -2\mathbf{X}_N' \end{pmatrix}, \tag{4.24}$$

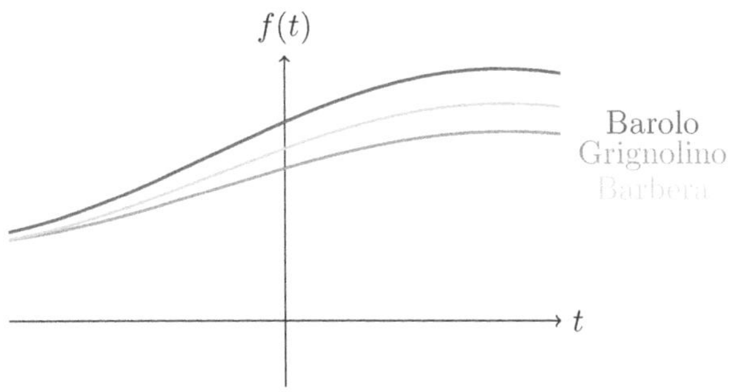

FIGURE 4.19: Andrews' curves for representative chemical measurements in the three wine cultivars. Barolo (top), Barbera (middle), Grignolino (bottom).

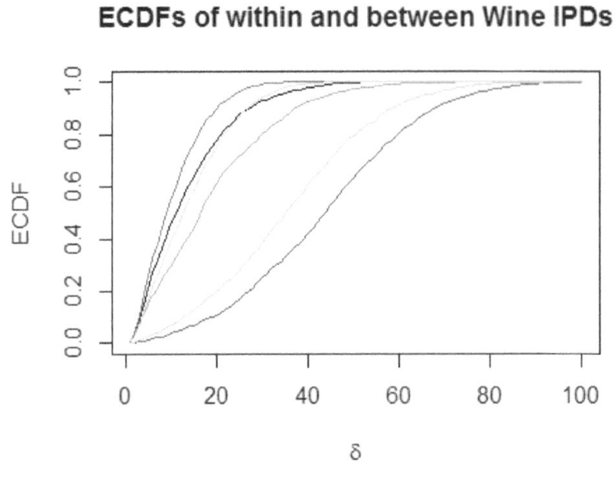

FIGURE 4.20: Empirical CDFs of the IPDs for three varieties of wine. The ECDFs from top to bottom are $\hat{H}_z$ (brown), $\hat{H}_y$ (black), $\hat{H}_{yz}$ (pink), $\hat{H}_x$ (red), $\hat{H}_{xz}$ (green), and $\hat{H}_{xy}$ (blue). The IPDs clearly support the alternative $H_a : F_z \le F_y \le F_x$.

$$\mathbf{V} = \begin{pmatrix} 1 & 1 & \cdots & 1 \\ \mathbf{X}_1'\mathbf{X}_1 & \mathbf{X}_2'\mathbf{X}_2 & \cdots & \mathbf{X}_N'\mathbf{X}_N \\ \mathbf{X}_1 & \mathbf{X}_2 & \cdots & \mathbf{X}_N \end{pmatrix}, \tag{4.25}$$

where $\{\mathbf{X}_1, \ldots, \mathbf{X}_N\}$ *is any ordering of the pooled observations* $\mathbb{X}$. *The last block row of* $\mathbf{V}$ *consists of* d *rows, with the* j*th column equal to the vector* $\mathbf{X}_j \in \mathbb{R}^d$. *Thus, the distance matrix containing all within and between-group squared IPDs,* $D(\mathbb{X}) \in \mathbb{R}^{N \times N}$, *admits the low-rank factorization* $D(\mathbb{X}) = \mathbf{UV}$. *Indeed, the* (i, j) *entry of* $\mathbf{UV}$ *is*

$$(\mathbf{UV})_{ij} = \mathbf{X}_i'\mathbf{X}_i + \mathbf{X}_j'\mathbf{X}_j - 2\mathbf{X}_i'\mathbf{X}_j = \|\mathbf{X}_i - \mathbf{X}_j\|^2,$$

which holds regardless of the group membership of $\mathbf{X}_i$ *and* $\mathbf{X}_j$.

The above representation shows that the rank of the full distance matrix is bounded by $d + 2$, independently of the number of groups k and the total sample size N. This low-rank structure is particularly useful for efficient and parallel computation of within- and between-group IPDs in high-dimensional settings.

4.13 Summary

This chapter presents a survey of graphical methods for the analysis of high-dimensional data, addressing the limitations of traditional visualization techniques when confronted with complex multivariate structures. It motivates the need for specialized high-dimensional graphics capable of revealing dependence, clustering, and distributional features that are otherwise obscured in low-dimensional projections. The chapter emphasizes visualization as a central component of exploratory data analysis and as a complement to formal statistical inference.

Foundational methods based on graph theory and linear algebra are introduced first. Minimum spanning trees are used to expose proximity relationships and latent clustering, while matrix decomposition techniques and principal component analysis, and multidimensional scaling provide principled dimension reduction. Biplots enable joint visualization of observations and variables, enhancing interpretability of reduced-dimensional representations. Projection-based and transformation-based displays, such as parallel coordinate plots, Andrews' plots, and Chernoff faces, and the Dutta–Ghosh transformation, are then developed as flexible tools for uncovering non-linear structure and multivariate patterns.

The chapter further focuses on graphical techniques for comparing distributions across two or multiple groups, with particular emphasis on IPD-based methods and CDF plots. Asymptotic properties of IPDs in high dimensions are discussed to support theoretical interpretation of the graphics. The methodology is illustrated through analyses of the Iris and Wine datasets, demonstrating how high-dimensional visualization facilitates group comparison, pattern recognition, and inferential insight in practical applications.

4.14 Exercises

Exercise 4.1 *Suppose $\lambda_1, \ldots, \lambda_d$ are the eigenvalues of a positive definite covariance matrix $\boldsymbol{\Sigma}$. Prove that*

$$(\sum_{i=1}^{d} \lambda_i^r)^2 \le \mathbb{P}(\sum_{i=1}^{d} \lambda_i^{2r})$$

with equality if and only if $\lambda_1 = \lambda_2 = \ldots = \lambda_d = \lambda$ for some constants λ and r.

Exercise 4.2 *Find the eigenvalues and eigenvectors of $\boldsymbol{\Sigma}$ where*

$$\boldsymbol{\Sigma} = \begin{pmatrix} b^2 + d & b & b \\ b & 1 + d & 1 \\ b & 1 & 1 + d \end{pmatrix}.$$

Exercise 4.3 *Suppose $\mathbf{X} = (X_1, \ldots, X_d)$ and $\mathbf{Y} = (Y_1, \ldots, Y_d)$ are independent vectors with zero means and covariance $\boldsymbol{\Sigma}$. Let*

$$f_{\mathbf{X}}(t) = \frac{X_1}{\sqrt{2}} + X_2 \sin t + X_3 \cos t + X_4 \sin 2t + X_5 \cos 2t + \ldots$$

for $-\pi < t < \pi$. Define $||f_{\mathbf{X}}(t) - f_{\mathbf{Y}}(t)||^2 = \int_{-\pi}^{\pi} (f_{\mathbf{X}}(t) - f_{\mathbf{Y}}(t))^2 dt$. Show that

$$||f_{\mathbf{X}}(t) - f_{\mathbf{Y}}(t)||^2 = \int_{-\pi}^{\pi} (f_{\mathbf{X}}(t) - f_{\mathbf{Y}}(t))^2 dt = \pi ||\mathbf{X} - \mathbf{Y}||^2.$$

and find $\mathbb{E}(||\mathbf{X} - \mathbf{Y}||^2)$.

Exercise 4.4 *Consider the symmetric Toeplitz covariance matrix*

$$\mathbf{C} = \begin{bmatrix} 1 & \rho & \rho^2 & \rho^3 \\ \rho & 1 & \rho & \rho^2 \\ \rho^2 & \rho & 1 & \rho \\ \rho^3 & \rho^2 & \rho & 1 \end{bmatrix}.$$

This matrix has the structure $C_{ij} = \rho^{|i-j|}$. Such matrices have known spectral properties. Obtain the spectral decomposition of $\mathbf{C}$.

Exercise 4.5 *Consider the covariance matrix*

$$\mathbf{C} = \begin{bmatrix} 1 & 0.5 & -0.5 & 0 \\ 0.5 & 1 & 0.707 & 0.5 \\ -0.5 & 0.707 & 1 & -0.5 \\ 0 & 0.5 & -0.5 & 1 \end{bmatrix}.$$

Obtain the variable biplot for this structure.

Exercise 4.6 *Suppose* $d = 10000$, *and* $s = 100$. *Generate* $m = n = 50$ *observations from* $\mathbb{N}(\mathbf{0}, \sigma_x^2 \mathbf{I}_d)$ *and* $\mathbb{N}(\boldsymbol{\mu}, \sigma_y^2 \mathbf{I}_d)$, *respectively. Replicate Figure 4.13 Panels (a) and (b).*

Exercise 4.7 *Consider the ECDFs of* $d_{(x)ij}$, $d_{(y)ij}$, $d_{(xy)ij}$ *given in* (4.16). *Write a program to replicate the ECDFs of sample IPDs for Iris data that appears in 4.17.*

Exercise 4.8 *Verify the conditions in section 4.10.*

Exercise 4.9 *Given projection vector* $V(t)$ *of the form,*

$$V(t) = \left(\frac{1}{\sqrt{2}}, \sin(t), \cos(t), \sin(2t), \cos(2t), \ldots\right)$$

where the terms alternate between sine and cosine functions. Find the norm of $V(t)$, *i.e.,*

$$\|V(t)\| = \sqrt{\left(\frac{1}{\sqrt{2}}\right)^2 + \sin^2(t) + \cos^2(t) + \sin^2(2t) + \cos^2(2t) + \ldots}.$$

5

Testing the Equality of High-Dimensional Distributions

5.1 Abstract

This chapter provides a systematic treatment of statistical tests for assessing the equality of high-dimensional distributions, with particular emphasis on HDLSS regimes. Classical multivariate tests often fail in high dimensions due to distance concentration, loss of neighborhood structure, and the breakdown of covariance-based methods. The chapter reviews a broad class of graph-based, rank-based, and IPD-based procedures designed to overcome these challenges, including the Friedman—Rafsky runs test, Rosenbaum's cross-match test, energy statistics, and several permutation-based frameworks. Both location- and dispersion-sensitive methods are examined, alongside traditional tests such as Hotelling's T^2 and Puri—Sen rank tests. Novel dissimilarity-based statistics are developed to address distance concentration and to retain power as dimensionality increases. Theoretical properties, including asymptotic behavior as dimension diverges, are complemented by simulation studies and extensive applications to benchmark datasets and gene expression data from cancer studies. Overall, the chapter offers a unified perspective on modern non-parametric testing methodologies for comparing high-dimensional distributions under practical and theoretical constraints.

5.2 Introduction

High-dimensional data poses unique challenges for analysis and interpretation due to several factors, including the concentration of Euclidean distances, data sparsity, computational complexity, and data visualization. High-dimensional data are now commonplace in microarrays, proteomics, brain images, videos, functional and financial data. In microarray analysis, the sample size N is usually in the order of tens or hundreds while the number of variables d is in the order of tens of thousands. In other applications, N is the order of thousands, but d is in the order of millions.

Testing for the equality of multivariate distributions is a cornerstone of statistical analysis. Interpoint distances (IPDs) are the building blocks of several multivariate techniques such as comparison of distributions, the k-nearest neighbors for supervised learning, and k-means clustering for unsupervised learning, depth functions, correspondence analysis, and multidimensional scaling. Comparing multivariate distributions is a challenging task due to the concentration of distances phenomenon. Angiulli [11] describes the concentration of the interpoint distances around a single value in high-dimensional spaces. Interpoint distances become almost indistinguishable in high-dimensional spaces and their coefficient of variation converges to zero as $d \to \infty$. Aggarwal et al. [47] prove that the relative difference of the distances of the minimum and maximum IPDs tends to zero as $d \to \infty$.

DOI: 10.1201/9781003789581-5

Distributional differences may be dense: small but spread across many coordinates, or sparse: large but confined to a few coordinates. Detectability thresholds differ substantially between these regimes. Sparse alternatives may be detectable even when $\|\mu\|_2$ is small, provided the sparsity level s satisfies $s \log d \ll n$. Beyond mean differences, in high dimensions, distributions may differ only in covariance structure, tail behavior, or dependence. For example, $F_1 = \mathbb{N}(\mathbf{0}, \mathbf{I}_d)$, $F_2 = \mathbb{N}(\mathbf{0}, \boldsymbol{\Sigma})$, with $\boldsymbol{\Sigma} \neq \mathbf{I}_d$. Detectability then depends on quantities such as $\|\boldsymbol{\Sigma} - \mathbf{I}_d\|_F^2 \gtrsim \frac{d}{n}$. Many classical tests lack power against such alternatives, motivating distance and kernel-based approaches.

Testing for the equality of distributions, Friedman and Rafsky [109] constructed a runs test based on the minimum spanning tree, and Henze and Penrose [147] showed that the test is asymptotically distribution-free and consistent. Bickel and Breiman [32] and Schilling [300] developed consistent and asymptotically distribution-free tests based on nearest-neighbor distances. Hall and Tajvidi [134] propose innovative permutation methods for equality of high-dimensional distributions. Chen and Qin [64], Huang et al. [158], Lopes et al. [203] offer tests for high-dimensional data. Székely and Rizzo [333] and, Baringhaus and Franz [21] and Biswas and Ghosh [35] propose permutation tests based on the IPDs. Rosenbaum [284] developed an exact distribution–free test using adjacency. Petrie [270] extends graph-based tests from the two-sample to the k-sample problem in high dimensions and proposes the use of orthogonal perfect matchings. For comparing means and variability of multivariate data, Marozzi et al. [216] propose tests based on IPDs. Modarres and Song [239] discuss IPDs, their applications and visualizations. Liu et al. [197] suggest Log-rank-type tests for the equality of high-dimensional distributions. Sarkar et al. [297] presents graph-theoretic methods of comparing high-dimensional distributions.

Puri-Sen test [274] is a rank-based counterpart of the Hotelling T^2 test [155]. Other non-parametric tests for the two-sample location problem appear in survey of Oja and Randles [256] and Oja [255]. Jurečková and Kalina [165] propose test statistics based on the ranks of the IPDs. These tests are distribution-free against some monotone alternatives of the Lehmann type and are finite-sample unbiased. Pesarin and Salmaso [269] discuss suitable transformations of the d-dimensional observations into univariate derived observations within the permutation testing framework. While it seems that there are a plethora of choices to test H_0, such is not the case when there are more variables than observations in each group. Many multivariate techniques such as nearest neighbors methods or Hotelling T^2 test become ineffective or lose statistical power in high dimensions. Methods of multivariate density estimation require very large sample sizes and suffer from the curse of dimensionality. Modarres [238] discusses testing the equality of high-dimensional distributions.

Pal et al. [261] and Modarres [234] show that classification methods based on the nearest distances will lose power as $d \to \infty$ due to the concentration of Euclidean distances. Mondal et al. [241] propose new multivariate two-sample tests based on nearest neighbor type coincidences. They show that several tests for the multivariate two-sample problem perform poorly for high-dimensional data, including Friedman and Rafsky [109] multivariate run test, Hall and Tajvidi's [134] test, test of Baringhaus and Franz [21], and the multivariate run test based on the shortest Hamiltonian path proposed by Biswas et al. [36]. Using dissimilarities introduced in chapter 3 , instead of Euclidean distances, we present modifications of the existing tests and novel methods of comparing distributions in $\mathbb{R}^d$ that take advantage of the concentration phenomena.

5.3 Friedman-Rafsky Runs Test

Consider samples of size m and n from distributions F_1 and F_2, both defined on $\mathbb{R}^d$. The null hypothesis $H_0 : F_1 = F_2$ is to be tested against general alternative hypotheses $H_a : F_1 \neq F_2$.

The Wald-Wolfowitz runs test (for $d = 1$) sorts the $N = m + n$ combined observations in ascending order. Each observation is then replaced by a label "X" or "Y" depending upon the sample to which it originally belonged. The test statistic is the total number of runs, R. A run is a consecutive sequence of identical labels. Rejection of H_0 is for small values of R. The null distribution of the test statistic can be derived by a straightforward combinatorial argument. For example, $X : 5, 3, 9, 2$ and $Y : 1, 7, 6$ produce the sorted list $1, 2, 3, 5, 6, 7, 9$, the sequence $YXXXYYX$ and $R = 4$. Asymptotically, the quantity $W = \frac{R-\mu_R}{\sigma_R}$ is $\mathbb{N}(0, 1)$ as m and n tend to ∞ such that $\lambda = m/n$ is bounded away from 0 or ∞. Here $\mu_r = 2mn/N+1$ and $\sigma_R^2 = \frac{2mn(2mn-N)}{N_3-N^2}$.

Friedman and Rafsky [109] multivariate runs test generalizes the Wald-Wolfowitz univariate runs test as follows.

(1) construct the MST of the pooled sample multivariate data points,

(2) remove all edges connecting different samples, and

(2) the test statistic R is the number of disjoint sub-trees. This will be one more than the number of edges deleted.

Rejection of H_0 is for a small number of sub-trees (runs). This generalization preserves the spirit of the Wald-Wolfowitz test since the sorted list is used primarily to link points that are close together in R^1. The multivariate MST links points that are close together in $\mathbb{R}^d$. Under H_0, the mean and variance of R can be calculated using indicator variables . Number the N - 1 edges of the MST arbitrarily and define $Z_i = 1$ if the ith edge links nodes from different samples, $1 \leq i \leq N - 1$. Otherwise, $Z_i = 0$. Hence,

$$R = \sum_{i=1}^{N-1} Z_i + 1.$$

To compute the mean and variance of R under H_0 note that $\mathbb{E}(Z_i) = \mathbb{P}(Z_i = 1)$, where $\mathbb{P}(Z_i = 1)$ is the probability of two nodes defining this edge are labeled X and Y or Y and X. Therefore,

$$\mathbb{P}(XY) = \mathbb{P}(YX) = \frac{mn}{N(N-1)}$$

$$\mathbb{P}(Z_i = 1) = 2\frac{mn}{N(N-1)},$$

and $\mathbb{E}(R) = 2mn/N + 1$.

$$\mathbb{V}ar(R) = \mathbb{V}ar(\sum_{i=1}^{N-1} Z_i + 1) = \sum_{i=1}^{N-1} \mathbb{V}ar(Z_i) + 2\sum_{i<j} \mathbb{C}ov(Z_i, Z_j). \tag{5.1}$$

It follows that

$$\sum_{i=1}^{N-1} \mathbb{V}ar(Z_i) = 2mn/N - \frac{4m^2n^2}{N^2(N-1)}$$

and

$$\mathbb{C}ov(Z_i, Z_j) = \mathbb{E}(Z_iZ_j) - (\mathbb{E}Z_i))^2.$$

Note that $\mathbb{E}(Z_iZ_j) = \mathbb{P}(Z_iZ_j = 1)$ and this probability depends on whether or not the ith and jth edges share a common node. Let C be the number of edge pairs that share a common node. $\mathbb{V}ar(R|C) = \frac{2mn}{N(N-1)}\{\frac{2mn-N}{N} + \frac{C-N+2}{(N-2)(N-3)}[N(N-1)-4mn+2]\}$. The value of C depends on the configuration of the MST and is determined by the node degrees. In the univariate case, these degrees are fixed independent of the sample: there are two nodes of degree one and N–2 nodes of degree two (string MST). In this case, $C = N - 2$ and $\mathbb{V}ar(R|C)$ reduces to the Wald-Wolfowitz result.

Figure 5.1 shows the full minimum spanning tree (MST) connecting all points from two groups. Blue triangles: $\mathbf{X} \sim \mathbb{N}(\mathbf{0}, \mathbf{I})$, 10 points. Red squares: $\mathbf{Y} \sim \mathbb{N}(\mathbf{31}, 0.5\ \mathbf{I})$, 15 points. The MST includes a cross-group edge. Figure 5.1 shows the MST after removing cross-group edges. Each connected component represents a run. Here, the MST splits into two components, so number of runs = 2. This illustrates how the Friedman–Rafsky test detects separation between distributions.

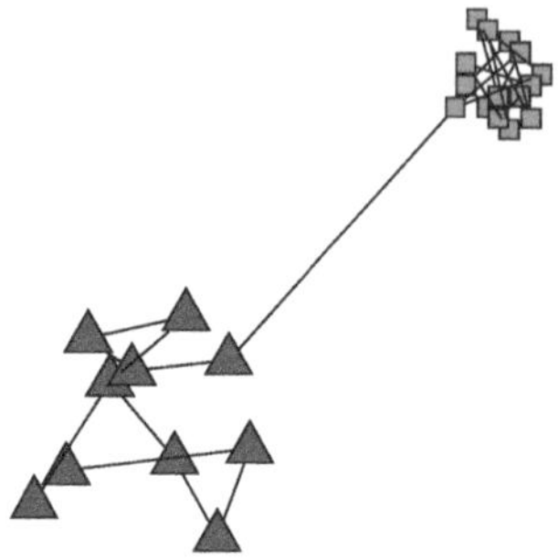

FIGURE 5.1: Full minimum spanning tree (MST) connecting all points from two groups. Blue triangles: $\mathbf{X} \sim \mathbb{N}(\mathbf{0}, \mathbf{I})$, 10 points. Red squares: $\mathbf{Y} \sim \mathbb{N}(\mathbf{31}, 0.5\ \mathbf{I})$, 15 points. The MST includes a cross-group edge.

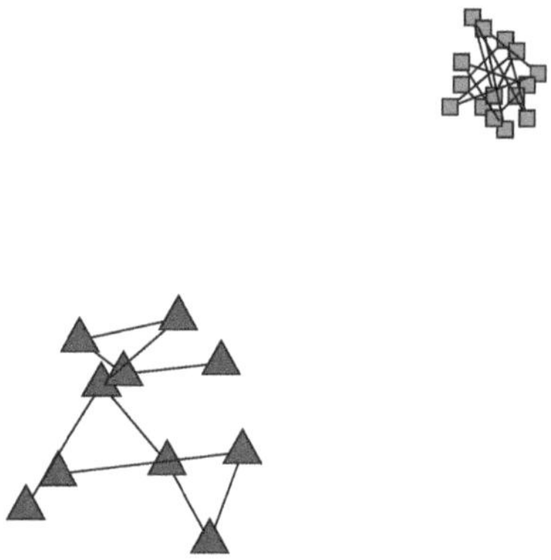

FIGURE 5.2: MST after removing the cross-group edge. Each connected component represents a run. Here, the MST splits into two components, so number of runs is 2. This illustrates how the Friedman–Rafsky test detects separation between distributions.

Example 5.3.1 *Suppose the minimum spanning tree of the combined sample of X and Y is a complete binary tree with 4 levels as follows.*

A run is a connected component consisting entirely of nodes with the same label after removing all X–Y edges. We first remove all edges that connect an X node to a Y node.

- *Root Y disconnects from both X children*
- *Every X node disconnects from its Y children*
- *Every Y node disconnects from its X children*

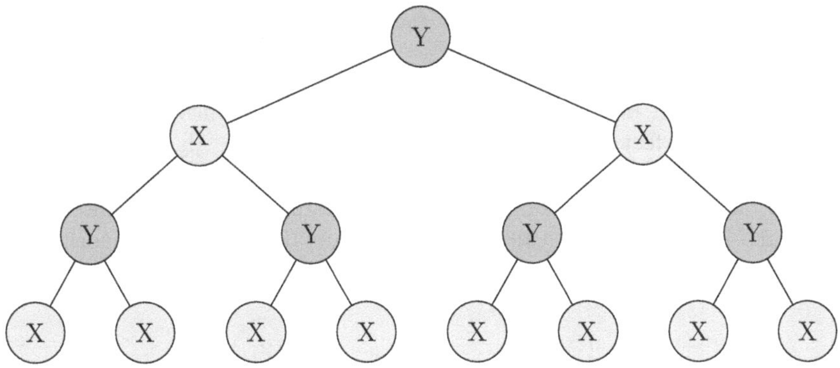

FIGURE 5.3: Minimum spanning tree as a complete binary tree with 4 levels.

Next, we identify Connected Components

- *Root* Y *is isolated* $\Rightarrow$ *1 run*
- *Each level-2* X *node is isolated* $\Rightarrow$ *2 runs*
- *Each level-3* Y *node is isolated* $\Rightarrow$ *4 runs*
- *Each level-4* X *node is isolated* $\Rightarrow$ *8 runs*

The total number of runs is $R = 1 + 2 + 4 + 8 = 15$. *Since all nodes alternate labels across levels, removing cross-sample edges completely disconnects the MST. This produces the maximum possible number of runs for this tree structure, consistent with strong mixing between samples. Under* H_0*, such behavior is typical; under* H_a*, fewer runs would be expected.*

5.4 Cross-Match Test of Rosenbaum

Rosenbaum's [284] cross-match test assesses whether two multivariate samples come from the same distribution by pairing nearest observations and counting how many of those pairs cross group labels. Under identical distributions, group labels have no bearing on distances, so roughly half the pairs should have mixed labels. If the distributions differ, within-group points tend to be closer, causing fewer cross-group pairs.

The test statistic A_1 is the number of cross-pairs in an optimal matching that minimizes the total distance between paired points. Let $N = m+n$ be the total number of observations.

This matching problem can be naturally represented as a bipartite graph. A bipartite graph $G = (U \cup W, E)$ is a graph whose vertices can be divided into two disjoint sets U and W such that every edge connects a vertex in U to a vertex in W:

$$V = U \cup W, \quad U \cap W = \emptyset, \quad E \subseteq U \times W.$$

In the cross-match test:

- U (blue) represents the observations from the first sample.
- W (red) represents the observations from the second sample.

- Each edge $(u, w) \in E$ is weighted by the distance between observation $u \in U$ and $w \in W$.

Finding the optimal matching that minimizes total distance is equivalent to finding a minimum-weight perfect matching in the bipartite graph. Algorithms for this include the Hungarian algorithm or general network flow methods. After computing the optimal matching, the number of cross-group pairs among matched edges is counted to obtain the test statistic A_1. Figure 5.4 shows an example of optimal pairing. An Algorithm for finding the value of A_1 follows.

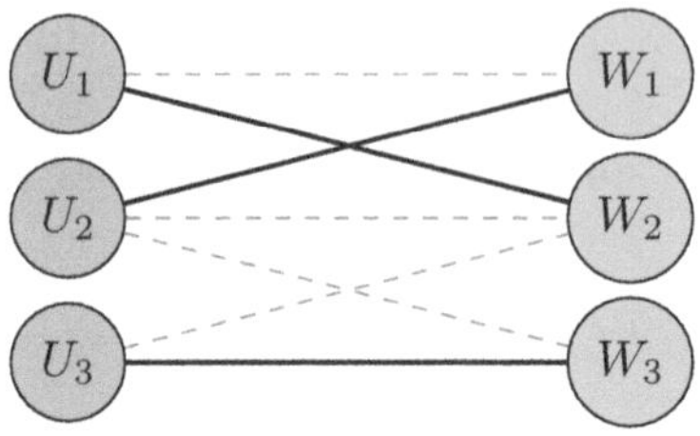

FIGURE 5.4: Blue nodes represent sample U, red nodes represent sample W. Dashed gray lines indicate all possible pairings. Thick black lines show the optimal matching that minimizes total distance between matched points, which is used to compute the Rosenbaum cross-match statistic A_1.

Algorithm 5.4.1 Cross-Match Test

Input: Two groups of observations in $\mathbb{R}^d$, total N observations
Output: Test statistic A_1 and decision to reject null hypothesis

Step 1: Compute distances
Pool all N observations and compute the complete graph with edge weights given by Euclidean distances δ_{ij}.

Step 2: Minimum-distance matching
Find a minimum-distance, non-bipartite matching into $I = \lfloor \frac{N}{2} \rfloor$ pairs.
Note: If N is odd, add a dummy "pseudo-subject," match $N + 1$ points, then discard the pair containing the pseudo-subject.

Step 3: Label pairs
Label pairs $1, \ldots, I$ and for $k = 0, 1, 2$, define

$$A_k = \#\{\text{pairs containing exactly } k \text{ observations from group 1}\}.$$

Step 4: Compute test statistic
The test statistic is A_1 = number of cross-match pairs (one from each group).
Step 5: Decision rule: Reject the null hypothesis for unusually small values of A_1.

In Step 3, because each pair has 0, 1, or 2 from group 1, $A_0 + A_1 + A_2 = I$, and because there are m observations in group 1, $2A_2 + A_1 = m$.

Example 5.4.1 *Compute Rosenbaum's cross-match statistic for testing whether two samples come from the same distribution. We work in* $\mathbb{R}^4$ *with* $m = 3$ *observations from Group 1 and* $n = 3$ *observations from Group 2. The six points are*

$$\text{Group 1 } (m=3): \mathbf{A} = (0,0,0,0), \quad \mathbf{B} = (1,0,0,0), \quad \mathbf{C} = (0,1,0,0),$$
$$\text{Group 2 } (n=3): \mathbf{X} = (5,5,5,5), \quad \mathbf{Y} = (5,5,5,6), \quad \mathbf{Z} = (5,5,6,5).$$

We order the pooled sample as $(\mathbf{A},\mathbf{B},\mathbf{C},\mathbf{X},\mathbf{Y},\mathbf{Z})$. *All pairwise Euclidean distances are square roots of integers. The distance matrix is*

$$D = (\delta_{ij}) = \begin{pmatrix} 0 & 1 & 1 & \sqrt{100} & \sqrt{111} & \sqrt{111} \\ 1 & 0 & \sqrt{2} & \sqrt{91} & \sqrt{102} & \sqrt{102} \\ 1 & \sqrt{2} & 0 & \sqrt{91} & \sqrt{102} & \sqrt{102} \\ \sqrt{100} & \sqrt{91} & \sqrt{91} & 0 & 1 & 1 \\ \sqrt{111} & \sqrt{102} & \sqrt{102} & 1 & 0 & \sqrt{2} \\ \sqrt{111} & \sqrt{102} & \sqrt{102} & 1 & \sqrt{2} & 0 \end{pmatrix}.$$

There are 15 *perfect matchings of six points. For each matching we compute the total cost (sum of the three pairwise distances) and* A_1, *the number of cross-pairs (pairs containing exactly one Group 1 observation).*

Matching	*Total cost*	(A_1)
$(\mathbf{A},\mathbf{B}),(\mathbf{C},\mathbf{X}),(\mathbf{Y},\mathbf{Z})$	$1+\sqrt{91}+\sqrt{2}$	(1)
$(\mathbf{A},\mathbf{C}),(\mathbf{B},\mathbf{X}),(\mathbf{Y},\mathbf{Z})$	$1+\sqrt{91}+\sqrt{2}$	(1)
$(\mathbf{A},\mathbf{B}),(\mathbf{C},\mathbf{Y}),(\mathbf{X},\mathbf{Z})$	$1+\sqrt{102}+1$	(1)
$(\mathbf{A},\mathbf{B}),(\mathbf{C},\mathbf{Z}),(\mathbf{X},\mathbf{Y})$	$1+\sqrt{102}+1$	(1)
$(\mathbf{A},\mathbf{C}),(\mathbf{B},\mathbf{Y}),(\mathbf{X},\mathbf{Z})$	$1+\sqrt{102}+1$	(1)
$(\mathbf{A},\mathbf{C}),(\mathbf{B},\mathbf{Z}),(\mathbf{X},\mathbf{Y})$	$1+\sqrt{102}+1$	(1)
$(\mathbf{A},\mathbf{X}),(\mathbf{B},\mathbf{C}),(\mathbf{Y},\mathbf{Z})$	$\sqrt{100}+\sqrt{2}+\sqrt{2}$	(1)
$(\mathbf{A},\mathbf{Y}),(\mathbf{B},\mathbf{C}),(\mathbf{X},\mathbf{Z})$	$\sqrt{111}+\sqrt{2}+1$	(1)
$(\mathbf{A},\mathbf{Z}),(\mathbf{B},\mathbf{C}),(\mathbf{X},\mathbf{Y})$	$\sqrt{111}+\sqrt{2}+1$	(1)
$(\mathbf{A},\mathbf{Y}),(\mathbf{B},\mathbf{Z}),(\mathbf{C},\mathbf{X})$	$\sqrt{111}+\sqrt{102}+\sqrt{91}$	(3)
$(\mathbf{A},\mathbf{Z}),(\mathbf{B},\mathbf{Y}),(\mathbf{C},\mathbf{X})$	$\sqrt{111}+\sqrt{102}+\sqrt{91}$	(3)
$(\mathbf{A},\mathbf{Y}),(\mathbf{B},\mathbf{X}),(\mathbf{C},\mathbf{Z})$	$\sqrt{111}+\sqrt{91}+\sqrt{102}$	(3)
$(\mathbf{A},\mathbf{Z}),(\mathbf{B},\mathbf{X}),(\mathbf{C},\mathbf{Y})$	$\sqrt{111}+\sqrt{91}+\sqrt{102}$	(3)
$(\mathbf{A},\mathbf{X}),(\mathbf{B},\mathbf{Y}),(\mathbf{C},\mathbf{Z})$	$\sqrt{100}+\sqrt{102}+\sqrt{102}$	(3)
$(\mathbf{A},\mathbf{X}),(\mathbf{B},\mathbf{Z}),(\mathbf{C},\mathbf{Y})$	$\sqrt{100}+\sqrt{102}+\sqrt{102}$	(3)

The minimum total cost is $1+\sqrt{91}+\sqrt{2}$, *achieved by the two matchings*

$$(\mathbf{A},\mathbf{B}),(\mathbf{C},\mathbf{X}),(\mathbf{Y},\mathbf{Z}) \quad \text{and} \quad (\mathbf{A},\mathbf{C}),(\mathbf{B},\mathbf{X}),(\mathbf{Y},\mathbf{Z}).$$

Each has $A_1 = 1$. *We select the first for concreteness. For the matching* $(\mathbf{A},\mathbf{B}),(\mathbf{C},\mathbf{X}),(\mathbf{Y},\mathbf{Z})$,

$$(\mathbf{A},\mathbf{B}): \text{within Group 1}, \qquad (\mathbf{C},\mathbf{X}): \text{cross-pair}, \qquad (\mathbf{Y},\mathbf{Z}): \text{within Group 2}.$$

Thus the observed statistic is $A_1^{obs} = 1$. *Under* H_0, *the matching is treated as fixed, and the* $m = 3$ *Group 1 labels are equally likely to occupy any* 3 *of the* 6 *positions. There are* $\binom{6}{3} = 20$ *equally likely labelings. The possible compositions across the three pairs are*

- $(1,1,1)$: *each pair is a cross-pair. Number of labelings:* $2^3 = 8$, *yielding* $A_1 = 3$.
- *permutations of* $(2,1,0)$: *There are* $3! = 6$ *assignments of counts, each with* $\binom{2}{2}\binom{2}{1}\binom{2}{0} = 2$ *labelings, for a total of* 12, *each yielding* $A_1 = 1$.

Thus,

$$\mathbb{P}(A_1 = 1) = \frac{12}{20} = \frac{3}{5}, \qquad \mathbb{P}(A_1 = 3) = \frac{8}{20} = \frac{2}{5}.$$

Rosenbaum's test rejects for unusually small values of A_1. *The one-sided p-value is* $p = \mathbb{P}(A_1 \leq 1) = \frac{3}{5}$. *Since* $p > 0.05$, *we fail to reject* H_0.

5.5 Test of Hall and Tajvidi

Hall and Tajvidi [134] proposed the following ingenious statistics for testing the equality of two distributions. Suppose we obtain independent random samples $\mathbb{X} = \{\mathbf{X}_1, \ldots, \mathbf{X}_m\}$ and $\mathbb{Y} = \{\mathbf{Y}_1, \ldots, \mathbf{Y}_n\}$ from distributions F_1 and F_2, respectively. We are interested in testing $H_0 : F_1 = F_2$. Let $D(\mathbf{X}, \mathbf{Y})$ denote the Euclidean distance between $\mathbf{X}$ and $\mathbf{Y}$. Let $\mathbb{Z} = \{\mathbf{Z}_1, \ldots, \mathbf{Z}_{m+n}\}$ denote the pooled sample $\mathbb{X} \cup \mathbb{Y}$. For integers $1 \leq i \leq m$ $(1 \leq i \leq n)$ and $1 \leq j \leq m + n - 1$, define $M_i(j)$ $(N_i(j))$ to equal the number of $\mathbf{Y}_k$'s ($\mathbf{X}_k$'s) that are included among that subset of $\mathbb{Y}$ for which $D(\mathbf{X}_i, \mathbf{Z}_k)$ $(D(\mathbf{Y}_i, \mathbf{Z}_k))$ assumes no greater value than its jth smallest among all $m + n - 1$ values computed for $Z_k \in \mathbb{X} \backslash \mathbf{X}_i$. From the continuity assumption, and identical support for F_1 and F_2, it follows with probability 1, $M_i(j)$ and $N_i(j)$ are well and uniquely defined. Ties would be broken either at random or by adding fractions to $M_i(j)$ and $N_i(j)$.

Hall and Tajvidi [134] show that, conditional on $\mathbb{Z}$, $M_i(j)$ and $N_i(j)$ have Hypergeometric distributions:

$$\begin{aligned} \mathbb{P}_{H_0}(M_i(j) = k|\mathbb{Z}) &= \binom{m-1}{j-k}\binom{n}{k} \Big/ \binom{m+n-1}{j}, \\ \mathbb{P}_{H_0}(N_i(j) = k|\mathbb{Z}) &= \binom{n-1}{j-k}\binom{m}{k} \Big/ \binom{m+n-1}{j} \end{aligned} \tag{5.2}$$

for $0 \leq k \leq j$. Furthermore,

$$\mathbb{E}_{H_0}(M_i(j)|\mathbb{Z}) = \frac{nj}{m+n-1}, \qquad \mathbb{E}_{H_0}(N_i(j)|\mathbb{Z}) = \frac{mj}{m+n-1}. \tag{5.3}$$

$$\mathbb{V}ar_{H_0}(M_i(j)) = \frac{(m-1)n}{m+n-2}\omega_j(1-\omega_j), \qquad \mathbb{V}ar_{H_0}(N_i(j)) = \frac{(n-1)m}{m+n-2}\omega_j(1-\omega_j), \tag{5.4}$$

where $\omega_j = j/(m+n-1)$. Test statistics for testing H_0 are

$$DM_i(j) = |M_i(j) - \frac{nj}{m+n-1}|, \qquad DN_i(j) = |N_i(j) - \frac{mj}{m+n-1}|. \tag{5.5}$$

A test statistic for $H_0 : F_1 = F_2$ using DM or DN rejects H_0 for large values.

The Hall and Tajvidi test examines whether points from the two samples $\mathbb{X}$ and $\mathbb{Y}$ are mixed together in space.

- For each $\mathbf{X}_i$, we consider the distances $D(\mathbf{X}_i, \mathbf{Z}_k)$ to all other points in the pooled sample $\mathbb{Z} \setminus \{\mathbf{X}_i\}$, and identify the j points closest to $\mathbf{X}_i$ (i.e., whose distance is no greater than the jth smallest distance).

- Among these j nearest points, $M_i(j)$ counts how many belong to the other sample $\mathbb{Y}$. Similarly, for each $\mathbf{Y}_i$, $N_i(j)$ counts how many of the j nearest points belong to $\mathbb{X}$.
- Under H_0, the neighbors are randomly distributed between groups, so $M_i(j)$ and $N_i(j)$ behave like Hypergeometric counts.
- If points cluster within their own sample, $M_i(j)$ and $N_i(j)$ tend to be smaller than expected, signaling a difference between F_1 and F_2.

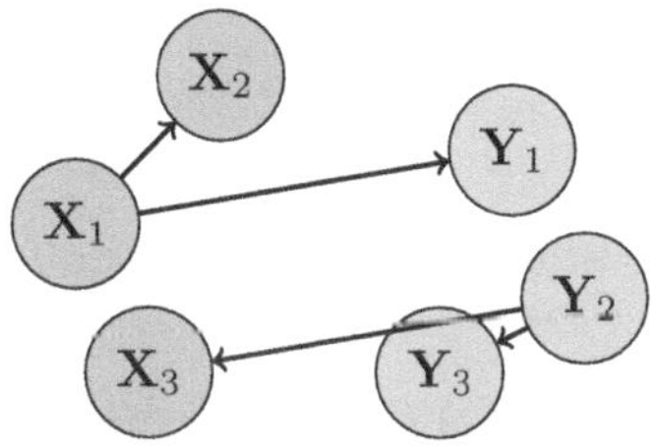

FIGURE 5.5: Blue nodes on the left represent sample $\mathbb{X}$ and red nodes on the right represent sample $\mathbb{Y}$. Arrows indicate nearest neighbors. For each point, we consider the j nearest neighbors (those whose distance is no greater than the jth smallest among all distances to other points) and count how many belong to the other sample. These counts define $M_i(j)$ for $\mathbf{X}_i$ and $N_i(j)$ for $\mathbf{Y}_i$. Under H_0, neighbors are randomly mixed; if points cluster within their own sample, these counts are smaller, signaling a difference between F_1 and F_2.

Example 5.5.1 *Hall and Tajvidi Test for $H_0 : F_1 = F_2$ with $m = n = 3$. For simplicity, we take 1-dimensional integers, $\mathbb{X} = \{1, 3, 5\}$, and $\mathbb{Y} = \{2, 4, 6\}$. The pooled sample is given by*

$$\mathbb{Z} = \mathbb{X} \cup \mathbb{Y} = \{1, 2, 3, 4, 5, 6\}.$$

For 1D Euclidean distance, we have $D(X_i, Z_k) = |X_i - Z_k|, \quad D(Y_i, Z_k) = |Y_i - Z_k|$.

Excluding the self-point, we obtain

$$D(X_1, Z.) = |1-2|, |1-3|, |1-4|, |1-5|, |1-6| = 1, 2, 3, 4, 5$$
$$D(X_2, Z.) = |3-1|, |3-2|, |3-4|, |3-5|, |3-6| = 2, 1, 1, 2, 3$$
$$D(X_3, Z.) = |5-1|, |5-2|, |5-3|, |5-4|, |5-6| = 4, 3, 2, 1, 1$$

Note that $M_i(j)$ = number of Y points among the j closest points to X_i (excluding X_i). For $X_1 = 1$, distances to others are $\{2, 3, 4, 5, 6\} \Rightarrow 1, 2, 3, 4, 5$

$$j = 1: \textit{closest} = \textit{2 (Y)} \Rightarrow M_1(1) = 1$$
$$j = 2: \textit{closest} = \textit{2,3} \Rightarrow Y's = \{2\} \Rightarrow M_1(2) = 1$$
$$j = 3: \textit{closest} = \textit{2,3,4} \Rightarrow Y's = \{2, 4\} \Rightarrow M_1(3) = 2$$
$$j = 4: \textit{closest} = \textit{2,3,4,5} \Rightarrow Y's = \{2, 4\} \Rightarrow M_1(4) = 2$$
$$j = 5: \textit{closest} = \textit{2,3,4,5,6} \Rightarrow Y's = \{2, 4, 6\} \Rightarrow M_1(5) = 3$$

For $X_2 = 3$, distances to others are $\{1, 2, 4, 5, 6\} \Rightarrow 2, 1, 1, 2, 3$
Sorted distances: 1 (Z=2, Y), 1 (Z=4, Y), 2 (Z=1,X), 2 (Z=5,X), 3 (Z=6, Y)

$$j = 1: \textit{closest} = \textit{2 (Y)} \Rightarrow M_2(1) = 1$$
$$j = 2: \textit{closest} = \textit{2,4 (Y, Y)} \Rightarrow M_2(2) = 2$$
$$j = 3: \textit{closest} = \textit{2,4,1} \Rightarrow Y's = \{2, 4\} \Rightarrow M_2(3) = 2$$
$$j = 4: \textit{closest} = \textit{2,4,1,5} \Rightarrow Y's = \{2, 4\} \Rightarrow M_2(4) = 2$$
$$j = 5: \textit{all points} \Rightarrow Y's = \{2, 4, 6\} \Rightarrow M_2(5) = 3$$

For $X_3 = 5$, *distances are* $\{1, 2, 3, 4, 6\} \Rightarrow 4, 3, 2, 1, 1$
Sorted distances: 1 (Z=4,Y), 1 (Z=6,Y), 2 (Z=3,X), 3 (Z=2,Y), 4 (Z=1,X).

$$j = 1 : M_3(1) = 1 \quad j = 2 : M_3(2) = 2$$
$$j = 3 : M_3(3) = 2 \quad j = 4 : M_3(4) = 3 \quad j = 5 : M_3(5) = 3$$

Table of $M_i(j)$,

i	$j=1$	$j=2$	$j=3$	$j=4$	$j=5$
1	1	1	2	2	3
2	1	2	2	2	3
3	1	2	2	3	3

Null distribution probabilities: Using the Hypergeometric formula,

$$\mathbb{P}_{H_0}(M_i(j) = k \mid \mathbb{Z}) = \frac{\binom{m-1}{j-k}\binom{n}{k}}{\binom{m+n-1}{j}}$$

For $m = n = 3$, $m + n - 1 = 5$, $j = 3$ *we have*

$$k = 0 : \quad \frac{\binom{2}{3-0}\binom{3}{0}}{\binom{5}{3}} = 0$$
$$k = 1 : \quad \frac{\binom{2}{2}\binom{3}{1}}{10} = 0.3$$
$$k = 2 : \quad \frac{\binom{2}{1}\binom{3}{2}}{10} = 0.6$$
$$k = 3 : \quad \frac{\binom{2}{0}\binom{3}{3}}{10} = 0.1$$

Expectation and Variance are

$$\omega_j = \frac{j}{m+n-1} = \frac{3}{5} = 0.6$$

$$\mathbb{E}_{H_0}[M_i(j)] = \frac{nj}{m+n-1} = \frac{3(3)}{5} = 1.8$$

$$\mathbb{V}ar_{H_0}[M_i(j)] = \frac{(m-1)n}{m+n-2}\omega_j(1-\omega_j) = \frac{2(3)}{4}(0.6)(0.4) = 0.36$$

The test statistics is $DM_i(j) = |M_i(j) - \mathbb{E}_{H_0}[M_i(j)]|$.
For $j = 3$ *we obtain*

$$DM_1(3) = |2 - 1.8| = 0.2, \quad DM_2(3) = |2 - 1.8| = 0.2, \quad DM_3(3) = |2 - 1.8| = 0.2.$$

For the p-value we obtain the probabilities under $H_0 : F_1 = F_2$ *as*

$$\mathbb{P}(M_i(3) = 1) = 0.3, \quad \mathbb{P}(M_i(3) = 2) = 0.6, \quad \mathbb{P}(M_i(3) = 3) = 0.1.$$

Observed deviation is $|M_i(3) - 1.8| = 0.2$ *and the two-sided p-value is*

$$p = \mathbb{P}(|M_i(3) - 1.8| \geq 0.2) = \mathbb{P}(M_i(3) = 1 \text{ or } 2 \text{ or } 3) = 1,$$

which indicates no evidence against H_0.

5.6 Test of Mondal-Biswas-Ghosh

The classical two-sample problem seeks to test $H_0 : F_1 = F_2$ vs. $H_a : F_1 \neq F_2$, where F_1 and F_2 are distributions in $\mathbb{R}^d$. Traditional tests often lose power in high dimensions due to the curse of dimensionality. Nearest-neighbor (NN) based tests are attractive because they avoid explicit density estimation and exploit local geometric structure in the data. The challenge is to adapt these tests to the high-dimensional regime where d may be comparable to or larger than sample sizes.

Mondal, Biswas, and Ghosh [241] propose a class of two-sample tests based on k-nearest neighbor (kNN) graphs and study their asymptotic properties in high-dimensional settings. Let $\{\mathbf{X}_1, \ldots, \mathbf{X}_m\} \sim F_1$ and $\{\mathbf{Y}_1, \ldots, \mathbf{Y}_n\} \sim F_2$ be two independent samples in $\mathbb{R}^d$, and let $\mathbf{Z}_1, \ldots, \mathbf{Z}_{m+n}$ denote the pooled sample, with labels

$$L(\mathbf{Z}_i) = \begin{cases} 0, & \mathbf{Z}_i \in \{\mathbf{X}_1, \ldots, \mathbf{X}_m\}, \\ 1, & \mathbf{Z}_i \in \{\mathbf{Y}_1, \ldots, \mathbf{Y}_n\}. \end{cases}$$

For each observation $\mathbf{Z}_i$, let $\mathbb{N}_k(i)$ denote the indices of its k nearest neighbors in the pooled sample (excluding $\mathbf{Z}_i$ itself), based on Euclidean distance. The directed kNN test statistic is defined as

$$T_{m,n}^{(k)} = \sum_{i=1}^{m+n} \sum_{j \in \mathbb{N}_k(i)} \mathbb{I}\{L(\mathbf{Z}_i) \neq L(\mathbf{Z}_j)\}.$$

Thus, $T_{m,n}^{(k)}$ counts the number of directed kNN edges that connect points from different samples. Under the null hypothesis $H_0 : F_1 = F_2$, the labels are exchangeable, and the null distribution of $T_{m,n}^{(k)}$ can be approximated either by permutation or, after appropriate centering and scaling, by asymptotic normality as $m, n \to \infty$ and $d \to \infty$ with k fixed.

Under the alternative hypothesis $H_a : F_1 \neq F_2$, observations tend to have nearest neighbors from the same distribution more frequently, leading to fewer cross-sample edges. Consequently, $T_{m,n}^{(k)}$ tends to be smaller than its null expectation. Mondal, Biswas, and Ghosh [241] show that the test is consistent, in the sense that

$$\mathbb{P}_{H_a}(\text{reject } H_0) \to 1 \quad \text{as } m, n \to \infty,$$

under fixed alternatives and certain local alternatives.

Example 5.6.1 *We illustrate the computation of $T_{m,n}^{(k)}$ using a toy dataset. Let $m = n = 3$ and $d = 5$, and define*

$$\mathbf{X}_1 = (0, 0, 0, 0, 0), \quad \mathbf{X}_2 = (1, 0, 0, 0, 0), \quad \mathbf{X}_3 = (0, 1, 0, 0, 0),$$
$$\mathbf{Y}_1 = (5, 0, 0, 0, 0), \quad \mathbf{Y}_2 = (5, 1, 0, 0, 0), \quad \mathbf{Y}_3 = (5, 0, 1, 0, 0).$$

The X observations form one cluster near the origin, while the Y observations form a separate cluster centered around $(5, 0, 0, 0, 0)$. For instance,

$$\|\mathbf{X}_1 - \mathbf{X}_2\|^2 = 1, \quad \|\mathbf{X}_1 - \mathbf{X}_3\|^2 = 1, \quad \|\mathbf{X}_1 - \mathbf{Y}_1\|^2 = 25,$$

and similar calculations show that all X points are much closer to each other than to any Y point, and vice versa. For $k = 1$, each observation has its nearest neighbor within the same group. Therefore, no cross-sample kNN edges occur, and $T_{3,3}^{(1)} = 0$. For $k = 2$, each observation again selects its two nearest neighbors from within the same group, so $T_{3,3}^{(2)} = 0$. In this example, the two samples are perfectly separated in $\mathbb{R}^5$, leading to a very small value of the test statistic, which provides strong evidence against H_0. In practice, the null distribution of $T_{m,n}^{(k)}$ is obtained via permutation.

5.7 Energy Statistic

The energy Statistic is a non-parametric test for the equality of two multivariate distributions, introduced by Székely and Rizzo [21] and further developed by Baringhaus and Franz [21]. It is based on the concept of energy distance, which measures the disparity between two distributions using IPDs. The energy distance between two distributions F_1 and F_2 of random vectors $\boldsymbol{X} \sim F_1$, $\boldsymbol{Y} \sim F_2$ in $\mathbb{R}^d$ is defined as

$$\mathcal{E}(F_1, F_2) = 2\mathbb{E}\|\boldsymbol{X} - \boldsymbol{Y}\| - \mathbb{E}\|\boldsymbol{X} - \boldsymbol{X}'\| - \mathbb{E}\|\boldsymbol{Y} - \boldsymbol{Y}'\|,$$

where $\boldsymbol{X}' \sim F_1$, $\boldsymbol{Y}' \sim F_2$ are i.i.d. copies.

The energy test measures whether two multivariate distributions F_1 and F_2 differ by comparing distances within and between samples. Figure 5.6 illustrates the energy statistic.

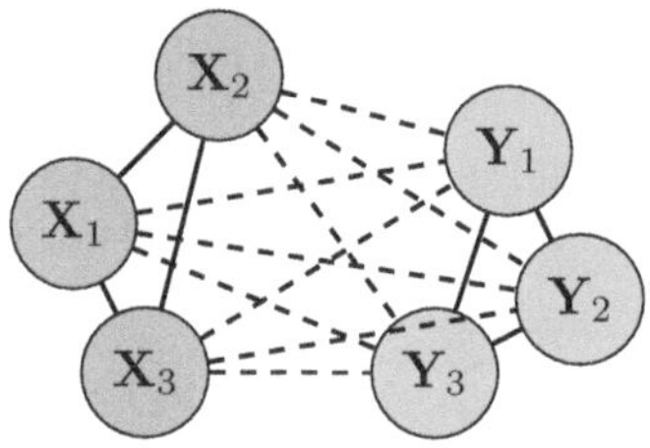

FIGURE 5.6: Illustration of the energy statistic. Blue nodes represent sample $\mathbb{X}$, red nodes represent sample $\mathbb{Y}$. Solid lines show distances within the same sample, and dashed lines show distances between samples. If the two distributions are the same, solid and dashed lines have roughly similar lengths. If the distributions differ, dashed lines are generally longer than solid lines, increasing the energy distance.

One can show that $\mathcal{E}(F_1, F_2) \geq 0$ with equality holding under H_0. Furthermore, Székely and Rizzo [333] and Baringhaus and Franz [21] prove that

$$\mathcal{E}(F_1, F_2) = 0 \quad \Longleftrightarrow \quad F_1 = F_2.$$

Let $\mathbf{X}_1, \ldots, \mathbf{X}_m$ and $\mathbf{Y}_1, \ldots, \mathbf{Y}_n$ be independent random samples from distributions F_1 and F_2, respectively, where each $\mathbf{X}_i, \mathbf{Y}_j \in \mathbb{R}^d$. The energy Statistic is defined as

$$EN = 2\bar{d}_{(xy)} - \bar{d}_{(x)} - \bar{d}_{(y)}.$$

Under the null hypothesis $H_0 : F_1 = F_2$, the statistic EN is expected to be small. A large value of EN suggests a departure from the null hypothesis, indicating that the distributions differ. It is noteworthy that EN uses the Euclidean distance to measure class separation. The energy Statistic provides a powerful, non-parametric method for testing the equality of two multivariate distributions. Its reliance on IPDs makes it particularly useful in high-dimensional settings where traditional parametric methods may be inadequate.

5.8 Biswas-Ghosh Test

Let $\mathbf{X}_1, \ldots, \mathbf{X}_m \overset{\text{i.i.d.}}{\sim} F_1$ and $\mathbf{Y}_1, \ldots, \mathbf{Y}_n \overset{\text{i.i.d.}}{\sim} F_2$, where $\mathbf{X}_i, \mathbf{Y}_j \in \mathbb{R}^d$ (possibly with d large relative to m, n). We wish to test $H_0 : F_1 = F_2 \quad$ vs. $\quad H_a : F_1 \neq F_2$, without parametric assumptions and in a way that remains applicable when d is large. Suppose $\mathbf{X}_1, \mathbf{X}_2$ are i.i.d. F_1 and $\mathbf{Y}_1, \mathbf{Y}_2$ are i.i.d. F_2. Assume $\mu_{F_1F_1} = \mathbb{E}(\|\mathbf{X}_1 - \mathbf{X}_2\|)$, $\mu_{F_2F_2} = \mathbb{E}(\|\mathbf{Y}_1 - \mathbf{Y}_2\|)$, and $\mu_{F_1F_2} = \mathbb{E}(\|\mathbf{X}_1 - \mathbf{Y}_1\|)$ exist. Biswas and Ghosh [35] show that

$$\mu_{F_1F_1} = \mu_{F_2F_2} = \mu_{F_1F_2} \iff F_1 = F_2.$$

Biswas and Ghosh [35] noticed that, when both $(\hat{\mu}_{F_1F_2} - \hat{\mu}_{F_1F_1})$ and $(\hat{\mu}_{F_1F_2} - \hat{\mu}_{F_2F_2})$ are significantly different from zero, they are likely to be of different signs. As a result, when they are added together, the energy statistic $EN = (\hat{\mu}_{F_1F_2} - \hat{\mu}_{F_1F_1}) + (\hat{\mu}_{F_1F_2} - \hat{\mu}_{F_2F_2})$ may take a value close to zero, and consequently, we may fail to reject H_0. They propose a test built from averages of interpoint Euclidean distances. Their Euclidean-distance statistic is

$$BG \;=\; \left(\bar{d}_{(xy)} - \bar{d}_{(x)}\right)^2 \;+\; \left(\bar{d}_{(xy)} - \bar{d}_{(y)}\right)^2. \tag{5.6}$$

Under H_0, all three expected pairwise distances coincide,

$$\mathbb{E}\,\|X - X'\| = \mathbb{E}\,\|Y - Y'\| = \mathbb{E}\,\|X - Y\|,$$

so the differences inside (5.6) concentrate near 0; under many alternatives, $\bar{d}_{(xy)}$ drifts away from the within-sample averages and (5.6) grows. One can show that $BG \geq 0$ with equality holding under H_0.

Example 5.8.1 *We illustrate the Biswas-Ghosh test with $m = n = 3$ in $\mathbb{R}^2$. Let*

$$\mathbf{X}_1 = (0,0),\ \mathbf{X}_2 = (1,0),\ \mathbf{X}_3 = (0,1), \qquad \mathbf{Y}_1 = (2,0),\ \mathbf{Y}_2 = (2,1),\ \mathbf{Y}_3 = (3,0).$$

We obtain

$$\textit{Within } X: \ \|\mathbf{X}_1 - \mathbf{X}_2\| = 1,\ \|\mathbf{X}_1 - \mathbf{X}_3\| = 1,\ \|\mathbf{X}_2 - \mathbf{X}_3\| = \sqrt{2},$$
$$\bar{d}_{(x)} = \tbinom{3}{2}^{-1}(1 + 1 + \sqrt{2}) = \frac{1}{3}(2 + \sqrt{2}) \approx 1.1381.$$
$$\textit{Within } Y: \ \|\mathbf{Y}_1 - \mathbf{Y}_2\| = 1,\ \|\mathbf{Y}_1 - \mathbf{Y}_3\| = 1,\ \|\mathbf{Y}_2 - \mathbf{Y}_3\| = \sqrt{2},$$
$$\bar{d}_{(y)} = \tbinom{3}{2}^{-1}(1 + 1 + \sqrt{2}) = \frac{1}{3}(2 + \sqrt{2}) \approx 1.1381.$$

We compute the 3×3 table $\|\mathbf{X}_i - \mathbf{Y}_j\|$,

	$\mathbf{Y}_1$	$\mathbf{Y}_2$	$\mathbf{Y}_3$
$\mathbf{X}_1$	2	$\sqrt{5}$	3
$\mathbf{X}_2$	1	$\sqrt{2}$	2
$\mathbf{X}_3$	$\sqrt{5}$	2	$\sqrt{10}$

Sum $= 2 + \sqrt{5} + 3 + 1 + \sqrt{2} + 2 + \sqrt{5} + 2 + \sqrt{10} \approx 19.0487$. *Hence*

$$\bar{d}_{(xy)} = (mn)^{-1} \sum_{i,j} \|\mathbf{X}_i - \mathbf{Y}_j\| = \frac{19.0487}{9} \approx 2.1165.$$

$$BG = \left(2.1165 - 1.1381\right)^2 + \left(2.1165 - 1.1381\right)^2 = 2 \times (0.9784)^2 \approx 1.916.$$

We would now obtain a permutation p-value by repeatedly shuffling the six labels, recomputing BG each time, and comparing to 1.916.

5.9 Matrix Valued Distributions

Matrix-valued distributions, such as the matrix normal, Wishart, and inverse Wishart, are widely used in statistics, machine learning, and applied sciences because they naturally model data with an inherent matrix structure (e.g., images, multivariate time series, or network data). In engineering and communications, for example, MIMO systems often model channel matrices using matrix normal distributions to capture correlations across antennas and time, while beamforming and array signal processing frequently employ Wishart or inverse Wishart distributions to analyze covariance of signal matrices. In finance, matrix-valued distributions are used for covariance modeling of asset returns over multiple time periods, and matrix normal priors are incorporated in Bayesian factor models to estimate dynamic relationships among variables.

Beyond these areas, matrix-valued distributions are increasingly applied in machine learning, bioinformatics, and physics due to their ability to capture complex dependencies across rows and columns. In machine learning, observed matrices such as ratings or images can be modeled as noisy draws from a matrix normal distribution, while spatio-temporal observations exploit Kronecker-structured covariance models to capture dependencies across both dimensions. Weight matrices in Bayesian neural networks are sometimes assigned matrix normal priors to reduce parameters. In bioinformatics, gene expression data and neuroimaging (fMRI, EEG) form natural matrices, where matrix normal distributions capture correlations across genes, conditions, or brain regions. In physics and engineering, random matrix theory uses matrix-valued distributions to describe energy levels or eigenvalue distributions of physical systems, and covariance matrices in state-space models are often modeled as Wishart or matrix normal, connecting statistical modeling with systems theory.

Suppose $\{\mathbf{X}_1, \cdots, \mathbf{X}_m\}$ are i.i.d. $d \times m$ random matrices drawn from distribution function F_1, and $\{\mathbf{Y}_1, \cdots, \mathbf{Y}_n\}$ are i.i.d. $d \times m$ random matrices drawn from distribution function F_2. We plan to test the null hypothesis $H_0 : F_1 = F_2$ versus the general alternative $H_a : F_1 \neq F_2$. Using the Frobenius norm, we define $\mu_{F_1F_1}, \mu_{F_2F_2}$ and $\mu_{F_1F_2}$ and estimate them with $\hat{\mu}_{F_1F_1}, \hat{\mu}_{F_2F_2}$ and $\hat{\mu}_{F_1F_2}$, respectively where, $\mu_{F_1F_1} = E\|\mathbf{X}_1 - \mathbf{X}_2\|_F$, $\mu_{F_2F_2} = E\|\mathbf{Y}_1 - \mathbf{Y}_2\|_F$, $\mu_{F_1F_2} = E\|\mathbf{X}_1 - \mathbf{Y}_1\|_F$,

$$\hat{\mu}_{F_1F_2} = \binom{m}{2}^{-1} \sum_{i=1}^{m-1} \sum_{j=1}^{m} \|\mathbf{X}_i - \mathbf{X}_j\|_F,$$

$$[4pt]\hat{\mu}_{F_2F_2} = \binom{n}{2}^{-1} \sum_{i=1}^{n-1} \sum_{j=1}^{n} \|\mathbf{Y}_i - \mathbf{Y}_j\|_F,$$

$$\hat{\mu}_{F_1F_2} = (mn)^{-1} \sum_{i=1}^{m} \sum_{j=1}^{n} \|\mathbf{X}_i - \mathbf{Y}_j\|_F.$$

Theorem 5.9.1 extends the equivalence of Maa, Pearl, and Bartoszyński [206] to matrix distributions. It follows that two matrix distributions are equal if and only if the mean Frobenius norm of the within and between distances is equal.

Theorem 5.9.1 *Let* $\mathbf{X}_1, \mathbf{X}_2, \mathbf{Y}_1, \mathbf{Y}_2$ *be independent* $d \times m$ *random matrices. Suppose* $\mathbf{X}_1, \mathbf{X}_2$ *are identically distributed with the matrix distribution* F_1*, and* $\mathbf{Y}_1, \mathbf{Y}_2$ *are identically distributed with the matrix distribution* F_2*. If the expectations* $E\|\mathbf{X}_1\|_F$ *and* $E\|\mathbf{Y}_1\|_F$ *are finite, we have* $\mu_{F_1F_1} = \mu_{F_2F_2} = \mu_{F_1F_2}$ *if and only if* $F_1 = F_2$.

Based on Theorem 5.9.1, Guo and Modarres [127] propose BG for testing equality of matrix distributions, where

$$BG = (\hat{\mu}_{F_1F_1} - \hat{\mu}_{F_1F_2})^2 + (\hat{\mu}_{F_2F_2} - \hat{\mu}_{F_1F_2})^2. \tag{5.7}$$

The following results extend the energy statistic to matrix setting.

Theorem 5.9.2 *Let $\mathbf{X_1}, \mathbf{X_2}, \mathbf{Y_1}, \mathbf{Y_2}$ be independent $d \times m$ random matrices. Suppose $\mathbf{X}_1, \mathbf{X}_2$ are i.i.d. from F_1, and $\mathbf{Y}_1, \mathbf{Y}_2$ are i.i.d. from F_2. If the expectations $\mathbb{E}\|\mathbf{X}_1\|_F$ and $\mathbb{E}\|\mathbf{Y}_1\|_F$ are finite, we have*

$$2\mathbb{E}\|\mathbf{X}_1 - \mathbf{Y}_1\|_F - \mathbb{E}\|\mathbf{X}_1 - \mathbf{Y}_2\|_F - \mathbb{E}\|\mathbf{X}_1 - \mathbf{Y}_2\|_F \geq 0,$$

with equality holding if and only if $F_1 = F_2$.

Guo and Modarres [127] extend the energy statistic for testing the equality of two matrix variate distributions.

$$EN = 2\hat{\mu}_{F_1F_2} - \hat{\mu}_{F_1F_1} - \hat{\mu}_{F_2F_2}, \tag{5.8}$$

For the matrix normal distribution Guo and Modarres [127] prove the following.

Lemma 5.9.1 *Suppose $\mathbf{X}_1, \mathbf{X}_2$ are i.i.d. random matrices from $\mathbb{MN}_{d,m}(M_1, \mathbf{\Sigma} \otimes \mathbf{\Psi})$, and $\mathbf{Y}_1, \mathbf{Y}_2$ are i.i.d. random matrices from $\mathbb{MN}_{d,m}(M_2, \Sigma \otimes \mathbf{\Psi})$. It follows that $2E\|\mathbf{X}_1 - \mathbf{Y}_1\|_F^2 - E\|\mathbf{X}_1 - \mathbf{X}_2\|_F^2 - E\|\mathbf{Y}_1 - \mathbf{Y}_2\|_F^2 \geq 0$, and the equality holds if and only if $M_1 = M_2$.*

Using the Frobenius distance as the IPD between two random matrices, Song and Modarres [319] test for a mixture of Wishart matrices,

$$H_0 : \mathbf{X} \sim W(m,\ V_1) \quad \text{against} \quad H_a : \mathbf{X} \sim \alpha W(m,\ V_1) + (1-\alpha) W(n,\ V_2). \tag{5.9}$$

They observe that the IPD-test for homogeneity of Wishart matrices maintains its nominal level and has good power. The test statistic shows sensitivity to the changing proportion of mixtures. Moreover, the testing results based on L_1 distance perform the best while the results based on L_2 distance are competitive.

5.10 Test of Li

Following Li [192] we consider the asymptotic normality of IPDs in the HDLSS settings and its application to two-sample testing. Let $\mathbf{X}_1, \ldots, \mathbf{X}_n \sim F_1$ and $\mathbf{Y}_1, \ldots, \mathbf{Y}_m \sim F_2$, be independent random samples in $\mathbb{R}^d$. For any $\mathbf{U}, \mathbf{V} \in \mathbb{R}^d$, define the squared Euclidean distance

$$\Delta(\mathbf{U}, \mathbf{V}) = \|\mathbf{U} - \mathbf{V}\|^2 = \sum_{k=1}^{d} (U_k - V_k)^2.$$

We consider testing $H_0 : F_1 = F_2$ versus $H_a : F_1 \neq F_2$, under the HDLSS asymptotic regime $d \to \infty$ with n, m fixed. We assume that for each coordinate k,

- $\mathbb{E}(X_{ik}) = \mu_{X,k}$ and $\mathbb{E}(Y_{jk}) = \mu_{Y,k}$;
- $\mathbb{V}ar(X_{ik})$ and $\mathbb{V}ar(Y_{jk})$ are finite;
- Uniformly bounded fourth moments, $\sup_k \mathbb{E}(X_{ik}^4) < \infty$, and $\sup_k \mathbb{E}(Y_{jk}^4) < \infty$.

For a fixed observation $\mathbf{X}_i$, let $\mathcal{F}_a^b(\mathbf{X}_i) = \sigma(X_{ik} : a \le k \le b)$ denote the σ-field generated by coordinates a through b. Define the strong mixing coefficients

$$\alpha(h) = \sup_k \sup_{A \in \mathcal{F}_1^k,\, B \in \mathcal{F}_{k+h}^\infty} |\mathbb{P}(A \cap B) - \mathbb{P}(A)\mathbb{P}(B)|.$$

Analogous definitions are assumed for $\mathbf{Y}_j$. Li [192] assumes that $\sum_{h=1}^\infty \alpha(h)^{\delta/(2+\delta)} < \infty$ for some $\delta > 0$. This summability condition ensures that long-range dependence between coordinates is asymptotically negligible.

The mixing condition ensures the application of the Central limit theorem for sums of weakly dependent quadratic coordinate contributions and the joint asymptotic normality for multiple such sums, even when they share observations. This provides the foundation for normal limits for the entire collection of IPDs.

Define the distance collections,

$$\mathcal{D}_{XX} = \{\Delta(\mathbf{X}_i, \mathbf{X}_j) : 1 \le i < j \le n\}, \quad \mathcal{D}_{YY} = \{\Delta(\mathbf{Y}_i, \mathbf{Y}_j) : 1 \le i < j \le m\},$$

and $\mathcal{D}_{XY} = \{\Delta(\mathbf{X}_i, \mathbf{Y}_j) : 1 \le i \le n, 1 \le j \le m\}$. Let

$$\mathbf{D}_d = (\mathcal{D}_{XX}, \mathcal{D}_{YY}, \mathcal{D}_{XY}) \in \mathbb{R}^K, \quad K = \binom{n}{2} + \binom{m}{2} + nm.$$

Each component of $\mathbf{D}_d$ can be expressed as a sum of d weakly dependent coordinate-wise terms. Under the stated moment and mixing conditions, Li [192] proves that there exist centering vectors $\boldsymbol{\mu}_d$ and scaling matrices $\boldsymbol{\Sigma}_d$ such that

$$\boldsymbol{\Sigma}_d^{-1/2}(\mathbf{D}_d - \boldsymbol{\mu}_d) \xrightarrow{d} N_K(0, \mathbf{I}_K), \quad d \to \infty.$$

This joint central limit theorem for the entire collection of within- and between-sample distances is the foundation of all subsequent test statistics. Let

$$\bar{\Delta}_{XX} = \frac{1}{\binom{n}{2}} \sum_{i<j} \Delta(\mathbf{X}_i, \mathbf{X}_j), \quad \bar{\Delta}_{YY} = \frac{1}{\binom{m}{2}} \sum_{i<j} \Delta(\mathbf{Y}_i, \mathbf{Y}_j),$$

and $\bar{\Delta}_{XY} = \frac{1}{nm} \sum_{i=1}^n \sum_{j=1}^m \Delta(\mathbf{X}_i, \mathbf{Y}_j)$. The location statistic is defined as

$$T_{\text{loc}} = \bar{\Delta}_{XY} - \frac{1}{2}(\bar{\Delta}_{XX} + \bar{\Delta}_{YY}).$$

Under H_0, $\mathbb{E}(T_{\text{loc}}) = 0$; under a location alternative, $\mathbb{E}(T_{\text{loc}})$ increases proportionally to $\|\mu_X - \mu_Y\|^2$.

Define the empirical variances of the distance collections,

$$\widehat{\mathbb{V}ar}(\mathcal{D}_{XX}) = \frac{1}{\binom{n}{2}} \sum_{i<j} (\Delta(\mathbf{X}_i, \mathbf{X}_j) - \bar{\Delta}_{XX})^2,$$

$$\widehat{\mathbb{V}ar}(\mathcal{D}_{YY}) = \frac{1}{\binom{m}{2}} \sum_{i<j} (\Delta(\mathbf{Y}_i, \mathbf{Y}_j) - \bar{\Delta}_{YY})^2,$$

$$\widehat{\mathbb{V}ar}(\mathcal{D}_{XY}) = \frac{1}{nm} \sum_{i=1}^n \sum_{j=1}^m (\Delta(\mathbf{X}_i, \mathbf{Y}_j) - \bar{\Delta}_{XY})^2.$$

The distance-based scale statistic is

$$T_{\text{scal}} = \widehat{\mathbb{V}ar}(\mathcal{D}_{XY}) - \frac{1}{2}\left[\widehat{\mathbb{V}ar}(\mathcal{D}_{XX}) + \widehat{\mathbb{V}ar}(\mathcal{D}_{YY})\right].$$

This statistic contrasts the variability of between-sample distances with the average variability of within-sample distances. Under H_0, $\mathbb{E}(T_{\text{scal}}) = 0$.

Using the joint asymptotic normality of $(T_{\text{loc}}, T_{\text{scal}})$, the two p-values may be combined using Fisher's method,

$$S = -2(\log p_{\text{loc}} + \log p_{\text{scal}}) \xrightarrow{d} \chi^2_4.$$

Example 5.10.1 *Consider $n = m = 25$, $d = 100$ and let*

$$\mathbf{X}_i \sim \mathbb{N}_{100}(0, I_{100}), \quad \mathbf{Y}_j \sim \mathbb{N}_{100}(\mu, \sigma^2 I_{100}),$$

with $\mu = (\delta, \delta, 0, \dots, 0) \in \mathbb{R}^{100}$. Assume that

$$|\mathcal{D}_{XX}| = \binom{25}{2} = 300, \quad |\mathcal{D}_{YY}| = 300, \quad |\mathcal{D}_{XY}| = 25 \cdot 25 = 625.$$

We obtain

$$E\|\mathbf{X}_i - \mathbf{X}_j\|^2 = 2d = 200, \quad E\|\mathbf{Y}_i - \mathbf{Y}_j\|^2 = 2\sigma^2 d, \quad E\|\mathbf{X}_i - \mathbf{Y}_j\|^2 = 2d + \|\mu\|^2 = 200 + 2\delta^2.$$

For the location test we get

$$T_{\text{loc}} = \bar{\Delta}_{XY} - \frac{1}{2}(\bar{\Delta}_{XX} + \bar{\Delta}_{YY}), \quad \mathbb{E}(T_{\text{loc}}) = 2\delta^2 \quad (\textit{if } \sigma^2 = 1).$$

Standard deviation of distances are

$$\mathbb{V}ar(\|\mathbf{X}_i - \mathbf{X}_j\|^2) = 8d = 800, \quad \text{sd}(T_{\text{loc}}) = O(\sqrt{d}) \approx 10\sqrt{8} \approx 28.3.$$

Signal-to-noise ratio is

$$\frac{\mathbb{E}(T_{\text{loc}})}{\text{sd}(T_{\text{loc}})} \approx \frac{2\delta^2}{28.3}.$$

For the scale test we compute empirical variances,

$$\widehat{\mathbb{V}ar}(\mathcal{D}_{XX}) = \frac{1}{300} \sum_{i<j} (\|\mathbf{X}_i - \mathbf{X}_j\|^2 - \bar{\Delta}_{XX})^2 \approx 800,$$

$$\widehat{\mathbb{V}ar}(\mathcal{D}_{YY}) \approx 8\sigma^4 d, \quad \widehat{\mathbb{V}ar}(\mathcal{D}_{XY}) \approx 8d + \textit{location contribution}.$$

Then

$$T_{\text{scal}} = \widehat{\mathbb{V}ar}(\mathcal{D}_{XY}) - \frac{1}{2}(\widehat{\mathbb{V}ar}(\mathcal{D}_{XX}) + \widehat{\mathbb{V}ar}(\mathcal{D}_{YY})).$$

If $\sigma^2 = 1$, $T_{\text{scal}} \approx 0$.
If $\sigma^2 > 1$, $\widehat{\mathbb{V}ar}(\mathcal{D}_{YY})$ inflates, yielding $T_{\text{scal}} > 0$.

5.11 Permutation Testing

Permutation testing is a non–parametric statistical method used to assess the significance of observed effects by estimating the sampling distribution of a test statistic under the null hypothesis. Unlike classical tests that often rely on specific distributional assumptions, permutation tests generate a reference distribution directly from the data by permuting labels or values. If the null hypothesis is true, then the observed grouping of data is arbitrary.

Therefore, any statistic calculated from the data should not depend on that grouping. By recalculating the test statistic after shuffling labels many times, we can estimate how extreme the observed statistic is, given the null hypothesis.

The main assumption of permutation testing is that observations are exchangeable under the null hypothesis. This means that permuting labels does not change the joint distribution of the data. In a two-sample test, this implies that under the null hypothesis, both samples come from the same distribution. We also assume that the observations are independent of one another. With m observations in the first group and n observations in the second group, there are $\binom{m+n}{n}$ assignments of labels to the two groups. When the number of possible permutations is very large, it becomes computationally infeasible to enumerate all permutations. Approximate permutation testing typically works with 5 to 10 thousand permutations. To compute the P-value of the test, one first obtains the values of test statistic for the observed data or T_{obs}. The permutation test works as follows.

Algorithm 5.11.1 Permutation Testing

Input: Observed data, test statistic T, number of permutations M

Output: P-value for the observed test statistic

Step 1: Compute the test statistic for the observed data: T_{obs}.

Step 2: Combine all data into a single pooled sample.

Step 3: Permute group labels

Randomly reassign group labels to simulate the null hypothesis H_0.

Step 4: Compute the test statistic T_i for this permutation.

Step 5: Repeat permutations

Repeat Steps 3–4 M times to build an empirical distribution of T under H_0.

The empirical cumulative distribution function (ECDF) of T is the reference set.

Step 6: Compute P-value $P_{\text{value}} = \frac{1}{M}\sum_{i=1}^{M}\mathbb{I}(|T_i| \geq |T_{obs}|)$

Here, M is the number of permutations, T_i is the test statistic from the i-th permutation, and $\mathbb{I}$ is the indicator function.

The permutation principle rests on the idea that if the null hypothesis is true, then the observed data labels do not carry any special meaning. Hence, they can be randomly permuted without affecting the underlying distribution of the data. This principle makes permutation testing distribution-free (non–parametric) and widely applicable, especially in small-sample settings or when parametric assumptions are questionable.

5.12 Hotelling T^2 Test

Consider the basic method of comparing two groups for the equality of their multivariate means. The classical parametric approach uses the two-sample Hotelling T^2 [155] under several assumptions, including multivariate normality and equal covariance matrices. It also assumes that $d < m + n - 2$. Now, $\mathbf{X}_i \sim \mathbb{N}_d(\boldsymbol{\mu}_x, \boldsymbol{\Sigma})$ for $1 \leq i \leq m$ and $\mathbf{Y}_j \sim \mathbb{N}_d(\boldsymbol{\mu}_y, \boldsymbol{\Sigma})$ for $1 \leq j \leq n$. Hence, $\bar{\mathbf{X}} \sim \mathbb{N}_d(\boldsymbol{\mu}_x, \boldsymbol{\Sigma}/m)$ and $\bar{\mathbf{Y}} \sim \mathbb{N}_d(\boldsymbol{\mu}_y, \boldsymbol{\Sigma}/n)$. Under $H_0 : \boldsymbol{\mu}_x = \boldsymbol{\mu}_y$,

Hotelling statistic

$$T^2 = \frac{mn}{m+n}(\bar{\mathbf{X}} - \bar{\mathbf{Y}})' S_d^{-1}(\bar{\mathbf{X}} - \bar{\mathbf{Y}}) \sim \frac{(m+n-2)d}{m+n-d-1} F(d, m+n-d-1) \tag{5.10}$$

is where F stands for the F distribution and the pooled covariance is $S_d = \frac{(m-1)S_x+(n-1)S_y}{m+n-2}$. The Hotelling T^2 statistic is invariance under affine transformations, has known exact and asymptotic distributions and is the uniformly most powerful invariant for testing the mean against two sided alternatives.

Hotelling T^2 [155] is a function of the likelihood ratio test of H_0. For a random sample of size n from d-variate normal distributions, the limiting distributions under the null hypothesis for the likelihood ratio tests of the means and covariance matrices follow a Chi-Square distribution as $n \to \infty$ while d remains fixed. When both d and n tend to infinity and their ratio $d/n \to\in (0, 1]$, Fan [93] shows that the likelihood ratio test statistics converge to a normal random variable. As the dimension d increases, even for simple one-sample testing on the mean of a normal with a known covariance matrix $\sigma^2\mathbf{I}$, the standard Wald, score, or likelihood ratio tests may have power that decreases to the Type I error rate as the departure from the null hypothesis increases.

5.13 Puri-Sen Rank Test for Location

The Puri-Sen test [274] is a non-parametric, rank-based alternative to the Hotelling's T^2 [155] test for comparing multivariate mean vectors of two independent samples. It is robust and does not require the assumption of multivariate normality. Let $\mathbf{X}_1, \ldots, \mathbf{X}_m$ and $\mathbf{Y}_1, \ldots, \mathbf{Y}_n$ be two independent samples from continuous d-dimensional distributions. The null hypothesis is $H_0 : \boldsymbol{\mu}_1 = \boldsymbol{\mu}_2$. To construct the test statistic

1. Combine all $N = m + n$ observations.
2. Rank each variable (component-wise) from 1 to N.
3. Let $\mathbf{R}_i = (R_{i1}, R_{i2}, \ldots, R_{id})'$ be the rank vector for observation i.
4. Compute average rank vectors for each group,

$$\bar{\mathbf{R}}_1 = \frac{1}{m}\sum_{i=1}^{m} \mathbf{R}_i, \quad \bar{\mathbf{R}}_2 = \frac{1}{n}\sum_{i=m+1}^{N} \mathbf{R}_i.$$

5. Compute the pooled rank covariance matrix $\mathbf{S}_R$.
6. Compute the test statistic,

$$Q = \frac{mn}{N}(\bar{\mathbf{R}}_1 - \bar{\mathbf{R}}_2)'\mathbf{S}_R^{-1}(\bar{\mathbf{R}}_1 - \bar{\mathbf{R}}_2).$$

Under H_0, for large m and n, $Q \sim \chi^2_d$. This test is applicable when $N > d$.

Example 5.13.1 *Consider two samples each with five observations in four dimensions.*

$$\textit{Group 1} = \begin{bmatrix} 2 & 3 & 7 & 1 \\ 3 & 2 & 6 & 2 \\ 4 & 4 & 5 & 3 \\ 5 & 6 & 4 & 4 \\ 6 & 5 & 3 & 5 \end{bmatrix}, \qquad \textit{Group 2} = \begin{bmatrix} 7 & 8 & 2 & 6 \\ 8 & 9 & 1 & 7 \\ 9 & 7 & 0 & 8 \\ 10 & 10 & -1 & 9 \\ 11 & 11 & -2 & 10 \end{bmatrix}.$$

Ranking each variable component-wise over the pooled sample gives

Obs	*1*	*2*	*3*	*4*	*5*	*6*	*7*	*8*	*9*	*10*
Group	*1*	*1*	*1*	*1*	*1*	*2*	*2*	*2*	*2*	*2*
Dim 1	*1*	*2*	*3*	*4*	*5*	*6*	*7*	*8*	*9*	*10*
Dim 2	*2*	*1*	*3*	*5*	*4*	*7*	*8*	*6*	*9*	*10*
Dim 3	*10*	*9*	*8*	*7*	*6*	*5*	*4*	*3*	*2*	*1*
Dim 4	*1*	*2*	*3*	*4*	*5*	*6*	*7*	*8*	*9*	*10*

Mean rank vectors are $\bar{\mathbf{R}}_1 = (3,3,8,3)'$, $\bar{\mathbf{R}}_2 = (8,8,3,8)'$ *and the overall mean rank vector is* $\bar{\mathbf{R}} = (5.5,5.5,5.5,5.5)'$. *Define* $\mathbf{d}_i = \mathbf{R}_i - \bar{\mathbf{R}}$. *The pooled rank covariance matrix is*

$$\mathbf{S}_R = \frac{1}{N-2}\sum_{i=1}^{10}\mathbf{d}_i\mathbf{d}_i' = \frac{1}{8}\begin{bmatrix} 82.5 & 77.5 & -77.5 & 77.5 \\ 77.5 & 82.5 & -77.5 & 77.5 \\ -77.5 & -77.5 & 82.5 & -77.5 \\ 77.5 & 77.5 & -77.5 & 82.5 \end{bmatrix}.$$

The Puri–Sen test statistic is $Q \approx 30.4$. *Under* H_0, *for large* m *and* n, $Q \sim \chi^2_4$. *At* $\alpha = 0.05$, *the critical value is* $\chi^2_{4,0.95} = 9.488$. *Since* $Q > 9.488$, *we reject* H_0, *concluding that the two groups differ significantly in multivariate location.*

The Puri-Sen test provides a powerful and robust alternative to Hotelling's T^2 test for multivariate data, especially when the assumption of normality is questionable.

5.14 Bartlett's Test for Dispersion

To describe the Bartlett's test for homogeneity of covariance matrices, let k independent samples be drawn from multivariate normal populations

$$\mathbf{X}_{ij} \sim \mathbb{N}_d(\boldsymbol{\mu}_i, \boldsymbol{\Sigma}_i), \quad j = 1,\ldots,n_i, \quad i = 1,\ldots,k.$$

We test the null hypothesis $H_0 : \boldsymbol{\Sigma}_1 = \boldsymbol{\Sigma}_2 = \cdots = \boldsymbol{\Sigma}_k = \boldsymbol{\Sigma}$.

Let $\mathbf{S}_i$ be the sample covariance matrix for group i,

$$\mathbf{S}_i = \frac{1}{n_i - 1}\sum_{j=1}^{n_i}(\mathbf{X}_{ij} - \bar{\mathbf{X}}_i)(\mathbf{X}_{ij} - \bar{\mathbf{X}}_i)'.$$

Define the pooled covariance matrix

$$\mathbf{S}_p = \frac{1}{N-k}\sum_{i=1}^{k}(n_i - 1)\mathbf{S}_i$$

where $N = \sum_{i=1}^{k} n_i$. Define the Bartlett test statistic as

$$B = (N-k)\ln|\mathbf{S}_p| - \sum_{i=1}^{k}(n_i - 1)\ln|\mathbf{S}_i|.$$

Apply the correction factor,

$$C = 1 - \frac{1}{3(d-1)}\left(\sum_{i=1}^{k}\frac{1}{n_i - 1} - \frac{1}{N-k}\right),$$

the corrected test statistic is $B^* = C \cdot B$. Under H_0, the statistic B^* approximately follows a chi-squared distribution, $B^* \sim \chi^2_{\frac{1}{2}d(d+1)(k-1)}$. Reject H_0 at significance level α if $B^* > \chi^2_{1-\alpha,\frac{1}{2}d(d+1)(k-1)}$. This test is applicable when $\min(n_i) > d$. Bartlett's test is widely recognized for its high sensitivity to departures from normality. Consequently, employing permutation testing is strongly recommended to derive more robust critical values for the test.

5.15 Puri–Sen Rank Test for Dispersion

The Puri–Sen rank-based approach for testing the equality of dispersion (scale) between two multivariate samples is explained below. We have two independent samples

$$\mathbf{X}_1, \ldots, \mathbf{X}_m \sim F_1 \subset \mathbb{R}^d, \qquad \mathbf{Y}_1, \ldots, \mathbf{Y}_n \sim F_2 \subset \mathbb{R}^d,$$

with pooled sample size $N = m + n$. We wish to test

$$H_0 : \text{ the two populations have equal dispersion}$$

versus a general alternative that the dispersions differ. The Puri–Sen approach is rank-based and non-parametric (coordinate-wise ranks) and proceeds as follows.

1. Pool the two samples into $\mathbf{Z}_1, \ldots, \mathbf{Z}_N \in \mathbb{R}^d$ where the first m rows correspond to group 1 and the next n rows to group 2.

2. For each coordinate $j = 1, \ldots, d$, replace the coordinate values $Z_{1j}, \ldots, Z_{Nj}$ by their pooled ranks $R_{1j}, \ldots, R_{Nj} \in \{1, \ldots, N\}$, using average ranks for ties if any.

3. Let $\mathbf{R}_i = (R_{i1}, \ldots, R_{id})'$ denote the d-vector of ranks for the i-th pooled observation. Let $\bar{\mathbf{R}} = \frac{1}{N}\sum_{i=1}^{N} \mathbf{R}_i$ be the overall mean rank vector (note: for a full rank range $1 \ldots N$ in a coordinate the coordinate mean is $(N+1)/2$).

4. Compute the total sum-of-products matrix

$$T = \sum_{i=1}^{N} (\mathbf{R}_i - \bar{\mathbf{R}})(\mathbf{R}_i - \bar{\mathbf{R}})' \in \mathbb{R}^{d\times d},$$

   and the between-group sum-of-products matrix

$$B = m(\bar{\mathbf{R}}_1 - \bar{\mathbf{R}})(\bar{\mathbf{R}}_1 - \bar{\mathbf{R}})' + n(\bar{\mathbf{R}}_2 - \bar{\mathbf{R}})(\bar{\mathbf{R}}_2 - \bar{\mathbf{R}})',$$

   where $\bar{\mathbf{R}}_1$ and $\bar{\mathbf{R}}_2$ are the mean rank vectors of group 1 and group 2 respectively.

5. Provided T is non–singular, the Puri–Sen test statistic for the two-sample case is

$$W_{PS} = (N-1)\,\mathrm{tr}\big(BT^{-1}\big).$$

Under mild regularity conditions (continuity and common distributional shape assumptions required in the rank framework) and under H_0, the statistic W_{PS} is asymptotically χ^2_d distributed. Thus for a significance level α we reject H_0 if $W_{PS} > \chi^2_{d,\,1-\alpha}$, or use the asymptotic p-value $p = \mathbb{P}\big(\chi^2_d \geq W_{PS}\big)$. This test is not applicable when $N > d$. The Puri–Sen tests are designed as extensions of classical rank tests to multivariate data, typically involving calculations that assume a fixed dimension while the sample sizes tend to infinity. When $N > d$, the covariance matrix or the rank covariance structure becomes singular (non-invertible), rendering classical MANOVA or rank-based dispersion tests unusable. The test is rank-based (hence robust) and is a multivariate extension of univariate rank dispersion tests (e.g., Ansari--Bradley, Mood scores). The asymptotic chi-square approximation is often acceptable for moderate sample sizes; permutation methods may be used to improve small-sample calibration.

Example 5.15.1 *Consider two random samples with $d = 5$, $m = 10$, $n = 10$, $N = 20$. After coordinate-wise pooled ranking (ranks $1, \ldots, 20$ in each coordinate), we compute*

$$\bar{\mathbf{R}} = (10.5, 10.5, 10.5, 10.5, 10.5), \quad \bar{\mathbf{R}}_1 = (14, 13, 12, 15, 13), \quad \bar{\mathbf{R}}_2 = (7, 8, 9, 6, 8).$$

Compute the between-group and total sum-of-products matrices,

$$B = m(\bar{\mathbf{R}}_1 - \bar{\mathbf{R}})(\bar{\mathbf{R}}_1 - \bar{\mathbf{R}})' + n(\bar{\mathbf{R}}_2 - \bar{\mathbf{R}})(\bar{\mathbf{R}}_2 - \bar{\mathbf{R}})', \quad T = \sum_{i=1}^{20} (\mathbf{R}_i - \bar{\mathbf{R}})(\mathbf{R}_i - \bar{\mathbf{R}})'.$$

Then the Puri–Sen test statistic is $W_{PS} = (N-1)\,\mathrm{tr}(BT^{-1}) \approx 198.3$. Under H_0, $W_{PS} \sim \chi_5^2$ asymptotically. The 5% critical value is $\chi^2_{5,0.95} \approx 11.07$. Since $W_{PS} = 198.3 > 11.07$, we reject H_0. There is strong evidence that the two groups differ in dispersion.

In practice,

- If the pooled coordinates contain ties, use average ranks (or another tie-handling rule). The rank-based test's validity depends on proper tie-handling.
- In very small samples the matrix T may be singular or near-singular. Options include using a generalized inverse, dimension reduction, or permutation testing for calibration (permutation of group labels yields a non-parametric null distribution).
- A recommended approach in practice is to compute W_{PS} on the observed data and then compute its permutation distribution by randomly permuting group labels many times and recomputing the statistic each time to get an empirical p-value.
- The Puri–Sen coordinate-wise rank approach is simple and robust but is not affine-invariant; other multivariate rank/sign methods (spatial ranks/signs, depth-based ranks) address some invariance concerns.

5.16 Triangle Test

Liu and Modarres [201] propose the triangle statistic as a method for comparing two high-dimensional distributions based on sample IPDs. Given two independent d-dimensional random samples, a triangle is formed by randomly selecting one observation from one sample and two observations from the other sample. The triangle statistic estimates the probability that the distance between the two observations from the same distribution is the largest, the middle, or the smallest among the three edges of the triangle. It can be shown that the test based on the triangle statistic is asymptotically distribution-free under the null hypothesis of equal but unknown distribution functions. The performance of the triangle test is evaluated and compared with other non-parametric tests through a simulation study.

The triangle statistic is well-defined even when the number of variables exceeds the number of observations ($d \geq n$), and its computational complexity is independent of d, making it particularly suitable for high-dimensional settings.

Suppose that $\mathbf{X} = \{\mathbf{X}_1, \ldots, \mathbf{X}_m\}$ and $\mathbf{Y} = \{\mathbf{Y}_1, \ldots, \mathbf{Y}_n\}$ are independent random samples of $\mathbb{R}^d$-valued random vectors, with $d \geq 1$, having distribution functions F_1 and F_2, respectively. The classical two-sample problem consists of testing the null hypothesis $H_0 : F_1 = F_2$ against the general alternative $H_a : F_1 \neq F_2$, where the distribution functions F_1 and F_2 are unknown. Testing H_0 has a long and well-established history and continues to be of substantial practical importance in a wide range of disciplines. In modern applications such as data mining and the analysis of microarray gene expression, challenges arise when the dimensionality of the data exceeds the sample size ($d \geq n$) or when observations take the form of functions rather than finite-dimensional vectors. In such contexts, procedures that rely on component-wise ranks are no longer applicable. An alternative approach is to utilize IPDs, which effectively reduce multivariate or functional data to a univariate representation. These distances can be employed to develop statistics for testing $H_0 : F_1 = F_2$ that are both consistent and distribution-free.

Alternative strategies utilize nearest-neighbor distances, as proposed by Bickel and Breiman [32], Henze [146], and Schilling [300], yielding tests that retain both consistency and desirable asymptotic properties. In addition, permutation-based procedures founded on the E-distance or energy distance has been introduced by Székely and Rizzo [333], Baringhaus and Franz [21], and Aslan and Zeoh [14]. Empirical investigations via simulation studies suggest that these methods perform effectively in detecting differences in location between distributions, while their power is comparatively limited when distinguishing differences in dispersion.

We introduce the triangle statistic as a method for testing the equality of two multivariate distribution functions. The construction of the statistic is motivated by geometric considerations and bears a connection to the goodness-of-fit test proposed by Bartoszyński, Pearl, and Lawrence [25]. We provide a formal definition of the triangle statistic and establish that the corresponding test is asymptotically distribution-free. The performance of the triangle statistic is examined through its power under various alternative hypotheses, and comparisons are made with existing procedures. Our findings indicate that the triangle statistic is particularly effective in detecting differences in dispersion matrices, as well as variations in skewness and kurtosis between multivariate distributions. The final section presents a discussion of the results and offers concluding remarks.

Motivation

Let $\delta(\cdot, \cdot)$ be any appropriately chosen distance function such that $\delta(x, y) = 0 \Leftrightarrow x = y$. Maa, Pearl, and Bartoskzyński [206] proved that, under mild conditions, $F_1 = F_2$ if and only if the IPD within samples and between samples have the same univariate distribution. That is, if $\mathbf{X}_1$ and $\mathbf{X}_2$ are independent and identically distributed (i.i.d.) with DF F_1 and $\mathbf{Y}_1$ and $\mathbf{Y}_2$ are i.i.d. with DF F_2,

$$\delta(\mathbf{X}_1, \mathbf{Y}_1) \stackrel{D}{=} \delta(\mathbf{X}_1, \mathbf{X}_2) \stackrel{D}{=} \delta(\mathbf{Y}_1, \mathbf{Y}_2) \qquad \Longleftrightarrow \qquad F_1 = F_2.$$

We form a triangle with randomly selected data points $\mathbf{X}_1 \sim F_1$ and $\mathbf{Y}_1, \mathbf{Y}_2 \sim F_2$. Let θ_1, θ_2 and θ_3 denote the probabilities of $\delta(\mathbf{Y}_1, \mathbf{Y}_2)$ being the largest, the middle, or the smallest side in the triangle, respectively. As shown in Figure 5.7 panel (a), for the general case in two dimensions, two circles centered at $\mathbf{Y}_1$ and $\mathbf{Y}_2$, respectively, with radius $\delta(\mathbf{Y}_1, \mathbf{Y}_2)$ partition the sample space into three regions: The lens area (Region 1), the lunar areas (Region 2) and the area outside the circles (Region 3). If $\mathbf{X}_1$ falls in Region 1, $\delta(\mathbf{Y}_1, \mathbf{Y}_2)$ is the largest side in the triangle; if $\mathbf{X}_1$ falls in Region 2, $\delta(\mathbf{Y}_1, \mathbf{Y}_2)$ is the middle side in the triangle, and if $\mathbf{X}_1$ falls in Region 3, $\delta(\mathbf{Y}_1, \mathbf{Y}_2)$ is the smallest side in the triangle. Therefore, θ_1, θ_2, and θ_3 denote the probabilities that $\mathbf{X}_1$ falls in the region 1, 2, and 3, respectively.

$$\delta(\mathbf{X}_1, \mathbf{Y}_1) \stackrel{D}{=} \delta(\mathbf{X}_1, \mathbf{Y}_2) \stackrel{D}{=} \delta(\mathbf{Y}_1, \mathbf{Y}_2) \qquad \Longrightarrow \qquad \theta_1 = \theta_2 = \theta_3 = 1/3.$$

Similarly, Figure 5.7 panel (b) is for a triangle with randomly selected data points $\mathbf{Y}_1 \sim F_2$ and $\mathbf{X}_1, \mathbf{X}_2 \sim F_1$ as vertices. We denote the probabilities of $\delta(\mathbf{X}_1, \mathbf{X}_2)$ being the largest, the middle, and the smallest side in the triangle as λ_1, λ_2, and λ_3, respectively. We have

$$\delta(\mathbf{Y}_1, \mathbf{X}_1) \stackrel{D}{=} \delta(\mathbf{Y}_1, \mathbf{X}_2) \stackrel{D}{=} \delta(\mathbf{X}_1, \mathbf{X}_2) \qquad \Longrightarrow \qquad \lambda_1 = \lambda_2 = \lambda_3 = 1/3.$$

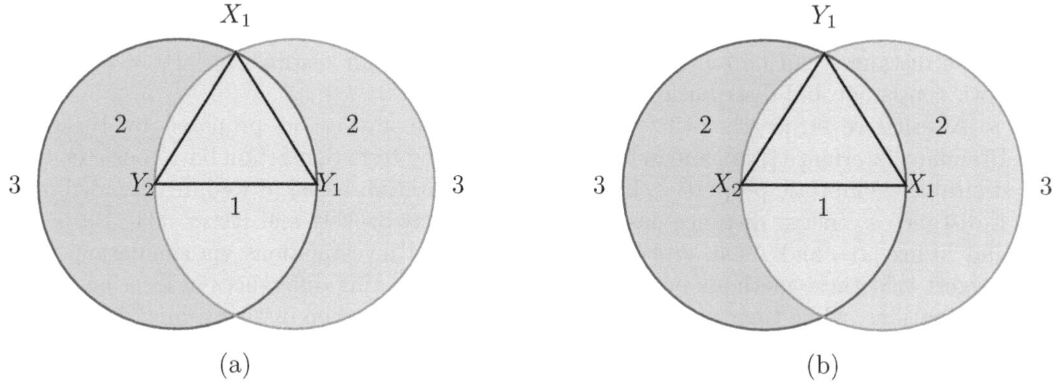

FIGURE 5.7: Regions defining θ_k and λ_k, $k = 1, 2, 3$.

Therefore, we have

$$F_1 = F_2 \qquad \Longrightarrow \qquad \begin{cases} \theta_1 = \theta_2 = \theta_3 = 1/3 \\ \lambda_1 = \lambda_2 = \lambda_3 = 1/3 \end{cases}$$

Triangle Statistics

Suppose $I_{1,k}[x_r; y_i, y_j] = 1$ if $\delta(y_i, y_j)$ is the largest ($k = 1$), the intermediate ($k = 2$), or the smallest ($k = 3$) side length in the triangle with vertices x_r, y_i, and y_j, and $I_{1,k}[x_r; y_i, y_j] = 0$ otherwise. Similarly, $I_{2,k}[y_{r'}; x_{i'}, x_{j'}] = 1$ if $\delta(x_{i'}, x_{j'})$ is the largest ($k = 1$), the intermediate ($k = 2$), or the smallest ($k = 3$) side length in the triangle with vertices $y_{r'}$, $x_{i'}$, and $x_{j'}$, and $I_{2,k}[y_{r'}; x_{i'}, x_{j'}] = 0$ otherwise. Define U_k and V_k with,

$$\begin{aligned} U_k &= \frac{1}{m\binom{n}{2}} \sum_{r=1}^{m} \sum_{i<j}^{n} I_{1,k}[x_r; y_i, y_j], \\ V_k &= \frac{1}{n\binom{m}{2}} \sum_{r'=1}^{n} \sum_{i'<j'}^{m} I_{2,k}[y_{r'}; x_{i'}, x_{j'}]. \end{aligned}$$

for $k = 1, 2, 3$, $r, i', j' = 1, \ldots, m$ and $r', i, j = 1, \ldots, n$. Here, U_k and V_k are two-sample U-statistics with kernel functions $I_{1,k}[x_r; y_i, y_j]$ and $I_{2,k}[y_{r'}; x_{i'}, x_{j'}]$, respectively. These statistics are unbiased estimators of θ_k and λ_k. For $k = 1, 2, 3$, under the null hypothesis H_0, it follows that

$$\mathbb{E}(U_k) = \theta_k = \tfrac{1}{3} \quad \text{and} \quad \mathbb{E}(V_k) = \lambda_k = \tfrac{1}{3}.$$

Hence, substantial deviations of U_k or V_k from 1/3 provide evidence against H_0 in favor of the alternative hypothesis $H_a : F_1 \neq F_2$. Accordingly, U_k and V_k may be used individually, or combined through suitable functions, to construct tests of $H_0 : F_1 = F_2$. Let $N = m + n$ be the total number of observations, and assume that as $m, n \to \infty$, $\frac{m}{N} \to \rho$, $\frac{n}{N} \to 1 - \rho$, and $0 < \rho < 1$. It follows from the theory of two-sample U-statistic (Puri and Sen [274]) that U_k and V_k are asymptotically normal.

Let $\Delta(\mathbf{X}_1, \mathbf{Y}_1, \mathbf{Y}_2)$ denote a triangle with vertices $\mathbf{X}_1$, $\mathbf{Y}_1$ and $\mathbf{Y}_2$. Note that θ_k is defined for $\Delta(\mathbf{X}_1, \mathbf{Y}_1, \mathbf{Y}_2)$, which is symmetric in $\mathbf{Y}_1$ and $\mathbf{Y}_2$. We define the following conditional probabilities for θ_k. Let

$$d_{1k}(x_1) = \mathbb{E}_{F_2,F_2}(I_{1,k}[x_1; \mathbf{Y}_1, \mathbf{Y}_2])$$

be the conditional probability that given x_1, $\delta(\mathbf{Y}_1, \mathbf{Y}_2)$ is LMS (the largest ($k = 1$), the intermediate ($k = 2$) and the smallest ($k = 3$)) in $\Delta(x_1, \mathbf{Y}_1, \mathbf{Y}_2)$). Let

$$q_{1k}(y_1) = \mathbb{E}_{F_1,F_2}(I_{1,k}[\mathbf{X}_1; y_1, \mathbf{Y}_2])$$

be the conditional probability that given y_1, $\delta(y_1, \mathbf{Y}_2)$ is LMS in $\Delta(\mathbf{X}_1, y_1, \mathbf{Y}_2)$. Similarly, we define the following conditional probabilities for λ_k. Let

$$d_{2k}(y_1) = \mathbb{E}_{F_1,F_1}(I_{2,k}[y_1; \mathbf{X}_1, \mathbf{X}_2])$$

be the conditional probability that given y_1, $\delta(\mathbf{X}_1, \mathbf{X}_2)$ is LMS in $\Delta(y_1, \mathbf{X}_1, \mathbf{X}_2)$, and

$$q_{2k}(\mathbf{X}_1) = \mathbb{E}_{F_1,F_2}(I_{2,k}[\mathbf{Y}_1; x_1, \mathbf{X}_2])$$

be the conditional probability that given x_1, $\delta(x_1, \mathbf{X}_2)$ is LMS in $\Delta(\mathbf{Y}_1, x_1, \mathbf{X}_2)$. We define the following parameters for the conditional probabilities, with $s, t = 1, 2, 3$,

$$\begin{aligned}
\alpha_{1st} &= \mathbb{C}ov(p_{1s}(\mathbf{X}_1), p_{1t}(\mathbf{X}_1)); & \alpha_{2st} &= \mathbb{C}ov(p_{2s}(\mathbf{Y}_1), p_{2t}(\mathbf{Y}_1));\\
\beta_{1st} &= \mathbb{C}ov(q_{1s}(\mathbf{Y}_1), q_{1t}(\mathbf{Y}_1)); & \beta_{2st} &= \mathbb{C}ov(q_{2s}(\mathbf{X}_1), q_{2t}(\mathbf{X}_1));\\
\gamma_{1st} &= \mathbb{C}ov(p_{1s}(\mathbf{X}_1), q_{2t}(\mathbf{X}_1)); & \gamma_{2st} &= \mathbb{C}ov(p_{2s}(\mathbf{Y}_1), q_{1t}(\mathbf{Y}_1)).
\end{aligned}$$

Since $U_1 + U_2 + U_3 = 1$, this relationship reduces the six covariances α_{1st}'s or β_{1st}'s to three linearly independent parameters (e.g., if s, t and k are distinct indices, then $\alpha_{1st} = \alpha_{1kk} - (\alpha_{1ss} + \alpha_{1tt})/2$). Similarly, $V_1 + V_2 + V_3 = 1$ reduces the six covariances α_{2st}'s or β_{2st}'s to three linearly independent parameters. These relationships reduce the six covariances γ_{1st}'s or γ_{2st}'s to three linearly independent parameters. Furthermore, under H_0: $F_1 = F_2$, for $k, s, t = 1, 2, 3$,

$$d_{1k}(\mathbf{X}_1) \stackrel{D}{=} p_{2k}(\mathbf{Y}_1), \text{ and } q_{1k}(\mathbf{Y}_1) \stackrel{D}{=} q_{2k}(\mathbf{X}_1),$$

where $\alpha_{1st} = \alpha_{2st}$, $\beta_{1st} = \beta_{2st}$, $\gamma_{1st} = \gamma_{2st}$.

For $k = 1, 2, 3$, Hájek projections of $U_k - \theta_k$ and $V_k - \lambda_k$ onto the sets of all linear functions of $\sum_{i=1}^{m} p_i(\mathbf{X}_i)$ and $\sum_{j=1}^{n} q_j(\mathbf{Y}_j)$ are given by

$$\begin{aligned}
\hat{U}_k &= \frac{1}{m}\sum_{i=1}^{m}(p_{1k}(\mathbf{X}_i) - \theta_k) + \frac{2}{n}\sum_{j=1}^{n}(q_{1k}(\mathbf{Y}_j) - \theta_k);\\
\hat{V}_k &= \frac{1}{n}\sum_{j=1}^{n}(p_{2k}(\mathbf{Y}_j) - \lambda_k) + \frac{2}{m}\sum_{i=1}^{m}(q_{2k}(\mathbf{X}_i) - \lambda_k).
\end{aligned}$$

Under condition (1), we obtain the following results, the proof of which appears at the end of the chapter.

Lemma 5.16.1 *For $k, s, t = 1, 2, 3$, $\hat{U}_k$ and $\hat{V}_k$ have asymptotically a multivariate normal distribution with zero mean vector and dispersion matrix with elements*

$$u_{st} = \mathbb{C}ov(\hat{U}_s, \hat{U}_t) = \frac{1}{m}\alpha_{1st} + \frac{4}{n}\beta_{1st};$$
$$v_{st} = \mathbb{C}ov(\hat{V}_s, \hat{V}_t) = \frac{1}{m}\alpha_{2st} + \frac{4}{n}\beta_{2st};$$
$$w_{st} = \mathbb{C}ov(\hat{U}_s, \hat{V}_t) = \frac{2}{m}\gamma_{1st} + \frac{2}{m}\gamma_{2ts}.$$

Lemma 5.16.2 *For $k = 1, 2, 3$,*

$$(U_k - \theta_k) - \hat{U}_k \xrightarrow{P} 0;$$
$$(V_k - \lambda_k) - \hat{V}_k \xrightarrow{P} 0.$$

Theorem 5.16.1 *For $k, s, t = 1, 2, 3$, $(U_k - \theta_k)$ and $(V_k - \lambda_k)$ asymptotically have a multivariate normal distribution with zero mean vector and dispersion matrix with elements*

$$u_{st} = \frac{1}{m}\alpha_{1st} + \frac{4}{n}\beta_{1st}; \quad v_{st} = \frac{1}{m}\alpha_{2st} + \frac{4}{n}\beta_{2st}; \quad w_{st} = \frac{2}{m}\gamma_{1st} + \frac{2}{n}\gamma_{2ts}.$$

Because the dispersion matrices are functions of the unknown distributions F_1 and F_2, we first derive consistent estimators for these matrices and then construct quadratic forms of the triangle statistics, thereby obtaining test statistics that are asymptotically distribution-free. Define

$$\hat{d}_{1k}(x_r) = \frac{1}{\binom{n}{2}}\sum_{i<j}^{n} I_{1k}[x_r; y_i, y_j]; \qquad \hat{q}_{1k}(y_i) = \frac{1}{m}\frac{1}{n-1}\sum_{r=1}^{m}\sum_{\substack{j=1\\ j\neq i}}^{m} I_{1k}[x_r; y_i, y_j];$$

$$\hat{d}_{2k}(y_r) = \frac{1}{\binom{m}{2}}\sum_{i<j}^{m} I_{2k}[y_r; x_i, x_j]; \qquad \hat{q}_{2k}(x_i) = \frac{1}{n}\frac{1}{m-1}\sum_{r=1}^{m}\sum_{\substack{j=1\\ j\neq i}}^{n} I_{2k}[y_r; x_i, x_j],$$

$$\hat{\alpha}_{1st} = \frac{1}{m-1}\sum_{r=1}^{m}[\hat{d}_{1s}(x_r) - U_s]\cdot[\hat{d}_{1t}(x_r) - U_t];$$

$$\hat{\beta}_{1st} = \frac{1}{n-1}\sum_{i=1}^{n}[\hat{q}_{1s}(y_i) - U_s]\cdot[\hat{q}_{1t}(y_i) - U_t];$$

$$\hat{\alpha}_{2st} = \frac{1}{n-1}\sum_{r=1}^{n}[\hat{d}_{2s}(y_r) - V_s]\cdot[\hat{d}_{2t}(y_r) - V_t];$$

$$\hat{\beta}_{2st} = \frac{1}{m-1}\sum_{i=1}^{m}[\hat{q}_{2s}(x_i) - V_s]\cdot[\hat{q}_{2t}(x_i) - V_t];$$

$$\hat{\gamma}_{1st} = \frac{1}{m-1}\sum_{r=1}^{m}[\hat{d}_{1s}(x_r) - U_s]\cdot[\hat{q}_{2t}(x_r) - V_t];$$

$$\hat{\gamma}_{2st} = \frac{1}{n-1}\sum_{r=1}^{n}[\hat{d}_{2s}(y_r) - V_s]\cdot[\hat{d}_{1t}(y_r) - U_t].$$

Lemma 5.16.3 *For* $s, t = 1, 2, 3,$

$$\hat{\alpha}_{1st} \xrightarrow{P} \alpha_{1st}; \qquad \hat{\beta}_{1st} \xrightarrow{P} \beta_{1st};$$
$$\hat{\alpha}_{2st} \xrightarrow{P} \alpha_{2st}; \qquad \hat{\beta}_{2st} \xrightarrow{P} \beta_{2st};$$
$$\hat{\gamma}_{1st} \xrightarrow{P} \gamma_{1st}; \qquad \hat{\gamma}_{1st} \xrightarrow{P} \gamma_{2st}.$$

One can use the statistics, U_k or V_k, $k = 1, 2, 3$, or a quadratic form based on U_k or V_k to test H_0: $F_1 = F_2$. Let $\mathbf{U} = (U_1 - 1/3, U_3 - 1/3)'$, $\mathbf{V} = (V_1 - 1/3, V_3 - 1/3)'$ and $\mathbf{W} = (U_1 - 1/3, U_3 - 1/3, V_1 - 1/3, V_3 - 1/3)'$, and denote

$$\hat{\alpha}_1 = \begin{pmatrix} \hat{\alpha}_{111} & \hat{\alpha}_{113} \\ \hat{\alpha}_{131} & \hat{\alpha}_{133} \end{pmatrix}, \qquad \hat{\alpha}_2 = \begin{pmatrix} \hat{\alpha}_{211} & \hat{\alpha}_{213} \\ \hat{\alpha}_{231} & \hat{\alpha}_{233} \end{pmatrix},$$

$$\hat{\beta}_1 = \begin{pmatrix} \hat{\beta}_{111} & \hat{\beta}_{113} \\ \hat{\beta}_{131} & \hat{\beta}_{133} \end{pmatrix}, \qquad \hat{\beta}_2 = \begin{pmatrix} \hat{\beta}_{211} & \hat{\beta}_{213} \\ \hat{\beta}_{231} & \hat{\beta}_{233} \end{pmatrix},$$

$$\hat{\gamma}_1 = \begin{pmatrix} \hat{\gamma}_{111} & \hat{\gamma}_{113} \\ \hat{\gamma}_{131} & \hat{\gamma}_{133} \end{pmatrix}, \qquad \hat{\gamma}_2 = \begin{pmatrix} \hat{\gamma}_{211} & \hat{\gamma}_{213} \\ \hat{\gamma}_{231} & \hat{\gamma}_{233} \end{pmatrix},$$

$$\hat{\mathbf{\Sigma}}_1 = \frac{1}{n}\hat{\alpha}_1 + \frac{4}{m}\hat{\beta}_1, \quad \hat{\mathbf{\Sigma}}_2 = \frac{1}{m}\hat{\alpha}_2 + \frac{4}{n}\hat{\beta}_2, \quad \hat{\mathbf{\Gamma}} = \frac{2}{n}\hat{\gamma}_1 + \frac{2}{m}\hat{\gamma}_2',$$

$$\hat{\mathbf{\Omega}} = \begin{pmatrix} \hat{\mathbf{\Sigma}} & \hat{\mathbf{\Gamma}} \\ \hat{\mathbf{\Gamma}}' & \hat{\mathbf{\Sigma}} \end{pmatrix}.$$

The test statistics $T_1 = \mathbf{U}'\hat{\mathbf{\Sigma}}_1^{-1}\mathbf{U}$, $T_2 = \mathbf{V}'\hat{\mathbf{\Sigma}}_2^{-1}\mathbf{V}$, and $Q = \mathbf{W}'\hat{\mathbf{\Omega}}^{-1}\mathbf{W}$ asymptotically have chi-squared distributions with 2, 2, and 4 degrees of freedom, respectively, under H_0: $F_1 = F_2$.

5.17 Comparison

Liu and Modarres [201] conducted a simulation study to examine the observed significance level and to compare the empirical power of the test statistics for testing $H_0 : F_1 = F_2$. All simulations are based on 1000 d-variate random samples with $m = n = 30$, and 50 from distributions that are specified in the Table. We used significance level $\alpha = 0.05$. Six other existing statistics for testing $H_0 : F_1 = F_2$ are included. The statistics are Friedman-Rafsky multivariate runs test (R), the nearest neighbor (NN) test (Henze [146]), Hotelling T^2 test, Puri-Sen rank test for location (PSL), Puri-Sen rank test for dispersion (PSS), and Bartlett's test (B) for dispersion. Liu and Modarres [201] demonstrate in their simulation study that all the tests attain approximately the nominal significance level α, with the notable exception of the B test when the ratio n/d is small. They further report that the NN test exhibits pronounced conservatism, and that the asymptotic chi-squared approximation for Bartlett's test is highly sensitive to departures from normality.

Moreover, Liu and Modarres [201] investigate the empirical power of these tests under location shift alternatives. Across all scenarios, the triangle-based tests are consistently the least powerful. Under normality, Hotelling's T^2 test [155] is the most powerful procedure for detecting location shifts, followed closely by the PSL test. Nevertheless, for shift alternatives, the power of all tests decreases as the dimension d increases. In contrast, for scale alternatives, Liu and Modarres [201] show that power is an increasing function of d. In this setting, the triangle statistics are most powerful, whereas the R and NN tests perform comparably poorly. For both location and scale alternatives, the NN test performs marginally better than the R test.

Table 5.1 displays alternative distributions that capture differences in skewness and kurtosis. The empirical powers of the tests are reported in Table 5.2. The runs test R and the nearest neighbor test NN exhibit comparable power across all alternatives, whereas the proposed tests consistently outperform both R and NN. Notably, although the multivariate Student's t distributions have heavy tails, the proposed tests achieve high power. In contrast, Hotelling's T^2 and the PSL test display very limited discriminatory ability. While the PSS test demonstrates some sensitivity, its performance is uniformly dominated by that of the proposed statistics.

TABLE 5.1: Four dimensional alternative distributions emphasizing skewness or kurtosis differences.

case	F_1	F_2
1	$\mathbb{N}(\mathbf{0},\mathbf{I})$	$C(\mathbf{0},\mathbf{I})$
2	$\mathbb{N}(\mathbf{0},\mathbf{I})$	$N_{\log}(\mathbf{0},\mathbf{I})$
3	$\mathbb{N}(\mathbf{0},\mathbf{I})$	$80\%\mathbb{N}(\mathbf{0},\mathbf{I}) + 20\%\mathbb{N}(\mathbf{0}, 0.2^2\mathbf{I})$
4	$\mathbb{N}(\mathbf{0},\mathbf{I})$	Student's t_2
5	$\mathbb{N}(\mathbf{0},\mathbf{I})$	Student's t_4
6	$U(\mathbf{0},\mathbf{I})$	$80\%U(\mathbf{0},\mathbf{I}) + 20\%\mathbb{N}(\mathbf{0.5}, 0.05^2\mathbf{I})$

TABLE 5.2: Empirical powers at $\alpha = 0.05$, calculated from 1000 pairs of samples, $d = 4$.

case	$n = m$	T^2	PSL	PSS	R	NN	T_1	T_2	Q
1	30	0.021	0.044	0.353	0.182	0.134	0.976	0.976	0.981
	50	0.031	0.049	0.657	0.359	0.345	1.000	1.000	1.000
2	30	0.972	0.879	0.145	0.461	0.562	0.739	0.229	0.660
	50	1.000	0.985	0.190	0.741	0.831	0.935	0.461	0.910
3	30	0.042	0.041	0.249	0.065	0.049	0.427	0.590	0.571
	50	0.054	0.063	0.361	0.065	0.054	0.670	0.770	0.760
4	30	0.034	0.051	0.151	0.053	0.039	0.850	0.729	0.836
	50	0.044	0.055	0.239	0.010	0.069	0.979	0.973	0.977
5	30	0.048	0.043	0.090	0.044	0.031	0.489	0.281	0.421
	50	0.042	0.049	0.110	0.056	0.031	0.739	0.571	0.659
6	30	0.051	0.047	0.551	0.101	0.064	0.779	0.783	0.828
	50	0.051	0.048	0.840	0.145	0.127	0.961	0.969	0.980

5.18 Discussion

For testing the equality of two multivariate distributions, triangle tests are particularly appealing because their computational complexity is independent of the dimension of the observations. Consequently, they are well suited for high-dimensional settings, especially in situations where $d \geq n$. The triangle tests have computational complexity $O(n^3)$, in contrast to the $O(n^2 \log n)$ complexity associated with the runs test and the NN test. While the runs and NN tests are most powerful against location shift alternatives, the triangle tests exhibit superior power for detecting scale alternatives as well as differences in skewness and kurtosis.

Bartoskzyński, Pearl, and Lawrence [206]) point out that tests based on sample IPDs are broadly applicable to discrete, continuous, multivariate, and even non-numerical data, provided that an appropriate distance measure between sample objects is available. The distance function δ plays a crucial role in determining the invariance properties of IPD-based tests. The triangle test remains invariant under any transformation that preserves the ordering of the IPDs. For instance, when δ is chosen, as in the simulation study considered here, the resulting test is invariant under location and/or scale transformations, as well as under orthogonal rotations. If invariance with respect to changes in location, scale, and/or permutations of the coordinates is desired, then any L_s metric of the form $\delta(\mathbf{x} - \mathbf{y}) = \|\mathbf{x} - \mathbf{y}\|_s$ provides an alternative. The parameter s may then be selected to balance power and computational efficiency.

Given a data point y_1 from distribution F_2, the conditional probability that $\delta(\mathbf{X}_1, \mathbf{X}_2)$ is the largest side of the triangle formed by the vertices $\mathbf{Y}_1$, $\mathbf{X}_1$, and $\mathbf{X}_2$ induces an ordering of observations from F_2 with respect to distribution F_1. The larger this probability, the closer the point y_1 is to distribution F_1. This probability can therefore serve as a useful alternative to data depth functions, which order multivariate observations from the center outward, as discussed by Liu, Parelius, and Singh [200]. The resulting triangle data depth has computational complexity $O(n^2)$ and is independent of the dimension d. As a result, it offers substantial computational simplicity and efficiency in high-dimensional settings when compared with existing notions such as Tukey's depth or simplicial depth. This depth notion can be further exploited to construct a multivariate Wilcoxon rank-sum test and to define a non-parametric classification rule.

5.19 Dissimilarity Tests

Let $\mathbf{X}_1$ and $\mathbf{X}_2$ be i.i.d. random vectors from distribution F_1, and let $\mathbf{Y}_1$ and $\mathbf{Y}_2$ be i.i.d. from F_2. Liu and Modarres [201] introduced the triangle statistic as a tool for comparing two distributions in high-dimensional spaces, noting that much of the existing literature concentrates on contrasts between the mean within-sample and between-sample IPDs. Székely and Rizzo [333] and Baringhaus and Franz [21] established that

$$2\,\mathbb{E}\|\mathbf{X}_1 - \mathbf{Y}_1\| - \mathbb{E}\|\mathbf{X}_1 - \mathbf{X}_2\| - \mathbb{E}\|\mathbf{Y}_1 - \mathbf{Y}_2\| \geq 0, \tag{5.11}$$

with equality if and only if $F_1 = F_2$, thereby motivating the energy test for two-sample comparison. Building on this line of research, Biswas and Ghosh [35] proposed an IPD-based procedure that improves upon the energy test in terms of power for detecting scale alternatives. From a geometric perspective, Hall et al. [133] showed that, as the dimension increases, observations within each sample tend to concentrate near the vertices of a regular simplex, with the observed variability largely attributable to random rotations of this configuration. They further documented the detrimental impact of distance concentration on Euclidean-distance-based measures and on several classification techniques. Ahn et al. [6] demonstrated that this geometric phenomenon persists under fairly general conditions. More recently, Sarkar and Ghosh [299] reported that disruptions in neighborhood structure substantially degrade the performance of clustering algorithms that rely on Euclidean distances.

Although IPD—based procedures are conceptually attractive for testing the equality of distributions in high dimensions, they encounter two fundamental obstacles. First, Euclidean distances tend to concentrate as the dimension increases, substantially diminishing their discriminatory power. Second, the explicit computation of IPDs becomes prohibitively expensive when both the sample size N and the dimension d are large. While these difficulties are intrinsic to high dimensional data analysis, we propose a new framework for comparing high-dimensional distributions that substantially reduces computational complexity while simultaneously enabling informative visualization.

High-dimensional datasets pose additional challenges, including sparsity, algorithmic scalability, and limited interpretability through standard visualization tools. To address these issues, we introduce a fast and effective methodology for the comparison and visualization of high-dimensional data. The proposed approach is built around a novel statistic that is motivated by IPDs yet circumvents their direct calculation. It is applicable to non-sparse high-dimensional settings and is capable of detecting changes in both the mean and the variance of the underlying variables.

Because of the pronounced concentration phenomenon in high-dimensional spaces, the Euclidean distance is generally ill-suited for direct use in such settings, and methods that rely explicitly on IPDs often suffer a loss of power. Rather than attempting to overcome distance concentration, our approach exploits this phenomenon through the construction of two dissimilarity indices. These indices are derived from lower bounds on Euclidean distances that depend only on sample means and variances, thereby avoiding the explicit evaluation of IPDs. In addition to reducing computational burden, the proposed methodology provides a joint visual representation of sample means and standard deviations, which enhances the separation between competing distributions.

The proposed framework is broadly applicable to a range of multivariate inference problems, including classification, clustering, change-point detection, and outlier identification, all of which can be formulated in terms of testing distributional equality. Nevertheless, the dissimilarity indices are subject to certain limitations. In particular, they cannot distinguish between multivariate distributions that share identical first and second moments. Consequently, while the method is effective for identifying differences in one-dimensional marginals through changes in means or variances, it lacks power when such marginals are identical across distributions. Despite this limitation, the indices remain sensitive to subtle discrepancies in the first two moments and are therefore well suited for a wide class of high-dimensional applications.

5.19.1 Test Based on δ

We introduce a new statistic for the comparison and visualization of high-dimensional datasets that is motivated by IPDs but circumvents their explicit computation. Owing to the well-known phenomenon of distance concentration, the Euclidean distance is generally ill-suited for high-dimensional settings. The proposed statistic is instead constructed from a high-dimensional dissimilarity index that explicitly leverages this concentration effect. In addition, we discuss a joint representation of sample means and standard deviations that facilitates visualization, aids in the identification of potential outliers, and improves class separation in the transformed space. The finite-sample convergence properties of the proposed dissimilarity indices are examined, eight competing statistics are compared under a range of distributional scenarios, and three real-data applications are presented.

Several existing procedures, including the energy test (Székely and Rizzo [330]) and the BG statistic of Biswas and Ghosh [35], are effective in high-dimensional regimes where the dimension d far exceeds the sample size N. Nevertheless, these methods face serious computational challenges when N is moderate to large. A primary computational bottleneck for many IPD-based algorithms is the evaluation of all pairwise distances. For a sample of size N, the cost of computing IPDs scales as $O(dN^2)$. While this remains manageable when N is on the order of a few hundred even for large d, it becomes prohibitive when N reaches the thousands in conjunction with high dimensionality.

The proposed test statistic is defined through the square of the limiting quantity $\tilde{\delta}(F_1, F_2)$ as $d \to \infty$. Modarres [233] establishes a connection between the eigen-structure of the two-sample distance matrix and tests for distributional equality, showing that the leading eigenvalues of the expected two-sample dissimilarity matrices based on δ and ρ converge to $\frac{mn}{N}\tilde{\delta}^2(F_1, F_2)$ and $\frac{mn}{N}\tilde{\rho}^2(F_1, F_2)$, respectively, as the dimension tends to infinity. Suppose the $m \times d$ data matrix of $\mathbb{X}$ values is used to find the mean and standard deviations of each row as follows,

$$\mathbb{X} = \begin{bmatrix} X_{11} & X_{12} & \dots & X_{1d} \\ \vdots & \vdots & \vdots & \vdots \\ X_{m1} & X_{m2} & \dots & X_{md} \end{bmatrix} \Longrightarrow \begin{bmatrix} \bar{X}_1^d & S_{\mathbf{X}_1}^d \\ \vdots & \vdots \\ \bar{X}_m^d & S_{\mathbf{X}_m}^d \end{bmatrix}.$$

In an analogous manner, we compute the row-wise means and corresponding standard deviations of the data matrix $\mathbb{Y}$. It is important to emphasize the conceptual distinction between these row-level summary statistics, which characterize individual observations, and the column-wise means and standard deviations, which describe the behavior of the variables themselves. To define the proposed statistic, let $\bar{x}$ denote the grand mean formed from the collection of observation means and standard deviations associated with the samples $\mathbb{X}$ and $\mathbb{Y}$.

$$\hat{\mu}_x^d = \frac{1}{m}\sum_{i=1}^{m} \bar{X}_i^d = \frac{1}{md}\sum_{i=1}^{m}\sum_{j=1}^{d} X_{ij}, \quad \hat{\sigma}_x^d = \frac{1}{m}\sum_{i=1}^{m} S_{\mathbf{X}_i}^d, \tag{5.12}$$

$$\hat{\mu}_y^d = \frac{1}{n}\sum_{i=1}^{n} \bar{Y}_i^d = \frac{1}{nd}\sum_{i=1}^{n}\sum_{j=1}^{d} Y_{ij}, \quad \hat{\sigma}_y^d = \frac{1}{n}\sum_{i=1}^{n} S_{\mathbf{Y}_i}^d. \tag{5.13}$$

To test $H_0 : F_1 = F_2$ the new test statistic is given by

$$A_\delta = (\hat{\mu}_x^d - \hat{\mu}_y^d)^2 + (\hat{\sigma}_x^d - \hat{\sigma}_y^d)^2. \tag{5.14}$$

The statistic A_δ follows the form of the limiting value of δ and aggregates the overall discrepancy between the group-wise observation means and variances. The null hypothesis H_0 is rejected for sufficiently large values of A_δ. Although its construction is motivated by the analysis of IPDS and the dissimilarity measure δ, the proposed statistic circumvents the explicit computation of IPDs. Consequently, A_δ admits a computational complexity of order $O(dN)$, in contrast to IPD-based procedures whose computational cost scales as $O(dN^2)$. The same principle may be extended by employing the limiting form of the dissimilarity measure, denoted by $\tilde{\rho}^2(F_1, F_2)$, to derive an alternative test statistic. Let $\hat{\boldsymbol{\mu}}_x$ and $\hat{\boldsymbol{\mu}}_y$ represent the estimators of the mean vectors, and let $\mathbf{S}_x$ and $\mathbf{S}_y$ denote the corresponding sample covariance matrices. Define

$$\hat{\sigma}_{F_1} = \frac{\mathrm{tr}(\mathbf{S}_x)}{d}, \qquad \hat{\sigma}_{F_2} = \frac{\mathrm{tr}(\mathbf{S}_y)}{d}, \qquad \hat{\nu}^2 = \frac{\|\hat{\boldsymbol{\mu}}_x - \hat{\boldsymbol{\mu}}_y\|}{d}.$$

We obtain the statistic

$$A_\rho = \frac{1}{N-2}[(n-1)|\sqrt{2}\hat{\sigma}_{F_1} - \sqrt{\hat{\sigma}_{F_1}^2 + \hat{\sigma}_{F_2}^2 + \hat{\nu}^2}| + (m-1)|\sqrt{\hat{\sigma}_{F_1}^2 + \hat{\sigma}_{F_2}^2 + \hat{\nu}^2} - \sqrt{2}\hat{\sigma}_{F_2}|]. \tag{5.15}$$

The null hypothesis H_0 is rejected for sufficiently large realizations of A_ρ. To approximate the sampling distributions of both A_ρ and A_δ, a permutation-based approach is employed.

This procedure is justified because, under H_0, all observations are identically distributed, rendering them exchangeable. At each permutation iteration, the pooled observations are randomly reassigned. After performing R such permutations, the empirical permutation distribution of the test statistic, denoted by $H_{A_\delta}(t)$, is constructed as

$$H_{A_\delta}(t) = \frac{1}{R}\sum_{r=1}^{R} \mathbb{I}\left(A_\delta^{(r)} \leq t\right),$$

where $\mathbb{I}(\cdot)$ is the indicator function and A_δ^r is the test statistic calculated for the r-th permutation sample. The p-value of the permutation test, denoted by p_{perm}, is given by

$$p_{\text{perm}} = 1 - H_{A_\delta}(A_\delta(\text{obs})) = \frac{1}{R}\sum_{r=1}^{R} \mathbb{I}(A_\delta^r > A_\delta(obs)),$$

where $A_\delta(obs)$ is the test statistic for the observed sample.

Theorem 5.19.1 *Under assumptions (B1)–(B3), the test statistic A_δ converges in probability to $\tilde{\delta}^2(F_1, F_2) \geq 0$ as $d \to \infty$.*

5.19.2 Dissimilarity Matrix

Outlier detection has long been a central problem in statistical research. An extensive review of both traditional and contemporary methodologies is presented in Hadi et al. [131]. Developments tailored to high-dimensional data settings are examined in Ro et al. [281]. More recently, Modarres ([232], [235]) proposed novel high-dimensional outlier detection procedures based on dissimilarity matrices derived from the observed samples. To demonstrate how the newly proposed dissimilarity matrix can be utilized for the visual identification of anomalous observations or potential change points, consider the $N \times N$ dissimilarity matrix corresponding to the samples $\mathbb{X}$ and $\mathbb{Y}$.

$$\begin{aligned} A_\delta(X_i, X_j) &= (\bar{X}_i^d - \bar{X}_j^d)^2 + (S_{\mathbf{X}_i}^d - S_{\mathbf{X}_j}^d)^2, && i,j = 1,\ldots,m, \quad (5.16)\\ A_\delta(Y_i, Y_j) &= (\bar{Y}_i^d - \bar{Y}_j^d)^2 + (S_{\mathbf{Y}_i}^d - S_{\mathbf{Y}_j}^d)^2, && i,j = 1,\ldots,n,\\ A_\delta(X_i, Y_j) &= (\bar{X}_i^d - \bar{Y}_j^d)^2 + (S_{\mathbf{X}_i}^d - S_{\mathbf{Y}_j}^d)^2, \quad i = 1,\ldots,m, && j = 1,\ldots,n. \end{aligned}$$

It is immediate that $A_\delta \geq 0$ and that its diagonal entries are identically zero. Moreover, one can show that, as the dimension $d \to \infty$, the dissimilarity A_δ converges to zero for pairs of observations originating from the same population, while for observations drawn from distinct populations, A_δ converges to a strictly positive limit $\tilde{\delta}(F_1, F_2) > 0$. Consequently, the resulting dissimilarity matrix provides a useful tool for identifying structural change points or detecting anomalous observations.

Let $S_r = \sum_{c=1}^{N}[A_\delta]_{rc}$ denote the sum of the entries in the rth row of the dissimilarity matrix A_δ, for $r = 1,\ldots,N$. Suppose that the observed sample contains a single outlier ($k = 1$) located at index t. Then, as $d \to \infty$, it follows that $S_r \to \tilde{\delta}(F_1, F_2)$ for all $r \neq t$, whereas $S_t \to (N-1)\tilde{\delta}(F_1, F_2)$. Hence, the observation indexed by t exhibits the largest aggregate dissimilarity and is therefore identified as a candidate outlier. In practice, plotting the values of S_r against the corresponding observation indices provides a simple graphical diagnostic for outlier detection.

Example 5.19.1 *We return to the Wine dataset of [16], which consists of 178 observations corresponding to the concentrations of 13 chemical constituents measured in Italian wines. The wines originate from three grape cultivars: Barolo, Grignolino, and Barbera with sample* *sizes of 59, 71, and 48 observations, respectively. Barolo wine is produced from Nebbiolo grapes, whereas the remaining two wine types are named directly after the grape varieties from which they are derived.*

Figure 5.8 displays the row sums of the dissimilarity matrix A_δ. The resulting plot indicates that the chemical measurements are recorded on markedly different scales and that the three wine categories are clearly separated. In particular, Barolo wines exhibit substantially higher Proline concentrations compared with both Barbera and Grignolino wines. Additionally, six Barolo observations appear on the left side of the plot with notably large row-wise dissimilarity values, suggesting atypical behavior relative to the remaining samples.

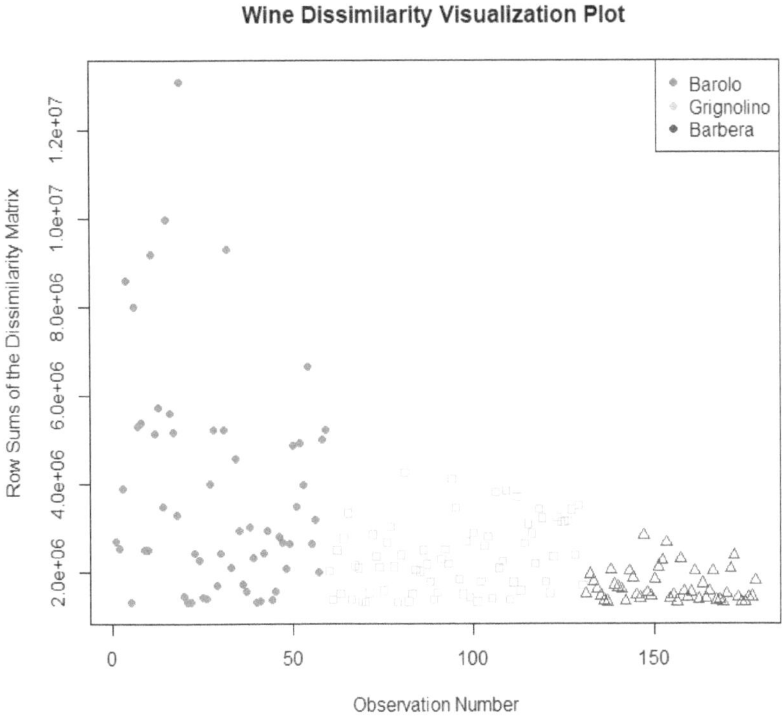

FIGURE 5.8: Row sums of the dissimilarities in Wine data. Barolo (left), Grignolino (middle), Barbera (right).

Example 5.19.2 *To examine the influence of a single outlying observation on the statistic A_δ, assume that the sample $\mathbb{Y}$ is obtained from $\mathbb{X}$ through a linear scaling and translation, namely $Y_{ij} = aX_{ij} + b, \quad i = 1, \ldots, m,\ j = 1, \ldots, d$, with $m = n$. Under this transformation, the empirical mean vectors and dispersion measures satisfy*

$$\hat{\mu}_y^d = a\hat{\mu}_x^d + b, \qquad S_{\mathbf{Y}_i}^d = aS_{\mathbf{X}_i}^d, \qquad \hat{\sigma}_y^d = a\hat{\sigma}_x^d.$$

Consequently, the transformed value of the test statistic is given by

$$\begin{aligned} A_\delta &= (\hat{\mu}_x^d - \hat{\mu}_y^d)^2 + (\hat{\sigma}_x^d - \hat{\sigma}_y^d)^2 = (\hat{\mu}_x^d - (a\hat{\mu}_x^d + b))^2 + (\hat{\sigma}_x^d - a\hat{\sigma}_x^d)^2 \\ &= b^2 + (1-a)^2\left([\hat{\mu}_x^d]^2 + [\hat{\sigma}_x^d]^2\right) - 2b(1-a)\hat{\mu}_x^d. \end{aligned} \tag{5.17}$$

The final three components quantify, respectively, the contributions of location shifts, scale changes, and their combined effect on the value of the test statistic. To investigate the impact of a single outlying observation, consider a transformation applied to only one data point. Specifically, assume that $Y_{1j} = aX_{1j} + b, \quad j = 1, \ldots, d$, *while all remaining rows of the observation matrix* $\mathbf{Y}$ *are left unchanged. Under this partial transformation, the corresponding sample mean and dispersion for the first observation satisfy*

$$\bar{Y}_1^d = a\bar{X}_1^d + b, \qquad \left(S_{\mathbf{Y}_1}^d\right)^2 = a^2\left(S_{\mathbf{X}_1}^d\right)^2.$$

Moreover, the new mean of the $\mathbf{Y}$ *sample is*

$$\begin{aligned}\hat{\mu}_{y(new)}^d &= \frac{1}{nd}\sum_{i=1}^{n}\sum_{j=1}^{d} Y_{ij} = \frac{1}{nd}\Big[\sum_{i=2}^{n}\sum_{j=1}^{d} Y_{ij} + \sum_{j=1}^{d}(aX_{1j}+b)\Big] \\ &= \frac{1}{nd}\Big[\sum_{i=2}^{n}\sum_{j=1}^{d} Y_{ij} + ap\bar{X}_1^d + db + \sum_{j=1}^{d} Y_{1j} - \sum_{j=1}^{d} Y_{1j}\Big] = \hat{\mu}_y^d + C_1,\end{aligned} \tag{5.18}$$

where $C_1 = \frac{1}{n}[a\bar{X}_1^d - \bar{Y}_1^d + b]$. *Similarly, the new standard deviation of the* $\mathbf{Y}$ *sample is*

$$\hat{\sigma}_{y(new)}^d = \frac{1}{n}\sum_{i=1}^{n} S_{\mathbf{Y}_i}^d = \frac{1}{n}[\sum_{i=2}^{n} S_{\mathbf{Y}_i}^d + S_{\mathbf{Y}_1}^d] = \frac{1}{n}[\sum_{i=2}^{n} S_{\mathbf{Y}_i}^d + |a|S_{\mathbf{X}_1}^d + S_{\mathbf{Y}_1}^d - S_{\mathbf{Y}_1}^d] = \hat{\sigma}_y^d + C_2,$$

where $C_2 = \frac{1}{n}(|a|S_{\mathbf{X}_1}^d - S_{\mathbf{Y}_1}^d)$. *Hence, the value of the test statistic is*

$$\begin{aligned}A_\delta(new) &= [\hat{\mu}_x^d - (\hat{\mu}_y^d + C_1)]^2 + [\hat{\sigma}_x^d - (\hat{\sigma}_y^d + C_2)]^2 \\ &= A_\delta + C_1^2 + C_2^2 - 2C_1(\hat{\mu}_x^d - \hat{\mu}_y^d) - 2C_2(\hat{\sigma}_x^d - \hat{\sigma}_y^d).\end{aligned} \tag{5.19}$$

Example 5.19.3 *We illustrate the effectiveness of the outlier visualization plot using a normal mixture model with heterogeneous covariance structures. A sample of* $N = 30$ *observations with* $d = 500$ *variables was generated from the mixture* $(1-q)\,\mathrm{N}_d(\mathbf{0}, \mathbf{\Sigma}_x) + q\,\mathrm{N}_d(\mathbf{0}, \mathbf{\Sigma}_y)$, *where* $q = 0.05$. *Here,* $\mathbf{\Sigma}_x$ *has unit variances and constant correlation* $\rho_x = 0.4$, *while* $\mathbf{\Sigma}_y$ *has variances* $1, 2, \ldots, d$ *and correlation* $\rho_y = 0.7$. *The resulting dataset includes three observations drawn from the* $\mathrm{N}_d(\mathbf{0}, \mathbf{\Sigma}_y)$ *component and 27 observations from the* $\mathrm{N}_d(\mathbf{0}, \mathbf{\Sigma}_x)$ *component. As shown in Figure 5.9, the three observations originating from the second mixture component are clearly distinguishable in the visualization, highlighting the utility of the plot for identifying outlying observations.*

5.19.3 Test for the Equality of Several Groups

Let $\mathbf{X}_i^{(k)}$ for $i = 1, \ldots, n_k$ denote the observations of the k-th group for $k = 1, \ldots, K$. Denote the mean and standard deviation of the groups with $\bar{X}_i^{d(k)}$ and $S_{\mathbf{X}_i}^{d(k)}$. Denote the grand mean and the grand standard deviation of the k-th group with

$$\hat{\mu}_k = \frac{1}{n_k}\sum_{i=1}^{n_i} \bar{X}_i^{d(k)} \text{ and } \hat{\sigma}_k = \frac{1}{n_k}\sum_{i=1}^{n_i} S_{\mathbf{X}_i}^{d(k)}. \tag{5.20}$$

The test statistic for comparing $K \geq 2$ groups is the sum of $K(K-1)/2$ two-way comparison and is given by

$$A_\delta = \sum_{t=1}^{K}\sum_{s>t=1}^{K} (\hat{\mu}_t - \hat{\mu}_s)^2 + (\hat{\sigma}_t - \hat{\sigma}_s)^2. \tag{5.21}$$

The sampling distribution of A_δ can be computed using permutations of the observations.

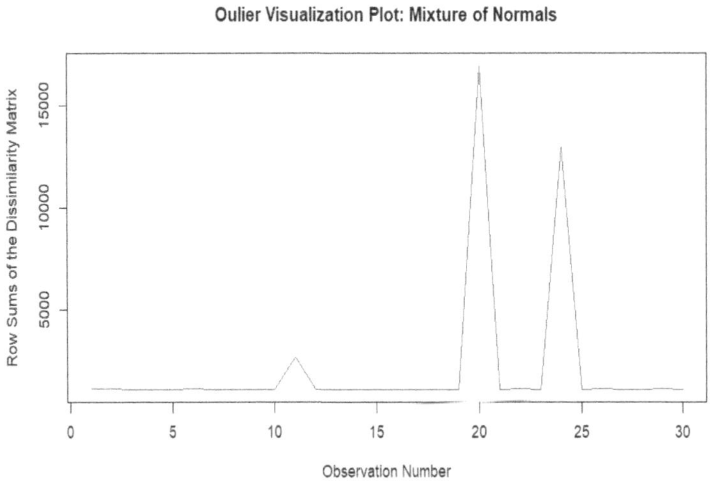

FIGURE 5.9: Row sums of the dissimilarities from the normal mixture with $d = 500$ variables.

5.19.4 Joint Display: OBS and VAR Plots

To obtain a joint display of observations means and variances from K groups we form a new $N \times d$ matrix

$$\mathbb{X}_{N\times d} = \begin{bmatrix} \mathbf{X}_i^{(1)}(n_1 \times d) \\ \vdots \\ \mathbf{X}_i^{(K)}(n_k \times d) \end{bmatrix}. \tag{5.22}$$

For the data matrix $\mathbb{X}$, we compute the row-wise means and standard deviations, denoted by $(\bar{X}_i^d, S_{\mathbf{X}_i}^d)$ for $i = 1, \ldots, N$. These quantities are then visualized by plotting the observation means against their corresponding standard deviations. To illustrate the joint behavior of observation means and variances, we conduct separate sampling experiments. In each experiment, the group sizes are equal with $n_1 = n_2 = n_3 = 30$, and the dimensionality is set to $d = 500$. The figures displayed on the left present the means and variances of the variables, whereas the figures on the right depict the means and standard deviations of the observations.

The left panel of Figure 5.10 shows the joint plot of means and standard deviations of the variables based on independent samples taken from Group 1: $\mathbb{N}(\mathbf{0}, 2\mathbf{I})$, Group 2: $\mathbb{N}(0.2\mathbf{1}_d, \mathbf{I})$ and Group 3, $\mathbb{N}(-0.2\mathbf{1}_d, \mathbf{I})$. The right panel of Figure 5.10 shows joint plot of means and standard deviations of the observations. Clearly, one sees a separation between the three groups in this Figure. The left panel of Figure 5.11 shows the joint plot of means and standard deviations of the variables based on independent samples taken from Group 1: $\chi^2_{(2)}$, Group 2: $\chi^2_{(3)}$, and Group 3: $\frac{1}{4}\chi^2_{(4)}$. The right panel of Figure 5.11 shows joint plot of means and standard deviations of the observations. The Figures show that displaying the means and standard deviations of high-dimensional datasets can provide insight into the behavior of the observations.

5.19.5 Analysis of OBS and VAR Plots

We consider two-group comparisons under different distributional settings. For each setting, we analyze two graphical summaries:

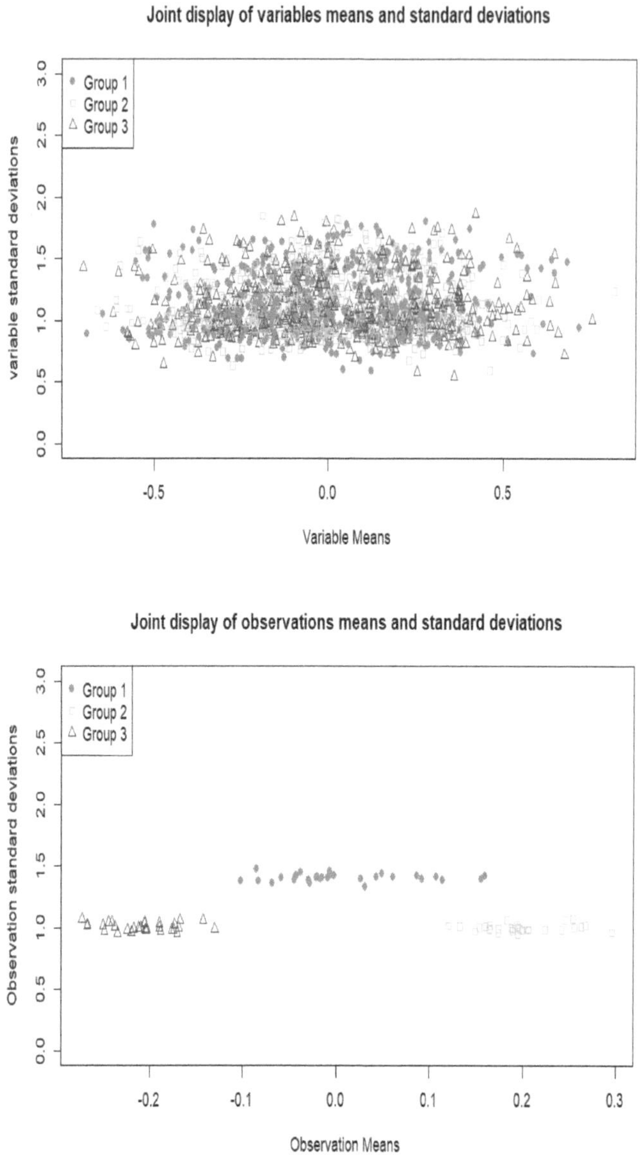

FIGURE 5.10: Left panel: Variable means and standard deviations for normal distributions. Right panel: Observation means and standard deviation for normal distributions.

VAR plot (variable-wise analysis). For a data matrix $\mathbf{X}$ of size $n_1 \times d$ and $\mathbf{Y}$ of size $n_2 \times d$, we compute column means and column standard deviations:

For group 1:

$$\bar{X}_{.j} = \frac{1}{n_1}\sum_{i=1}^{n_1} X_{ij}, \qquad S_{X_{.j}} = \sqrt{\frac{1}{n_1}\sum_{i=1}^{n_1}(X_{ij} - \bar{X}_{.j})^2}.$$

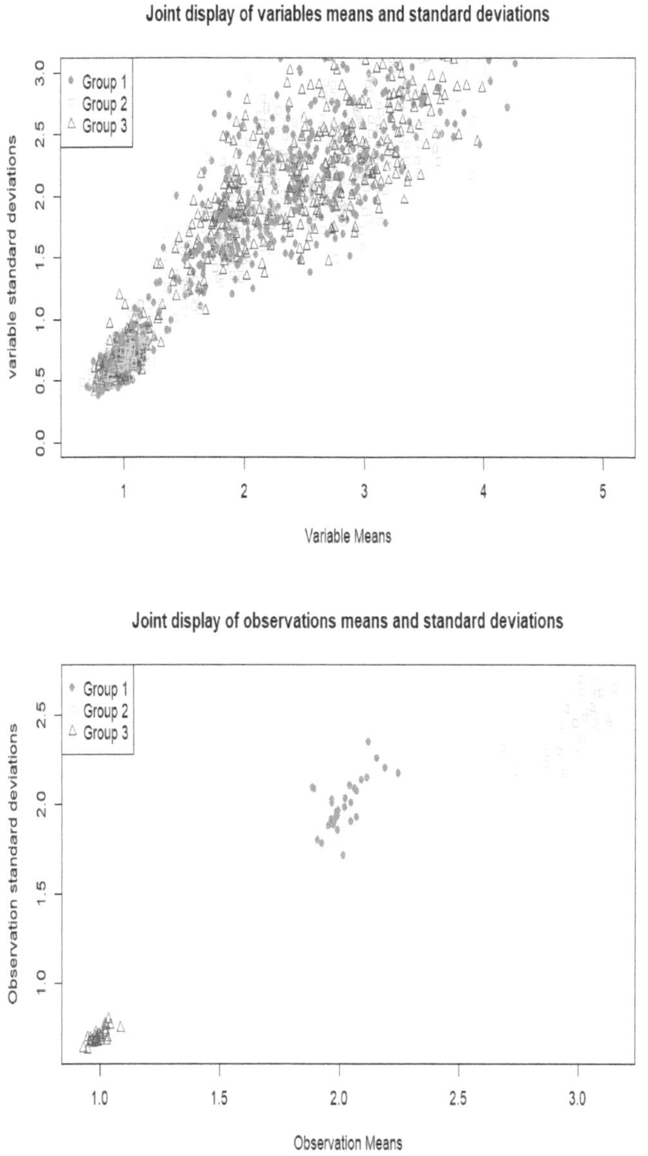

FIGURE 5.11: Left panel: Variable means and standard deviations for χ^2 distributions. Right panel: Observation means and standard deviations: χ^2 distributions.

For group 2:

$$\bar{Y}_{\cdot j} = \frac{1}{n_2}\sum_{i=1}^{n_2} Y_{ij}, \qquad S_{Y_{\cdot j}} = \sqrt{\frac{1}{n_2}\sum_{i=1}^{n_2}(Y_{ij} - \bar{Y}_{\cdot j})^2}.$$

The VAR plot consists of the $2d$ points $(\bar{X}_{\cdot j}, S_{X_{\cdot j}})$, and $(\bar{Y}_{\cdot j}, S_{Y_{\cdot j}})$.

OBS plot (observation-wise analysis). Row-wise mean and standard deviation for group 1:

$$\bar{X}_i^d = \frac{1}{d}\sum_{j=1}^{d} X_{ij}, \qquad S_{\mathbf{X}_i}^d = \sqrt{\frac{1}{d}\sum_{j=1}^{d}(X_{ij} - \bar{X}_i^d)^2}.$$

Similarly for group 2:

$$\bar{Y}_i^d = \frac{1}{d}\sum_{j=1}^{d} Y_{ij}, \qquad S_{\mathbf{Y}_i}^d = \sqrt{\frac{1}{d}\sum_{j=1}^{d}(Y_{ij} - \bar{Y}_i^d)^2}.$$

The OBS plot contains $n_1 + n_2$ points $(\bar{X}_i^d, S_{\mathbf{X}_i}^d)$ and $(\bar{Y}_i^d, S_{\mathbf{Y}_i}^d)$.

We look for the separation in the OBS plot for four different cases. Note that with large d the VAR plot will have Kd pairs of observations, where K is the number of groups, and thus cluttered, whereas the OBS plot has N pairs to plot regardless of d or K.

We generated $n_1 = 20$ and $n_2 = 30$ observations with $d = 500$ from

- $F_1 = \mathbb{N}(\mathbf{0}, \mathbf{I})$ and $F_2 = \mathbb{N}(0.3\ \mathbf{1}, \mathbf{I})$,
- $F_1 = \chi_3^2$ and $F_2 = \chi_6^2$,
- $F_1 = \beta(2, 5)$ and $F_2 = \beta(3, 5)$,
- $F_1 = \beta(2, 5)$ and $F_2 = \beta(5, 5)$.

For the normal shift model the column statistics are,

$$\bar{X}_{\cdot j} \sim \mathbb{N}\left(0, \frac{1}{n_1}\right), \qquad \bar{Y}_{\cdot j} \sim \mathbb{N}\left(\mu, \frac{1}{n_2}\right).$$

Standard deviations converge to 1. Separation in VAR plot (difference between the means after scale adjustment) is governed by $\text{Sep}_{VAR} = \mu\sqrt{n}$. The row statistics are

$$\bar{X}_i^d \sim \mathbb{N}\left(0, \frac{1}{d}\right), \qquad \bar{Y}_i^d \sim \mathbb{N}\left(\mu, \frac{1}{d}\right).$$

Thus, separation in OBS plot is governed by $\text{Sep}_{OBS} = \mu\sqrt{d}$. If $d \to \infty$ with N fixed, then $\mu\sqrt{d} \to \infty$, and we get perfect separation in OBS plot. Hence, OBS plot detects mean shifts in high dimension. Figure 5.12 shows the observation and variable plots for the distributions.

For the chi-squared distributions, means and variances are $\mathbb{E}(X) = 3$, $\mathbb{V}ar(X) = 6$, $\mathbb{E}(Y) = 6$, and $\mathbb{V}ar(Y) = 12$. OBS plot separates when $df_x \neq df_y$ as $d \to \infty$. Figure 5.13 shows the observation and variable plots for the distributions.

For the beta distribution with mean shift, row mean variance scales as $1/d$. Thus, if $d \to \infty$, OBS separates the groups. Figure 5.14 shows the observation and variable plots for the distributions.

For the beta shift model, row mean variance scales as $1/d$. Signal strength larger than the previous case. Figure 5.15 shows the observation and variable plots for the distributions.

The separation in OBS plots stand out and the groups are completely visible even though $d = 500$. The phenomenon is driven by variance reduction of $1/d$ for row means versus $1/n$ for column means.

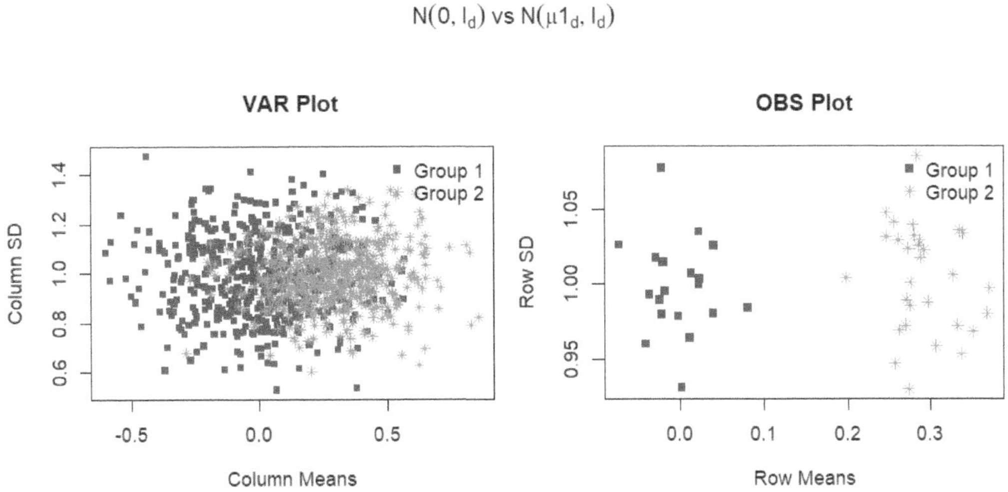

FIGURE 5.12: VAR and OBS plots for the means and standard deviations of the normal distribution.

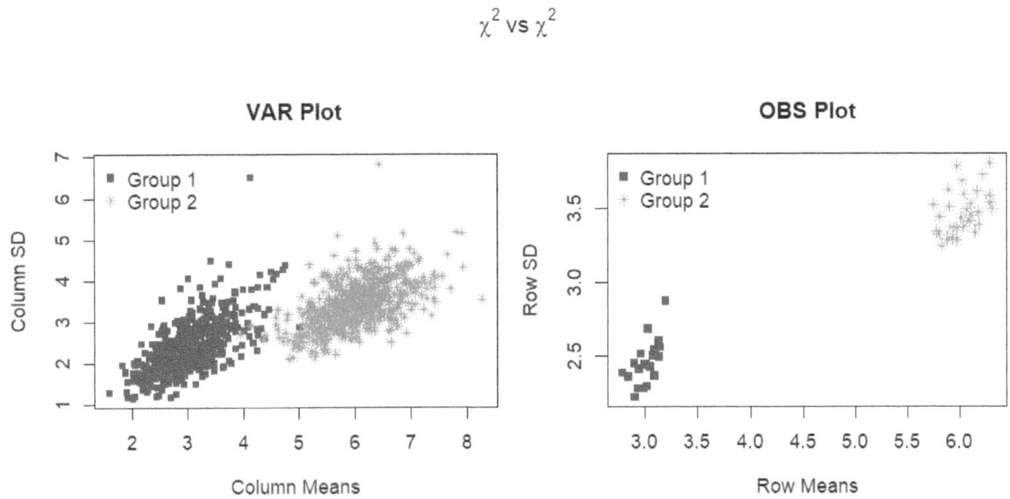

FIGURE 5.13: VAR and OBS plots for the means and standard deviations of χ^2 distribution.

Example 5.19.4 *Limiting Dissimilarity Measures: $\tilde{\delta}(F_1, F_2)$ and $\tilde{\rho}(F_1, F_2)$. Let f and g be two distributions in $\mathbb{R}^d$,*

$$f(\mathbf{x}) = \alpha\,\mathbb{N}(\mu_1 \mathbf{1}_d, \sigma_1^2 \mathbf{I}_d) + (1-\alpha)\,\mathbb{N}(\mu_2 \mathbf{1}_d, \sigma_2^2 \mathbf{I}_d),$$
$$g(\mathbf{y}) = p\,\mathbb{N}(\mu_3 \mathbf{1}_d, \sigma_3^2 \mathbf{I}_d) + (1-p)\,\mathbb{N}(\mu_4 \mathbf{1}_d, \sigma_4^2 \mathbf{I}_d).$$

Define the mixture means and variances by

$$\bar{\mu}_x = \alpha\mu_1 + (1-\alpha)\mu_2, \quad \bar{\mu}_y = p\mu_3 + (1-p)\mu_4, \quad \nu^2 = (\bar{\mu}_x - \bar{\mu}_y)^2,$$
$$\sigma_{F_1}^2 = \alpha\sigma_1^2 + (1-\alpha)\sigma_2^2 + \alpha(1-\alpha)(\mu_1 - \mu_2)^2,$$
$$\sigma_{F_2}^2 = p\sigma_3^2 + (1-p)\sigma_4^2 + p(1-p)(\mu_3 - \mu_4)^2.$$

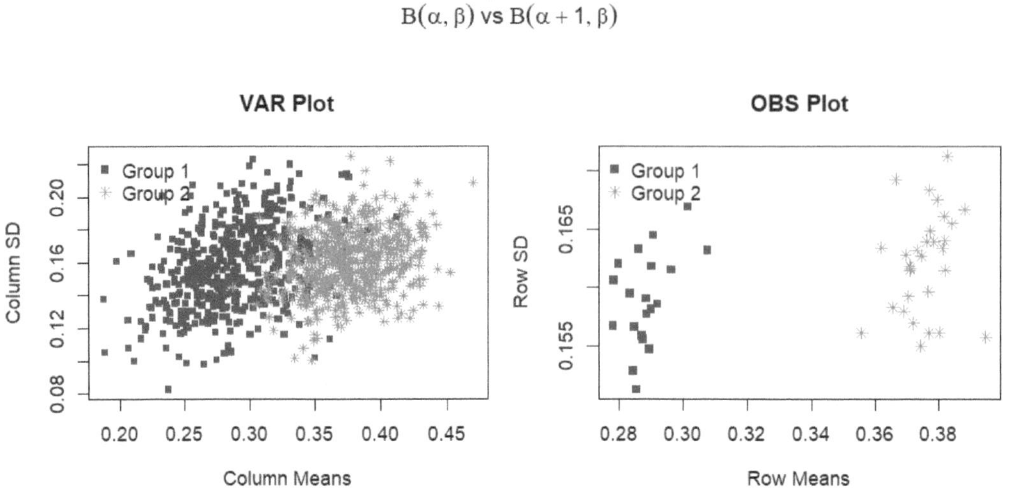

FIGURE 5.14: VAR and OBS plots for the means and standard deviations of the beta distribution with α shift.

Limiting Dissimilarity Measures $\tilde{\delta}(F_1, F_2)$ *is*

$$\tilde{\delta}(F_1, F_2) = \sqrt{\nu^2 + (\sigma_{F_1} - \sigma_{F_2})^2}$$

Assuming $m = n$*, so* $N = 2m$*, we get*

$$\tilde{\rho}(F_1, F_2) = \frac{m-1}{2m-2}\left(\left|\sqrt{2}\sigma_{F_1} - \sqrt{\sigma_{F_1}^2 + \sigma_{F_2}^2 + \nu^2}\right| + \left|\sqrt{\sigma_{F_1}^2 + \sigma_{F_2}^2 + \nu^2} - \sqrt{2}\sigma_{F_2}\right|\right)$$

Equal Means, Unequal Variances. Assume $\mu_1 = \mu_2 = \mu_3 = \mu_4 = \mu$*, so* $\bar{\mu}_x = \bar{\mu}_y = \mu$*, and* $\nu^2 = 0$*. We have* $\tilde{\delta}(F_1, F_2) = |\sigma_{F_1} - \sigma_{F_2}|$, $\sigma_{F_2}^2 = p\sigma_3^2 + (1-p)\sigma_4^2$ *and*

$$\tilde{\delta}(F_1, F_2) = |\sigma_{F_1} - \sigma_{F_2}|\,.$$

$$\tilde{\rho}(F_1, F_2) = \frac{m-1}{2m-2}\left(\left|\sqrt{2}\sigma_{F_1} - \sqrt{\sigma_{F_1}^2 + \sigma_{F_2}^2}\right| + \left|\sqrt{\sigma_{F_1}^2 + \sigma_{F_2}^2} - \sqrt{2}\sigma_{F_2}\right|\right)$$

Equal Variances, Unequal Means. Assume $\sigma_1^2 = \sigma_2^2 = \sigma_3^2 = \sigma_4^2 = \sigma^2$*. We have*

$$\sigma_{F_1}^2 = \sigma^2 + \alpha(1-\alpha)(\mu_1 - \mu_2)^2, \quad \sigma_{F_2}^2 = \sigma^2 + p(1-p)(\mu_3 - \mu_4)^2,$$

$$\tilde{\delta}(F_1, F_2) = \sqrt{\nu^2 + (\sigma_{F_1} - \sigma_{F_2})^2},$$
$$\tilde{\rho}(F_1, F_2) = \frac{m-1}{2m-2}\left(\left|\sqrt{2}\sigma_{F_1} - \sqrt{\sigma_{F_1}^2 + \sigma_{F_2}^2 + \nu^2}\right| + \left|\sqrt{\sigma_{F_1}^2 + \sigma_{F_2}^2 + \nu^2} - \sqrt{2}\sigma_{F_2}\right|\right)$$

Figure 5.16 displays the VAR and OBS plots of the above two mixtures with

$$f(\mathbf{x}) = 0.5\,\mathbb{N}(0\mathbf{1}_d, \mathbf{I}_d) + 0.5\,\mathbb{N}(1\mathbf{1}_d, \mathbf{I}_d),$$

$$g(\mathbf{y}) = 0.4\,\mathbb{N}(2\mathbf{1}_d, \mathbf{I}_d) + (1-p)\,\mathbb{N}(3\mathbf{1}_d, 4\mathbf{I}_d).$$

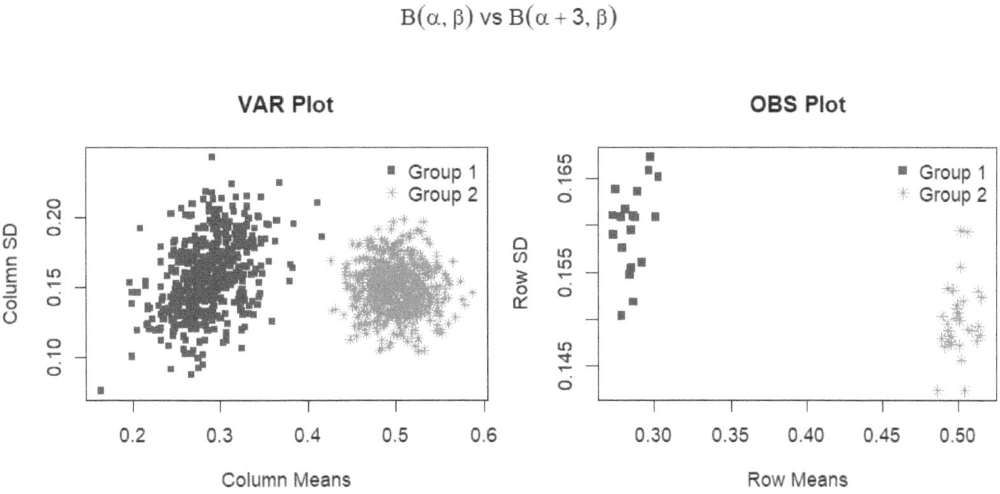

FIGURE 5.15: VAR and OBS plots for the means and standard deviations of the beta distribution with a larger α shift.

While the VAR plot shows two clusters corresponding to the mixtures, the OBS plot shows four clusters that are the mixture components, two in group 1 and two in group 2.

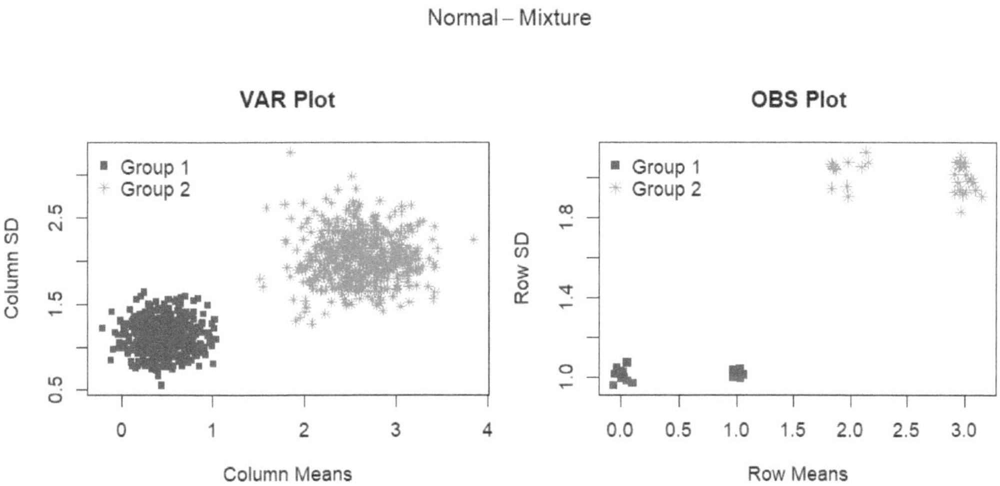

FIGURE 5.16: VAR and OBS plots for the means and standard deviations of the mixture of two normal distributions. The OBS plot correctly identifies the components of each mixture.

5.19.6 New Statistics EN_δ and BG_δ

The energy statistic of Székely and Rizzo [330] and the test proposed by Biswas and Ghosh [35] are widely recognized as powerful and commonly used methods for testing the equality of high-dimensional distribution functions. These test statistics are formulated using IPDs and remain applicable even when the dimension d far exceeds the sample sizes. In order to compare their statistical power, we concentrate on high-dimensional tests that are based on IPDs. Moreover, Székely and Rizzo [333], Baringhaus and Franz ([21], and Biswas and

Ghosh [35] establish necessary and sufficient conditions for the equality of vector-valued distributions. To test $H_0 : F_1 = F_2$ Székely and Rizzo [333], Baringhaus and Franz [21] propose the statistic

$$EN_E = 2\bar{d}_{(xy)} - \bar{d}_{(x)} - \bar{d}_{(y)}, \tag{5.23}$$

where $\bar{d}_{(x)}$, $\bar{d}_{(y)}$ and $\bar{d}_{(xy)}$ are defined in (2.7), (2.11), and (2.15), respectively. We reject H_0 for large values of EN_E. This statistic is powerful against shift alternatives. Biswas and Ghosh [35] proposed the statistic

$$BG_E = (\bar{d}_{(xy)} - \bar{d}_{(x)})^2 + (\bar{d}_{(xy)} - \bar{d}_{(y)})^2. \tag{5.24}$$

Note that both EN_E and BG_E statistics use the Euclidean distance to measure dissimilarity.

To obtain the BG and EN statistics using the dissimilarity measure δ, we compute

$$\bar{\delta}_{(x)} = \frac{2}{m(m-1)} \sum_{i=1}^{m-1} \sum_{j=i+1}^{m} \delta(\mathbf{X}_i, \mathbf{X}_j), \quad \bar{\delta}_{(y)} = \frac{2}{n(n-1)} \sum_{i=1}^{n-1} \sum_{j=i+1}^{n} \delta(\mathbf{Y}_i, \mathbf{Y}_j), \tag{5.25}$$

and $\bar{\delta}_{(xy)} = \frac{1}{mn} \sum_{i=1}^{m} \sum_{j=1}^{n} \delta(\mathbf{X}_i, \mathbf{Y}_j)$. The new statistics are

$$EN_\delta = 2\bar{\delta}_{(xy)} - \bar{\delta}_{(x)} - \bar{\delta}_{(y)}, \tag{5.26}$$

$$BG_\delta = (\bar{\delta}_{(xy)} - \bar{\delta}_{(x)})^2 + (\bar{\delta}_{(xy)} - \bar{\delta}_{(y)})^2. \tag{5.27}$$

The statistics EN_δ, and BG_δ are extensions of existing statistics EN and BG that use the dissimilarity δ. They are introduced to make comparisons with A_δ. To obtain the BG and EN statistics using dissimilarity measure ρ, we compute

$$\bar{\rho}_{(x)} = \frac{2}{m(m-1)} \sum_{i=1}^{m-1} \sum_{j=i+1}^{m} \rho(\mathbf{X}_i, \mathbf{X}_j), \quad \bar{\rho}_{(y)} = \frac{2}{n(n-1)} \sum_{i=1}^{n-1} \sum_{j=i+1}^{n} \rho(\mathbf{Y}_i, \mathbf{Y}_j), \tag{5.28}$$

and $\bar{\rho}_{(xy)} = \frac{1}{mn} \sum_{i=1}^{m} \sum_{j=1}^{n} \rho(\mathbf{X}_i, \mathbf{Y}_j)$. The test statistics are

$$EN_\rho = 2\bar{\rho}_{(xy)} - \bar{\rho}_{(x)} - \bar{\rho}_{(y)}, \quad BG_\rho = (\bar{\rho}_{(xy)} - \bar{\rho}_{(x)})^2 + (\bar{\rho}_{(xy)} - \bar{\rho}_{(y)})^2. \tag{5.29}$$

The test statistics EN_ρ and BG_ρ are proposed by Sarkar and Ghosh [299].

5.19.7 Analysis of Simulated Datasets

We have previously established that the time complexity of A_δ is $O(dN)$, whereas methods relying on the computation of IPDs exhibit a time complexity of $O(dN^2)$. Consequently, for large N, A_δ offers a computational advantage. Indeed, the competing methods were originally designed for HDLSS scenarios. To compare performance, we generated 1000 samples of size $m = n = 20$ with dimension $d = 100$, and evaluated the following methods

(a) the energy statistics EN_E, EN_δ, and EN_ρ;

(b) the BG statistics BG_E, BG_δ, and BG_ρ;

(c) the limiting form of the δ statistic, A_δ defined in (5.21); and

(d) the limiting form of the ρ statistic, A_ρ.

Although A_δ can be computed efficiently, we restricted the experiments to small sample sizes due to the substantial computational demands of the IPD-based methods. The following plots illustrate the empirical power of the statistics under a shift model. Random permutations were employed to compute the significance levels and power of the tests, and the observed alpha levels were all approximately 0.05. In this experiment, the dimension d is fixed at 100, and the observations are shifted by μ under the alternative hypothesis. In

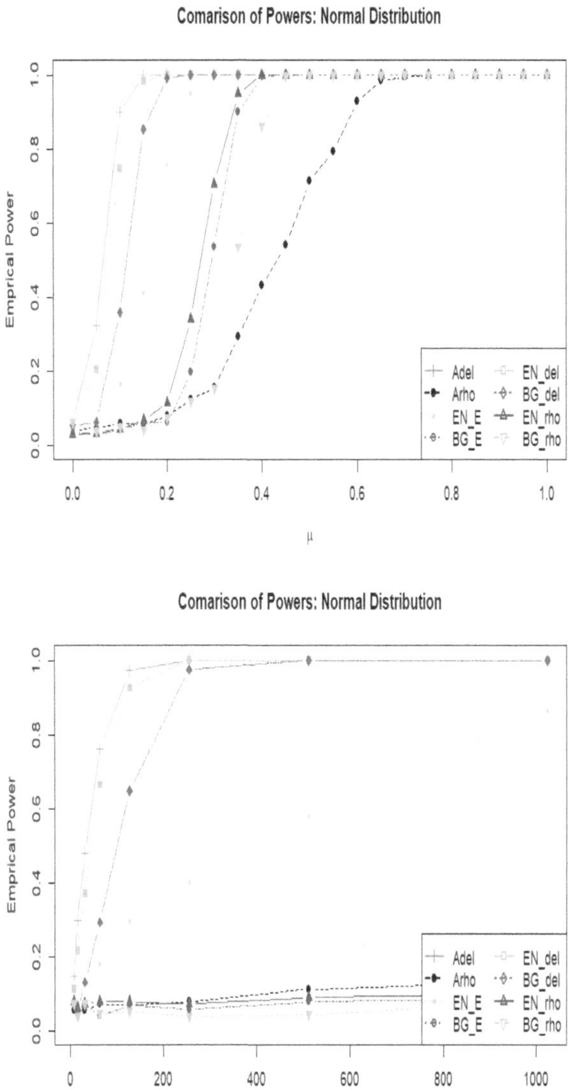

FIGURE 5.17: Comparison of estimated power. Left panel: $\mathbf{X} = \mathbb{N}(\mathbf{0}, \mathbf{I})$, $\mathbf{Y} = \mathbb{N}(\mu\ \mathbf{1}, \mathbf{I})$, Right Panel: $\mathbf{X} \sim \mathbb{N}(\mathbf{0}, \mathbf{I})$, $\mathbf{Y} \sim \mathbb{N}(0.1\ \mathbf{1}, \mathbf{I})$.

the left panel of Figure 5.17 you see the power of the eight statistics based on independent samples taken from $\mathbf{X} \sim \mathbb{N}(\mathbf{0}, \mathbf{I})$ and $\mathbf{Y} \sim \mathbb{N}(\mu\ \mathbf{1}_d, \mathbf{I})$ where $\mu = \{0, 0.05, 0.10, \ldots, 1\}$. The power ordering is $A_\delta > EN_\delta > BG_\delta > EN_E > EN_\rho < BG_E < BG_\rho < A_\rho$. Using independent samples from $\mathbf{X} \sim U(0,1)$ and $\mathbf{Y} \sim U(0,1) + \mu$ for $\mu = \{0, 0.01, 0.02, \ldots, 0.15\}$, the power ordering remains the same.

In the next experiment we are interested in the power of these statistics for fixed alternatives as $d \to \infty$. We generated 1000 samples of sizes $m = n = 20$ from the above distributions and computed the power for a fixed shift of 0.1 when $d = 2^i$, $3 \leq i \leq 10$. In the right panel of Figures 5.17 you see the power of the statistics for increasing d. The plots show that the power ordering is $A_\delta > EN_\delta > BG_\delta > EN_E$. The other statistics do not show much power as $d \to \infty$.

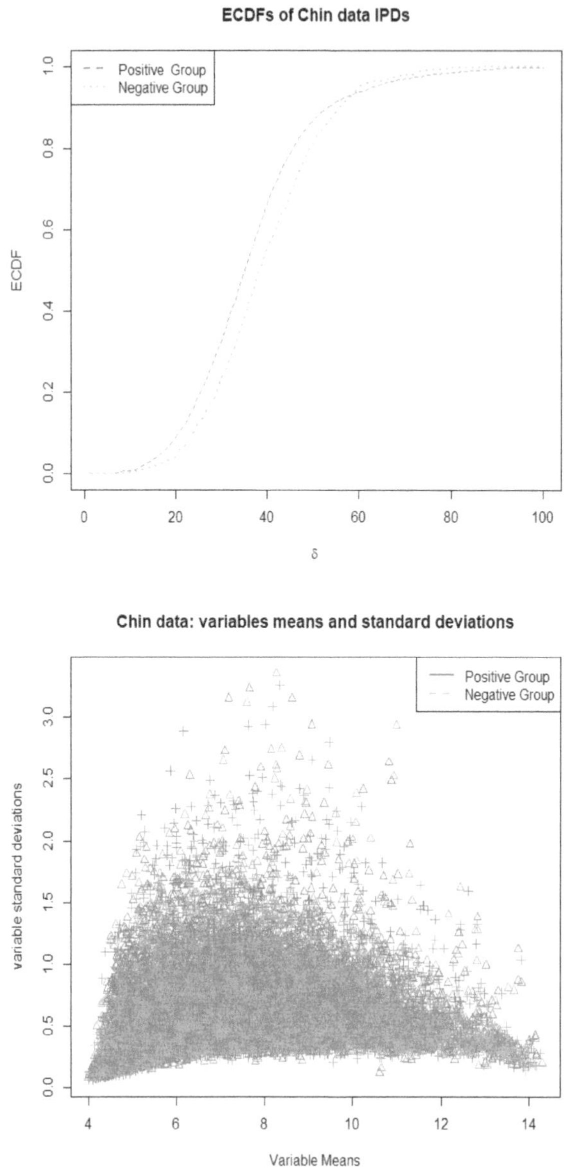

FIGURE 5.18: Left panel: Empirical distribution function of Chin IPDS. Right panel: Means and standard deviations of the variables for Chin data.

5.19.8 Analysis of Chin Gene Expression Data for Breast Cancer

Chin et al. [66] present gene expression data pertaining to breast cancer. The dataset comprises 118 samples, including 75 positive and 63 negative cases, across 22,215 variables. The data can be accessed through the microarray library in R. The left panel of Figure 5.18 displays the empirical cumulative distribution functions (ECDFs) of the mean inter-point distances (IPDs) for the two groups. The close alignment of these ECDFs suggests that the underlying distributions are similar but shifted relative to one another.

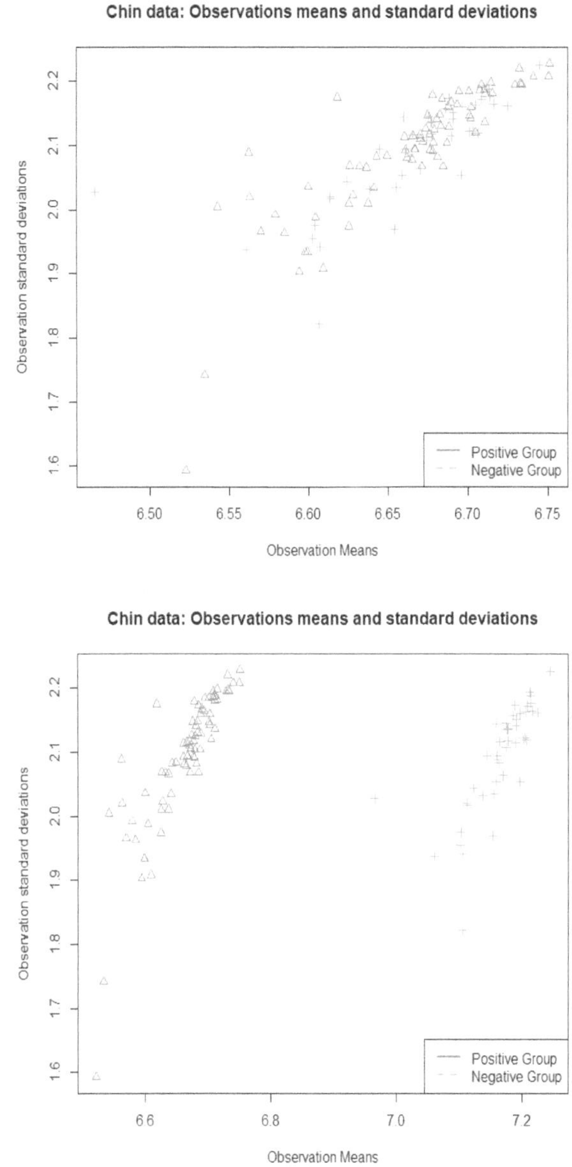

FIGURE 5.19: Left panel: Means and standard deviations of the observations for Chin data. Right panel: Means and standard deviations of the observations for Chin after 0.5 shift of the data.

To formally assess whether the distributions of the positive and negative groups are equivalent, we computed eight distinct test statistics using the observed data. P-values were estimated via 1,000 random permutations for each statistic. All eight resulting p-values were zero, providing strong evidence to reject the hypothesis of equality between the distributions. The right panel of Figure 5.18 shows the means and standard deviations of the variables in the Chin dataset. The left panel of Figure 5.19 illustrate the means and standard deviations of the observations and highlights three potential outliers located on the left side of the plot. We further hypothesized that the distribution of the negative group represents a shifted version of the positive group. We see in the right panel of Figure 5.19 the means and standard deviations after the negative group has been adjusted by a shift of 0.5, corroborating the presence of a consistent shift separating the two distributions.

Modarres [238] demonstrates that, for the Chin dataset, the relative variance diminishes toward zero as $d \to \infty$, indicating that the distances in this high-dimensional dataset exhibit concentration phenomena.

5.20 Summary

This chapter provides a comprehensive treatment of statistical tests for assessing the equality of high-dimensional distributions, a problem that arises frequently in modern data analysis where classical multivariate tests often fail due to dimensionality, sparsity, or non-normality. The chapter begins by motivating the need for non-parametric and geometry-based approaches that remain effective when the dimension is comparable to or exceeds the sample size. Emphasis is placed on distribution-free methods that uses IPDs, graph structures, and rank-based ideas to detect differences in location, scale, and overall distributional shape.

A broad class of graph and distance based tests is introduced, including the Friedman–Rafsky runs test, Rosenbaum's cross-match test, and test of Hall–Tajvidi, Mondal–Biswas–Ghosh, and Biswas–Ghosh. Energy statistics and related tests are presented as powerful omnibus procedures capable of detecting general distributional differences. The chapter also discusses permutation-based inference as a unifying framework for calibrating these tests in finite samples. Classical procedures such as Hotelling's T^2, Puri–Sen rank tests for location and dispersion, and Bartlett's test are revisited to highlight their limitations and adaptations in high-dimensional contexts.

The latter part of the chapter introduces novel methodologies, including tests for matrix-valued distributions, the triangle test and a general class of dissimilarity-based tests. These methods are extended to multi-sample settings and accompanied by new test statistics, joint graphical displays, and theoretical justification. Simulation studies and real-data applications illustrate the comparative performance of the proposed procedures. Collectively, the chapter establishes a unifying perspective on high-dimensional two- and multi-sample testing, balancing theoretical rigor, computational feasibility, and practical effectiveness.

5.21 Proofs

Proof of Lemma 5.16.1

Since $\mathbf{X}_i, i = 1, \ldots, m$, are i.i.d. and $\mathbf{Y}_j, j = 1, \ldots, n$, are i.i.d., $d_{1k}(\mathbf{X}_i), i = 1, \ldots, m$, are bounded, i.i.d. with mean θ_k and variance α_{1kk}. Similarly, $q_{1k}(\mathbf{Y}_j), j = 1, \ldots, n$, are bounded, i.i.d. with mean θ_k and variance β_{1kk}. Under condition (1), for $k = 1, 2, 3$, by the central limit theorem under Lindeberg-Levy condition,

$$\frac{1}{m}\sum_{i=1}^{m}(p_{1k}(\mathbf{X}_i) - \theta_k) \xrightarrow{D} \mathbb{N}(0, \frac{\alpha_{1kk}}{m}),$$

$$\frac{1}{n}\sum_{j=1}^{n}(q_{1k}(\mathbf{Y}_j) - \theta_k) \xrightarrow{D} \mathbb{N}(0, \frac{\beta_{1kk}}{m}).$$

Thus, $\hat{U}_k$ is asymptotically normal with mean 0 and variance $\frac{\alpha_{1kk}}{m} + \frac{\beta_{1kk}}{m}$. Similarly, we can show that $\hat{V}_k$ is asymptotically normal with mean 0 and variance $\frac{\alpha_{2kk}}{m} + \frac{\beta_{2kk}}{m}$. Furthermore, any vector with linearly independent components of $\hat{U}_k$'s and $\hat{V}_k$'s are asymptotically multivariate normal by the multivariate version of the Lindeberg-Levy central limit theorem. We also obtain the elements of dispersion matrix, v_{st} and w_{st},

$$\begin{aligned}
u_{st} &= \mathbb{C}ov(\hat{U}_s, \hat{U}_t) \\
&= \mathbb{C}ov(\frac{1}{m}\sum_{i=1}^{m}(p_{1s}(\mathbf{X}_i) - \theta_s), \frac{1}{m}\sum_{i=1}^{m}(p_{1t}(\mathbf{X}_i) - \theta_t)) \\
&+ \mathbb{C}ov(\frac{2}{n}\sum_{j=1}^{n}(q_{1s}(\mathbf{Y}_j) - \theta_s), \frac{2}{n}\sum_{j=1}^{n}(q_{1t}(\mathbf{Y}_j) - \theta_t)) \\
&= \frac{1}{m^2}\sum_{i=1}^{m}\mathbb{C}ov(p_{1s}(\mathbf{X}_i) - \theta_s, p_{1t}(\mathbf{X}_i) - \theta_t) + \frac{4}{n^2}\sum_{j=1}^{n}\mathbb{C}ov(q_{1s}(\mathbf{X}_i) - \theta_s, q_{1t}(\mathbf{X}_i) - \theta_t) \\
&= \frac{1}{m^2}\sum_{i=1}^{m}\alpha_{1st} + \frac{4}{n^2}\sum_{j=1}^{n}\beta 1st = \frac{1}{m}\alpha_{1st} + \frac{4}{n}\beta_{1st},
\end{aligned}$$

$$v_{st} = \mathbb{C}ov(\hat{V}_s, \hat{V}_t) = \frac{1}{m}\alpha_{2st} + \frac{4}{n}\beta_{2st}, \qquad w_{st} = \mathbb{C}ov(\hat{U}_s, \hat{V}_t) = \frac{2}{m}\gamma_{1st} + \frac{2}{m}\gamma_{2ts}.$$

Proof of Lemma 5.16.2

To show that $(U_k - \theta_k) - \hat{U}_k \xrightarrow{P} 0$, we first obtainvspace*4pt

$$\begin{aligned}
\mathbb{C}ov(U_s, U_t) &= \mathbb{C}ov(\frac{1}{m\binom{n}{2}}\sum_{r=1}^{m}\sum_{i<j}^{n} I_{1,s}[x_r; y_i, y_j], \frac{1}{m\binom{n}{2}}\sum_{r'=1}^{m}\sum_{i'<j'}^{n} I_{1,t}[x_{r'}; y_{i'}, y_{j'}]) \\
&= A_1 + A_2 + A_3 + A_4 + A_5,
\end{aligned}$$

where A_1 collects the covariance terms of $I_{1,s}[x_r; y_i, y_j]$ and $I_{1,t}[x_{r'}; y_{i'}, y_{j'}]$ that has only x_r in common,

$$\begin{aligned}
A_1 &= \frac{m\binom{n}{2}\binom{n-2}{2}}{m^2\binom{n}{2}^2}\mathbb{C}ov(I_{1,s}[x_r; y_i, y_j], I_{1,t}[x_r; y_{i'}, y_{j'}]) \\
&= \frac{\binom{n-2}{2}}{m\binom{n}{2}}\mathbb{E}I_{1,s}[x_r; y_i, y_j] - \theta s) \cdot (I_{1,t}[x_r; y_{i'}, y_{j'}] - \theta_t) \\
&= \frac{\binom{n-2}{2}}{m\binom{n}{2}}\mathbb{E}\mathbb{E}((I_{1,s}[x_r; y_i, y_j] - \theta s) \cdot (I_{1,t}[x_r; y_{i'}, y_{j'}] - \theta_t)|x_r)) \\
&= \frac{\binom{n-2}{2}}{m\binom{n}{2}}\mathbb{E}p_{1s}(x_r) - \theta s) \cdot (p_{1t}(x_r) - \theta_t) = \frac{\binom{n-2}{2}}{m\binom{n}{2}}\alpha_{1st} = \frac{1}{m}\alpha_{1st} + o(m^{-1}),
\end{aligned}$$

A_2 collects the covariance terms of $I_{1,s}[x_r; y_i, y_j]$ and $I_{1,t}[x_{r'}; y_{i'}, y_{j'}]$ that has only one of y_i or y_j in common, say y_{i*},

$$A_2 = \frac{2m\binom{n}{2}(m-1)(n-2)}{m^2\binom{n}{2}^2}\mathbb{C}ov(I_{1,s}[x_r; y_i, y_j], I_{1,t}[x_{r'}; y_i, y_{j'}])$$

$$= \frac{2(m-1)(n-2)}{m\binom{n}{2}}\mathbb{E}I_{1,s}[x_r; y_{i*}, y_j] - \theta s) \cdot (I_{1,t}[x_{r'}; y_{i*}, y_{j'}] - \theta_t)$$

$$= \frac{2(m-1)(n-2)}{m\binom{n}{2}}\mathbb{E}\mathbb{E}((I_{1,s}[x_r; y_i, y_j] - \theta s) \cdot (I_{1,t}[x_r; y_{i'}, y_{j'}] - \theta_t)|y_{i*}))$$

$$= \frac{2(m-1)(n-2)}{m\binom{n}{2}}\mathbb{E}q_{1s}(y_{i*}) - \theta s) \cdot (q_{1t}(y_{i*}) - \theta_t)$$

$$= \frac{2(m-1)(n-2)}{m\binom{n}{2}}\beta_{1st} = \frac{1}{n}\beta_{1st} + o(n^{-1}),$$

A_3 collects the covariance terms of $I_{1,s}[x_r; y_i, y_j]$ and $I_{1,t}[x_{r'}; y_{i'}, y_{j'}]$ that has x_r and one of y_i or y_j in common, say y_{i*},

$$A_3 = \frac{2m\binom{n}{2}(n-2)}{m^2\binom{n}{2}^2}\mathbb{C}ov(I_{1,s}[x_r; y_{i*}, y_j], I_{1,t}[x_r; y_{i*}, y_{j'}]) = o(m^{-1}),$$

A_4 collects the covariance terms of $I_{1,s}[x_r; y_i, y_j]$ and $I_{1,t}[x_{r'}; y_{i'}, y_{j'}]$ that has both y_i and y_j in common,

$$A_4 = \frac{m\binom{n}{2}(m-1)}{m^2\binom{n}{2}^2}\mathbb{C}ov(I_{1,s}[x_r; y_i, y_j], I_{1,t}[x_{r'}; y_i, y_j]) = o(m^{-1}),$$

A_5 collects the covariance terms of $I_{1,s}[x_r; y_i, y_j]$ and $I_{1,t}[x_{r'}; y_{i'}, y_{j'}]$ that has exactly the same x_r, y_i, and y_j in common,

$$A_5 = \frac{m\binom{n}{2}}{m^2\binom{n}{2}^2}\mathbb{C}ov(I_{1,s}[x_r; y_i, y_j], I_{1,t}[x_{r'}; y_i, y_j]) = o(m^{-1}).$$

Therefore, $\mathbb{C}ov(U_s, U_t) = \frac{1}{m}\alpha_{1st} + \frac{4}{n}\beta_{1st} + o(m^{-1})$. Also, we obtain

$$\begin{aligned}
&\mathbb{E}(\hat{U}_k(U_k - \theta_k)) \\
=&\mathbb{E}((\frac{1}{m}\sum_{i=1}^{m}(p_{1k}(\mathbf{X}_i) - \theta_k) + \frac{2}{n}\sum_{j=1}^{n}(q_{1k}(\mathbf{Y}_j) - \theta_k)) \cdot (U_k - \theta_k)) \\
=&\mathbb{E}((\frac{1}{m}\sum_{i=1}^{m}(p_{1k}(\mathbf{X}_i) - \theta_k)) \cdot (U_k - \theta_k)) + \mathbb{E}((\frac{2}{n}\sum_{j=1}^{n}(q_{1k}(\mathbf{Y}_j) - \theta_k)) \cdot (U_k - \theta_k)) \\
=&\mathbb{E}\mathbb{E}((\frac{1}{m}\sum_{i=1}^{m}(p_{1k}(\mathbf{X}_i) - \theta_k)) \cdot (U_k - \theta_k))|x_i) + \mathbb{E}\mathbb{E}(((\frac{2}{n}\sum_{j=1}^{n}(q_{1k}(\mathbf{Y}_j) - \theta_k)) \cdot (U_k - \theta_k))|y_j) \\
=&\mathbb{E}(\frac{1}{m}\sum_{i=1}^{m}((p_{1k}(\mathbf{X}_i) - \theta_k) \cdot \frac{1}{m}(p_{1k}(\mathbf{X}_i) - \theta_k)) + \mathbb{E}(\frac{2}{n}\sum_{j=1}^{n}(q_{1k}(\mathbf{Y}_j) - \theta_k \cdot \frac{2}{n}(q_{1k}(\mathbf{Y}_j) - \theta_k)) \\
=&\frac{1}{m^2}\sum_{i=1}^{m}((p_{1k}(\mathbf{X}_i) - \theta_k) \cdot (p_{1k}(\mathbf{X}_i) - \theta_k)) + \frac{4}{n^2}\sum_{j=1}^{n}((q_{1k}(\mathbf{Y}_j) - \theta_k) \cdot (q_{1k}(\mathbf{Y}_j) - \theta_k)) \\
=&\frac{1}{m}\alpha_{1kk} + \frac{4}{n}\beta_{1kk}
\end{aligned}$$

Since $\mathbb{E}(\hat{U}_k^2) = \frac{1}{m}\alpha_{1kk} + \frac{4}{n}\beta_{1kk}$ and $\mathbb{E}U_k - \theta_k)^2 = \frac{1}{m}\alpha_{1kk} + \frac{4}{n}\beta_{1kk} + o(m^{-1})$,

$$\begin{aligned}\mathbb{E}((U_k - \theta_k) - \hat{U}_k)^2 &= \mathbb{E}U_k - \theta_k)^2 - 2\mathbb{E}(\hat{U}_k(U_k - \theta_k)) + \mathbb{E}(\hat{U}_k^2)\\ &= \frac{1}{m}\alpha_{1kk} + \frac{4}{n}\beta_{1kk} + o(m^{-1}) - 2(\frac{1}{m}\alpha_{1kk} + \frac{4}{n}\beta_{1kk}) + \frac{1}{m}\alpha_{1kk} + \frac{4}{n}\beta_{1kk}\\ &= o(m^{-1}).\end{aligned}$$

Thus, by the Chebyshev's Lemma, $(U_k - \theta_k) - \hat{U}_k \xrightarrow{P} 0$. Similarly, we can show $(V_k - \lambda_k) - \hat{V}_k \xrightarrow{P} 0$.

Proof of Theorem 5.16.1

Since $\hat{U}_k$ is asymptotically normal with mean 0 and variance $\frac{\alpha_{1kk}}{m} + \frac{\beta_{1kk}}{m}$ and $(U_k - \theta_k) - \hat{U}_k \xrightarrow{P} 0$, by Slusky's Theorem, we obtain that $U_k - \theta_k$ is asymptotically normal with mean 0 and variance $\frac{\alpha_{1kk}}{m} + \frac{\beta_{1kk}}{m}$. Similarly, we have $V_k - \lambda_k$ is asymptotically normal with mean 0 and variance $\frac{\alpha_{2kk}}{m} + \frac{\beta_{2kk}}{m}$. The asymptotic multivariate normality for any vector with linearly independent components of $U_k - \theta_k$'s and $V_k - \lambda_k$'s follows from the multivariate version of Slusky's Theorem.

Proof of Lemma 5.16.3

To show the consistency of $\hat{\alpha}_{1st}$, we first obtain

$$\begin{aligned}\hat{\alpha}_{1st} &= \frac{1}{m-1}\sum_{r=1}^{m}[\hat{d}_{1s}(x_r) - U_s] \cdot [\hat{d}_{1t}(x_r) - U_t]\\ &= \frac{1}{m-1}\sum_{r=1}^{m}[\hat{d}_{1s}(x_r) - \theta_s + \theta_s - U_s] \cdot [\hat{d}_{1t}(x_r) - \theta_t + \theta_t - U_t]\\ &= \frac{1}{m-1}\sum_{r=1}^{m}[\hat{d}_{1s}(x_r) - \theta_s] \cdot [\hat{d}_{1t}(x_r) - \theta_t] - \frac{1}{m-1}\sum_{r=1}^{m}[\hat{d}_{1s}(x_r) - \theta_s] \cdot (U_t - \theta_t)\\ &- \frac{1}{m-1}\sum_{r=1}^{m}(U_s - \theta_s) \cdot [\hat{d}_{1t}(x_r) - \theta_t] + \frac{1}{m-1}\sum_{r=1}^{m}(U_s - \theta_s) \cdot (U_t - \theta_t)\\ &= \frac{1}{m-1}\sum_{r=1}^{m}[\hat{d}_{1s}(x_r) - \theta_s] \cdot [\hat{d}_{1t}(x_r) - \theta_t] - (U_t - \theta_t) \cdot \frac{1}{m-1}\sum_{r=1}^{m}[\hat{d}_{1s}(x_r) - \theta_s])\\ &- (U_s - \theta_s) \cdot \frac{1}{m-1}\sum_{r=1}^{m}[\hat{d}_{1t}(x_r) - \theta_t] + \frac{m}{m-1}\sum_{r=1}^{m}(U_s - \theta_s) \cdot (U_t - \theta_t)\\ &= \frac{1}{m-1}\sum_{r=1}^{m}[\hat{d}_{1s}(x_r) - \theta_s] \cdot [\hat{d}_{1t}(x_r) - \theta_t] - (U_t - \theta_t) \cdot \frac{1}{m-1} \cdot m(U_s - \theta_s)\\ &- (U_s - \theta_s) \cdot \frac{1}{m-1} \cdot m(U_t - \theta_t) + \frac{m}{m-1}(U_s - \theta_s) \cdot (U_t - \theta_t)\\ &= \frac{1}{m-1}\sum_{r=1}^{m}[\hat{d}_{1s}(x_r) - \theta_s] \cdot [\hat{d}_{1t}(x_r) - \theta_t] - \frac{m}{m-1}(U_s - \theta_s) \cdot (U_t - \theta_t).\end{aligned}$$

Note that $\hat{d}_{1k}(x_r) = \frac{1}{\binom{n}{2}} \sum_{i<j}^{n} I_{1,k}[x_r; y_i, y_j]$. Thus,

$$\begin{aligned}
&\mathbb{E}([\hat{d}_{1s}(x_r) - \theta_s] \cdot [\hat{d}_{1t}(x_r) - \theta_t]) \\
=&\mathbb{E}((\frac{1}{\binom{n}{2}} \sum_{i<j}^{n} I_{1,s}[x_r; y_i, y_j] - \theta_s) \cdot (\frac{1}{\binom{n}{2}} \sum_{i'<j'}^{n} I_{1,t}[x_r; y_{i'}, y_{j'}] - \theta_t) \\
=&\frac{1}{\binom{n}{2}^2} \sum_{i<j} \mathbb{E} I_{1,s}[x_r; y_i, y_j] - \theta_s) \cdot \frac{1}{\binom{n}{2}} \sum_{i'<j'}^{n} I_{1,t}[x_r; y_{i'}, y_{j'}] - \theta_t) = \qquad B_1 + B_2 + B_3,
\end{aligned}$$

where B_1 collects the terms of $I_{1,s}[x_r; y_i, y_j]$ and $I_{1,t}[x_r; y_{i'}, y_{j'}]$ that has only x_r in common,

$$\begin{aligned}
B_1 &= \frac{1}{\binom{n}{2}} \mathbb{E}((I_{1,s}[x_r; y_i, y_j] - \theta_s) \cdot (I_{1,t}[x_r; y_{i'}, y_{j'}] - \theta_t) \cdot \binom{n-2}{2})) \\
&= \frac{\binom{n-2}{2}}{\binom{n}{2}} \cdot \alpha_{1st} = \alpha_{1st} + o(n^{-1}),
\end{aligned}$$

B_2 collects the terms of $I_{1,s}[x_r; y_i, y_j]$ and $I_{1,t}[x_r; y_{i'}, y_{j'}]$ that has x_r and one of y_i or y_j in common, say y_{i*},

$$B_1 = \frac{1}{\binom{n}{2}} \mathbb{E}((I_{1,s}[x_r; y_{i*}, y_j] - \theta_s) \cdot (I_{1,t}[x_r; y_{i*}, y_{j'}] - \theta_t) \cdot 2(n-2)) = o(n^{-1}),$$

B_3 collects the terms of $I_{1,s}[x_r; y_i, y_j]$ and $I_{1,t}[x_r; y_{i'}, y_{j'}]$ that has exactly the same x_r, y_i and y_j in common,

$$B_3 = \frac{1}{\binom{n}{2}} \mathbb{E}((I_{1,s}[x_r; y_i, y_j] - \theta_s) \cdot (I_{1,t}[x_r; y_i, y_j] - \theta_t)) = o(n^{-1}),$$

Therefore,

$$\mathbb{E}([\hat{d}_{1s}(x_r) - \theta_s] \cdot [\hat{d}_{1t}(x_r) - \theta_t]) = \alpha_{1st} + o(n^{-1}) \tag{5.30}$$

Since $([\hat{d}_{1s}(x_r) - \theta_s] \ldots [\hat{d}_{1t}(x_r) - \theta_t]), r = 1, 2, \ldots, m$, are bounded and exchangeable, by the weak law of large numbers for exchangeable random variables (Hong and Lee [153]), we can get

$$\frac{1}{m-1} \sum_{r=1}^{m} [\hat{d}_{1s}(x_r) - \theta_s] \cdot [\hat{d}_{1t}(x_r) - \theta_t] \xrightarrow{P} \alpha_{1st}$$

Also, using Lemma 2, we can show that $U_s - \theta_s \xrightarrow{P} 0$ and $U_t - \theta_t \xrightarrow{P} 0$. Therefore, $\hat{\alpha}_{1st} \xrightarrow{P} \alpha_{1st}$. Similarly, we can show that under condition (1), for $s, t = 1, 2, 3$,

$$\hat{\beta}_{1st} \xrightarrow{P} \beta_{1st}; \quad \hat{\alpha}_{2st} \xrightarrow{P} \alpha_{2st}; \quad \hat{\beta}_{2st} \xrightarrow{P} \beta_{2st}; \quad \hat{\gamma}_{1st} \xrightarrow{P} 2\gamma_{1st}; \quad \hat{\gamma}_{1st} \xrightarrow{P} 2\gamma_{2st}.$$

Proof of Theorem 5.19.1: Proof: Let $\bar{X}^d = \frac{1}{md} \sum_{i=1}^{m} \sum_{t=1}^{d} X_{it}$, and $\bar{Y}^d = \frac{1}{nd} \sum_{i=1}^{n} \sum_{t=1}^{d} Y_{it}$. One can show that

$$\mathbb{E}(\hat{\mu}_x^d) = \frac{1}{md} \sum_{i=1}^{m} \sum_{t=1}^{d} \mathbb{E}(X_{it}) = \frac{1}{d} \mathbf{1}'_d \boldsymbol{\mu}_x, \tag{5.31}$$

where $\mathbf{1}_d$ is a d-dimensional vector of 1s. It is not difficult to show that

$$\begin{aligned} \mathbb{V}ar(\hat{\mu}_x^d) &= \mathbb{V}ar(\frac{1}{md}\sum_{i=1}^{m}\sum_{t=1}^{d} X_{it}) = \frac{1}{m^2d^2}\sum_{i=1}^{m}\sum_{t=1}^{d}\sum_{r=1}^{d}\mathbb{C}ov(X_{it}, X_{ir}), \\ &= \frac{1}{m^2d^2}\sum_{i=1}^{m}[\sum_{t=1}^{d}\mathbb{V}ar(X_{it}) + \sum_{t=1}^{d}\sum_{r\neq t=1}^{d}\mathbb{C}ov(X_{it}, X_{ir})] = \frac{1}{md^2}tr(\mathbf{\Sigma}_x) + \frac{o(d^2)}{md^2}, \end{aligned} \tag{5.32}$$

where the last line follows from Assumption (B3). The same expressions hold for the $\mathbb{Y}$ sample with obvious change of notation. It follows that $\mathbb{V}ar(\hat{\mu}_x^d)$ and $\mathbb{V}ar(\hat{\mu}_y^d)$ converge to 0 in probability as $d \to \infty$. By the weak law of large numbers, $\hat{\mu}_x^d$ and $\hat{\mu}_y^d$ converge in probability to $\frac{1}{d}\mathbf{1}_d'\boldsymbol{\mu}_x$ and $\frac{1}{d}\mathbf{1}_d'\boldsymbol{\mu}_y$, respectively, as $d \to \infty$. Furthermore, since the rows of of the data matrix are independent, it follows that

$$\begin{aligned} \mathbb{E}(\hat{\sigma}_x^d)^2 &= \frac{1}{m^2}\mathbb{E}(\sum_{i=1}^{m} S_{\mathbf{X}_i}^d)^2 \\ &= \frac{1}{m^2}\big[\mathbb{E}(\sum_{i=1}^{m}(S_{\mathbf{X}_i}^d)^2) + 2\sum_{1\leq i<j\leq m}\mathbb{E}((S_{\mathbf{X}_i}^d)(S_{\mathbf{X}_j}^d))\big] \\ &= \frac{1}{m^2}(m(\sigma_x^d)^2 + m(m-1)(\sigma_x^d)^2) = (\sigma_x^d)^2 \end{aligned} \tag{5.33}$$

and $\mathbb{E}(\hat{\sigma}_y^d)^2 = (\sigma_y^d)^2$, similarly. Moreover,

$$\begin{aligned} \mathbb{V}ar\big((\hat{\sigma}_x^d)^2\big) &= \frac{1}{m^2}\mathbb{V}ar\big(\sum_{i=1}^{m} S_{\mathbf{X}_i}^d\big)^2 \\ &= \frac{1}{m^2d^2}\sum_{i=1}^{m}\sum_{l=1}^{d}\sum_{l'=1}^{d}\mathbb{C}ov\big((\boldsymbol{X}_{il} - \bar{X}_i^d)^2, (\boldsymbol{X}_{il'} - \bar{X}_i^d)^2\big) \\ &= \frac{1}{m^2d^2}\sum_{i=1}^{m}\sum_{l=1}^{d}\mathbb{V}ar\big((\boldsymbol{X}_{il} - \bar{X}_i^d)^2\big) + \frac{o(d^2)}{md^2} \\ &= \frac{1}{m^2d^2}\sum_{i=1}^{m}\sum_{l=1}^{d}\Big[\mathbb{E}\big((\boldsymbol{X}_{il} - \bar{X}_i^d)^4\big) - \sum_{i=1}^{m}\sum_{l=1}^{d}\big(\mathbb{V}ar(\boldsymbol{X}_{il})\big)^2\Big] + o(1). \end{aligned} \tag{5.34}$$

One obtains the same expressions with obvious change of notation for $\mathbb{V}ar\big((\hat{\sigma}_y^d)^2\big)$. By Assumptions (B2)–(B3), $\mathbb{V}ar\big((\hat{\sigma}_x^d)^2\big)$ and $\mathbb{V}ar\big((\hat{\sigma}_y^d)^2\big)$ converge to 0 in probability as $d \to \infty$. By the weak law of large numbers, $\hat{\sigma}_x^d$ and $\hat{\sigma}_y^d$ converge in probability to σ_x^d and σ_y^d, respectively. It follows from Slutsky's theorem that $A_\delta \to \tilde{\delta}^2(F_1, F_2) \geq 0$ as $d \to \infty$. □

5.22 Exercises

Exercise 5.1 *Suppose* $\mathbf{X} = \{x_1, \ldots, x_m\}$ *and* $\mathbf{Y} = \{y_1, \ldots, y_n\}$ *are independent random samples from distribution functions* F_1 *and* F_2 *in* $\mathbb{R}^d$*, respectively. We would like to test* $H_0 : F_1 = F_2$ *against* $H_a : F_1 \neq F_2$ *using Friedman-Rafsky multivariate runs test.*

a) Define a minimum spanning tree.

b) Describe the Friedman-Rafsky multivariate runs test.

c) *Suppose the minimum spanning tree of the combined sample of X and Y is a complete binary tree with 4 levels where all terminal nodes are labeled "Y" and all non-terminal nodes are labeled "X". Find the number of runs and identify the runs that are formed.*

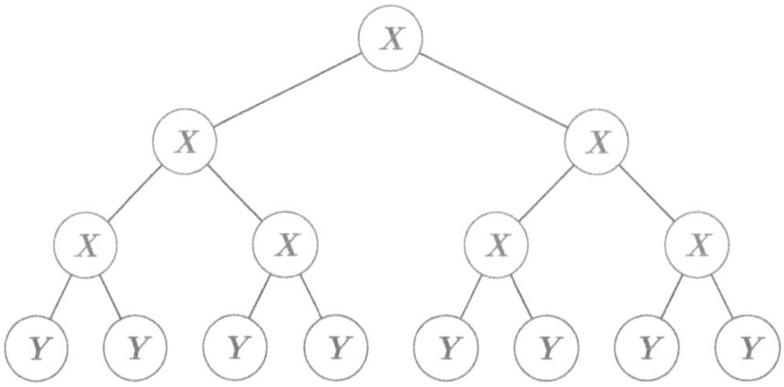

Exercise 5.2 *Provide the R code to construct the minimum spanning tree and the Friedman-Rafsky test. How do you perform an approximate permutation test and compute p-value?*

Exercise 5.3 *Suppose* $\mathbf{X} = \{x_1, x_2\}$ *and* $\mathbf{Y} = \{y_1, y_2, y_3\}$ *are independent random samples from distribution functions* F_1 *and* F_2 *in* $\mathbb{R}^10$*, respectively. We would like to test* $H_0 : F_1 = F_2$ *against* $H_a : F_1 \neq F_2$ *using the statistic*

$$T = \sum_{k=1}^{3} (U_k - \frac{1}{3})^2 + (V_k - \frac{1}{3})^2. \tag{5.35}$$

Suppose further that $x_1 = i\mathbf{1}_d$*,* $y_j = j\mathbf{1}_d$ *for* $i = 1, 2$ *and* $j = 1, 2, 3$*.*

a) *Obtain the values of* U_k *and* V_k *for the triangle statistics and compute the observed value of* T*.*

b) *Show that there are 10 atoms in the reference set. Obtain the sampling distribution of* T *using the permutation testing algorithm in 5.11.1. Calculate the* P_{value}*.*

Exercise 5.4 *Biswas and Ghosh [35] proposed the statistic in* (5.23)*.*

a) *Prove that in* (5.24) $2BG_E = EN_E^2 + (\bar{d}_{(x)} - \bar{d}_{(y)})^2$*.*

b) *Show that while the term* $(\bar{d}_{(x)} - \bar{d}_{(y)})^2$ *in* EN_E *provides information about scale differences it becomes noise under shift alternatives.*

Exercise 5.5 *Provide the complete computation of ranks,* $\bar{\mathbf{R}}_1$*,* $\bar{\mathbf{R}}_2$*,* $\mathbf{S}_R$*, and* Q *in Example 5.13.1.*

Exercise 5.6 *Testing different means in bivariate normal distributions. Generate* $\mathbf{X}_1, \ldots, \mathbf{X}_{50}$ *from* $\mathbb{N}_2(\mathbf{0}, \mathbf{I}_2)$ *and* $\mathbf{Y}_1, \ldots, \mathbf{Y}_{50}$ *from* $\mathbb{N}_2(\mathbf{1}, \mathbf{I}_2)$*. Test whether* $F_1 = F_2$ *using the Friedman-Rafsky test.*

Exercise 5.7 *Testing different means in bivariate normal distributions, same mean, different covariance matrices. Let* $\mathbf{X}_1, \ldots, \mathbf{X}_{50} \sim \mathbb{N}_2(\mathbf{0}_2, \mathbf{I}_2)$ *and* $\mathbf{Y}_1, \ldots, \mathbf{Y}_{50}$ *be taken from* $\mathbb{N}_2(\mathbf{0}_2, 2\,\mathbf{I}_2)$*. Test whether* $F_1 = F_2$ *using the Friedman-Rafsky statistic.*

Exercise 5.8 *Suppose* $m = 3$ *and* $n = 2$ *observations are taken from* F_1 *and* F_2*, respectively in the test of Hall and Tajvidi. Enumerate the sampling distribution of* $M_i(j)$ *and* $N_i(j)$ *under* $H_0 : F_1 = F_2$*.*

Exercise 5.9 *Use (d=4, $m = 3, n = 3$). Using the same six points in Example 5.4.1, do the following:*

1. *Compute the complete distance matrix.*

2. *List candidate optimal matchings and their total costs.*

3. *Identify the matching with minimum total distance.*

4. *Determine A_1 and compute the one-sided p-value.*

Exercise 5.10 *Suppose $\mathbf{X}_1, \mathbf{X}_2$ are i.i.d. F_1 and $\mathbf{Y}_1, \mathbf{Y}_2$ are i.i.d. F_2. Assume $\mu_{F_1} = \mathbb{E}(\|\mathbf{X}_1 - \mathbf{X}_2\|)$, $\mu_{F_2} = \mathbb{E}(\|\mathbf{Y}_1 - \mathbf{Y}_2\|)$, and $\mu_{F_1F_2} = \mathbb{E}(\|\mathbf{X}_1 - \mathbf{Y}_1\|)$ exist. Prove that $\mu_{F_1} = \mu_{F_2} = \mu_{F_1F_2}$ if and only if $F_1 = F_2$.*

6

High-Dimensional Classification of Categorical Data

6.1 Abstract

This chapter develops a comprehensive framework for classification in high-dimensional settings, with particular emphasis on discrete and categorical data where the number of variables may far exceed the sample size. Classical parametric classifiers often deteriorate in such regimes due to covariance singularity, distance concentration, and overfitting. To address these challenges, the chapter introduces non-parametric, test-based classification methods built on IPDs and permutation principles, providing distribution-free decision rules with enhanced robustness. Modifications based on relative distance measures are proposed to improve stability and interpretability in high dimensions. A broad range of classification paradigms is examined, including depth-based methods, support vector machines, nearest-neighbor rules, diagonal linear discriminant analysis, and dissimilarity-based classifiers, alongside energy- and rank-based tests. The methodology is extended to multisample classification problems and evaluated through extensive simulation studies under multivariate Bernoulli, multinomial, and Poisson models. Comparative analyses highlight strengths and limitations across competing approaches. Applications to gene expression data, including prostate tumor studies, demonstrate practical relevance. The chapter concludes with methodological recommendations for effective high-dimensional classification under diverse distributional and structural conditions.

6.2 Introduction

Classification is a central task in multivariate analysis, focused on assigning new observations to one of several pre-defined groups based on their feature profiles. In the context of high dimensional discrete data, we introduce a novel classification approach based on the IPDs, which serve as measures of similarity between observations. The proposed methodology is non-parametric and uses a test-based framework, wherein classification decisions are guided by permutation testing to assess the statistical significance of proximity–based group membership.

To enhance robustness and interpretability, we introduce a modification to the traditional test-based rule by incorporating relative distance measures, computed with respect to the training sample baseline. We evaluate the performance of this rule through a comprehensive comparative study involving both parametric and non-parametric alternatives. Specifically, we benchmark it against classical parametric techniques such as the likelihood ratio rule and modified linear discriminant analysis, as well as established non-parametric

DOI: 10.1201/9781003789581-6

methods including support vector machines, k-nearest neighbors, and depth-based classifiers. Empirical assessments are conducted under three discrete multivariate frameworks, including multivariate Bernoulli, multinomial, and multivariate Poisson distributions, highlighting the versatility and competitiveness of the proposed method in challenging high-dimensional settings.

Suppose F_1 and F_2 are distribution functions associated with the d-dimensional discrete random vectors $\mathbf{X}$ and $\mathbf{Y}$ under populations F_1 and F_2, respectively. Given training samples $\mathbf{X}_1, \ldots, \mathbf{X}_m$ from F_1, and $\mathbf{Y}_1, \ldots, \mathbf{Y}_n$ from F_2 with known class memberships, we would like to define a classification rule to allocate a new observation into F_1 or F_2. Nominal categorical data are abundant in practice and include gender, race, eye color, hair color, marital status, political affiliation, age, or income level, to name a few. The classification statistics discussed in this chapter are based on the IPDs of the sample observations and can be applied to datasets of nominal variables. For example, the multivariate Bernoulli (MB) distribution is the building block of discrete multivariate distributions (Marshall and Olkin [217]), with diverse applications in DNA profiling (Wilber et al. [354]) and diagnosis of cardiac images of normal and abnormal patients based on binary feature patterns (Kurgan et al. [179]), among others. Erlich et al. [90] discuss methods of clustering based on the binary representation of datasets.

In a wide range of contemporary applications, spanning data mining, communication networks, epidemiological modeling, and clustering, data analysts frequently encounter discrete or categorical data in high-dimensional settings, where the number of variables surpasses the number of available observations. Such scenarios introduce significant methodological challenges for traditional classification techniques. In particular, the classical linear discriminant function is known to exhibit degraded performance in high-dimensional regimes, primarily due to instability and singularity of the sample covariance matrix (Bickel and Levina [33]). When the dimensionality exceeds the sample size, the sample covariance matrix becomes singular, rendering the discriminant function ill-defined.

To address this issue, a common strategy involves replacing the full covariance matrix with its diagonal approximation, thereby eliminating the need to estimate off-diagonal elements and reducing sensitivity to over fitting. In this context, Huang et al. [157] proposed an unbiased diagonal discriminant rule specifically tailored for high-dimensional classification problems. This approach maintains computational tractability while improving classification performance under certain distributional assumptions. In the comparison section, we undertake a detailed evaluation of this method and compare its performance against a suite of alternative classification strategies, both parametric and non-parametric, within various high-dimensional discrete data settings. Johnson and Wichern [162] discuss a method to assign a new observation to one of two multivariate normal distributions by minimizing the expected cost of misclassification (ECM). Let p_x and p_y denote the prior probability of F_1 and F_2, respectively, and $\mathbb{P}(2|1)$ be the conditional probability of classifying a F_1 observation as F_2. Similarly, we define $\mathbb{P}(1|2)$. Therefore,

$$\begin{aligned} \mathbb{P}(\text{an observation is misclassified as } F_1) &= \mathbb{P}(1|2)p_y, \\ \mathbb{P}(\text{an observation is misclassified as } F_2) &= \mathbb{P}(2|1)p_x. \end{aligned} \tag{6.1}$$

We also define the cost of misclassification as $c(1|2)$ when we misclassify a F_2 observation as F_1, while $c(2|1)$ stands for classifying a F_1 observation as F_2. The expected cost of misclassification is

$$ECM = c(2|1)\mathbb{P}(2|1)p_x + c(1|2)\mathbb{P}(1|2)p_y. \tag{6.2}$$

An optimal classification rule should have an ECM as small as possible. Let $f_x(\mathbf{z})$ and $f_y(\mathbf{z})$ be the probability function of $\mathbf{z}$ under F_1 and F_2, respectively. Johnson and Wichern [162] show that ECM is minimized if we assign $\mathbf{z}$ to F_1 when $\frac{f_x(\mathbf{z})}{f_y(\mathbf{z})} \leq \frac{c(1|2)p_y}{c(2|1)p_x}$. We could also assign

$\mathbf{z}$ to the population with the largest posterior probability $\mathbb{P}(F_1|\mathbf{z})$ or $\mathbb{P}(F_2|\mathbf{z})$. By Bayes's rule, the posterior probabilities are

$$\begin{aligned} \mathbb{P}(F_1|\mathbf{z}) &= \frac{p_x f_x(\mathbf{z})}{p_x f_x(\mathbf{z}) + p_y f_y(\mathbf{z})}, \\ \mathbb{P}(F_2|\mathbf{z}) &= \frac{p_y f_y(\mathbf{z})}{p_x f_x(\mathbf{z}) + p_y f_y(\mathbf{z})}. \end{aligned} \tag{6.3}$$

One can show that $\mathbb{P}(F_1|\mathbf{z}) \leq \mathbb{P}(F_2|\mathbf{z})$ is equivalent to $\frac{f_x(\mathbf{z})}{f_y(\mathbf{z})} \leq \frac{p_y}{p_x}$, which proves that under equal misclassification costs, the Bayes's classifier and ECM-minimized classifier are equivalent. Based on the form of the classifier, we refer to it as the likelihood ratio (LR) method.

6.3 Discrete Distributions

Following Guo and Modarres [125], we will discuss some non-parametric methods and compare them with LR method under high dimensional discrete distributions. The distributions we consider are the multivariate Bernoulli, multinomial, and multivariate Poisson. Each distribution represents a different type of dataset in real life: finite-category datasets (multinomial), binary datasets (multivariate Bernoulli) and datasets with over-dispersed margins (multivariate Poisson). Below, we provide a brief description of each distribution. For other distributions, see section 1.6. A multinomial distribution is sparse when the size n of observations is less than the number of categories d. Sparse multinomial distribution often occurs in sample surveys with numerous possible responses. For sparse multinomial samples, the likelihood ratio statistic cannot be computed and as a solution, categories with zero counts are replaced with a small constant such as 0.5. Feinberg [97] discussed several problems that arise when the number of categories is increased in response to an increasing sample size. One can obtain the Poisson distribution by taking the limit of a binomial distribution as the number of trials n tends to infinity, probability of success θ tends to zero and $n\theta$ approaches the Poisson mean λ. Following Holgate [151] construction, suppose $T, T_1, \ldots, T_d$ are mutually independent Poisson random variables with means λ_0 and $\lambda_1, \ldots, \lambda_d$, respectively. Let $W_i = T_i + T$ for $i = 1, \ldots, d$. Hence, $\mathbf{W} = (W_1, \ldots, W_d)$ has a multivariate Poisson distribution, denoted as MP($\boldsymbol{\lambda}$), with mean vector $\boldsymbol{\lambda} = (\lambda_0 + \lambda_1 \ldots, \lambda_0 + \lambda_d)'$.

One can show that the marginal distributions of $W_1, \ldots, W_d$ follow a Poisson distributions with means $\lambda_1 + \lambda_0, \ldots, \lambda_d + \lambda_0$, and $\mathbb{C}ov(W_i, W_j) = \lambda_0$ for all $i \neq j$. Hence, $\mathbf{W}$ follows a multivariate Poisson distribution with a constant positive correlation matrix. Let $m = \min(w_1, \ldots, w_d)$. The PMF of $\mathbf{W}$ is given by

$$\mathbb{P}(\mathbf{W}) = exp(-\sum_{t=1}^{d} \lambda_t) \prod_{t=1}^{d} \frac{\lambda_t^{w_t}}{w_t!} \sum_{i=0}^{m} \prod_{j=1}^{d} \binom{w_j}{i} i! (\frac{\lambda_0}{\prod_{k=1}^{d} \lambda_k})^i.$$

6.4 Classification Methods

6.4.1 Test-based Classification

Liao and Akritas [194] propose a test-based classification procedure. To assign a new observation $\mathbf{z}$ to one of the populations F_1 or F_2, we first assume that $\mathbf{z}$ belongs to F_1 and test

$H_0 : F_1 = F_2$ using $m+1$ observations from F_1 and n observations from F_2. We denote this test by $\mathbf{T}_1$. Next, we assume that $\mathbf{z}$ belongs to F_2 and perform $H_0 : F_1 = F_2$ using m observations from F_1 and $n+1$ observations from F_2, denoting this test as $\mathbf{T}_2$.

A crucial insight is that if $\mathbf{z}$ indeed arises from F_1, including it with the F_2 sample tends to diminish the distinction between the two populations. Let PV_1 and PV_2 be the p-values of $\mathbf{T}_1$ and $\mathbf{T}_2$, respectively. Now, $\mathbf{z}$ is allocated to F_1 if $PV_1 < PV_2$; otherwise, it is assigned to F_2. This approach performs well when the theoretical p-values are accurately known and directly comparable. Given a new observation $\mathbf{z}$, we wish to allocate it to either F_1 or F_2 using classification via two-sample testing algorithm.

Algorithm: Classification via Two-Sample Testing

Input: Samples $X = \{X_1, \ldots, X_m\} \sim F_1$, $Y = \{Y_1, \ldots, Y_n\} \sim F_2$, new observation $\mathbf{z}$
Output: Class label of $\mathbf{z}$
Assume $\mathbf{z} \in F_1$ and construct $X^* = X \cup \{\mathbf{z}\}$
Test $H_0 : F_1 = F_2$ using $(m+1, n)$ samples
Let PV_1 be the p-value
Assume $\mathbf{z} \in F_2$ and construct $Y^* = Y \cup \{\mathbf{z}\}$
Test $H_0 : F_1 = F_2$ using $(m, n+1)$ samples
Let PV_2 be the p-value
if $PV_1 < PV_2$ **then**
    Assign $\mathbf{z}$ to F_1
else
    Assign $\mathbf{z}$ to F_2
end if

Example 6.4.1 *We have two d-dimensional populations, F_1 and F_2 with samples*

$$\mathbf{X}_1, \ldots, \mathbf{X}_m \overset{i.i.d.}{\sim} F_1, \quad \mathbf{Y}_1, \ldots, \mathbf{Y}_n \overset{i.i.d.}{\sim} F_2.$$

We use Hotelling's T^2 statistic, $T^2 = \frac{mn}{m+n}(\bar{\mathbf{X}} - \bar{\mathbf{Y}})' S_p^{-1} (\bar{\mathbf{X}} - \bar{\mathbf{Y}})$, where $S_p = \frac{(m-1)S_X + (n-1)S_Y}{m+n-2}$ is the pooled sample covariance matrix. The statistic is converted to an F-statistic,

$$F = \frac{m+n-d-1}{d(m+n-2)} T^2 \quad \sim \quad F_{d,\, m+n-d-1}.$$

We take

$$m = 3, \quad n = 4, \quad d = 3, \quad \boldsymbol{\mu} = (1.8, 1.8, 1.8)',$$
$$F_1 \sim \mathbb{N}(\mathbf{0}, \mathbf{I}_3), \quad F_2 \sim \mathbb{N}(\boldsymbol{\mu}, \mathbf{I}_3).$$

The simulated samples are

$$X = \begin{bmatrix} -0.2154 & 0.4821 & -0.7612 \\ 0.4738 & -0.3594 & 0.1285 \\ -1.0247 & 0.2569 & 0.7841 \end{bmatrix}, \quad Y = \begin{bmatrix} 2.1453 & 1.7259 & 1.8932 \\ 1.5648 & 2.2043 & 1.6157 \\ 2.2214 & 1.7328 & 2.4576 \\ 1.9875 & 2.4152 & 1.9584 \end{bmatrix}.$$

Two new observations are drawn,

$$\mathbf{z}_x = (0.3521, -0.2845, 0.4179), \quad \mathbf{z}_y = (1.9463, 2.2857, 2.1124).$$

Classification for $\mathbf{z}_x$ (true F_1 sample), $\mathbf{T}_1 : X \cup \{\mathbf{z}_x\}$ vs Y

$$T^2 = 46.1824, \quad F = 30.7883, \quad PV_1 = 0.0012.$$

$$\mathbf{T}_2 : X \text{ vs } Y \cup \{\mathbf{z}_x\},$$
$$T^2 = 71.9441, \quad F = 37.6139, \quad PV_2 = 0.0007.$$

Since $PV_1 < PV_2$, $\mathbf{z}_x$ is allocated to F_1 (correct).

Classification for $\mathbf{z}_y$ (true F_2 sample), $\mathbf{T}_1 : X \cup \{\mathbf{z}_y\}$ vs Y,

$$T^2 = 90.5729, \quad F = 60.3819, \quad PV_1 < 0.0001.$$

$$\mathbf{T}_2 : X \text{ vs } Y \cup \{\mathbf{z}_y\}$$

$$T^2 = 38.9472, \quad F = 20.7855, \quad PV_2 = 0.0025.$$

Since $PV_1 > PV_2$, $\mathbf{z}_y$ is allocated to F_2 (correct).

When the populations F_1 and F_2 are strongly separated, estimated p-values for the two tests can both be extremely small. If the new observation $\mathbf{z}$ truly belongs to F_1, the statistic $\mathbf{T}_1$ is expected to take a large value because it quantifies the already pronounced separation between F_1 and F_2, yielding a very low PV_1. Similarly, computing $\mathbf{T}_2$ by temporarily associating $\mathbf{z}$ with the F_2 sample still results in two distinguishable groups, albeit slightly less separated, which produces a comparably small PV_2.

Simulation results indicate that when p-values are derived from permutation distributions, both PV_1 and PV_2 frequently reach zero in cases of well-separated populations. This scenario highlights a limitation of the original procedure and motivates an adjustment to the test-based classification criterion.

An alternative to directly comparing p-values is to base classification on the relative impact of the new observation on the test statistic. Let T_0 denote the value of the test statistic computed from the original training samples $\mathbf{X}$ and $\mathbf{Y}$. The effect of incorporating the new observation $\mathbf{z}$ into the $\mathbf{X}$ sample is measured by $|T_1 - T_0|$, while the corresponding effect for inclusion in F_2 is given by $|T_2 - T_0|$.

Under the assumption of equal prior probabilities, the classification rule can be reformulated by assigning $\mathbf{z}$ to F_1 if $\frac{|T_1-T_0|}{|T_2-T_0|} < 1$ and to F_2 otherwise. When prior probabilities are unequal or misclassification costs differ, the decision criterion is adjusted to $\frac{|T_1-T_0|}{|T_2-T_0|} < \frac{c(2|1)}{c(1|2)} \cdot \frac{p_x}{p_y}$, where $c(j|i)$ and p_x, p_y are the prior probabilities of the respective populations.

6.4.2 Interpoint Distances

Many classical multivariate techniques, such as the Hotelling T^2 test [155], become ineffective or lose statistical power in high dimensions. We discuss the IPDs for testing the equality of two groups. Let $\mathbb{X} = \{\mathbf{X}_i\}$ and $\mathbb{Y} = \{\mathbf{Y}_j\}$, for $i = 1, \ldots m$ and $j = 1, \ldots, n$. Consider two points $\mathbf{X}_i = (X_{1i}, \ldots, X_{di})$ and $\mathbf{X}_j = (X_{1j}, \ldots, X_{dj})$. Notation for IPDs appears in section 2.5. Given a new observation $\mathbf{z}$, let $x \cup z$ denotes the set containing the $\mathbf{X}$ observations and $\mathbf{z}$. The updated average IPDs are

$$\begin{aligned}
\bar{d}_{(x\cup z)y} &= \frac{mn\bar{d}_{(xy)} + \sum_{j=1}^{n} \left[(\mathbf{z}-\mathbf{Y}_j)'(\mathbf{z}-\mathbf{Y}_j)\right]^{1/2}}{(m+1)n}, \\
\bar{d}_{x\cup z} &= \frac{(m-1)m\bar{d}_{(x)} + \sum_{i=1}^{m} \left[(\mathbf{z}-\mathbf{X}_i)'(\mathbf{z}-\mathbf{X}_i)\right]^{1/2}}{(m+1)m}.
\end{aligned} \tag{6.4}$$

Similarly, we define, $\bar{d}_{x(y\cup z)}$ and $\bar{d}_{y\cup z}$.

6.4.2.1 Energy Test

Barighaus and Franz [21] and Székely and Rizzo [329] proved that $2E\|\mathbf{X}_1 - \mathbf{Y}_1\| - E\|\mathbf{X}_1 - \mathbf{X}_2\| - E\|\mathbf{Y}_1 - \mathbf{Y}_2\| \geq 0$ with equality holding if and only if $F_1 = F_2$. By replacing the expectations with the average IPDs, they obtain the statistic $EN = 2\bar{d}_{(xy)} - \bar{d}_{(x)} - \bar{d}_{(y)}$.

The closer EN is to zero, the more likely that $F_1 = F_2$. We can use EN as the test statistic for the null hypothesis $H_0 : F_1 = F_2$, and reject the hypothesis with large EN. Using the test-based method with EN we define

$$\begin{aligned} T_1 &: EN_1 = 2\bar{d}_{(x\cup z)y} - \bar{d}_{x\cup z} - \bar{d}_{(y)}, \\ T_2 &: EN_2 = 2\bar{d}_{x(y\cup z)} - \bar{d}_{(x)} - \bar{d}_{y\cup z}. \end{aligned} \tag{6.5}$$

Let PV_1 be the p-value of EN_1, and PV_2 be the p-value of EN_2. If the prior probabilities are equal, $\mathbf{z}$ should be classified in F_1 if $PV_1 < PV_2$, otherwise, classify $\mathbf{z}$ in F_2. If the prior probability are not equal, the classification rule is modified to assign $\mathbf{z}$ to F_1, if $\frac{PV_2}{PV_1} > \frac{1-p_x}{1-p_y}$, where p_x and p_y are the prior probability of F_1 and F_2.

We use $|EN_1 - EN_0|$ to measure the change in the value of EN when the new observation is added to the $\mathbf{X}$ sample. Similarly, the influence of adding the new observation to F_2 is given by $|EN_2 - EN_0|$. Therefore, if the prior probabilities are equal, the classification rule is modified by assigning $\mathbf{z}$ to F_1 if $\frac{|EN_1 - EN_0|}{|EN_2 - EN_0|} < 1$, otherwise, classify $\mathbf{z}$ in F_2. When we consider unequal prior probability and cost, we assign $\mathbf{z}$ to F_1 if

$$\frac{|EN_1 - EN_0|}{|EN_2 - EN_0|} < \left(\frac{c(2|1)}{c(1|2)}\right)\left(\frac{p_x}{p_y}\right). \tag{6.6}$$

6.4.2.2 Biswas and Ghosh Statistic

Consider bivariate vector $(\|\mathbf{X}_1 - \mathbf{X}_2\|, \|\mathbf{X}_1 - \mathbf{Y}_1\|)$ with bivariate distribution M_1 and $(\|\mathbf{X}_1 - \mathbf{Y}_1\|, \|\mathbf{Y}_1 - \mathbf{Y}_2\|)$ with distribution M_2. Let $\boldsymbol{\mu}_{M_1}$ and $\boldsymbol{\mu}_{M_2}$ denote the mean vectors of M_1 and M_2, respectively. Biswas and Ghosh [35] show that $\boldsymbol{\mu}_{M_1} = \boldsymbol{\mu}_{M_2} \Leftrightarrow \mu_{F_1F_1} = \mu_{F_1F_2} = \mu_{F_2F_2}$. Hence, instead of testing $H_0 : F_1 = F_2$, one can test an equivalent null hypothesis $H_0' : \boldsymbol{\mu}_{M_1} = \boldsymbol{\mu}_{M_2}$ against the alternative $H_a' : \boldsymbol{\mu}_{M_1} \neq \boldsymbol{\mu}_{M_2}$. The test statistic is

$$BG = \|\hat{\boldsymbol{\mu}}_{M_1} - \hat{\boldsymbol{\mu}}_{M_2}\|^2 = (\bar{d}_{(xy)} - \bar{d}_{(x)})^2 + (\bar{d}_{(xy)} - \bar{d}_{(y)})^2,$$

and H_0 is rejected for larger value of BG. Using the test-based method of classification, we introduce the classification rule using BG statistic.

Let BG_i be the test statistic. One can show that

$$\begin{aligned} T1 &: BG_1 = (\bar{d}_{(x\cup z)y} - \bar{d}_{(x\cup z)})^2 + (\bar{d}_{(x\cup z)y} - \bar{d}_{(y)})^2, \\ T2 &: BG_2 = (\bar{d}_{x(y\cup z)} - \bar{d}_{(y\cup z)})^2 + (\bar{d}_{x(y\cup z)} - \bar{d}_{(x)})^2. \end{aligned} \tag{6.7}$$

Suppose BG_0 is the test statistic that only involves training samples. If the prior probabilities are equal, $\mathbf{z}$ should be classified in F_1 if $\frac{|BG_1 - BG_0|}{|BG_2 - BG_0|} < 1$, otherwise, classify $\mathbf{z}$ in F_2. If we consider unequal prior probability and cost, we assign $\mathbf{z}$ to F_1 if

$$\frac{|BG_1 - BG_0|}{|BG_2 - BG_0|} < \left(\frac{c(2|1)}{c(1|2)}\right)\left(\frac{p_x}{p_y}\right). \tag{6.8}$$

6.4.2.3 Euclidean Distance Statistic

The Euclidean distance (ED) statistic is defined as $ED = |\bar{d}_{(x)} - \bar{d}_{(y)}|$, which quantifies the absolute difference between the centers of the two groups. Let ED_1 denote the ED statistic obtained when the new observation $\mathbf{z}$ is added to the $\mathbf{X}$ sample; then $ED_1 = |\bar{d}_{(x\cup z)} - \bar{d}_{(y)}|$. Similarly, if $\mathbf{z}$ is incorporated into the $\mathbf{Y}$ sample, the corresponding statistic ED_2 is given by $ED_2 = |\bar{d}_{(x)} - \bar{d}_{(y\cup z)}|$. Let ED_0 denote the Euclidean distance computed from the original

$\mathbf{X}$ and $\mathbf{Y}$ samples. Under equal prior probabilities, a natural classification rule assigns $\mathbf{z}$ to F_1 if $\frac{|ED_1 - ED_0|}{|ED_2 - ED_0|} < 1$, and to F_2 otherwise. In the presence of unequal priors or differing misclassification costs, the decision criterion can be generalized as

$$\frac{|ED_1 - ED_0|}{|ED_2 - ED_0|} < \frac{c(2|1)}{c(1|2)} \cdot \frac{p_x}{p_y},$$

where $c(j|i)$ represents the cost of assigning an observation from F_i to F_j, and p_x, p_y denote the respective prior probabilities.

6.4.3 Depth-based Method

Another method of classification is based on the closeness of the new observation to center of the training samples. A data depth function is a function $D(\mathbf{t}; F)$ that measures the closeness of a point $\mathbf{t} \in \mathbb{R}^d$ with respect to the center of a probability distribution F. If $\mathbf{t}_1$ is more central than $\mathbf{t}_2$ with respect to distribution F, a depth function should imply that $D(\mathbf{t}_1; F) > D(\mathbf{t}_2; F)$. There are many existing depth functions, such as Mahalanobis depth (Mahalanobis [207]), Tukey depth (Tukey [342]), simplicial depth (Liu [199]), and convex hull peeling depth (Mosler [243]), among others. Liu and Modarres [201] introduced the lens depth function. Suppose $\mathbf{X}_1$ and $\mathbf{X}_2$ are two random vectors in $\mathbb{R}^d$; the lens of $\mathbf{X}_1$ and $\mathbf{X}_2$ is defined as

$$L(\boldsymbol{X}_1, \boldsymbol{X}_2) = B(\boldsymbol{X}_1, \|\boldsymbol{X}_1 - \boldsymbol{X}_2\|) \cap B(\boldsymbol{X}_2, \|\boldsymbol{X}_1 - \boldsymbol{X}_2\|),$$

where $B(\mathbf{c}, r)$ is the closed ball of radius r centered at $\mathbf{c}$. The lens is the intersection of the closed balls with radius $\|\mathbf{X}_1 - \mathbf{X}_2\|$, centered at $\mathbf{X}_1$ and $\mathbf{X}_2$, respectively. The probability $\mathbb{P}(\boldsymbol{t} \in L(\boldsymbol{X}_1, \boldsymbol{X}_2))$ can be viewed as a statistic to measure the closeness of $\mathbf{t}$ with respect to F. Liu and Modarres [201] show that

$$LD(\mathbf{t}; F) = \mathbb{P}(\boldsymbol{t} \in L(\boldsymbol{X}_1, \boldsymbol{X}_2)),$$

is a data depth function for a vector $\mathbf{t} \in \mathbb{R}^d$ with respect to distribution F, where $\mathbf{X}_1$ and $\mathbf{X}_2$ are i.i.d. random variables from F.

$$\text{If } \|\mathbf{X}_1 - \mathbf{X}_2\| > \max(\|\mathbf{t} - \mathbf{X}_1\|, \|\mathbf{t} - \mathbf{X}_2\|), \text{ then } \boldsymbol{t} \in L(\boldsymbol{X}_1, \boldsymbol{X}_2).$$

Therefore, we can build a classifier using the lens depth function.

In contrast to the parametric likelihood ratio (LR) approach, the lens depth (LD) classification method is non-parametric and does not require any assumptions regarding the underlying distribution of the data. Let $X = \{\mathbf{X}_i\}$ and $Y = \{\mathbf{Y}_i\}$ denote samples drawn from populations F_1 and F_2, respectively, with equal prior probabilities. To classify a new observation $\mathbf{z}$, we first compute its lens data depth relative to X.

For each pair of vectors $(\mathbf{X}_i, \mathbf{X}_j)$ in X, define $dis_0 = \|\mathbf{X}_i - \mathbf{X}_j\|$, $dis_1 = \|\mathbf{z} - \mathbf{X}_i\|$, and $dis_2 = \|\mathbf{z} - \mathbf{X}_j\|$. The observation $\mathbf{z}$ is considered to fall within the "lens" formed by $\mathbf{X}_i$ and $\mathbf{X}_j$ if both $dis_1 < dis_0$ and $dis_2 < dis_0$. Repeating this procedure for all pairs in X and counting the frequency with which $\mathbf{z}$ lies within the corresponding lens regions yields an estimate of its lens depth with respect to X, denoted LD_1. Analogously, the lens depth of $\mathbf{z}$ relative to Y is computed and denoted LD_2. The classification decision is then made by comparing these depths: $\mathbf{z}$ is assigned to F_1 if $LD_1 > LD_2$ and to F_2 otherwise. This procedure effectively assigns $\mathbf{z}$ to the population whose points collectively form a lens region closer to it. See Example 8.5.19 for an illustration.

6.4.4 Support Vector Machine

Support Vector Machines (SVMs), introduced by Cortes and Vapnik [70], are non-parametric supervised learning methods used for binary classification and regression problems. The central idea of SVM is to find an optimal separating hyperplane between two classes of data such that the margin, i.e., the distance between the hyperplane and the nearest data points from either class is maximized. These nearest data points are called support vector. Meyer and Wien [224] developed an R package, `e1071`, that implements SVMs and their extensions for practical applications.

Suppose we have a labeled dataset $\{(\mathbf{x}_i, y_i) : \mathbf{x}_i \in \mathbb{R}^d,\ y_i \in \{-1, +1\},\ i = 1, \ldots, n\}$. We would like to find a linear decision rule $f(\mathbf{x}) = \text{sign}(\mathbf{w}'\mathbf{x} + b)$, where $\mathbf{w} \in \mathbb{R}^d$ is the normal vector to the separating hyperplane, and $b \in \mathbb{R}$ is the intercept.

For perfectly separable data, the goal is to find the hyperplane $\mathbf{w}'\mathbf{x} + b = 0$, such that $y_i(\mathbf{w}'\mathbf{x}_i + b) \geq 1, \quad i = 1, \ldots, n$. The margin, defined as the perpendicular distance between the hyperplane and the closest data points, equals $\frac{2}{\|\mathbf{w}\|}$. Maximizing the margin is equivalent to minimizing $\|\mathbf{w}\|^2$, which leads to the following quadratic optimization problem

$$\min_{\mathbf{w},b} \frac{1}{2}\|\mathbf{w}\|^2 \quad \text{subject to} \quad y_i(\mathbf{w}'\mathbf{x}_i + b) \geq 1,\ i = 1, \ldots, n.$$

Introducing Lagrange multipliers $\alpha_i \geq 0$, the dual optimization problem is

$$\max_{\alpha} \sum_{i=1}^{n} \alpha_i - \frac{1}{2}\sum_{i=1}^{n}\sum_{j=1}^{n} \alpha_i\alpha_j y_i y_j \mathbf{x}_i'\mathbf{x}_j,$$

subject to $\sum_{i=1}^{n} \alpha_i y_i = 0, \quad \alpha_i \geq 0$. The solution vector is $\mathbf{w} = \sum_{i=1}^{n} \alpha_i y_i \mathbf{x}_i$. Only the points with $\alpha_i > 0$ contribute to $\mathbf{w}$, and these are the support vectors.

For non–separable data, slack variables $\xi_i \geq 0$ are introduced, $y_i(\mathbf{w}'\mathbf{x}_i + b) \geq 1 - \xi_i, \quad \xi_i \geq 0$. The optimization problem becomes $\min_{\mathbf{w},b,\xi} \ \frac{1}{2}\|\mathbf{w}\|^2 + C\sum_{i=1}^{n} \xi_i$, where $C > 0$ is a tuning parameter balancing margin maximization and classification error.

To handle non-linear classification, SVMs use kernels $K(\mathbf{x}_i, \mathbf{x}_j)$, replacing the inner product $\mathbf{x}_i'\mathbf{x}_j$. Common kernels include, linear: $K(\mathbf{x}, \mathbf{z}) = \mathbf{x}'\mathbf{z}$, polynomial: $K(\mathbf{x}, \mathbf{z}) = (\mathbf{x}'\mathbf{z} + c)^d$, and radial basis function: $K(\mathbf{x}, \mathbf{z}) = \exp(-\gamma\|\mathbf{x} - \mathbf{z}\|^2)$.

Consider two classes, X and Y. If the classes are linearly separable, many hyperplanes can separate them. SVM chooses the hyperplane that maximizes the distance to the nearest points from either class. These closest points are called support vectors, and they are critical because moving them would change the hyperplane. SVM focuses only on the points near the decision boundary (support vectors) and ignores points that are far away, making it robust and efficient.

Figure 6.1 illustrates the main elements of an SVM. Class X (blue squares) and Class Y (red triangles) are the two sets of training points. The solid black line is the optimal separating hyperplane found by the SVM. This line divides the two classes. The dashed lines represent the margins, which are parallel to the hyperplane and pass through the support vectors. Support vectors are the critical points lying on the margin boundaries. They are highlighted with thicker markers. The position of these points determines the hyperplane. The black dot is a new observation. Its classification depends on which side of the hyperplane it falls: if it is above the hyperplane, it is classified as Class Y; otherwise, it is Class X. Points that are far from the hyperplane do not affect its position; only support vectors matter. This visual summary shows why SVM is robust: the hyperplane is determined by only a few points (support vectors), and the margin maximization ensures good separation between classes.

Example 6.4.2 *Consider a two-dimensional dataset with points from two classes,*

$$\mathbf{x}_1 = (2, 2),\ y_1 = +1, \quad \mathbf{x}_2 = (4, 4),\ y_2 = +1, \quad \mathbf{x}_3 = (4, 0),\ y_3 = -1, \quad \mathbf{x}_4 = (0, 0),\ y_4 = -1.$$

The optimal separating hyperplane can be derived using the dual problem, yielding approximately $\mathbf{w} = (1, 1), \quad b = -3$. *Thus, the decision boundary is* $x_1 + x_2 - 3 = 0$. *Points satisfying* $x_1 + x_2 > 3$ *are classified as* $+1$, *and those with* $x_1 + x_2 < 3$ *are classified as* -1.

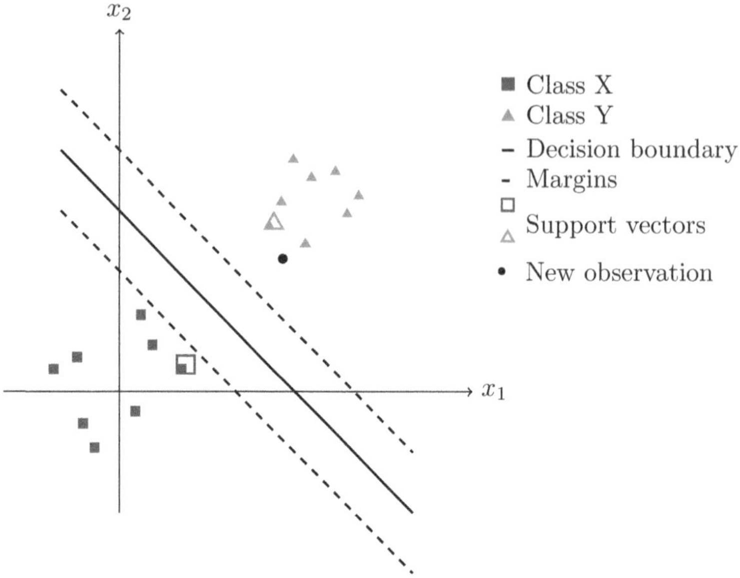

FIGURE 6.1: **SVM Illustration.** Blue squares and red triangles represent two classes of data. The solid black line is the optimal separating hyperplane, and the dashed lines show the margins. Support vectors are highlighted with thicker markers. The black dot represents a new observation, which will be classified based on which side of the hyperplane it falls.

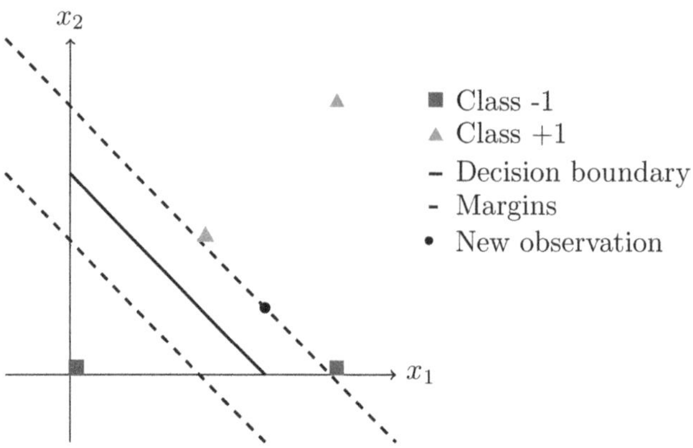

FIGURE 6.2: **SVM Plot for the Example Points.** Blue squares and red triangles are the two classes. The solid black line is the decision boundary $x_1 + x_2 - 3 = 0$. Dashed lines are the margins. Support vectors (thicker markers) determine the hyperplane. The black dot is a new observation classified based on the side of the hyperplane.

SVM Plot for the Example appears in Figure 6.2. Blue squares are class -1 *points:* $(0, 0)$ *and* $(4, 0)$. *Red triangles are class* $+1$ *points:* $(2, 2)$ *and* $(4, 4)$. *The solid black line is the SVM decision boundary:* $x_1 + x_2 - 3 = 0$. *The dashed lines show the margins, parallel to the decision boundary. Support vectors are highlighted:* $(0, 0)$ *and* $(2, 2)$ *in this example. These points define the hyperplane. The black dot represents a new observation; it will be classified as* $+1$ *because it is above the decision boundary.*

This illustrates the SVM principle: only the support vectors determine the separating hyperplane; other points are irrelevant.

6.5 Multisample classification

6.5.1 Test-based Classification

Similar to its two-sample counterpart, the key observation in the multisample case is that allocation of the new observation to the correct class implies that the multisample test statistic for $H_0 : F_1 = F_2 = \ldots = F_s$ will be more rejective. We define within-group and between-group average IPDs. For the training sample G_k with n_k observations, $k = 1, \ldots, s$, the within-group IPD and the between-group average IPD are

$$\begin{aligned}\bar{d}_{(G_k)} &= \frac{1}{n_k(n_k-1)} \sum_{\mathbf{X}_i, \mathbf{X}_j \in G_k} \|\mathbf{X}_i - \mathbf{X}_j\|, \\ \bar{d}_{(G_k G_k^c)} &= \frac{1}{n_k n_{k^c}} \sum_{\mathbf{X}_i \in G_k, \mathbf{X}_j \in G_k^c} \|\mathbf{X}_i - \mathbf{X}_j\|,\end{aligned} \tag{6.9}$$

respectively, where G_k^c is the complement of sample group G_k. For example, the EN statistics for multi-sample becomes $EN = \sum_{k=1}^{s} \bar{d}_{(G_k G_k^c)} - \sum_{k=1}^{s} \bar{d}_{(G_k)}$. When $s = 2$, we obtain the two-sample EN statistic.

Let $G_k \cup \mathbf{z}$ denotes the set containing the sample G_k and the new observation $\mathbf{z}$. The updated average IPDs are

$$\begin{aligned}\bar{d}_{(G_k \cup \mathbf{z})} &= \frac{(n_k-1)n_k \bar{d}_{(G_k)} + \sum_{\mathbf{X}_i \in G_k} \|\mathbf{z} - \mathbf{X}_i\|}{n_k(n_k+1)}, \\ \bar{d}_{(G_k \cup \mathbf{z})G_k^c} &= \frac{n_k n_{k^c} \bar{d}_{(G_k G_k^c)} + \sum_{\mathbf{X}_i \in G_k^c} \|\mathbf{z} - \mathbf{X}_i\|}{n_{k^c}(n_k+1)}, \\ \bar{d}_{(G_t)G_t^c} &= \frac{n_{G_t} n_{G_t^c} \bar{d}_{(G_t G_t^c)} + \sum_{\mathbf{X}_i \in G_t} \|\mathbf{z} - \mathbf{X}_i\|}{n_{G_t^c+1}(n_{G_t})} \quad \text{for } t \neq k.\end{aligned}$$

Using the updated average IPDs, the EN statistics for the kth class becomes

$$EN_k = \sum_{i \neq k} \bar{d}_{(G_i G_i^c)} + \bar{d}_{(G_k \cup \mathbf{z})} - \sum_{i \neq k} \bar{d}_{(G_i)} - \bar{d}_{(G_k \cup \mathbf{z})G_k^c}.$$

We next obtain the difference between EN_k and EN_0, where $EN_0 = \sum_{k=1}^{s} \bar{d}_{(G_k G_k^c)} - \sum_{k=1}^{s} \bar{d}_{(G_k)}$ depends only on the training samples. We allocate the new observation $\mathbf{z}$ to the class with minimum $|EN_k - EN_0|$ for $k = 1, \ldots, s$.

The multisample BG test statistic becomes $BG = \sum_{i=1}^{s} (\bar{d}_{(G_i G_i^c)} - \bar{d}_{(G_i)})^2$. Given a new observation $\mathbf{z}$, we calculate BG_k for kth class,

$$BG_k = \sum_{i \neq k} (\bar{d}_{(G_i G_i^c)} - \bar{d}_{(G_i)})^2 + (\bar{d}_{(G_k \cup \mathbf{z})G_k^c} - \bar{d}_{(G_k \cup \mathbf{z})})^2.$$

Note that BG_0 is the test statistic that only involves training samples. Hence, $\mathbf{z}$ is allocated to the class with minimum $|BG_k - BG_0|$. The ED statistic for the multi-sample case is defined as $ED = \sum_{i=1}^{s} |\bar{d}_{(G_i)} - \bar{d}_{(G_i^c)}|$ with ED_0 based on the training samples. Since

$$ED_k = \sum_{i \neq k} |\bar{d}_{(G_i)} - \bar{d}_{(G_i^c)}| + |\bar{d}_{(G_k \cup \mathbf{z})} - \bar{d}_{(G_i^c)}|,$$

we assign $\mathbf{z}$ to the class with minimum $|ED_k - ED_0|$.

6.5.2 Depth-based Method

Ghosh and Chaudhuri [116] offer a depth-based approach to classification. Lange, Mosler and Mozharovskyi [182] discuss a non-parametric classification rule based on zonoid depth. Given a new observation $\mathbf{z}$, the lens depth of $\mathbf{z}$ with respect to k-th class is defined as $LD(\mathbf{z}; F_k) = \mathbb{P}(t \in L(\boldsymbol{X}_{k1}, \boldsymbol{X}_{k2}))$, where $\boldsymbol{X}_{k1}, \boldsymbol{X}_{k2} \overset{iid}{\sim} F_k$. Suppose $\{\boldsymbol{X}_{ki}\}$ are samples from population F_k, for any vector pair $(\boldsymbol{X}_{ki}, \boldsymbol{X}_{kj})$, define $dis_0 = \|\boldsymbol{X}_{ki} - \boldsymbol{X}_{kj}\|$, $dis_1 = \|\boldsymbol{z} - \boldsymbol{X}_{ki}\|$, and $dis_2 = \|\boldsymbol{z} - \boldsymbol{X}_{kj}\|$. If both dis_1 and dis_2 are less than dis_0, then $\boldsymbol{z}$ falls in the lens area of the two points. One repeats the process for all pairs in kth sample and counts the frequency that $\boldsymbol{z}$ falls in the lens area. We assign $\mathbf{z}$ to the class with highest data depth.

6.6 Performance

In this section, we study the performance of the IPD classifiers (EN, BG and ED) under multivariate Bernoulli, multinomial, and multivariate Poisson distributions and compare them with LR, ED, Lens, LDA, SVM and NN methods. One important way of evaluating the performance of any classification procedure is to calculate the apparent error rate

$$\text{AER} = (n_{1M} + n_{2M})/(m+n),$$

where n_{1M} is the number of F_1 observations misclassified as F_2 items, and n_{2M} is the number of F_2 observations misclassified as F_1 items.

However, the apparent error rate always underestimate the misclassification rate (Anderson [10]). This optimistic evaluation happens because we use the same data to obtain the classification rule and to evaluate it. To modify the evaluation procedure, one can split the total sample into two parts: training sample and validation sample. One uses the training sample to construct the classification models, and the validation sample to evaluate the rule. Using Lachenbruch and Mickey [180] holdout procedure, we start with the data from F_1, omit one observation at a time, recalculate the classification based on the remaining $m-1$ and n observations. Then, we use the classification rule to classify the holdout observation.

After performing the procedure for every observation in F_1, let $n_{1M}^{(H)}$ to represent the number of holdout observations misclassified in F_1 group. Similarly, use $n_{2M}^{(H)}$ to represent the misclassification number for F_2. An estimate of the expected actual error rate is

$$\hat{\mathbb{E}}(AER) = (n_{1M}^{(H)} + n_{2M}^{(H)})/(m+n).$$

In this section, we use $\hat{\mathbb{E}}(AER)$ to evaluate and compare different classification procedures.

Assuming the population of different classes are all normally distributed with mean $\boldsymbol{\mu}_k$ and covariance matrix $\boldsymbol{\Sigma}$, the linear discriminant analysis (LDA) rule is: $\boldsymbol{z}$ is assigned to the class that minimizes $(\boldsymbol{z} - \boldsymbol{\mu}_k)'\boldsymbol{\Sigma}^{-1}(\boldsymbol{z} - \boldsymbol{\mu}_k)$. Despite the normal assumption, LDA is widely used in practice and works fine for many other continuous distribution, especially for real data whose distribution is not clearly known. However, classic LDA will loss power with high-dimensional data. The singularity of the estimated sample covariance matrix will make LDA inapplicable when the dimension is larger than sample size. Huang et al. [157] proposed a modified LDA rule for high-dimensional classification based on diagonal approximation of the covariance matrix. Suppose $\hat{\sigma}_i^2, i = 1, \ldots, d$, are the variance estimates, the discriminate score for kth class is

$$\sum_{i=1}^{d} (z_i - \hat{\mu}_{ki})^2/\hat{\sigma}_i^2 - 2lnp_k,$$

where p_k denotes the prior probability of class k. They show that this discriminate score is biased and modify it to a bias-corrected rule, given by

$$\frac{n-s-2}{n-s}\sum_{i=1}^{d}(z_i-\hat{\mu}_{k_i})^2/\hat{\sigma}_i^2-\frac{d}{n_k}-2lnp_k,$$

where s is the number of classes, n_k is the number of observations in each class and n is the total number of observations.

6.6.1 Multivariate Bernoulli Samples

We compare the performance of the classification methods for various (d, n) values. The sampling experiment, with 1000 replications, consists of generating $d = (2, 10, 50)$ dimensional multivariate Bernoulli random vectors of sizes $m = n \in \{10, 20, 30, 100\}$ with specified $(\mathbf{p}_x, \mathbf{\Sigma}_x)$ and $(\mathbf{p}_y, \mathbf{\Sigma}_y)$, respectively. The covariance matrices $\mathbf{\Sigma}_x$ and $\mathbf{\Sigma}_y$ are constructed as constant correlation matrices with $\rho = (0, 0.2, 0.5)$.

Due to the unavailability of the form of the distribution, except in the exchangeable case, and large number of parameters involved, it is not practical to use the likelihood function to perform classification for general MB distribution. For this reason, many practitioners assume independence among the components. Hand and Yu [138] even show that the Bayes rule based on false assumption of independence can be more effective than sophisticated rules in some situations. Suppose F_1 and F_2 have multivariate Bernoulli distributions with independent components. Given a new observation $\mathbf{z} = (z_1, \ldots, z_d)$, the likelihood of $\mathbf{z}$ under F_1 is given as $f_x(\mathbf{z}) = \prod_{i=1}^{d} p_i^{z_i}(1-p_i)^{1-z_i}$, where d_i is the marginal probability for ith component in F_1. Similarly, one can obtain the likelihood of $\mathbf{z}$ under F_2 as $f_y(\mathbf{z}) = \prod_{j=1}^{d} p_j^{z_j}(1-p_j)^{(1-z_j)}$, where d_j is the marginal probability for jth component in F_2. We consider both independent and equally correlated Bernoulli vectors.

Let R_1 and R_2 be the regions where we assign F_1 and F_2 observations. Assuming equal priors and misclassification costs, the LR classification rule is determined by R_1 and R_2 such that

$$\begin{aligned} R_1 &: \prod_{i=1}^{d}\left(\frac{p_{xi}}{p_{yi}}\right)^{z_i}\left(\frac{1-p_{xi}}{1-p_{yi}}\right)^{1-z_i} \geq 1, \\ R_2 &: \prod_{i=1}^{d}\left(\frac{p_{xi}}{p_{yi}}\right)^{z_i}\left(\frac{1-p_{xi}}{1-p_{yi}}\right)^{1-z_i} < 1. \end{aligned} \tag{6.10}$$

The maximum likelihood estimate of $\boldsymbol{p}_x$ and $\boldsymbol{p}_y$ have components $\widehat{p}_{xi} = \frac{1}{n}\sum_{j=1}^{n} x_{ij}$ and $\widehat{p}_{yi} = \frac{1}{n}\sum_{j=1}^{n} y_{ij}$, respectively, for $i = 1, \ldots, d$.

When F_1 and F_2 have identical populations a 95% confidence interval for the proportion of the observation from F_1 that are correctly classified is (0.469, 0.530). Guo and Modarres [125] demonstrate in their experiment that nearly all the proportion are close to 0.5 and fall in the confidence interval. They show that the proportion of correct classification when the components are independents with unequal means and variances. Their simulations show that a simple test based on the distance between the population centers outperforms more sophisticated methods under the MB distribution.

6.6.2 Multinomial Samples

Suppose the two populations F_1 and F_2 follow $MT(n, \mathbf{p}_x)$ and $MT(n, \mathbf{p}_y)$ respectively. Given a new observations $\mathbf{z} = (z_1, \ldots, z_d)$, the likelihood of $\mathbf{z}$ for F_1 is written as $f_x(\mathbf{z}) =$

$n!\prod_{i=1}^{d}\frac{p_{xi}^{z_i}}{z_i!}$, where d_{xi} is the ith element of $\mathbf{p}_x, i = 1,\ldots,d$. Similarly, the likelihood of $\mathbf{z}$ for F_2 is $f_y(\mathbf{z}) = n!\prod_{i=1}^{d}\frac{1}{z_i!}p_{yi}^{z_i}$, where d_{yi} is the ith element of $\mathbf{p}_y, i = 1,\ldots,d$. The maximum likelihood estimate of $\boldsymbol{p}_x$ and $\boldsymbol{p}_y$ have components $\widehat{p}_{xk} = \sum_{j=1}^{n} x_{kj}/n$ and $\widehat{p}_{yk} = \sum_{j=1}^{n} y_{kj}/n$. Assuming equal priors and misclassification costs, the LR classification rule is determined by R_1 and R_2 such that

$$\begin{aligned} R_1 &: \sum_{i=1}^{d} z_i(\ln\sum_{j=1}^{n} x_{ij} - \ln(n) - \ln\sum_{j=1}^{n} y_{ij} + \ln(n)) \geq 0, \\ R_2 &: \sum_{i=1}^{d} z_i(\ln\sum_{j=1}^{n} x_{ij} - \ln(n) - \ln\sum_{j=1}^{n} y_{ij} + \ln(n)) < 0. \end{aligned} \tag{6.11}$$

To evaluate the classification rules under multinomial samples, we draw $m = (10, 20, 50)$ observations from multinomial distribution with $n = (5, 50)$ trials, and $d = (6, 36, 60)$ categories. Similar to the multivariate Bernoulli samples, we use F_1 and F_2 to represent the same populations. Guo and Modarres [125] show that under identical distributions the proportion of the observations from F_1 that are correctly classified is about 0.5.

Guo and Modarres [125] show that increasing the number of trials increases the discriminant power in all classifiers. Dutta and Ghosh [83] show the adverse effects of high dimensions on the performance of the classic nearest neighbor classifier. This is also seen in the experiments of Guo and Modarres [125]. They report the power ordering $LR \simeq LDA > (EN \simeq BG) > Lens > (ED \simeq SVM \simeq NN)$.

6.6.3 Multivariate Poisson Samples

For multivariate Poisson, we consider the samples with constant correlations.

Suppose two populations F_1 and F_2 follow MP($\boldsymbol{\lambda}_x$) and MP($\boldsymbol{\lambda}_y$), respectively. The likelihood of a new observation $\mathbf{z}$ for F_1 is

$$\mathbb{P}(\mathbf{z}|\boldsymbol{\lambda}_x) = \exp(-\sum_{t=1}^{d}\lambda_{xt})\prod_{t=1}^{d}\frac{\lambda_{xt}^{z_t}}{z_t!}\sum_{i=0}^{m}\prod_{j=1}^{d}\binom{z_j}{i}i!(\frac{\lambda_{x0}}{\prod_{k=1}^{d}\lambda_{xk}})^i,$$

where $\lambda_{x0}, \lambda_{x1}, \ldots, \lambda_{xd}$ are the components of $\boldsymbol{\lambda}_x$ and $m = \min(z_1, \ldots, z_d)$. Similarly, we can get the likelihood of $\mathbf{z}$ under F_2 as $\mathbb{P}(\mathbf{z}|\boldsymbol{\lambda}_y)$. One can obtain the likelihood ratio criteria as

$$LR(\mathbf{z}) = \frac{\mathbb{P}(\mathbf{z}|\boldsymbol{\lambda}_x)}{\mathbb{P}(\mathbf{z}|\boldsymbol{\lambda}_y)} = \frac{\sum_{i=0}^{m}\prod_{j=1}^{d}\binom{z_j}{i}i!c_x^i}{\sum_{i=0}^{m}\prod_{j=1}^{d}\binom{z_j}{i}i!c_y^i}\exp(\sum_{t=1}^{d}\lambda_{yt} - \sum_{t=1}^{d}\lambda_{xt})\prod_{t=1}^{d}(\frac{\lambda_{xt}}{\lambda_{yt}})^{z_t},$$

where $c_x = \lambda_{x0}/\prod_{k=1}^{d}\lambda_{xk}$, $c_y = \lambda_{y0}/\prod_{k=1}^{d}\lambda_{yk}$.

The estimation of $\boldsymbol{\lambda}_x$ can be computed as

$$\hat{\lambda}_{x0} = \binom{d}{2}^{-1}\sum_{s=1}^{d}\sum_{h=s+1}^{d}\frac{1}{m-1}\sum_{i=1}^{m}(X_{si} - \bar{X}_s)(X_{hi} - \bar{X}_h)$$

and $\hat{\lambda}_{xt} = \bar{X}_t - \hat{\lambda}_{x0}$ for $t = 1, \ldots, d$, where X_{si} is the sth component of $\mathbf{X}_i$ and $\bar{X}_s$ is the mean of the sth components. Similarly, we can get the estimation of $\boldsymbol{\lambda}_y$. Assuming equal priors and misclassification costs, the LR classification rule is determined by R_1 and R_2 such that $R_1 : LR(\mathbf{z}) \geq 1$ and $R_2 : LR(\mathbf{z}) < 1$.

Using $d = (3, 10, 50)$ dimensional Poisson random vectors of sizes $m = n \in \{10, 20, 30\}$, Guo and Modarres [125] show that under identical distributions the proportion of the observation from F_1 that are correctly classified is about 0.5. They also report on the proportion of the correct classification when the Poisson samples are from different populations. Their sampling experiments show that the discriminant power increases with d and m, especially when the correlation parameter in the two distributions is different they obtain the power ordering $(EN \simeq SVM \simeq NN) > BG > (Lens \simeq LDA) > ED > LR$.

6.6.4 Comparison of Methods

The main concern is the ability to classify discrete observations in high dimensions when the number of variables exceeds the number of observations. How do the methods we discuss compare and relate? In terms of assumptions, LDA assumes $\mathbb{P}(\mathbf{X} = \mathbf{X}|C = j)$ follows a normal distribution with mean $\boldsymbol{\mu}_j$ and covariance $\boldsymbol{\Sigma}$, for $j = 1, 2$. For LDA, the log odds is linear because of the equal covariance assumption. Popular with practitioners, logistic regression models the log odds as a linear function of the predictors. It does not assume a form for the distribution of $\mathbf{X}$. Both LDA and logistic regression assume a linear decision boundary, but the methods of estimation and the estimated coefficients are different. Hastie, Tibshirani, and Friedman [141] note that LDA and logistic regression tend to perform similarly in a variety of situations in practice.

ED measures the separation of the two groups by computing the distance between the centroids. Using the spectral decomposition of $\boldsymbol{\Sigma} = \mathbf{U}\boldsymbol{\Lambda}U'$, one can show that a new observation $\mathbf{X}$ is first transformed to $\mathbf{z} = \boldsymbol{\Lambda}^{-\frac{1}{2}}U'\mathbf{X}$, before it is assigned to the class with the nearest centroid $\boldsymbol{\mu}_{z(i)} = \boldsymbol{\Lambda}^{-\frac{1}{2}}\boldsymbol{\mu}_i$ in the transformed space. That is, we assign $\mathbf{X}$ to the class i for which $\|\mathbf{z} - \boldsymbol{\mu}_{z(i)}\|^2 - log(p_i)$ is smallest.

Interpoint distance methods do not assume a form for the distribution of the observations. Their asymptotic theory is based on U-statistics. Let $\sigma^2_{F_1} + \sigma^2_{F_2} + v^2$. Method of BG assumes uniformly bounded fourth moments for the distributions and weak dependence among the component variables to show that the following converge in probability as $d \to \infty$

$$d^{-1/2}\|\mathbf{X}_i - \mathbf{X}_j\| \to \sqrt{2}\sigma_{F_1}, \quad d^{-1/2}\|\mathbf{Y}_i - \mathbf{Y}_j\| \to \sqrt{2}\sigma_{F_2}, \quad d^{-1/2}\|\mathbf{X}_i - \mathbf{Y}_j\| \to \Delta^{1/2}$$

where $\text{trace}(\boldsymbol{\Sigma}_1)/d \to \sigma_{F_1}$, $\text{trace}(\boldsymbol{\Sigma}_2)/d \to \sigma_{F_2}$ and $\|\boldsymbol{\mu}_1 - \boldsymbol{\mu}_2\|^2/d \to v^2$.

The group boundary built by a test-based classifier is the set of all data points for which the null hypothesis of the equality of the distributions is not rejected. For a new observation $\mathbf{z}$ from class 1,

$$d^{-1/2}\|\mathbf{z} - \mathbf{X}_i\| \to \sqrt{2}\sigma_{F_1}, \quad d^{-1/2}\|\mathbf{z} - \mathbf{Y}_j\| \to \Delta^{1/2}$$

as $d \to \infty$. The parameters $\sigma^2_{F_1}$, $\sigma^2_{F_2}$ and v^2 identify the group boundary. A new observation $\mathbf{z}$ from class 1 is correctly classified by the NN classifier if $v^2 > \sigma^2_{F_1} - \sigma^2_{F_2}$. Similarly, if $\mathbf{z}$ is from class 2, it is correctly classified if $v^2 > \sigma^2_{F_2} - \sigma^2_{F_1}$. The NN rule correctly classifies all new observations if $v^2 > |\sigma^2_{F_1} - \sigma^2_{F_2}|$. See section 4.10. In high dimensional setting, NN is influenced by non–informative variables and suffers from the curse of dimensionality (Aggarwal et al. [47]). Pal et al. [261] show that, in high-dimensional settings, NN classifier misclassified about half of the observations. NN can become ill defined because all pairwise distances concentrate around a single value. Beyer et al. [29] show that distance concentration can occur with as few as 15 dimensions. See Dutta and Ghosh [83] and Hall et al. [133].

6.7 High-Dimensional Classification

We consider the non-parametric classification of HDLSS data where the classical discrimination methods break down due to the singularity of the sample covariance matrix. Using dissimilarity indices,we build powerful classifiers, and compare their behavior with existing methods. We illustrate the difficulties with the Euclidean nearest neighbor (NN) method and prove that dissimilarity-based classifiers produce misclassification rates that tend to zero as $d \to \infty$. We present test-based classifiers in the HDLSS setting. A simulation study compares the misclassification rates of diagonal linear discriminant analysis with twelve other non-parametric classifiers. The methods are applied to microarray data for classification of prostate cancer.

With a long history in Statistics, classification is a multivariate technique that is concerned with allocating new observations to two or more groups. Fisher [101] introduced the linear discriminant analysis as a rule for assigning a new observation to one of two normally distributed groups. One can assign a new observation to populations F_1 or F_2 by minimizing the expected cost of misclassification ECM.

The literature on classification is extensive, encompassing a wide range of methods including linear and quadratic classifiers, logistic regression, the naive Bayes classifier, support vector machines, kernel-based classifiers, k-nearest neighbors, boosting, decision trees, random forests, neural networks, and various ensemble methods. Despite this diversity, there exist relatively few non-parametric classification procedures specifically designed for high-dimensional data, for which the misclassification rate asymptotically approaches zero as the number of variables $d \to \infty$. Lange et al. [182] proposed a non-parametric classification method based on the zonoid depth. Several authors have also suggested modifications to the linear and quadratic discriminant functions in situations where the covariance matrix is singular. Under the assumption that the populations of two classes are normally distributed with a common positive-definite covariance matrix, the linear discriminant analysis (LDA) rule is widely used and performs satisfactorily for many non-normal distributions as well. However, classical LDA loses effectiveness in high-dimensional settings and becomes infeasible when d exceeds the sample size N. To address this, Xu et al. [361] introduced two modified LDA approaches tailored for HDLSS data and compared their performance with other popular classification algorithms. Furthermore, to render LDA applicable when the covariance matrix is singular, many authors have proposed a variant that employs a diagonal approximation of the covariance matrix.

The nearest-neighbor (NN) method relies on Euclidean distances, which makes it vulnerable to the concentration of distances phenomenon in high-dimensional settings. In this setting, the sparsity of the observations has unfavorable effects on the performance of NN-based classifiers. Angiulli [11] shows that the interpoint distances concentrate around a single value in high-dimensional spaces. The concentration phenomenon describes the tendency for distances to become almost indistinguishable in high-dimensional spaces. The coefficient of variation of the IPDs tends to zero as $d \to \infty$. The relative difference of the distances of the closest and farthest observations of an independently selected data point tends to zero as $d \to \infty$. Hence, the concept of the NN of an observation is no longer meaningful.

Hall et al. [134] argue that in the HDLSS setting, observations in each group tend to lie at the vertices of a regular simplex and randomness of the observations is the result of the random rotation of the simplex. They explore the curse-of-dimensionality on the Euclidean distance in the HDLSS setting and identify its negative effects on several classifiers, including the nearest neighbor classifier. On a similar note, Chan and Hall [56] identified the shortcomings of the nearest neighbor and average distance classifiers, and proposed some "scale-adjustments" to rectify them. Ahn et al. [6] show that the HDLSS geometric representation holds under mild conditions. Sarkar and Ghosh [299] show that violation of the neighborhood structure is responsible for poor performance of clustering algorithms in high dimensions based on the Euclidean distance.

Recently, Roy et al. [289] use a generalized class of distance functions to define modifications of NN and DS classifiers. In fact, both the T_ρ and $T(NN_\rho)$ classifiers in this work have been thoroughly studied by Roy et al. [289]. We compare them against classifiers based on a new dissimilarity index T_δ and $T(NN_\delta)$. Roy et al. [288] offer fast and robust classifiers for HDLSS data. The shortcomings of the Euclidean distance in high dimensions have been explored in the literature, e.g., Beyer et al. [29] and Radovanovic et al. [275]. Ting et al. [338], Aryal et al. [13], and Chakraborty and Zhang [54] use of different distance functions than the Euclidean distance.

We introduce a powerful non-parametric classification method based on the dissimilarity matrix of the observations. As a fundamental building block in many multivariate techniques, a dissimilarity matrix is an $N \times N$ matrix that encodes all pairwise dissimilarities among the N sample points. Such matrices are central to the implementation of widely used classifiers, including k-means, k-nearest neighbors, and support vector machines. Dissimilarity measures tailored for high-dimensional observations play a pivotal role in numerous applications, ranging from DNA microarray analysis and proteomics to computer vision, machine learning, data mining, classification, and natural language processing. For microarray and proteomics datasets (West et al. [353]), the sample size is often on the order of tens, whereas the number of variables can reach thousands or even hundreds of thousands. In such HDLSS settings, the dissimilarity matrix offers substantial data reduction, transforming the original $N \times d$ data matrix into a more tractable $N \times N$ representation, while preserving essential information for subsequent analysis.

We consider new NN classifiers of HDLSS data based on dissimilarity indices. The new classifiers perform much better than NN classifiers based on the Euclidean distance. Since the dissimilarity indices are asymptotic in d, the classifiers based on these indices exhibit remarkable performances, producing misclassification rates that tend to zero as $d \to \infty$. To proof that misclassification rates tend to zero as $d \to \infty$ we assume that classification relevant information grows systematically at a rate proportional to d. The Euclidean distance only uses two observations to measure the distance between them. It does utilize the neighborhood structure provided by the other $N-2$ observations. We use the dissimilarities (3.6) and (3.13) for classification.

We generated a random sample of size $m = 15$ from a skew-normal distribution (F_1) with location parameter $\boldsymbol{\mu}_x$, scale parameter $\boldsymbol{\omega}_x$ and shape parameter $\boldsymbol{\alpha}_x$. We also generated $n = 15$ random observations from a skew-normal distribution(F_2) with location parameter $\boldsymbol{\mu}_y$, scale parameter $\boldsymbol{\omega}_y$ and shape parameter $\boldsymbol{\alpha}_y$. We set $\boldsymbol{\mu}_x = 0.81$ and $\boldsymbol{\mu}_y = \mathbf{0}$, $\boldsymbol{\omega}_x = \boldsymbol{\omega}_y = \mathbf{I}$ and the scale parameters $\boldsymbol{\alpha}_x = \boldsymbol{\alpha}_y = \mathbf{1}$. We estimated the misclassification rates of 13 classifiers in 1000 simulation experiments. For each experiment, we generated a new observation $\mathbf{z}$ from F_1 or F_2 with probability of 1/2. Figure 6.3 shows the estimated misclassification rates of the classifiers for $d = 2^i$, for $i = 1, 2, \ldots, 10$. Two classifiers stand out in as their misclassification rates do not vanish as $d \to \infty$. T_E is based on the total Euclidean distances and the NN estimator NN_E is based on the nearest neighbor classifier using the Euclidean distance. The misclassification rates of the remaining classifiers approach zero at different rates as $d \to \infty$. In particular, classifiers using the dissimilarity δ in outperform the same classifiers using the dissimilarity ρ. These classifiers are defined in the following sections.

We discuss dissimilarity indices for the classification of high-dimensional data and describe their asymptotic properties as $d \to \infty$. We present high dimensional nearest neighbors methods based on the new dissimilarity indices. Using the new indices, we also discuss test-based classifiers, adopting tests of equality of distributions to the classification in the HDLSS setting. Since the dissimilarity indices are asymptotic in d, we prove that the classifiers based on these indices produce misclassification rates that tend to zero as $d \to \infty$. We report on a simulation study that compares the misclassification rates of diagonal LDA with twelve other classifiers. We apply the methods to a real dataset and the last section is devoted to concluding remarks and recommendations.

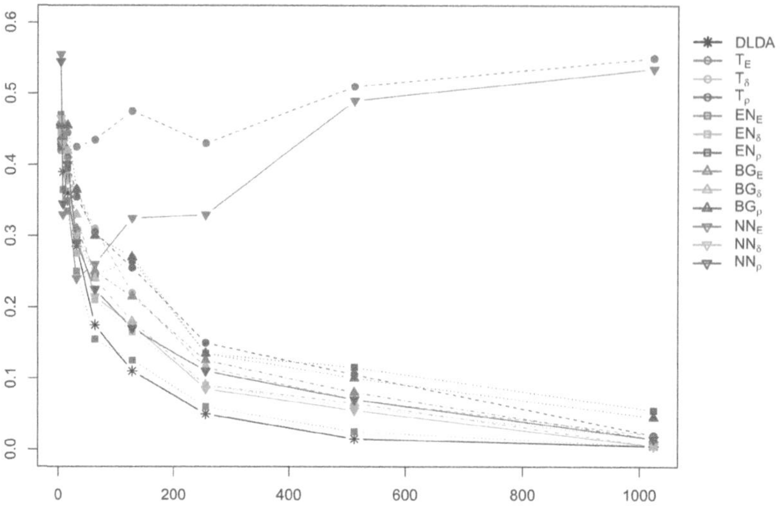

FIGURE 6.3: Estimated misclassification rates against d. Groups F_1 and F_2 follow the skew-normal distributions. T_E and NN_E are the worst performers.

6.8 Classification Methods

A generalization of the distance between vectors $\mathbf{x}_i$ and $\mathbf{x}_j$ in Euclidean d-dimensional space, the most commonly used proximity measure for continuous measurements is the Minkowski metric L_s. Marozzi ([215], [216]) discusses powerful multivariate tests based on IPDs for various value of s. Modarres and Song [239] survey recent developments on IPDs, discuss their properties, applications, and present their distributions for several probability models. Suppose $\mathbf{X}$ and $\mathbf{Y}$ are independent d-variate vectors taken independently from distributions F_1 and F_2 with means $\boldsymbol{\mu}_x$, and $\boldsymbol{\mu}_y$ and covariance matrices $\boldsymbol{\Sigma}_x$ and $\boldsymbol{\Sigma}_y$, respectively. Let $\mathbb{X} = \{\mathbf{X}_i\}_{i=1}^m$ and $\mathbb{Y} = \{\mathbf{Y}_j\}_{j=1}^n$ be i.i.d. copies of $\mathbf{X}$ and $\mathbf{Y}$, respectively. Consider i.i.d. vectors $\mathbf{X}_i$ and $\mathbf{X}_j$ from F_1 and i.i.d. vectors $\mathbf{Y}_i$ and $\mathbf{Y}_j$ from F_2. We use the dissimilarity indices ρ of (3.6) and δ of (3.13) in this chapter.

We discuss how the test statistics of Baringhaus and Franz [21], Biswas and Ghosh [35]for testing the equality of distribution function can be applied as discriminant functions using the test-based approach. We also discuss the nearest neighbor classification using the Euclidean distance and dissimilarity measures ρ and δ. Other classifiers that are used for comparison include the dissimilarity classifiers and a modified linear discriminant rule based on diagonal approximation of the covariance matrix.

We will use the test-based classification with Euclidean distance and dissimilarity indices ρ and δ to construct non–parametric classifiers. Let $\mathbf{X} = \{\mathbf{X}_i\}$ and $\mathbf{Y} = \{\mathbf{Y}_j\}$, for $i = 1, \ldots m$ and $j = 1, \ldots, n$. Consider two points $\mathbf{X}_i = (X_{1i}, \ldots, X_{di})$ and $\mathbf{X}_j = (X_{1j}, \ldots, X_{dj})$. Consider the IPDs $d_{(x)ij}$, $d_{(y)ij}$ and $d_{(xy)ij}$ defined in (2.5), (2.9), and (2.13), respectively. The average IPDs are defined in (2.7), (2.11), and (2.15), respectively. Given a new observation $\mathbf{z}$, let $x \cup z$ denotes the set containing the $\mathbf{X}$ observations and $\mathbf{z}$. The updated average IPDs are

$$\bar{d}_{(x\cup z)y} = \frac{mn\bar{d}_{(xy)} + \sum_{j=1}^n \|\mathbf{z} - \mathbf{Y}_j\|}{(m+1)n}, \quad \bar{d}_{x\cup z} = \frac{(m-1)m\bar{d}_{(x)} + \sum_{i=1}^m \|\mathbf{z} - \mathbf{X}_i\|}{(m+1)m}. \tag{6.12}$$

Similarly, we define, $\bar{d}_{x(y\cup z)}$ and $\bar{d}_{y\cup z}$.

6.8.1 Energy Test

To test $H_0 : F_1 = F_2$ Székely and Rizzo [329] and Barighaus and Franz [21] prove that $2E\|\mathbf{X}_1 - \mathbf{Y}_1\| - E\|\mathbf{X}_1 - \mathbf{X}_2\| - E\|\mathbf{Y}_1 - \mathbf{Y}_2\| \geq 0$ with equality holding if and only if $F_1 = F_2$ and propose the statistic $EN = 2\bar{d}_{(xy)} - \bar{d}_{(x)} - \bar{d}_{(y)}$. Applying test-based classification, we obtain

$$T_1(EN_E) = 2\bar{d}_{(x\cup z)y} - \bar{d}_{x\cup z} - \bar{d}_{(y)}, \quad T_2(EN_E) = 2\bar{d}_{x(y\cup z)} - \bar{d}_{(x)} - \bar{d}_{y\cup z}. \tag{6.13}$$

To obtain a classification rule based on dissimilarity index ρ, we first compute the matrix $\boldsymbol{\rho}$. The average IPDs of $\mathbf{X}$ and $\mathbf{Y}$ are

$$\bar{\rho}_{(x)} = \frac{2}{m(m-1)} \sum_{i=1}^{m-1} \sum_{j=i+1}^{m} \rho(\mathbf{X}_i, \mathbf{X}_j), \quad \bar{\rho}_{(y)} = \frac{2}{n(n-1)} \sum_{i=1}^{n-1} \sum_{j=i+1}^{n} \rho(\mathbf{Y}_i, \mathbf{Y}_j), \tag{6.14}$$

respectively. The average IPD between $\mathbf{X}$ and $\mathbf{Y}$ is $\bar{\rho}_{(xy)} = \frac{1}{mn} \sum_{i=1}^{m} \sum_{j=1}^{n} \rho(\mathbf{X}_i, \mathbf{Y}_j)$. We obtain

$$T_1(EN_\rho) = 2\bar{\rho}_{(x\cup z)y} - \bar{\rho}_{x\cup z} - \bar{\rho}_y, \quad T_2(EN_\rho) = 2\bar{\rho}_{x(y\cup z)} - \bar{\rho}_x - \bar{\rho}_{y\cup z}. \tag{6.15}$$

To obtain classification rule based on dissimilarity index δ, we first compute the $N \times N$ matrix $\boldsymbol{\delta}$. The average IPD of $\mathbf{X}$ and $\mathbf{Y}$ are

$$\bar{\delta}_{(x)} = \frac{2}{m(m-1)} \sum_{i=1}^{m-1} \sum_{j=i+1}^{m} \delta(\mathbf{X}_i, \mathbf{X}_j), \quad \bar{\delta}_{(y)} = \frac{2}{n(n-1)} \sum_{i=1}^{n-1} \sum_{j=i+1}^{n} \delta(\mathbf{Y}_i, \mathbf{Y}_j), \tag{6.16}$$

respectively. The average IPD between $\mathbf{X}$ and $\mathbf{Y}$ is $\bar{\delta}_{(xy)} = \frac{1}{mn} \sum_{i=1}^{m} \sum_{j=1}^{n} \delta(\mathbf{X}_i, \mathbf{Y}_j)$. We obtain

$$T_1(EN_\delta) = 2\bar{\delta}_{(x\cup z)y} - \bar{\delta}_{x\cup z} - \bar{\delta}_y, \quad T_2(EN_\delta) = 2\bar{\delta}_{x(y\cup z)} - \bar{\delta}_x - \bar{\delta}_{y\cup z}. \tag{6.17}$$

In each case, the classification rule assigns $\mathbf{z}$ to F_1 if $T_1 < T_2$; otherwise, it is assigned to F_2. We refer to these methods as $T(EN_E)$, $T(EN_\rho)$ and $T(EN_\delta)$.

Lemma 6.8.1 *The misclassification rates of $T(EN_\delta)$ under the assumptions (B1)-(B3), and $T(EN_\rho)$ under assumptions (A1)-(A3), tend to zero as $d \to \infty$.*

6.8.2 Biswas and Ghosh Test

To test $H_0 : F_1 = F_2$ Biswas and Ghosh [35] present the statistic $BG = (\bar{d}_{(xy)} - \bar{d}_{(x)})^2 + (\bar{d}_{(xy)} - \bar{d}_{(y)})^2$, and reject H_0 for larger value of BG. We can use BG as a discriminant function. We obtain

$$\begin{aligned}
T_1(BG_E) &= (\bar{d}_{(x\cup z)y} - \bar{d}_{(x\cup z)})^2 + (\bar{d}_{(x\cup z)y} - \bar{d}_{(y)})^2, \\
T_2(BG_E) &= (\bar{d}_{x(y\cup z)} - \bar{d}_{(y\cup z)})^2 + (\bar{d}_{x(y\cup z)} - \bar{d}_{(x)})^2, \\
T_1(BG_\rho) &= (\bar{\rho}_{(x\cup z)y} - \bar{\rho}_{(x\cup z)})^2 + (\bar{\rho}_{(x\cup z)y} - \bar{\rho}_{(y)})^2, \\
T_2(BG_\rho) &= (\bar{\rho}_{x(y\cup z)} - \bar{\rho}_{(y\cup z)})^2 + (\bar{\rho}_{x(y\cup z)} - \bar{\rho}_{(x)})^2, \\
T_1(BG_\delta) &= (\bar{\delta}_{(x\cup z)y} - \bar{\delta}_{(x\cup z)})^2 + (\bar{\delta}_{(x\cup z)y} - \bar{\delta}_{(y)})^2, \\
T_2(BG_\delta) &= (\bar{\delta}_{x(y\cup z)} - \bar{\delta}_{(y\cup z)})^2 + (\bar{\delta}_{x(y\cup z)} - \bar{\delta}_{(x)})^2.
\end{aligned} \tag{6.18}$$

The classification rule assigns $\mathbf{z}$ to F_1 if $T_1 < T_2$; otherwise, it is assigned to F_2. We will refer to these as $T(BG_E)$, $T(BG_\rho)$, and $T(BG_\delta)$.

Lemma 6.8.2 *Under assumptions (B1)-(B3), the misclassification rates of $T(BG_\delta)$ tends to zero as $d \to \infty$. Under assumptions (A1)-(A3), the misclassification rate of $T(BG_\rho)$ tends to zero as $d \to \infty$.*

6.8.3 Nearest Neighbor Classification

The k-nearest neighbor rule (Fix and Hodges [102]; Henze [146]) is a popular, flexible and intuitive method of non-parametric classification. While this method is commonly used in the machine learning and statistics communities, less is known about its misclassification rate when $d \to \infty$. To count the number of the nearest neighbors of a new observation z let $\mathbb{I}_j^{\mathbb{E}}(\mathbf{X}_i, \mathbf{z})$, $\mathbb{I}_j^{\delta}(\mathbf{X}_i, \mathbf{z})$ and $\mathbb{I}_j^{\rho}(\mathbf{X}_i, \mathbf{z})$ be indicator variables that equal 1 if $\mathbf{X}_i$ is the j-th nearest neighbor of $\mathbf{z}$ using Euclidean distance, dissimilarity indices δ and ρ, respectively. Similarly, define $\mathbb{I}_j^{\mathbb{E}}(\mathbf{Y}_i, \mathbf{z})$, $\mathbb{I}_j^{\delta}(\mathbf{Y}_i, \mathbf{z})$ and $\mathbb{I}_j^{\rho}(\mathbf{Y}_i, \mathbf{z})$ as indicator variables that equal 1 if $\mathbf{Y}_i$ is the j-th nearest neighbor of $\mathbf{z}$. We fix the number of nearest neighbors at k. We obtain

$$\begin{aligned} T_1(NN_E) &= \sum_{i=1}^{m}\sum_{j=1}^{k} \mathbb{I}_j^{\mathbb{E}}(\mathbf{X}_i, \mathbf{z}), \quad T_2(NN_E) = \sum_{i=1}^{n}\sum_{j=1}^{k} \mathbb{I}_j^{\mathbb{E}}(\mathbf{Y}_i, \mathbf{z}), \\ T_1(NN_\delta) &= \sum_{i=1}^{m}\sum_{j=1}^{k} \mathbb{I}_j^{\delta}(\mathbf{X}_i, \mathbf{z}), \quad T_2(NN_\delta) = \sum_{i=1}^{n}\sum_{j=1}^{k} \mathbb{I}_j^{\delta}(\mathbf{Y}_i, \mathbf{z}), \\ T_1(NN_\rho) &= \sum_{i=1}^{m}\sum_{j=1}^{k} \mathbb{I}_j^{\rho}(\mathbf{X}_i, \mathbf{z}), \quad T_2(NN_\rho) = \sum_{i=1}^{n}\sum_{j=1}^{k} \mathbb{I}_j^{\rho}(\mathbf{Y}_i, \mathbf{z}). \end{aligned} \tag{6.19}$$

We assign the new observation $\mathbf{z}$ to F_1 if $T_1 > T_2$; otherwise, $\mathbf{z}$ is assigned to F_2. We refer to these methods as $T(NN_E)$, $T(NN_\delta)$ and $T(NN_\rho)$.

Lemma 6.8.3 *The misclassification rates of $T(NN_\delta)$ and $T(NN_\rho)$, under assumptions (A1)–(A3) and (B1)–(B3), respectively, tend to zero as $d \to \infty$.*

6.8.4 Dissimilarity Classifiers

We compute

$$\begin{aligned} T_1(\delta) &= \sum_{i=1}^{m} \delta(\mathbf{z}, \mathbf{X}_i), \quad T_2(\delta) = \sum_{j=1}^{n} \delta(\mathbf{z}, \mathbf{Y}_j), \\ T_1(\rho) &= \sum_{i=1}^{m} \rho(\mathbf{z}, \mathbf{X}_i), \quad T_2(\rho) = \sum_{j=1}^{n} \rho(\mathbf{z}, \mathbf{Y}_j), \\ T_1(E) &= \sum_{i=1}^{m} \|\mathbf{z} - \mathbf{X}_i\|, \quad T_2(E) = \sum_{j=1}^{n} \|\mathbf{z} - \mathbf{Y}_j\|. \end{aligned} \tag{6.20}$$

Statistics $T_1(E)$ and $T_2(E)$ measure the total dissimilarity of $\mathbf{z}$ from the F_1 and F_2 groups using L_1 norm, respectively. We assign $\mathbf{z}$ to the closest group. That is, we assign $\mathbf{z}$ to F_1 if $T_1 < T_2$; otherwise, $\mathbf{z}$ is assigned to F_2. We refer to these methods as T_δ, T_ρ, and T_E.

Lemma 6.8.4 *The misclassification rates of T_ρ and T_δ, under assumptions (A1)–(A3) and (B1)–(B3), respectively, tend to zero as $d \to \infty$.*

6.8.5 Diagonal LDA

Huang et al. [157] proposed a modified LDA rule for high-dimensional classification based on diagonal approximation of the covariance matrix. Suppose $\hat{\sigma}_i^2, i = 1, \ldots, d$, are the variance estimates, the discriminate score for kth class is $\sum_{i=1}^{d}(y_i - \hat{\mu}_{ki})^2/\hat{\sigma}_i^2 - 2\ln p_k$. Huang et al. [157] show that this discriminate score is biased and modify it to a bias-corrected rule, which is

$$\frac{N-4}{N-2}\sum_{i=1}^{d}(y_i - \hat{\mu}_{k_i})^2/\hat{\sigma}_i^2 - \frac{d}{n_k} - 2\ln p_k, \tag{6.21}$$

where n_k is the number of the observations in each class and N is the total number of observations. A new observation $\mathbf{z}$ is assigned to the class with maximum discriminant score. We will refer to this method as DLDA.

6.9 Comparisons

In this section, we provide the results of a sampling experiment to assess the misclassification rates of diagonal LDA, with the following discriminant functions. More extensive comparisons appear in Modarres [234].

- Dissimilarity classifiers (DS): T_E, T_δ, T_ρ,
- Nearest neighbor classifiers (NN): $T(NN_E)$, $T(NN_\delta)$, $T(NN_\rho)$,
- EN classifiers (EN): $T(EN_E)$, $T(EN_\delta)$, $T(EN_\rho)$,
- BG classifiers (BG): $T(BG_E)$, $T(BG_\delta)$, $T(BG_\rho)$.

We generated random samples of sizes $m = n = 20$ from d-dimensional log-normal distributions $L\mathbb{N}_d(\mathbf{0}_d, \mathbf{I}_d)$ and $L\mathbb{N}_d(\mu_y \mathbf{1}_d, \mathbf{I}_d)$ and estimated the misclassification rates of the classifiers in 1000 simulation experiments. For each experiment, we generated a new observation $\mathbf{z}$ from F_1 or F_2 with probability of 1/2. The following Table shows the rates for 13 classifiers when $\mu_y = 0.45$ and $d = 2^i$, for $i = 1, 2, \ldots, 10$. Two classifiers stand out in the Table as their misclassification rates do not vanish as $d \to \infty$. They are $DS(E)$ the NN estimator $NN(E)$ the Euclidean distance. The misclassification rates of the remaining classifiers approach zero at different rates as $d \to \infty$. In particular, classifiers using the dissimilarity δ outperform the same classifiers using the dissimilarity ρ.

TABLE 6.1: Misclassification rates for two multivariate log-normal distributions.

d	DLDA	T_E	T_ρ	T_δ	EN_E	EN_δ	EN_ρ	BG_E	BG_δ	BG_ρ	NN_E	NN_δ	NN_ρ
2	0.372	0.431	0.454	0.462	0.365	0.372	0.373	0.407	0.429	0.421	0.430	0.391	0.408
4	0.361	0.408	0.389	0.419	0.337	0.361	0.351	0.379	0.407	0.409	0.379	0.398	0.381
8	0.351	0.498	0.432	0.428	0.345	0.337	0.371	0.338	0.418	0.438	0.406	0.318	0.377
16	0.282	0.472	0.338	0.352	0.281	0.319	0.321	0.312	0.328	0.357	0.338	0.302	0.322
32	0.181	0.498	0.289	0.311	0.187	0.223	0.269	0.275	0.259	0.308	0.358	0.213	0.271
64	0.138	0.501	0.214	0.237	0.130	0.151	0.211	0.231	0.168	0.264	0.437	0.176	0.217
128	0.069	0.487	0.109	0.151	0.051	0.059	0.148	0.129	0.069	0.159	0.425	0.051	0.15
256	0.023	0.512	0.065	0.101	0.039	0.051	0.117	0.087	0.048	0.121	0.523	0.038	0.074
512	0.003	0.498	0.014	0.039	0.003	0.013	0.063	0.039	0.017	0.064	0.49	0.014	0.034
1024	0.000	0.516	0.003	0.013	0.000	0.004	0.016	0.009	0.004	0.014	0.535	0.002	0.011

Modarres [234] displays the misclassification rates of the methods under several distributions. Under a shift model, all methods reach a misclassification rate of zero as μ_y increases for fixed d. However, classifiers that use dissimilarity δ provide more discriminatory power at smaller values of μ_y. For some parameter values, discriminant using index δ show misclassification rates less close to zero whereas the other discriminant show higher rates.

6.10 Analysis of Prostate Tumor Gene Expression Data

The Prostate Tumor Gene Expression dataset is described and studied in Chung and Keles [69]. The dataset has two components: the gene expression matrix with 102 rows and 6033 columns, and the class index. The prostate dataset consists of 52 prostate tumor and 50 normal samples. Normal and tumor classes are coded in 0 and 1, respectively. We applied the classifiers discussed in this chapter to the dataset with the first d variables. The misclassification rates appear in Figure 6.4. One observes that for the entire dataset $d = 6033$, classifiers based on δ index show zero misclassification rates while other classifiers show much larger rates. For $d = 128$, the classifiers based on δ index show very small misclassification rates. A good runner up is DLDA.

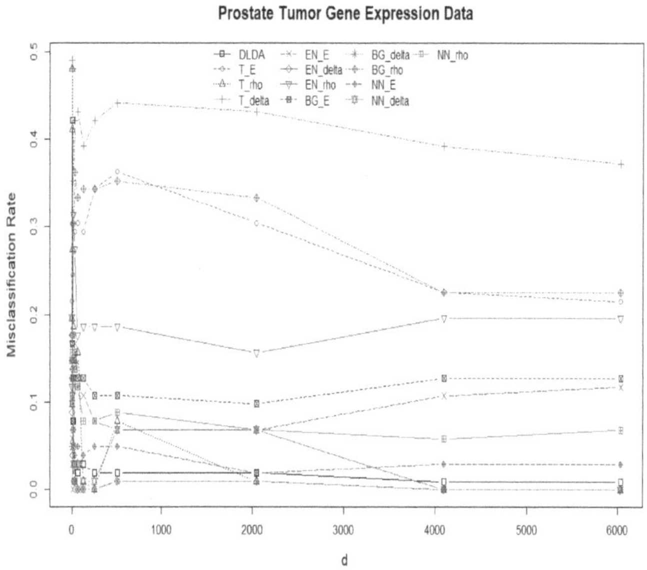

FIGURE 6.4: Estimated misclassification rates against d for prostate tumor data.

6.11 Summary

We investigate classification methodologies for high-dimensional discrete data, with particular emphasis on regimes in which conventional classifiers encounter structural or inferential limitations. Three canonical multivariate discrete distributions, multivariate Bernoulli, multinomial, and multivariate Poisson, are examined, as they capture prevalent data-generating mechanisms in a wide range of applied settings. IPDs form the basis of a novel non-parametric, test-based classification rule tailored to high-dimensional discrete observations. The proposed classifier is bench-marked against the likelihood ratio (LR) rule, modified linear discriminant analysis, support vector machines (SVM), k-nearest neighbors (NN), and depth-based classifiers. Simulation results demonstrate that classification accuracy is highly sensitive to the underlying data distribution, with no single method uniformly dominating across all scenarios. Consistent with the findings of Fernández-Delgado et al. [98], SVM exhibits suboptimal performance for discrete data, particularly for multivariate Bernoulli samples. NN performs competitively for high-dimensional multinomial and multivariate Poisson data but deteriorates in low-dimensional settings and for Bernoulli distributions. Euclidean distance–based methods are most effective for binary data, whereas LR, energy-based (EN), and NN classifiers are preferable for sparse multinomial data. EN-based classifiers are especially well suited to multivariate Poisson data with over-dispersed marginal distributions.

We further address the problem of classification in HDLSS settings, where classical parametric and distance-based approaches often fail due to singular covariance structures and the concentration of distances. The emphasis is on non-parametric classification procedures that remain effective as the dimension increases while the sample size remains fixed. It is shown that Euclidean distance–based nearest neighbor classifiers are severely affected by the curse of dimensionality, thereby motivating the use of alternative dissimilarity measures that more effectively capture class separation in high-dimensional spaces. Two such high-dimensional dissimilarity indices, denoted by ρ and δ, are introduced and studied as foundational tools for constructing robust classifiers. A general classification framework is developed by integrating these dissimilarity measures with several classification strategies, including nearest neighbor rules, dissimilarity-based classifiers, and test-based classifiers derived from the energy (EN) and Biswas–Ghosh (BG) statistics. These approaches are compared with Euclidean-based classifiers and diagonal linear discriminant analysis (DLDA). Theoretical results establish the asymptotic behavior of the proposed dissimilarity measures under mild dependence conditions, demonstrating that within-class dissimilarities converge to zero while between-class dissimilarities converge to positive constants as $d \to \infty$. Consequently, classifiers based on ρ and δ achieve misclassification rates that converge to zero in the HDLSS regime, in stark contrast to Euclidean distance–based methods. The relative merits of ρ and δ are also examined, with δ-based classifiers exhibiting superior discriminatory power for alternatives close to the null hypothesis. The chapter concludes with extensive comparative studies and an application to a real gene expression dataset involving prostate tumor samples.

6.12 Proofs

Proof: Lemma 6.8.3: Assuming (B1)-(B3) hold, if $\mathbf{z}$ is from F_1, then $\mathbb{I}_j^{\delta}(\mathbf{X}_i) \to 1$ and $\mathbb{I}_j^{\delta}(\mathbf{Y}_j) \to 0$ so that $T_1(NN_\delta) \to k > 0$ while $T_2(NN_\delta) \to 0$ with probability 1 as $d \to \infty$. Hence, $T_1(NN_\delta) > T_2(NN_\delta)$ and $\mathbf{z}$ will be assigned to F_1 with probability 1 as $d \to \infty$. Similarly, if $\mathbf{z}$ is from F_2, then $\mathbb{I}_j^{\delta}(\mathbf{X}_i) \to 0$ and $\mathbb{I}_j^{\delta}(\mathbf{Y}_j) \to 1$ so that $T_1(NN_\delta) \to 0$ while $T_2(NN_\delta) \to k > 0$ with probability 1 as $d \to \infty$. Hence, $T_1(NN_\delta) < T_2(NN_\delta)$ and $\mathbf{z}$ will be assigned to F_1 with probability 1 as $d \to \infty$. Replacing δ with ρ establishes that as $d \to \infty$, the NN method with index ρ has zero misclassification rate as $d \to \infty$. □

Proof: Lemma 6.8.4: Assuming (B1)-(B3) hold, if $\mathbf{z}$ is from F_1, then $T_1(\delta) \to 0$ while $T_2(\delta) \to m\tilde{\delta}(F_1, F_2) > 0$ with probability 1 as $d \to \infty$. Hence, $T_1(\delta) < T_2(\delta)$ and $\mathbf{z}$ will be assigned to F_1 with probability 1 as $d \to \infty$ If $\mathbf{z}$ is from F_2, then $T_1(\delta) \to m\tilde{\delta}(F_1, F_2) > 0$ while $T_2(\delta) \to 0$ with probability 1 as $d \to \infty$. Hence, $T_1(\delta) > T_2(\delta)$ and $\mathbf{z}$ will be assigned to F_2 with probability 1 as $d \to \infty$.

If $\mathbf{z}$ is from F_1, then $T_1(\rho) \to 0$ while $T_2(\rho) \to m\tilde{\rho}(F_1, F_2) > 0$ with probability 1 as $d \to \infty$. Hence, $T_1(\rho) < T_2(\rho)$ and $\mathbf{z}$ will be assigned to F_1 with probability 1 as $d \to \infty$ If the new observation $\mathbf{z}$ is from F_2, then $T_1(\rho) \to m\tilde{\rho}(F_1, F_2) > 0$ while $T_2(\rho) \to 0$ with probability 1 as $d \to \infty$. Hence, $T_1(\rho) > T_2(\rho)$ and $\mathbf{z}$ will be assigned to F_2 with probability 1 as $d \to \infty$. □

Proof: Lemma 6.8.1: We obtain $T_1(EN_\delta) = 2\bar{\delta}_{(x\cup z)y} - \bar{\delta}_{x\cup z} - \bar{\delta}_{y}$. and $T_2(EN_\delta) = 2\bar{\delta}_{x(y\cup z)} - \bar{\delta}_x - \bar{\delta}_{y\cup z}$. Assuming (B1)-(B3) hold, if $\mathbf{z}$ is from F_1 and assigned to F_1, then there are $m+1$ observations in $x \cup z$ and n observations in the $\mathbf{Y}$ sample. Hence, $\bar{\delta}_{x\cup z} \to 0$, $\bar{\delta}_{(y)} \to 0$, and $\bar{\delta}_{(x\cup z)y} \to \tilde{\delta}(F_1, F_2)$ as $d \to \infty$. Hence, $T_1(EN_\delta)$ converges to $2\tilde{\delta}(F_1, F_2) > 0$ in probability as $d \to \infty$. If the new observation $\mathbf{z}$ is from F_1 and assigned to F_2, then there are m observations in $\mathbf{X}$ sample and $n+1$ observations in the $y \cup z$ sample. Hence, $\delta_x \to 0$, $\bar{\delta}_{y\cup z} \to \frac{2}{n+1}\tilde{\delta}(F_1, F_2)$, and $\bar{\delta}_{x(y\cup z)} \to \frac{n}{n+1}\tilde{\delta}(F_1, F_2)$. Hence, $T_2(EN_\delta)$ converges to $\frac{2(n-1)}{n+1}\tilde{\delta}(F_1, F_2)$ in probability as $d \to \infty$. Since $T_1(EN_\delta) > T_2(EN_\delta)$, $\mathbf{z}$ will be assigned to F_1 with probability 1 as $d \to \infty$.

Assuming (B1)-(B3) hold, if $\mathbf{z}$ is from F_2 and assigned to F_1, then there are $m+1$ observations in $x \cup z$ and n observations in the $\mathbf{Y}$ sample. Hence, $\bar{\delta}_{x\cup z} \to \frac{2}{m+1}\tilde{\delta}(F_1, F_2)$, $\bar{\delta}_{(y)} \to 0$, and $\bar{\delta}_{(x\cup z)y} \to \frac{m}{m+1}\tilde{\delta}(F_1, F_2)$ as $d \to \infty$. Now, $T_1(EN_\delta)$ converges to $\frac{2(m-1)}{m+1}\tilde{\delta}(F_1, F_2) > 0$ in probability as $d \to \infty$. If $\mathbf{z}$ is from F_2 and assigned to F_2, then there are m observations in $\mathbf{X}$ sample and $n+1$ observations in the $y \cup z$ sample. Hence, $\delta_x \to 0$, $\bar{\delta}_{y\cup z} \to 0$, and $\bar{\delta}_{x(y\cup z)} \to \frac{n}{n+1}\tilde{\delta}(F_1, F_2)$. Hence, $T_2(EN_\delta)$ converges to $2\tilde{\delta}(F_1, F_2)$ in probability as $d \to \infty$. Since $T_1(EN_\delta) < T_2(EN_\delta)$, $\mathbf{z}$ will be assigned to F_2 with probability 1 as $d \to \infty$. □

Proof: Lemma 6.8.2: We obtain $T_1(BG_\delta) = (\bar{\delta}_{(x\cup z)y} - \bar{\delta}_{(x\cup z)})^2 + (\bar{\delta}_{(x\cup z)y} - \bar{\delta}_{(y)})^2$ and $T_2(BG_\delta) = (\bar{\delta}_{x(y\cup z)} - \bar{\delta}_{(y\cup z)})^2 + (\bar{\delta}_{x(y\cup z)} - \bar{\delta}_{(x)})^2$. Assuming (B1)-(B3) hold, if $\mathbf{z}$ is from F_1 and assigned to F_1, then there are $m+1$ observations in $x \cup z$ and n observations in the $\mathbf{Y}$ sample. Hence, $\bar{\delta}_{x\cup z} \to 0$, $\bar{\delta}_{(y)} \to 0$, and $\bar{\delta}_{(x\cup z)y} \to \tilde{\delta}(F_1, F_2)$ as $d \to \infty$. Hence, $T_1(EN_\delta)$ converges to $2\tilde{\delta}^2(F_1, F_2) > 0$ in probability as $d \to \infty$. If $\mathbf{z}$ is from F_1 and assigned to F_2, then there are m observations in $\mathbf{X}$ sample and $n+1$ observations in the $y\cup z$ sample. Hence, $\delta_x \to 0$, $\bar{\delta}_{y\cup z} \to \frac{2}{n+1}\tilde{\delta}(F_1, F_2)$, and $\bar{\delta}_{x(y\cup z)} \to \frac{n}{n+1}\tilde{\delta}(F_1, F_2)$. Hence, $T_2(EN_\delta)$ converges to $\frac{(n-2)^2+n^2}{(n+1)^2}\tilde{\delta}^2(F_1, F_2)$ in probability as $d \to \infty$. Since $T_1(BG_\delta) > T_2(BG_\delta)$, $\mathbf{z}$ is assigned to F_1 with probability 1 as $d \to \infty$.

Assuming (B1)-(B3) hold, if $\mathbf{z}$ is from F_2 and assigned to F_1, then there are $m+1$ observations in $x\cup z$ and n observations in the $\mathbf{Y}$ sample. Hence, $\bar{\delta}_{x\cup z} \to \frac{2}{m+1}\tilde{\delta}(F_1, F_2)$, $\bar{\delta}_{(y)} \to 0$, and $\bar{\delta}_{(x\cup z)y} \to \frac{m}{m+1}\tilde{\delta}(F_1, F_2)$ as $d \to \infty$. Hence, $T_1(EN_\delta)$ converges to $\frac{(m-2)^2+m^2}{(m+1)^2}\tilde{\delta}^2(F_1, F_2) > 0$ in probability as $d \to \infty$. If $\mathbf{z}$ is from F_2 and assigned to F_2, then there are m observations in $\mathbf{X}$ sample and $n+1$ observations in the $y \cup z$ sample. Hence, $\delta_x \to 0$, $\bar{\delta}_{y\cup z} \to 0$, and $\bar{\delta}_{x(y\cup z)} \to \frac{m}{m+1}\tilde{\delta}(F_1, F_2)$. Hence, $T_2(EN_\delta)$ converges to $\frac{2m^2}{(m+1)^2}\tilde{\delta}^2(F_1, F_2)$ in probability as $d \to \infty$. Since $T_1(BG_\delta) < T_2(BG_\delta)$, $\mathbf{z}$ is assigned to F_2 with probability 1 as $d \to \infty$. □

6.13 Exercises

Exercise 6.1 *See Sarkar et al. [297]. For fixed k, the nearest neighbor (NN) test (Schilling [300]; Henze [146]) uses the statistic*

$$T = \frac{1}{Nk}\Big(\sum_{i=1}^{m}\sum_{r=1}^{k}\mathbb{I}_r(\mathbf{X}_i) + \sum_{j=1}^{n}\sum_{r=1}^{k}\mathbb{I}_r(\mathbf{Y}_j)\Big) \tag{6.22}$$

where $\mathbf{X}_1, \ldots, \mathbf{X}_m \sim F_1$, $\mathbf{Y}_1, \ldots, \mathbf{Y}_n \sim F_2$, and $\mathbb{I}_r(\mathbf{Z})$ is an indicator variable that is 1 if $\mathbf{Z}$ and its r-th nearest neighbor, using Euclidean distances, arise from the same distribution. The null hypothesis $H_0 : F_1 = F_2$ is rejected for large values of T. Under what conditions, does the NN method fail to reject H_0 under the alternative? Explain.

Exercise 6.2 *Compute the nearest neighbor test statistic T,*

$$T = \frac{1}{Nk}\left(\sum_{i=1}^{N}\sum_{r=1}^{k}\mathbb{I}_r(\mathbf{Z}_i)\right)$$

Let $N = 8$, $k = 2$, and the following points in $\mathbb{R}^3$,

$$\begin{aligned}
&\mathbf{Z}_1 = (0,0,0), \quad \mathbf{Z}_2 = (0.1,0,0), \quad \mathbf{Z}_3 = (0,0.1,0), \quad \mathbf{Z}_4 = (0,0,0.1),\\
&\mathbf{Z}_5 = (1,1,1), \quad \mathbf{Z}_6 = (1,1,2), \quad \mathbf{Z}_7 = (2,1,1), \quad \mathbf{Z}_8 = (1,2,1).
\end{aligned}$$

Exercise 6.3 *Classification with equal covariance matrices, different means. Suppose $F_1 = \mathbb{N}(\boldsymbol{\mu}_1, \boldsymbol{\Sigma})$ and $F_2 = \mathbb{N}(\boldsymbol{\mu}_2, \boldsymbol{\Sigma})$. Derive the likelihood ratio test and the decision boundary for classifying a new observation $\mathbf{z}$.*

Exercise 6.4 *Diagonal covariances, unequal variances. Let $F_1 = \mathbb{N}(\boldsymbol{\mu}_1, \boldsymbol{\Sigma}_1)$ and $F_2 = \mathbb{N}(\boldsymbol{\mu}_2, \boldsymbol{\Sigma}_2)$, with $\boldsymbol{\Sigma}_1 = diag(\sigma_1^2, \ldots, \sigma_d^2)$ and $\boldsymbol{\Sigma}_2 = diag(\tau_1^2, \ldots, \tau_d^2)$. Derive the log-likelihood ratio for classifying $\mathbf{z}$.*

Exercise 6.5 *Full covariances, quadratic discriminant. Assume $F_1 = \mathbb{N}(\boldsymbol{\mu}_1, \boldsymbol{\Sigma}_1)$ and $F_2 = \mathbb{N}(\boldsymbol{\mu}_2, \boldsymbol{\Sigma}_2)$, where $\boldsymbol{\Sigma}_1 \neq \boldsymbol{\Sigma}_2$ are general positive definite matrices. Derive the form of the quadratic discriminant function and explain how to classify $\mathbf{z}$.*

Exercise 6.6 *Suppose $F_1 = Bernoulli(p_1)$ and $F_2 = Bernoulli(p_2)$. Derive the likelihood ratio test to classify $z \in \{0, 1\}$.*

Exercise 6.7 *Given $F_1 = Mult(n, \theta^{(1)})$, $F_2 = Mult(n, \theta^{(2)})$, derive the log-likelihood ratio test.*

Exercise 6.8 *Let $\mathbf{X} = \{\mathbf{X}_i\}$ and $\mathbf{Y} = \{\mathbf{Y}_j\}$, for $i = 1, j \ldots n$ be samples from populations F_1 and F_2, respectively. Let $\mathbf{X}_0$ be a new sample item to be classified and $\mathbf{S}_p = \frac{1}{2}(\mathbf{S}_1 + \mathbf{S}_2)$ denote the pooled covariance matrix obtained from the two training samples. Let T_1^2 and T_2^2 be Hotelling test statistic for Tests 1 and 2. Thus,*

$$T_1^2 = \left[\frac{n\bar{\mathbf{X}}_1 + \mathbf{X}_0}{n+1} - \bar{\mathbf{X}}_2\right]'\left[\left(\frac{1}{n+1} + \frac{1}{n}\right)\mathbf{S}_p\right]^{-1}\left[\frac{n\bar{\mathbf{X}}_1 + \mathbf{X}_0}{n+1} - \bar{\mathbf{X}}_2\right],$$
$$T_2^2 = \left[\bar{\mathbf{X}}_1 - \frac{n\bar{\mathbf{X}}_2 + \mathbf{X}_0}{n+1}\right]'\left[\left(\frac{1}{n} + \frac{1}{n+1}\right)\mathbf{S}_p\right]^{-1}\left[\bar{\mathbf{X}}_1 - \frac{n\bar{\mathbf{X}}_2 + \mathbf{X}_0}{n+1}\right],$$

Let $PV_1(\mathbf{X}_0)$ be the p-value of T_1^2, and $PV_2(\mathbf{X}_0)$ be the p-value of T_2^2. Test-based classification method suggests that, if the prior probabilities are equal, $\mathbf{x}_0$ be classified as F_1 if $PV_1(\mathbf{X}_0) < PV_2(\mathbf{X}_0)$. Prove that

$$PV_1(\mathbf{X}_0) < PV_2(\mathbf{X}_0) \Longleftrightarrow T_1^2 > T_2^2.$$

Show that the above is equivalent to

$$(\bar{\mathbf{X}}_1 - \bar{\mathbf{X}}_2)'\mathbf{S}_p^{-1} - \frac{1}{2}(\bar{\mathbf{X}}_1 - \bar{\mathbf{X}}_2)'\mathbf{S}_p^{-1}(\bar{\mathbf{X}}_1 + \bar{\mathbf{X}}_2).$$

Therefore, the test-based classification rule on the pooled T^2 statistic, is equivalent to the linear discriminant (LDA) rule.

Exercise 6.9 *Use the following training data to find a separating hyperplane with the SVM approach.*

$$\begin{aligned}
&\mathbf{X}_1 = (1,1), \quad y_1 = 1, \quad \mathbf{X}_2 = (2,2), \quad y_2 = 1\\
&\mathbf{X}_3 = (1,3), \quad y_3 = -1, \quad \mathbf{X}_4 = (2,4), \quad y_4 = -1
\end{aligned}$$

7

IPD of Multivariate Power Series Distributions

7.1 Abstract

This chapter develops a unified theoretical framework for the distribution of squared IPDs within the broad class of multivariate power series distributions (MPSDs). Encompassing important models such as the multinomial, negative multinomial, multivariate Poisson, and multivariate logarithmic distributions, MPSDs provide a natural foundation for analyzing high-dimensional discrete and count data. The chapter derives explicit distributional results for IPDs under a variety of settings, including independent and non-identically distributed vectors, mixture models, and distances from fixed reference points. These results establish analytical tools for understanding IPD behavior in high dimensions. Building on these developments, the chapter formulates IPD-based tests of homogeneity for mixture models with both known and estimated parameters, complemented by bootstrap density estimation. Comparative studies demonstrate the effectiveness of IPD-based procedures relative to competing homogeneity tests, with applications ranging from synthetic mixtures to real data, including the Old Faithful geyser dataset. By providing closed-form expressions and rigorous proofs, this chapter underscores the central role of IPDs in high-dimensional inference, classification, clustering, and model discrimination for discrete multivariate data.

7.2 Multivariate Power Series Interpoint Distances

The primary objective of this chapter is to derive the distributions of squared Euclidean IPDs within the framework of multivariate power series distributions (MPSDs). The work of Modarres and Song [240] focuses on the multivariate power series IPDs. This broad and important class of discrete multivariate distributions includes, as notable members, the multinomial (MN), negative multinomial (NGMN), multivariate Poisson (MP), and multivariate logarithmic (ML) distributions. The IPD is a fundamental metric in multivariate analysis and arises naturally in numerous statistical methodologies. Its relevance spans diverse fields, as it underpins procedures such as the comparison of multivariate distributions, cluster analysis, classification algorithms, correspondence analysis, and multidimensional scaling, among others.

IPDs have proven indispensable in the analysis of spatial and high-dimensional data. For instance, they play a crucial role in the study of point patterns, such as those explored in Ripley's seminal work [280], and are integral to methods for identifying spatial disease clusters via minimum spanning trees in epidemiological surveillance. Moreover, they serve as a foundation for tests assessing distributional homogeneity and are central to the

DOI: 10.1201/9781003789581-7

formulation of statistical depth functions. Modarres and Song [240] discuss multivariate power series IPDs and Song and Modarres [319] consider IPD test of homogeneity for multivariate mixture models.

It is worth emphasizing that, while not all tests of distributional homogeneity rely on IPDs, those that do frequently exhibit enhanced performance, especially in high-dimensional contexts where classical procedures may lose power or become infeasible due to the singularity of covariance matrices. For instance, Kolesnik [171] derives the distribution of IPDs between two independent telegraph processes, thereby highlighting the relevance of IPDs in stochastic process modeling. In a related vein, Modarres [228] studies IPDs in the setting of high-dimensional multivariate Bernoulli distributions, and Lukens [205] utilizes IPD densities arising from mixtures of multivariate normal distributions to uncover structure and to identify outliers in high-dimensional data. Within the field of image analysis, Osada et al. [260] demonstrate the effectiveness of IPDs for shape discrimination, a line of inquiry further advanced by Berrendero et al. [28], who connect IPD distributions to the identification of geometric forms.

Categorical data, frequently encountered in disciplines such as biology, ecology, physics, and genomics, naturally give rise to multivariate count models, with MPSDs providing a robust statistical framework for their analysis. These distributions are especially relevant in contingency table analysis and in scenarios involving the modeling of multiple count outcomes. For example, Shin [308] applies multinomial change point theory to health care diagnostics, while Waller and Zelterman [344]) propose log-linear models under negative multinomial sampling schemes. Royle and Link [291] extend these applications by introducing a general class of multinomial mixture models tailored for analyzing anuran calling survey data.

Conventional multivariate techniques often falter in high-dimensional regimes when the number of variables exceeds the number of observations. Classical inferential methods that rely on non–singular covariance matrices become infeasible, and density estimation methods demand prohibitively large sample sizes due to the curse of dimensionality. In contrast, methodologies based on IPDs offer a compelling alternative. Their key advantage lies in their dimensional invariance: regardless of the number of variables, IPDs remain scalar-valued, which enables robust application in high-dimensional settings.

The applicability of IPDs further extends to hypothesis testing, where they have been instrumental in the construction of powerful procedures for the general two-sample problem. Seminal contributions in this domain include those of Friedman and Rafsky [109], Henze and Penrose [147], Hall and Tajvidi [134], Baringhaus and Franz [21], Liu and Modarres [201] Biswas and Ghosh [35], and Székely and Rizzo [333]. Further IPD-based testing methodologies are presented in Rosenbaum [284], Jurečková and Kalina [165], and Marozzi [214], and [215].

Another prominent application of IPDs arises in classification. Because both the expected value and the distributional form of IPDs can vary substantially across competing models, they provide a powerful basis for model discrimination. Silverman and Brown [311] employ IPDs to differentiate among stationary point processes, whereas Guo and Modarres [125] develop IPD-based methodologies for the classification of high-dimensional discrete distributions. Additional studies, including those by Shurygin [309], Zhang et al. [368], and Hall et al. [135], further demonstrate the adaptability and effectiveness of IPDs within a variety of classification frameworks. IPDs also constitute a fundamental building block of many clustering procedures. IPDs likewise underpin a broad class of statistical depth functions. For example, Lok and Lee [202] introduce a depth function tailored to multimodal data that integrates both the distribution function and the mean of IPDs computed from fixed reference points. Such depth-based approaches yield robust, non-parametric instruments for the analysis of multivariate distributions, particularly in settings where standard assumptions regarding data structure or distributional form are violated.

In this section, we derive closed-form expressions and distributional results for IPDs, their norms, and their distances from fixed points for random vectors following various members of the MPSD family. These derivations provide the analytical foundation necessary for using IPDs in a wide array of multivariate applications, particularly those involving high dimensional or discrete data.

7.3 Multivariate Power Series Distributions

Noack [252] first introduces a class of discrete power series distribution. Sibuya et al. [310] characterized the NGMN distribution and Patil [262] define the Multivariate analogue of the power series distribution. Joshi and Patil [163] find certain structural properties of the power series distributions. Suppose $\mathbf{X} = (X_1, X_2, \ldots, X_d)'$ is a d-dimensional random vector and $\boldsymbol{\theta} = (\theta_1, \ldots, \theta_d)'$ is a parameter vector. Define $\boldsymbol{\Theta}$ to be the d-dimensional parameter space of $\boldsymbol{\theta}$. Let χ be a countable subset without any limit point of a d-fold Cartesian product of the set $\mathbb{R}$ of non–negative integers. Set χ can be written as $\chi = \{(x_1, x_2, \ldots, x_d)'\} \subseteq \mathbb{N}_0^d$, where $\mathbb{N}_0$ is the set of non–negative integers. Now, define the series function

$$f(\boldsymbol{\theta}) = \sum_{\mathbf{x}\in\chi} a(x_1, x_2, \ldots, x_d)\theta_1^{x_1}\theta_2^{x_2}\ldots\theta_d^{x_d},$$

where the summation extends over χ, and $\theta_i \geq 0$ with $\boldsymbol{\theta} \in \boldsymbol{\Theta}$, so that $f(\boldsymbol{\theta})$ is finite and differentiable. Also, define the coefficient function $a(\mathbf{x})$ to be a positive-valued function of $\mathbf{x}$. The random vector $\mathbf{X}$ is said to have multivariate power series distribution (MPSD) with range χ, parameter $\boldsymbol{\theta}$ and series function $f(\boldsymbol{\theta})$, if it has PMF

$$\mathbb{P}(\mathbf{X} = \mathbf{x}\,; \boldsymbol{\theta}) = \frac{a(x_1, \ldots, x_d)\theta_1^{x_1}\ldots\theta_d^{x_d}}{f(\theta_1, \ldots, \theta_d)}, \text{ given } (x_1, \ldots, x_d)' \in \chi \text{ and } f(\theta_1, \ldots, \theta_d) \neq 0, \tag{7.1}$$

where $a(\mathbf{x}) > 0$, if $\mathbf{x} \in \chi$, and $a(\mathbf{x}) = 0$, otherwise. We denote it as $\mathbf{X} \sim \text{MPSD}(\boldsymbol{\theta}_x, f_x(\boldsymbol{\theta}_x))$.

The MPSD is a discrete multivariate exponential type distribution with marginal PMF for each component X_j, $j = 1, \ldots, d$ as

$$\mathbb{P}(X_j = x_j) = \frac{\sum_{\{x_1,\ldots,x_d\}\setminus x_j}\left(a(x_1, \ldots, x_d)\prod_{i=1}^{d}\theta_i^{x_i}\right)}{f(\boldsymbol{\theta})} = \frac{a_j(\theta_1, \ldots, x_j, \ldots, \theta_d)\theta_j^{x_j}}{f(\boldsymbol{\theta})}, \tag{7.2}$$

where $\{x_1, \ldots, x_d\} \setminus x_j$ represents the relative complement of $\{x_j\}$ in $\{x_1, \ldots, x_d\}$ and the coefficient function is

$$a_j(\theta_1, \ldots, x_j, \ldots, \theta_d) = \sum_{\{x_1,\ldots,x_d\}\setminus x_j}\left(a(x_1, \ldots, x_d)\prod_{i\neq j}\theta_i^{x_i}\right).$$

Hence, X_j has a univariate power series distribution with the series function $f(\boldsymbol{\theta})$ as expanded in powers of θ_j, other θ's are treated as known. It is not difficult to obtain the mean vector and covariance matrix by using the r-th factorial moment of $\mathbf{X}$ (Patil [262]).

By taking different series function and reparametrization, we can obtain different multivariate distributions. For the purpose of this section, we illustrate the MN, NGMN, MP and ML distributions. Patil [262] describes a certain population where each individual has one distinct character out of $(d + 1)$ different possible characters, the $d + 1$ possible responses are considered for the categorical variables, while only d categories are assumed in the rest of the chapter. Let the population size be N, and let N_i be the number of individuals in the population having character i for $i = 0, 1, 2, \ldots, d$. Therefore, $\sum_{i=0}^{d} N_i = N$. Let $d_i = \frac{N_i}{N}$ be the proportion of individuals in the population having character i so that $\sum_{i=0}^{d} p_i = 1$. Now, a sample of individuals from the population may be drawn with replacement or without replacement. Further, sampling with replacement may be direct or inverse. Depend on the nature of the sampling procedure used, the random vector $\mathbf{X}$ is known to have an MN or NGMN distribution. MP and ML distributions can be obtained by taking limits and truncations of the first two distributions. Table 7.1 shows the corresponding series functions of these four distributions.

TABLE 7.1: MPSD with different parameters and series functions.

Distribution	Reparameterization	Series function	Coefficient function
MN	$\theta_i = \frac{p_i}{1-\sum_{i=1}^{d} p_i}$	$\left(1 + \sum_{i=1}^{d} \theta_i\right)^n$	$\frac{n!}{\prod_{i=1}^{d} x_i!(n-\sum_{i=1}^{d} x_i)!}$
NGMN	$\theta_i = p_i$	$\left(1 - \sum_{i=1}^{d} \theta_i\right)^{-k}$	$\frac{\Gamma(\sum_{i=1}^{d} x_i+k)}{\prod_{i=1}^{d} x_i!\Gamma(k)}$
MP	$\theta_i = \lambda_i$	$e^{\sum_{i=1}^{d} \theta_i}$	$1/\prod_{i=1}^{d} x_i!$
ML	$\theta_i = p_i$	$-log\left(1 - \sum_{i=1}^{d} \theta_i\right)$	$\frac{(\sum_{i=1}^{d} x_i-1)!}{\prod_{i=1}^{d} x_i!}$

By using equation (7.2), one can show that the marginal distributions, respectively, are

$$\begin{aligned}
B(n, p_t) &: \binom{n}{x_t} p_t^{x_t} (1-p_t)^{n-x_t}, \\
NB(k, p_t) &: \binom{k+x_t-1}{x_t} p_t^{x_t} (1-p_t)^{k}, \\
Poi(\lambda_t) &: \frac{e^{-\lambda_t}\lambda_t^{x_t}}{x_t!}, \\
L(p_t) &: \frac{log\left(1-\sum_{s=1}^{d} p_s + p_t\right)}{log\left(1-\sum_{s=1}^{d} p_s\right)} \cdot \mathbf{1}_{\{x_t=0\}}, \\
L(P,t) &: \frac{1}{-x_t log\left(1-\sum_{s=1}^{d} p_s\right)} \left(\frac{p_t}{1-\sum_{s=1}^{d} p_s + p_t}\right)^{x_t} \cdot \mathbf{1}_{\{x_t \geq 1\}}.
\end{aligned} \tag{7.3}$$

Since all the series functions in Table 7.1 can be written in the form of $f(\boldsymbol{\theta}) = g\left(\sum_{t=1}^{d} \theta_t\right)$, then the MN, NGMN, MP and ML distributions have the following properties, the proofs of which appear in Song [320].

Lemma 7.3.1 *Suppose the random vector $\boldsymbol{X}$ has the MPSD distribution with parameters $\boldsymbol{\theta}$ and series function $g\left(\sum_{t=1}^{d} \theta_t\right)$, then the random vector $\boldsymbol{X}^* = (X_{i_1}, \ldots, X_{i_d})'$ has the MPSD with parameters $(\theta_{i_1}, \ldots, \theta_{i_d})$ and series function $g\left(\sum_{t=1}^{d} \theta_{i_t}\right)$, where $(i_1, \ldots, i_d)$ is a permutation of $(1, 2, \ldots, d)$.*

Lemma 7.3.2 *Suppose* $\boldsymbol{X}$ *has the MPSD distribution with parameters* $\boldsymbol{\theta}$ *and series function* $g\left(\sum_{t=1}^{d}\theta_t\right)$, *then the random vector* $\boldsymbol{X}^*|(x_{r+1},\ldots,x_d)$ *has the MPSD with parameters* $(\theta_1,\ldots,\theta_r)$ *and series function* $h\left(\sum_{t=1}^{r}\theta_r\right)$, *where* $\boldsymbol{X} = (X_1,\ldots,X_r)'$, $h\left(\sum_{t=1}^{r}\theta_r\right)$ *depends on the form of* $g\left(\sum_{t=1}^{d}\theta_t\right)$ *and the value of* $\sum_{t=r+1}^{d}x_t$, *but it is independent of* $(\theta_{r+1},\ldots,\theta_d)$.

Lemma 7.3.3 *Suppose* $\boldsymbol{X}$ *has the MPSD distribution with parameters* $\boldsymbol{\theta}$ *and series function* $g\left(\sum_{t=1}^{d}\theta_t\right)$, *then* $\boldsymbol{X}^* = (X_1,\ldots,X_r)'$ *has the MPSD with parameters* $(\theta_1,\ldots,\theta_r)$ *and series function* $g^*\left(\sum_{t=1}^{r}\theta_t\right)$, *where* $g^*(a) = g\left(a + \sum_{t=r+1}^{d}\theta_t\right)$.

Definition 7.3.1 *Suppose* $\boldsymbol{X}$ *has the MPSD with parameter* $\boldsymbol{\theta}$ *and series function* $f(\boldsymbol{\theta})$. *Suppose further that*

$$\theta_t = \frac{p_t}{1-\sum_{t=1}^{d}p_t} \quad \textit{with} \quad 0 \leq \sum_{t=1}^{d}p_t < 1, 0 \leq p_t < 1$$

for $t = 1,\ldots,d$, *and* $n \in \mathbb{N}_0$. *Random vector* $\boldsymbol{X}$ *is said to have the d-dimensional MN distribution if the coefficient function is*

$$a(\boldsymbol{x}) = \frac{n!}{\prod_{t=1}^{d}x_t!\left(n-\sum_{t=1}^{d}x_t\right)!}$$

and series function is $f(\boldsymbol{\theta}) = \left(1+\sum_{t=1}^{d}\theta_t\right)^n$. *It has PMF*

$$\mathbb{P}(\boldsymbol{X}=\boldsymbol{x}) = \frac{n!}{\prod_{t=1}^{d}x_t!\cdot x_0!}\prod_{t=1}^{d}p_t^{x_t}\cdot p_0^{x_0}, \tag{7.4}$$

where $x_0 = n - \sum_{t=1}^{d}x_t$, $p_0 = 1-\sum_{t=1}^{d}p_t$ *and* $x_t \in \mathbf{N}_0$, $\sum_{t=1}^{d}x_t \leq n$. *Alternatively, the distribution of* $\boldsymbol{X}$ *can be called a* $(d+1)$*-dimensional singular MN distribution with parameters* $(n, p_0, p_1, \ldots, p_d)$.

Conventionally, the definition of the d-dimensional MN distribution is singular, so it is in fact $d-1$ dimensional non–singular MN distribution. Patil approaches this form in another direction, his $d+1$ dimensional MN is singular, so he defines a d-dimensional MN distribution. We use a d-dimensional singular MN distribution in the remaining sections and denoted by $\mathbf{X} \sim MN_d(n,\mathbf{p})$. For example, if $\mathbf{X} = (X_1, X_2)$ has a 2-dimensional singular MN distribution, it has a binomial distribution for X_2. To interpret MN, suppose sample size is n, let X_t be the number of individuals in the sample having character t. Then $\mathbf{X} = (X_1,\ldots,X_d)'$ is a random observation vector with the MN distribution.

Example 7.3.1 *Suppose* $\boldsymbol{X} \sim MN(n,\boldsymbol{p}$, *where* $\boldsymbol{p} = (p_1,\ldots,p_d)'$. *From (7.2), the marginal distribution for each component of* $\boldsymbol{X}$, *say* X_i *is* $B(n,p_t)$ *for* $t = 1,\ldots,d$. *The descending r th factorial moment of* $\boldsymbol{X}$ *is*

$$\mathbb{E}\left(\prod_{t=1}^{d}X_t^{(r_t)}\right) = n^{\left(\sum_{t=1}^{d}r_t\right)}\prod_{t=1}^{d}p_t^{r_t}. \tag{7.5}$$

One can use (7.5) *to find* $\mathbb{C}\mathrm{ov}(\mathrm{X_s}, \mathrm{X_t}) = -\mathrm{np_s p_t}$ *for* $s, t = 1, \ldots, d, s \neq t$, *so that the correlation is* $\mathbb{C}or(X_s, X_t) = -\sqrt{\frac{p_s p_t}{(1-p_s)(1-p_t)}}$. *Hence, counts in different categories are negatively correlated. The probability generating function is*

$$G_{\boldsymbol{X}}(\boldsymbol{z}) = \left(\sum_{t=1}^{d} p_t z_t\right)^n. \tag{7.6}$$

The covariance matrix of $\boldsymbol{X}$ *is singular with rank* $d-1$, *of the form* $\boldsymbol{\Sigma}_x = \mathbb{C}\mathrm{ov}(\boldsymbol{X}) = \mathrm{n}\,(\mathrm{diag}(\boldsymbol{p}) - \boldsymbol{p}\boldsymbol{p}')$, *where diag*$(\boldsymbol{p})$ *is the diagonal matrix formed by* $\boldsymbol{p} = (p_1, \ldots, p_d)'$.

Definition 7.3.2 *Suppose the random vector* $\boldsymbol{X}$ *has the MPSD with parameter* $\boldsymbol{\theta}$ *and series function* $f(\boldsymbol{\theta})$. *Suppose further that* $\theta_t = p_t$ *with* $0 \leq \sum_{t=1}^{d} p_t < 1$, $0 \leq p_t < 1$ *for* $t = 1, \ldots, d$, *and* $k \in \mathbb{N}_0$. *Random vector* $\boldsymbol{X}$ *is said to have d-dimensional NGMN distribution if the coefficient function is* $a(\boldsymbol{x}) = \frac{\Gamma(\sum_{t=1}^{d} x_t + k)}{\prod_{t=1}^{d} x_t! \cdot \Gamma(k)}$ *and series function is* $f(\boldsymbol{\theta}) = \left(1 - \sum_{t=1}^{d} \theta_t\right)^{-k}$. *It has PMF*

$$\mathbb{P}(\boldsymbol{X} = \boldsymbol{x}) = \frac{\Gamma\left(\sum_{t=1}^{d} x_t + k\right)}{\prod_{t=1}^{d} x_t! \cdot \Gamma(k)} \prod_{t=1}^{d} p_t^{x_t} \cdot p_0^k, \tag{7.7}$$

where $p_0 = 1 - \sum_{t=1}^{d} p_t$ *and* $x_t \in \mathbf{N}_0$. *We shall use notation* $\boldsymbol{X} \sim NGMN(k, \boldsymbol{p})$ *to specify the distribution.*

To interpret the NGMN distribution, suppose the sampling is continued until k individuals of character 0 are observed. Let X_t be the number of individuals in the sample having character t. Hence, $\mathbf{X} = (X_1, \ldots, X_d)'$ is a random observation vector with a negative multinomial distribution.

Example 7.3.2 *Suppose* $\boldsymbol{X} \sim NGMN(k, \boldsymbol{p})$, *where* $\boldsymbol{p} = (p_1, \ldots, p_d)'$. *The marginal distribution for each component of* $\boldsymbol{X}$, *say* X_t *is* $NB\left(k, \frac{p_t}{1-\sum_{s\neq t} p_s}\right)$, *for* $t = 1, \ldots, d$. *The descending* r *th factorial moment of* $\boldsymbol{X}$ *is*

$$\mathbb{E}\left(\prod_{t=1}^{d} X_t^{(r_t)}\right) = k^{[\sum_{t=1}^{d} r_t]} \prod_{t=1}^{d} \left(\frac{p_t}{p_0}\right)^{r_t}, \tag{7.8}$$

where $a^{[b]} = a(a+1)\ldots(a+b-1)$ *and* $p_0 = 1 - \sum_{t=1}^{d} p_t$. *The mean and variance for each component of* $\boldsymbol{X}$, *say* X_t *are*

$$\mathbb{E}(X_t) = \frac{kp_t}{p_0}, \quad \mathbb{V}ar(X_t) = \frac{kp_0(p_t + p_0)}{p_0^2}.$$

One can use (7.8) *to find* $\mathbb{C}\mathrm{ov}(\mathrm{X_s}, \mathrm{X_t}) = \frac{\mathrm{kp_s p_t}}{\mathrm{p_0^2}}$, *for* $s, t = 1, \ldots, d, s \neq t$, *so that the correlation is* $\mathbb{C}or(X_s, X_t) = \sqrt{\frac{p_s p_t}{(p_0+p_s)(p_0+p_t)}}$. *It is interesting to notice that this correlation is positive, while for most discrete multivariate distributions the correlations are negative.*

7.4 Distribution of MPSD IPDs

Let random vectors $\mathbf{X}$ and $\mathbf{Y}$ be sampled from discrete multivariate distributions. In this setting, the squared IPD $d^2_{(xy)}$ can assume only non-negative integer values, and consequently the IPD $d_{(xy)}$ itself is restricted to the square roots of those integers lying in the support of the squared IPD. Accordingly,

$$\mathbb{P}(d_{(xy)} = \sqrt{v}) = \mathbb{P}\left(d^2_{(xy)} = v\right),$$

where v ranges over the non-negative integers contained in the support of $d^2_{(xy)}$, as determined by the joint distributions of $\mathbf{X}$ and $\mathbf{Y}$. In light of this one-to-one correspondence, our analysis focuses exclusively on the distribution of the squared IPD for discrete multivariate distributions.

The core of the ensuing theoretical results is grounded in the observation that

$$\mathbb{P}(\|\mathbf{X} - \mathbf{Y}\|^2 = v) = \sum_{\mathbf{a}\in D}\sum_{\mathbf{b}\in E} \mathbb{P}(\mathbf{X} = \mathbf{a})\,\mathbb{P}(\mathbf{Y} = \mathbf{b}), \tag{7.9}$$

where D and E denote the supports of the distributions of $\mathbf{X}$ and $\mathbf{Y}$, respectively. The summation extends over all pairs $(\mathbf{a}, \mathbf{b}) \in D \times E$ satisfying $\|\mathbf{a} - \mathbf{b}\|^2 = v$. By specifying different choices of the support sets D and E, the results presented in the following theorems are obtained.

In the following section, we determine the IPD distribution between i.i.d. MN vectors (Theorem 7.4.1), the difference between i.i.d. MPSD vectors except MN (Lemma 7.4.1), IPD between i.n.i.d. MN vectors (Theorem 7.4.3), IPD between i.n.i.d. MPSD vectors except MN (Theorem 7.4.4), IPD between a mixture of MN vectors (Theorem 7.4.5), difference between MPSD mixture vectors except MN (Lemma 7.4.3), and IPD between a mixture of MPSD vectors except MN (Theorem 7.4.6). We also discuss three related corollaries for NGMN vectors. We consider IPDs arising from the MN distribution separately and prior to other prominent members of the MPSD family, as the MN distribution is characterized by inherent constraints on the range of attainable values.

Suppose $\mathbf{X}$ and $\mathbf{X}^*$ are d-dimensional random vectors drawn independently from the MN distribution with parameters n and $\mathbf{p}$, so that $\sum_{t=1}^d X_t = \sum_{t=1}^d X_t^* = n$ and the maximum possible value of the IPD is $2n^2$. Moreover, we can express the MN IPD as

$$\begin{aligned} d^2_{(x)} &= \sum_{t=1}^{d}(X_t - X_t^*)^2 = \sum_{t=1}^{d-1}(X_t - X_t^*)^2 + \left(n - \sum_{t=1}^{d-1} X_t - n + \sum_{t=1}^{d-1} X_t^*\right)^2 \\ &= 2\sum_{t=1}^{d-1}(X_t - X_t^*)^2 + 2\sum_{s=1}^{d-2}\sum_{t=s+1}^{d-1}(X_s - X_s^*)(X_t - X_t^*), \end{aligned} \tag{7.10}$$

which means the squared IPDs from MN can only be non–negative even numbers. However, not all even numbers are included in its range. For example, when $d = 3$ and $n = 2$, the range of squared IPD is $\{0, 2, 6, 8\}$ with 4 excluded. We define the set $\mathbf{H}_r = \left\{v \mid v = \sum_{t=1}^d v_t^2,\ \sum_{t=1}^d v_t = 0,\ v_t \in \{-n, \ldots, 0, \ldots, n\}\right\}$ as the range of the squared MN IPD. The subscript of $\mathbf{H}_r$ emphasizes the fact that the underlying distribution has restricted support. The same applies for sets $\mathbf{J}_r$ and $\mathbf{O}_r$ that we define later. To obtain set $\mathbf{H}_r$, we perform a full grid search for $v_t \in \{-n, \ldots, 0, \ldots, n\}$ that satisfy $\sum_{t=1}^d v_t = 0$ and $\sum_{t=1}^d v_t^2 = v \in \{0, \ldots, 2n^2\}$.

7.4.1 IPD Between i.i.d. MN Vectors

Theorem 7.4.1 *Suppose $\boldsymbol{X}$ and $\boldsymbol{X}^*$ are random vectors that are drawn independently from the $MN(n, \boldsymbol{p})$ distribution. Let $\Phi = \{0, 1, \ldots, n\}$ and*

$$\boldsymbol{V}_r = \left\{\boldsymbol{v} = (v_1, \ldots, v_d)' \mid v_1^2 + \ldots + v_d^2 = v, v \in \boldsymbol{H}_r\right\}.$$

The PMF of the IPD between $\boldsymbol{X}$ *and* $\boldsymbol{X}^*$ *is given by*

$$\mathbb{P}\left(d^2_{(x)} = v\right) = \sum_{v \in \boldsymbol{V}_r} \sum_{\phi_1 \in \Phi} \cdots \sum_{\phi_d \in \Phi} \mathbb{P}\left(X_1 = \phi_1 + v_1\right) \cdot \mathbb{P}\left(X_1^* = \phi_1\right) \cdot \mathbf{1}_{\{\sum_{t=1}^d \phi_t = n, \phi_1 + v_1 \geq 0\}}$$
$$\times \prod_{t=2}^{d} \mathbb{P}\left(X_t = \phi_t + v_t | x_1, \ldots, x_{t-1}\right) \cdot \mathbb{P}\left(X_t^* = \phi_t | x_1^*, \ldots, x_{t-1}^*\right) \cdot \mathbf{1}_{\{\phi_t + v_t \geq 0\}}, \tag{7.11}$$

where $\mathbf{1}_{\{\sum_{t=1}^d \phi_t = n,\ \phi_1 + v_1 \geq 0\}}$ *equals to* 1 *if* $\sum_{t=1}^d \phi_t = n$ *and* $\phi_1 + v_1 \geq 0$, *and equals* 0 *otherwise. The distribution of* $X_t | x_1, \ldots, x_{t-1}$ *and* $X_t^* | x_1^*, \ldots, x_{t-1}^*$ *are*

$$X_1 \sim B(n, p_1), \text{ and } X_t | x_1, \ldots, x_{t-1} \sim B\left(n - \sum_{r=1}^{t-1} x_r, \frac{p_t}{1 - \sum_{r=1}^{t-1} p_r}\right), \text{ for } t=2, \ldots, d;$$
$$X_1^* \sim B(n, p_1), \text{ and } X_t^* | x_1^*, \ldots, x_{t-1}^* \sim B\left(n - \sum_{r=1}^{t-1} x_r^*, \frac{p_t}{1 - \sum_{r=1}^{t-1} p_r}\right), \text{ for } t=2, \ldots, d,$$

where $B(n, p_1)$ *represents the binomial distribution with parameters* n *and* d_1. *The expected value of the IPD is* $\mathbb{E}\left(d^2_{(x)}\right) = 2 \sum_{t=1}^d n(p_t - p_t^2)$.

Suppose $\mathbf{X}$ and $\mathbf{X}^*$ are d-dimensional random vectors that are drawn independently from MPSD($\boldsymbol{\theta}, f(\boldsymbol{\theta})$) except the MN distribution, with mean $\boldsymbol{\mu}_x$ and covariance matrix $\boldsymbol{\Sigma}_x$. Let $T_{(x)}(t) = X_t - X_t^*$ for $t = 1, \ldots, d$ and $\mathbf{T}_{(x)} = \mathbf{X} - \mathbf{X}^* = \left(T_{(x)}(1), \ldots, T_{(x)}(d)\right)$. The squared IPD can be expressed as

$$d^2_{(x)} = \sum_{t=1}^{d} (X_t - X_t^*)^2 = \sum_{t=1}^{d} T^2_{(x)}(t).$$

Because $\mathbf{X}$ has integer valued support $T_{(x)}(t)$ has integer valued support $\mathbb{Z}$, where $\mathbb{Z}$ is the set of all integers. Hence, $T^2_{(x)}(t) \in \{0^2, 1^2, 2^2, 3^2, \ldots\}$, the set of squared integers. Therefore, the range of $d^2_{(x)}$ is the set that contains all possible sums of squared integers up to dimension d. We define set $\mathbf{H}_u = \{v \mid v = v_1^2 + \ldots + v_d^2, v_t \in \mathbb{N}_0\}$ as the range of squared IPD for the MPSD family except the MN distribution. The set $\mathbf{H}_u$ contains values of $v = v_1^2 + \ldots + v_d^2$ based on all partitions of $v \geq 0$ into $d \geq 1$ parts $v_1^2, \ldots, v_d^2$.

Lemma 7.4.1 *The PMF of* $T^2_{(x)}(t)$ *for* $t = 1, \ldots, d$ *can be written as*

$$\mathbb{P}\left(T^2_{(x)}(t) = v_t^2\right) = \left(1 + \mathbf{1}_{\{v_t \neq 0\}}\right) \sum_{\psi \in \Psi} \mathbb{P}\left(X_t = \psi + v_t\right) \mathbb{P}\left(X_t^* = \psi\right), \; v_t \in \mathbb{N}_0 \tag{7.12}$$

where Ψ *is the support of* X_t *for the corresponding marginal distribution,* $\mathbf{1}_{\{v_t \neq 0\}} = 1$ *when* $v_t \neq 0$ *and* $\mathbf{1}_{\{v_t \neq 0\}} = 0$, *otherwise.*

7.4.2 IPD Between i.i.d. MPSD vectors except MN

Theorem 7.4.2 *Define the set* $\boldsymbol{V}_u = \left\{\boldsymbol{v} = (v_1, \ldots, v_d)' \mid v_1^2 + \ldots + v_d^2 = v, v \in \boldsymbol{H}_u\right\}$. *The PMF of IPD between observations* $\boldsymbol{X}$ *and* $\boldsymbol{X}^*$ *is given by*

$$\mathbb{P}\left(d^2_{(x)} = v\right) = \sum_{v \in \boldsymbol{V}_u} \prod_{t=1}^{d} \left(1 + \mathbf{1}_{\{v_t \neq 0\}}\right) \sum_{\psi \in \Psi} \frac{a_t(\theta_1, \ldots, \psi, \ldots, \theta_d)\theta_t^{\psi}}{f(\boldsymbol{\theta})} \frac{a_t(\theta_1, \ldots, \psi + v_t, \ldots, \theta_d)\theta_t^{\psi + v_t}}{f(\boldsymbol{\theta})}, \tag{7.13}$$

where $\sum_{\boldsymbol{v}\in \boldsymbol{V}_u}$ is the sum taken over all vectors $\boldsymbol{v}$ in set $\boldsymbol{V}_u$. Let $\boldsymbol{\Sigma}_x$ be the covariance of the MPSD vectors. The expected value of the IPD is $\mathbb{E}\left(d^2_{(x)}\right) = 2tr\left(\boldsymbol{\Sigma}_x\right)$.

A grid search can be used to compute $\mathbb{P}\left(d^2_{(x)} = v\right)$. Since $v = v_1^2 + \ldots + v_d^2$ and $v_t \leq \sqrt{v}$ for $t = 1, \ldots, d$, then a full grid search would have time complexity $O\left(v^{d/2}\right)$. It is time consuming to use a full grid search when the dimension d is large. The number of compositions of v into d parts is well studied and the number of such partition is $\binom{v+d-1}{v}$, which is the number of ways to place $v \geq 0$ indistinguishable balls in $d \geq 0$ labeled cells. Nijenhuis and Wilf [250] illustrate an algorithm called NEXTCOM to list all such partitions. However, not all parts of each partition have integer square root. Therefore, one can use the algorithm NEXTCOM to find all partitions of v into d parts, then admit the partitions whose parts are perfect squares.

We use I.N.I.D. to denote independently, but not identically distributed random variables or vectors. In a manner similar to the IPDs between the i.i.d. MPSD random vectors, we consider IPD between the MN vectors first, followed by IPDs between vectors from the other three distributions.

7.4.3 IPD Between i.n.i.d. MN vectors

Theorem 7.4.3 *Suppose $\boldsymbol{X}$ is a d-dimensional vector that is drawn from $MN(n, \boldsymbol{p}_x)$ distribution. Suppose $\boldsymbol{Y}$ is a d-dimensional vector that is drawn from $MN(n, \boldsymbol{p}_y)$ distribution independent with $\boldsymbol{X}$. Let $\Phi = \{0, 1, \ldots, n\}$. The PMF of the IPD between $\boldsymbol{X}$ and $\boldsymbol{Y}$ is given by*

$$\mathbb{P}\left(d^2_{(xy)} = v\right) = \sum_{\boldsymbol{v}\in \boldsymbol{V}_r} \sum_{\phi_1\in\Phi} \cdots \sum_{\phi_d\in\Phi} \mathbb{P}\left(X_1 = \phi_1 + v_1\right) \cdot \mathbb{P}\left(Y_1 = \phi_1\right) \cdot \mathbf{1}_{\{\sum_{t=1}^{d} \phi_t = n, \phi_1 + v_1 \geq 0\}}$$
$$\times \prod_{t=2}^{d} \mathbb{P}\left(X_t = \phi_t + v_t | x_1, \ldots, x_{t-1}\right) \cdot \mathbb{P}\left(Y_t = \phi_t | y_1, \ldots, y_{t-1}\right) \cdot \mathbf{1}_{\{\phi_t + v_t \geq 0\}}, \tag{7.14}$$

where the distribution of $X_t|x_1, \ldots, x_{t-1}$ and $Y_t|y_1, \ldots, y_{t-1}$ are

$$X_1 \sim B(n, p_{(x)1}), \text{ and } X_t|x_1, \ldots, x_{t-1} \sim B\left(n - \sum_{r=1}^{t-1} x_r, \frac{p_{(x)t}}{1 - \sum_{r=1}^{t-1} p_{(x)r}}\right), \text{ for } t=2, \ldots, d;$$

$$Y_1 \sim B(n, p_{(y)1}), \text{ and } Y_t|y_1, \ldots, y_{t-1} \sim B\left(n - \sum_{r=1}^{t-1} y_r, \frac{p_{(y)t}}{1 - \sum_{r=1}^{t-1} p_{(y)r}}\right), \text{ for } t=2, \ldots, d.$$

The expected value of IPD is $E\left(d^2_{(xy)}\right) = \sum_{t=1}^{d} n\left(p_{(x)t} + p_{(y)t} - 2p_{(x)t}p_{(y)t}\right)$.

Suppose $\mathbf{X}$ and $\mathbf{Y}$ are d-dimensional random vectors that are drawn independently from MPSD$(\boldsymbol{\theta}_x, f_x(\boldsymbol{\theta}_x))$ and MPSD$(\boldsymbol{\theta}_y, f_y(\boldsymbol{\theta}_y))$, respectively, except the MN distribution. The IPD between vectors $\mathbf{X}$ and $\mathbf{Y}$ can be expressed as $d^2_{(xy)} = \sum_{t=1}^{d}\left(X_t - Y_t\right)^2 = \sum_{t=1}^{d} T^2_{(xy)}(t)$, where $\mathbf{T}_{(xy)} = \mathbf{X} - \mathbf{Y} = \left(T_{(xy)}(1), \ldots, T_{(xy)}(d)\right)'$. Because both $\mathbf{X}$ and $\mathbf{Y}$ have integer valued support and there are no constraints on the range of $\mathbf{X}$ and $\mathbf{Y}$, it follows that $T_{(x)}(t) \in \mathbb{Z}$.

Lemma 7.4.2 *Let $\boldsymbol{T}_{(xy)} = \boldsymbol{X} - \boldsymbol{Y} = (X_1 - Y_1, \ldots, X_d - Y_d)' = \left(T_{(xy)}(1), \ldots, T_{(xy)}(d)\right)'$. The PMF of $T^2_{(xy)}(t)$ for $t = 1, \ldots, d$ can be written as*

$$\mathbb{P}\left(T^2_{(xy)}(t) = v_t^2\right) = \begin{cases} \sum_{\psi\in\Psi}\left[\eta_y(\psi) \cdot \zeta_x(\psi + v_t) + \zeta_x(\psi) \cdot \eta_y(\psi + v_t)\right], & \text{if } v_t \neq 0, \\ \sum_{\psi\in\Psi}\left[\zeta_x(\psi) \cdot \eta_y(\psi)\right], & \text{if } v_t = 0, \end{cases} \tag{7.15}$$

where

$$\zeta_x(\psi) = \frac{a_{(x)t}\left(\theta_{(x)1}, \ldots, \psi, \ldots, \theta_{(x)d}\right)\theta_{(x)t}^{\psi}}{f_x(\boldsymbol{\theta}_x)},$$
$$\eta_y(\psi) = \frac{a_{(y)t}\left(\theta_{(y)1}, \ldots, \psi, \ldots, \quad \theta_{(y)d}\right)\theta_{(y)t}^{\psi}}{f_y(\boldsymbol{\theta}_y)}, \tag{7.16}$$

and Ψ *is the common support of* $\boldsymbol{X}$ *and* $\boldsymbol{Y}$.

Theorem 7.4.4 *To obtain the IPD between i.n.i.d. MPSD vectors except MN note that the PMF of the squared Euclidean IPD between observations* $\boldsymbol{X}$ *and* $\boldsymbol{Y}$ *is given by*

$$\mathbb{P}\left(d^2_{(xy)} = v\right) = \sum_{v \in \boldsymbol{V}_u} \left(\prod_{\{t|v_t=0\}} \mathbb{P}\left(X_t - Y_t = 0\right) \times \prod_{\{t|v_t \neq 0\}} \left[\mathbb{P}\left(X_t - Y_t = v_t\right) + \mathbb{P}\left(X_t - Y_t = -v_t\right)\right] \right). \tag{7.17}$$

Computing the probabilities $\mathbb{P}(X_t - Y_t = 0)$, $\mathbb{P}(X_t - Y_t = v_t)$ *and* $\mathbb{P}(Y_t - X_t = v_t)$ *are left as an exercise.*

We discuss the distribution of the squared IPD between vectors that are drawn from a mixture of MPSD distributions in the following theorems. We derive the PMF in two different cases, one for the MN distribution and the other one for the rest of distributions in the MPSD family. Suppose $\mathbf{X}$ is a d-dimensional random vector that is drawn from a mixture distribution of k MPSD distributions with PMF $g_1(\cdot)$ to $g_k(\cdot)$, where $k \in \{1, 2, \ldots\}$. Suppose the means and covariance matrices for distribution $g_1(\cdot)$ to $g_k(\cdot)$ are $\boldsymbol{\mu}_1, \ldots, \boldsymbol{\mu}_k$ and $\boldsymbol{\Sigma}_1, \ldots, \boldsymbol{\Sigma}_k$, respectively. The PMF of $\mathbf{X}$ is $g_{\mathbf{X}}(\mathbf{x}) = \sum_{i=1}^{k} \alpha_i g_i(\mathbf{x})$, where $0 \leq \alpha_i \leq 1$ and $\sum_{i=1}^{k} \alpha_i = 1$. Note that the mixture distribution means $\mathbf{X}$ is drawn from $g_i(\cdot)$ with probability α_i for $i = 1, \ldots, k$. Suppose $\mathbf{X}^*$ is an independent d-dimensional random vector that is drawn from the same distribution as $\mathbf{X}$. The following identities are useful in the proofs of mixture IPD theorems.

$$\mathbb{P}\left(d^2_{(x)} = v\right) = \sum_{s} \mathbb{P}\left(d^2_{(x)} = v \mid S = s\right) \cdot \mathbb{P}\left(S = s\right), \tag{7.18}$$

$$\mathbb{E}\left(d^2_{(x)}\right) = \sum_{s} \mathbb{E}\left(d^2_{(x)} \mid S = s\right) \cdot \mathbb{P}\left(S = s\right), \tag{7.19}$$

where S is a categorical variable representing different cases in the mixture models.

7.4.4 IPD Between a mixture of MN vectors

Theorem 7.4.5 *Suppose* $\boldsymbol{X}$ *and* $\boldsymbol{X}^*$ *are* d*-dimensional random vectors that are drawn from a mixture of the* k *MN distributions with* $\boldsymbol{X}, \boldsymbol{X}^* \sim \sum_{i=1}^{k} \alpha_i MN(n, \boldsymbol{p}_i)$. *The PMF of the* $\|\boldsymbol{X} - \boldsymbol{X}^*\|^2$ *is given by,*

$$\mathbb{P}\left(d^2_{(x)} = v\right) = \sum_{i=1}^{k} \alpha_i^2 \sum_{v \in \boldsymbol{V}_r} \sum_{\phi_1, \ldots, \phi_d \in \Phi} h_{it}\left(\phi_t + v_t\right) h_{it}\left(\phi_t\right) \cdot \prod_{t=1}^{d} \mathbf{1}_{\{\phi_t + v_t \geq 0\}}$$
$$+ \sum_{i=1}^{k-1} \sum_{j=i+1}^{k} \alpha_i \alpha_j \sum_{v \in \boldsymbol{V}_r} \sum_{\phi_1, \ldots, \phi_d \in \Phi} h_{it}\left(\phi_t + v_t\right) h_{jt}\left(\phi_t\right) \cdot \prod_{t=1}^{d} \mathbf{1}_{\{\phi_t + v_t \geq 0\}}$$

$$+\sum_{i=1}^{k-1}\sum_{j=i+1}^{k}\alpha_i\alpha_j\sum_{v\in \boldsymbol{V}_r}\sum_{\phi_1,\ldots,\phi_d\in\Phi} h_{jt}\left(\phi_t+v_t\right)h_{it}\left(\phi_t\right)\cdot\prod_{t=1}^{d}\mathbf{1}_{\{\phi_t+v_t\geq 0\}}, \tag{7.20}$$

where $\sum_{\phi_1,\ldots,\phi_d\in\Phi}$ *is the sum taken over all* $\phi_1,\ldots,\phi_d$ *such that* $\sum_{t=1}^{d}\phi_t = n$ *and* $\Phi = \{0,1,\ldots,n\}$*. For* $i = 1,\ldots,k$*, functions* $h_{it}(\phi_t+v_t)$ *are the PMF of* $X_{it}|x_{i1},\ldots,x_{i(t-1)}$ *that is evaluated at* ϕ_t+v_t *for* $t = 2,\ldots,d$*, which is the PMF of binomial distribution with parameters* $n-\sum_{r=1}^{t-1}(\phi_r+v_r)$ *and* $\frac{p_{(i)t}}{1-\sum_{r=1}^{t-1}p_{(i)r}}$*. For* $t = 1$*, function* $h_{i1}(\phi_1+v_1)$ *is the PMF of* X_{i1} *that is evaluated at* ϕ_1+v_1*, which is the PMF of binomial distribution with parameters* n *and* d_{i1}*. The expected value of the IPD is,*

$$\mathbb{E}\left[d_{(x)}^2\right] = 2\sum_{i=1}^{k}\alpha_i^2\sum_{t=1}^{d}n\left(p_{(i)t}-p_{(i)t}^2\right) \tag{7.21}$$
$$+2\sum_{i=1}^{k-1}\sum_{j=i+1}^{k}\alpha_i\alpha_j\sum_{t=1}^{d}n\left[\left(p_{(i)t}-p_{(i)t}^2\right)+\left(p_{(j)t}-p_{(j)t}^2\right)+n\left(p_{(i)t}-p_{(j)t}\right)^2\right].$$

Lemma 7.4.3 *Suppose* $\boldsymbol{X}$ *and* $\boldsymbol{X}^*$ *are d-dimensional random vectors that are drawn independently from a mixture of MPSD distributions except MN. Let* $\boldsymbol{T}_{(x)} = \boldsymbol{X}-\boldsymbol{X}^*$*. The PMF of* $T_{(x)}^2(t)$ *for* $t = 1,\ldots,d$ *can be written as*

$$\mathbb{P}\left(T_{(x)}(t)=v_t\right) = \alpha_i^2\sum_{\psi\in\Psi}l_i(\psi+v_t)\cdot l_i(\psi)+\alpha_i\alpha_j\sum_{\psi\in\Psi}l_i(\psi+v_t)\cdot l_j(\psi)$$
$$+\alpha_i\alpha_j\sum_{\psi\in\Psi}l_j(\psi+v_t)\cdot l_i(\psi), \tag{7.22}$$

where Ψ *is the common support of* $l_1(\cdot)$ *to* $l_k(\cdot)$ *and* $l_i(\cdot)$ *is the marginal PMF of* $g_i(\cdot)$*.*

7.4.5 IPD Between a Mixture of MPSD Vectors Except MN

Theorem 7.4.6 *The PMF of the IPD between* $\boldsymbol{X}$ *and* $\boldsymbol{X}^*$ *is given by*

$$\mathbb{P}\left(d_{(x)}^2=v\right) = \sum_{v\in \boldsymbol{V}_u}\prod_{t=1}^{d}\left(1+\mathbf{1}_{\{v_t\neq 0\}}\right)\mathbb{P}\left(T_{(x)}(t)=v_t\right), \tag{7.23}$$

where $P\left(T_{(x)}(t)=v_t\right)$ *has expression in* (7.22)*. The mean of the IPD is*

$$\mathbb{E}\left[d_{(x)}^2\right] = 2\sum_{i=1}^{k}\alpha_i^2 tr\left(\boldsymbol{\Sigma}_i\right)+2\sum_{i=1}^{k-1}\sum_{j=i+1}^{k}\alpha_i\alpha_j\left[tr\left(\boldsymbol{\Sigma}_i+\boldsymbol{\Sigma}_j\right)+\left(\boldsymbol{\mu}_i-\boldsymbol{\mu}_j\right)'\left(\boldsymbol{\mu}_i-\boldsymbol{\mu}_j\right)\right], \tag{7.24}$$

where $\boldsymbol{\mu}_i$ *and* $\boldsymbol{\Sigma}_i$ *are the mean vectors and covariance matrices of the i-th component in the mixture model.*

7.4.6 IPD Between i.i.d. NGMN Vectors

Corollary 7.4.1 *Suppose* $\boldsymbol{X}$ *and* $\boldsymbol{X}^*$ *are independent d-dimensional random vectors that are drawn from the NGMN distribution with parameters* k *and* $\mathbf{p}$ *defined in Table 7.1. The PMF of the squared Euclidean IPD between* $\boldsymbol{X}$ *and* $\boldsymbol{X}^*$ *is given by*

$$\mathbb{P}\left(d_{(x)}^2=v\right) = \sum_{v\in \boldsymbol{V}_u}\prod_{t=1}^{d}\left(1+\mathbf{1}_{\{v_t\neq 0\}}\right)\sum_{\psi=0}^{\infty}NB\left(v_t+\psi;k,\frac{p_t}{p_0}\right)\cdot NB\left(\psi;k,\frac{p_t}{p_0}\right), \tag{7.25}$$

where $NB\left(v_t+\psi;k,\frac{p_t}{p_0}\right)$ *is the negative binomial distribution with parameters* k *and* $\frac{p_t}{p_0}$ *that is evaluated at* $v_t+\psi$ *and* $d_0=1-\sum_{t=1}^{d}p_t$. *The expected value of the IPD is*

$$E\left(d_{(x)}^2\right)=2k\sum_{t=1}^{d}\frac{p_t}{p_0}\left(1+\frac{p_t}{p_0}\right).$$

Similarly, by using Theorem 7.4.2 and the definition of MP and ML distributions, the distribution of IPDs between i.i.d. MP and ML vectors can be obtained. Modarres and Song [240] and Song [320] show the PMF of IPDs between vectors from MPSD family of distributions.

7.4.7 IPD Between i.n.i.d. NGMN Vectors

Corollary 7.4.2 *Suppose* $\boldsymbol{X}$ *and* $\boldsymbol{Y}$ *are independent d-dimensional vectors that are drawn from* $NGMM(k,\boldsymbol{p}_x)$ *and* $NGMN(k,\boldsymbol{p}_y)$ *defined in Table 7.1. The PMF of the squared Euclidean IPD between* $\boldsymbol{X}$ *and* $\boldsymbol{Y}$ *is given by*

$$\begin{aligned}\mathbb{P}\left(d_{(xy)}^2=v\right)=\sum_{v\in \boldsymbol{V}_u}&\left[\prod_{t|v_t=0}\left\{\sum_{\psi=0}^{\infty}NB\left(\psi;k,\frac{p_{(x)t}}{p_{(x)0}}\right)\cdot NB\left(\psi;k,\frac{p_{(y)t}}{p_{(y)0}}\right)\right\}\right.\\&\cdot\prod_{t|v_t\neq 0}\sum_{\psi=0}^{\infty}\left\{NB\left(v_t+\psi;k,\frac{p_{(x)t}}{p_{(x)0}}\right)\cdot NB\left(\psi;k,\frac{p_{(y)t}}{p_{(y)0}}\right)\right.\\&\left.\left.+NB\left(v_t+\psi;k,\frac{p_{(y)t}}{p_{(y)0}}\right)\cdot NB\left(\psi;k,\frac{p_{(x)t}}{p_{(x)0}}\right)\right\}\right],\end{aligned}\tag{7.26}$$

where $d_{(x)0}=1-\sum_{t=1}^{d}p_{(x)t}$ *and* $d_{(y)0}=1-\sum_{t=1}^{d}p_{(y)t}$. *The expected value of* $d_{(xy)}^2$ *is*

$$\mathbb{E}\left(d_{(xy)}^2\right)=\sum_{t=1}^{d}\left\{k(k+1)\left(\left(\frac{p_{(x)t}}{p_{(x)0}}\right)^2+\left(\frac{p_{(y)t}}{p_{(y)0}}\right)^2\right)-2k^2\frac{p_{(x)t}p_{(y)t}}{p_{(x)0}^2}+k\left(\frac{p_{(x)t}}{p_{(x)0}}+\frac{p_{(y)t}}{p_{(y)0}}\right)\right\}.\tag{7.27}$$

Similarly, by using Theorem 7.4.4 and the definition of MP and ML distributions, the distribution of IPDs between i.n.i.d. MP and ML vectors can be obtained.

The plot of PMF of IPDs between i.n.i.d. vectors from MN distributions appears in Figure 7.1. Vectors have parameter $n=5$ with $\mathbf{p}=(0.05,0.15,0.25,0.15,0.15,0.25)'$ and $\mathbf{p}=(0.14,0.27,0.19,0.12,0.21,0.07)'$. The PMF is skewed with a short tail on the right. The expected value of the IPD is 8.38. Figure 7.2 displays the PMF of IPDs between i.n.i.d. vectors from NGMN distributions. Both vectors have parameter $k=5$, one has $\mathbf{p}_x=(0.05,0.15,0.25,0.10,0.15,0.05)'$ and the other one has $\mathbf{p}_y=(0.04,0.17,0.19,0.09,0.21,0.07)'$. The PMF is skewed to the right with a long tail and the CDF converge to 1 slowly. The expected value of the IPD is 57.21.

7.4.8 IPD Between a Mixture of the NGMN Vectors

Corollary 7.4.3 *Suppose* $\boldsymbol{X}$ *and* $\boldsymbol{X}^*$ *are independent d-dimensional random vectors that are drawn from a mixture of the NGMN distributions, where one component has parameters* k *and* $\boldsymbol{p}_1$, *and the other component has* k *and* $\boldsymbol{p}_2$. *The PMF of the IPD between* $\boldsymbol{X}$ *and* $\boldsymbol{X}^*$

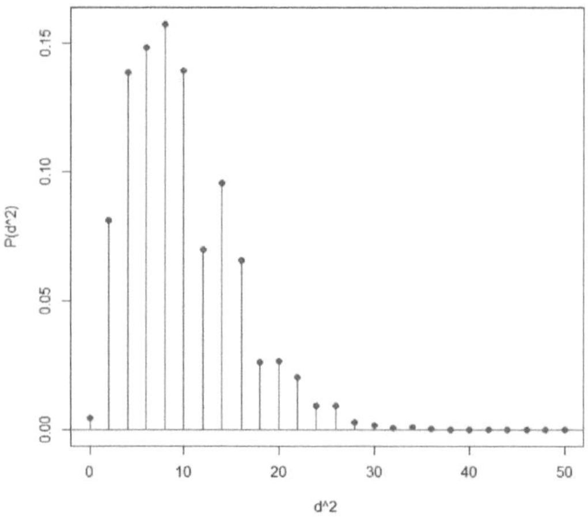

FIGURE 7.1: MN IPD of i.n.i.d. vectors.

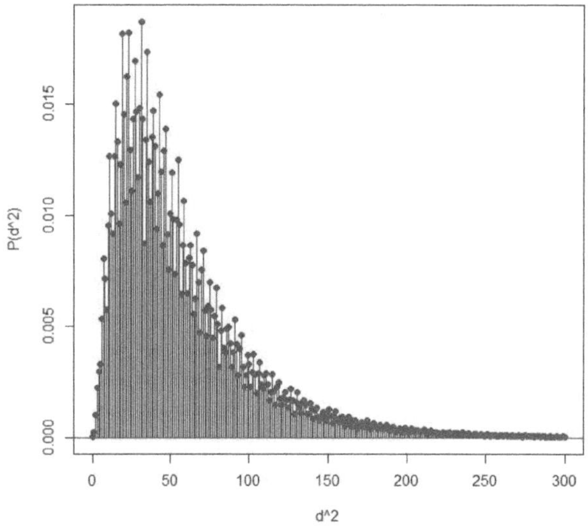

FIGURE 7.2: NGMN IPD of i.n.i.d. vectors.

is given by

$$\mathbb{P}\left(d^2_{(x)} = v\right) = \sum_{v \in \boldsymbol{V}_u} \prod_{t=1}^{d} \left(1 + \mathbf{1}_{\{v_t \neq 0\}}\right) \Bigg[\alpha^2 \sum_{\psi \in \Psi} NB\left(\psi + v_t; k, \frac{p_{(1)t}}{p_{(1)0}}\right) \cdot NB\left(\psi; k, \frac{p_{(1)t}}{p_{(1)0}}\right)$$
$$+ (1-\alpha)^2 \sum_{\psi \in \Psi} NB\left(\psi + v_t; k, \frac{p_{(2)t}}{p_{(2)0}}\right) \cdot NB\left(\psi; k, \frac{p_{(2)t}}{p_{(2)0}}\right)$$
$$+ \alpha(1-\alpha) \sum_{\psi \in \Psi} NB\left(\psi + v_t; k, \frac{p_{(1)t}}{p_{(1)0}}\right) \cdot NB\left(\psi; k, \frac{p_{(2)t}}{p_{(2)0}}\right)$$
$$+ \alpha(1-\alpha) \sum_{\psi \in \Psi} NB\left(\psi + v_t; k, \frac{p_{(2)t}}{p_{(2)0}}\right) \cdot NB\left(\psi; k, \frac{p_{(1)t}}{p_{(1)0}}\right)\Bigg], \quad (7.28)$$

where $0 \leq \alpha \leq 1$ *and* $NB\left(\psi + v_t, k, \frac{p_{(1)t}}{p_{(1)0}}\right)$ *is the negative binomial distribution with parameters* k *and* $\frac{p_{(1)t}}{p_{(1)0}}$ *that is evaluated at* $\psi + v_t$. *The expected value of* $d_{(x)}^2$ *is*

$$\mathbb{E}\left(d_{(x)}^2\right) = 2k\sum_{t=1}^{d}\left[k\alpha(1-\alpha)\left(\frac{p_{(1)t}}{p_{(1)0}} - \frac{p_{(2)t}}{p_{(2)0}}\right)^2 + \alpha\frac{p_{(1)t}}{p_{(1)0}}\left(1 + \frac{p_{(1)t}}{p_{(1)0}}\right) + (1-\alpha)\frac{p_{(2)t}}{p_{(2)0}}\left(1 + \frac{p_{(2)t}}{p_{(2)0}}\right)\right]. \tag{7.29}$$

The IPD PMF plot between i.i.d. vectors from a mixture of MN distributions appears in Song [320].

7.4.9 Distribution of MPSD Norm

We find the distribution of MPSD norm and distance from a fixed vector in this section. There are two cases to consider for MN. The Euclidean norm has applications in vector median filtering (Barni et al. [24]). Suppose $\mathbf{X} \in \mathbb{R}^d$ is random vector from MPSD$(\boldsymbol{\theta}, f(\boldsymbol{\theta}))$, with mean $\boldsymbol{\mu}_x$ and covariance matrix $\boldsymbol{\Sigma}_x$. We define $\|\mathbf{X}\|^2 = \mathbf{X}'\mathbf{X} = \sum_{t=1}^{d} X_t^2$. Let $\eta_t \in \{0, \ldots, n\}$ and $\boldsymbol{\eta} = (\eta_1, \ldots, \eta_d)'$. The range of the squared MN norm is $\mathbf{J}_r = \left\{\boldsymbol{\eta} \mid \eta = \sum_{t=1}^{d} \eta_t^2, \sum_{t=1}^{d} \eta_t = n, \right\}$, and the range for the squared norm from the other three distribution is $\mathbf{J}_u = \left\{\boldsymbol{\eta} \mid \eta = \sum_{t=1}^{d} \eta_t^2\right\}$. Define $\mathbf{N}_r = \left\{\boldsymbol{\eta} \mid \eta = \sum_{t=1}^{d} \eta_t^2, , \eta \in \mathbf{J}_r\right\}$, and $\mathbf{N}_u = \left\{\boldsymbol{\eta} \mid \eta = \sum_{t=1}^{d} \eta_t^2, , \eta \in \mathbf{J}_u\right\}$.

Theorem 7.4.7 *The PMF of the squared Euclidean norm of* $\boldsymbol{X}$ *is*

$$\mathbb{P}\left(\|\boldsymbol{X}\|^2 = \eta\right) = \begin{cases} \sum_{\boldsymbol{\eta} \in \boldsymbol{N}_r} B\left(\eta_1; n, p_1\right) \cdot \prod_{t=1}^{d} B\left(\eta_t; n - \sum_{r=1}^{t-1} \eta_r, \frac{p_t}{1 - \sum_{r=1}^{t-1} p_r}\right), \\ \text{if } \boldsymbol{X} \sim MN(n, \boldsymbol{p}); \\ \sum_{\boldsymbol{\eta} \in \boldsymbol{N}_u} \prod_{t=1}^{d} \frac{a_t(\theta_1, \ldots, \eta_t, \ldots, \theta_d)\theta_t^{\eta_t}}{f(\boldsymbol{\theta})}, \\ \text{if } \boldsymbol{X} \text{ is drawn from MPSD except MN,} \end{cases}$$

where $f(\boldsymbol{\theta})$ *is the series function for different types of MPSD according to Table 7.1 and* $a_t(\theta_1, \ldots, \eta_t, \ldots, \theta_d) = \sum_{\{x_1, \ldots, x_d\} \setminus x_t}\left(a(x_1, \ldots, \eta_t, \ldots, x_d)\prod_{s \neq t} \theta_s^{x_s}\right)$. *The expected value of the squared norm is* $\mathbb{E}\left(\|\boldsymbol{X}\|^2\right) = tr\left(\boldsymbol{\Sigma}_x\right) + \boldsymbol{\mu}_x'\boldsymbol{\mu}_x$.

Proof:
Similarly to the proof of Theorems 7.4.1 and 7.4.2 by choosing $\mathbf{Y} = \mathbf{0}$. By using the general form of the mean of IPD and taking matrix $\mathbf{A} = \mathbf{I}_d$, we obtain the expected value of $\|\mathbf{X}\|^2$. □ Song [320] displays the PMF of squared norm of MN distributions.

7.4.10 Distance of MPSD Vectors from a Fixed Vector in $\mathbb{Z}^d$

Suppose $\mathbf{X}$ is a d-dimensional random vector drawn from distribution MPSD$(\boldsymbol{\theta}, f(\boldsymbol{\theta}))$ and $\mathbf{q}$ is a fixed d-dimensional vector in $\mathbb{Z}^d$. The range of the squared Euclidean distance from the fixed vector depends on the support of X_t because $q_t's$ are fixed, for $t = 1, \ldots, d$. We define the set

$$\mathbf{O}_r = \left\{ w \mid w = \sum_{t=1}^{d} (w_t - q_t)^2, \sum_{t=1}^{d} w_t = n, w_t \in \{0, \ldots, n\} \right\}$$

be the range of $\|\mathbf{X} - \mathbf{q}\|^2$ when $\mathbf{X}$ is drawn from the MN distribution. Define the set $\mathbf{O}_u = \left\{ w \mid w = \sum_{t=1}^{d} (w_t - q_t)^2, w_t \in \{0, \ldots, n\} \right\}$ be the range of $\|\mathbf{X} - \mathbf{q}\|^2$ when $\mathbf{X}$ is drawn from MPSD except MN distribution. Let

$$\mathbf{W}_r = \left\{ \boldsymbol{w} = (w_1, \ldots, w_d)' \mid w_1^2 + \ldots + w_d^2 = w, w \in \mathbf{O}_r \right\},$$

and similarly, $\mathbf{W}_u = \left\{ \boldsymbol{w} \mid w_1^2 + \ldots + w_d^2 = w, w \in \mathbf{O}_u \right\}$.

Theorem 7.4.8 *The PMF of squared MPSD Euclidean distance from the fixed vector* $\boldsymbol{q}$ *to* $\boldsymbol{X}$*, is given by*

$$\mathbb{P}\left(\|\boldsymbol{X} - \boldsymbol{q}\|^2 = w\right) = \begin{cases} \sum_{\boldsymbol{w} \in \boldsymbol{W}_r} B\left(w_1 + q_1; n, p_1\right) \\ \times \prod_{t=2}^{d} B\left(w_t + q_t; n - \sum_{r=1}^{t-1} (w_r + q_r), \frac{p_t}{1 - \sum_{r=1}^{t-1} p_r}\right), \\ \text{if } \boldsymbol{X} \sim MN(n, \boldsymbol{p}); \\ \sum_{\boldsymbol{w} \in \boldsymbol{W}_u} \left(\prod_{t \mid w_t = 0} \frac{a_t(\theta_1, \ldots, q_t, \ldots, \theta_d) \theta_t^{q_t}}{f(\boldsymbol{\theta})} \right. \\ \left. \times \prod_{t \mid w_t \neq 0} \left[\frac{a_t(\theta_1, \ldots, q_t + w_t, \ldots, \theta_d) \theta_t^{q_t + w_t}}{f(\boldsymbol{\theta})} + \frac{a_t(\theta_1, \ldots, q_t - w_t, \ldots, \theta_d) \theta_t^{q_t - w_t}}{f(\boldsymbol{\theta})} \right] \right), \\ \text{if } \boldsymbol{X} \text{ is in MPSD family except MN distribution,} \end{cases}$$

where $f(\boldsymbol{\theta})$ *is the series function for different types of MPSD according to Table 7.1 and* $a_t(\theta_1, \ldots, q_t, \ldots, \theta_d) = \sum_{\{x_1, \ldots, x_d\} \setminus x_t} \left(a(x_1, \ldots, q_t, \ldots, x_d) \prod_{s \neq t} \theta_s^{x_s} \right)$. *The expected value of the squared distance from* $\boldsymbol{q}$ *is*

$$\mathbb{E}\left(\|\boldsymbol{X} - \boldsymbol{q}\|^2\right) = tr\left(\boldsymbol{\Sigma}_x\right) + \left(\boldsymbol{\mu}_x - \boldsymbol{q}\right)' \left(\boldsymbol{\mu}_x - \boldsymbol{q}\right). \tag{7.30}$$

Proof: Similar to the proof of Theorems 7.4.1 and 7.4.2 by choosing $\mathbf{Y} = \mathbf{q}$. □

Song [320] states similar corollaries for other members of the MPSD family and provides plots of the PMF of $\|\mathbf{X} - \mathbf{q}\|^2$, where $\mathbf{X}$ has a NGMN distribution.

7.5 IPD Test of Homogeneity for Mixture Models

Populations comprising a finite collection of homogeneous subgroups are commonly represented through finite mixture distributions. Following Pearson's [267] pioneering analysis of crab morphology data from the Bay of Naples using a mixture of two normal distributions, finite mixture models have become a central component of statistical modeling in a wide array of fields, including astronomy, biology, genetics, economics, and engineering. Their appeal lies in the fact that many continuous distributions can be effectively approximated by mixtures of normal densities with a common variance, rendering these models particularly useful for density estimation. In this context, Priebe [272] showed that log-normal distributions may be accurately approximated using mixtures with far fewer components than those typically required by kernel density estimation methods.

Comprehensive theoretical treatments and applied developments of finite mixture models are presented in works such as Lindsay [195], and Böhning [39]. McLachlan and Peel [222] further expanded the role of multivariate mixture models within the framework of model-based clustering. More recently, Nagode [245] investigated the estimation and classification performance of mixtures of multivariate normal distributions. A substantial portion of the literature presumes that mixture components belong to known parametric families, with parameter estimation commonly carried out via the expectation–maximization (EM) algorithm. Within this setting, EM-type test statistics have been proposed and studied by Chen and Li [62], Li et al. [193] , and Niu et al. [251]. For inferential purposes, Chen and Chen [59] introduced a likelihood ratio test for evaluating component homogeneity in mixture models, employing bootstrap techniques to approximate the null distribution. Although their methodology is well established for univariate normal mixtures, analogous developments for the multivariate case remain relatively underexplored.

IPDs constitute a fundamental component of a broad spectrum of multivariate statistical methodologies, including clustering, classification, correspondence analysis, and multidimensional scaling. They also occupy a central position in spatial statistics, where they underpin the analysis of point patterns, the construction of minimum spanning trees, the detection of spatial disease clusters, tests of distributional homogeneity, and the formulation of statistical depth functions. Bonetti et al. [40] highlight the role of IPDs in public health surveillance systems, while Lukens [205] utilizes IPDs to reveal high-dimensional data structure and to identify groups of outliers. In the context of image analysis, Osada et al. [260] and Berrendero et al. [28] employ IPD distributions for shape discrimination and geometric identification. Marozzi ([214] [215]) further extends IPD-based multivariate testing procedures to HDLSS settings, with particular emphasis on medical imaging applications. Additionally, Baringhaus and Franz [21], Rosenbaum [284], Jurečková and Kalina [165], and Guo and Modarres [125] incorporate IPDs into the development of robust two-sample tests and classification methods for high-dimensional categorical data.

Multivariate categorical and count data arise frequently in a wide range of fields, including economics, business analytics, biostatistics, ecology, and environmental science. Finite mixture models offer a highly adaptable framework for the analysis of such data, as they can effectively accommodate over-dispersion, capture latent heterogeneity, and model complex dependence structures. As an illustration, Li and Zha [191] propose a two-way multivariate Poisson mixture model for the joint classification of documents and clustering of words, highlighting its applicability to text mining problems. In a related application, Brijs et al. [45] employ Poisson mixture models to segment supermarket shoppers, while Karlis and Meligkotsidou [166] use multivariate Poisson mixtures to model multiple types of crime incidents across different geographic regions.

In addition to Poisson-based mixtures, mixtures of multinomial distributions are widely used to model categorical count data. For example, Royle and Link [291] introduce a multinomial mixture model to estimate amphibian population abundance from call survey data, a standard approach in ecological monitoring. To account for excess variability and correlation in count data, Morel and Nagaraj [242] propose finite mixtures of multinomial random variables, providing a practical strategy for addressing over-dispersion and latent group dependence in the modeling of discrete data.

A comprehensive and unified framework for modeling such data is provided by the Multivariate Power Series Distribution (MPSD) family. This family includes several key multivariate discrete distributions, such as the multinomial (MN), negative multinomial (NGMN), multivariate Poisson (MP), and multivariate logarithmic (ML) distributions. The critical issue of identifiability, essential for valid statistical inference in mixture models, is addressed by Sapatinas [296], who establishes sufficient conditions under which finite mixtures of MPSD distributions are identifiable, thereby guaranteeing that distinct parameter configurations correspond to distinct distributions.

Despite the versatility and broad applicability of finite mixture models, testing for the homogeneity of mixture components-that is, determining whether the observed data originate from a single component or a mixture-remains a statistically challenging and comparatively underexplored problem. We introduce a novel, non-parametric methodology for assessing homogeneity in finite mixtures, grounded in the distribution of IPDs. Our approach derives the theoretical distribution of IPDs under the assumption that the data are generated from multivariate normal or MPSD distributions. For other multivariate distributions, particularly those lacking closed-form analytical expressions, we employ a bootstrap procedure to estimate the distribution of IPDs empirically.

The resulting IPD-based homogeneity test provides a compelling alternative to classical approaches, such as the EM test, particularly in high-dimensional or discrete-data settings where parametric assumptions may be difficult to justify or computationally intensive. Additionally, the proposed methodology accommodates a hierarchical testing framework, allowing for comparisons between nested models, for example, testing whether a dataset is better described by a mixture with k components versus $k+1$ components. Through extensive simulation studies, we show that the IPD-based test exhibits competitive performance in terms of both power and robustness, making it a valuable tool for validating and assessing the fit of finite mixture models in contemporary multivariate data analysis.

7.5.1 IPD Test for Mixtures

Suppose $\{\mathbf{X}\}_{i=1}^{m}$ is a sample of d-dimensional random vectors that are drawn independently from a mixture of k distributions with distribution $f(\mathbf{x}) = \sum_{i=1}^{k} \alpha_i f_i(\mathbf{x})$, where $0 \leq \alpha_i \leq 1$ and $\sum_{i=1}^{k} \alpha_i = 1$. The quantities α_i are the mixing proportions and the $f_i(\mathbf{x})$ are the component densities or PMF of the mixture model. The test of mixture of k distributions can be built hierarchically. The first hypothesis is for $k = 2$,

$$H_0 : f(\mathbf{x}) = f_1(\mathbf{x}; \boldsymbol{\theta}_1) \text{ against } H_a : f(\mathbf{x}) = \alpha_1 f_1(\mathbf{x}; \boldsymbol{\theta}_1) + \alpha_2 f_2(\mathbf{x}; \boldsymbol{\theta}_2), \tag{7.31}$$

where $\alpha_1 + \alpha_2 = 1$. In general, we test the hypothesis of

$$H_0 : f(\mathbf{x}) = \sum_{i=1}^{k} \alpha_i f_i(\mathbf{x}; \boldsymbol{\theta}_i) \text{ against } H_a : f(\mathbf{x}) = \sum_{j=1}^{k+1} \beta_j f_j(\mathbf{x}; \boldsymbol{\theta}_j), \tag{7.32}$$

where $\sum_{i=1}^{k} \alpha_i = 1$, $\sum_{j=1}^{k+1} \beta_j = 1$ and $\boldsymbol{\theta}_i' s$ are parameters of density or PMF f_i. Measures of center and dispersion, such as mean, median, and standard deviation are often stated in terms of s-norms or L_s-norms, which are defined by $\|\mathbf{X}\|_s = \left(\sum_{t=1}^{d} |X_t|^s\right)^{\frac{1}{s}}$. The squared

IPD, henceforth IPD, between any of the two random vectors $\mathbf{X}_i$ and $\mathbf{X}_j$ is defined by $d^2_{(x)ij} = \|\mathbf{X}_i - \mathbf{X}_j\|^2 = \sum_{t=1}^{d}(X_{it} - X_{jt})^2$, the Euclidean distance throughout this chapter. Marozzi [214] compares the IPD tests based on different distances and demonstrates that the test based on the Euclidean distance performs well for normal and light-tailed distributions, whereas the test based on the minimum distance performs well for heavy-tailed distributions. We construct the IPD-test in two cases with different assumptions and based on different distances. The following discussion is based on the work of Song and Modarres [319].

7.5.2 IPD-test with Known Proportions and Parameters

In this sub-section, we assume that the form of the components, the proportions and parameters of the mixture models are known. An algorithm for finding the empirical critical value is discussed next. Throughout the chapter, we use a nominal level of 5% for testing hypothesis.

Algorithm 7.5.1 Algorithm A: Critical Value Estimation under H_0 with Known Parameters

Input: Observations $\{\mathbf{X}\}_{i=1}^{m}$, number of intervals K, number of repetitions R
Output: 95th percentile $C_{0.95}$ of χ^2 under H_0

Step 1: Find IPD distribution under H_0
Compute the distribution of interpoint distances (IPDs) $d^2_{(x)}$ of $\{\mathbf{X}\}_{i=1}^{m}$ under H_0, using either theoretical or bootstrap-estimated distributions.

Step 2: Divide into intervals and compute expected frequencies
Divide the range of IPDs into K intervals and compute cumulative probabilities p_j in each interval.
Let $\mathbb{E}_j = \binom{m}{2}p_j$ be the expected frequency in the j-th interval, $j = 1, \ldots, K$.

Step 3: Generate sample and observed frequencies
Generate a random sample $\{\mathbf{x}\}_{i=1}^{m}$ under H_0 and compute the $\binom{m}{2}$ IPDs.
Let O_j be the observed frequency in the j-th interval, $j = 1, \ldots, K$.

Step 4: Compute chi-squared statistic

$$\chi^2 = \sum_{j=1}^{K} \frac{(O_j - \mathbb{E}_j)^2}{\mathbb{E}_j}$$

Denote ${}^{T}\chi^2$ and ${}^{B}\chi^2$ as χ^2 values using theoretical and bootstrap densities, respectively.

Step 5: Repeat and estimate critical value
Repeat Steps 3–4 R times.
Let $C_{0.95}$ be the 95th percentile of the resulting χ^2 values.

For a given alternative hypothesis, the empirical power is computed as follows.

Algorithm 7.5.2 Algorithm B: Empirical Power Estimation under H_a

Input: Mixture distribution under H_a, number of intervals K, expected frequencies $\mathbb{E}_j$ from Algorithm A, critical value $C_{0.95}$, number of repetitions R

Output: Empirical power estimate

Step 1: Generate sample and compute IPDs

Generate a random sample of size m from the mixture distribution under H_a.

Compute the $\binom{m}{2}$ interpoint distances (IPDs) within the sample.

Let O_j be the observed frequency in each interval $j = 1, \ldots, K$.

Step 2: Compute test statistic $\chi^2 = \sum_{j=1}^{K} \frac{(O_j - \mathbb{E}_j)^2}{\mathbb{E}_j}$ where $\mathbb{E}_j$ are defined in Step 2 of Algorithm A.

Step 3: Repeat sampling

Repeat Steps 1–2 a total of R times.

Step 4: Estimate power

Let $\mathbb{N}$ be the number of times $\chi^2 > C_{0.95}$.

Estimate empirical power as Power $= \frac{\mathbb{N}}{R}$.

Deriving the distribution of IPDs based on L_1, $\|\mathbf{X}\|_1 = \sum_{t=1}^{d} |X_t|$, and L_∞, $\|\mathbf{X}\|_\infty = \max_{1 \leq t \leq d} |X_t|$, norms is challenging. Accordingly, we employ a parametric bootstrap to approximate these distributions and utilize them in testing the homogeneity of mixture models with these distance metrics. The choice of the number of intervals K for the IPD test can affect the features captured in the data and is primarily determined by the range of the IPDs. For data that are approximately normally distributed, we adopt Sturges' formula [325], $K = \lceil \log_2 m \rceil + 1$, to determine the number of intervals. For non-normal data, we apply Doane's modification [77] of Sturges' formula, $K = 1 + \log_2(m) + \log_2 \left(1 + \frac{|g_1|}{\sigma_{g_1}}\right)$, where g_1 denotes the estimated third-moment skewness of the IPD distribution, and

$$\sigma_{g_1} = \sqrt{\frac{6(m-2)}{(m+1)(m+3)}},$$

to improve interval selection and better capture features in skewed or non-normal data.

Song [320] applies the IPD-test for homogeneity of NGMN mixture as there does exist a test for this class of mixture models in the literature. He obtains the theoretical PMF of IPDs between NGMN mixture distributions. He also uses parametric bootstrap to obtain the distribution of NGMN IPD with L_1 and L_∞. Consider testing the following hypotheses,

$$\begin{aligned} H_0 &: \mathbf{X} \sim NGMN(r, \mathbf{p}_1), \\ H_a &: \mathbf{X} \sim \alpha NGMN(r, \mathbf{p}_1) + (1-\alpha) NGMN(r, \mathbf{p}_2), \end{aligned} \tag{7.33}$$

where r is the number of failures before the process is halted. Using fixed parameter values, he compares the powers of the IPD-test with L_2, L_1 and L_∞ distances, respectively and shows that IPD-test has stable type I errors and the power values that are sensitive to changing proportion and parameter values with the three distances. The power with L_∞ distance is the highest in all cases and is significantly larger than the power with the other two distances. The power with L_1 distance is smaller than the power with L_2 distance in many cases.

Currently, there is no existing test in the literature for assessing homogeneity of negative multinomial (NGMN) mixture models. Chen and Li [62] propose an EM-based test statistic for univariate normal mixture models; however, their methodology has not been extended to multivariate normal mixtures, and EM-type tests are generally challenging to implement in high-dimensional settings. Song and Modarres[319] apply the IPD-based test to multivariate normal samples that serve as approximations to the NGMN samples. Following Childs and Balakrishnan [65], the distribution of the NGMN can be approximated by a multivariate normal distribution with the same mean vector and covariance matrix for large sample sizes. They compare testing H_0 in (7.33) to testing

$$\begin{aligned} H_0 &: \phi(\mathbf{x}) = \phi_1(\mathbf{x}; \boldsymbol{\mu}_1, \boldsymbol{\Sigma}_1), \\ H_a &: \phi(\mathbf{x}) = \alpha\phi_1(\mathbf{x}; \boldsymbol{\mu}_1, \boldsymbol{\Sigma}_1) + (1-\alpha)\phi_2(\mathbf{x}; \boldsymbol{\mu}_2, \boldsymbol{\Sigma}_2), \end{aligned} \tag{7.34}$$

where ϕ is the p.d.f. of multivariate normal distribution. The parameters $\boldsymbol{\mu}_1, \boldsymbol{\Sigma}_1, \boldsymbol{\mu}_2$ and $\boldsymbol{\Sigma}_2$ are obtained by using formulas $\boldsymbol{\mu} = \frac{r\mathbf{p}}{p_0}$ and

$$\boldsymbol{\Sigma}_{(ts)} = \begin{cases} \frac{rp_0(p_{1t}+p_{01})}{p_0^2} \text{ if } t = s, \\ \frac{rp_t p_s}{p_0^2} \text{ if } t \neq s, \end{cases} \tag{7.35}$$

where r, $\mathbf{p} = (p_1, \ldots, p_d)'$ and $d_0 = 1 - \sum_{t=1}^{d} p_t$ are the parameters for the NGMN distributions in (7.33). Using the multivariate normal mixture models to approximate the NGMN mixture models, their simulation experiments show that the IPD-test maintains the nominal level under the null hypothesis. The testing powers for the NGMN models are higher than the testing powers of the approximated models if the mixing proportion $(1-\alpha)$ is large, whereas the later one performs better when the mixing proportion is small.

Song and Modarres [319] further consider testing

$$\begin{aligned} &H_0 : \mathbf{X} \sim \alpha_1 NGMN(k, \mathbf{p}_1) + \alpha_2 NGMN(k, \mathbf{p}_2) \text{ against} \\ &H_a : \mathbf{X} \sim \beta_1 NGMN(k, \mathbf{p}_1) + \beta_2 NGMN(k, \mathbf{p}_2) + \beta_3 NGMN(k, \mathbf{p}_3). \end{aligned} \tag{7.36}$$

They show that the IPD-test maintains the nominal level. The power is sensitive to the changes in proportions since the power increases when the mixture proportion $(1-\beta_2)$ increases under the alternatives. The IPD-test based on L_∞ distance performs better than the other two distances for NGMN mixture model.

7.5.3 IPD-test with Estimated Proportions and Parameters

The reported experiments by Song and Modarres [319] assume that the form of components, the proportions and parameters of the mixture model are known. Here, we only assume that the form of the components is known. The next algorithm estimates the critical values under H_0 with estimated parameters. The algorithm for power estimation remains the same as Algorithm B.

Algorithm 7.5.3 Algorithm C: Critical Value Estimation under H_0 with Estimated Parameters

Input: Random sample $\{\mathbf{x}\}_{i=1}^{m}$, number of intervals K, number of repetitions R
Output: 95th percentile $C_{0.95}$ of ${}^{E}\chi^2$ under H_0

Step 1: Estimate parameters from sample
Compute the sample mean $\bar{\mathbf{x}}$ and sample covariance matrix

$$\hat{\Sigma} = \frac{1}{m-1}\sum_{i=1}^{m}(\mathbf{x}_i - \bar{\mathbf{x}})(\mathbf{x}_i - \bar{\mathbf{x}})'.$$

Step 2: Compute IPD distribution with estimated parameters
Find the distribution of interpoint distances (IPDs) $d^2_{(x)}$ of $\{\mathbf{X}\}_{i=1}^{m}$ using the estimated parameters.

Step 3: Compute expected frequencies
Compute the expected frequency $\mathbb{E}_j$ in each interval $j = 1, \ldots, K$ for the IPD distribution obtained in Step 2.

Step 4: Generate sample and compute observed frequencies
Generate another random sample $\{\mathbf{x}^*\}_{i=1}^{m}$ using the estimated parameters.
Compute the $\binom{m}{2}$ IPDs and count the observed frequency O_j in each interval $j = 1, \ldots, K$.

Step 5: Compute test statistic ${}^{E}\chi^2 = \sum_{j=1}^{K} \frac{(O_j - \mathbb{E}_j)^2}{\mathbb{E}_j}$.

Step 6: Repeat and estimate critical value
Repeat Steps 4–5 R times.
Let $C_{0.95}$ be the 95th percentile of the resulting ${}^{E}\chi^2$ values.

Song [320] reports that the IPD-test with estimated proportion and parameters has competitive power to the same test by assuming proportions and parameters are all known. The estimated proportion under the null hypothesis is also quite accurate by using Matteson-James [221] method. While this method performs well for location shift alternatives, it does not estimate accurately in the cases of scale differences for multivariate normal mixtures. We recommend this method for location shift alternatives when sample sizes are relatively large.

7.5.4 Bootstrap Estimated Density

While we presented the distribution of IPDs from MPSD and multivariate normal mixtures the distribution of IPD is difficult to derive for other families and one may resort to its asymptotic distribution (Li [192]). When the distribution of IPDs are unknown, as noted in Algorithm A, one uses parametric bootstrap to find the distribution of IPDs (Liu and Modarres [201]). The procedure of bootstrapping is shown below.

Algorithm 7.5.4 Algorithm D: IPD Density Estimation

Input: Target distribution under H_0, sample size m, number of repetitions R
Output: Estimated probabilities of interpoint distances (IPDs)

Step 1: Generate sample
Generate a random sample of size m from the target distribution under H_0.

Step 2: Compute IPDs
Find the $\binom{m}{2}$ interpoint distances (IPDs) within the generated sample.

Step 3: Repeat sampling
Repeat Steps 1–2 a total of R times and record frequencies for the $\binom{m}{2} \times R$ total IPDs.

Step 4: Estimate IPD probabilities
Compute the estimated probabilities of IPDs by dividing the frequencies by $\binom{m}{2} \times R$.
These estimated probabilities are used in Step 1 of Algorithm A.

The above algorithm works well for estimating the distribution of IPDs from mixture of discrete multivariate distributions. For the continuous multivariate case, we need a larger number of simulations because the distribution of the IPD is continuous. Hence, one can find the frequencies of IPDs within many small intervals or simply round all IPDs to integers to cumulate frequencies. For d-dimensional random vectors that are drawn from Multivariate Normal, MN, NGMN, and MP distributions Song [320] shows that the empirical distribution of IPDS converge to the true distributions under the bootstrap.

7.5.5 Analysis of Old Faithful Geyser

Old Faithful is a cone geyser located in Yellowstone National Park, Wyoming, USA. Its eruptions expel between 14,000 and 32,000 liters of boiling water to heights ranging from 32 to 56 meters, with durations of 1.5 to 5 minutes. The intervals between eruptions vary from 45 to 125 minutes, averaging 66.5 minutes in 1939 and gradually increasing to approximately 90 minutes today. The Old Faithful dataset, available in R, consists of 272 observations on two variables: the waiting time between eruptions and the duration of each eruption. While Figure 7.3 suggests the presence of two elliptical clusters, we formally assess homogeneity using the IPD-based test.

Assuming the data follow a bivariate normal distribution, we compute the sample mean and covariance matrix from all observations and implement steps 1—5 of Algorithm C to bootstrap the distribution of the IPD under the null hypothesis of homogeneity. The observed value of the IPD-test is then calculated from the data. Using the estimated null distribution and the observed statistic, we obtain a p-value of 0.017. Because this p-value is substantially smaller than the nominal level of 0.05, we reject the null hypothesis of homogeneity and conclude that the Old Faithful data are better described by a mixture of bivariate normal distributions.

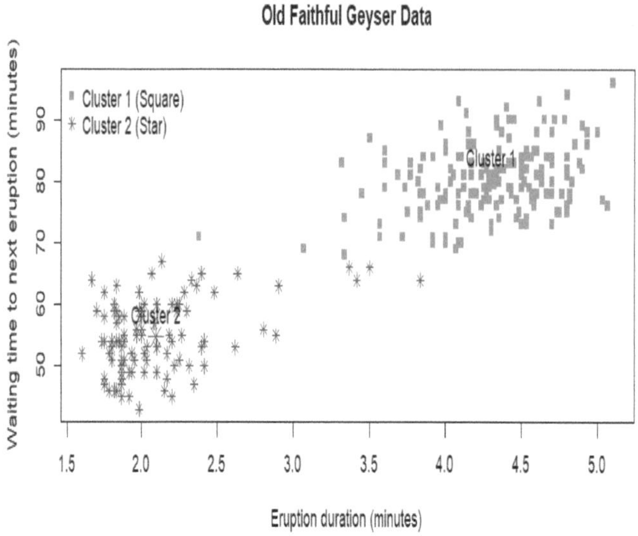

FIGURE 7.3: Old Faithful Geyser scatter plot suggests two clusters.

7.6 Summary

This chapter develops a unified and rigorous framework for analyzing IPDs within the class of multivariate power series distributions (MPSDs), encompassing important models such as the multinomial, negative multinomial, multivariate Poisson, and related families. Explicit distributional results for squared IPDs are derived under a variety of settings, including independent and non-identically distributed observations, mixtures of distributions, and distances from fixed vectors in $\mathbb{Z}^d$. These results provide a comprehensive understanding of IPD behavior in high-dimensional discrete spaces and establish IPDs as effective low-dimensional summaries of complex multivariate structure.

Building on these theoretical developments, the chapter proposes IPD-based tests of homogeneity for mixture models, applicable when mixture proportions and component parameters are either known or estimated. The distribution of the test statistic is obtained using both theoretical results and parametric bootstrap approximations. Compared with likelihood-based procedures such as the EM test and the modified likelihood ratio test (MLRT), the IPD-test avoids fundamental difficulties including non-standard null distributions, unbounded likelihoods, infinite Fisher information, and singular covariance matrices. Since IPDs are inherently one-dimensional regardless of the ambient dimension, the proposed test remains computationally stable and effective in high-dimensional and low-sample-size settings.

The finite-sample performance of the IPD-test is examined through simulation studies and applications to mixture models. These studies show that the IPD-test maintains good control of type I error and competitive power relative to EM and MLRT procedures, particularly in high-dimensional contexts where likelihood-based methods tend to be conservative or unstable. An application to the Old Faithful geyser dataset further illustrates the practical utility of the methodology. The chapter establishes IPDs as a powerful and flexible tool for testing homogeneity in multivariate mixture models, with broad relevance to high-dimensional inference, clustering, and model discrimination.

7.7 Proofs

Proof of Theorem 7.4.1:

For any $v \in \mathbf{V}_r$, we write

$$\begin{aligned}
\mathbb{P}\left(d^2_{(x)} = v\right) &= \sum_{\mathbf{v}\in\mathbf{V}_r} \mathbb{P}\left(\mathbf{X} - \mathbf{X}^* = (v_1, \ldots, v_d)'\right) = \sum_{\mathbf{v}\in\mathbf{V}_r} \mathbb{P}\left(X_1 - X_1^* = v_1, \ldots, X_d - X_d^* = v_d\right)\\
&= \sum_{\mathbf{v}\in\mathbf{V}_r} \sum_{\phi_1\in\Phi} \mathbb{P}\left(X_1 - X_1^* = v_1, \ldots, X_d - X_d^* = v_d \mid X_1^* = \phi_1\right) \mathbb{P}\left(X_1^* = \phi_1\right)\\
&= \sum_{\mathbf{v}\in\mathbf{V}_r} \sum_{\phi_1\in\Phi} \mathbb{P}\left(X_1 = v_1 + \phi_1, \ldots, X_d - X_d^* = v_d\right) \mathbb{P}\left(X_1^* = \phi_1\right)\\
&\ \ \vdots\\
&= \sum_{\mathbf{v}\in\mathbf{V}_r} \sum_{\phi_1\in\Phi} \cdots \sum_{\phi_d\in\Phi} \mathbb{P}\left(X_1 = v_1 + \phi_1, \ldots, X_d = v_d + \phi_d\right) \cdot \mathbb{1}_{\left\{\sum_{t=1}^d \phi_t = n\right\}}\\
&\quad \times \prod_{t=1}^{d} \mathbb{P}\left(X_t^* = \phi_t\right) \cdot \mathbb{1}_{\{\phi_t + v_t \geq 0\}}\\
&= \sum_{\mathbf{v}\in\mathbf{V}_r} \sum_{\phi_1\in\Phi} \cdots \sum_{\phi_d\in\Phi} \mathbb{P}\left(X_1 = \phi_1 + v_1\right) \cdot \mathbb{P}\left(X_1^* = \phi_1\right) \cdot \mathbb{1}_{\left\{\sum_{t=1}^d \phi_t = n, \phi_1 + v_1 \geq 0\right\}}\\
&\quad \times \prod_{t=2}^{d} \mathbb{P}\left(X_t = \phi_t + v_t | x_1, \ldots, x_{t-1}\right) \cdot \mathbb{P}\left(X_t^* = \phi_t | x_1^*, \ldots, x_{t-1}^*\right) \cdot \mathbb{1}_{\{\phi_t + v_t \geq 0\}},
\end{aligned} \tag{32}$$

where the distribution of X_1 and X_1^* follow from the marginal distribution of the MN distribution. The distributions of $X_t|x_1, \ldots, x_{t-1}$ and $X_t^*|x_1^*, \ldots, x_{t-1}^*$ can be obtained by using the Grouping and Conditioning properties of the MPSD family.

Proof of Lemma 7.4.1:

For $t = 1, \ldots, d$ one can write,

$$\begin{aligned}
\mathbb{P}\left(T^2_{(x)}(t) = v_t^2\right) &= \begin{cases} \mathbb{P}\left(T_{(x)}(t) = v_t\right) + \mathbb{P}\left(T_{(x)}(t) = -v_t\right), & \text{if } v_t \neq 0;\\ \mathbb{P}\left(T_{(x)}(t) = 0\right), & \text{if } v_t = 0, \end{cases}\\
&= \begin{cases} 2\mathbb{P}\left(T_{(x)}(t) = v_t\right), & \text{if } v_t \neq 0;\\ \mathbb{P}\left(T_{(x)}(t) = 0\right), & \text{if } v_t = 0, \end{cases}\\
&= \left(1 + \mathbb{1}_{\{v_t \neq 0\}}\right) \mathbb{P}\left(T_{(x)}(t) = v_t\right)\\
&= \left(1 + \mathbb{1}_{\{v_t \neq 0\}}\right) \sum_{\psi\in\Psi} \mathbb{P}(X_t - X_t^* = v_t \mid X_t^* = \psi) \cdot \mathbb{P}(X_t^* = \psi)\\
&= \left(1 + \mathbb{1}_{\{v_t \neq 0\}}\right) \sum_{\psi\in\Psi} \mathbb{P}(X_t = \psi + v_t) \cdot \mathbb{P}(X_t^* = \psi),
\end{aligned}$$

by using the marginal distribution of MPSD from equation (2).

Proof of Theorem 7.4.2:

By using Lemma 3.1, one can write

$$
\begin{aligned}
\mathbb{P}\left(d^2_{(x)} = v\right) &= \mathbb{P}\left(\sum_{t=1}^{d} T^2_{(x)}(t) = v\right) = \mathbb{P}\left(T^2_{(x)}(1) + \sum_{t=2}^{d} T^2_{(x)}(t) = v\right) \\
&= \sum_{\mathbf{v}\in\mathbf{V}_u} \mathbb{P}\left(T^2_{(x)}(1) + \sum_{t=2}^{d} T^2_{(x)}(t) = v \mid T^2_{(x)}(1) = v_1^2\right) \cdot \mathbb{P}\left(T^2_{(x)}(1) = v_1^2\right) \\
&= \sum_{\mathbf{v}\in\mathbf{V}_u} \mathbb{P}\left(\sum_{t=2}^{d} T^2_{(x)}(t) = v - v_1^2\right) \cdot \mathbb{P}\left(T^2_{(x)}(1) = v_1^2\right) \\
&= \sum_{\mathbf{v}\in\mathbf{V}_u} \mathbb{P}\left(\sum_{t=3}^{d} T^2_{(x)}(t) = v - v_1^2 - v_2^2\right) \cdot \mathbb{P}\left(T^2_{(x)}(2) = v_2^2\right) \cdot \mathbb{P}\left(T^2_{(x)}(1) = v_1^2\right) \\
&\vdots \\
&= \sum_{\mathbf{v}\in\mathbf{V}_u} \prod_{t=1}^{d} \mathbb{P}\left(T^2_{(x)}(t) = v_t^2\right) \\
&= \sum_{\mathbf{v}\in\mathbf{V}_u} \prod_{t=1}^{d} \left(1 + \mathbb{1}_{\{v_t \neq 0\}}\right) \sum_{\psi\in\Psi} \mathbb{P}(X_t = \psi + v_t) \cdot \mathbb{P}(X_t^* = \psi) \\
&= \sum_{\mathbf{v}\in\mathbf{V}_u} \prod_{t=1}^{d} \left(1 + \mathbb{1}_{\{v_t \neq 0\}}\right) \sum_{\psi\in\Psi} \frac{a_t(\theta_1, \ldots, \psi, \ldots, \theta_d)\theta_t^{\psi}}{f(\boldsymbol{\theta})} \cdot \frac{a_t(\theta_1, \ldots, \psi + v_t, \ldots, \theta_d)\theta_t^{\psi+v_t}}{f(\boldsymbol{\theta})}.
\end{aligned}
\tag{33}
$$

Since $\mathbb{E}\left(\mathbf{X} - \mathbf{X}^*\right) = \mathbf{0}$ and $\mathbb{C}ov\left(\mathbf{X} - \mathbf{X}^*\right) = 2\boldsymbol{\Sigma}_x$, we obtain the expected value of IPD by using equation (4).

Proof of Theorem 7.4.3:

The proof follows from the proof of Theorem 3.1 by changing $\mathbf{X}^*$ to $\mathbf{Y}$ with different parameters. The distribution of $X_t|x_1, \ldots, x_{t-1}$ and $Y_t|y_1, \ldots, y_{t-1}$ can be obtained by using the Grouping and Conditioning properties of the MPSD family

Proof of Lemma 7.4.2:

For $t = 1, \ldots, d$, we write,

$$
\mathbb{P}\left(T^2_{(xy)}(t) = v_t^2\right) = \begin{cases} \mathbb{P}\left(X_t - Y_t = v_t\right) + \mathbb{P}\left(X_t - Y_t = -v_t\right), & \text{if } v_t \neq 0, \\ \mathbb{P}\left(X_t - Y_t = 0\right), & \text{if } v_t = 0. \end{cases}
\tag{34}
$$

We also write that

$$
\begin{aligned}
\mathbb{P}\left(X_t - Y_t = v_t\right) &= \sum_{\psi\in\Psi} \mathbb{P}(X_t - Y_t = v_t \mid Y_t = \psi) \cdot \mathbb{P}(Y_t = \psi) \\
&= \sum_{\psi\in\Psi} \mathbb{P}(X_t = \psi + v_t) \cdot \mathbb{P}(Y_t = \psi),
\end{aligned}
\tag{35}
$$

and similarly,

$$\begin{aligned}
\mathbb{P}\left(Y_t - X_t = v_t\right) &= \sum_{\psi \in \Psi} \mathbb{P}(Y_t = \psi + v_t) \cdot \mathbb{P}(X_t = \psi), \\
\mathbb{P}\left(X_t - Y_t = 0\right) &= \sum_{\psi \in \Psi} \mathbb{P}(X_t = \psi) \cdot \mathbb{P}(Y_t = \psi).
\end{aligned} \tag{36}$$

Substitute equations (35) and (36) in (34) to get the PMF shown in expressions (14) and all the marginal distributions can be expressed explicitly by using equation (2).

Proof of Theorem 7.4.4:

In a manner similar to the proof of Theorem 7.4.2, we find

$$\begin{aligned}
\mathbb{P}\left(d^2_{(xy)} = v\right) &= \mathbb{P}\left(\sum_{t=1}^{d} T^2_{(xy)}(t) = v\right) = \mathbb{P}\left(T^2_{(xy)}(1) + \sum_{t=2}^{d} T^2_{(xy)}(t) = v\right) \\
&= \sum_{\mathbf{v} \in \mathbf{V}_u} \mathbb{P}\left(T^2_{(xy)}(1) + \sum_{t=2}^{d} T^2_{(xy)}(t) = v \mid T^2_{(xy)}(1) = v_1^2\right) \cdot \mathbb{P}\left(T^2_{(xy)}(1) = v_1^2\right) \\
&= \sum_{\mathbf{v} \in \mathbf{V}_u} \mathbb{P}\left(\sum_{t=2}^{d} T^2_{(xy)}(t) = v - v_1^2\right) \cdot \mathbb{P}\left(T^2_{(xy)}(1) = v_1^2\right) \\
&= \sum_{\mathbf{v} \in \mathbf{V}_u} \mathbb{P}\left(\sum_{t=3}^{d} T^2_{(xy)}(t) = v - v_1^2 - v_2^2\right) \cdot \prod_{t=1}^{2} \mathbb{P}\left(T^2_{(xy)}(t) = v_t^2\right) \\
&\;\;\vdots \\
&= \sum_{\mathbf{v} \in \mathbf{V}_u} \prod_{t=1}^{d} \mathbb{P}\left(T^2_{(xy)}(t) = v_t^2\right),
\end{aligned} \tag{37}$$

where $\mathbb{P}\left(T^2_{(xy)}(t) = v_t^2\right)$ can be expressed by (14). Because $\mathbb{E}\left(\mathbf{X} - \mathbf{Y}\right) = \boldsymbol{\mu}_x - \boldsymbol{\mu}_y$ and $\mathbb{C}ov\left(\mathbf{X} - \mathbf{Y}\right) = \boldsymbol{\Sigma}_x + \boldsymbol{\Sigma}_y$, one can find the expected value by using equation (7).

Proof of Theorem 7.4.5:

By definition of the mixture distribution, we obtain the following cases.

$$\begin{cases}
\text{1. } \mathbf{X} \text{ and } \mathbf{X}^* \text{ are both drawn from distribution with PMF } g_i(\cdot) \text{ w.p. } \alpha_i^2, \\
\text{2. } \mathbf{X} \text{ and } \mathbf{X}^* \text{ are drawn from distributions with PMF } g_i(\cdot) \text{ and } g_j(\cdot) \text{ w.p. } \alpha_i\alpha_j, \\
\text{3. } \mathbf{X} \text{ and } \mathbf{X}^* \text{ are drawn from distributions with PMF } g_j(\cdot) \text{ and } g_i(\cdot) \text{ w.p. } \alpha_i\alpha_j,
\end{cases}$$

where w.p. denotes for with probability. We find the PMF of IPD for each case by using the same argument as equation (32), where $h_{it}(\cdot)$ is the PMF of $X_{it}|x_{i1}, \ldots, x_{i(t-1)}$ for $t = 2, \ldots, d$ and h_{1t} is the PMF of X_{i1}. Apply identity (16) to the cases listed above and substitute equation (32) to get the final expression in (18). The expected value follows from equation (7) and identity (17).

Proof of Lemma 7.4.3:

$$\mathbb{P}\left(T_{(x)}(t) = v_t\right) = \sum_{\psi \in \Psi} \mathbb{P}\left(X_t - X_t^* = v_t | X^* = \psi\right) \cdot \mathbb{P}\left(X_t^* = \psi\right)$$

$$= \sum_{\psi \in \Psi} \mathbb{P}(X_t = \psi + v_t) \cdot \mathbb{P}(X_t^* = \psi)$$

$$= \begin{cases} \sum_{\psi \in \Psi} l_i(\psi + v_t) \cdot l_i(\psi), & \text{w.p } \alpha_i^2, \\ \sum_{\psi \in \Psi} l_i(\psi + v_t) \cdot l_j(\psi), & \text{w.p } \alpha_i \alpha_j, \\ \sum_{\psi \in \Psi} l_j(\psi + v_t) \cdot l_i(\psi), & \text{w.p } \alpha_i \alpha_j. \end{cases} \tag{38}$$

Hence, one can rewrite it as expression (20).

Proof of Theorem 7.4.6:

Note that equation (21) can be obtained by using the same argument of expression (33) and substituting $P\left(T_{(x)}(t) = v_t\right)$ from (20). The expected value follows from equation (7) and identity (17).

7.8 Exercises

Exercise 7.1 *Also called the Hamming distance in information theory (Hamming [136]), the squared IPDs of Bernoulli data are often used to examine the multivariate structure of Bernoulli data in high dimensions. Suppose* $\mathbf{X} = \{\mathbf{X}_i\}$ *for* $i = 1, \ldots, m$ *is a sample of independent and identically distributed random vectors in* $\{0,1\}^d$ *drawn from a MB distribution with mean vector* $\mathbf{P}$*, covariance matrix* $\mathbf{\Sigma}$ *and distribution function* F*.*

a) Define $d^2_{(x)ij}$*, the IPD between MB vectors* $\mathbf{X}_i$ *and* $\mathbf{X}_j$*.*

b) Show that $d^2_{(x)ij}$ *is a proper distance function and satisfies the triangle inequality.*

c) Let $\mathbf{T}_{(x)ij} = (T_{(x)ij1}, \ldots, T_{(x)ijd})$ *be a vector in* $\{0,1\}^d$ *corresponding to the disagreements between observations* $\mathbf{X}_i$ *and* $\mathbf{X}_j$*. Show that* $\mathbf{T}_{(x)ij}$ *has a MB(*$\boldsymbol{\theta}_{(x)ij}, \mathbf{\Gamma}_{(x)ij,ij}$*). Specify the mean and variance of* $\mathbf{T}_{(x)ij}$*.*

d) When are the two distances $d^2_{(x)ij}$ *and* $d^2_{(x)kh}$ *dependent?*

e) Find $\mathbb{E}d^2_{(x)ij})$ *and* $\mathbb{V}ar(d^2_{(x)ij})$*.*

Exercise 7.2 *Obtain the probability distribution of* $d^2_{(x)ij}$*, the squared distance between two randomly selected vectors* $\mathbf{X}_i$ *and* $\mathbf{X}_j$*,* $i \neq j$*, drawn from a MB(*$\mathbf{P}, \mathbf{\Sigma}$*). Since the* L_2*-norm is the same as the* L_1*-norm for Bernoulli variables, the distribution of* $d^2_{(x)ij}$ *and* $d_{(x)ij}$ *are the same. Assume* $\mathbf{\Sigma}$ *is diagonal.*

Exercise 7.3 *Suppose* $\mathbf{X}_i$ *are d-dimensional multivariate Bernoulli, for* $i = 1, \ldots, m$*, with covariance that follows an exchangeable structure so that* $\mathbf{P} = p\mathbf{1}_d$ *and* $\rho_{rs} = \rho$*. Obtain the mean, variance and covariance of the IPDs.*

Exercise 7.4 *The squared IPD between* $\mathbf{X}_i$ *and* $\mathbf{Y}_j$ *(independent groups) is defined as*

$$d^2_{(xy)ij} = (\mathbf{X}_i - \mathbf{Y}_j)'(\mathbf{X}_i - \mathbf{Y}_j) = \sum_{r=1}^{d} (X_{ir} - Y_{jr})^2 = \sum_{r=1}^{d} T_{(xy)ijr}. \tag{7.37}$$

for $1 \leq i \leq m$ *and* $1 \leq j \leq n$*, where* $\mathbf{X}_i$ *and* $\mathbf{Y}_j$ *have MB distributions with means* $p\mathbf{1}_d$ *and* $p^*\mathbf{1}_d$*, and covariances* $\sigma_{rs} = \mathbb{C}ov(X_{ir}, X_{is})$*, and* $\sigma^*_{rs} = \mathbb{C}ov(Y_{ir}, Y_{is})$ *. Obtain the mean, variance and covariance of IPDs.*

Exercise 7.5 *Suppose* $\{\mathbf{X}_i\}$ *for* $i = 1, \ldots, m$ *and* $\{\mathbf{Y}_i\}$ *for* $i = 1, \ldots, n$ *are independent samples drawn from* $MB(\mathbf{P}, \mathbf{\Sigma})$ *and* $MB(\mathbf{P}^*, \mathbf{\Sigma}^*)$*, respectively. When* $\mathbf{\Sigma}$ *and* $\mathbf{\Sigma}^*$ *are diagonal, obtain the distribution of* $d^2_{(xy)}$.

Exercise 7.6 *Consider a sequence of Bernoulli random variables* $X_1, \ldots, X_n$. *Suppose* $\mathbb{P}(X_i = 1) = 1 - \mathbb{P}(X_i = 0) = p$, *for* $i = 1, \ldots, n,$. *Let* $\mathbb{P}(X_i = 1|x_{i-1} = 1) = \lambda$, *for* $i = 2, \ldots, n$.

a) *Find* $\mathbb{P}(X_i = 1|X_{i-1} = 0)$ *and* $\mathbb{P}(X_i = 0|X_{i-1} = 0)$.

b) *Find* $a \leq b$ *such that* $a \leq \lambda \leq b$ *for* $0 \leq p \leq 1$.

Exercise 7.7 *Suppose* $\mathbf{\Sigma} = (\sigma_{ij})$ *is the covariance matrix of a d-dimensional random with eigenvalue and eigenvector pairs* (λ_i, e_i) *for* $i = 1, \ldots d$ *where* $\lambda_1 \geq \lambda_2 \geq \ldots \geq \lambda_d > 0$. *Let* $\mathbf{\Sigma} = (\sigma_{ij})$ *where* $\sigma_{ii} = p_i(1 - p_i)$, *for* $i = 1, \ldots, d$ *and* $0 < p_i < 1$. *Prove that*

a) $0 < \sigma_{ii} \leq 1/4$ *and* $-1/4 \leq \sigma_{ij} \leq 1/4$ *where* $i \neq j = 1, \ldots, d$.

b) $\sum_{i=1}^d \lambda_i \leq d/4$ *and* $\prod_{i=1}^d \lambda_i \leq 4^{-d}$. *The joint CDF must obey*

$$\max(0, \mathbb{P}(X_i = 1) + \mathbb{P}(X_{i-1} = 1) - 1) \leq \mathbb{P}(X_i = 1, X_{i-1} = 1) \leq \min(\mathbb{P}(X_i = 1), \mathbb{P}(X_{i-1} = 1)).$$

Exercise 7.8 *Find the PMF of* $d^2_{(x)ij}$ *in Exercise 7.1 when* $d = 2$ *and* $d = 3$.

Exercise 7.9 *Let* Y_1, Y_2 *be correlated Bernoulli random variables such that* $\mathbb{E}Y_i) = p_i$ *for* $i = 1, 2$. *Find the PMF of* $Y_{(1)} = min(Y_1, Y_2)$ *and* $Y_{(2)} = max(Y_1, Y_2)$.

Exercise 7.10 *Obtain simultaneous plots of the within and between IPDs for samples from multivariate Bernoulli distributions with independent components and probability vectors* $\mathbf{p}_x = p\mathbf{1}$ *and* $\mathbf{p}_y = q\mathbf{1}$. *Let d=10000,* $m = n = 50$ *and obtain plots when a)* $(p = 0.5, q = 0.5)$, *b)* $(p = 0.5, q = 0.8)$, *and c)* $(p = 0.8, q = 0.5)$. *What are the distributions of within and between squared IPDs from multivariate Bernoulli distributions when the components are independent?*

Exercise 7.11 *Suppose* $\mathbf{X} \in \{0, 1\}^d$: *A multivariate Bernoulli random vector with mean vector* $\mathbb{E}[\mathbf{X}] = \mathbf{p} = (p_1, \ldots, p_d)'$. *Further, assume that* $\mathbf{Y} \in \{0, 1\}^d$: *A random vector independent of* X, *with mean vector* $\mathbb{E}[\mathbf{Y}] = \mathbf{q} = (q_1, \ldots, q_d)'$. *Find the mean and variance* $Z = \mathbf{X}'\mathbf{Y} = \sum_{i=1}^d X_i Y_i$ *Assuming the components of* $\mathbf{X}$ *and* $\mathbf{Y}$ *are independent, find the distribution of* Z.

Exercise 7.12 *Let* $\mathbf{X}, \mathbf{Y} \in \{0, 1\}^d$ *be independent multivariate Bernoulli random vectors, such that*

$$\mathbb{E}[\mathbf{X}] = \mathbf{p}, \quad \mathbb{C}ov(\mathbf{X}) = \mathbf{\Sigma}_X, \quad \mathbb{E}[\mathbf{Y}] = \mathbf{q}, \quad \mathbb{C}ov(\mathbf{Y}) = \mathbf{\Sigma}_Y$$

- *Find the distribution, mean, and variance of* $z_x = \|\mathbf{X}\|^2$.
- *Find the distribution, mean, and variance of* $W = \|\mathbf{X} + \mathbf{Y}\|^2$.
- *Simplify the solutions when* $\mathbf{p} = p\mathbf{1}_d$ *and* $\mathbf{q} = q\mathbf{1}_d$.

Exercise 7.13 *Suppose* $\mathbf{X} \sim$ *Multinomial*$(m, \mathbf{p})$ *and* $\mathbf{Y} \sim$ *Multinomial*$(n, \mathbf{q})$ *are two independent* $d \times 1$ *multinomial vectors with probability vectors* $\mathbf{p}$ *and* $\mathbf{q}$ *respectively. Define* $\mathbf{W} = \mathbf{X} + \mathbf{Y}$. *Find the distribution, mean, covariance, of* $\mathbf{W}$ *and its squared norm. Simplify when* $\mathbf{p} = \mathbf{q}$. *What are asymptotic distributions of* $\mathbf{W}$ *and* $\|\mathbf{W}\|^2$ *as* $m, n \to \infty$*?*

Exercise 7.14 *Suppose* $\boldsymbol{X}$ *has multivariate power series distribution as defined in* (7.1). *Show that its probability generating function is*

$$G_{\mathbf{X}}(\boldsymbol{z}) = \frac{f(\theta_1 z_1, \ldots, \theta_d z_d)}{f(\theta_1, \ldots, \theta_d)}. \tag{7.38}$$

Exercise 7.15 *Suppose* $\boldsymbol{X}$ *and* $\boldsymbol{X}^*$ *are* 6*-dimensional vectors that are drawn from the MN distribution with parameters* $n = 5$ *and* $\mathbf{p} = (0.10, 0.15, 0.30, 0.10, 0.15, 0.20)'$.

- *Write R code to compute and plot the PMF of the simulated IPD* $\mathrm{d}^2_{(x)}$.
- *Write R code to compute and plot the PMF of the theoretical IPD* $\mathrm{d}^2_{(x)}$.
- *Obtain their Q-Q plot.*

8

General Notation of Data Depth

8.1 Abstract

Data depth, initially introduced by Tukey [342] and systematically developed by Liu [199], provides a comprehensive non-parametric framework for quantifying the centrality of observations in $\mathbb{R}^d$ through center-outward ordering. A depth function $D(\mathbf{t}; F)$ assigns a non-negative measure of centrality to a point $\mathbf{t}$ relative to a distribution F, satisfying affine invariance, maximality at the center, monotonicity with respect to the deepest point, and vanishing at infinity. Classical depth notions—including Mahalanobis, half-space (Tukey), simplicial, projection, spatial, spherical, convex hull peeling, interpoint, lens, and zonoid depths—support robust multivariate medians, ranking, classification, outlier detection, and functional data analysis. Recent advances exploit graph-based constructions, such as minimum spanning trees (MST), β-skeletons, path and RUNT depths, and geodesic convexity, offering computational efficiency in high-dimensional contexts while preserving robustness and distribution-free properties. Depth contours, level sets, and depth–depth plots facilitate geometric visualization of multivariate structure, whereas MST and graph-based depths reduce exponential complexity associated with multivariate medians. Applications include multivariate ranking, robust inference, rank-based testing, confidence region estimation, and centrality-based classification in skewed or non-elliptical distributions. Collectively, these methodologies establish data depth as a foundational tool for modern multivariate statistics, high dimensional data analysis, and robust non-parametric inference.

8.2 Introduction

The notion of data depth, as described by the seminal works of Liu [199] and Tukey [342], provides a systematic non-parametric approach for defining quantitative and graphical multivariate distributional characteristics and developing inferential methods based on center-outward ranking of observations in $\mathbb{R}^d$. A data depth function is any function $D(\mathbf{t}; F)$ that measures the closeness or centrality of a point $\mathbf{t} \in \mathbb{R}^d$ with respect to a distribution function F or its empirical counterpart F_m. Thus, a depth function assigns to each $x \in \mathbb{R}^d$ a non–negative score as its center-outward depth with respect to F. If a rank order is defined with respect to F, then observations which are further from the center of F have low ranks whereas observations close to the center receive higher ranks.

Hotelling [155] characterized the univariate median as the point which minimizes the maximum number of observations on one of its sides, $D(t; F) = \min(F(t), 1 - F(t))$. This notion was generalized by Tukey [342], giving rise to the definition of half-space or Tukey's multivariate depth (Donoho [79]). Small [317] provides an excellent treatment of multidimensional medians using geodesic convexity on graphs. Depth functions have been used

DOI: 10.1201/9781003789581-8

to define measures of multivariate location, scale, and skewness, to conduct multivariate rank tests, and to visualize multivariate observations with depth contour plots. Several of the existing depth-based multivariate medians pose formidable computational challenges as their time complexities grow exponentially with d. Multivariate medians based on the MST are especially attractive as their time complexities are independent of the dimension.

8.3 Applications of Depth Functions

1. Outlier Detection

Statistical depth functions provide a natural and robust framework for identifying outliers in multivariate data. Unlike traditional Mahalanobis proximity–based approaches, depth functions such as the Tukey depth or projection depth do not rely on parametric assumptions and remain effective even in the presence of skewed or heavy-tailed distributions. Points with low depth values are considered outliers, as they lie far from the "center" defined by the depth function. This technique is particularly valuable in high-dimensional settings where visual inspection is not feasible.

2. Multivariate Ranking and Centrality Depth functions enable the extension of univariate concepts such as medians and quantiles to the multivariate case. For example, the multivariate median can be defined as the point with the highest depth. This centrality measure is useful in applications such as environmental monitoring and finance, where determining a robust center of multivariate distributions is essential for summarization and risk assessment. Rankings based on depth values allow practitioners to order multivariate observations in a meaningful way.

3. Classification Depth-based classification methods have gained popularity as non-parametric alternatives to traditional techniques. The basic idea is to assign a new observation to the class in which it achieves the highest depth. This approach is distribution-free and robust, making it suitable for high dimensional and skewed data. For instance, Guo and Modarres [125] developed a depth-based classifier that performs competitively against support vector machines and nearest neighbor methods, especially when the underlying distributions are not well separated or are non–elliptical.

4. Functional Data Analysis In functional data analysis (FDA), statistical depth functions are used to define centrality and variability among curves or functions. Depth notions such as band depth and modified band depth allow for the construction of functional box-plots, which generalize the univariate box-plot to infinite-dimensional settings. These are widely used in medical imaging, weather forecasting, and spectroscopic data to detect atypical patterns and summarize temporal trends.

5. Robust Statistical Inference Depth functions underpin robust versions of statistical inference procedures, including hypothesis testing and confidence region estimation. Because depth-based methods are less sensitive to deviations from model assumptions and outliers, they are particularly useful in applications where data contamination is expected. For example, depth-based confidence regions can be constructed without assuming multivariate normality, offering a robust alternative to ellipsoidal regions defined by covariance matrices.

8.4 Statistical Depth Functions

Zuo and Serfling [372] proposed a formal framework for statistical depth functions based on the following four properties. Let F be the class of distributions on the Borel set on $\mathbb{R}^d$

and $F_{\mathbf{X}}$ be the distribution function of a given random vector $\mathbf{X}$. Let $D : \mathbb{R}^d \times \mathrm{F} \longmapsto R$ be bounded, non–negative, and satisfy the following four conditions

- Affine invariance: $D(A\mathbf{t} + \mathbf{b}; F_{A\mathbf{X}+\mathbf{b}}) = D(\mathbf{t}; F_{\mathbf{X}})$ holds for any random vector $\mathbf{X}$, any $d \times d$ non–singular matrix A, and any vectors $\mathbf{t}$ and $\mathbf{b}$ in $\mathbb{R}^d$;
- Maximality at center: $D(\theta; F) = sup_{\mathbf{t}\in\mathbb{R}^d} D(\mathbf{t}; F)$ holds for any $F \in \mathrm{F}$ with center θ_F (a point of symmetry);
- Monotonicity relative to deepest point: For any $F \in \mathrm{F}$ with center θ, $D(\mathbf{t}; F) \leq D(\theta + \alpha(\mathbf{t} - \theta); F)$ holds for $\alpha \in [0, 1]$; and
- Vanishing at infinity: $D(\mathbf{t}; F) \to 0$ as $\| \mathbf{t} \| \to \infty$, for each $F \in \mathrm{F}$.

A sample version of $D(\cdot; F)$ is denoted by $D(\cdot; F_m)$, where F_m is the sample or empirical distribution function. Several novel depth functions have appeared in the literature, including Mahalanobis (MD) (Mahalanobis [207]), half-space (HSD) or Tukey's depth (Tukey [342]), convex hull peeling (CH) (Mosler [243], simplicial volume (SV) or Oja's Depth (Oja [254]), majority (MJD) (Singh [314]), simplicial ($\mathrm{SD_{simp}}$) (Liu [199]), spatial ($DPTD$) (Vardi and Zhang [343], projection (PD) (Stahel [323] and Donoho and Gasko [78]), zonoid (ZD) (Koshevoy and Mosler [172] and Mosler [243]), spherical (SPD) (Elmore et al. [88], lens (LD) (Liu and Modarres [201]) depth functions, interpoint depth (ID) (Lok and Lee [202]). Some of these depth functions appear below. Depth functions are not only useful for ranking multivariate observations, they can also be used to define level sets. The sets $\{\mathbf{t} \in \mathbb{R}^d : D(\mathbf{t}, F) \geq \alpha, \alpha \geq 0\}$ are called the level sets of depth function D with respect to distribution function F.

8.5 Depth Functions and Contours

8.5.1 Mahalanobis Depth

Let F be a multivariate distribution function in $\mathbb{R}^d, d \geq 1$ and $\mathbf{X} = \{\mathbf{X}_1, \ldots, \mathbf{X}_m\}$ be a random sample from F. The Mahalanobis depth $MD(\mathbf{t}; F)$ is defined as

$$MD(\mathbf{t}; F) = \left[1 + (\mathbf{t} - \mu_F)\mathbf{\Sigma}_F^{-1}(\mathbf{t} - \mu_F)\right]^{-1},$$

where μ_F and $\mathbf{\Sigma}_F$ are the mean vector and dispersion matrix of F, respectively. The sample version of $MD(\mathbf{t}; F)$ is obtained by replacing μ_F and $\mathbf{\Sigma}_F$ with their sample estimates. The computational complexity of Mahalanobis depth is $O(md^2)$. To visualize the depth functions we plot the contour of Mahalanobis Depth for bivariate normal distribution $\mathbb{N}(\mathbf{0}, \mathbf{I}_2)$ in Figure 8.1 and plot the contour for bivariate uniform distribution $U(0, 1)$ in Figure 8.2.

8.5.2 Half Space Depth

Let $\mathbf{X} \in \mathbb{R}^d$ and let F be a probability distribution on $\mathbb{R}^d$. The half-space depth $HSD(\mathbf{t}; F)$ (Tukey [342]) is defined as

$$\begin{aligned} HSD(\mathbf{t}; F) &= \inf_H \{\mathbb{P}(H) : H \text{ is a closed half-space in } R^d \text{ and } \mathbf{t} \in H\} \\ &= \inf_{\mathbf{u}\in\mathbb{R}^d} \mathbb{P}_F(\{\mathbf{X} : \mathbf{u}^T\mathbf{X} \geq \mathbf{u}^T\mathbf{t}\}) = \inf_{\|\mathbf{u}\|=1} \mathbb{P}_F(\{\mathbf{X} : \mathbf{u}^T\mathbf{X} \geq \mathbf{u}^T\mathbf{t}\}). \end{aligned}$$

The sample version of $HSD(\mathbf{t}; F)$ is $HSD(\mathbf{t}; F_m) = \frac{1}{m} \min_{\|\mathbf{u}\|=1} \#\{i : \mathbf{u}^T\mathbf{X}_i \geq \mathbf{u}^T\mathbf{t}\}$. Consider a cloud of m point in R^2. Fix a point $\mathbf{t}$. For every line through $\mathbf{t}$, count the

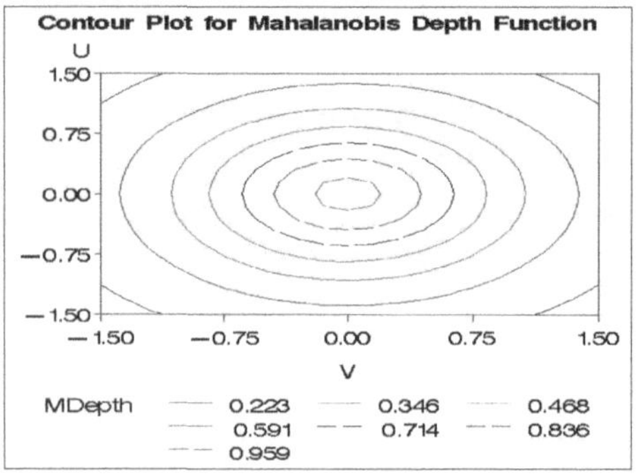

FIGURE 8.1: Mahalanobis: $\mathbb{N}(0, \mathbf{I}_2)$.

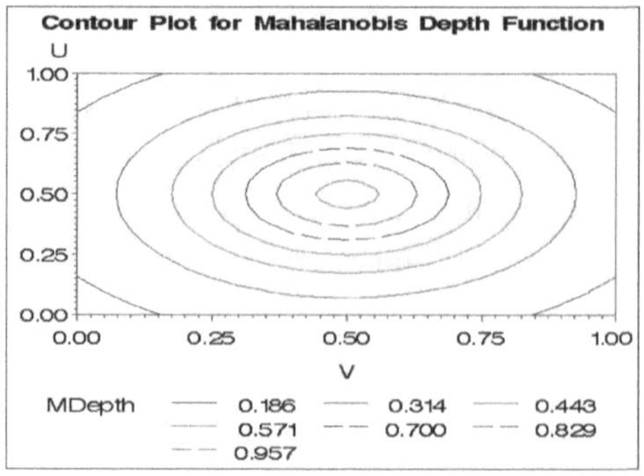

FIGURE 8.2: Mahalanobis: $U(0, 1)$.

number of points above and below the line. Find the minimum number over all lines and you obtain the $HSD(\mathbf{t}; F_m)$. This represents the smallest probability mass over all closed halfspaces that contain the point $\mathbf{X}$. It reflects how deep or central $\mathbf{X}$ is with respect to the distribution F. Some properties of the halfspace depth are

- $\text{HSD}(\mathbf{X}; F) \in [0, 1]$.
- Maximum depth occurs at the Tukey median (center point with maximum halfspace depth).
- $\text{HSD}(\mathbf{X}; F) \to 0$ as $\|\mathbf{X}\| \to \infty$.
- Halfspace depth is affine-invariant.

Example 8.5.1 *When $d = 1$ a halfspace is a ray. The halfspace depth becomes*

$$HSD(x; F) = \min\{F(x), 1 - F(x)\}.$$

Let F be the standard normal distribution. Then,$HSD(0; F) = \min\{\Phi(0), 1 - \Phi(0)\} = 0.5$. As $x \to \infty$, $HSD(x; F) \to 0$.

Example 8.5.2 *When $d = 2$ let F be the uniform distribution over the unit disk in $\mathbb{R}^2$. The halfspace depth is*

- *Maximal at the center (origin), where every halfspace through the point has at least half the probability mass.*
- *Decreasing as $\mathbf{X}$ moves toward the boundary.*

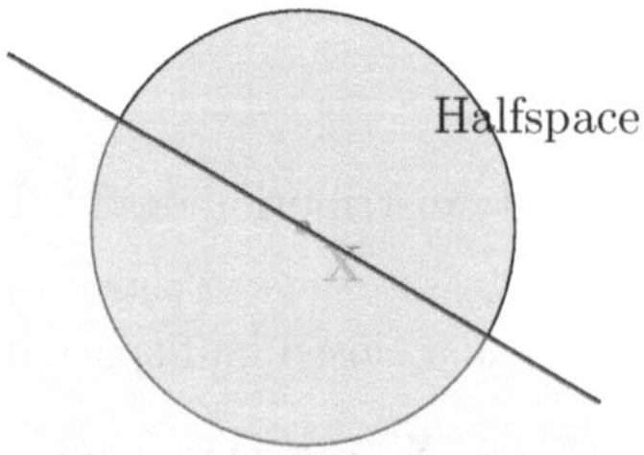

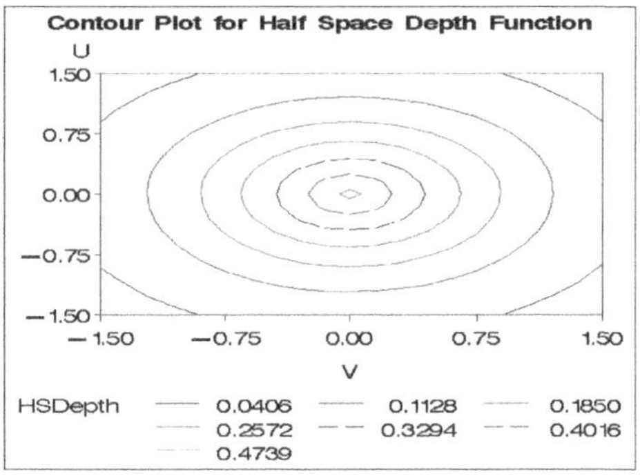

FIGURE 8.3: Halfspace: $\mathbb{N}(0, \mathbf{I}_2)$.

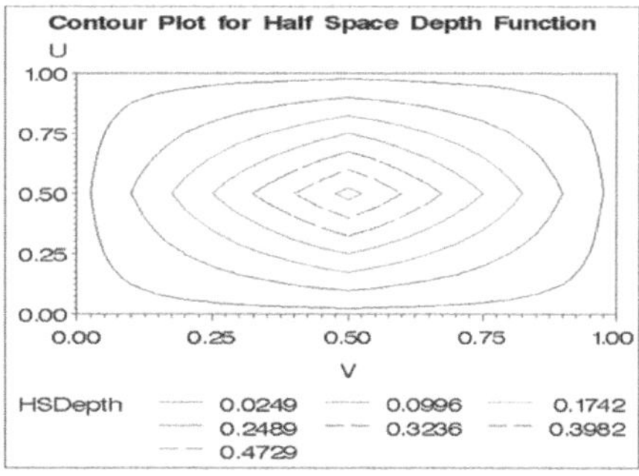

FIGURE 8.4: Halfspace: $U(0, 1)$.

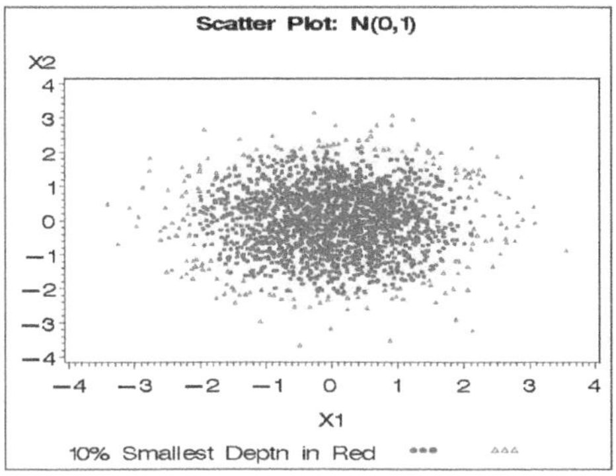

FIGURE 8.5: Normal Depth and outliers.

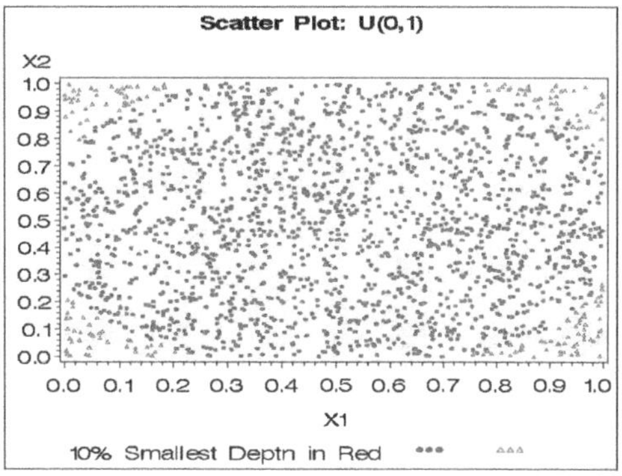

FIGURE 8.6: Uniform Depth and outliers.

The contour plot of half-space depth for bivariate normal distribution $\mathbb{N}(\mathbf{0}, \mathbf{I}_2)$ appears in Figure 8.3 and the half-space depth of bivariate uniform distribution $U(0, 1)$ in Figure 8.4.

Halfspace depth provides a robust, non-parametric measure of centrality in multivariate data. It is particularly useful in defining multivariate medians, constructing confidence regions, and in classification. Points with high-halfspace depth are centrally located, while those with low depth may be considered outliers. The computational complexity of halfspace depth is $O(m^{d-1} \log m)$.

The scatter plot for bivariate the bivariate normal distribution $\mathbb{N}(\mathbf{0}, \mathbf{I}_2)$ appears in Figure 8.5. The scatter plot for the bivariate uniform distribution appears in Figure 8.6. The scatter plot for the bivariate exponential distribution appears in Figure 8.7. In each Figure the red triangles constitute the 5% extreme values.

8.5.3 Simplicial Volume Depth

Let $\mathbf{X} = \{X_1, \ldots, X_m\} \subset \mathbb{R}^d$ be a data set in general position, and let $\mathbf{x} \in \mathbb{R}^d$. For any $(d+1)$-tuple $(X_{i_1}, \ldots, X_{i_{d+1}})$, denote by $\Delta(X_{i_1}, \ldots, X_{i_{d+1}})$ the simplex spanned by these points, and by $\mathrm{Vol}(\cdot)$ its d-dimensional volume.

The simplicial volume depth of $\mathbf{x}$ with respect to $\mathbf{X}$ is defined as

$$D_{\mathrm{SV}}(\mathbf{x}; \mathbf{X}) = \frac{\sum_{i_1 < \cdots < i_{d+1}} \mathrm{Vol}\big(\Delta(X_{i_1}, \ldots, X_{i_{d+1}})\big)\, \mathbb{I}\{\mathbf{x} \in \Delta(X_{i_1}, \ldots, X_{i_{d+1}})\}}{\sum_{i_1 < \cdots < i_{d+1}} \mathrm{Vol}\big(\Delta(X_{i_1}, \ldots, X_{i_{d+1}})\big)}.$$

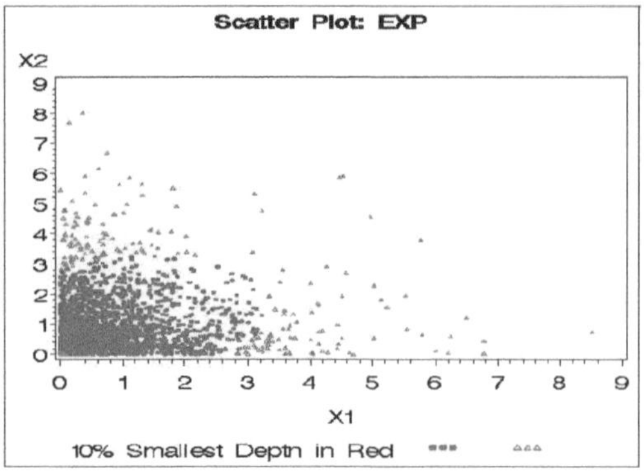

FIGURE 8.7: Exp. Depth and outliers.

Properties of simplicial volume depth are

1. For all $\mathbf{x} \in \mathbb{R}^d$, $0 \leq D_{\text{SV}}(\mathbf{x}; \mathbf{X}) \leq 1$.
2. Affine invariance. For any non-singular affine transformation $\mathbf{T}$, $D_{\text{SV}}(\mathbf{Tx}; \mathbf{TX}) = D_{\text{SV}}(\mathbf{x}; \mathbf{X})$. As a consequence of affine invariance, simplicial volume depth is invariant under translations, rotations, reflections, and uniform scalings.
3. For fixed $\mathbf{X}$, $\lim_{\|\mathbf{x}\| \to \infty} D_{\text{SV}}(\mathbf{x}; \mathbf{X}) = 0$.
4. The function $\mathbf{x} \mapsto D_{\text{SV}}(\mathbf{x}; \mathbf{X})$ is continuous on $\mathbb{R}^d$.
5. For any $\alpha \in [0, 1]$, the region $R_\alpha = \{\mathbf{x} : D_{\text{SV}}(\mathbf{x}; \mathbf{X}) \geq \alpha\}$ is convex. The depth regions $\{R_\alpha\}$ are nested: if $\alpha_1 > \alpha_2$, then $R_{\alpha_1} \subseteq R_{\alpha_2}$.
6. There exists at least one point $\mathbf{x}^*$ such that $D_{\text{SV}}(\mathbf{x}^*; \mathbf{X}) = \sup_{\mathbf{x} \in \mathbb{R}^d} D_{\text{SV}}(\mathbf{x}; \mathbf{X})$.
7. If the distribution is centrally symmetric about $\boldsymbol{\theta}$, then $\boldsymbol{\theta}$ is a deepest point. Let $\mathbf{x}^*$ be a deepest point. Along any ray starting from $\mathbf{x}^*$, the depth is non-increasing.
8. The breakdown point of the simplicial volume depth median is $\frac{1}{d+1}$, which is the same as Tukey and simplicial depth.
9. Computational complexity. A naive computation requires $O(m^{d+1})$ simplex volume evaluations, which is computationally prohibitive for large m or d.
10. A population simplicial volume depth can be defined via expectations over $(d+1)$-tuples. The empirical depth converges almost surely to its population counterpart.
11. If all simplices are given equal weight, D_{SV} reduces to classical simplicial depth. Thus, D_{SV} is a volume-weighted refinement of simplicial depth.

Example 8.5.3 *Let $\mathbf{X}_4 = \{(0,0), (1,0), (0,1), (1,1)\} \subset \mathbb{R}^2$ and let $\mathbf{z} = (0.5, 0.5)$. In $\mathbb{R}^2$, simplicial volume depth is based on all 2-simplices (triangles) formed by 3 data points.*
There are $\binom{4}{3} = 4$ distinct triangles,

$$T_1 = \text{conv}\{(0,0), (1,0), (0,1)\}, \quad T_2 = \text{conv}\{(0,0), (1,0), (1,1)\},$$
$$T_3 = \text{conv}\{(0,0), (0,1), (1,1)\}, \quad T_4 = \text{conv}\{(1,0), (0,1), (1,1)\}.$$

Each triangle is a right triangle with area

$$\text{Area}(T_k) = \frac{1}{2}, \qquad k = 1, \ldots, 4.$$

A direct geometric check shows that $\mathbf{z} = (0.5, 0.5)$ *lies in the interior of all four triangles. Therefore, the simplicial volume depth of* $\mathbf{z}$ *is*

$$D_{\text{SV}}(\mathbf{z}; \mathbf{X}_4) = \frac{\sum_{k=1}^{4} \text{Area}(T_k)\, \mathbb{I}\{\mathbf{z} \in T_k\}}{\sum_{k=1}^{4} \text{Area}(T_k)}[1ex] \quad = \frac{4 \cdot \frac{1}{2}}{4 \cdot \frac{1}{2}} = 1.$$

Hence, $\mathbf{z}$ *attains the maximum possible simplicial volume depth with respect to* $\mathbf{X}_4$.

Example 8.5.4 *Let* $\mathbf{X}_5 = \{X_1, \ldots, X_5\} \subset \mathbb{R}^2$ *be five points in general position lying on a circle centered at the origin. Consider the simplicial volume depth of* $\mathbf{z} = \mathbf{0}$.

In $\mathbb{R}^2$, *simplicial volume depth is based on all triangles formed by triples of data points. By symmetry, for any triangle* $\Delta(X_i, X_j, X_k)$ *that contains the origin, there exists a congruent triangle with the same area. Moreover, triangles containing the origin tend to have relatively large area, while triangles not containing the origin contribute zero to the depth.*

Hence, the numerator of

$$D_{\text{SV}}(\mathbf{0}; \mathbf{X}_5) = \frac{\sum_{i<j<k} \text{Area}(\Delta_{ijk})\, \mathbb{I}\{\mathbf{0} \in \Delta_{ijk}\}}{\sum_{i<j<k} \text{Area}(\Delta_{ijk})}$$

is maximized at the origin by rotational symmetry. Hence, $\mathbf{0}$ *is a deepest point with respect to simplicial volume depth, while points sufficiently far from the origin have depth zero.*

8.5.4 Oja Median

Given a data set $\mathbf{X}_m = \{\mathbf{X}_1, \ldots, \mathbf{X}_m\} \subset \mathbb{R}^d$, the Oja objective function is defined as

$$O(\mathbf{z}; \mathbf{X}_m) = \sum_{1 \le i_1 < \cdots < i_d \le m} \text{vol}_d(\text{conv}(\mathbf{z}, \mathbf{X}_{i_1}, \ldots, \mathbf{X}_{i_d})),$$

where $\text{vol}_d(\cdot)$ denotes the d-dimensional volume of a simplex.

Any point $\mathbf{z}^* \in \mathbb{R}^d$ that minimizes $O(\mathbf{z}; \mathbf{X}_m)$ is called an Oja median.

Example 8.5.5 *Let* $\mathbf{X}_3 = \{(0,0), (1,0), (0,1)\}$. *Any point inside triangle has a finite sum. Let* $\mathbf{z} = (z_1, z_2)$. *The area of a triangle with vertices* $(x_i, y_i), (x_j, y_j), (x_k, y_k)$ *is*

$$\mathit{Area} = \frac{1}{2}\left|\det\begin{pmatrix} x_i & y_i & 1 \\ x_j & y_j & 1 \\ x_k & y_k & 1 \end{pmatrix}\right|.$$

$$\mathit{Area}(\triangle(\mathbf{z}, \mathbf{x}_1, \mathbf{x}_2)) = \frac{1}{2}\left|\det\begin{pmatrix} z_1 & z_2 & 1 \\ 0 & 0 & 1 \\ 1 & 0 & 1 \end{pmatrix}\right| = \frac{1}{2}|z_2|,$$

$$\mathit{Area}(\triangle(\mathbf{z}, \mathbf{x}_1, \mathbf{x}_3)) = \frac{1}{2}\left|\det\begin{pmatrix} z_1 & z_2 & 1 \\ 0 & 0 & 1 \\ 0 & 1 & 1 \end{pmatrix}\right| = \frac{1}{2}|z_1|,$$

$$\mathit{Area}(\triangle(\mathbf{z}, \mathbf{x}_2, \mathbf{x}_3)) = \frac{1}{2}\left|\det\begin{pmatrix} z_1 & z_2 & 1 \\ 1 & 0 & 1 \\ 0 & 1 & 1 \end{pmatrix}\right| = \frac{1}{2}|z_1 + z_2 - 1|.$$

Thus the Oja sum is $O(\mathbf{z}) = \frac{1}{2}\left(|z_1| + |z_2| + |z_1 + z_2 - 1|\right)$.

Minimizing this expression inside the triangle formed by the points, we find that the Oja median for the three points $(0,0),\ (1,0),\ (0,1)$ *is*

$$\mathbf{z} = \frac{\mathbf{x}_1 + \mathbf{x}_2 + \mathbf{x}_3}{3} = \left(\frac{0+1+0}{3}, \frac{0+0+1}{3}\right) = \left(\frac{1}{3}, \frac{1}{3}\right).$$

8.5.5 Simplicial Depth

Let $\mathbf{X} \in \mathbb{R}^d$ and let F be a probability distribution on $\mathbb{R}^d$. The simplicial depth (Liu [199]) of $\mathbf{X}$ with respect to F is defined as the probability that $\mathbf{X}$ lies inside a random d-dimensional simplex formed by $d+1$ independent samples from F,

$$\mathrm{SD}_{\mathrm{simp}}(\mathbf{X}; F) = \mathbb{P}\left(\mathbf{X} \in \mathrm{conv}(\mathbf{X}_1, \ldots, \mathbf{X}_{d+1})\right),$$

where $\mathrm{conv}(\cdot)$ denotes the convex hull, and $\mathbf{X}_1, \ldots, \mathbf{X}_{d+1}$ are i.i.d. random vectors from F. For empirical data, let $\mathbf{X}_1, \ldots, \mathbf{X}_m$ be m sample points in $\mathbb{R}^d$. The empirical simplicial depth of $\mathbf{X}$ is given by

$$\mathrm{SD}_{\mathrm{simp}}(\mathbf{X}; \hat{F}_m) = \frac{1}{\binom{m}{d+1}} \sum_{1 \le i_1 < \ldots < i_{d+1} \le m} \mathbb{I}\left(\mathbf{X} \in \mathrm{conv}(\mathbf{X}_{i_1}, \ldots, \mathbf{X}_{i_{d+1}})\right).$$

Example 8.5.6 *In one dimension, a simplex is just an interval defined by two points. The simplicial depth of* x *is*

$$SD_{simp}(x; F) = \mathbb{P}\left(\min(X_1, X_2) \le x \le \max(X_1, X_2)\right) = 2F(x)(1 - F(x)),$$

where F *is the cumulative distribution function (CDF) of the univariate distribution. Let* F *be the uniform distribution on* $[0,1]$. *It follows that* $SD_{simp}(x; F) = 2x(1-x), \quad x \in [0,1]$.

Example 8.5.7 *When* $d = 2$ *let* $\mathbf{X}$ *follow the uniform distribution over the unit square* $[0,1]^2$. *To compute the simplicial depth of a point* $\mathbf{X} \in \mathbb{R}^2$ *empirically, we sample many triplets* $(\mathbf{X}_i, \mathbf{X}_j, \mathbf{X}_k)$ *from the data and count the proportion of triangles* $conv(\mathbf{X}_i, \mathbf{X}_j, \mathbf{X}_k)$ *that contain* $\mathbf{X}$. *The depth is highest near the geometric center of the distribution (around* $(0.5, 0.5)$*) and decreases toward the boundaries.*

Some properties of simplicial depth are

- $\mathrm{SD}_{\mathrm{simp}}(\mathbf{X}; F) \in [0,1]$ and the depth decays to 0 as $\|\mathbf{X}\| \to \infty$.
- The depth is maximized at points deep inside the distribution.
- Simplicial depth is affine-invariant.

Simplicial depth measures the likelihood that a point lies inside random simplices formed by the data. It provides a robust, non-parametric notion of data depth useful in classification, and outlier detection. Unlike some other depths, it reflects the geometric structure of the data directly. For example, given m points in R^2, the SD of a data point $\mathbf{t}$ is the number of triangles that contain t divided by $\binom{m}{3}$. For $d \ge 4$, there is no known algorithm faster than $O(m^{d+1})$ brute force approach requiring generation of all simplices and calculating the proportion that contain a given $\mathbf{t}$. The computational complexity of simplicial depth is $O(m^{d+1})$. The contour plot of the simplicial depth for $N(0, \mathbf{I}_2)$ appears in Figure 8.8.

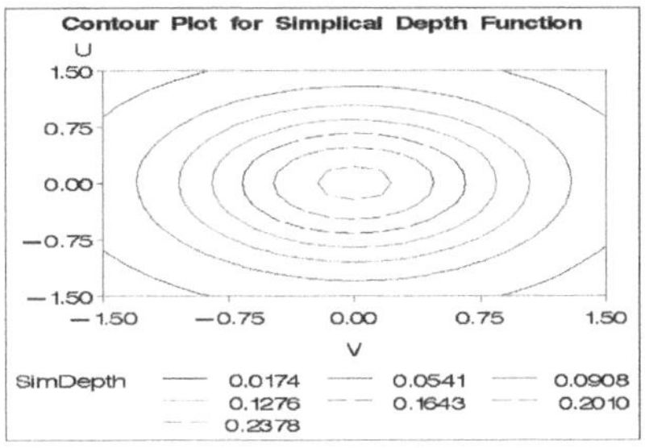

FIGURE 8.8: Simplicial Depth: $N(0, \mathbf{I}_2)$.

8.5.6 Projection Depth

Let $\mathbf{X} \in \mathbb{R}^d$ and let F be a probability distribution on $\mathbb{R}^d$. The projection depth of $\mathbf{X}$ with respect to F is defined as

$$\text{PD}(\mathbf{X}; F) = \left(1 + \sup_{\mathbf{u} \in \mathbb{S}^{d-1}} \frac{|\mathbf{u}'\mathbf{X} - \text{Med}(\mathbf{u}'\mathbf{X})|}{\text{MAD}(\mathbf{u}'\mathbf{X})}\right)^{-1},$$

where,

- $\mathbb{S}^{d-1}$ is the unit sphere in $\mathbb{R}^d$,
- Med$(\cdot)$ is the univariate median,
- MAD$(\cdot)$ is the median absolute deviation.

This measures how far a point lies from the center of the data in the most extreme projected direction, scaled by dispersion (Zuo and Serfling [372]).

Example 8.5.8 *In one dimension, projection depth becomes*

$$PD(x; F) = \left(1 + \frac{|x - Med(X)|}{MAD(X)}\right)^{-1}.$$

Let F be the standard normal distribution. Then, we have

$$Med(X) = 0, \quad MAD(X) \approx 0.6745, \quad PD(x; F) = \left(1 + \frac{|x|}{0.6745}\right)^{-1}.$$

Example 8.5.9 *Let $d = 2$. To compute $PD(\mathbf{X}; F)$ for F on $\mathbb{R}^2$,*

1. *Project all data and the point $\mathbf{X}$ onto a direction $\mathbf{u}$.*
2. *Compute the univariate median and MAD in that direction.*
3. *Calculate the standardized distance of the projection of $\mathbf{X}$.*
4. *Maximize this over all directions $\mathbf{u}$.*

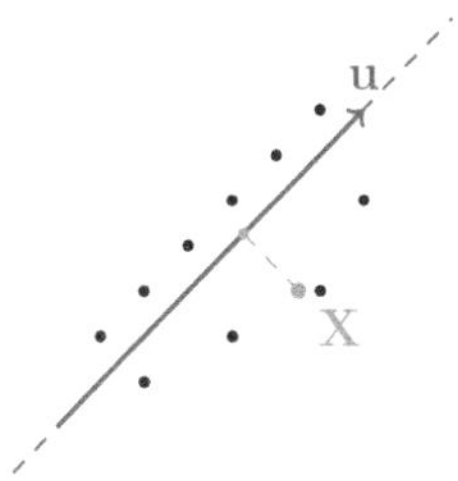

Some properties of projection depth are

- $0 < \text{PD}(\mathbf{X}; F) \leq 1$.
- Maximum depth is achieved at the projection median.
- Projection depth is affine-invariant.
- Sensitive to both position and scale outliers.

Projection depth provides a robust measure of how central a point is by looking at its worst standardized deviation in all directions. It is computationally more intensive but highly informative in identifying both position and dispersion anomalies. Useful in robust estimation, outlier detection, and data visualization.

- Each projection of m data points and x onto a direction u takes $O(nd)$.
- Computing median and MAD each takes $O(m)$.
- A brute-force or grid-based approximation of the unit sphere with t directions requires repeating the process t times.
- The exact supremum over all directions is computationally intractable for large d, so t random or deterministic directions (e.g., using spherical caps) are sampled in practice.
- The approximation improves with increasing t, which can be a function of d.
- The method is robust and computationally feasible with moderate m, d, and t.

8.5.7 Spatial Depth

Let $\mathbf{x} \in \mathbb{R}^d$ be a point and let F be a probability distribution on $\mathbb{R}^d$. The spatial depth (Vardi, and Zhang [343]) of $\mathbf{x}$ with respect to F is defined as

$$\text{SD}(\mathbf{X}; F) = 1 - \left\| E_F \left[\frac{\mathbf{X} - \mathbf{x}}{\|\mathbf{X} - \mathbf{x}\|} \right] \right\|,$$

where the expectation is taken over the distribution of the random vector $\mathbf{X} \sim F$, and the norm is the Euclidean norm. This measure reflects the centrality of the point $\mathbf{X}$ with respect to the distribution F. Points near the center of F will have higher depth, while those farther away will have lower depth. Some properties of spatial depth are

- $0 \leq \text{SD}(\mathbf{X}; F) \leq 1$ and $\text{SD}(\mathbf{X}; F) \to 0$ as $\|\mathbf{X}\| \to \infty$
- $\text{SD}(\mathbf{X}; F) = 1$ if and only if $\mathbf{X}$ is the spatial median of F.

Example 8.5.10 *When $d = 1$ let F be the standard normal distribution $\mathbb{N}(0, 1)$. Then the spatial depth of a point $x \in \mathbb{R}$ is*

$$SD(x; F) = 1 - \left| E_F \left[\frac{X - x}{|X - x|} \right] \right| = 1 - |\mathbb{P}(X > x) - \mathbb{P}(X < x)|.$$

Since F is symmetric around 0, the depth is maximized at $x = 0$,

$$SD(0; F) = 1 - |0.5 - 0.5| = 1.$$

At $x = 2$, $\mathbb{P}(X > 2) \approx 0.0228$, and $\mathbb{P}(X < 2) \approx 0.9772$,

$$SD(2; F) = 1 - |0.0228 - 0.9772| = 1 - 0.9544 = 0.0456.$$

Example 8.5.11 *When $d = 2$ let F be the uniform distribution over the unit circle in $\mathbb{R}^2$, i.e., $\mathbf{X} \sim Unif(\{\mathbf{X} \in \mathbb{R}^2 : \|\mathbf{X}\| \leq 1\})$. At the origin $\mathbf{0}$,*

$$E\left[\frac{\mathbf{X}}{\|\mathbf{X}\|}\right] = \mathbf{0}, \quad so \quad SD(\mathbf{0}; F) = 1.$$

At $\mathbf{X} = (0.5, 0)$,

$$SD(\mathbf{X}; F) = 1 - \left\| E\left[\frac{\mathbf{X} - \mathbf{X}}{\|\mathbf{X} - \mathbf{X}\|}\right] \right\| < 1.$$

The closer $\mathbf{X}$ is to the boundary of the support, the lower the spatial depth.

Spatial depth measures how deep or central a point is within a distribution. It generalizes quantiles and medians to multivariate settings and is used in robust statistics, classification, and multivariate data visualization. The computational complexity of the spatial depth is md.

- Spatial depth is computationally efficient and suitable for high-dimensional data.
- Unlike halfspace or simplicial depths, which have higher complexities (e.g., $O(m^d)$ or $O(m^3)$ in 2D), spatial depth is easy to compute.
- It is robust to outliers but not as deeply robust as some other depth functions.

8.5.8 Spherical Depth

Let $\mathbf{X} = \{\mathbf{X}_1, \mathbf{X}_2, \ldots, \mathbf{X}_m\} \subset \mathbb{R}^d$ be a set of points. The spherical depth (Elmore, Hettmansperger and Xuan [88]) of a point $\mathbf{x} \in \mathbb{R}^d$ with respect to $\mathbf{X}$ is defined as the proportion of all closed hyperspheres (in d-dimensional space) with diameter determined by pairs of points $(\mathbf{X}_i, \mathbf{X}_j)$ such that $\mathbf{x}$ lies within the hypersphere. Formally, the spherical depth is defined as

$$DPTD(\mathbf{x}; \mathbf{X}) = \frac{2}{m(m-1)} \sum_{1 \leq i < j \leq m} \mathbb{I}\left\{\|\mathbf{x} - \frac{1}{2}(\mathbf{X}_i + \mathbf{X}_j)\| \leq \frac{1}{2}\|\mathbf{X}_i - \mathbf{X}_j\|\right\},$$

where $\mathbb{I}\{\cdot\}$ is the indicator function. Each term checks whether $\mathbf{x}$ lies in the closed hypersphere having $\mathbf{X}_i$ and $\mathbf{X}_j$ as endpoints of a diameter. Properties of spherical depth are

1. Boundedness. For all $\mathbf{x} \in \mathbb{R}^d$, $\quad 0 \leq DPTD(\mathbf{x}; \mathbf{X}) \leq 1$.
2. Translation invariance. For any $\mathbf{a} \in \mathbb{R}^d$, $\quad DPTD(\mathbf{x} + \mathbf{a}; \mathbf{X} + \mathbf{a}) = DPTD(\mathbf{x}; \mathbf{X})$.
3. Orthogonal invariance. For any orthogonal matrix $\mathbf{Q}$, $\quad DPTD(\mathbf{Q}\mathbf{x}; \mathbf{Q}\mathbf{X}) = DPTD(\mathbf{x}; \mathbf{X})$.
4. Scale invariance. For any scalar $c > 0$, $\quad DPTD(c\mathbf{x}; c\mathbf{X}) = DPTD(\mathbf{x}; \mathbf{X})$.
5. Vanishing at infinity. For fixed $\mathbf{X}$, $\quad \lim_{\|\mathbf{x}\| \to \infty} DPTD(\mathbf{x}; \mathbf{X}) = 0$.
6. For $\alpha \in [0, 1]$, the region $R_\alpha = \{\mathbf{x} : DPTD(\mathbf{x}; \mathbf{X}) \geq \alpha\}$ is convex.
7. Existence of deepest points. There exists at least one point $\mathbf{x}^*$ maximizing $DPTD(\mathbf{x}; \mathbf{X})$.
8. If $\mathbf{X}$ is centrally symmetric about $\boldsymbol{\theta}$, then $\boldsymbol{\theta}$ is a deepest point.
9. Monotonicity from the center. Let $\mathbf{x}^*$ be a deepest point. Along any ray starting at $\mathbf{x}^*$, the depth is non-increasing.

10. A direct evaluation of $DPTD(\mathbf{x}; \mathbf{X})$ requires $O(m^2 d)$ operations.

11. The empirical spherical depth converges almost surely to its population counterpart.

Example 8.5.12 *Consider the points $\mathbf{X} = \{-2, 0, 2\} \subset \mathbb{R}$. Let $x = 0$. The pairs of points are*

- $(-2, 0)$*: midpoint* -1*, radius* $1 \Rightarrow 0$ *not in interval* $[-2, 0]$
- $(-2, 2)$*: midpoint* 0*, radius* $2 \Rightarrow 0$ *in interval* $[-2, 2]$
- $(0, 2)$*: midpoint* 1*, radius* $1 \Rightarrow 0$ *not in interval* $[0, 2]$

So $DPTD(0; \mathbf{X}) = \frac{1}{3}$.

Example 8.5.13 *Consider $d = 2$. Let $\mathbf{X} = \{(0,0), (2,0), (1, \sqrt{3})\}$ and $x = (1, 1)$. Consider the disks defined by all pairs*

- $(0,0)$ *and* $(2,0)$*: center* $(1,0)$*, radius* $1 \Rightarrow x$ *lies in circle since distance* $\sqrt{1} < 1$
- $(0,0)$ *and* $(1, \sqrt{3})$*: center* $(0.5, \sqrt{3}/2)$*, radius* $1 \Rightarrow x$ *lies in circle*
- $(2,0)$ *and* $(1, \sqrt{3})$*: similar result* $\Rightarrow x$ *lies in circle*

So $DPTD(\mathbf{x}; \mathbf{X}) = 1$*. See Figure 8.9.*

To color the triangle vertices blue, the query point x red, and the circles black, here is the modified LaTeX code:

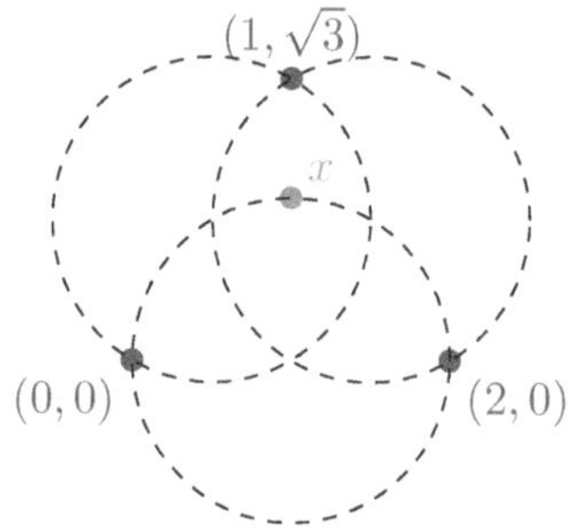

FIGURE 8.9: Illustration of Spherical Depth in 2D.

Computing spherical depth involves evaluating $\binom{m}{2} = \frac{m(m-1)}{2}$ pairs of points to determine if x lies in the corresponding ball. Hence, the time complexity is $O(m^2)$ for a single point x. Efficient implementations or approximate versions can be used in practice to handle larger datasets.

8.5.9 Convex Hull Peeling Depth

Let $\mathbf{X} = \{\mathbf{X}_1, \mathbf{X}_2, \ldots, \mathbf{X}_m\} \subset \mathbb{R}^d$ be a set of points. The convex hull peeling depth (Barnett [22]) of a point $\mathbf{x} \in \mathbb{R}^d$ is defined as the number of layers of convex hulls required to reach $\mathbf{x}$ in a recursive peeling process. In each step, the outermost convex hull of the remaining points is removed. Formally, define $\mathbf{X}^{(0)} = \mathbf{X}$, and recursively

$$\text{Hull}^{(k)} = \text{ConvexHull}(\mathbf{X}^{(k)})$$

$$\mathbf{X}^{(k+1)} = \mathbf{X}^{(k)} \setminus \text{Hull}^{(k)}.$$

The peeling depth of $\mathbf{x}$ is the smallest k such that $\mathbf{x} \in \text{Hull}^{(k)}$. If $\mathbf{x} \notin \text{ConvexHull}(\mathbf{X})$, then $D_{\text{peel}}(\mathbf{x}; \mathbf{X}) = 0$. Properties of convex hull peeling depth are

1. Integer-valued depth. $D_{\text{CH}}(\mathbf{x}; \mathbf{X})$ takes values in $\{0, 1, 2, \ldots, K\}$, where K is the total number of peeling layers. Points outside $\text{conv}(\mathbf{X})$ have depth 0.
2. Affine invariance. For any non-singular affine transformation T, $D_{\text{CH}}(T\mathbf{x}; T\mathbf{X}) = D_{\text{CH}}(\mathbf{x}; \mathbf{X})$.
3. Convex hull peeling is invariant under translations, rotations, reflections, and uniform scaling.
4. Vanishing at infinity. If $\|\mathbf{x}\| \to \infty$, then $D_{\text{CH}}(\mathbf{x}; \mathbf{X}) = 0$.
5. Convexity of depth regions. For $k \geq 1$, the region $R_k = \{\mathbf{x} : D_{\text{CH}}(\mathbf{x}; \mathbf{X}) \geq k\}$ is a convex set equal to the k-th peeled convex hull $H^{(k)}$.
6. The depth regions are nested: $H^{(1)} \supseteq H^{(2)} \supseteq \cdots \supseteq H^{(K)}$.
7. There exists at least one point attaining maximal depth K. Such points lie in the innermost convex hull.
8. Maximality at center of symmetry. If the data set is centrally symmetric about $\boldsymbol{\theta}$, then $\boldsymbol{\theta}$ lies in the deepest layer.
9. Moving away from a deepest point along any ray, the depth does not increase.
10. The breakdown point of the deepest peeling estimator is low: in $\mathbb{R}^d$, it is of order $1/m$. Thus, convex hull peeling is not robust. A single extreme outlier can significantly change the outer hull and hence the peeling structure.
11. For all $\mathbf{x}$, $D_{\text{HSD}}(\mathbf{x}; \mathbf{X}) \leq D_{\text{CH}}(\mathbf{x}; \mathbf{X})$, so convex hull peeling provides an upper bound on Tukey depth.

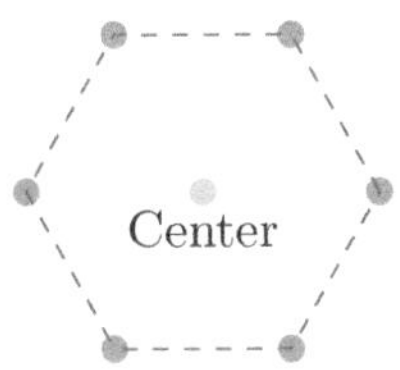

FIGURE 8.10: Convex Hull Peeling: Outer layer (dashed) and center point.

Example 8.5.14 *Let X be 7 points: 6 vertices of a regular hexagon and the center point. See Figure 8.10. The center is not on the convex hull, but all 6 other points form the first convex hull. After removing those, the center remains and forms the next hull. Thus,*

- *All 6 outer points have depth 1.*
- *The center point has depth 2.*

Example 8.5.15 *Consider a set X of 20 random points in $\mathbb{R}^2$. The outermost layer of points forms the convex hull (depth 1). Remove those, and repeat until all points are assigned a depth. A point at the center of the cluster might have depth 3 or 4.*

Each convex hull can be computed in $O(m \log m)$ time in 2D. In the worst case, if only one point is removed per layer, this process could take up to n layers. Thus, the total complexity is $O(m^2 \log m)$ in the worst case, though typically fewer layers are needed.

8.5.10 IPD Depth

Lok and Lee [202] define a statistical depth function based on IPDs, which accommodates multimodal distributions. Let $\delta(\cdot,\cdot)$ be a measurable distance function (e.g., Euclidean). Define

$$\Lambda(t|\mathbf{X},F) = \mathbb{P}_F\left(\delta(\mathbf{X},\mathbf{X}) \le t\right), \quad \text{where } \mathbf{X} \sim F.$$

Let $\mathbf{X},\mathbf{Y} \sim F$ be independent. Define the IPD distribution function as

$$\Lambda^*(t|F) = \mathbb{P}_F\left(\delta(\mathbf{X},\mathbf{Y}) \le t\right) = \mathbb{E}_F\left[\Lambda(t|\mathbf{X},F)\right].$$

Let Λ^{-1} and $(\Lambda^*)^{-1}$ be the respective generalized inverses. The IPD depth is defined as

$$ID(\mathbf{X};F) = \left\{1 + \frac{\Lambda^{-1}(\gamma|\mathbf{X},F)}{(\Lambda^*)^{-1}(1-\epsilon|F)}\right\}^{-1},$$

where $\gamma, \epsilon \in (0,1)$ control the centrality and scaling.

The empirical version is

$$ID_m(\mathbf{X}) = \left(\frac{1 + \inf\left\{t \,\middle|\, \frac{1}{m}\sum_{i=1}^m \mathbf{1}_{\{\delta(\mathbf{X},\mathbf{X}_i)\le t\}} \ge \gamma\right\}}{\inf\left\{t \,\middle|\, \frac{1}{\binom{m}{2}}\sum_{1\le i<j\le m} \mathbf{1}_{\{\delta(\mathbf{X}_i,\mathbf{X}_j)\le t\}} \ge 1-\epsilon\right\}}\right)^{-1}.$$

Example 8.5.16 *Let F be a mixture of two 2D normals centered at $(0,0)$ and $(5,5)$, a bimodal density. The IPD depth correctly assigns low depth to points in the low-density region between the modes and high depth near each mode. See Figure 8.11.*

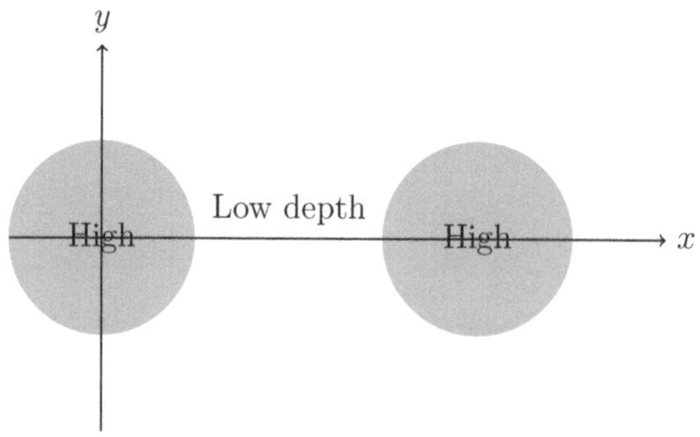

FIGURE 8.11: Illustration of IPD Depth in Bimodal Setting.

Example 8.5.17 *An annulus is the region between two concentric circles. Let F be uniform over an annulus centered at origin. A point at the center of the annulus has low IPD depth, despite being geometrically central, due to the low density in its neighborhood.*

Example 8.5.18 *Consider the dataset $\mathbf{X} = \{(0,0),(2,0),(0,2),(2,2),(1,1)\} \subset \mathbb{R}^2$, and let the test point be $\mathbf{z} = (1,1)$. We use Euclidean distance and parameters $\gamma = 0.6$, $\epsilon = 0.2$.*

To compute the distances from $\mathbf{z}$ to sample points,

$$\begin{aligned}
&\|\mathbf{z}-\mathbf{X}_1\| = \sqrt{2} \approx 1.414, \quad \|\mathbf{z}-\mathbf{X}_2\| = \sqrt{2} \approx 1.414,\\
&\|\mathbf{z}-\mathbf{X}_3\| = \sqrt{2} \approx 1.414, \quad \|\mathbf{z}-\mathbf{X}_4\| = \sqrt{2} \approx 1.414,\\
&\|\mathbf{z}-\mathbf{X}_5\| = 0
\end{aligned}$$

Sorted distances: $0, 1.414, 1.414, 1.414, 1.414$. *For* $\gamma = 0.6$*, 3rd smallest distance is* $t_\gamma \approx 1.414$

Pairwise distances in the sample are

$$\|\mathbf{X}_1 - \mathbf{X}_2\| = 2, \qquad \|\mathbf{X}_1 - \mathbf{X}_3\| = 2, \quad \|\mathbf{X}_1 - \mathbf{X}_4\| = \sqrt{8} \approx 2.828,$$
$$\|\mathbf{X}_1 - \mathbf{X}_5\| = \sqrt{2} \approx 1.414, \qquad \ldots$$

Sorted pairwise distances: $1.414, 1.414, 1.414, 1.414, 2, 2, 2, 2.828, 2.828, \ldots$
For $1 - \epsilon = 0.8$*, the 8th smallest distance is* $t_{1-\epsilon} \approx 2.828$.

Computing the empirical IPD,

$$ID_5(\mathbf{z}) = \left(\frac{1 + t_\gamma}{t_{1-\epsilon}}\right)^{-1} = \frac{t_{1-\epsilon}}{1 + t_\gamma} \approx \frac{2.828}{1 + 1.414} \approx 0.854$$

The central point $(1, 1)$ *has high IPD relative to the sample, reflecting its centrality. Points near the edges would have smaller IPD values due to larger relative distances.*

Let m be the number of data points in dimension d. Computing $\delta(\mathbf{X}, \mathbf{X}_i)$ for all i requires $O(md)$. Computing all $\binom{m}{2}$ IPDs requires $O(m^2 d)$. Sorting the distances to obtain quantile-based thresholds: $O(m \log m)$ (for each individual or pairwise set). Hence, The total time complexity of IPD depth is $O(m^2 d)$ due to the pairwise computations.

Some properties of IPD depth are

- Affine Invariance: Yes, if δ is chosen appropriately (e.g., Mahalanobis distance).
- Robust to Multimodality: Effectively captures local density.
- No Center Bias: Unlike Mahalanobis or Tukey depth.
- Scale Invariance: Controlled by denominator quantile $(\Lambda^*)^{-1}(1 - \epsilon)$.
- Computationally Intensive: Due to $O(n^2)$ pairwise distances.

8.5.11 Lens Depth

For a continuous distribution F on $\mathbb{R}^d$, let $L(\mathbf{X}_1, \mathbf{X}_2)$ denote the hyper-lens defined by the intersection of two identical closed hyper-spheres centered at $\mathbf{X}_1$ and $\mathbf{X}_2$, respectively, with radius $\|\mathbf{X}_1 - \mathbf{X}_2\|$. The hyper-lens of any two i.i.d. random vectors $\mathbf{X}_1$ and $\mathbf{X}_2$ in $\mathbb{R}^d$ is defined as $L(\mathbf{X}_1, \mathbf{X}_2) = B(\mathbf{X}_1, \|\mathbf{X}_1 - \mathbf{X}_2\|) \cap B(\mathbf{X}_2, \|\mathbf{X}_1 - \mathbf{X}_2\|)$ where $B(\mathbf{c}, r)$ is the closed ball of radius r centered at $\mathbf{c}$.

Suppose $\mathbf{X}_1$ and $\mathbf{X}_2$ are two i.i.d. random vectors from a continuous distribution function F in $\mathbb{R}^d$. Consider two intersecting closed balls with radius $\|\mathbf{X}_1 - \mathbf{X}_2\|$, centered at $\mathbf{X}_1$ and $\mathbf{X}_2$, respectively. Denote this intersection with $L(\mathbf{X}_1, \mathbf{X}_2)$. That is,

$$L(\mathbf{X}_1, \mathbf{X}_2) = B(\mathbf{X}_1, \|\mathbf{X}_1 - \mathbf{X}_2\|) \cap B(\mathbf{X}_2, \|\mathbf{X}_1 - \mathbf{X}_2\|),$$

where $B(\mathbf{c}, r)$ is the closed ball of radius r centered at $\mathbf{c}$. For bivariate distributions, Figure 8.12 illustrates how these two closed circles partition the two-dimensional Euclidean space into three regions: The lens area (Region 1), the lunar areas (Region 2) and the area outside the discs (Region 3). Note that $L(\mathbf{X}_1, \mathbf{X}_2)$ reduces to an interval when $d = 1$.

Consider the triangle $\Delta(\mathbf{t}, \mathbf{X}_1, \mathbf{X}_2)$ with lengths $\|\mathbf{X}_1 - \mathbf{X}_2\|$, $\|\mathbf{t} - \mathbf{X}_1\|$ and $\|\mathbf{t} - \mathbf{X}_2\|$ where $\mathbf{t} \in \mathbb{R}^d$. One can verify that $\|\mathbf{X}_1 - \mathbf{X}_2\|$ is the longest, middle and shortest side in $\Delta(\mathbf{t}, \mathbf{X}_1, \mathbf{X}_2)$ if and only if $\mathbf{t}$ falls in Region 1, Region 2 and Region 3, respectively.

For any vector $\mathbf{t} \in \mathbb{R}^d$, the probability $\mathbb{P}(\mathbf{t} \in L(\mathbf{X}_1, \mathbf{X}_2))$ provides a measure of closeness of $\mathbf{t}$ with respect to F since the points inside $L(\mathbf{X}_1, \mathbf{X}_2)$ are close to both $\mathbf{X}_1$ and $\mathbf{X}_2$; hence, close to F. Naturally, the higher the probability, the closer $\mathbf{t}$ is with respect to the center of F.

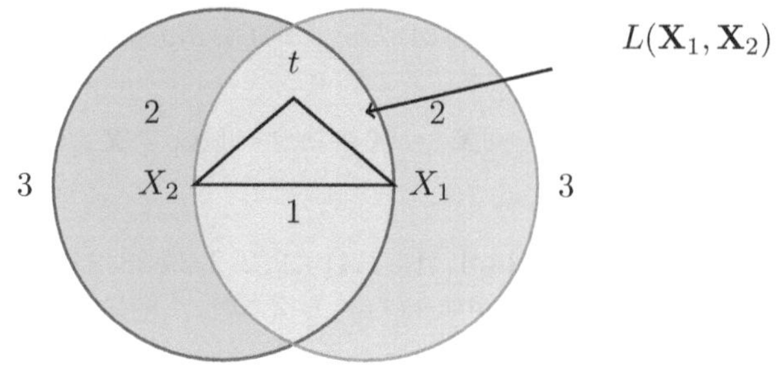

FIGURE 8.12: Regions defining $L(\mathbf{X}_1, \mathbf{X}_2)$ in $\mathbb{R}^2$.

Definition 8.5.1 *: The lens data depth function $LD(\mathbf{t}; F)$ (Liu and Modarres [201]) of a vector $\mathbf{t} \in \mathbb{R}^d$, with respect to a distribution function F on $\mathbb{R}^d$, is the probability that $\mathbf{t}$ is contained in the random hyper-lens $L(\mathbf{X}_1, \mathbf{X}_2)$. That is,*

$$LD(\mathbf{t}; F) = \mathbb{P}(\mathbf{t} \in L(\mathbf{X}_1, \mathbf{X}_2)),$$

where $\mathbf{X}_1$ and $\mathbf{X}_2$ are i.i.d. from F.

Definition 8.5.2 *: Let $\mathbf{X}_1, \ldots, \mathbf{X}_m$ be an i.i.d. random sample from a continuous distribution function F on $\mathbb{R}^d$. The sample $LD(\mathbf{t}; F)$ is the proportion of $L(\mathbf{X}_i, \mathbf{X}_j), 1 \leq i < j \leq m$, that contain $\mathbf{t}$. That is,*

$$LD(\mathbf{t}; F_m) = \frac{1}{\binom{m}{2}} \sum_{i<j}^{m} \mathbb{I}[\mathbf{t} \in L(\mathbf{X}_i, \mathbf{X}_j)],$$

where $\mathbb{I}(A)$ is the indicator function for event A, i.e. $\mathbb{I}(A) = 1$ if event A occurs and $\mathbb{I}(A) = 0$ otherwise.

For example the sample minimum (or maximum) is included in the $m - 1$ lenses that it makes with the other observations so the lens depth of the minimum (or maximum) is 2/m. The maximum value of the sample lens depth is 1 and is reached by any sample point that is contained in all $\binom{m}{2}$ lenses. Since $LD(\mathbf{t}; F)$ is a probability, we have $0 \leq LD(\mathbf{t}; F) \leq 1$. Note that $LD(\mathbf{t}; F_m)$ is an order-two U-statistic with the kernel function

$$h(\mathbf{X}_1, \mathbf{X}_2) = \mathbb{I}[\mathbf{t} \in L(\mathbf{X}_1, \mathbf{X}_2)] = \mathbb{I}[\|\mathbf{X}_1 - \mathbf{X}_2\| > \max(\|\mathbf{t} - \mathbf{X}_1\|, \|\mathbf{t} - \mathbf{X}_2\|)].$$

Hence, $LD(\mathbf{t}; F)$ is the probability that $\|\mathbf{X}_1 - \mathbf{X}_2\|$ is the longest side of the triangle $\Delta(\mathbf{t}, \mathbf{X}_1, \mathbf{X}_2)$. This observation is important for computing the sample $LD(\mathbf{t}; F_m)$ and proving some of the properties of the $LD(\mathbf{t}; F)$. One can find accurate estimates of $LD(\mathbf{t}; F)$ by its parametric bootstrap or the proportion of random hyper-lenses $L(\mathbf{X}_i, \mathbf{X}_j)$'s that contain $\mathbf{t}$ where $\mathbf{X}_i$ and $\mathbf{X}_j$ are simulated repeatedly from F.

Example 8.5.19 *Consider the six points in $\mathbb{R}^3$,*

$$\begin{aligned} \mathbf{X}_1 &= (0,0,0), & \mathbf{X}_4 &= (0,0,1), \\ \mathbf{X}_2 &= (1,0,0), & \mathbf{X}_5 &= (1,1,1), \\ \mathbf{X}_3 &= (0,1,0), & \mathbf{X}_6 &= (2,2,2) \end{aligned}$$

The lens depth of a point $\mathbf{t}$ with respect to $\mathbb{X} = \{\mathbf{X}_1, \ldots, \mathbf{X}_6\}$ is

$$LD(\mathbf{t}; \mathbb{X}) = \frac{1}{\binom{6}{2}} \sum_{1 \leq i < j \leq 6} \mathbb{I}\big[\mathbf{t} \in L(\mathbf{X}_i, \mathbf{X}_j)\big],$$

where $\mathbf{t} \in L(\mathbf{X}_i, \mathbf{X}_j)$ *if and only if* $\max(\|\mathbf{t}-\mathbf{X}_i\|, \|\mathbf{t}-\mathbf{X}_j\|) < \|\mathbf{X}_i - \mathbf{X}_j\|$.

We count the number of lenses containing each point (excluding lenses where the point is an endpoint),

- $\mathbf{X}_1$*: lenses* $(2,3),(2,4),(3,4) \to$ *3 lenses*
- $\mathbf{X}_2$*: lenses* $(1,3),(1,4),(3,4) \to$ *3 lenses*
- $\mathbf{X}_3$*: lenses* $(1,2),(1,4),(2,4) \to$ *3 lenses*
- $\mathbf{X}_4$*: lenses* $(1,2),(1,3),(2,3) \to$ *3 lenses*
- $\mathbf{X}_5$*: 0 lenses*
- $\mathbf{X}_6$*: 0 lenses*

Hence, the lens depth values are

$$LD(\mathbf{X}_1) = LD(\mathbf{X}_2) = LD(\mathbf{X}_3) = LD(\mathbf{X}_4) = \frac{3}{15} = 0.2,$$
$$LD(\mathbf{X}_5) = LD(\mathbf{X}_6) = 0.$$

Points near the center" of the cloud (like $\mathbf{X}_1$*–*$\mathbf{X}_4$*) have moderate lens depth, while peripheral points such as* $\mathbf{X}_5$ *and* $\mathbf{X}_6$ *have minimal depth.*

8.5.11.1 Properties of Lens Depth

The following theorems show that $LD(\mathbf{t};F)$ satisfies many desirable properties for a statistical depth function as defined by Liu [199], Zuo and Serfling [372].

Theorem 8.5.1 *:* $LD(\mathbf{t};F)$ *is (a) orthogonally invariant using Euclidean distance and (b) affine invariant using Mahalanobis distance.*

Theorem 8.5.2 *: For any c.d.f* F *on* $\mathbb{R}^d$ *and* $\mathbf{t} \in \mathbb{R}^d$, $LD(\mathbf{t};F)$ *vanishes at infinity, i.e.* $\sup_{\|\mathbf{t}\|\geq M} LD(\mathbf{t};F) \to 0$ *as* $M \to \infty$.

Clearly, as one moves to any point where the probability density approaches zero, the probability of being inside any lens approaches zero and LD vanishes.

Theorem 8.5.3 *: If* F *is an absolutely continuous distribution function on* $\mathbb{R}^d$*, then* $LD(\mathbf{t};F)$ *is continuous with respect to* $\mathbf{t}$.

Theorem 8.5.4 *: For any absolutely continuous distribution function* F *on* $\mathbb{R}^d$*, the maximal value of* $LD(\mathbf{t};F)$ *is* $\geq 1/2$.

A random vector $\mathbf{X}$ with a distribution function F is said to be centrally symmetric about its center $\boldsymbol{\mu}$ if $\mathbf{X} - \boldsymbol{\mu} \stackrel{D}{=} \boldsymbol{\mu} - \mathbf{X}$. Under the class of centrally symmetric distributions, we can show the following two properties for $LD(\mathbf{t};F)$.

Theorem 8.5.5 *: If* F *is centrally symmetric about its center* $\boldsymbol{\mu}$*, then* $LD(\mathbf{t};F)$ *is symmetric about* $\boldsymbol{\mu}$*. That is,* $LD(\mathbf{t}+\boldsymbol{\mu};F) = LD(\boldsymbol{\mu}-\mathbf{t};F)$.

Example 8.5.20 *Suppose* $\mathbf{X} \sim \mathrm{N}_d(\boldsymbol{\mu}, \mathbf{I})$*. Without loss of generality, assume that the center* $\boldsymbol{\mu} = 0$*. Since* $\mathbf{X}_1$ *and* $\mathbf{X}_2$ *are i.i.d. multivariate normal vectors,* $\|\mathbf{t}-\mathbf{X}_1\|$ *and* $\|\mathbf{t}-\mathbf{X}_2\|$ *are independently and identically distributed as a non–central* χ^2 *random variable with* d *degrees of freedom and non–centrality parameter* $\boldsymbol{\theta} = \|\mathbf{t}\|^2$ *(Anderson [10]). Since the* *distribution function of a non-central χ^2 random variable with d degrees of freedom is a decreasing function of $\boldsymbol{\theta}$, $\|\mathbf{t}-\mathbf{X}_1\|$ and $\|\mathbf{t}-\mathbf{X}_2\|$ become stochastically larger when the non-centrality parameter $\boldsymbol{\theta}$ increases (i.e., $\mathbf{t}$ moves away from the center $\boldsymbol{\mu}$). Therefore, $LD(\mathbf{t};F)=\mathbb{P}(\|\mathbf{X}_1-\mathbf{X}_2\|>\max(\|\mathbf{t}-\mathbf{X}_1\|,\|\mathbf{t}-\mathbf{X}_2\|))$ is (a) maximized at the center $\boldsymbol{\mu}$ or when $\boldsymbol{\theta}=\|\mathbf{t}\|^2=0$ and (b) monotonically decreasing as $\mathbf{t}$ moves away from $\boldsymbol{\mu}$.*

Since $\mathbb{I}[\mathbf{t}\in L(\mathbf{X}_i,\mathbf{X}_j)]$ is a measurable symmetric function in $\mathbf{X}_i$ and $\mathbf{X}_j$, and $LD(\mathbf{t};F)\le 1$, one can use the law of large numbers for U-processes (Theorem 3.1, Arcones and Giné [12], to show $LD(\mathbf{t};F_m)$ is uniformly consistent for $LD(\mathbf{t};F)$, i.e., as $n\to\infty$,

$$\sup_{\mathbf{t}\in\mathbb{R}^d}|TD(\mathbf{t};F_m)-TD(\mathbf{t};F)|\xrightarrow{a.s}0.$$

For any $\mathbf{t}\in\mathbb{R}^d$, $LD(\mathbf{t};F_m)$ is an unbiased estimator for $LD(\mathbf{t};F)$ with variance

$$\mathbb{V}ar(LD(\mathbf{t};F_m))=\frac{\alpha^2+2(m-2)\beta^2}{\binom{m}{2}},$$

where $\alpha^2=\mathbb{V}ar(\mathbb{I}[\mathbf{t}\in L(\mathbf{X}_1,\mathbf{X}_2)])$ and $\beta^2=\mathbb{V}ar(\mathbb{P}(\mathbf{t}\in L(\mathbf{X}_i,\mathbf{X}_j)|\mathbf{X}_i))$. When $d=1$, $\mathbb{E}(LD(t;F_m))=2F(t)(1-F(t))$, and

$$\mathbb{V}ar(LD(t;F_m))=\frac{2F(t)(1-F(t))(1-2F(t)+2F(t)^2)}{\binom{m}{2}}+\frac{2(m-2)F(t)(1-F(t))(1-4F(t)+4F(t)^2)}{\binom{m}{2}}.$$

Theorem 8.5.6 : *For any given $\mathbf{t}$, let $\mathbb{P}(\mathbf{X}_i)=\mathbb{P}(\mathbf{t}\in L(\mathbf{X}_i,\mathbf{X}_j)|\mathbf{X}_i)$, i.e., $\mathbb{P}(\mathbf{X}_i)$ denotes the conditional probability that $\mathbf{t}$ is contained in the random hyper-lens $L(\mathbf{X}_i,\mathbf{X}_j)$ given $\mathbf{X}_i$, for $i,j=1,\ldots,m$. One can show*

$$m^{1/2}(LD(\mathbf{t};F_m)-LD(\mathbf{t};F))/(4\beta^2)\xrightarrow{D}\mathbb{N}(0,1),$$

where β^2 denotes the variance of $\mathbb{P}(\mathbf{X}_i)$.

The asymptotic normality of $LD(\mathbf{t};F_m)$ is useful for comparing the efficiencies of different depth functions. It is also very helpful for outlier detection in the multivariate settings. Figure 8.13 shows the relationship between lens and spherical depth functions.

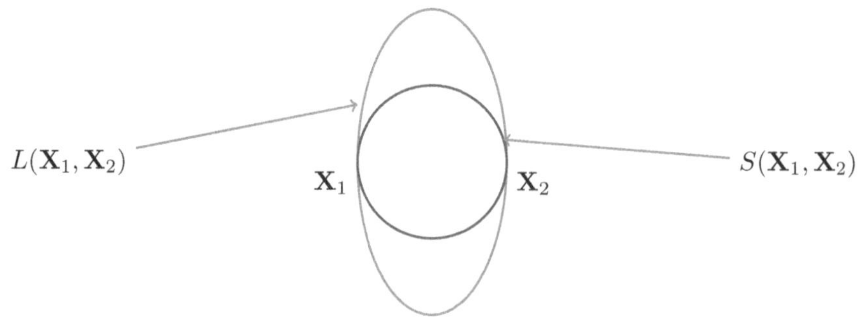

FIGURE 8.13: Relationship between Lens and Spherical depth functions.

8.5.11.2 Lens Depth: Maximality at the Center

One essential property of statistical depth functions is maximality at the center. If F is centrally symmetric about $\boldsymbol{\theta}$, then $D(\boldsymbol{\theta}) = \sup_{\mathbf{t}\in\mathbb{R}^d} D(\mathbf{t})$. For lens depth, [201] claimed this holds for the lens depth function. However, Kleindessner and von Luxburg [169] point out a gap in the proof of the Theorem. We explain the difficulty and offer a solution for the multivariate elliptical family of distributions.

The population lens depth can be written as:

$$LD(\mathbf{t}; F)(\mathbf{t}) = \mathbb{E}_{\mathbf{X}_1}\Big[\mathbb{P}\big(\mathbf{t} \in \text{Lens}(\mathbf{X}_1, \mathbf{X}_2) \mid \mathbf{X}_1\big)\Big].$$

The incorrect argument assumes that for every fixed $\mathbf{X}_1$, the function

$$\mathbf{t} \mapsto \mathbb{P}\big(\mathbf{t} \in \text{Lens}(\mathbf{X}_1, \mathbf{X}_2) \mid \mathbf{X}_1\big)$$

is radially decreasing as $\mathbf{t}$ moves away from the center. If this were true, then integrating over $\mathbf{X}_1$ would preserve monotonicity, proving maximality. The conditional probability depends on the geometry of the lens determined by $\mathbf{X}_1$ and $\mathbf{X}_2$. For certain $\mathbf{X}_1$, moving $\mathbf{t}$ away from the center does not necessarily decrease the conditional probability. Thus, pointwise monotonicity in $\mathbf{t}$ is not guaranteed. Therefore, monotonicity of the integral is not guaranteed. Hence, maximality at the center is unproven.

The difficulty is that the lens regions depend on pairs of random points. Conditioning on one point destroys global symmetry. Symmetry arguments must operate at the joint level. Similar difficulty appears in spherical depth [88]. Despite the missing proof no counterexample for symmetric distributions has been found. Empirical simulations support maximality and geometric intuition strongly suggests the center maximizes lens inclusion probability. Thus, the conjecture remains open but plausible.

A proof is possible under stronger assumptions such as spherically symmetric about $\mathbf{0}$ and density of the form $f(\mathbf{x}) = g(\|\mathbf{x}\|)$, where g is non-increasing in the radius (radially unimodal).

Example 8.5.21 *The lens $L(\mathbf{X}_1, \mathbf{X}_2)$ reduces to the interval (X_1, X_2) or (X_2, X_1) when $d = 1$. Therefore, $LD(t; F) = \mathbb{P}(t \in L(X_1, X_2)) = 2 \cdot F(t)(1 - F(t))$.*

Note that $\frac{\partial LD(t;F)}{\partial F(t)} = 2(1 - 2F(t))$ and $\frac{\partial^2 LD(t;F)}{\partial F(t)^2} < 0$. Hence, $LD(t; F)$ is maximized at t such that $F(t) = \frac{1}{2}$, which is the usual univariate median μ, and $LD(t; F)$ decreases monotonically to zero as t moves away from μ.

Example 8.5.21 shows the exact characterization of the lens when $d = 1$. If F has atoms, the formula $D(t) = 2F(t)(1 - F(t))$ still holds. The function remains maximized at any median. Under central symmetry, $\boldsymbol{\theta}$ is a median. Thus maximality still holds without continuity assumptions. Hence in one dimension, the conjecture is completely resolved under minimal assumptions.

Why is the multivariate case fundamentally harder? In $\mathbb{R}^d$ with $d \geq 2$, the lens is no longer the interval between the two points. Instead, it is a non-linear intersection of balls. In one dimension

$$t \in \text{Lens}(\mathbf{X}_1, \mathbf{X}_2) \iff t \text{ lies between } \mathbf{X}_1 \text{ and } \mathbf{X}_2.$$

In higher dimensions,

$$\mathbf{t} \in \text{Lens}(\mathbf{X}_1, \mathbf{X}_2) \iff \mathbf{t} \text{ lies in a non-linear, non-convex region.}$$

Consequently, when $d = 1$ the depth reduces to a function of $F(t)$. In higher dimensions no such scalar reduction exists. The event depends jointly and non-linearly on $(\mathbf{X}_1, \mathbf{X}_2)$.

8.5.11.3 Multivariate Elliptical Distributions

Let $\mathbf{X} \in \mathbb{R}^d$ follow an elliptical distribution with center $\boldsymbol{\mu} \in \mathbb{R}^d$ and positive-definite covariance $\boldsymbol{\Sigma}$, denoted $\mathbf{X} \sim \text{Elliptical}(\boldsymbol{\mu}, \boldsymbol{\Sigma}, g)$, where $g : [0, \infty) \to [0, \infty)$ is a decreasing generator function. By definition, the density (if it exists) has the form

$$f_{\mathbf{X}}(\mathbf{x}) = c\, g\big((\mathbf{x} - \boldsymbol{\mu})' \boldsymbol{\Sigma}^{-1} (\mathbf{x} - \boldsymbol{\mu})\big),$$

where c is a normalization constant. Examples include multivariate normal, with $g(r) = \exp(-r/2)$ and t-distributions.

Theorem 8.5.7 *Let $\mathbf{X} \in \mathbb{R}^d$ follow an elliptical distribution with center $\boldsymbol{\mu}$ and positive-definite covariance $\boldsymbol{\Sigma}$, with generator g non-increasing. Then the lens depth is maximized at the center, $LD(\mu; F) = \sup_{\mathbf{t} \in \mathbb{R}^d} LD(\mathbf{t}; F)$.*

The proof relies only on affine invariance and radial monotonicity; no assumptions on continuity of g are needed beyond integrability. The general conjecture for non-elliptical centrally symmetric distributions remains open.

8.5.11.4 Lens Median

Based on the notion of LD, one can define a multivariate analogue of the univariate median, LM, as the point or sets of points where the lens data depth is maximized. That is, $LM = arg\max_{\mathbf{t}} LD(\mathbf{t}; F)$. Given a random sample $\mathbf{X} = \{\mathbf{X}_1, \ldots, \mathbf{X}_m\}$ from F, the sample lens median LM_m is defined as

$$LM_m = arg\max_{\mathbf{t}} LD(\mathbf{t}; F_m).$$

A robust optimization method such as the method of Nelder and Mead [246] can be employed to obtain the point or the region that maximizes the sample lens data depth. This downhill simplex optimization algorithm searches for a new vector by extrapolating the lens depths measured at vertices of a $(d + 1)$ simplex and then replaces one of these vertices with the new vector before the next iteration. The simulation results show that the function *optim* in *R* package works well for finding the point with maximal sample lens depth. This function provides optimization methods based on Nelder-Mead, quasi-Newton and conjugate-gradient algorithms. Methods of constrained optimization and simulated annealing are also implemented. For a good treatment of numerical optimization, see Nocedal and Wright [253]. The implementation of Nelder and Mead [246] algorithm in *R* package works well with non–differentiable functions and is the default method. Other gradient-based methods fail due to non–smoothness of the sample lens data depths. To avoid convergence to a local maximum, we suggest to use several random starting points for optimization.

It is easy to show that LM and LM_m are both orthogonally invariant using the invariance property stated in Theorem 8.5.1 for $LD(\mathbf{t}; F)$. Note that we can get an affine invariant version of $LD(\mathbf{t}; F)$ and associated LM, and LM_m using the Mahalanobis distance as a measure of the difference in two multivariate observations. Furthermore, different transformation-retransformation techniques discussed in Serfling [305] can be employed to obtain affine invariant versions of $LD(\mathbf{t}; F)$ and associated LM as well as affine invariant, robust and computationally efficient versions of LM_m. Consistency and the asymptotic normality of LM_m follow from the maximality at the center. Consistency of $LD(\mathbf{t}; F_m)$ follows using the same arguments of Arcones and Giné [12].

We will next study the breakdown point and finite sample relative efficiency of LM_m as an unbiased estimator for the center of a centrally symmetric distribution.

8.5.11.5 Breakdown Point of $\mathbf{LM_m}$

The breakdown point of an estimator is a measure of its robustness. It measures the proportion of the data that can be contaminated without causing the estimator to become arbitrarily large. Let $\mathbf{X}^m = \{\mathbf{X}_1, \ldots, \mathbf{X}_m\}$ be a given random sample of size m and denote the estimator of interest with $\theta(\mathbf{X}^m)$. Consider adjoining $\mathbf{X}^m$ with c arbitrary vectors $\mathbf{X}^c$ and denote the contaminated sample by $\mathbf{X}_m^{m+c}$. If $\|\theta(\mathbf{X}_m^{m+c}) - \theta(\mathbf{X}^m)\|$ becomes arbitrarily large, then the estimator breaks down under a contamination fraction of $c/(m+c)$. Donoho and Gasko [78] defined the **additive** breakdown point $\epsilon^\star(\theta, \mathbf{X}^m)$ as the smallest contamination fraction under which the estimator breaks down

$$\epsilon^\star(\boldsymbol{\theta}, \mathbf{X}^m) = \min\{\frac{c}{m+c} : \sup_{\mathbf{X}_m^{m+c}} \|\theta(\mathbf{X}^m) - \theta(\mathbf{X}_m^{m+c})\| = \infty\}$$

and among translational-invariant location estimators, the best achievable breakdown point is 1/2 (Donoho [79]).

Let the m uncontaminated observations be such that

$$LD(\mathbf{X}_1; F_m) > LD(\mathbf{X}_2; F_m) > \ldots > LD(\mathbf{X}_m; F_m).$$

That is, $\mathbf{X}_1$ is the lens median LM_m and there are no ties in depth in the dataset. Let $L(\mathbf{X}_1; F_m)$ denote the number of lenses that contain $\mathbf{X}_1$. It is not difficult to show that $L(\mathbf{X}_1; F_m) \leq \frac{(m+1)^2}{4}$.

For the lower bound of the breakdown point, let the c contaminated observations $\mathbf{X}_{m+1} = \ldots = \mathbf{X}_{m+c}$ be at infinity such that $LD(\mathbf{X}_{m+1}; F_m) = 0$. The lens median breaks down when $L(\mathbf{X}_{m+1}; F_{m+c})$ exceeds $L(\mathbf{X}_1; F_{m+c})$. That is, $mc + \binom{c}{2} \geq \frac{(m+c+1)^2}{4}$. Solving this inequality gives $c \geq m(-1+\sqrt{2}) + o(m)$. Therefore,

$$\epsilon^\star(\boldsymbol{\theta}, \mathbf{X}^m) = \frac{c}{m+c} = \frac{m(-1+\sqrt{2}) + o(m)}{\sqrt{2}m + o(m)}.$$

As $m \to \infty$, $\epsilon^\star \to \frac{\sqrt{2}-1}{\sqrt{2}} = 0.2929$. Therefore, asymptotically at least 29.29% of the observations should be contaminated in order for LM_m to break down. The breakdown point for the lens median is the same as the breakdown point for the univariate Hodges-Lehmann estimator, which is defined as the median of $\frac{m(m+1)}{2}$ values of $(x_{[j]} + x_{[k]})/2$, where $x_{[j]}$ and $x_{[k]}$ are jth and kth order statistics in the ordered sample and $j \leq k$.

8.5.11.6 Finite Sample Relative Efficiency

To study the finite sample relative efficiency of LM_m, we generate $s = 1000$ random samples from the bivariate standard normal distribution $\mathbb{N}(\mathbf{0}, \mathbf{I}_2)$ as well as a heavy-tail bivariate t-distribution $t_2(\mathbf{0}, \mathbf{I}_2)$ with 2 degrees of freedom (DF) for several sample sizes m. We include other depth-associated medians such as the half space median (HM), simplicial median (SM), Oja median (OM), spatial median (SPM), spherical median ($SPHM$) and component-wise median (CM) as well as the sample mean vector (M) for comparison. We compute sample LM as well as $SPHM$ using the function **optim** in R to find the vector that maximizes the sample lens or spherical depth values. Sample medians HM, SM, OM, SPM and CM are computed using the "depth" package in R. The empirical mean squared error ($EMSE$) for an estimator $\hat{\boldsymbol{\theta}}$ can be calculated as $\frac{1}{s}\sum_{i=1}^{s} \|\hat{\theta}_i - \boldsymbol{\theta}\|^2$, where $\boldsymbol{\theta} = (0,0)'$ and $\hat{\theta}_i$ is the estimate from ith sample. The relative efficiency (RE) of the estimator can be obtained as the ratio of $EMSE$ for the sample mean vector to its $EMSE$.

The simulation results for bivariate normal show that the sample mean vector is most efficient for estimating the center of a bivariate normal distribution and least efficient for heavy-tail bivariate t-distribution. Among all non-parametric estimators, sample SPM, HM and OM are most efficient, followed by sample LM, $SPHM$ and CM. Sample simplicial depth median SM appears to be least efficient among all non-parametric estimators considered in this simulation study. There are no efficient algorithms for computing sample HM, SM and OM in high dimensions.

8.6 Depth-Depth Plot

Depth-depth plots (DD-plots) (Liu, Parelius and Singh [200]) provide a graphical means of comparing two multivariate distributions by computing depth values of points with respect to each distribution. A depth function $D(\mathbf{t}; F)$ assigns to each point $\mathbf{t} \in \mathbb{R}^d$ a value indicating its centrality relative to distribution F.

Given two distributions F_1 and F_2, the DD-plot is the collection of ordered pairs,

$$\mathrm{DD}(F_1, F_2) = \left\{(D(\mathbf{t}; F_1), D(\mathbf{t}; F_2)) : \mathbf{t} \in \mathbb{R}^d\right\}. \tag{8.1}$$

There are two sample versions:

- Two-sample version
$$\left\{(D(\mathbf{t}; F_{1n}), D(\mathbf{t}; F_{2m})) : \mathbf{t} \in \mathbb{R}^d\right\}, \tag{8.2}$$
where F_{1n} and F_{2m} are empirical distributions.

- One-sample vs. theoretical distribution
$$\left\{(D(\mathbf{t}; F_{1n}), D(\mathbf{t}; F_1)) : \mathbf{t} \in \mathbb{R}^d\right\}. \tag{8.3}$$

Example 8.6.1 *Using $m = 3$, $n = 2$ in $\mathbb{R}^4$, and Mahalanobis depth to obtain the DD-plot.*

***Group** F_1: $\mathbf{X}_1 = (1, 0, 0, 0)$, $\mathbf{X}_2 = (0, 1, 0, 0)$, $\mathbf{X}_3 = (0, 0, 1, 0)$.*

***Group** F_2: $\mathbf{Y}_1 = (2, 2, 2, 2)$, $\mathbf{Y}_2 = (3, 3, 3, 3)$.*

For Mahalanobis depth computation let $\bar{X} = (1/3, 1/3, 1/3, 0)$ and sample covariance matrix S_X:

$$S_X = \frac{1}{2}\sum_{i=1}^{3}(\mathbf{X}_i - \bar{X})(\mathbf{X}_i - \bar{X})'$$

Similarly for F_{2m}.

For each point $\mathbf{t}$ in $\{\mathbf{X}_1, \mathbf{X}_2, \mathbf{X}_3, \mathbf{Y}_1, \mathbf{Y}_2\}$, compute Mahalanobis depth with respect to both F_{1n} and F_{2m}, $D(\mathbf{t}; F_{1n}) = \left[1 + (\mathbf{t} - \bar{X})' S_X^{-1} (\mathbf{t} - \bar{X})\right]^{-1}$ and for F_{2m}. Resulting DD-plot points (rounded for brevity), $\mathbf{X}_1$: $(0.75, 0.05)$, $\mathbf{X}_2$: $(0.75, 0.04)$, $\mathbf{X}_3$: $(0.75, 0.03)$, $\mathbf{Y}_1$: $(0.02, 0.88)$, $\mathbf{Y}_2$: $(0.01, 0.85)$.

Liu, Parelius, and Singh [200] proposed the use of statistical depth functions to provide robust, non-parametric descriptions of multivariate data. Their work introduced

- The concept of data depth contours to visualize the multivariate "shape" of a dataset.
- The use of depth-based ranks for multivariate hypothesis testing.
- DD-plots as a tool for distribution comparison, analogous to Q-Q plots in univariate settings.

These tools are especially helpful in identifying location and shape differences between distributions and for assessing goodness-of-fit in multivariate problems.

Example 8.6.2 *We revisit the Wine dataset containing three varieties: Barolo, Grignolino, and Barbera. For each pair of cultivars, we compute Mahalanobis depth values for equal-sized subsamples ($n = 48$). Depth–Depth (DD) plots compare the centrality of each observation across two distributions:*

$$\mathrm{DD}(F_1, F_2) = \{(D(\mathbf{t}; F_1), D(\mathbf{t}; F_2)) : \mathbf{t} \in \mathbb{R}^d\}.$$

If the distributions are similar, the points cluster near the diagonal $y = x$. Figure 8.14 shows three DD plots. For Barolo vs Grignolino: Points lie below $y = x$, showing Barolo samples are more central within their distribution; clear class separation. For Barolo 7vs Barbera: Even stronger downward pattern: Barolo wines have higher depths relative to Barbera, indicating strong distinction. For Grignolino vs Barbera: Points cluster closer to the diagonal, suggesting higher overlap between these two cultivars.

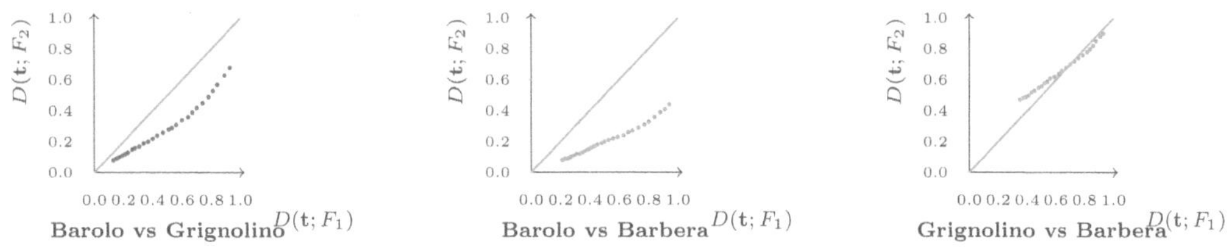

FIGURE 8.14: Depth–Depth (DD) plots comparing wine cultivars using Mahalanobis depth (scaled 0–1). Axes are labeled cleanly with tick marks at 0.0–1.0. The red line $y = x$ indicates equal depth across groups; points below the line show higher centrality in the first cultivar.

8.7 Computational Complexity

The lens depth function proposed in this chapter is based on the concept of geometric neighborhood. It satisfies many desirable properties of a statistical depth function such as affine and orthogonal invariance, maximality at the center of symmetry, monotonicity relative to the center (deepest point) and vanishing at infinity. The sample lens depth is uniformly consistent and asymptotically normal, and it enjoys computational simplicity in high dimensions.

Table 8.1 compares six depth functions in terms of the time complexity in big-O notation for computing the depth function and the bivariate median. We also compare the maximal depth values and the asymptotic breakdown points of sample medians. The first three rows in Table 8.1 correspond to the half-space, simplicial and Oja depth functions. These pose formidable computational challenges for all but the small values of d. For example, when $d = 2$ there are several efficient algorithms for Oja, simplicial and half-space depth functions and medians whose time complexities appear in column 3 of the Table. Aloupis [8] provides a good discussion on computations associated with these depth functions. Lopuhaa and Rousseeuw [204] discuss the breakdown point of various estimators. The last three rows correspond to spatial, spherical and lens depth functions whose time complexity for depth computation grows linearly in d. The listed time complexity of $O(m^2 d)$ for spherical and lens depth functions is based on a brute force method that considers all $\binom{m}{2}$ cases. It is unlikely that the computational complexity of the lens median can be reduced beyond $O(m^2)$.

TABLE 8.1: Comparison of Some Existing Depth Functions.

Depth Function	Depth Time Complexity	Bivariate Median Time Complexity	Maximal Depth	Median Breakdown
Oja	$O(m^d)$	$O(m^3 \log m)$	1	0
Halfspace	$O(m^{d-1})\log(m)$	$O(m \log^3 m)$	$(1/(d+2), 1/2)$	$1/3$
Spherical	$O(m^2 d)$	$O(m^2)$	$1/2$	$\frac{\sqrt{2}-1}{\sqrt{2}}$
Simplicial	$O(m^{d+1})$	$O(m^4)$	$1/2^d$	$< 1/3$
Lens	$O(m^2 d)$	$O(m^2)$	$\geq 1/2$	$\frac{\sqrt{2}-1}{\sqrt{2}}$
Spatial	$O(md)$	$O(m)$	1	$1/2$

We show that the sample lens median has a breakdown point of $\frac{\sqrt{2}-1}{\sqrt{2}}$ and a fairly good relative efficiency compared to other medians for estimating the center of a centrally symmetric multivariate distribution.

The range of the lens depth in one dimension is $(0, 1/2)$ and it increases to 1 as d increases. Hence, the maximal depth of the lens depth function is at least $1/2$ whereas the maximal depth of the spherical depth is always $1/2$. As a result, center-outward ranking based on the sample lens depth yields fewer ties than SD, HSD, and SPD.

In practice, we recommend depth functions whose time complexity grows only linearly in d and whose associated median is robust with a high breakdown point such as LD and SPD. Such depth functions are especially important for non-parametric analysis in high-dimensional settings.

8.8 Behavior of Depth Functions as $d \to \infty$

Let $\mathbf{X}_1, \ldots, \mathbf{X}_n \in \mathbb{R}^d$ be a random sample from a distribution F, and consider the behavior of various statistical depth functions as $d \to \infty$, possibly with fixed n or $n = o(d)$. Many depth functions exhibit degeneracy, concentration, or computational instability in this regime.

- Mahalanobis Depth: The sample covariance $\mathbf{S}$ becomes singular when $d \geq n$. As $d \to \infty$, Mahalanobis depth tends to zero for all non–central points, leading to a loss of discrimination. One can use regularized covariance estimators (e.g., shrinkage or sparse inverse covariance) or perform dimensionality reduction prior to using Mahalanobis when $d \geq m$.

- Half-space Depth: For $d \gg n$, almost all points lie on the convex hull, and HSD becomes zero for most interior points. Depth contours degenerate. One remedy is to compute this depth in lower-dimensional random projections.

- Convex Hull Peeling Depth: In high dimensions, most data points lie on the convex hull of the data; the first layer already includes all observations, so depth is 1 or 0.

- Simplicial Volume Depth: The number of d-simplices becomes computationally infeasible ($\binom{n}{d} \to \infty$ or undefined when $d > n$). Volumes of high-dimensional simplices converge to zero. One solution is to perform random subsampling or projection to reduce dimensionality.

- Majority Depth: See Exercise 8.11. It becomes undefined or unstable when most hyperplanes do not separate any majority due to high sparsity.
- Simplicial Depth: The probability of $\mathbf{X}$ lying in any random simplex goes to zero as $d \to \infty$, since most simplices are empty in high dimensions.
- Spatial Depth: More robust in high dimensions since it does not rely on covariance matrix. Depth decreases more gradually but can still suffer from concentration of distances.
- Spherical Depth: It degenerates in high dimensions due to sparsity and concentration of spheres away from central points.
- Lens Depth: It suffers from similar degeneracies as spherical depth in high dimensions; most points lie outside of any lens.
- Interpoint Depth: Interpoint distances become nearly equal in high dimensions due to concentration of measure.

Many classical depth functions degenerate or lose discriminatory power as $d \to \infty$. Projection depth, spatial depth, and interpoint-based depths tend to perform better in high dimensions. Random projection, dimensionality reduction, regularization, and functional depth formulations are necessary for practical use in high dimensions.

8.9 Depth-Based Graphs

We represent the observations of a random sample in $\mathbb{R}^d$ as vertices of a complete weighted graph. We denote a random data matrix by $\mathbf{X} = \{\mathbf{X}_1, \ldots, \mathbf{X}_n\}$, which is also used to represent the set of vertices in a graph G, where $\mathbf{X}_i$ is an independent d-dimensional random vector from F. A weighted and connected simple graph G has a set of vertices $v(G) = \mathbf{X}$ and a set of edges $e(G) = \{(\mathbf{X}_i, \mathbf{X}_j)|1 \leq i < j \leq n\} \subset v \times v$. We also associate a weight $w(G)$ with every edge of G. The weight of each edge can be the Euclidean distance, the Mahalanobis distance, or a measure of discrepancy such as the Kullback-Leibler distance between the vertices. A proximity graph has the same vertex set $v(G)$, but an edge set that includes only those edges $(\mathbf{X}_i, \mathbf{X}_j)$ for which a neighborhood condition holds.

We discuss proximity graphs, expose their connection to spherical and lens depth functions, and define a family of depth functions on the $\beta-$skeleton graph. We discuss MST and propose new depth functions based on the MST. We examine the properties of MST-based depth functions, compare their efficiency in estimating the median and present an example. The last section is devoted to summary and concluding remarks.

8.10 Proximity Graphs

Different notions of closeness or neighborhood give rise to different proximity graphs. For example, the r-ball graph places an edge between vertices $\mathbf{X}_i$ and $\mathbf{X}_j$ if $\mathbf{X}_j \in B(\mathbf{X}_i, r)$

where $B(\mathbf{c}, r)$ is a ball with center $\mathbf{c}$ and radius r. The nearest neighbor graph NNG(K) places an edge between vertices $\mathbf{X}_i$ and $\mathbf{X}_j$ if $\mathbf{X}_j$ is one of the K nearest neighbors of $\mathbf{X}_j$. The minimum spanning tree (MST), the Delaunay Triangulation (DT), the relative neighborhood graph (RNG) (Toussaint [340], and the Gabriel Graph (GG) (Gabriel and Sokal [113]) are prominent representatives of proximity graphs. For any set of points X in the plane

$$\mathrm{NN}(1) \subseteq \mathrm{MST}(X) \subseteq \mathrm{RNG}(X) \subseteq \mathrm{GG}(X) \subseteq \mathrm{DT}(X)$$

holds for proximity graphs (Toussaint [340]).

We assume that a connected graph G is given with a set of vertices $v(G) = \mathbf{X}$ and a set of edges $e(G)$. We associate a weight $w(G) : e(G) \mapsto \mathbb{R}^+$ with every edge of G. The weights are the Euclidean or Mahalanobis distances between the vertices A connected graph with no cycles is a tree. The total weight of a tree is the sum of the weights of its edges. The MST of a graph G is a weighted and connected tree that contains all the vertices in the graph G such that the total weight of the tree is minimized. The MST of a graph G is unique if the set of weights of G contains no ties. Minimum spanning trees have been used in numerous applications, including outlier detection (Rohlf [283]) and Caroni and Prescott ([51] [52]), two-sample problems (Friedman and Rafsky [109]) and depth functions. Patil and Taillie [265] define upper level set trees and use them for detection of arbitrary shaped hotspots on a graph.

Prim [273] and Kruskal [176] provide two distinct algorithms for building an MST. Prim's Algorithm [273]: Given a vertex set $\mathbf{X}$ and an edge set E, this algorithm constructs the MST by starting with an empty vertex set $\mathbf{X}^*$ and edge set E^*. We start by moving an arbitrary vertex from $\mathbf{X}$ to $\mathbf{X}^*$. Maintaining a tree at all stages, the algorithm proceeds to identify the shortest edge from $\mathbf{X}$ to $\mathbf{X}^*$. This edge is next added to E^*. The corresponding vertex is removed from $\mathbf{X}$ and added to $\mathbf{X}^*$. The algorithm terminates when all vertices are in $\mathbf{X}^*$. Kruskal Algorithm [176]: Given a vertex set $\mathbf{X}$ and an edge set E, this algorithm construct the MST by iteratively adding the next shortest edge that does not create a cycle to the tree. The algorithm ends when all $n-1$ edges are found. One major difference between the two algorithms is that Prim's maintains a connected tree in all stages of construction while Kruskals' algorithm deals with a forest that eventually merges into a single tree.

A spanning tree provides a path between every two vertices of a graph and is a compact method of displaying and summarizing observations in $\mathbb{R}^d$. An MST contains as a subgraph the nearest neighbor graph. If any edge of an MST is deleted, thereby dividing the graph into two disjoint connected subgraphs, and thus dividing the points into two disjoint subsets, the deleted edge weight is the smallest IPD between the two subsets. By deleting an edge of the MST, we create two clusters of the observations. If we delete k edges of the MST, we create $k+1$ clusters. The following discussion on β-skeleton graphs and multidimensional medians on minimum spanning tree appears in Yang and Modarres [363] and Yang and Modarres [364], respectively.

8.10.1 β-Skeleton

Kirkpatrick and Radke [168] considered a parameterized family of neighborhood graphs called the β-skeletons. The shape parameter $\beta = 1$ corresponds to GG(random hypersphere) and $\beta = 2$ corresponds to RNG (random hyper-lens) formed by two random observations. Edges of a neighborhood graph are defined by the influence region (IR) between the vertices. Given a pair of points $\mathbf{X}_i$ and $\mathbf{X}_j$ in $\mathbf{X}$, an edge is placed in the neighborhood graph if and only if $\mathrm{IR}(\mathbf{X}_i, \mathbf{X}_j) \cap \mathbf{X} = \phi$.

Let $\mathrm{Skel}_\beta(\mathbf{X}_i, \mathbf{X}_j)$ denote the IR of β-skeleton where $0 \leq \beta < \infty$. Note that

$$\mathrm{Skel}_\beta(\mathbf{X}_i, \mathbf{X}_j) = B\left(\left(1 - \frac{\beta}{2}\right)\mathbf{X}_i + \frac{\beta}{2}\mathbf{X}_j, \frac{\beta}{2}r\right) \bigcap B\left(\frac{\beta}{2}\mathbf{X}_i + \left(1 - \frac{\beta}{2}\right)\mathbf{X}_j, \frac{\beta}{2}r\right). \quad (8.4)$$

One can show that $\beta = 0$ produces the line segment between $\mathbf{X}_i$ and $\mathbf{X}_j$. Since for $\beta > 2$ the β-skeleton may become disconnected, we will only consider $0 \le \beta \le 2$.

8.10.2 β-Skeleton Depth

For any $\mathbf{t}$ in $\mathbb{R}^d$, we define the β-skeleton depth function as $\text{SkelD}_\beta(\mathbf{t}; F) = \mathbb{P}(t \in \text{Skel}_\beta(\mathbf{X}_i, \mathbf{X}_j))$. The β-skeleton depth provides a measure of closeness of $\mathbf{t}$ with respect to F since points inside $\text{Skel}_\beta(\mathbf{X}_i, \mathbf{X}_j)$ are close to both $\mathbf{X}_i$ and $\mathbf{X}_j$. The β-skeletons are nested and provide a nested family of depth functions, each with its own multivariate median. One can readily verify that

$$\{\mathbf{t} : \mathbf{t} \in \text{Skel}_{\beta_1}(\mathbf{X}_i, \mathbf{X}_j)\} \subset \{\mathbf{t} : \mathbf{t} \in \text{Skel}_{\beta_2}(\mathbf{X}_i, \mathbf{X}_j)\}$$

implies $\text{SkelD}_{\beta_1}(\mathbf{t}; F) = \mathbb{P}(\mathbf{t} \in \text{Skel}_{\beta_1}(\mathbf{X}_i, \mathbf{X}_j))$ is less than or equal to $\text{SkelD}_{\beta_2}(\mathbf{t}; F) = \mathbb{P}(\mathbf{t} \in \text{Skel}_{\beta_2}(\mathbf{X}_i, \mathbf{X}_j))$ when $\beta_1 \le \beta_2$. Let $\mathbb{I}(A) = 1$ if event A occurs and $\mathbb{I}(A) = 0$, otherwise. We define the sample β-skeleton depth of vertex $\mathbf{t}$ by

$$\text{SkelD}_\beta(\mathbf{t}; F_n) = \frac{1}{\binom{n}{2}} \sum_{i<j}^{n} \mathbb{I}[\mathbf{t} \in \text{Skel}_\beta(\mathbf{X}_i, \mathbf{X}_j)]. \tag{8.5}$$

It is not difficult to obtain estimates of β-skeleton depth by parametric bootstrap. For spherical and lens depth functions (GG and RNG graphs) it has been shown that $\text{SkelD}_\beta(\mathbf{t}; F_n)$ is an order-two U-statistic. The β-skeleton graph has a kernel function $k(\mathbf{X}_i, \mathbf{X}_j) = \mathbb{I}[\mathbf{t} \in \text{Skel}_\beta]$, which can be used to show that β-skeleton medians have asymptotic multivariate normal distributions. The spherical depth $SPD(\mathbf{t}; F)$ (Elmore, Hettmansperger and Xuan [88]) is defined on the 1-skeleton graph as the probability that $\mathbf{t}$ is contained in the unique, closed random hypersphere, $S(\mathbf{X}_i, \mathbf{X}_j)$, formed by two random points $\mathbf{X}_i$ and $\mathbf{X}_j$, which are i.i.d. with DF F. That is, $SPD(\mathbf{t}; F) = \mathbb{P}(\mathbf{t} \in S(\mathbf{X}_i, \mathbf{X}_j))$ where $S(\mathbf{X}_i, \mathbf{X}_j)$ is the unique d-dimensional ball passing through $\mathbf{X}_i$ and $\mathbf{X}_j$ with diameter $\|\mathbf{X}_i - \mathbf{X}_i\|$ and center $\frac{1}{2}(\mathbf{X}_i + \mathbf{X}_j)$. The sample version of the spherical depth at a point $\mathbf{t}$ is defined as

$$SPD(\mathbf{t}; F_n) = \frac{1}{\binom{n}{2}} \sum_{i<j}^{n} \mathbb{I}[\mathbf{t} \in S(\mathbf{X}_i, \mathbf{X}_j)]. \tag{8.6}$$

The spherical depth function is based on the Gabriel graph which has an edge between vertices $\mathbf{X}_i$ and $\mathbf{X}_j$ if there are no other vertices in the influence region $\text{IR}(\mathbf{X}_i, \mathbf{X}_j) = B(\frac{1}{2}(\mathbf{X}_i + \mathbf{X}_j), \frac{1}{2}\|\mathbf{X}_i - \mathbf{X}_j\|)$. The lens data depth function $LD(\mathbf{t}; F)$ of a vector $\mathbf{t} \in \mathbb{R}^d$ is defined on the 2-skeleton graph as the probability that $\mathbf{t}$ is contained in the random hyperlens $L(\mathbf{X}_i, \mathbf{X}_j)$ formed at the intersection of two closed balls centered at random points $\mathbf{X}_i$ and $\mathbf{X}_j$, which are i.i.d. from F with radius $r = \frac{1}{2}\|\mathbf{X}_i - \mathbf{X}_j\|$. That is, $LD(\mathbf{t}; F) = \mathbb{P}(\mathbf{t} \in L(\mathbf{X}_i, \mathbf{X}_j))$, where $\mathbf{X}_i$ and $\mathbf{X}_j$ are i.i.d. from F and $L(\mathbf{X}_i, \mathbf{X}_j) = B(\mathbf{X}_i, r) \cap B(\mathbf{X}_j, r)$. One can show $0 \le LD(\mathbf{t}; F) \le 1$. The sample lens depth is defined as

$$LD(\mathbf{t}; F_n) = \frac{1}{\binom{n}{2}} \sum_{i<j}^{n} \mathbb{I}[\mathbf{t} \in L(\mathbf{X}_i, \mathbf{X}_j)]. \tag{8.7}$$

The lens depth function is based on another underlying proximity graph. A relative neighbor graph places an edge between $\mathbf{X}_i$ and $\mathbf{X}_j$, if $L(\mathbf{X}_i, \mathbf{X}_j)$, the lune of $(\mathbf{X}_i, \mathbf{X}_j)$, does not contain any other vertices, except $\mathbf{X}_i$ and $\mathbf{X}_j$. That is, RNG places an edge between vertices $\mathbf{X}_i$ and $\mathbf{X}_j$ if and only if for any other point $\mathbf{X}_t$ we have $\|\mathbf{X}_i - \mathbf{X}_j\| \le \max\{\|\mathbf{X}_i - \mathbf{X}_t\|, \|\mathbf{X}_j - \mathbf{X}_t\|\}$. Since RNG is a planar graph we have, in terms of the edge set, $n - 1 \le |\text{RNG}(E)| \le 3n - 6$ where the lower bound is reached by the MST (Marchette [210]).

Figure 8.15 shows β-skeleton depth functions and depicts the connection between proximity graphs and two existing depth functions, when β increases the influence region of each point becomes larger and the β-skeleton becomes sparser.

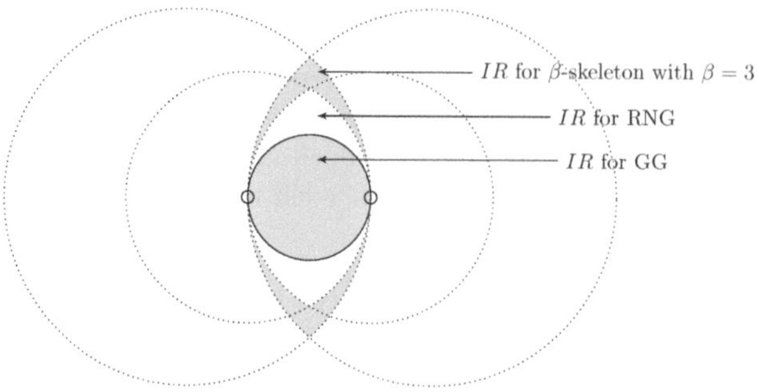

FIGURE 8.15: β-skeleton depth functions: For spherical depth $\beta = 1$ and for lens depth $\beta = 2$.

8.10.3 MST Revisited

Given a weighted and connected simple graph G, the degree of a vertex is the number of edges that connects that vertex to other vertices. A path between vertices $\mathbf{X}_i$ and $\mathbf{X}_j$ is denoted by Path$(\mathbf{X}_i, \ldots, \mathbf{X}_j)$ and is a sequence of adjacent vertices that connect $\mathbf{X}_i$ to $\mathbf{X}_j$. The length of a path is the number of edges on that path. A cycle is a path with the same initial and final vertices. A connected graph with no cycles is a tree. The MST of a graph G is a weighted and connected tree that contains all the vertices in graph G such that the total weight of the tree is minimized.

A sequence of vertices $\mathbf{X}_i, \ldots, \mathbf{X}_j$ is a geodesic path from $\mathbf{X}_i$ to $\mathbf{X}_j$, if it has minimum length among all paths from $\mathbf{X}_i$ to $\mathbf{X}_j$. If k is the number of edges on this path, then the geodesic distance from $\mathbf{X}_i$ to $\mathbf{X}_j$ is $d(\mathbf{X}_i, \mathbf{X}_j) = k$. Moreover, $d(\mathbf{X}_i, \mathbf{X}_j) = 0$, if and only if $\mathbf{X}_i = \mathbf{X}_j$. The MST is unique when the edge weights are unique. Furthermore, all paths are geodesic if and only if there are no duplicated vertices. These conditions are satisfied with probability 1 when the underlying distribution is absolutely continuous.

Minimum spanning trees have been used in numerous applications, including two-sample problems (Friedman and Rafsky [109]), outliers detection (Rohlf [283]) and Caroni and Prescott ([51] [52]). Patil et al. [264] propose a method to identify spatially constrained clusters (hotspots) using upper level sets. Patil and Taillie [265] define upper level set trees and use them for detection of arbitrary shaped hotspots on a graph. Assuncão et al. [15] use the connection between the upper level sets and the MST to suggest methods for fast detection of arbitrarily shaped disease clusters. Boruvka [43], Prim [273] and Kruskal [176] provide three distinct algorithm for building an MST.

A spanning tree provides a path between every two vertices of a graph and is a compact method of displaying and summarizing observations in $\mathbb{R}^d$. Depth functions defined on the MST allow for the use of colors for visualization purposes, identify those vertices in the outskirts of the sample or discriminate among several groups. We study the following MST depth functions and their associated medians

- Path depth (PD),
- Path depth of length at most r ($\text{PD}_{\leq r}$),
- RUNT depth (RUNTD),

- MST eccentricity depth (ECD),
- MST peeling depth (PLD),
- Geodesic convexity depth (GCD).

We assume that the vertex whose depth we seek is already present on the MST; otherwise, we find its depth on the MST after it is added to the complete graph.

Let $h(\mathbf{t}; \mathbf{X}_1, \mathbf{X}_2, \ldots, \mathbf{X}_\gamma)$ be some bounded, non–negative function that measures how close the point $\mathbf{t}$ is to the set of points $\{\mathbf{X}_1, \mathbf{X}_2, \ldots, \mathbf{X}_\gamma\}$ based on their relationship on the MST. The type A depth functions on the MST is defined as $D(\mathbf{t}; F) = E\big[h(\mathbf{t}; \mathbf{X}_1, \mathbf{X}_2, \ldots, \mathbf{X}_\gamma)\big]$, where the expectation is taken over all possible MSTs with vertices that are i.i.d. DF F. The following three depth functions are type A depth functions on the MST.

8.11 MST-based Depth Functions

8.11.1 Path Depth

We define the path depth function $PD(\mathbf{t}; F)$ of a vector $\mathbf{t} \in \mathbb{R}^d$ with respect to a DF F to be the probability that $\mathbf{t}$ is contained on a path between two randomly selected observations $\mathbf{X}_i$ and $\mathbf{X}_j$ from an MST with vertices that are n i.i.d. random vectors with DF F. One can verify that by setting $\gamma = 2$ and $h(\mathbf{t}; \mathbf{X}_1, \mathbf{X}_2) = \mathbf{I}[\mathbf{t} \in \text{Path}(\mathbf{X}_1, \ldots, \mathbf{X}_2)]$, PD is a type A depth function on the MST. The empirical path depth is

$$PD(\mathbf{t}; F_n) = \frac{1}{\binom{n}{2}} \sum_{i<j}^{n} \mathbb{I}[\mathbf{t} \in Path(\mathbf{X}_i, \ldots, \mathbf{X}_j)]. \tag{8.8}$$

We define connectivity of a node by the number of paths that contain that node in the graph. For example, the number of paths that contain a terminal node in a string tree (a linear list of nodes whose degrees are 2, except the first and the last on the list that have a degree of 1) of n nodes is $n-1$. A vertex that is accessible by the most number of vertices maximizes the PD and provides the path median (PM). Clearly, the central nodes are more accessible than terminal nodes as there are more paths that connect a node in the center to other nodes than there are paths that connect a terminal node to other nodes. Theorem 8.11.2 can be used for fast computation of PD and PM. The median vertex provides access to the greatest number of vertices on the MST and has the highest path depth value whereas the terminal vertices have $PD(\mathbf{t}; F_n) = \frac{n-1}{\binom{n}{2}} = 2/n$ because any terminal vertex $\mathbf{t}$ is on $n-1$ paths with other vertices. We provide an example of path depth below.

Example 8.11.1 ***Computation of path depth.*** *We consider the path depths of vertices in Figure 8.16. One observes that the graph is still a tree after removing $\mathbf{X}_4$. Hence, any path between $\mathbf{X}_i$ and $\mathbf{X}_j$, $i \neq j \in \{1, 2, 3, 5, 6, 7\}$ does not contain $\mathbf{X}_4$. The only paths containing $\mathbf{X}_4$ are the six paths between $\mathbf{X}_4$ and $\mathbf{X}_i$, $i \neq 4$. The same may be said for other terminal vertices $\mathbf{X}_5$, $\mathbf{X}_6$ and $\mathbf{X}_7$. One can show that*

$$PD(\mathbf{X}_4, F_n) = PD(\mathbf{X}_5, F_n) = PD(\mathbf{X}_6, F_n) = PD(\mathbf{X}_7, F_n) = \frac{6}{\binom{7}{2}} = \frac{2}{7}.$$

One may also observe the following.

- *The vertex* $\mathbf{X}_1$ *is contained on paths between* $(\mathbf{X}_3, \mathbf{X}_2)$, $(\mathbf{X}_3, \mathbf{X}_5)$, $(\mathbf{X}_3, \mathbf{X}_6)$, $(\mathbf{X}_3, \mathbf{X}_7)$, $(\mathbf{X}_4, \mathbf{X}_2)$, $(\mathbf{X}_4, \mathbf{X}_5)$, $(\mathbf{X}_4, \mathbf{X}_6)$, $(\mathbf{X}_4, \mathbf{X}_7)$ *and the other 6 paths between* $\mathbf{X}_1$ *and* $\mathbf{X}_i$, $i \neq 1$. $PD(\mathbf{X}_1, F_n) = \frac{14}{\binom{7}{2}} = \frac{2}{3}$.

- *The vertex* $\mathbf{X}_2$ *is contained on paths between* $(\mathbf{X}_1, \mathbf{X}_5)$, $(\mathbf{X}_1, \mathbf{X}_6)$, $(\mathbf{X}_1, \mathbf{X}_7)$, $(\mathbf{X}_3, \mathbf{X}_5)$, $(\mathbf{X}_3, \mathbf{X}_6)$, $(\mathbf{X}_3, \mathbf{X}_7)$, $(\mathbf{X}_4, \mathbf{X}_5)$, $(\mathbf{X}_4, \mathbf{X}_6)$, $(\mathbf{X}_4, \mathbf{X}_7)$ *and the other 6 paths between* $\mathbf{X}_2$ *and* $\mathbf{X}_i$, $i \neq 2$. $PD(\mathbf{X}_2, F_n) = \frac{15}{\binom{7}{2}} = \frac{5}{7}$.

- *The vertex* $\mathbf{X}_3$ *is contained on paths between* $(\mathbf{X}_4, \mathbf{X}_1)$, $(\mathbf{X}_4, \mathbf{X}_2)$, $(\mathbf{X}_4, \mathbf{X}_5)$, $(\mathbf{X}_4, \mathbf{X}_6)$, $(\mathbf{X}_4, \mathbf{X}_7)$ *and the other 6 paths between* $\mathbf{X}_3$ *and* $\mathbf{X}_i$, $i \neq 3$. $PD(\mathbf{X}_2, F_n) = \frac{11}{\binom{7}{2}} = \frac{11}{21}$. □

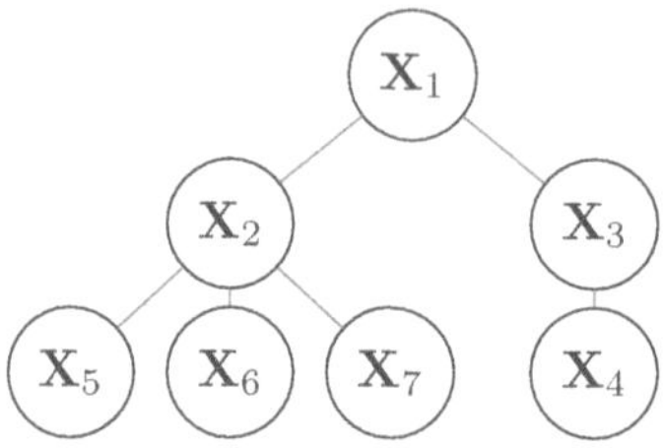

FIGURE 8.16: Construction of path depth.

When $d = 1$ the MST connects the order statistics $X_{(i)}$ of the sample and $PD(X_{(i)}; F_n) = \frac{i(n-i+1)-1}{\binom{n}{2}}$. The depth of the median is $PD(X_{(\frac{n+1}{2})}; F) = \frac{(n+1)^2/4-1}{\binom{n}{2}}$ when n is odd and the average of the two vertices with maximum depth values when n is even. In both cases, $PD(X_M; F)$ approaches $\frac{1}{2}$ as n $\to \infty$. For $d \geq 2$, maximum depth value is 1 as can be verified from a star tree with 6 nodes, five of which are terminal nodes. However, the maximum is not always attainable. For example, when n=8, the maximum degree of a vertex in a planar graph cannot be greater than 6 due to the upper bound by the Hadwiger or the Kissing number (Martini and Swanepoel [219] and Small [317]).

Figure 8.17 shows a star-shaped tree with 11 nodes: 5 nodes on the left, 5 nodes on the right, and 1 center node C. We want to compute the path depth of the center node. Let the total number of vertices be $n = 11$. The total number of paths between distinct vertex pairs is $\binom{11}{2} = 55$. There are 25 paths connecting left nodes to right nodes that pass through the center node C. There are 10 paths from the center to each leaf that include the center. Hence, the center lies on $25 + 10 = 35$ paths, giving an empirical path depth $PD(\text{center}; F_n) = \frac{35}{55} = \frac{7}{11}$. This is the maximum path depth among all nodes. Terminal nodes each lie on $n - 1 = 10$ paths, so their path depth is $PD(\text{terminal}; F_n) = \frac{10}{55} = \frac{2}{11}$.

8.11.2 Path Depth of Length of at Most r

The path depth of vertex $\mathbf{t}$ receives contributions from every path in which it is a vertex, no matter how long the length of that path is. We define $PD_{\leq\delta}$ to denote the path depth of length at most δ, for fixed $1 \leq \delta \leq \alpha$, where α is the diameter or the length of the longest path on the MST. This path depth function places more emphasis on shorter paths; thus, lowering the depth of more extreme vertices in the outskirts of the distribution. One can show that $PD_{\leq\alpha} = PD$. Different values of δ generate a family of depth functions on the MST, each with their own multivariate median. Similar to the path depth, $\text{PD}_{\leq\delta}$ is a type A depth function on the MST with $\gamma = 2$ and $h(\mathbf{t}; \mathbf{X}_1, \mathbf{X}_2) = \mathbf{I}[\mathbf{t} \in Path_{\leq\delta}(\mathbf{X}_1, \ldots, \mathbf{X}_2)]$.

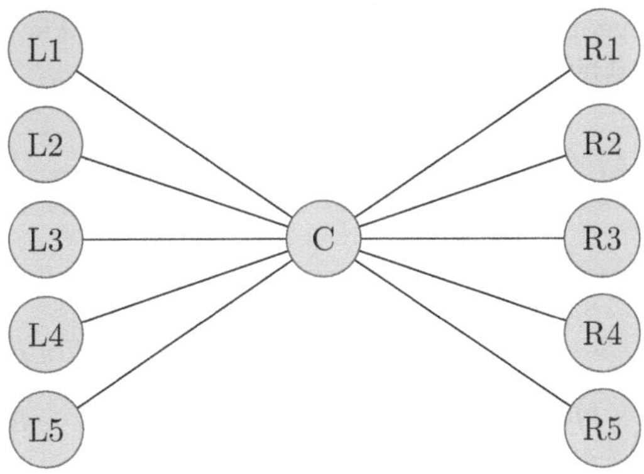

FIGURE 8.17: Star-shaped tree with 11 nodes: 5 nodes on the left, 5 nodes on the right, and 1 center node C. The red edges connect all nodes to the center. The center node has maximum path depth (path median) because it lies on all paths connecting left nodes to right nodes and on paths to each leaf.

We define the empirical $PD_{\leq\delta}$ of vertex $\mathbf{t}$ as

$$PD_{\leq\delta}(\mathbf{t}; F_n) = \frac{1}{\binom{n}{2}} \sum_{i<j}^{n} \mathbb{I}[\mathbf{t} \in Path_{\leq\delta}(\mathbf{X}_i, \dots, \mathbf{X}_j)], \tag{8.9}$$

where $\text{Path}_{\leq\delta}(\mathbf{X}_i, \dots, \mathbf{X}_j)$ is a path of length at most δ, $1 \leq \delta \leq \alpha$, that contains $\mathbf{t}$. One can show $P\mathbf{D}_\delta(\mathbf{X}_i, F_n) = PD_{\leq\delta}(\mathbf{X}_i; F_n) - PD_{\leq\delta-1}(\mathbf{X}_i; F_n)$ for $1 < \delta \leq \alpha$. Since only paths of at most length δ contribute to the depth of vertices, $PD_{\leq\delta}$ is sensitive only to the local behavior of the data cloud. For example $PD_1(\mathbf{X}_i; F_n) = \text{Degree}(\mathbf{X}_i)$. Local connectivity is closely related to the nearest neighbor graphs. In fact, the number of paths of length one provides the number of nearest neighbors of each vertex on the tree and $PD_{\leq\delta}$ measures how well a node is locally, within a path length of δ, connected. We provide an example of $PD_{\leq\delta}$ in the Exercises for this chapter.

Let $\mathcal{C}$ be a class of vertex sets on the MST. The type D depth functions on the MST is defined as $D(\mathbf{t}; F) = \inf_{\mathbf{t}\in C\in\mathcal{C}} \mathbb{P}(C) = \inf_{\mathbf{t}\in C\in\mathcal{C}} E\big[\mathbb{I}[\mathbf{X}_1 \in C]\big]$, where the expectation is taken over all possible MST with vertices that are i.i.d. random vectors with DF F. The next two depth functions are type D depth functions on the MST.

8.11.3 All Path Probability Depth

While $PD_{\leq\delta}$ may be used to investigate the local behaviors it may be difficult to determine the proper value of δ. The all paths probability depth (APPD) takes all paths into account while longer paths contribute less to depth values than shorter paths. We assign a weight $\mathbb{P}(\mathbf{X}_i)$ to each vertex such that $\sum_1^n \mathbb{P}(\mathbf{X}_i) = 1$ and define the depth of $\mathbf{t}$ to be

$$APPD(\mathbf{t}; F_n) = \sum_{i<j}^{n} \prod_{\mathbf{X}_k \in Path(\mathbf{X}_i,\dots,\mathbf{X}_j)} \mathbb{P}(\mathbf{X}_k)\mathbb{I}[\mathbf{t} \in Path(\mathbf{X}_i, \dots, \mathbf{X}_j)].$$

In the special case where $\mathbb{P}(\mathbf{X}_i) = \frac{1}{n}\ \forall\, i \in \{1, 2, \dots, n\}$, let $\gamma = 2$ and

$$h(\mathbf{t}; \mathbf{X}_1, \mathbf{X}_2) = \mathbb{I}[\mathbf{t} \in Path(\mathbf{X}_1, \dots, \mathbf{X}_2)]/n^{d(\mathbf{X}_1,\mathbf{X}_2)+1},$$

where $d(\mathbf{X}_1, \mathbf{X}_2)$ is the geodesic distance between $\mathbf{X}_1$ and $\mathbf{X}_2$. The APPD is a type A depth functions on the MST. The sample APPD is defined as

$$APPD(\mathbf{t}; F_n) = \sum_{i<j}^{n} \frac{\mathbb{I}[\mathbf{t} \in Path(\mathbf{X}_i, \ldots, \mathbf{X}_j)]}{n^{1+d(\mathbf{X}_i, \mathbf{X}_j)}},$$

where $d(\mathbf{X}_i, \mathbf{X}_j)$ is the geodesic distance between $\mathbf{X}_i$ and $\mathbf{X}_j$.

Example 8.11.2 *We show how to construct all path probability depth for Figure 8.16.*

- *There are two paths of length one, five paths of length two, four paths of length three and three paths of length four containing* $\mathbf{X}_1$. *Hence,*

$$APPD(\mathbf{X}_1, F_n) = \frac{2}{7^2} + \frac{5}{7^3} + \frac{4}{7^4} + \frac{3}{7^5} \approx 0.0572.$$

- *There are four paths of length one, four paths of length two, four paths of length three and three paths of length four containing* $\mathbf{X}_2$. *Hence,*

$$APPD(\mathbf{X}_2, F_n) = \frac{4}{7^2} + \frac{4}{7^3} + \frac{4}{7^4} + \frac{3}{7^5} \approx 0.0951.$$

- *There are two paths of length one, two paths of length two, four paths of length three and three paths of length four containing* $\mathbf{X}_3$. *Hence,*

$$APPD(\mathbf{X}_3, F_n) = \frac{2}{7^2} + \frac{2}{7^3} + \frac{4}{7^4} + \frac{3}{7^5} \approx 0.0485.$$

- *There are a paths of length one, a paths of length two, a paths of length three and three paths of length four containing* $\mathbf{X}_4$. *Hence,*

$$APPD(\mathbf{X}_4, F_n) = \frac{1}{7^2} + \frac{1}{7^3} + \frac{1}{7^4} + \frac{3}{7^5} \approx 0.0239.$$

- *There are a paths of length one, three paths of length two, a paths of length three and a paths of length four containing* $\mathbf{X}_5$. *Hence,*

$$APPD(\mathbf{X}_5, F_n) = \frac{1}{7^2} + \frac{3}{7^3} + \frac{1}{7^4} + \frac{1}{7^5} \approx 0.0296.$$

- *There are a path of length one, three paths of length two, a path of length three and a path of length four containing* $\mathbf{X}_6$. *Hence,*

$$APPD(\mathbf{X}_6, F_n) = \frac{1}{7^2} + \frac{3}{7^3} + \frac{1}{7^4} + \frac{1}{7^5} \approx 0.0296.$$

- *There are two paths of length one, three paths of length two, a path of length three and a path of length four containing* $\mathbf{X}_7$. *Hence,*

$$APPD(\mathbf{X}_7, F_n) = \frac{1}{7^2} + \frac{3}{7^3} + \frac{1}{7^4} + \frac{1}{7^5} \approx 0.0296.$$

8.11.4 RUNT Depth

The RUNT (Hartigan and Mohanty [140]) is a test statistic for detection of bimodality of a distribution based on the asymptotics of single linkage clusters. The RUNT of a node on the dendrogram is the smaller number of leaves of each of the two subtrees joined at that node. Stuetzle [324] associates a RUNT size with each edge of the MST for cluster detection. We define the depth of a terminal node as zero and the depth of a non–terminal node $\mathbf{t}$ as $1/n$ times the number of vertices in the smallest subtree formed in the forest generated after removing $\mathbf{t}$ from the MST. That is,

$$RUNTD(\mathbf{t}; F_n) = \frac{1}{n}\min(n_1, n_2, \ldots, n_k), \tag{8.10}$$

where $n_1, n_2, \ldots, n_k$ are the number of vertices in subtrees of vertex $\mathbf{t}$. The RUNT or the smallest cluster represents the direction through which a vertex is most exposed and RUNT depth measures the extent of this exposure. A large RUNT depth means that there are many vertices further out even on the most exposed direction. Hence, the vertex is deep or more central in the data cloud. The RUNT median is the vertex that maximizes the RUNT depth. The maximum value of RUNTD is $\frac{n-1}{2n}$ for odd n and approaches $1/2$ as n grows large. It should be noted that RUNT median may not be unique.

Figure 8.18 shows an MST with four subclusters of sizes 2, 3, 4, and 5. Nodes are shown in blue and edges in red. To find the runt of node C_3^3 the total number of vertices is 14. Removing node C_3^3 produces three connected components:

$$\begin{aligned}\text{Component 1 (left cluster)} &: 2 \text{ nodes},\\ \text{Component 2 (upper part)} &: 5 \text{ nodes},\\ \text{Component 3 (lower part)} &: 6 \text{ nodes}.\end{aligned}$$

Hence, $\text{Runt}(C_3^3) = \min\{2, 5, 6\} = 2$.

Let $h(\mathbf{t}; \mathbf{X}_1, \mathbf{X}_2, \ldots, \mathbf{X}_\gamma)$ be some unbounded, non–negative function that measures the distance between the point $\mathbf{t}$ and the set of points $\{\mathbf{X}_1, \mathbf{X}_2, \ldots, \mathbf{X}_\gamma\}$ based on their relationship on the MST. The type B depth functions on the MST is defined as $D(\mathbf{t}; F) = \big(1+E\big[h(\mathbf{t}; \mathbf{X}_1, \mathbf{X}_2, \ldots, \mathbf{X}_\gamma)\big]\big)^{-1}$, where the expectation is taken over all possible MST with vertices that are i.i.d. random vectors with DF F. The next depth function is a type B depth function on the MST.

8.11.5 Eccentricity Depth

Friedman and Rafsky [109] define a ranking scheme on trees based on the eccentricity of a vertex $\mathbf{X}_i$, $EC(\mathbf{X}_i)$, which is the maximum geodesic distance between $\mathbf{X}_i$ and any other vertex in the tree. The median is the vertex with the minimum eccentricity. To be consistent with other MST depth functions where the maximum depth is assigned to the central vertex, we define the eccentricity depth of any vertex in the MST as

$$ECD(\mathbf{t}; F_n) = 1 - \frac{1}{n-1}EC(\mathbf{t}). \tag{8.11}$$

The deepest vertex on the MST is the point with the largest ECD. In a string MST, where all nodes have degree 2 except the first and the last, the minimum value of ECD is zero and is reached at the two terminal nodes since the longest path that includes a terminal node is $n-1$ for this MST. The maximum value of ECD at the median is $\frac{n-3}{n-1}$, which approaches 1 as n grows large. This may also be verified for a tree of n vertices, n-1 of which are

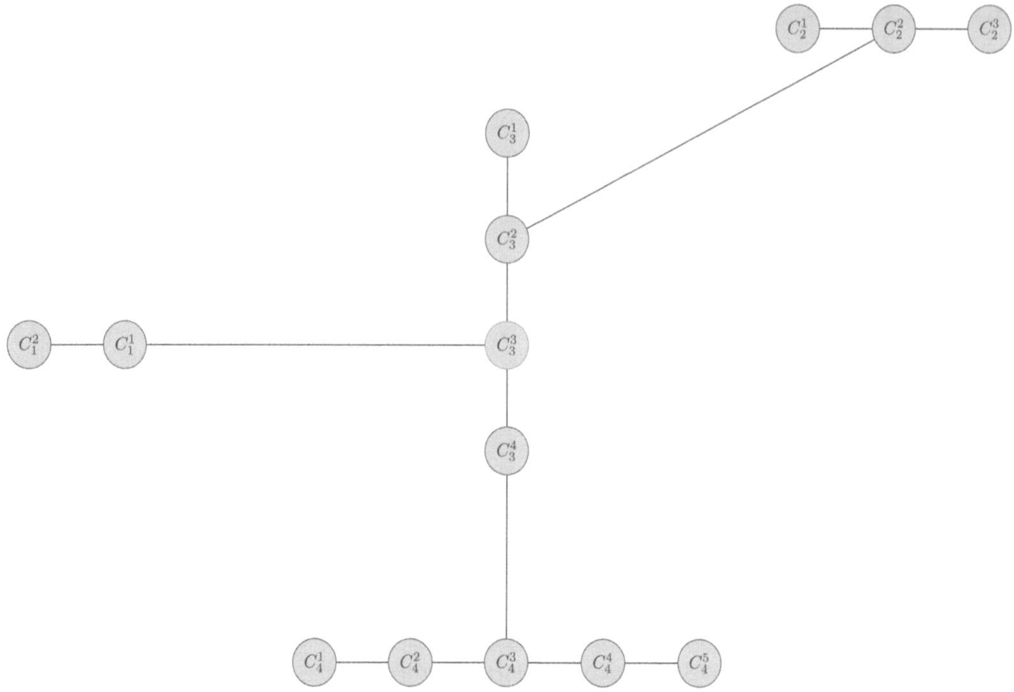

FIGURE 8.18: Minimum spanning tree with four subclusters of sizes 2, 3, 4, and 5. Node C_3^3 (shown in red) is removed when computing its runt. Removing C_3^3 produces three connected components of sizes 2, 5, and 6. Hence the runt of C_3^3 equals 2.

terminal nodes. When $d = 1$ one can show that $ECD(\mathbf{X}_i; F_n) = 1 - \frac{1}{n-1}\max(i-1, n-i)$ for $i = 1, \ldots, n$. Lemma 8.11.1 can be used for fast calculation of ECM.

Let $O(\mathbf{t}; \mathbf{X}_1, \mathbf{X}_2, \ldots, \mathbf{X}_\gamma)$ be some unbounded, non–negative function that measures the outlyingness of the point $\mathbf{t}$ with respect to the set of points $\{\mathbf{X}_1, \mathbf{X}_2, \ldots, \mathbf{X}_\gamma\}$ based on their relationship on the MST. The type C depth functions on the MST is defined as $D(\mathbf{t}; F) = \left(1 + E\left[O(\mathbf{t}; \mathbf{X}_1, \mathbf{X}_2, \ldots, \mathbf{X}_\gamma)\right]\right)^{-1}$, where the expectation is taken over all possible MST with vertices that are i.i.d. random vectors with DF F. The next depth function is a type C depth function on the MST.

8.11.6 Peeling Depth

Similar to convex hull peeling, peeling the MST provides an intuitive method of defining the depth of a vertex on the MST. We call this depth function peeling depth (PLD). Suppose we remove all vertices of degree 1 and their incident edges from the MST. These vertices have depth $1/n$. We repeat the process, and at stage k, after removing all vertices of degree 1 and their incident edge, we assign the rank of k/n to all such vertices. The process is terminated when we reach a single vertex or two vertices joined by an edge, in which case average depth is assigned to each vertex. It may be shown using a string MST that the maximum value of PLD is $\frac{n+1}{2}$ for odd n and $\frac{n}{2}$ for even n.

8.11.7 Geodesic Convexity Depth

Assuming that the data form the graph through their minimum spanning tree and the empirical distribution places mass $1/n$ on each vertex or data point, Small [317] introduced

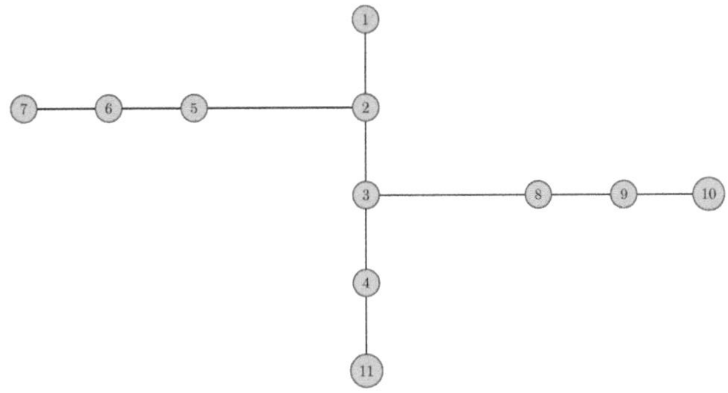

FIGURE 8.19: Tree with $n = 11$ vertices. Nodes are shown in blue and edges in red. Removing vertex 3 produces connected components of sizes 4, 3, and 3, yielding the maximum geodesic convexity depth.

the geodesic convexity depth and median that are applicable to distributions and data on graphs. If we remove $\mathbf{t}$ from the MST, the tree is divided into k clusters with $n_1, \ldots, n_k$ vertices, respectively. Small [317] defines the GCD at a vertex $\mathbf{t}$ as

$$GCD(\mathbf{t}; F_n) = 1 - \frac{1}{n}\max(n_1, n_2, \ldots, n_k). \tag{8.12}$$

The GCD median (GCM) is the vertex that maximizes the geodesic convexity depth. The vertices of minimum depth are the leaves and have a GCD of $1/n$. Small [317] shows that the level sets $\{x_i \in \mathbf{X} : GCD(x_i; F_n) \geq \alpha\}$, for $\alpha \geq 0$, are convex sets and the GCM is essentially unique as a consequence of the convexity.

Figure 8.19 shows an MST and we look for the GCD of vertex 3. Removing vertex 3 gives components of sizes 4, 3, 3. Thus, $GCD(3; F_{11}) = 1 - \frac{4}{11} = \frac{7}{11}$, yielding the maximum geodesic convexity depth.

Removing vertex 11 leaves one component of size 10, so $GCD(11; F_{11}) = 1 - \frac{10}{11} = \frac{1}{11}$. Vertex 3 maximizes the GCD and is therefore the geodesic convexity median.

8.11.8 Properties

Interpoint distances are the building blocks of proximity graphs. The IPDs remain invariant after an affine (orthogonal) transformation, depending on whether the edge weights use the Mahalanobis or the Euclidean distances. In particular, the transformations leave β-skeleton depth functions, including spherical and lens depth functions invariant. We state the result for lens depth.

Theorem 8.11.1 : *$LD(\mathbf{t}; F_n)$ is orthogonally invariant using the Euclidean distance and affine invariant using the Mahalanobis distance.*

Theorem 8.11.2 : *Given a random sample $\mathbf{X} = \{\mathbf{X}_1, \mathbf{X}_2, \ldots, \mathbf{X}_n\}$ from a continuous distribution F on $\mathbb{R}^d$, we construct an MST with vertex set $\mathbf{X}$ and edge set E. By removing $\mathbf{X}_i$, the MST splits into a forest of k trees $T_1, T_2, \ldots, T_k$. Let V_t and E_t be the vertex set and edge set of T_t, $t \in \{1, 2, \ldots, k\}$. If m_t is the cardinality of V_t, then the path depth of $\mathbf{X}_i$ is given by*

$$PD(\mathbf{X}_i; F_n) = \frac{1}{\binom{n}{2}}\{(n-1) + \sum_{1 \leq t < h}^{k} m_t m_h\}. \tag{8.13}$$

Proof: Since MST is connected, the vertex $\mathbf{X}_i$ is on a path with any other vertex $\mathbf{X}_j$, $i \neq j$ and there are $n-1$ such paths. Once the vertex $\mathbf{X}_i$ is removed, consider any other two vertices $\mathbf{X}_\alpha$ and $\mathbf{X}_\beta$, $\alpha \neq \beta$. If $\mathbf{X}_\alpha$ and $\mathbf{X}_\beta$ belong to the same subtree T_t, $t = 1, \ldots, k$, then $\mathbf{X}_i$ cannot be on any path between them within T_t. On the other hand, if $\mathbf{X}_\alpha$ and $\mathbf{X}_\beta$ belong to different subtrees T_t and T_h, $t \neq h$, then $\mathbf{X}_i$ is a bridge vertex joining T_t to T_h. Hence, $\mathbf{X}_i$ is on $m_t m_h$ paths that contain X_i. The result follows by adding the number of paths in all such subtrees. □

The computation of ECD median may be costly, as finding the longest path originating from a certain vertex X_i involves a depth first traversal of the tree. However, for finding ECD median, the following lemma helps us reduce the computing time.

Lemma 8.11.1 *Suppose* $\mathbf{X}_i$ *is any arbitrary vertex in the MST and* $\mathbf{X}_M$ *is the EC median. If* $Path(\mathbf{X}_i, \ldots, \mathbf{X}_j)$ *is the longest path between* $\mathbf{X}_i$ *and terminal node* $\mathbf{X}_j$*, then* $\mathbf{X}_M \in Path(\mathbf{X}_i, \ldots, \mathbf{X}_j)$.

Proof: We exclude the trivial case by assuming $\mathbf{X}_i \neq \mathbf{X}_M$, $\mathbf{X}_j \neq \mathbf{X}_M$. Suppose $\mathbf{X}_M \notin Path(\mathbf{X}_i, \ldots, \mathbf{X}_j)$. By the connectivity of MST, there must exists $Path(\mathbf{X}_M, \ldots, \mathbf{X}_i, \ldots, \mathbf{X}_j)$. However, the geodesic distance $d(\mathbf{X}_i, \mathbf{X}_j) < d(\mathbf{X}_M, \mathbf{X}_j)$. Hence, $\mathbf{X}_M$ does not have the highest ECD and cannot be ECM. □

Utilizing the above lemma to find the ECM, we start from any arbitrary vertex $\mathbf{X}_i$, then proceed to find the longest path originating from $\mathbf{X}_i$ and the ECD of all vertices on this path. The ECD median is the vertex with the highest depth on this path. Note that in string MST, this method is the same as calculating the ECD of all vertices. Addario-Berry et al. [1] show that expected diameter length of an MST is of $O(n^{1/3})$. Since the time complexity of a depth-first traversal is of $O(n)$ it follows that the time complexity of ECM is of $O(n^{4/3})$.

Due to its peeling nature, PLD is monotonically decreasing moving away from PLM. One may also show that ECD is monotonically decreasing moving away from ECM using Lemma 8.11.1.

8.11.9 Breakdown Point and Computational Complexity

We consider a random sample of size n denoted by $\mathbf{X} = \{\mathbf{X}_1, \mathbf{X}_2, \ldots, \mathbf{X}_n\}$. Let $\mathbf{X}^{n+m}$ be the contaminated sample obtained by appending $\mathbf{X}^* = \{\mathbf{X}_1^*, \mathbf{X}_2^*, \ldots, \mathbf{X}_m^*\}$ to $\mathbf{X}$ where $\mathbf{X}^*$ is m arbitrary points. The medians break down when the Euclidean distance between the medians of $\mathbf{X}$ and $\mathbf{X}^{n+m}$ becomes arbitrarily large. Donoho and Gasko [78] defined the additive breakdown point as the smallest contamination fraction $\frac{m}{n+m}$ under which the estimator breaks down. Without losing generality, we assume the two clusters are connected through the edge $(\mathbf{X}_1, \mathbf{X}_1^*)$.

As noted by Small [317], geodesic distance on a finite graph is bounded, which implies robustness of depth functions based on geodesic distance. When less than half of the vertices in $\mathbf{X}$ are relocated to ∞, removing any of these relocated vertices and their edges will still leave us with a connected subtree of unmoved vertices. Consequently, geodesic and path depths of the moved vertices will be less than $1/2$. The corresponding medians remain to be one of the unmoved vertices and can achieve a breakdown of 50 percent.

For PLM breakdown, we consider the case where $\mathbf{X}_1^*$ is the only vertex left when $\mathbf{X}$ is completely peeled. This condition is equivalent to having the longest path length between $\mathbf{X}_1^*$ and $\{\mathbf{X}_2^*, \mathbf{X}_3^*, \ldots, \mathbf{X}_m^*\}$ being greater than or equal to the longest path length between $\mathbf{X}_1^*$ and $\{\mathbf{X}_1, \mathbf{X}_2, \ldots, \mathbf{X}_n\}$. Breakdown point is found by finding the smallest m and largest

n for which the above holds. The smallest m is found when $\mathbf{X}^*$ forms a string tree. The longest path between $\mathbf{X}_1^*$ and $\{\mathbf{X}_2^*, \mathbf{X}_3^*, \ldots, \mathbf{X}_m^*\}$ is $m-1$ in this case. We note that the maximum degree of a vertex in the MST is bounded above by the Kissing number. The largest n may be found by considering a tree where all non–terminal nodes are of maximum degree while the longest path length between $\mathbf{X}_1$ and any terminal node is also $m-1$. Let k be the maximum degree then

$$n = 1 + k + k(k-1) + k(k-1)^2 + \ldots + k(k-1)^{\lfloor \frac{m-1}{2} \rfloor - 1} = 1 + k\frac{(k-1)^{\lfloor \frac{m-1}{2} \rfloor} - 1}{k-2}.$$

A similar argument shows ECM has the same breakdown point as PLM.

We summarize the computation complexities and breakdown points in Table 8.2. We consider the time complexity for the specified depth medians, assuming that the MST is given. The breakdown point of $PM_{\leq r}$ is listed as $\leq \frac{1}{2}$ in the table because it increases as r increases. When r equals the diameter of the MST, the breakdown point of $PM_{\leq r}$ is that of PM. PLM and ECM shares the same breakdown point where α is the diameter of the tree. We conclude the section by pointing out that RUNT median is not robust. It may be verified by considering a tree with one non–terminal node in the center that is connected to $n-1$ terminal nodes. In that case all vertices would have RUNT depth of either $1/n$ or zero. The median can breakdown by adding a string tree of 3 arbitrary points.

TABLE 8.2: The computation complexity and breakdown points of MST depth medians.

	PM	PM$_{\leq r}$	APPM	RUNTM	PLM	GCM	ECM
Computation Complexity	$O(n)$	$O(n)$	$O(n^2)$	$O(n)$	$O(n)$	$O(n)$	$O(n^{4/3})$
Breakdown Point	$\frac{1}{2}$	$\leq \frac{1}{2}$	$\leq \frac{1}{2}$	0	$\frac{\alpha}{n+\alpha}$	$\frac{1}{2}$	$\frac{\alpha}{n+\alpha}$

8.11.10 Finite Sample Relative Efficiency

To study the relative efficiency of using MST medians to estimate the center of a distribution, we consider the following estimators.

- Path depth median (PM), Path median of length at most r (PM$_{\leq r}$),
- All paths probability median (APPM), RUNT median (RUNTM),
- Eccentricity median (ECM), Peeling median (PLM),
- Geodesic convexity median (GCM), Spherical median (SPM), Lens median (LM).

Yang, Modarres and Guo [364] describe a sampling experiment using random samples from the bivariate standard normal distribution $\mathbb{N}(\mathbf{0}, \mathbf{I}_2)$, the heavy-tail bivariate t-distribution $t_2(\mathbf{0}, \mathbf{I}_2)$ with 2 degrees of freedom and standard bivariate Log-normal for several sample sizes n. They also considered the case $n \leq d$ for the multivariate normal distribution. They found that the averages of the medians using different depth functions are very close to the population medians of the distributions considered. The maximum deviation is less than 0.001 for all cases studied. The β-skeleton depth functions SPM and LM outperform the MST-based medians. However, when $d > n$, MST-based medians outperform SPM and have similar performance as LM. Overall, PM and GCM are recommended for the finding multidimensional medians on the MST. We also note that the MSE of $PM_{r\leq 2}$ and $PM_{r\leq 3}$ fluctuate with respect to sample size. This behavior is expected since they measure local connectivity in the tree. The RUNT depth of a new terminal node attached to the median immediately changes the depth of the median to $1/n$ and the new median will be in one of the sub-clusters. This explains why RUNT depth produces larger MSEs in some cases.

8.12 Depth of Ordinal Data

Let $\mathbf{O}$ be a finite set of objects. Instead of access to a numerical dissimilarity function $d : \mathbf{O} \times \mathbf{O} \to \mathbb{R}$, we are only given ordinal information in the form of comparisons

$$d(A, B) < d(C, D), \quad A, B, C, D \in \mathbf{O},$$

possibly for a subset of all quadruples. Such data arise in human similarity judgments, preference learning, machine learning, and non-metric multidimensional scaling.

We examine which classical and modern depth notions can be meaningfully computed, approximated, or interpreted under this ordinal information model, without relying on numerical distances, inner products, or affine structure. We expect any depth function based on data ordering and graph-based structures to work with ordinal data. These depths rely on rank information and neighborhood relations, making them invariant under monotone transformations of distance.

Halfspace Depth

Halfspace depth fundamentally relies on linear ordering induced by projections and affine geometry. Determining whether a point lies in a halfspace requires access to linear inequalities, which cannot be recovered from ordinal distance comparisons alone. Ordinal comparisons do not encode directional or affine information. Even if an ordinal embedding exists, multiple non-affinely equivalent configurations may satisfy the same comparisons. For example, two point clouds related by a non-linear monotone transformation preserve all distance orderings but yield different halfspace depths.

Simplicial Volume Depth

Simplicial volume depth depends on volumes of simplices spanned by subsets of points. Volume comparison is not invariant under monotone transformations of distances. Volume is a higher-order geometric quantity requiring metric consistency across dimensions. Ordinal distance data does not allow volume comparison even up to ordering.

Simplicial Depth

Simplicial depth depends only on whether a query point lies inside simplices formed by data points. Hence, it may be possible to only use ordinal information for computing this depth under strong assumptions. If ordinal data admit an embedding preserving orientation and convexity (e.g., exact Euclidean embedding), simplex containment can be inferred. For example, in one dimension, ordinal comparisons suffice to determine interval containment and thus simplicial depth.

Projection Depth

Projection depth relies on univariate depth after projecting data onto directions. If projections correspond to rankings along latent axes, ordinal information may suffice to estimate relative outlyingness. However, identifying meaningful projection directions requires more than distance comparisons.

Spatial Depth

Spatial depth depends on averaging normalized difference vectors. Vector subtraction and normalization are undefined without numerical coordinates or angles. Ordinal distance data lacks directional information.

Spherical Depth

Spherical depth is based on containment within metric balls. Ball containment can be determined from comparisons of the form $d(X, Y) \leq d(X, Z)$, which are directly available in ordinal data. For example, a point lies in the ball with diameter (A, B) if it is closer to both A and B than A and B are to each other, a purely ordinal condition.

Convex Hull Peeling

Convex hull peeling iteratively removes extreme points. Convex hull membership is not invariant under ordinal embeddings. Different embeddings consistent with the same comparisons may yield different hulls.

IPD Depth

IPD depth is defined on general metric spaces and relies on relative distance ordering. The construction uses only pairwise distance comparisons and does not require vector space structure. Depth rankings remain unchanged under any monotone transformation of the metric.

Lens Depth

Lens depth counts the number of lenses containing a query point. Lens membership depends solely on comparisons such as $d(X, A) \leq d(A, B)$ and $d(X, B) \leq d(A, B)$. Lens depth is used in comparison-based similarity search and ordinal embedding contexts.

β-Skeleton Depth

β-skeleton depth generalizes lens depth by varying the lens shape. For a fixed β, skeleton inclusion is determined entirely by ordinal distance comparisons.

RUNT Depth on the MST

RUNT depth is defined via the minimum spanning tree (MST). The MST depends only on the ordering of edge weights, not their numerical values. For example, Kruskal's algorithm operates entirely on ordinal edge comparisons.

Geodesic Convexity Depth

Geodesic convexity depth is defined on general metric spaces with geodesics. Identifying geodesics requires metric interpolation, which is not available from ordinal data.

Table 8.3 summarizes the structural requirements of the discussed depth functions. Depth functions relying exclusively on relative distance comparisons such as lens-based, skeleton-based, and MST-based depths are naturally compatible with ordinal data. In contrast, depths requiring affine structure, volumes, directions, or convex hulls are fundamentally incompatible with purely ordinal information.

TABLE 8.3: Structural requirements of depth functions. "Ordinal" indicates compatibility with distance comparisons only; "Metric" requires numerical distances; "Affine" requires vector space or affine structure.

Depth Function	**Ordinal**	**Metric**	**Affine**
Halfspace Depth	No	No	Yes
Simplicial Volume Depth	No	Yes	Yes
Simplicial Depth	Partial	Yes	Yes
Projection Depth	Partial	Yes	Yes
Spatial Depth	No	Yes	Yes
Spherical Depth	Yes	Yes	No
Convex Hull Peeling	No	Yes	Yes
Lok and Lee	Yes	Yes	No
Lens Depth	Yes	Yes	No
β-Skeleton Depth	Yes	Yes	No
RUNT Depth (MST)	Yes	Yes	No
Geodesic Convexity Depth	No	Yes	No

8.13 Analysis of University Rankings

Johnson and Wichern [162] analyze a dataset consisting of 25 U.S. universities described by six academic and institutional variables: the average SAT score of entering freshmen, the percentage of freshmen ranked in the top 10% of their high school class, the acceptance rate, the student-to-faculty ratio, estimated annual expenses, and the graduation rate. The objective is to visualize the observations, identify a central node, determine the terminal nodes of the minimum spanning tree (MST), and rank the universities relative to this center.

To this end, the universities are represented as vertices in $\mathbf{R}^6$, and an MST is constructed to facilitate the exploration of MST-based depth functions. Figure 8.20 displays the depth values of 25 universities across eight different depth functions. Each row corresponds to a university, labeled along the vertical axis, while the horizontal axis represents the depth values (scaled by a factor of 10). Each depth function is represented by a distinct marker and color, allowing for easy comparison across functions. This layout allows one to quickly identify patterns, such as universities with consistently high- or low-depth values across multiple functions, and facilitates visual comparisons between depth functions for a given university. The grid lines improve readability across the value range.

The MST of the underlying graph appears in Figure 8.21. Across all MST depth functions except $PD_{\leq r}$, Northwestern University attains the highest-depth value. This indicates that Northwestern is the most centrally located observation within the data cloud and may be regarded as the median node. Terminal nodes are characterized by a RUNT depth of zero; notably, all terminal nodes receive identical depth values, with the exception of those

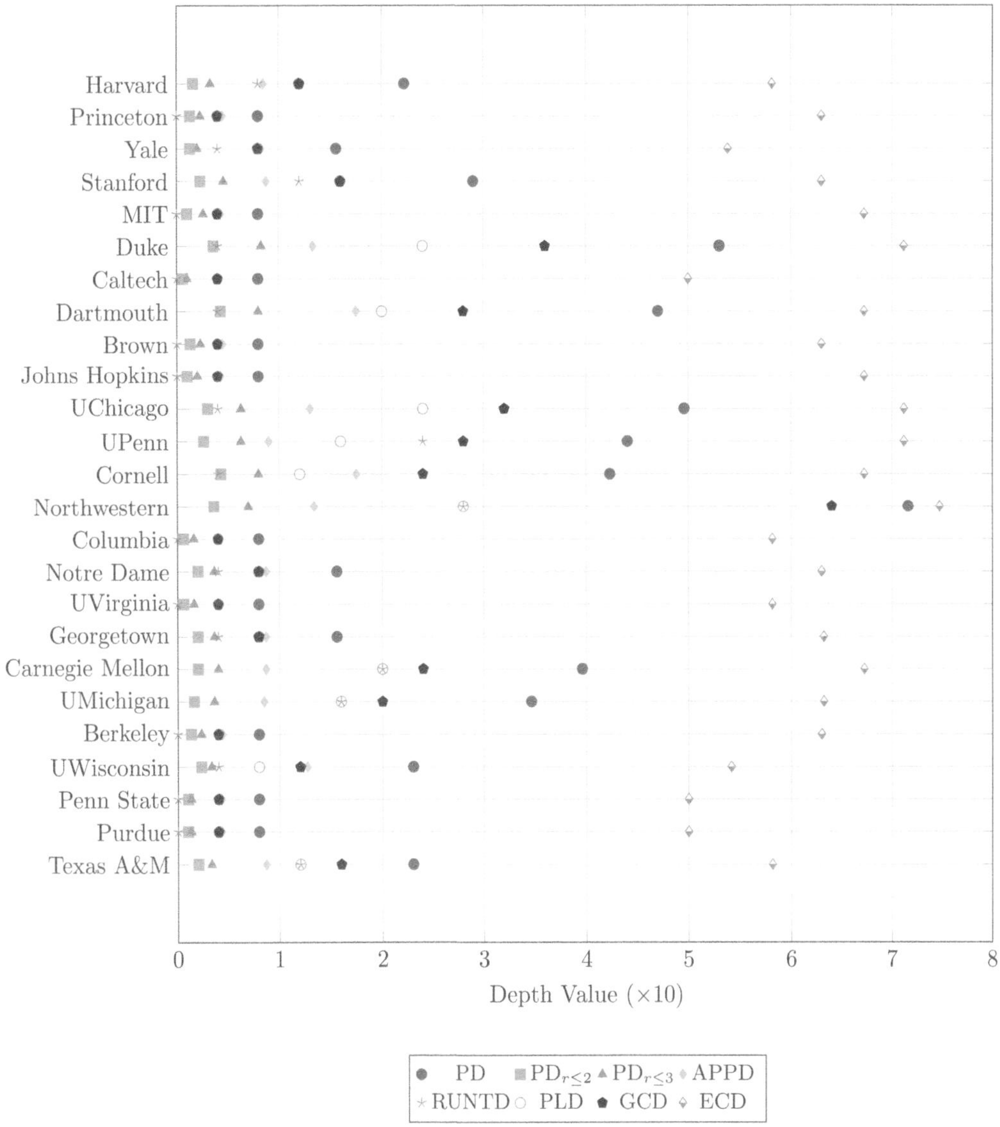

FIGURE 8.20: Depth values for all 25 universities across eight depth functions. Each row represents a university (vertical axis) and each colored marker represents a different depth function (see legend). This layout allows visual comparison of depth values across universities and depth functions.

defined by the ECD and $PD_{\leq r}$ depth functions. The two largest IPDs along the MST are associated with the California Institute of Technology and Pennsylvania State University. Furthermore, large state universities tend to cluster together as descendants of the Carnegie Mellon node, while most private institutions are situated on the upper left-hand side of the Northwestern node.

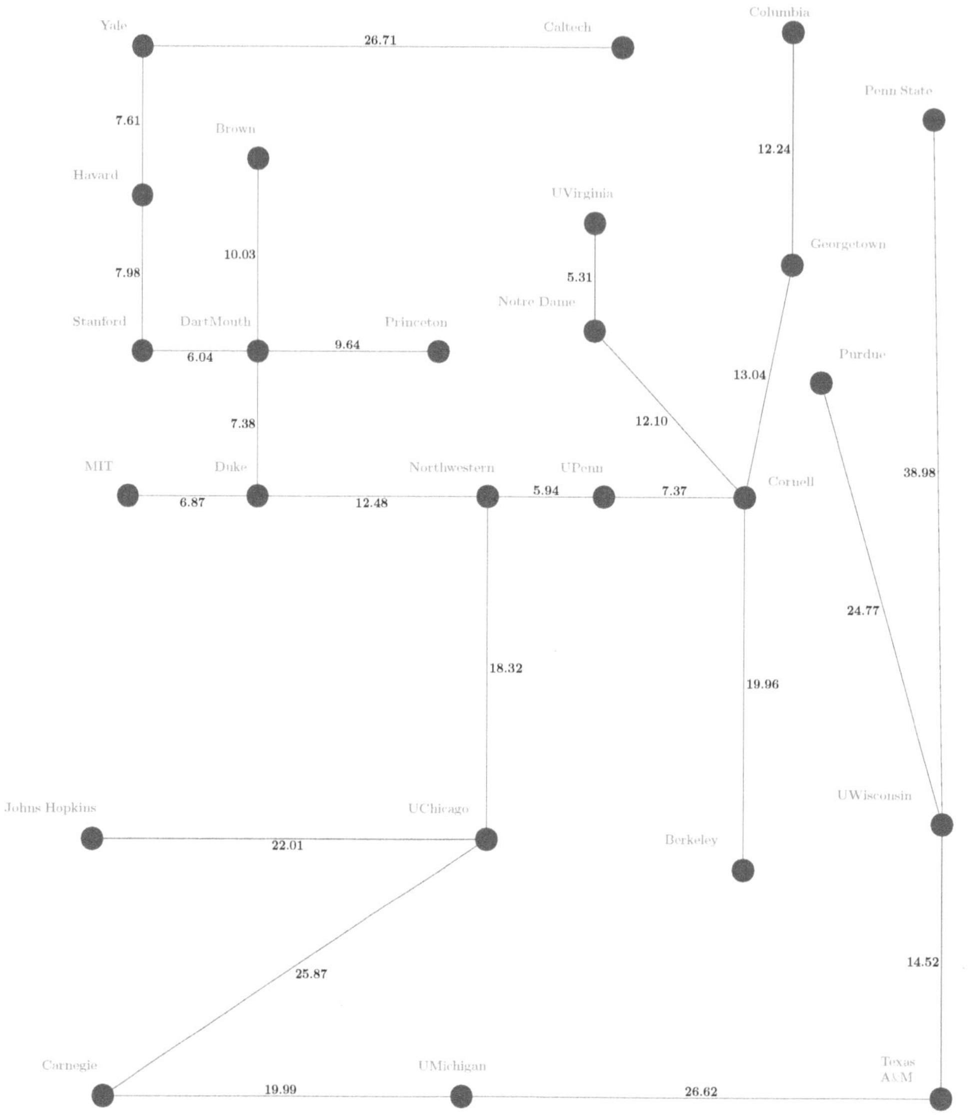

FIGURE 8.21: Depth values of 25 universities on the MST-based on six variables. The edge values are the Euclidean distances between vertices.

8.14 Summary

This chapter offers an overview of data depth as a foundational concept for non-parametric multivariate analysis. Data depth provides a principled framework for measuring the centrality of observations in $\mathbb{R}^d$ through center–outward ordering, thereby extending univariate notions of ranking and quantiles to higher dimensions. The chapter reviews the defining properties of statistical depth functions, such as affine invariance, monotonicity, maximality at the center, and vanishing at infinity, and surveys a broad class of classical depth notions, including Mahalanobis, half-space, simplicial, projection, spatial, spherical, convex hull peeling, interpoint, and lens depths. These depth functions support a wide range of statistical tasks, including robust estimation of multivariate location, ranking, classification, outlier detection, and visualization via depth contours and level sets.

A central theme of the chapter is the interaction between depth functions, geometry, and computational complexity, particularly in high-dimensional settings. While many traditional depth-based medians suffer from exponential computational cost as the dimension increases, graph-based constructions offer robust alternatives. The chapter develops depth notions based on proximity graphs and minimum spanning trees (MSTs), including β-skeleton depth, path-based depths, RUNT depth, eccentricity depth, peeling depth, and geodesic convexity depth. These graph-based depths retain desirable robustness and distribution-free properties while achieving computational efficiency that is largely independent of the ambient dimension, making them particularly attractive for high dimensional data analysis.

The chapter further explores theoretical and practical aspects of depth-based methods, including depth–depth plots for comparing multivariate distributions, proximity graph representations, and the asymptotic behavior of depth functions as the dimension grows. Properties such as breakdown point, finite-sample relative efficiency, and computational feasibility are examined to guide practical implementation. An applied analysis of university ranking data illustrates how depth-based methods provide interpretable, robust, and flexible tools for multivariate ranking and inference. The chapter establishes data depth as a unifying and versatile paradigm for modern multivariate statistics, offering both theoretical insight and practical methodologies for high-dimensional and non-elliptical data.

8.15 Proofs

Proof of Theorem 8.11.1: is left as an exercise.

Proof of Theorem 8.5.2: For any given $\mathbf{t} \in \mathbb{R}^d$, we observe that

$$
\begin{aligned}
\{\mathbf{t} \in L(\mathbf{X}_1, \mathbf{X}_2)\} &\subseteq \{\|\mathbf{t}\| \le \|\mathbf{X}_2\| + \|\mathbf{t} - \mathbf{X}_2\|\} \cup \{\|\mathbf{t}\| \le \|\mathbf{X}_1\| + \|\mathbf{t} - \mathbf{X}_1\|\} \\
&\subseteq \{\|\mathbf{t}\| \le \|\mathbf{X}_2\| + \|\mathbf{X}_1 - \mathbf{X}_2\|\} \cup \{\|\mathbf{t}\| \le \|\mathbf{X}_1\| + \|\mathbf{X}_1 - \mathbf{X}_2\|\} \\
&\subseteq \{\|\mathbf{t}\| \le \|\mathbf{X}_1\| + 2\|\mathbf{X}_2\|\} \cup \{\|\mathbf{t}\| \le 2\|\mathbf{X}_1\| + \|\mathbf{X}_2\|\} \\
&\subseteq \{\|\mathbf{t}\| \le 3\|\mathbf{X}_1\|\} \cup \{\|\mathbf{t}\| \le 3\|\mathbf{X}_2\|\} = \cup_{i=1,2}\{\|\mathbf{X}_i\| \ge \frac{\|\mathbf{t}\|}{3}\},
\end{aligned}
$$

$$
LD(\mathbf{t}; F) = \mathbb{P}(\mathbf{t} \in L(\mathbf{X}_1, \mathbf{X}_2)) \le \sum_{i=1,2} \mathbb{P}(\{\|\mathbf{X}_i\| \ge \frac{\|\mathbf{t}\|}{3}\}) = 2\mathbb{P}(\{\|\mathbf{X}_1\| \ge \frac{\|\mathbf{t}\|}{3}\}).
$$

Since $\mathbb{P}(\{\|\mathbf{X}_1\| \ge \frac{\|\mathbf{t}\|}{3}\})$ is decreasing in $\|\mathbf{t}\|$, it follows that as $M \longrightarrow \infty$,

$$
\sup_{\|\mathbf{t}\| \ge M} LD(\mathbf{t}; F) \le \sup_{\|\mathbf{t}\| \ge M} 2\mathbb{P}(\{\|\mathbf{X}_1\| \ge \frac{\|\mathbf{t}\|}{3}\}) \le 2\mathbb{P}(\{\|\mathbf{X}_1\| \ge \frac{M}{3}\}) \longrightarrow 0.
$$

Proof of Theorem 8.5.3: We prove the theorem for $d = 2$. A more general proof for any dimension d follows analogously. Let $\mathbf{X}_1$, $\mathbf{X}_2$ be i.i.d. with a bivariate c.d.f F. Let $\mathbf{t}_k$ be a sequence in $\mathbb{R}^2$ such that $\mathbf{t}_k \to \mathbf{t}$ and we will show that

$$
|LD(\mathbf{t}_k; F) - LD(\mathbf{t}; F)| \le \mathbb{P}(\partial(L(\mathbf{X}_1, \mathbf{X}_2)) \cap \overline{\mathbf{t}\mathbf{t}_k}) \to 0
$$

as $k \to \infty$, where $\partial(L(\mathbf{X}_1, \mathbf{X}_2))$ denotes the boundary of $L(\mathbf{X}_1, \mathbf{X}_2)$, " $\cap$ " denotes the intersection of a line or arc segment, and $\overline{\mathbf{t}\mathbf{t}_k}$ denotes the line segment connecting $\mathbf{t}$ and $\mathbf{t}_k$. $LD(\mathbf{t}_k; F) \neq LD(\mathbf{t}; F)$ only if $L(\mathbf{X}_1, \mathbf{X}_2)$ contains one of the points $\mathbf{t}$ or $\mathbf{t}_k$ but not the other. This implies that $\partial(L(\mathbf{X}_1, \mathbf{X}_2))$ must intersect the line segment $\overline{\mathbf{t}\mathbf{t}_k}$. Let $A = \{\mathbf{t} \in L(\mathbf{X}_1, \mathbf{X}_2)\}$ and $B = \{\mathbf{t}_k \in L(\mathbf{X}_1, \mathbf{X}_2)\}$.

$$\begin{aligned} LD(\mathbf{t}; F) - LD(\mathbf{t}_k; F) &= \mathbb{P}(A) - \mathbb{P}(B) \leq \mathbb{P}(A) - \mathbb{P}(A \cap B) \leq \mathbb{P}(A \cap B^c) \\ &= \mathbb{P}(\mathbf{t} \in L(\mathbf{X}_1, \mathbf{X}_2) \cap \mathbf{t}_k \notin L(\mathbf{X}_1, \mathbf{X}_2)) \leq \mathbb{P}(\partial(L(\mathbf{X}_1, \mathbf{X}_2)) \cap \mathbf{t}\bar{\mathbf{t}}_k). \end{aligned}$$

Similarly, We can show that $LD(\mathbf{t}_k; F) - LD(\mathbf{t}; F) \leq \mathbb{P}(\partial(L(\mathbf{X}_1, \mathbf{X}_2)) \cap \mathbf{t}\bar{\mathbf{t}}_k)$. Thus, $|LD(\mathbf{t}_k; F) - LD(\mathbf{t}; F)| \leq \mathbb{P}(\partial(L(\mathbf{X}_1, \mathbf{X}_2)) \cap \mathbf{t}\bar{\mathbf{t}}_k) \to 0$.

Let $A_k = \{\partial(L(\mathbf{X}_1, \mathbf{X}_2)) \cap \mathbf{t}\bar{\mathbf{t}}_k\}$, $\forall k$.

$$\limsup_k A_k = \cap_{k=1}^{\infty}[\cup_{j=k}^{\infty} A_j] = \{\mathbf{t} \in \partial(L(\mathbf{X}_1, \mathbf{X}_2))\}.$$

By Fatu's Lemma (Billingsley [34]),

$$\limsup_k \mathbb{P}(A_k) \leq \mathbb{P}(\limsup_k A_k) = \mathbb{P}(\mathbf{t} \in \partial(L(\mathbf{X}_1, \mathbf{X}_2))).$$

By conditioning on $\mathbf{X}_1$, for any $\mathbf{t} \in \partial(L(\mathbf{X}_1, \mathbf{X}_2))$, we observe that $\mathbf{X}_2$ must fall on the curve passing through a', a, t, b', b ($C(a', a, t, b', b)$). Since F is absolutely continuous, $\mathbb{P}(X_2 \in C(a', a, t, b', b)|\mathbf{X}_1) = 0$. Thus,

$$\mathbb{P}(\mathbf{t} \in \partial(L(\mathbf{X}_1, \mathbf{X}_2))) = \int_{\mathbf{X}_1} \mathbb{P}(X_2 \in C(a', a, t, b', b)|\mathbf{X}_1) d\mathbf{X}_1 = 0,$$

$$\limsup_{k\to\infty} |LD(\mathbf{t}_k; F) - LD(\mathbf{t}; F)| \leq \limsup_{k\to\infty} \mathbb{P}(A_k) \leq \mathbb{P}(\mathbf{t} \in \partial(L(\mathbf{X}_1, \mathbf{X}_2))) = 0.$$

Proof of Theorem 8.5.4: Elmore, Hettmansperger and Xuan [88] show that SPD has a maximal depth 1/2 for any absolutely continuous distribution function F on $\mathbb{R}^d$. It is easy to see that $\mathbf{t} \in S(\mathbf{X}_1, \mathbf{X}_2) \subset \mathbf{t} \in L(\mathbf{X}_1, \mathbf{X}_2)$, where $S(\mathbf{X}_1, \mathbf{X}_2)$ is the closed sphere of radius $\frac{\|\mathbf{X}_1 - \mathbf{X}_2\|}{2}$ centered at $\frac{\mathbf{X}_1 + \mathbf{X}_2}{2}$. Thus, $SPD(\mathbf{t}; F) = \mathbb{P}(\mathbf{t} \in S(\mathbf{X}_1, \mathbf{X}_2)) \leq \mathbb{P}(\mathbf{t} \in L(\mathbf{X}_1, \mathbf{X}_2)) = LD(\mathbf{t}; F)$. Therefore, LD has a maximal depth $\geq 1/2$ for any absolutely continuous distribution function F on $\mathbb{R}^d$.

Proof of Theorem 8.5.5: Without loss of generality, assume that $\boldsymbol{\mu} = \mathbf{0}$. We observe that $\|-\mathbf{t} - \mathbf{X}_i\| = \|\mathbf{t} + \mathbf{X}_i\| = \|\mathbf{t} - (-\mathbf{X}_i)\|$ for $i = 1, 2$. Since F is centrally symmetric about the origin, $\mathbf{X}_1 \overset{D}{=} -\mathbf{X}_1$ and $\mathbf{X}_2 \overset{D}{=} -\mathbf{X}_2$, we have $LD(\mathbf{t}; F) = \mathbb{P}(\|\mathbf{X}_1 - \mathbf{X}_2\| > \max(\|\mathbf{t} - \mathbf{X}_1\|, \|\mathbf{t} - \mathbf{X}_2\|)) = \mathbb{P}(\|\mathbf{X}_1 - \mathbf{X}_2)\| > \max(\|-\mathbf{t} - \mathbf{X}_1)\|, \|-\mathbf{t} - \mathbf{X}_2)\|)) = LD(-\mathbf{t}; F)$.

Proof of Theorem 8.5.7: We denote the lens depth of a point $\mathbf{t} \in \mathbb{R}^d$ as $LD(\mathbf{t}; F) = \mathbb{P}(t \in \text{Lens}(\mathbf{X}_1, \mathbf{X}_2))$, with $\mathbf{X}_1, \mathbf{X}_2$ independent copies of $\mathbf{X}$. We wish to show that the center $\boldsymbol{\mu}$ maximizes $LD(\mathbf{t}; F)$. Define $\mathbf{Y} = \boldsymbol{\Sigma}^{-1/2}(\mathbf{X} - \boldsymbol{\mu})$, so that $\mathbf{Y}$ is spherically symmetric about $\mathbf{0}$ in $\mathbb{R}^d$: $f_{\mathbf{Y}}(\mathbf{y}) = c\,g(\|\mathbf{y}\|^2)$. Under this transformation, the lens depth is preserved under affine transformations,

$$t \mapsto LD(\mathbf{t}; F) \quad \Longleftrightarrow \quad \mathbf{y} = \boldsymbol{\Sigma}^{-1/2}(\mathbf{t} - \boldsymbol{\mu}) \mapsto LD(\mathbf{y}; F).$$

Thus it suffices to study the spherically symmetric case in $\mathbb{R}^d$ with center 0. For two points $\mathbf{Y}_1, \mathbf{Y}_2 \in \mathbb{R}^d$, the lens is

$$\text{Lens}(\mathbf{Y}_1, \mathbf{Y}_2) = B(\mathbf{Y}_1, \|\mathbf{Y}_1 - \mathbf{Y}_2\|) \cap B(\mathbf{Y}_2, \|\mathbf{Y}_1 - \mathbf{Y}_2\|),$$

where $B(x, r)$ is the Euclidean ball of radius r. The depth at $\mathbf{y} \in \mathbb{R}^d$ is $LD(\mathbf{y}; F) = \mathbb{P}(\mathbf{y} \in \text{Lens}(\mathbf{Y}_1, \mathbf{Y}_2))$. Because $\mathbf{Y}_1, \mathbf{Y}_2$ are spherically symmetric about $\mathbf{0}$ we have

- $(\mathbf{Y}_1, \mathbf{Y}_2) \stackrel{d}{=} (-\mathbf{Y}_1, -\mathbf{Y}_2)$. Therefore, $LD(\mathbf{y}; F) = LD(-\mathbf{y}; F)$.
- Depth is radial, $LD(\mathbf{y}; F) = LD(\|\mathbf{y}\|; F)$

To show the monotonicity of lens depth with radius $r = \|\mathbf{y}\|$ consider the lens of two points $\mathbf{Y}_1, \mathbf{Y}_2$. The lens contains the midpoint $m = (\mathbf{Y}_1 + \mathbf{Y}_2)/2$. The probability that a point lies in the lens decreases as the point moves radially outward from the origin, because

$$g(\|\mathbf{y}\|^2) \text{ is non-increasing in } \|\mathbf{y}\|.$$

Hence, $LD(\mathbf{y}; F) \le LD(\mathbf{0}; F) \quad$ for all $\mathbf{y} \in \mathbb{R}^d$. Transforming back we get

$$t = \boldsymbol{\Sigma}^{1/2}\mathbf{y} + \boldsymbol{\mu} \quad \Longrightarrow \quad LD(\mathbf{t}; F) = LD(\mathbf{y}; F) \le LD(\mathbf{0}; F) = LD(\boldsymbol{\mu}; F).$$

Hence the lens depth is maximized at the center $\boldsymbol{\mu}$.

Proof of Theorem 8.5.6: Let $U = \frac{2}{m}\sum_{i=1}^m (\mathbb{P}(\mathbf{X}_i) - LD(\mathbf{t}; F))$. Since $\mathbf{X}_i, i = 1, \ldots, m$, are i.i.d., $\mathbb{P}(\mathbf{X}_i)$ are bounded, i.i.d. with mean $LD(\mathbf{t}; F)$ and variance β^2. By the central limit theorem under Lindeberg-Levy condition, $m^{1/2}U/(4\beta^2) \xrightarrow{D} \mathbb{N}(0, 1) \quad \text{as} \quad m \to \infty$. We obtain

$$\begin{aligned}
\mathbb{C}ov(LD(\mathbf{t}; F_m), U) &= \frac{2}{m}\sum_{i=1}^m \mathbb{E}((LD(\mathbf{t}; F_m) - LD(\mathbf{t}; F)) \cdot (\mathbb{P}(\mathbf{X}_i) - LD(\mathbf{t}; F))) \\
&= \frac{4}{m^2}\sum_{i=1}^m \mathbb{E}(\mathbb{P}(\mathbf{X}_i) - LD(\mathbf{t}; F) \cdot (\mathbb{P}(\mathbf{X}_i) - LD(\mathbf{t}; F))) \\
&= \frac{4}{m^2}\sum_{i=1}^m \beta^2 = \frac{4}{m}\beta^2, \\
\mathbb{V}ar(LD(\mathbf{t}; F_m) - U) &= \mathbb{V}ar(LD(\mathbf{t}; F_m)) - 2\mathbb{C}ov(LD(\mathbf{t}; F_m), U) + \mathbb{V}ar(U) \\
&= \frac{4}{m}\beta^2 + o(m^{-1}) - 2\frac{4}{m}\beta^2 + \frac{4}{m}\beta^2 = o(m^{-1}).
\end{aligned}$$

By the Chebyshev's Lemma, $(LD(\mathbf{t}; F_m) - LD(\mathbf{t}; F)) - U \xrightarrow{P} 0$. Since U is asymptotically normal with mean 0 and variance $\frac{4\beta^2}{m}$, by Slusky's Theorem, we obtain $LD(\mathbf{t}; F_m) - LD(\mathbf{t}; F)$ is asymptotically normal with mean 0 and variance $\frac{4\beta^2}{m}$.

8.16 Exercises

Exercise 8.1 *The Mahalanobis depth* $MD(\mathbf{t}; F)$ *of a vector* $\mathbf{t}$ *with respect to a distribution function* F *with mean vector* $\boldsymbol{\mu}_F$ *and dispersion matrix* $\boldsymbol{\Sigma}_F$ *is defined in* (8.5.1).

a) *Write a program to generate a sample of size* n *from a bivariate normal distribution* (F) *with mean* $\boldsymbol{\mu}_F = (0, 0)$, *unit variances and correlation 0.5. If you code in R, see the function mahalanobis(x, center, cov). Let* $n = 10$. *Print the observations, their mean and covariance, and the depth of each data point. Find and print the bivariate median; i.e., the bivariate vector that maximizes the depth function.*

b) *Generate* $n = 1000$ *bivariate observations from* F. *Plot all observations and identify the lowest 5 percent of the data in red.*

Exercise 8.2 *Prove Theorem 8.11.1.*

Exercise 8.3 *Prove that the kernel function of* SkelD_β *can be written in terms of IPDs as*

$$\begin{aligned} h(\mathbf{X}_1, \mathbf{X}_2) &= \mathbb{I}[\mathbf{t} \in \text{Skel}_\beta(\mathbf{X}_1, \mathbf{X}_2)] \\ &= \mathbb{I}\big[\frac{\beta}{2}||\mathbf{X}_1 - \mathbf{X}_2|| \geq \max\{||\mathbf{t} - \mathbf{C}_1||, ||\mathbf{t} - \mathbf{C}_2||\}\big], \end{aligned} \tag{8.14}$$

where $\mathbf{C}_1 = (1 - \frac{\beta}{2})\mathbf{X}_1 + \frac{\beta}{2}\mathbf{X}_2$, $\mathbf{C}_2 = \frac{\beta}{2}\mathbf{X}_1 + (1 - \frac{\beta}{2})\mathbf{X}_2$ *and* $1 \leq \beta < \infty$.

Exercise 8.4 *Consider the kernel function of* β*-skeleton depth defined in Equation* (8.3). *Prove that for* $1 \leq \beta < \infty$, *the* β*-skeleton depth is orthogonally invariant using the Euclidean distance and affine invariant under the Mahalanobis distance.*

Exercise 8.5 *Prove that for a continuous distribution* F *in* $\mathbb{R}^d$, *the half-space depth is maximized at the Tukey median, i.e., the point with the highest depth.*

Exercise 8.6 *Prove that the half-space depth of a point* x *in one dimension is* $\min(F(x), 1 - F(x))$, *where* F *is the empirical distribution function.*

Exercise 8.7 *Prove that convex hull peeling depth assigns depth* k *to points in the* k*-th convex layer of the data, and that it is not affine invariant.*

Exercise 8.8 *Compute the CH (convex hull peeling depth) of each point in the dataset*

$$\{(0, 0), (1, 0), (0, 1), (1, 1), (0.5, 0.5)\}.$$

Exercise 8.9 *Let* $D_{SV}(x) = \mathbb{E}[vol(S) \cdot \mathbb{I}\{x \in S\}]$ *over all simplices* S *formed by* $d+1$ *points from a dataset in* $\mathbb{R}^d$. *Show that* D_{SV} *is affine invariant.*

Exercise 8.10 *Explain how the simplicial volume depth of a point is affected by the volume of simplices it forms with other data points.*

Exercise 8.11 *Let* $\mathbf{X}_1, \ldots, \mathbf{X}_m$ *be a sample from a distribution* F *in* $\mathbb{R}^d$, *and let* $x \in \mathbb{R}^d$ *be a point. The majority depth (Singh [314]) of* x *with respect to* F *is defined as the probability that* x *lies in the majority half-space determined by a randomly chosen pair* $(\mathbf{X}_i, \mathbf{X}_j)$ *of data points. That is,*

$$D_{MJD}(x; F) = \mathbb{P}_{\mathbf{X}_i, \mathbf{X}_j}\left(x \in H^+(\mathbf{X}_i, \mathbf{X}_j)\right)$$

where $H^+(\mathbf{X}_i, \mathbf{X}_j)$ *contains the majority of the sample and* $H^+(\mathbf{X}_i, \mathbf{X}_j)$ *is the closed half-space bounded by the hyperplane through* $\mathbf{X}_i$ *and* $\mathbf{X}_j$ *such that it contains at least* $\lceil (n-2)/2 \rceil$ *of the remaining data points* $\{\mathbf{X}_k : k \neq i, j\}$.

In the population version, we define

$$D_{MJD}(x; F) = \mathbb{P}\left(x \in H^+(X,Y) \text{ and } F(H^+(X,Y)) \geq 1/2\right),$$

where (X, Y) *are independent random vectors drawn from* F. *Let* $x \in \mathbb{R}^d$. *Show that the majority depth is given by the proportion of hyperplanes formed by data pairs such that* x *lies on the side containing the majority of other points.*

Exercise 8.12 *Show that the majority depth of a point in 2D with respect to a dataset is proportional to the number of lines passing through pairs of points that have the same number of data points on each side.*

Exercise 8.13 *Show that the simplicial depth is maximal at the center of a symmetric distribution in* $\mathbb{R}^2$.

Exercise 8.14 *Show that the simplicial depth is affine invariant.*

Exercise 8.15 *Prove that DPTD is maximal at the spatial median and derive the estimating equation for the spatial median.*

Exercise 8.16 *Let* $\mathbf{X}_1, \ldots, \mathbf{X}_n$ *be data in* $\mathbb{R}^d$. *Show that the spatial depth of* $\mathbf{x}$ *is given by* $D_{DPTD}(\mathbf{x}) = \mathbf{1} - \left\| \frac{\mathbf{1}}{\mathbf{n}} \sum_{\mathbf{i}=\mathbf{1}}^{\mathbf{n}} \frac{\mathbf{X}_\mathbf{i} - \mathbf{x}}{\|\mathbf{X}_\mathbf{i} - \mathbf{x}\|} \right\|$.

Exercise 8.17 *Let* $\mathbf{X}_1, \ldots, \mathbf{X}_n$ *be in* $\mathbb{R}^d$. *Show that the projection depth of a point* $\mathbf{x}$ *is defined as*

$$D_{PD}(\mathbf{x}) = \left(\mathbf{1} + \sup_{\|\mathbf{u}\|=\mathbf{1}} \frac{|\mathbf{u}'\mathbf{x} - Med(\mathbf{u}'\mathbf{X})|}{MAD(\mathbf{u}'\mathbf{X})} \right)^{-1}.$$

Prove that this depth is affine invariant.

Exercise 8.18 *Explain why the projection depth of a point is robust against outliers.*

Exercise 8.19 *Let* $x \in \mathbb{R}^d$. *The spherical depth of* x *is the proportion of all pairs* $(\mathbf{X}_i, \mathbf{X}_j)$ *such that* $\mathbf{x}$ *lies in the closed ball with diameter* $\mathbf{X}_i\mathbf{X}_j$. *Prove that for any such pair,* $\mathbf{x}$ *lies in the ball if and only if* $\|\mathbf{x} - (\mathbf{X}_i + \mathbf{X}_j)/2\| \leq \|\mathbf{X}_i - \mathbf{X}_j\|/2$.

Exercise 8.20 *Define the spherical depth and compute it for* $x = 0$ *with data points* $\{1, -1, 2, -2\}$ *in* $\mathbb{R}$.

Exercise 8.21 *Let* $\mathbf{X}_1, \ldots, \mathbf{X}_n$ *be in a metric space. Define the lens depth of a point* $\mathbf{x}$ *as the proportion of pairs* $(\mathbf{X}_i, \mathbf{X}_j)$ *such that* $\mathbf{x}$ *lies in the lens region* $L(\mathbf{X}_i, \mathbf{X}_j) = B(\mathbf{X}_i, r) \cap B(\mathbf{X}_j, r)$ *with* $r = \delta(\mathbf{X}_i, \mathbf{X}_j)$. *Show that this depth is invariant under isometries, which preserve distances, of the metric space.*

Exercise 8.22 *Describe the lens depth function and explain how it generalizes Tukey's depth.*

Exercise 8.23 *Use* $m = 3$, $n = 2$ *in* $\mathbb{R}^4$, *and Mahalanobis depth to obtain the DD-plot.*
Group F_1 *sample from* $\mathbb{N}_4(\mathbf{0}, \mathbf{I}_4)$
Group F_2 *sample from* $\mathbb{N}_4((1, 1, 0, 0), \mathbf{I}_4)$,

- $\mathbf{X}_1 = (0.1, -0.3, 0.2, -0.1)$, $\mathbf{X}_2 = (0.3, 0.1, -0.4, 0.2)$
- $\mathbf{X}_3 = (-0.2, 0.2, 0.0, -0.3)$, $\mathbf{Y}_1 = (1.2, 1.1, -0.1, 0.0)$, $\mathbf{Y}_2 = (0.9, 0.8, 0.2, -0.1)$.

Exercise 8.24 *Prove that the nearest neighborhood graph $NNG(\mathbf{X})$ of the vertex set $\mathbf{X}$ is a proximity graph where an edge is placed between vertices $\mathbf{X}_i$ and $\mathbf{X}_j$, $i \in \{1, 2, \ldots, n\}, j \in \{1, 2, \ldots, n\}$, $i \neq j$, if there is no other vertex closer to $\mathbf{X}_i$ than $\mathbf{X}_j$.*

Exercise 8.25 *Prove that the β-skeleton IR's are nested. That is,*

$$\text{Skel}_{\beta_1}(\mathbf{X}_1, \mathbf{X}_2) \subseteq \text{Skel}_{\beta_2}(\mathbf{X}_1, \mathbf{X}_2)\ \forall\ \mathbf{X}_1 \textit{ and } \mathbf{X}_2 \in \mathbb{R}^d \textit{ and } 1 \leq \beta_1 < \beta_2 < \infty.$$

Exercise 8.26 *Prove that for $\beta = 1$ and $\beta = 2$, the kernel in Equation (8.3) reduces to the kernels of spherical depth and lens depth, respectively.*

Exercise 8.27 *Prove that for a fixed point $\mathbf{t} \in \mathbb{R}^d$ and any DF F defined over $\mathbb{R}^d$, the β-skeleton depth of $\mathbf{t}$ is non-decreasing in β.*

Exercise 8.28 *Prove that for $1 \leq \beta < \infty$, the β-skeleton depth $\text{SkelD}_\beta(\mathbf{t}; F)$ is symmetric about the center $\boldsymbol{\mu}$ provided F is centrally symmetric about $\boldsymbol{\mu}$.*

Exercise 8.29 *For $d = 1$ and $X_1 < X_2$, show that*

$$Skel_\beta(X_1, X_2) = \begin{cases} [X_1, X_2], & X_1 < X_2, \\ [X_2, X_1], & X1 > X_2. \end{cases} \tag{8.15}$$

Hence, the β-skeleton IR is independent of β in one dimension.

Exercise 8.30 *Let $\mathbf{X}_1$ and $\mathbf{X}_2$ be two random vectors in $\mathbb{R}^d$, $d \geq 2$. Define $\mathbf{C}_1 = (1 - \frac{\beta}{2})\mathbf{X}_1 + \frac{\beta}{2}\mathbf{X}_2$ and $\mathbf{C}_2 = \frac{\beta}{2}\mathbf{X}_1 + (1 - \frac{\beta}{2})\mathbf{X}_2$, where $1 < \beta < \infty$. Prove that the following statements are equivalent.*

- *$\mathbf{C}_1$ and $\mathbf{C}_2$ are both on the surface of some ball $B(\mathbf{t}, \frac{\beta}{2}r)$, denoted by $\partial B(\mathbf{t}, \frac{\beta}{2}r)$, centered at $\mathbf{t}$ with radius $r = ||\mathbf{X}_1 - \mathbf{X}_2||$,*
- *$\mathbf{X}_1$ and $\mathbf{X}_2$ are both on $\partial B(\mathbf{t}, \sqrt{\frac{\beta}{2}}r)$.*

Exercise 8.31 *Given an absolutely continuous and centrally symmetric DF F on $\mathbb{R}^d$ with center $\boldsymbol{\mu}$, any constants $0 \leq \alpha_1 < \alpha_2$ and vector $\mathbf{t} \in \mathbb{R}^d$,*

$$\text{SkelD}_\beta(\boldsymbol{\mu} + \alpha_2\mathbf{t}; F) \leq \text{SkelD}_\beta(\boldsymbol{\mu} + \alpha_1\mathbf{t}; F),\ 1 \leq \beta < \infty.$$

Exercise 8.32 *When $d = 1$, show that $SkelD_\beta$ is maximized at $t = F^{-}(1/2)$ for $1 \leq \beta < \infty$.*

Exercise 8.33 *Let F be a DF on $\mathbb{R}^d$ with center $\boldsymbol{\mu}$. We say F is spherically symmetric about $\boldsymbol{\mu}$ if $\mathbf{A}(\mathbf{X} - \boldsymbol{\mu}) \overset{d}{=} \boldsymbol{\mu} - \mathbf{X}$ for $\mathbf{X} \sim F$ and $d \times d$ orthogonal matrices $\mathbf{A}$. Prove that if F is absolutely continuous and spherical symmetric, then the level set of β-skeleton depth, $D^\alpha = \{\mathbf{t} : \text{SkelD}_\beta(\mathbf{t}; F) \geq \alpha\}$, is a convex set*

Exercise 8.34 *Given n i.i.d. random vectors $\mathbf{X} = \{\mathbf{X}_1, \mathbf{X}_2, \ldots, \mathbf{X}_n\} \sim F$ in $\mathbb{R}^d$, the sample β-skeleton depth of a point $\mathbf{t} \in \mathbb{R}^d$ has mean and variance given by*

$$\begin{aligned} \mathbb{E}\big[\text{SkelD}_\beta(\mathbf{t}; F_n)\big] &= \text{SkelD}_\beta(\mathbf{t}; F), \\ \mathbb{V}ar\big(\text{SkelD}_\beta(\mathbf{t}; F_n)\big) &= \frac{1}{\binom{n}{2}}\big[a + 2(n-2)b^2\big], \end{aligned} \tag{8.16}$$

where $a = \mathrm{SkelD}_\beta(\mathbf{t}; F)(1 - \mathrm{SkelD}_\beta(\mathbf{t}; F))$ *and*

$$b^2 = \mathbb{P}\big(\mathbf{t} \in \bigcap_{i=2,3} \mathrm{Skel}_\beta(\mathbf{X}_1, \mathbf{X}_i)\big) - \mathbb{P}\big((\mathbf{t} \in \mathrm{Skel}_\beta(\mathbf{X}_1, \mathbf{X}_2)\big)^2. \tag{8.17}$$

Exercise 8.35 *Construct the RUNT depth for Figure 8.16.*

Exercise 8.36 *Write an R function to implement the Lens depth.*

9

Hotelling T^2 Test and Wilks Outlier Method

9.1 Abstract

This chapter addresses the challenges of applying the Hotelling T^2 test [155] in HDLSS settings, where the conventional test suffers from singularity of the sample covariance matrix and loss of power when $d \gg N$. To overcome these limitations, we propose a block-partitioning approach using the union-intersection principle, computing T^2 statistics within each block of variables to enhance power while maintaining robustness. We extend this methodology to Wilks' outlier detection, enabling identification of multivariate outliers in blocks of variables with improved sensitivity and controlled significance levels. The chapter reviews testing the equality of multivariate means, including IPD-based methods and permutation tests. Applications to microarray gene expression data illustrate the practical utility of these approaches in HDLSS contexts. Theoretical properties, including invariance, asymptotic distributions, and power functions, are discussed. Collectively, these developments provide a framework for high-dimensional hypothesis testing and robust multivariate outlier detection, addressing critical limitations of traditional Hotelling T^2 and Wilks procedures.

9.2 Introduction

In this chapter, we consider the Hotelling T^2 test [155] in low-sample-size, high-dimensional setting. We partition the d variables into $b > 1$ blocks of d/b variables and use the union-intersection principle to propose a testing procedure that computes the T^2 test in each block. We show that the proposed method is more powerful than Hotelling T^2 test. We also consider the Wilks method of outlier detection and use the union-intersection principle to search for outliers in blocks of variables. The significance level and the power function of the new test are investigated. We show that the new outlier detection method produces more power compared to the Wilks [356] test.

Methods of testing for the equality of multivariate distributions have a long history in Statistics. Consider the basic method of comparing two groups of m and n independent observations for the equality of their multivariate means using the two-sample Hotelling T^2 test [155] with d observations where $d < N - 2$ and $N = m + n$. The T^2 statistic has many advantages such as invariance under affine transformations, known exact and asymptotic distributions and being the uniformly most powerful invariant for testing the mean against two sided alternatives. It is powerful when d is small compared with N. However, it has two severe practical drawbacks, including lack of power when d is close to N and being undefined due to the singularity of the covariance matrix when $d \geq N - 2$.

DOI: 10.1201/9781003789581-9

Suppose $\mathbf{X}$ and $\mathbf{Y}$ are independent d-variate vectors taken independently from normal distributions with means $\boldsymbol{\mu}_x$, and $\boldsymbol{\mu}_y$ and covariance matrices $\boldsymbol{\Sigma}_x$ and $\boldsymbol{\Sigma}_y$, respectively. Let $\mathbf{X}_i$ and $\mathbf{Y}_j$, $i = 1, \dots m$, $j = 1, \dots, n$ be i.i.d. copies of $\mathbf{X}$ and $\mathbf{Y}$ with mean vectors $\bar{\mathbf{X}}$ and $\bar{\mathbf{Y}}$ and sample covariances $\mathbf{S}_x$ and $\mathbf{S}_y$, respectively. We consider the one-sample and two-sample Hotelling T^2 test in high-dimensional low-sample-size (HDLSS) setting where the Hotelling T^2 test suffers from a singularity problem in the covariance matrix estimation and is undefined. Many authors have suggested modifications to the Hotelling T^2 test or other alternatives because testing for the mean vector is fundamental to inference and applications.

One can use the Euclidean distance instead of the Mahalanobis distance to measure separation between the two mean vectors and construct the T^2 test assuming the covariance is the identity matrix. Chen and Qin [64] use the Euclidean distance for testing in HDLSS setting where $d \gg N$. A second approach uses a shrinkage parameter (Chen et al. [63]; Shen et al. [307]).

To side-step the issue of covariance singularity, Sinha and Wieland [316] partition the sample in b block and present a union-intersection test for the mean vector in the one sample case where $\mathbf{X}_i$ of the i-th blocks is distributed as $\mathbb{N}(\boldsymbol{\mu}_i, \sigma_i^2\mathbf{I})$. Wu, Genton and Stefanski [359] suggest a similar approach by using the diagonal of the sample covariance matrix. In this section we extend their one sample Hotelling T^2 test to $\mathbf{X}_i \sim \mathbb{N}(\boldsymbol{\mu}_i, \boldsymbol{\Sigma}_i)$. Srivastava and Du [322] and Srivastava [321] offer similar test statistics in this category. This approach also loses information that exists in the covariance matrix by setting all of them to zero. While this approach produces high powers in many cases it loses many of the optimal properties of the Hotelling T^2 test, including a known distribution and affine invariance. Huang et al. [158] presents an excellent overview of tests on high-dimensional means. Li et al. [188] investigate the usage of spectral regularization in high-dimensional linear hypothesis. Lopes et al. [203] consider the method of random projection. He et al. [144] considers a screening method using the likelihood ratio test under the context of linear regression. To test the null hypothesis $H_0 : \boldsymbol{\mu}_x = \boldsymbol{\mu}_0$ against $H_a : \boldsymbol{\mu}_x \neq \boldsymbol{\mu}_0$, the Hotelling T^2 statistic [155] is defined as

$$T^2 = m(\bar{\mathbf{X}} - \boldsymbol{\mu}_0)'\mathbf{S}_x^{-1}(\bar{\mathbf{X}} - \boldsymbol{\mu}_0) \tag{9.1}$$

Under H_0, T^2 has $\frac{(m-1)d}{m-d}F(d, m-d)$ distribution. Hotelling T^2 test is invariant with respect to the group of affine transformations. It is uniformly most powerful against shift alternatives under multivariate normality. Moreover, its χ^2 asymptotic distribution under H_0 does not depend on normality as the sample sizes tend to infinity. A limitation of the Hotelling T^2 test is its reliance on the normality assumption, which restricts its applicability. Everitt [91] studies the robustness of Hotelling's one- and two-sample T^2 tests for samples from uniform, exponential, and lognormal distributions. Everitt [91] shows that the two-sample test is noticeably far more robust to normality than the one-sample test, which is badly affected by extreme skewness. Many investigators use Box-Cox transformation to induce normality (Freeman and Modarres [106]). Simaika [313] shows that T^2 test is uniformly most powerful in the class of tests whose power functions depend on $(\boldsymbol{\mu}_x, \boldsymbol{\Sigma}_x)$ only through $\boldsymbol{\mu}_x'{\boldsymbol{\Sigma}_x}^{-1}\boldsymbol{\mu}_x$. Let $t_\alpha = \frac{d}{m-d}F_\alpha(d, m-d)$ be the critical value and define the non–centrality parameter $\delta = m(\boldsymbol{\mu}_x - \boldsymbol{\mu}_0)'{\boldsymbol{\Sigma}_x}^{-1}(\boldsymbol{\mu}_x - \boldsymbol{\mu}_0)$. The power function of an α significance level T^2 test is given by Anderson [10],

$$Power(T^2) = \mathbb{P}(F(d, m-d, \delta) \geq t_\alpha), \tag{9.2}$$

The one-sample Hotelling T^2 test is very important for paired comparisons where we hypothesize that the mean vector of paired differences is zero (Marozzi [214]). Using the maximization lemma, one can show that

$$T^2 = \max_{\mathbf{a}} t_a^2 = \frac{m(\mathbf{a}'(\bar{\mathbf{X}} - \boldsymbol{\mu}))}{\mathbf{a}'\mathbf{S}_x\mathbf{a}}, \quad \mathbf{a} \in \mathbb{R}^d, \quad \|\mathbf{a}\| = 1.$$

Iwashita [160] considers the asymptotic null and non-null distribution of the T^2 test under the elliptical distributions.

Modarres [236] proposes a different strategy when the sample covariance is singular, based on conducting T^2 test in blocks of variables where the block covariances are no longer singular. Modarres [230] studies the blocks of IPDs, their distributions, correlations, independence and the homogeneity of their total variances. The Hotelling T^2 statistic exhibits low power when the sample covariance is close to being singular, and it is not well-defined when $d \geq m$. Consider the one-sample T^2 test where $\mathbf{S}_x$ has unit variances and constant correlation $\frac{-1}{d-1} < r < 1$. The sample covariance matrix is singular when $r \leq \frac{-1}{d-1}$. However, using $b > 1$ blocks with $d_i = d/b < m$ variables produces $r > \frac{-1}{d_i - 1}$ so that the lower-order sample covariance matrices are no longer singular.

We assume that the blocks are independent of each other under multivariate normality. That allows us to study each block of variables in the appropriate subspace where T^2 test is applicable. Performing T^2 test in appropriate subspaces retains the optimal properties of the Hotelling T^2 test in those subspaces, including unbiasedness, affine invariance, and most importantly a known distribution. If $d = 5000$ and $m = 100$, one may consider $b = 100$ blocks of $d_i = 50$ variables with $m = 100$ observations. The hypothesis of block independence has long been of interest in classical multivariate analysis. Suppose $\mathbf{X}$ has a d variate distribution with covariance matrix $\boldsymbol{\Sigma} = (\boldsymbol{\Sigma}_{ij})$ and b blocks of sizes $d_i, i = 1, \ldots b$. The null hypothesis is $H_0 : \boldsymbol{\Sigma}_{ij} = \mathbf{0}$ for $i \neq j = 1, \ldots, b$. The likelihood ratio test statistic is $V = \frac{|\mathbf{R}|}{\prod_{i=1}^{b} |\mathbf{R}_{ii}|}$, where $\mathbf{R}$ is the sample correlation matrix and $\mathbf{R}_{ii}$ are the blocks correlation matrices. Under H_0, Anderson [10] provides a χ^2 approximation for V and its extension to elliptically contoured distributions. The blocks' components and the number of blocks are specified by the researcher. Computer simulation provides a more accurate estimate of the sampling distribution of V.

Testing for sparse alternatives is at the heart of many applications such as DNA microarray, computer vision, machine learning, data mining, and classification. Sparsity is a common assumption in high-dimensional inference. The covariance matrix must be sparse in certain pattern in order to test the existence of independent blocks. It is well known that in a high-dimensional classification problem, the vast majority of the variables may prove to be noise and should be discarded before classification since they usually deteriorate the performance of any classifier. Working with a much smaller subset of the discriminatory variables improves the classification accuracy.

Let $\mathbf{X}$ and $\mathbf{Y}$ be independent d-variate vectors from multivariate normal distributions $\mathbb{N}(\mathbf{0}, \mathbf{I})$ and $\mathbb{N}(\boldsymbol{\mu}, \mathbf{I})$, where $\mu = (\mu, \ldots, \mu, 0, \ldots, 0)$ and $\mu > 0$ is in the first d_1 components. Depending on the size of the shift μ and the value of d_1, the hypothesis $H_0 : \boldsymbol{\mu}_x = \boldsymbol{\mu}_0$ is not detected. However, if we consider testing in b blocks of d_i variables, the discriminatory variables in the first block will lead to the rejection of H_0. Hence, testing in blocks is a useful tool for variable selection. This is especially true for ultra high-dimensional data. Furthermore, it is common in analysis of HDLSS data to assume that the covariance matrix is sparse. Hall et al. [133] have stated similar assumptions to study the asymptotic behavior of HDLSS techniques.

The chapter is organized as follows. We first discuss regularized T^2 statistic. We discuss the one-sample and two-sample T^2 test when d is larger than the sample size. We show that even when d is close to the sample size, Hotelling T^2 test loses power, whereas the subspace strategy shows more power. We describe an approach whereby we partition the vector of observations into $b > 1$ blocks of d_i variables, where the block covariances are no longer singular. One examines the null hypothesis in $b > 1$ subspaces of dimension d_i, where rejection of any block will result in the rejection of the null hypothesis. An inclusion-exclusion formulation allows one to compute the level of significance. We discuss Wilks [356] outlier test and its connection to Hotelling T^2 test. We show that the Wilks test also loses power as d approaches the sample size. We also propose a subspace strategy to perform Wilks outlier test.

9.3 Regularized Hotelling T^2 Tests

Testing equality of multivariate mean vectors is a fundamental problem in statistics. The classical Hotelling T^2 test [154] extends the univariate t-test to d-dimensional data by incorporating the inverse sample covariance matrix and is optimal under multivariate normality when d is fixed and smaller than the sample size. In modern high-dimensional settings, however, the dimension d may be comparable to or larger than the total sample size m, causing the sample covariance matrix to become ill-conditioned or singular. In such regimes the classical Hotelling statistic becomes undefined when $d \geq m$, and even when d approaches m from below it exhibits severe instability and inflated Type I error. Bai and Saranadasa [19] developed a high-dimensional test that avoids covariance inversion entirely and established asymptotic theory when $d/n \to \gamma > 0$. Ledoit and Wolf [183] proposed shrinkage covariance estimators to improve conditioning in large dimensions. Chen et al. [63] introduced a ridge-regularized Hotelling statistic that stabilizes the inverse covariance directly while preserving the geometric interpretation of the quadratic form. This document develops the regularized Hotelling T^2 statistic and explicitly distinguishes the regimes in which the classical and regularized procedures are defined.

Let $\mathbf{X}_1, \ldots, \mathbf{X}_m \in \mathbb{R}^d$ be i.i.d. with mean $\boldsymbol{\mu}$ and covariance $\boldsymbol{\Sigma}$. We test $H_0 : \boldsymbol{\mu} = \boldsymbol{\mu}_0$. Define

$$\bar{\mathbf{X}} = \frac{1}{m}\sum_{i=1}^{m}\mathbf{X}_i, \quad \mathbf{S} = \frac{1}{m-1}\sum_{i=1}^{m}(\mathbf{X}_i - \bar{\mathbf{X}})(\mathbf{X}_i - \bar{\mathbf{X}})^T.$$

Hotelling T^2 statistic 9.1 is defined only if $d < m$. If $d \geq m$, then $\mathbf{S}$ is singular and T^2 is undefined. For the regularized Hotelling T^2 we introduce ridge regularization by defining

$$\mathbf{S}_\lambda = \mathbf{S} + \lambda\mathbf{I}_d, \quad \lambda > 0.$$

and

$$T_\lambda^2 = m(\bar{\mathbf{X}} - \boldsymbol{\mu}_0)^T(\mathbf{S} + \lambda\mathbf{I}_d)^{-1}(\bar{\mathbf{X}} - \boldsymbol{\mu}_0).$$

This statistic is well-defined for all d, including $d \gg m$.

For the two-sample setting, let $\mathbf{X}_i \sim (\boldsymbol{\mu}_1, \boldsymbol{\Sigma})$ be independent of $\mathbf{Y}_j \sim (\boldsymbol{\mu}_2, \boldsymbol{\Sigma})$. We Test $H_0 : \boldsymbol{\mu}_1 = \boldsymbol{\mu}_2$. Define $\mathbf{D} = \bar{\mathbf{X}} - \bar{\mathbf{Y}}$ and the pooled covariance

$$\mathbf{S}_p = \frac{(m-1)\mathbf{S}_1 + (n-1)\mathbf{S}_2}{m+n-2}.$$

The classic Hotelling two-sample test is

$$T^2 = \frac{mn}{m+n}\mathbf{D}^T\mathbf{S}_p^{-1}\mathbf{D}.$$

It is defined only if $d < m + n - 2$.

The regularized two-sample statistic is defined by

$$T_\lambda^2 = \frac{mn}{m+n}\mathbf{D}^T(\mathbf{S}_p + \lambda\mathbf{I}_d)^{-1}\mathbf{D}.$$

This remains well-defined even when $d \geq m+n-2$

9.3.1 Spectral Decomposition and Rank Structure

Consider $\mathbf{S}_p = \mathbf{U}\mathbf{\Lambda}\mathbf{U}^T$, where $\mathbf{\Lambda} = \text{diag}(\hat{\lambda}_1, \ldots, \hat{\lambda}_r, 0, \ldots, 0)$, and $r = \text{rank}(\mathbf{S}_p) \leq m+n-2$. If $d > m+n-2$, then $d-r$ eigenvalues are zero. The inverse $\mathbf{S}_p^{-1}$ exists only if $r = d$. Hence classical T^2 requires full rank. After ridge regularization we obtain

$$\mathbf{S}_p + \lambda\mathbf{I}_d = \mathbf{U}(\mathbf{\Lambda} + \lambda\mathbf{I}_d)\mathbf{U}^T,$$

with eigenvalues $\hat{\lambda}_1 + \lambda, \ldots, \hat{\lambda}_r + \lambda, \lambda, \ldots, \lambda$.

Thus even when $d \gg m+n$, all eigenvalues are strictly positive and inversion is stable. The statistic becomes

$$T_\lambda^2 = \sum_{k=1}^{r} \frac{(\mathbf{u}_k^T\mathbf{Z})^2}{\hat{\lambda}_k + \lambda} + \sum_{k=r+1}^{d} \frac{(\mathbf{u}_k^T\mathbf{Z})^2}{\lambda}.$$

Therefore, the classical T^2 cannot be computed when $d \geq m+n$. and regularized T_λ^2 remains computable even if $d \gg m+n$. To see the behavior in different dimensional regimes for the one-sample Hotelling T^2 statistic, consider the following cases.

1. when $d < m$: Both classical and regularized statistics are defined.
2. when $d \to m$ from below: Classical T^2 becomes unstable because the smallest eigenvalue of $\mathbf{S}_p$ approaches zero. Regularized T_λ^2 remains stable.
3. when $d \ggg m$: Classical T^2 is undefined. Regularized T_λ^2 is well-defined.

Two-sample regularized T_λ^2 remains computable because ridge lifts all zero eigenvalues. However, The sample covariance has rank at most $m+n-2$ and most directions are noise-dominated. Its power depends on how the mean difference scales with d. Stronger regularization is typically required. Thus the regime $d \gg m+n$ applies only to the regularized statistic.

9.3.2 Asymptotic Normality in High Dimensions

We formalize the limiting behavior of the regularized Hotelling T_λ^2 in the high-dimensional regime, including $d \gg m$.

Theorem 9.3.1 (Asymptotic Normality of T_λ^2) *Let $\mathbf{X}_1, \ldots, \mathbf{X}_m \in \mathbb{R}^d$ be i.i.d. with mean $\boldsymbol{\mu}$ and covariance $\mathbf{\Sigma}$. Consider*

$$T_\lambda^2 = m(\bar{\mathbf{X}} - \boldsymbol{\mu}_0)^T(\mathbf{S} + \lambda\mathbf{I}_d)^{-1}(\bar{\mathbf{X}} - \boldsymbol{\mu}_0),$$

with $\lambda > 0$. Assume as $m \to \infty$ that $\frac{d}{m} \to \gamma \in (0,\infty)$ or $d \gg m$, and that $\frac{1}{d}\text{tr}(\mathbf{\Sigma})$ and $\frac{1}{d}\text{tr}(\mathbf{\Sigma}^2)$ remain bounded. Under $H_0 : \boldsymbol{\mu} = \boldsymbol{\mu}_0$,

$$\frac{T_\lambda^2 - dM_\lambda}{\sqrt{dv_\lambda}} \xrightarrow{d} \mathbb{N}(0,1),$$

where

$$M_\lambda = \frac{1}{d}\text{tr}\left((\mathbf{\Sigma} + \lambda\mathbf{I}_d)^{-1}\mathbf{\Sigma}\right), \quad v_\lambda = \frac{2}{d}\text{tr}\left((\mathbf{\Sigma} + \lambda\mathbf{I}_d)^{-1}\mathbf{\Sigma}(\mathbf{\Sigma} + \lambda\mathbf{I}_d)^{-1}\mathbf{\Sigma}\right).$$

9.4 Testing Sub-Hypothesis

9.4.1 One-sample Test

When $d \geq m-2$ we present the following approach based on partitioning $\mathbf{X}$ into $b > 1$ blocks of d_i variables such that $\sum_{i=1}^{b} d_i = d$. Let

$$\mathbf{X} = (\mathbf{X}^{(1)}, \mathbf{X}^{(2)}, \ldots, \mathbf{X}^{(b)}), \quad \boldsymbol{\mu}_x = (\boldsymbol{\mu}_x{}^{(1)}, \boldsymbol{\mu}_x{}^{(2)}, \ldots, \boldsymbol{\mu}_x{}^{(b)}),$$

and $\bar{\mathbf{X}} = (\bar{\mathbf{X}}^{(1)}, \bar{\mathbf{X}}^{(2)}, \ldots, \bar{\mathbf{X}}^{(b)})$. Partition $\boldsymbol{\Sigma}$ and $\mathbf{S}_x$ as

$$\boldsymbol{\Sigma} = Diag(\boldsymbol{\Sigma}_1, \boldsymbol{\Sigma}_2, \ldots, \boldsymbol{\Sigma}_b), \quad \mathbf{S}_x = Diag(\mathbf{S}_1, \mathbf{S}_2, \ldots, \mathbf{S}_b),$$

where $\boldsymbol{\Sigma}_i$ and $\mathbf{S}_i$ are $d_i \times d_i$ symmetric matrices. Variables in blocks i and j are independent of each other since $\boldsymbol{\Sigma}_{ij} = \mathbf{0}$ for $i \neq j = 1, \ldots, d$ under multivariate normality. Let

$$T_i^2 = m(\bar{\mathbf{X}}^{(i)} - \boldsymbol{\mu}_0^{(i)})'\mathbf{S}_i^{-1}(\bar{\mathbf{X}}^{(i)} - \boldsymbol{\mu}_0^{(i)})$$

be the Hotelling T^2 test in block i with d_i variables and m observations. It follows that

$$T_i^2 \quad \sim \quad \frac{(m-1)d_i}{m-d_i} F(d_i, m-d_i), \quad T^2 = \sum_{i=1}^{b} T_i^2.$$

If $F \sim F(d_i, m-d_i)$, then

$$\begin{aligned}
\mathbb{E}(F) &= \frac{m-d_i}{m-d_i-2}, & \mathbb{V}ar(F) &= \frac{2(d_i-2)^2(m-2)}{d_i(m-d_i-2)^2(m-d_i-4)}, \\
\mathbb{E}(T^2) &= \sum_{i=1}^{b} \frac{m-d_i}{m-d_i-2}, & \mathbb{V}ar(T^2) &= \sum_{i=1}^{b} \frac{2(m-1)^2(m-2)d_i}{(m-d_i-2)^2(m-d_i-4)}.
\end{aligned} \tag{9.3}$$

For the null hypothesis H_0, a family of tests adheres to the union-intersection principle (Roy [290]) provided that H_0 is true if and only if the null hypotheses of all b blocks of variables are true. The null hypothesis H_0 is rejected if and only if the null hypothesis of at least one of the blocks of variables is rejected. The intersection part of the union-intersection principle states that if the null hypothesis H_0 is true, then the null hypotheses of all the blocks of variables are true. Therefore, if even for one block of variables null hypothesis is false, then H_0 is also false (Olkin and Tomsky [257]). Instead of testing H_0 we propose to test an equivalent hypothesis. Let

$$K_{0i} : \boldsymbol{\mu}_x{}^{(i)} = \boldsymbol{\mu}_0{}^{(i)}, \quad \text{and} \quad K_{ai} : \boldsymbol{\mu}_x{}^{(i)} \neq \boldsymbol{\mu}_0{}^{(i)}.$$

We test the hypothesis

$$K_0 : \cap_{i=1}^{b} K_{0i}, \quad \text{against} \quad K_a : \cup_{i=1}^{b} K_{ai}.$$

We reject K_0 when any of the blocks rejects K_{0i}, for $i = 1, \ldots, b$. We assume $\alpha_i = \alpha^*$ is the common probability that K_{0i} based on d_i variables is rejected, $i = 1, \ldots, b$. We use $d_i = d/b = d^*$ for $i = 1, \ldots b$. The decision to reject K_0 if any of the blocks rejects K_{0i} is equivalent to rejecting K_0 when

$$F = \max_{1 \leq i \leq b} T_i > \frac{(m-1)d^*}{m-d*} F_{(\alpha^*, d^*, (m-1)d^*)}.$$

Note that H_0 and K_0 have the same type I error. All sub-hypotheses are conducted at the same level and rejected for the maximum value of the test statistic in all blocks where $d_i = d/b$ and $\alpha_i = \alpha^*$ for $i = 1, \dots b$. Hence, we obtain $\alpha^* = 1 - (1 - \alpha)^{1/b}$. When $b = 2$, at $\alpha = 0.05$ we obtain $\alpha^* = 0.02532$. For $b = 3$, we get $\alpha^* = 0.01676$. We refer to the test based on blocks as sub-T^2.

It is noteworthy that the T^2 formed in block i has the property that

$$T_i^2 = \max_{\mathbf{a}_i} t_{a_i}^2 = \frac{m(\mathbf{a}_i'(\bar{\mathbf{X}}^{(i)} - \boldsymbol{\mu}^{(i)}))}{\mathbf{a}_i'\mathbf{S}_i\mathbf{a}_i}$$

where $\mathbf{a}_i \in \mathbb{R}^{d_i}$ for $i = 1, \dots, b$. Using this method one can build simultaneous confidence intervals for the means of the blocks. An important aspect of testing H_0 in blocks using the union-intersection principle is that it identifies the component blocks that are responsible for rejection of H_0. Hence, we get a clearer picture of the nature of deviation from H_0. The block of variables that is responsible for rejection of H_0 provides a basis for variable selection before classification of high-dimensional data (Fan, Samworth, and Wu [94]).

We now present some examples comparing the powers of Hotelling T^2 and sub-T^2 test statistics. When comparing two testing methods, it is crucial to consider both the size of the test and its power. In all simulations that produced the resulting graphs the α levels are well within the margin of error and very close to the nominal level of $\alpha = 0.05$.

Example 9.4.1 *We generated $m = 20$ observations on $d = 18$ variables from $\mathbb{N}(\mathbf{0}, \boldsymbol{\Sigma})$ where $\boldsymbol{\Sigma}$ has 1 on the main diagonal and $\rho = 1/2$, everywhere else. We set $\alpha = 0.05$ and $b = 2$ with $d_1 = 9$ and d_2=9. We repeated the experiment 1000 times. The left panel in Figure 9.1 shows the observed power of testing $H_0 : \boldsymbol{\mu}_x = \mathbf{0}$ using Hotelling T^2 test at level α. The left panel in Figure 9.1 also shows the power of testing $K_0 : \boldsymbol{\mu}_x^{(1)} = \mathbf{0}$ and $\boldsymbol{\mu}_x^{(2)} = \mathbf{0}$. The observed significance level of the Hotelling T^2 test and sub-T^2 test are* 0.047 *and* 0.045, *respectively. It is evident that sub-T^2 test is more powerful than Hotelling T^2 test.*

Example 9.4.2 *To study the effects of increasing dimensions on the two test statistics we generated $d = 2^i$ variables with $m = d + 2$ for $i = 1, \dots, 10$ from a d-variate normal distribution with zero mean and constant covariance structure with correlation 1/2. We set $\alpha = 0.05$ and $b = 2$ with $d_1 = d_2 = d^{i-1}$. The right panel of Figure 9.1 shows the observed power of testing $H_0 : \boldsymbol{\mu}_x = \mathbf{0}$ using Hotelling T^2 test at level α. The right panel of Figure 9.1 also shows the power of testing $K_0 : \boldsymbol{\mu}_x^{(1)} = \mathbf{0}$ and $\boldsymbol{\mu}_x^{(2)} = \mathbf{0}$. Clearly, sub-$T^2$ dominates the Hotelling T^2 test as d becomes large.*

The sub-T^2 statistic offers applicability over T^2 when $m < d$ as we are able to calculate T_i^2 for each block and test K_0. The subspace approach of using sub-T^2 to test the subspaces separately offers more power over T^2 when d is less than, but close to m. The reason Hotelling's test loses power as d approaches m is the inaccuracy of the estimator of the covariance matrix. Assuming that the population covariance matrix is $\mathbf{I}_d$, the eigenvalues of the sample covariance matrix $\mathbf{S}_x$ for $d < m$ must converge to unity. However, as shown by Silverstein [312], the ratio of the largest and the smallest eigenvalues of $\mathbf{S}_x$ converge to

$$\frac{(1 + \sqrt{y_m})^2}{(1 - \sqrt{y_m})^2}, \quad y_m = d/m \to y \in (0, 1).$$

In particular, when y_m is close to 1, $\mathbf{S}_x$ has many small eigenvalues, leading to many large eigenvalues for $\mathbf{S}_x^{-1}$. Bai and Saranadasa [19] discuss the effects of high dimension on the Hotelling T^2 test as $d < n \to \infty$. Marozzi [215] discusses further drawbacks of the Hotelling T^2 test and recommends test statistics based on IPDs.

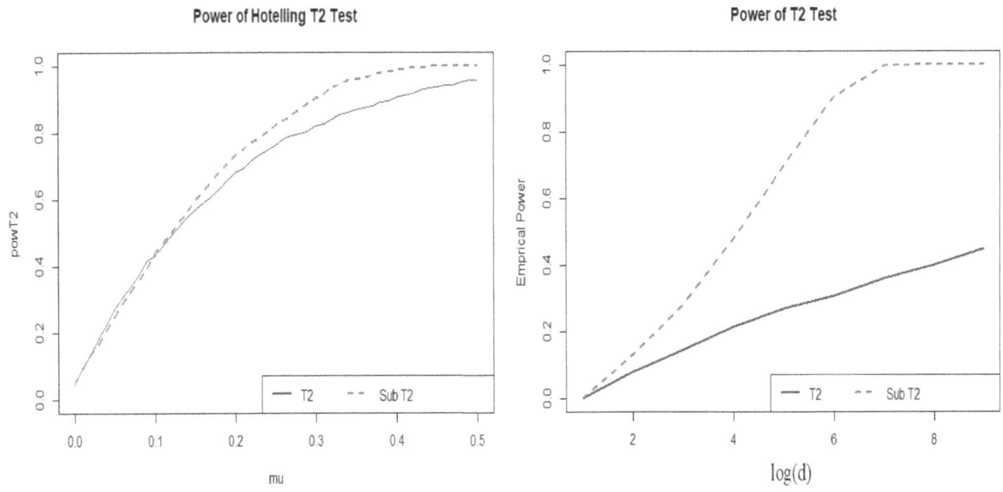

FIGURE 9.1: Left panel: Power of testing H_0 and K_0. Right panel: Effects of increasing dimensions.

Under H_0, $\frac{m-d}{d(m-1)}T^2$ has an $F(d, m-d)$ distribution. When d is fixed and $n \to \infty$, the limiting distribution is $\chi^2(d)$. Under H_0, using the Central Limit Theorem, it is well-known that T^2 test is asymptotically normal, as $n \to \infty$. However, the limiting power of the Hotelling's test for d close to m is α.

9.4.2 Two-sample Test

Let $\mathbf{S}_d = \frac{1}{N-2}((m-1)\mathbf{S}_x + (n-1)\mathbf{S}_y)$ be the pooled covariance of $\boldsymbol{\Sigma}_x = \boldsymbol{\Sigma}_y = \boldsymbol{\Sigma}$ and

$$T^2 = \frac{mn}{N}(\bar{\mathbf{X}} - \bar{\mathbf{Y}})'\mathbf{S}_d^{-1}(\bar{\mathbf{X}} - \bar{\mathbf{Y}}) \tag{9.4}$$

denote the Hotelling two-sample T^2 statistic for testing the null hypothesis $H_{02} : \boldsymbol{\mu}_x = \boldsymbol{\mu}_y$ against $H_{a2} : \boldsymbol{\mu}_x \neq \boldsymbol{\mu}_y$. Under H_{02}, T^2 has $\frac{(N-2)d}{N-d-1}F(d, N-d-1)$ distribution. The power function of an α significance level, two-sample T^2 test is given by

$$Power(T^2) = \mathbb{P}(F(d, N-d-1, \delta) \geq t_\alpha), \tag{9.5}$$

where $t_\alpha = \frac{d}{N-d-1}F_\alpha(d, N-d-1)$ is the critical value and the non–centrality parameter is $\delta = n(\boldsymbol{\mu}_x - \boldsymbol{\mu}_y)'\boldsymbol{\Sigma}^{-1}(\boldsymbol{\mu}_x - \boldsymbol{\mu}_y)$.

Similar to the $\mathbf{X}$ sample, when $d \geq N-2$, we partition $\mathbf{Y}$ into $b > 1$ blocks of d_i variables such that $\sum_{i=1}^{b} d_i = d$. Let $\mathbf{Y} = (\mathbf{Y}^{(1)}, \mathbf{Y}^{(2)}, \ldots, \mathbf{Y}^{(b)})$, $\boldsymbol{\mu}_y = ({\boldsymbol{\mu}_y}^{(1)}, {\boldsymbol{\mu}_y}^{(2)}, \ldots, {\boldsymbol{\mu}_y}^{(b)})$. We partition $\boldsymbol{\Sigma}_y$, $\mathbf{Y}$ and $\mathbf{S}_y$, similarly. Variables in blocks i and j are independent of each other since $\boldsymbol{\Sigma}_{xij} = \boldsymbol{\Sigma}_{yij} = \mathbf{0}$ for $i \neq j = 1, \ldots, d$ under multivariate normality. Suppose $d/b \geq N$ and let

$$T_i^2 = \frac{mn}{N}(\bar{\mathbf{X}}^{(i)} - \bar{\mathbf{Y}}^{(i)})'{\mathbf{S}_d^{(i)}}^{-1}(\bar{\mathbf{X}}^{(i)} - \bar{\mathbf{Y}}^{(i)}), \tag{9.6}$$

where

$$\mathbf{S}_d^{(i)} = \frac{1}{N-2}((m-1)\mathbf{S}_x^{(i)} + (n-1)\mathbf{S}_y^{(i)})$$

is the pooled covariance of the i-th block. It follows that

$$T_i^2 \sim \frac{(N-2)d_i}{(N-d_i-1)} F(d_i, N-d_i-1), \quad \text{and } T^2 = \sum_{i=1}^{b} T_i^2.$$

Instead of testing H_{02}, we propose to test an equivalent hypothesis. Let

$$L_{0i} : \boldsymbol{\mu}_x{}^{(i)} = \boldsymbol{\mu}_y{}^{(i)} \text{ and } L_{ai} : \boldsymbol{\mu}_x{}^{(i)} \neq \boldsymbol{\mu}_y{}^{(i)}. \tag{9.7}$$

We test the null hypothesis

$$L_0 : \cap_{i=1}^{b} L_{0i} \quad \text{against } L_a : \cup_{i=1}^{b} L_{ai}. \tag{9.8}$$

We reject L_0 when any of the blocks rejects L_{0i}, for $i = 1, \ldots, b$. We assume $\alpha_i = \alpha^*$ is the common probability that L_{0i} based on d_i variables is rejected, $i = 1, \ldots, b$, where α is the level of significance of H_{02}. When $d_i = d/b = d^*$, the decision to reject L_0 if any of the blocks rejects L_{0i} is equivalent to rejecting L_0 when

$$F = \max_{1 \le i \le b} T_i > \frac{(N-2)d^*}{(N-d^*-1)} F_{(\alpha^*, d^*, N-d^*-1)} \text{ where } \alpha^* = 1-(1-\alpha)^{1/b}. \tag{9.9}$$

Example 9.4.3 *We generated $m = n = 100$ observations from $\mathbb{N}(\mathbf{0}, \boldsymbol{\Sigma})$ and $\mathbb{N}(0.5\mathbf{1}, \boldsymbol{\Sigma})$, respectively, with $d = 2, 3, \ldots, 99$ variables. The covariance $\boldsymbol{\Sigma}$ has 1 on the main diagonal and ρ, everywhere else. We used $\rho = 0.5$ and $\rho = 0.9$ with 5000 simulations. The left panel in Figure 9.2 shows the observed power of testing $H_{02} : \boldsymbol{\mu}_x = \boldsymbol{\mu}_y$ using Hotelling T^2 test at level $\alpha = 0.05$. The power of T^2 test tends to its α level as d approaches N.*

Example 9.4.4 *We generated $m = 30$ observations from $\mathbb{N}(\mathbf{0}, \boldsymbol{\Sigma})$ and $n = 30$ observations from $\mathbb{N}(\mu_y \mathbf{1}, \boldsymbol{\Sigma})$ with $d = 14$, $d_1 = d_2 = 7$ variables and $b = 2$ where $\mu_y = 0, 0.1, 0.2, \ldots, 1$. The covariance $\boldsymbol{\Sigma}$ has 1 on the main diagonal and $\rho = 0.5$ everywhere else. The right panel in Figure 9.2 shows the observed power of testing $H_{02} : \boldsymbol{\mu}_x = \boldsymbol{\mu}_y$ using Hotelling T^2 and sub-T^2 tests at level $\alpha = 0.05$ with 5000 simulations. The T^2 and sub-T^2 tests show estimated significance levels of 0.066 and 0.053, respectively, at $\mu_y = 0$. The T^2 is less powerful than sub-T^2, even when $d < n$.*

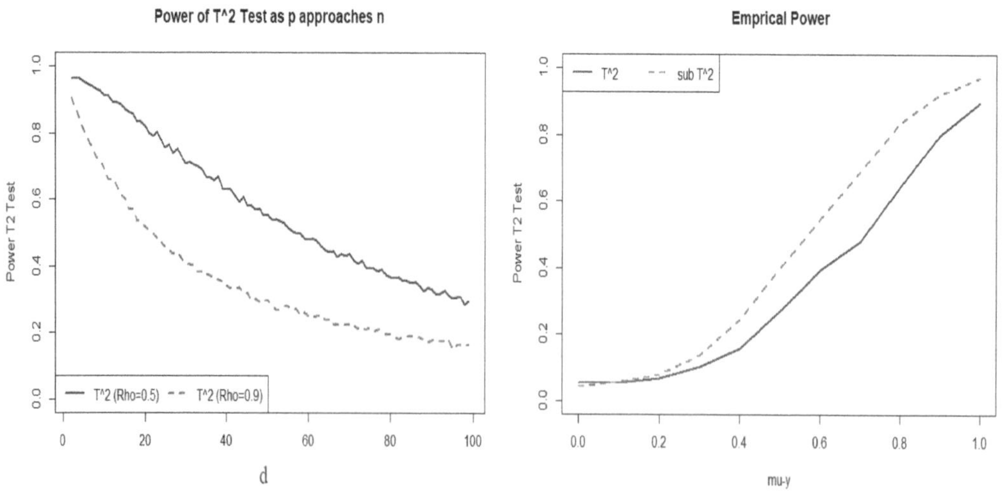

FIGURE 9.2: Left panel: Power of T^2 test. Right panel: Powers of T^2 and sub-T^2 tests.

Example 9.4.5 *We consider a two-sample Hotelling T^2 procedure using blockwise partitioning. Each group has $m = n = 5$ observations in $d = 9$ dimensions. The variables are grouped into $J = 3$ disjoint blocks, each of dimension $d = 3$.*

We test the global hypothesis

$$H_0^{global} : \boldsymbol{\mu}_1 = \boldsymbol{\mu}_2 \quad vs \quad H_a^{global} : \boldsymbol{\mu}_1 \neq \boldsymbol{\mu}_2$$

by testing the blockwise hypotheses

$$H_{0j} : \boldsymbol{\mu}_{1j} = \boldsymbol{\mu}_{2j}, \qquad j = 1, 2, 3,$$

using Hotelling's two-sample T^2 statistic within each block.

Block definitions Block 1: $(\mathbf{X}_1, \mathbf{X}_2, \mathbf{X}_3)$*, Block 2:* $(\mathbf{X}_4, \mathbf{X}_5, \mathbf{X}_6)$*, Block 3:* $(\mathbf{X}_7, \mathbf{X}_8, \mathbf{X}_9)$.

Group 1 data

ID	$\mathbf{X}_1$	$\mathbf{X}_2$	$\mathbf{X}_3$	$\mathbf{X}_4$	$\mathbf{X}_5$	$\mathbf{X}_6$	$\mathbf{X}_7$	$\mathbf{X}_8$	$\mathbf{X}_9$
G1_1	*0.497*	*-0.138*	*0.648*	*1.523*	*-0.234*	*-0.234*	*1.579*	*0.767*	*-0.469*
G1_2	*0.543*	*-0.463*	*-0.466*	*0.242*	*-1.913*	*-1.725*	*-0.562*	*-1.013*	*0.314*
G1_3	*-0.908*	*-1.412*	*1.466*	*-0.226*	*0.068*	*-1.425*	*-0.544*	*0.111*	*-1.151*
G1_4	*0.376*	*-0.601*	*-0.292*	*-0.602*	*1.852*	*-0.013*	*-1.058*	*0.823*	*-1.221*
G1_5	*0.209*	*-1.960*	*-1.328*	*0.197*	*0.738*	*0.171*	*-0.116*	*-0.301*	*-1.479*

Group 2 data

ID	$\mathbf{X}_1$	$\mathbf{X}_2$	$\mathbf{X}_3$	$\mathbf{X}_4$	$\mathbf{X}_5$	$\mathbf{X}_6$	$\mathbf{X}_7$	$\mathbf{X}_8$	$\mathbf{X}_9$
G2_1	*-0.720*	*-0.461*	*1.057*	*1.844*	*-0.263*	*1.824*	*-0.385*	*-0.677*	*0.612*
G2_2	*1.031*	*0.931*	*-0.839*	*1.191*	*1.831*	*2.476*	*-0.479*	*-0.186*	*-1.106*
G2_3	*-1.196*	*0.813*	*1.356*	*1.428*	*2.504*	*1.862*	*-0.645*	*0.361*	*1.538*
G2_4	*-0.036*	*1.565*	*-2.620*	*2.322*	*1.587*	*1.201*	*0.092*	*-1.988*	*-0.220*
G2_5	*0.357*	*1.478*	*-0.518*	*0.692*	*0.998*	*2.415*	*0.329*	*-0.530*	*0.513*

Blockwise sample means
Block 1,

$$\bar{\boldsymbol{X}}_1 = (0.143, -0.915, 0.006), \quad \bar{\boldsymbol{Y}}_1 = (-0.113, 0.865, -0.313),$$
$$\bar{\boldsymbol{d}}_1 = (0.256, -1.780, 0.319).$$

Block 2,

$$\bar{\boldsymbol{X}}_2 = (0.227, 0.102, -0.645), \quad \bar{\boldsymbol{Y}}_2 = (1.495, 1.731, 1.956),$$
$$\bar{\boldsymbol{d}}_2 = (-1.268, -1.629, -2.601).$$

Block 3,

$$\bar{\boldsymbol{X}}_3 = (-0.140, 0.078, -0.801), \quad \bar{\boldsymbol{Y}}_3 = (-0.218, -0.604, 0.267),$$
$$\bar{\boldsymbol{d}}_3 = (0.078, 0.682, -1.068).$$

Let S_j denote the pooled covariance matrix for block j,

$$S_j = \frac{(m-1)S_{1j} + (n-1)S_{2j}}{m+n-2}.$$

The blockwise Hotelling statistics are

$$T_j^2 = \frac{mn}{m+n} \bar{\boldsymbol{d}}_j' S_j^{-1} \bar{\boldsymbol{d}}_j.$$

Block	*Variables*	T_j^2
1	X_1, X_2, X_3	*20.27*
2	X_4, X_5, X_6	***50.51***
3	X_7, X_8, X_9	*5.83*

For $d = 3$,

$$F_j = \frac{m+n-d-1}{d(m+n-2)} T_j^2 \sim F_{d,\, m+n-d-1}.$$

With $m + n = 10$, *this reduces to* $F_j = \frac{1}{4}T_j^2$.

- *Block 1:* $F = 5.07$, *Block 2:* $F = 12.63$, *Block 3:* $F = 1.46$

The critical value at $\alpha = 0.05$ *is* $F_{3,6}(0.95) \approx 4.76$.

- *Block 1: Reject* H_{01}, *Block 2: Reject* H_{02}, *Block 3: Do not reject* H_{03}

Using a blockwise decision rule that rejects H_0^{global} *whenever at least one block is significant, we reject the global null hypothesis. Block 2 shows the strongest evidence of a mean difference.*

9.5 Improving Wilks Outlier Test

Given n independent vectors $\mathbf{X}_1, \mathbf{X}_2, \ldots, \mathbf{X}_n$ in $\mathbb{R}^d$ from a distribution $\mathbb{N}$ with mean $\boldsymbol{\mu}$ and covariance matrix $\boldsymbol{\Sigma}$, researchers are interested in testing $H_0 : \mathbf{X}_i \sim F(\boldsymbol{\mu}, \boldsymbol{\Sigma})$, $i = 1, \ldots, n$, against the alternative hypothesis $H_a : \mathbf{X}_i \sim F(\boldsymbol{\mu}, \boldsymbol{\Sigma})$, for some i, and $\mathbf{X}_j \sim F(\boldsymbol{\mu}+\mathbf{b}, \boldsymbol{\Sigma})$, for all $j \neq i$. The slippage parameter $\mathbf{b}$, the mean vector $\boldsymbol{\mu}$, and the covariance $\boldsymbol{\Sigma}$ matrix are unknown, so are the indices j of the outliers.

9.5.1 Wilks Outlier Method

The classical test of H_0 under the multivariate normal distribution is Wilks [356] method. Let $\bar{\mathbf{X}} = \frac{1}{n}\sum_{i=1}^n \mathbf{X}_i$ and $\mathbf{S} = \frac{1}{n-1}\sum_{i=1}^n (\mathbf{X}_i - \bar{\mathbf{X}})(\mathbf{X}_i - \bar{\mathbf{X}})'$ denote the sample mean and covariance matrix, respectively. The Wilks statistic is the likelihood ratio test statistics for $H_0 : \mathbf{X}_i \sim \mathbb{N}(\boldsymbol{\mu}, \boldsymbol{\Sigma})$, $i = 1, \ldots, n$, against the alternative hypothesis $H_a : \mathbf{X}_i \sim \mathbb{N}(\boldsymbol{\mu}, \boldsymbol{\Sigma})$ and $\mathbf{X}_j \sim \mathbb{N}(\boldsymbol{\mu}+\mathbf{b}, \boldsymbol{\Sigma})$ for $i \neq j$ and $\mathbf{b} \neq \mathbf{0}$. To find a single outlier one computes the scatter ratio $R_j = \frac{|\mathbf{A}^{(j)}|}{|\mathbf{A}|}$, where $|\mathbf{A}|$ is the determinant of $\mathbf{A} = (n-1)\mathbf{S}$ and $\mathbf{A}^{(j)}$ is the matrix of the sums of squares and cross-products with $\mathbf{X}_j$ removed from the sample. The suspect outlier is the observation $\mathbf{X}_t$, the removal of which provides the largest reduction in $|\mathbf{A}|$. The Wilks' statistic is defined as Caroni and Prescott [51],

$$D_t = \min_j R_j = \frac{|\mathbf{A}^{(t)}|}{|\mathbf{A}|} = 1 - \frac{n}{n-1}(\mathbf{X}_t - \bar{\mathbf{X}})'\mathbf{A}^{-1}(\mathbf{X}_t - \bar{\mathbf{X}}). \tag{9.10}$$

Let $T_j^2 = n(\mathbf{X}_j - \bar{\mathbf{X}})'\mathbf{S}^{-1}(\mathbf{X}_j - \bar{\mathbf{X}})$ denote the Hotelling T^2 statistic for the j-th observation. Using the fact that $\frac{n-d-1}{d(n-2)}T_j^2 \sim F(d, n-d-1)$, Krzanowski [177] proves that

$$R_j = (1 + \frac{T_j^2}{n-2})^{-1} \sim \beta(\frac{n-d-1}{2}, \frac{d}{2}). \tag{9.11}$$

Example 9.5.1 *We take dimension $d = 4$, sample size $n = 6$. Five observations are from $\mathbb{N}_4(\mathbf{0}, \mathbf{I}_4)$ and one is shifted by $\mu = 3$ in every coordinate.*

$$\mathbf{X}_1 = (-0.20,\ 0.48,\ -0.52,\ -0.56), \quad \mathbf{X}_2 = (1.97,\ 1.39,\ 0.09,\ 0.28),$$
$$\mathbf{X}_3 = (0.77,\ 1.25,\ 1.01,\ -1.30), \quad \mathbf{X}_4 = (0.27,\ 0.23,\ 1.35,\ 0.89),$$
$$\mathbf{X}_5 = (-2.00,\ -0.37,\ 1.67,\ -0.44), \quad \mathbf{X}_6 = (2.46,\ 3.48,\ 6.25,\ 1.98) \ \textit{(shifted)}.$$

We obtain $\bar{\mathbf{X}} = (0.54,\ 1.08,\ 1.64,\ 0.14)$, $A = (n-1)S$, $|A| \approx 275.95$. To compute the Wilks ratio, for each case j, compute $R_j = \frac{|A^{(j)}|}{|A|}$, and $D_t = \min_j R_j$.

j	1	2	3	4	5	6
R_j	0.203	0.464	0.050	0.222	0.259	**0.002**

The minimum occurs at $j = 6$, giving $D_t = 0.002$, which correctly identifies the shifted observation $\mathbf{X}_6$ as the outlier.

Bonferroni bounds are used to approximate the percentage points of D_1 as the lower $100\alpha/n$ percentage points of the beta distribution. Wilks and Hotelling tests are related by (9.11). Modarres [236] shows that the power of the Wilks outlier test decreases when d approaches n. The Wilks test is non–robust and its power approaches α as d increases. Bhandary [30] shows that the Wilks statistic remains the same within the larger class of elliptically symmetric distributions.

9.5.2 Why Wilks Test Breaks Down

Hu and Bai [156] study the effects of high dimension on the power of the one-sample Hotelling test statistic when $y_n = d/n \to y \in (0, 1)$ to ensure the existence of the test. They prove that the limiting power of the Hotelling's test is slowly increasing for y_n close to 1, as the non–central parameter increases. Consider testing the null hypothesis

$$K_0 : \boldsymbol{\mu} = \boldsymbol{\mu}_0 \quad \text{against} \quad K_a : \boldsymbol{\mu} \neq \boldsymbol{\mu}_0.$$

Let $T^2 = n(\bar{\mathbf{X}} - \boldsymbol{\mu}_0)'\mathbf{S}^{-1}(\bar{\mathbf{X}} - \boldsymbol{\mu}_0)$ and $\delta = \boldsymbol{\Sigma}^{-1/2}(\boldsymbol{\mu} - \boldsymbol{\mu}_0)$. Under K_0, $\frac{n-d}{d(n-1)}T^2$ has an $F(d, n-d)$ distribution. When d is fixed and $n \to \infty$, the limiting distribution is $\chi^2(d)$. Under K_0, using the Central Limit Theorem, Hu and Bai [156] show that for the one-sample T^2 test, as $n \to \infty$,

$$\sqrt{\frac{(1-y_n)^3}{2ny_n}}(T^2 - \frac{ny_n}{1-y_n} - \frac{n\|\delta\|^2}{1-y_n}) \xrightarrow{D} \mathbb{N}(0, 1). \tag{9.12}$$

Using the limiting normal distribution (9.12), the limiting power of the Hotelling T^2 test is

$$\Phi(-\eta_\alpha + \sqrt{\frac{n(1-y)}{2y}}\|\delta\|^2), \tag{9.13}$$

where η_α is the upper α quantile of the standard normal random variable. Note that (9.13) is slowly increasing for y close to 1 as $\|\delta\|^2$ increases. Since Wilks and Hotelling tests are related, the power of the Wilks outlier test also decreases when d/n is close to 1.

Example 9.5.2 *To see the effects of power loss, the left panel in Figure 9.3 shows the power function of the Wilks one-outlier test for fixed $n = 50$ as d ranges in $(2, 3, \ldots, 28)$*

for $\rho = 0.0$ and 0.9. Under $H_0 : \mathbf{X}_i \sim \mathbb{N}_d(\mathbf{0}_d, \boldsymbol{\Sigma})$ for $i = 1, \ldots n$ and under H_a, one of the observations come from $\mathbb{N}_d(6 \times \mathbf{1}_d, \boldsymbol{\Sigma})$. The covariance matrix has 1 on the main diagonal and ρ on the off diagonal. The Figure shows that as d approaches n the Wilks test loses power rapidly and converges to $\alpha = 0.05$.

Instead of testing H_0 we test an equivalent hypothesis. Let

$$K_{0i} : \mathbf{X}^{(i)} \sim \mathbb{N}(\boldsymbol{\mu}^{(i)}, \boldsymbol{\Sigma}^{(i)}),\ i = 1, \ldots, b$$

denote the null hypothesis and

$$K_{ai} : \mathbf{X}^{(i)} \sim \mathbb{N}(\boldsymbol{\mu}^{(i)}, \boldsymbol{\Sigma}) \text{ and } \mathbf{X}_j \sim \mathbb{N}(\boldsymbol{\mu}^{(i)} + \mathbf{b}, \boldsymbol{\Sigma}^{(i)})$$

for $i \neq j$ and $\mathbf{b} \neq \mathbf{0}$ denote the alternative hypothesis. We test the null hypothesis $K_0 : \cap_{i=1}^{b} K_{0i}$ against the alternative $K_a : \cup_{i=1}^{b} K_{ai}$. We apply the Wilks statistic to each of the $b > 1$ blocks of d_i variables. We reject K_0 when any of blocks rejects K_{0i}, for $i = 1, \ldots, b$.

Example 9.5.3 *We generated $n = 30$ observations from $\mathbb{N}(\mathbf{0}, \mathbf{I}_d)$ and created an outlier by shifting one observation by 2.5 units, where $d = 4, 6, 8, \ldots, 48$. We used the Wilks test to find the outlier with n observations and d variables. We split the d variables into two groups of d/2 variables and used the Wilks test with n observations and d/2 variables twice. An outlier was declared when it was found in either group. We repeated the experiment 5000 times. The right panel in Figure 9.3 shows the estimated powers of the Wilks test with one outlier at $\alpha = 0.05$ using all d variables and the sub-Wilks test with d/2 variables. The power of Wilks test tends to its α level as d approaches n while the power of the sub-Wilks method increases.*

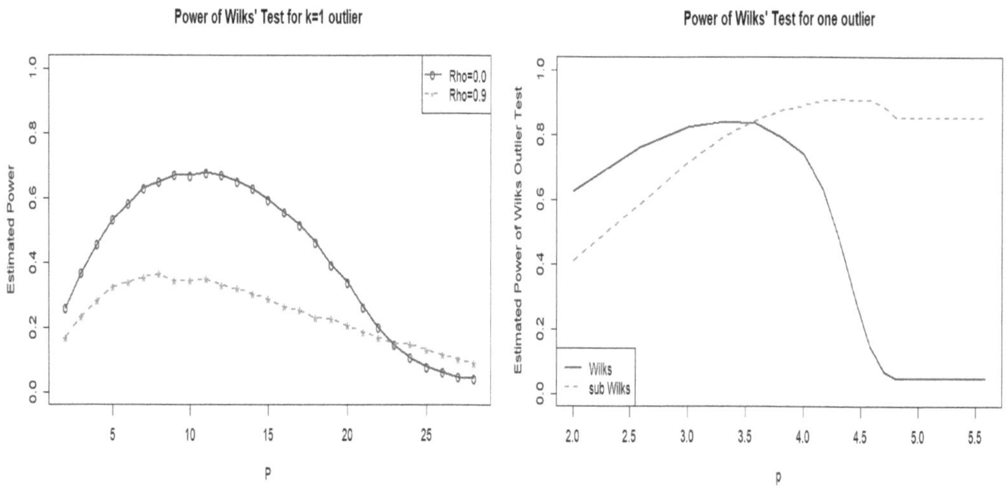

FIGURE 9.3: Left panel: Power of Wilks test for $\rho = 0.0$ and $\rho = 0.9$. Right panel: Powers of Wilks and sub-Wilks tests.

Example 9.5.4 *We generated $n = 45$ observations and $d = 40$ variables from $\mathbb{N}(\boldsymbol{\mu}, \mathbf{I})$ and created an outlier by shifting one observation by 2μ units where $\boldsymbol{\mu} = \mu\mathbf{1}$ and $\mu = 0, 0.1, \ldots, 2.0$. We used Wilks test to find the outlier with n observations and d variables. We also split the d variables into two groups of d/2 variables and used the sub-Wilks test*

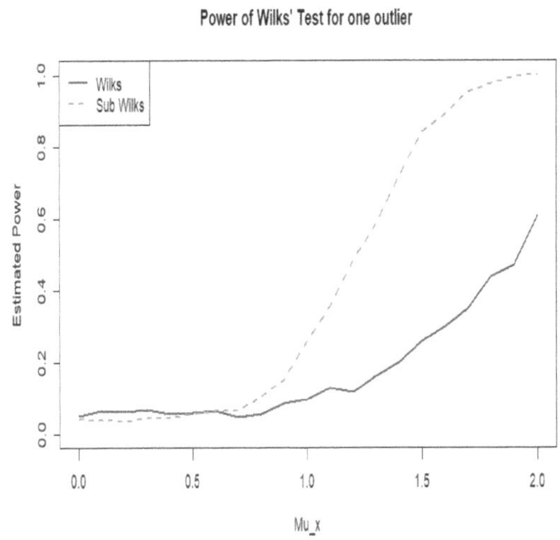

FIGURE 9.4: Powers of Wilks and sub-Wilks tests.

with n observations and $d/2$ variables. An outlier was declared when it was found in either groups. We repeated the experiment 5000 times. Figure 9.4 shows the estimated powers of Wilks test with one outlier at $\alpha = 0.05$ using all d variables and sub-Wilks method with $d/2$ variables. Wilks test has an estimated type I error of 0.067, *whereas the sub-Wilks test has an estimated type I error of* 0.051.

9.6 Christensen Microarray Data Revisited

For a high-dimensional application, we revisit the microarray data of Christensen et al. [67] who analyzed 217 non–pathologic human tissues from 10 anatomic sites at 1,413 autosomal CpG loci associated with 773 genes to investigate tissue-specific differences in DNA methylation. The data are available publicly in "datamicroarray" library of R. The dataset contains 217 observations with 1413 variables. The observation labels are obtained for "blood" (**X** group with 85 observations), "other" (**Y** group with 113 observations) and "placenta" (**Z** group with 19 observations). We selected the first 1400 observations in **X** and **Y** groups and tested for the equality of their means using sub-T^2 method.

We first partition the sample into $b = 14$ blocks of $d_i = 100$ variables so that $m+n-1 > d_i$. We test the null hypothesis $K_0 : \cap_{i=1}^{b} \boldsymbol{\mu}_x{}^{(i)} = \boldsymbol{\mu}_y{}^{(i)}$. We reject K_0 when any of the blocks rejects K_{0i}, for $i = 1, \ldots, b$. Solving $\alpha^* = 1 - (1-\alpha)^{1/b}$ for $\alpha = 0.05$ with $b = 14$ we obtain $\alpha^* = 0.003657$. The threshold value is obtained by

$$c_{\alpha^*} = ((m-n)d_i)/(m+n-d_i-1)\ F^{-1}(\alpha^*, d_i, m+n-d_i-1)$$

where F^{-1} is the quantile function of the F distribution. The hypothesis K_0 is rejected at level α^* in all sub-blocks. Solving $\alpha^* = 1 - (1-\alpha)^{1/b}$ for $\alpha = 0.05$ with $b = 140$ we obtain $\alpha^* = 0.0003663$. With $b = 140$ blocks of $d_i = 10$ variables, the hypothesis K_0 is again rejected at level α^* in all sub-blocks.

To determine if the two groups have the same distribution we analyzed the group $\mathbf{X}$ with 85 and group $\mathbf{Y}$ with 113 observations and $d = 1413$ variables using two statistics designed for high-dimensional data. We used the energy statistic (Székely and Rizzo [333]); Baringhaus and Franz [21]) and the test of Biswas and Ghosh [35]. The average IPD of $\mathbf{X}$ and $\mathbf{Y}$ groups using Euclidean distances are defined in (2.7), (2.11), and (2.15), respectively. The energy statistic is given by $E = 2\bar{d}_{(xy)} - \bar{d}_{(x)} - \bar{d}_{(y)}$. The test of Biswas and Gosh [35] is given by $BG = (\bar{d}_{(xy)} - \bar{d}_{(x)})^2 + (\bar{d}_{(xy)} - \bar{d}_{(y)})^2$. We obtained the distribution of E and BG under the null hypothesis using 1000 random permutations. The equality of distribution functions is rejected by both test statistics. The ECDFs plots of the average IPDs for the three groups show distinct differences, indicating that they arise from different distributions.

9.7 Summary

This chapter revisits the classical Hotelling T^2 test in the challenging HDLSS regime, where the standard procedure becomes undefined due to singularity of the sample covariance matrix and suffers from severe loss of power as $d \gg n$. To address these limitations, the chapter develops a block-based testing strategy grounded in the union–intersection principle. By partitioning variables into mutually independent blocks under multivariate normality, Hotelling T^2 statistics can be computed within lower-dimensional subspaces where their distributional properties remain valid. This approach preserves key optimal features of the classical test, including affine invariance, unbiasedness, and known null distributions, while restoring feasibility and improving power in high-dimensional settings. The methodology is developed for both one-sample and two-sample testing problems, with explicit derivations of significance levels and power functions using inclusion–exclusion arguments.

Building on this framework, the chapter addresses multivariate outlier detection through a critical reexamination of Wilks' outlier test. It is shown that Wilks' procedure deteriorates rapidly as the dimension approaches the sample size, leading to diminished power and unreliable detection. A blockwise adaptation of Wilks' method is proposed, which searches for outlying structure within subsets of variables rather than the full space. The resulting procedure exhibits improved sensitivity and controlled type I error, particularly in HDLSS contexts. Analytical results and simulation studies confirm that the proposed block-based outlier test outperforms the classical Wilks method across a range of dimensional regimes.

In practice, the assumption of multivariate normality of the observations is an impediment to using Hotelling T^2 test. We also explore resampling-based alternatives. Permutation versions of the two-sample T^2 test and bootstrap-based approaches for the one-sample case are discussed as practical tools for enhancing robustness under non-normality. These approximate procedures are shown to support and complement the blockwise T^2 methodology in high dimensions. An application to microarray gene expression data demonstrates the practical relevance of the proposed methods for detecting mean differences and outliers in modern high-throughput datasets. Overall, the chapter provides a coherent framework for extending Hotelling's T^2 test and Wilks' outlier detection to high-dimensional settings, highlighting both theoretical foundations and practical implications for multivariate inference.

9.8 Proofs

Proof: Theorem 9.3.1: The proof is broken down into the following six steps.

1. Eigen-decomposition of the regularized covariance. Let $\mathbf{S}+\lambda\mathbf{I}_d = \mathbf{U}(\mathbf{\Lambda}+\lambda\mathbf{I}_d)\mathbf{U}^T$, where $\mathbf{U}$ is orthogonal and $\mathbf{\Lambda} = \text{diag}(\hat{\lambda}_1, \ldots, \hat{\lambda}_r, 0, \ldots, 0)$ with $r = \text{rank}(\mathbf{S}) \leq m-1$. Then
$$(\mathbf{S} + \lambda\mathbf{I}_d)^{-1} = \mathbf{U}(\mathbf{\Lambda} + \lambda\mathbf{I}_d)^{-1}\mathbf{U}^T.$$

2. Express T_λ^2 in the eigen basis. Let $\mathbf{Z} = \sqrt{m}(\bar{\mathbf{X}} - \boldsymbol{\mu}_0)$. Then
$$T_\lambda^2 = \mathbf{Z}^T(\mathbf{S} + \lambda\mathbf{I}_d)^{-1}\mathbf{Z} = \sum_{k=1}^{d} \frac{(\mathbf{u}_k^T\mathbf{Z})^2}{\hat{\lambda}_k + \lambda}.$$

3. Distribution of the projected components. Conditional on $\mathbf{S}$, each component $(\mathbf{u}_k^T\mathbf{Z})$ is approximately normal with mean 0 and variance $\mathbf{u}_k^T\mathbf{\Sigma}\mathbf{u}_k$, since $\mathbf{Z}$ is the sum of m i.i.d. vectors scaled by $\sqrt{m}$.

4. Linearization. Write
$$T_\lambda^2 = \sum_{k=1}^{d} \frac{(\mathbf{u}_k^T\mathbf{Z})^2 - \mathbf{u}_k^T\mathbf{\Sigma}\mathbf{u}_k}{\hat{\lambda}_k + \lambda} + \sum_{k=1}^{d} \frac{\mathbf{u}_k^T\mathbf{\Sigma}\mathbf{u}_k}{\hat{\lambda}_k + \lambda} =: R_1 + R_2.$$

   R_2 is the mean part: as $d \to \infty$, by the law of large numbers, $R_2 \approx d \cdot M_\lambda$.

   R_1 is a sum of centered independent terms. Its variance is
$$\mathbb{V}ar(R_1) = 2\sum_{k=1}^{d} \frac{(\mathbf{u}_k^T\mathbf{\Sigma}\mathbf{u}_k)^2}{(\hat{\lambda}_k + \lambda)^2} \approx d \cdot v_\lambda.$$

5. Apply the Central Limit Theorem. Since R_1 is a sum of a large number of weakly dependent terms (the dependence is negligible in high dimensions under standard regularity conditions), the classical Lyapunov or Lindeberg-Feller CLT gives
$$\frac{R_1}{\sqrt{dv_\lambda}} \xrightarrow{d} \mathbb{N}(0, 1).$$

6. We combine the centered mean $R_2 - dM_\lambda$ with the standardized R_1 to obtain
$$\frac{T_\lambda^2 - dM_\lambda}{\sqrt{dv_\lambda}} \xrightarrow{d} \mathbb{N}(0, 1).$$

Importantly, $\lambda > 0$ ensures all denominators $\hat{\lambda}_k + \lambda$ are strictly positive, making the statistic well-defined even when $d \geq m$ or $d \gg m$. Classical T^2 cannot be computed in these regimes because $\mathbf{S}$ is singular. □

The theorem is valid in ultra-high-dimensional settings as long as $\lambda > 0$. The mean M_λ and variance v_λ account for the effect of ridge regularization. This result justifies using the standard normal approximation for p-values in high-dimensional or $d \gg m$ applications.

9.9 Exercises

Exercise 9.1 *Show that Hotelling's T^2 test statistic is invariant under non–singular linear transformations of the data.*

Exercise 9.2 *Prove that when $d = 1$, Hotelling's T^2 test reduces to the classical two-sample t-test.*

Exercise 9.3 *The Wilks' outlier statistic is defined in (9.10). Let $T_j^2 = n(\mathbf{X}_j - \bar{\mathbf{X}})'\mathbf{S}^{-1}(\mathbf{X}_j - \bar{\mathbf{X}})$ denote the Hotelling T^2 statistic for the j-th observation. Using the fact that $\frac{n-d-1}{d(n-2)}T_j^2 \sim F(d, n-d-1)$, prove that*

$$R_j = \left(1 + \frac{T_j^2}{n-2}\right)^{-1} \sim Beta\left(\frac{n-d-1}{2}, \frac{d}{2}\right). \tag{9.14}$$

Exercise 9.4 *Let $\frac{n-d}{d(n-1)}T^2 \sim F(d, n-d)$, where T^2 is the Hotelling's statistic. Prove that as $n \to \infty$ with fixed d, the distribution of T^2 converges to a chi-square distribution with d degrees of freedom.*

Exercise 9.5 *What is the limiting behavior of the Hotelling T^2 one-sample test statistic as the dimension $d \to \infty$, with n fixed and growing with d?*

Exercise 9.6 *Blockwise Hotelling T^2 Test in Stroke Rehabilitation Study*
We compare two rehabilitation programs for stroke patients.

- *Group 1 (G1): Cognitive training*
- *Group 2 (G2): Physical training*

Patients are evaluated on 8 standardized outcomes,

- *Block 1 (Cognitive outcomes): Memory, Attention, Executive Function, Verbal Fluency*
- *Block 2 (Motor/Physical outcomes): Grip Strength, Gait Speed, Balance, Coordination*

We have 6 patients per group. The goal is to test

$$H_0 : \boldsymbol{\mu}_1 = \boldsymbol{\mu}_2 \quad vs \quad H_a : \boldsymbol{\mu}_1 \neq \boldsymbol{\mu}_2$$

using Hotelling's T^2 on blocks of size 4.
Group 1 (Cognitive Training).

Mem	*Att*	*Exec*	*VF*	*Grip*	*Gait*	*Bal*	*Coord*
48	62	58	44	22	1.5	5.1	7.8
45	65	60	47	24	1.7	4.9	8.0
50	64	63	49	23	1.6	5.3	8.2
46	63	59	45	21	1.4	5.0	7.9
49	66	62	48	23	1.8	5.2	8.1
47	61	57	43	22	1.5	5.0	7.7

Group 2 (Physical Training).

Mem	*Att*	*Exec*	*VF*	*Grip*	*Gait*	*Bal*	*Coord*
42	58	54	40	28	2.0	6.1	8.7
41	60	55	41	27	1.9	6.0	8.6
44	59	57	43	29	2.2	6.2	8.9
43	57	56	42	28	2.1	6.3	9.0
45	61	58	44	30	2.3	6.5	9.1
42	58	55	41	27	2.0	6.0	8.8

Exercise 9.7 *Let $\mathbb{X}^{(k)} = \{\mathbf{X}_i^{(k)}\}_{i=1}^{n_i}$ denote the i.i.d. observations of the k-th group that arise from $\mathbb{N}(\boldsymbol{\mu}_k, \boldsymbol{\Sigma})$ for $k = 1, \ldots, K$. We are interested in testing*

$$H_0 : \boldsymbol{\mu}_1 = \boldsymbol{\mu}_2 = \ldots = \boldsymbol{\mu}_K, \tag{9.15}$$

against $H_a : \boldsymbol{\mu}_r \neq \boldsymbol{\mu}_t$ for at least one $r \neq t = 1, \ldots, K$. Extend the sub-T^2 case to $k > 2$ samples and conduct a multivariate analysis of variance when $d \gg n_i$.

10

High-Dimensional Outlier Detection Methods

10.1 Abstract

This chapter provides an overview of high-dimensional outlier detection methods, addressing the challenges posed when $d \gg N$. Traditional techniques, including Mahalanobis distance-based approaches, are sensitive to outliers and often fail in high-dimensional settings due to singularity of the covariance matrix and masking or swamping effects. To overcome these limitations, robust estimators such as the minimum covariance determinant, minimum volume ellipsoid, and component-wise medians are discussed. Proximity-based methods, including IPDs, nearest-neighbor approaches, and kernel-based outlier factors, offer distribution-free alternatives while retaining sensitivity to anomalies. The chapter also examines generalized dispersion measures, r-th nearest neighbor methods, and bootstrap-based procedures for assessing statistical significance. Performance studies under normal, t_4, skew-normal, and mixture distributions illustrate the comparative effectiveness of these techniques. Practical applications demonstrate the utility of high-dimensional outlier detection in real-world contexts, including finance, health monitoring, and data mining. Collectively, these methodologies provide a robust, computationally feasible framework for identifying anomalous observations in complex multivariate datasets.

10.2 Introduction

Methods of outlier detection have a long history in Statistics (Bernoulli [27]; Barnett & Lewis [23]; Hawkins[143]). Hadi et al. [131]present an overview of the methods of outlier detection. Broadly speaking, outliers are those observations that exhibit different characteristics than the majority of the observations in the dataset. Anomalous observations can adversely affect model selection and parameter estimation. There are numerous applications of outlier methods in early warning systems, public health, fraud detection, finance, geographic information systems, and data mining.

Multivariate methods of outlier detection include model-based and proximity–based techniques. Model-based methods specify a model and label observations that do not fit the assumed model as outliers. Hence, outliers are those observations that have a low probability of occurrence under the specified model (Hawkins [143]). Proximity techniques exploit the spatial proximity of the observations to detect outliers. The main approaches of the proximity methods are distanced-based, density-based and depth-based methods.

DOI: 10.1201/9781003789581-10

A proximity–based method of outlier detection uses the Mahalanobis distance (Mahalanobis [207]),

$$M(\mathbf{X}) = n(\mathbf{X} - \bar{\mathbf{X}})'\mathbf{S}^{-1}(\mathbf{X} - \bar{\mathbf{X}}),$$

where $\mathbf{X}$ is a d-variate vector of the observations, $\bar{\mathbf{X}}$ is the vector of the means and $\mathbf{S}$ is the sample covariance matrix based on n observations. Assuming that the observations are taken from a multivariate normal distribution,

$$\frac{n}{(n-1)^2} M(\mathbf{X}) \sim Beta(\frac{d}{2}, \frac{n-p-1}{2}).$$

As $n \to \infty$, the Mahalanobis distance has a $\chi^2(d)$ distribution with d degrees of freedom. While its asymptotic distribution is known, it is the difficulty of using robust parameter estimates that causes difficulties. The method requires that $n > d$ so that the inverse of the sample covariance exists. Its exact distribution is unknown under other multivariate distributions.

The sample mean vector and covariance matrix are infamous for their sensitivity to outliers. Outliers adversely affect estimates of the mean and the covariance matrix and cause two difficulties, outlier masking (not all outliers are identified) and outlier swamping (non–outlying observations are incorrectly labeled as outliers). Robust estimates of the parameters can often alleviate these problems. Minimum covariance determinant and the minimum volume ellipsoid (Rousseeuw [286]) are popular methods of obtaining robust estimators of the mean and the covariance. Hadi [130] estimates the center by a vector of component-wise medians and the covariance matrix for those observations with the smallest Mahalanobis distance. Penny and Jolliffe [268] provide a modification this method. Peña and Prieto [266] search for outliers along data projections that maximize or minimize the kurtosis coefficient. Most robust techniques of outlier detection are computationally expensive and encounter difficulties with high-dimensional data.

Approaches to outlier detection with the use of the bootstrap (Efron and Tibshirani [86]) include the work of Singh and Xie [315] on the bootstrap-based outlier detection plot as a non-parametric graphical tool for detecting outliers in a dataset. Martin and Roberts [218] propose a bootstrap approach to constructing critical points for use in outlier detection in the context of least-squares Studentized residuals. Beath [26] uses the bootstrap likelihood ratio test to determine if there are outliers present by comparing models with and without outliers. Amado et al. [9] discuss the bootstrap when outliers are present in the data and suggest resampling with different probabilities so that outliers have smaller probabilities of selection. Based on the empirical distribution of the IPDs, Song and Modarres [319] construct a bootstrap test of homogeneity for multivariate finite mixture models. Frees [108] discusses density estimation of functions of observations using the kernel density function and proves large sample properties of the resampling estimator of the kernel density. The empirical kernel density function is a U-statistic with a bounded kernel and has several well-known properties. In particular, one can resample the empirical kernel density function of the IPDs.

The number of outliers in a dataset of size n can never exceed half the observations as one cannot distinguish between the outlier and the non–outlier groups, otherwise. In search of the outliers, one can order the observations in the univariate case and plot the observations when $d = 2$ or 3. However, these methods fail for large d as it is possible for an observation to be an outlier in d-dimensional space, but not one in lower dimensions. Multivariate outliers pose special challenges, especially when $d \gg n$. Following Hall et al. [133] we assume that $d \to \infty$ with n fixed. A proximity–based method, IPDs have proved useful in many high-dimensional problems. Modarres and Song [239] present a recent survey of IPDs, and Guo and Modarres [125] use IPDs for high-dimensional classification. Modarres

[229] proposes methods for visualization and comparison of high-dimensional observations. Being one-dimensional, IPDs circumvent difficulties of the high-dimensional space and serve as the building blocks of many multivariate techniques.

10.3 Outlier Detection with Dissimilarity Indices

Following Modarres [232] suppose that k outliers are identified and the sample is divided into non–outlier (main) group $\mathbb{X} = \{\mathbf{X}_1, \ldots, \mathbf{X}_m\}$ with $m = n - k$ observations and outlier group $\mathbb{Y} = \{\mathbf{Y}_1, \ldots, \mathbf{Y}_k\}$ with k observations. This assumption is only for ease of the presentation and the method we describe automatically detects the outliers. Let $\boldsymbol{\mu}_x$, $\boldsymbol{\mu}_y$, $\boldsymbol{\Sigma}_x$, and $\boldsymbol{\Sigma}_y$ denote the mean vectors and the covariance matrices of the groups $\mathbb{X}$ and $\mathbb{Y}$, respectively. Let $\mathbb{Z}_n = \mathbb{X} \cup \mathbb{Y}$ be the set of all the observations from the main and the outlier groups, where $\mathbf{Z}_i = \mathbf{X}_i$ for $i = 1, \ldots, m$ and $\mathbf{Z}_i = \mathbf{Y}_{i-m}$ for $i = m+1, \ldots, n$. We use the dissimilarity measure $\rho(\mathbf{X}, \mathbf{Y})$ in (3.6) to detect outliers.

Let $S_r = \sum_{c=1}^{n} [\boldsymbol{\rho}]_{rc}$ be the sum of the r-th row of the dissimilarity matrix $\boldsymbol{\rho}$ for $r = 1, \ldots n$. Suppose the observed dataset of n observations has $k = 1$ outlier at position t. Using (3.16), one observes that as $d \to \infty$, S_r, $r \neq t = 1, \ldots, n$, approaches $\tilde{\rho}(F_1, F_2)$ whereas S_t approaches $(n-1)\tilde{\rho}(F_1, F_2)$. It follows that observation t holds the maximum dissimilarity and is a suspect outlier. One can plot S_r values against observation numbers in order to identify the outliers.

Example 10.3.1 *We generated a sample of* $n = 100$ *observations from* $\mathbb{N}_d(\mathbf{0}, \boldsymbol{\Sigma})$ *with* $d = 5$ *variables. The covariance matrix has unit variances and constant correlation* $\rho = 0.5$*. We next shifted* $k = 3$ *observations at positions* $20, 40$ *and* 60 *by* $5, 10$ *and* 15 *units, respectively. Figure 10.1 shows the outlier detection plot where the horizontal axis represents the observation number and the vertical axis displays the row sums of the dissimilarity*

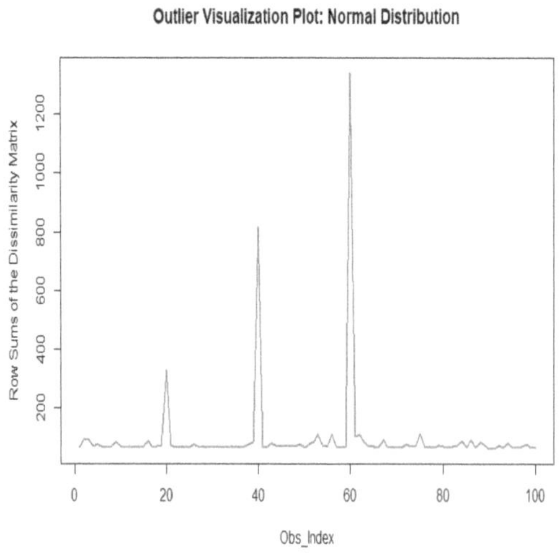

FIGURE 10.1: Row sums of the dissimilarities from a sample of $n = 100$ observations from $\mathbb{N}_d(\mathbf{0}, \boldsymbol{\Sigma})$ with $d = 5$ variables. The three outliers at positions 20, 40, 60 show large dissimilarities, whereas the remaining observations show small dissimilarities.

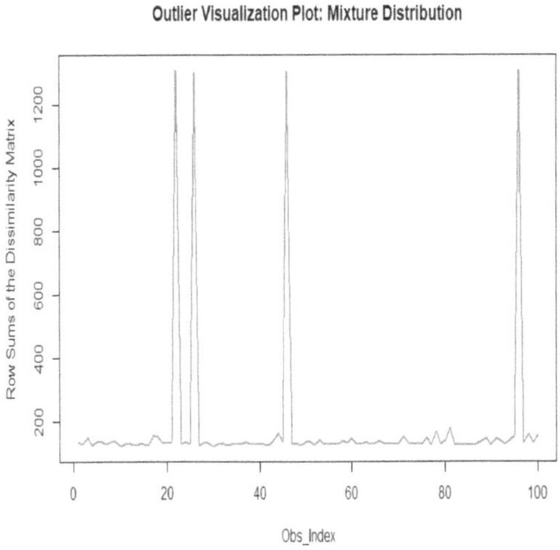

FIGURE 10.2: Row sums of the dissimilarities for $n = 100$ observations from the mixture $.95\mathbb{N}_d(\mathbf{0}, \mathbf{\Sigma}_x) + .05\mathbb{N}_d(\mathbf{0}, \mathbf{\Sigma}_y)$ and $d = 2000$ variables. The four observations from the second component of the mixture distribution appear at positions 22, 26, 46, and 96.

matrix. The three outliers stand out for visual detection. While Lemma 1 requires $d \to \infty$ for convergence of $\rho(\mathbf{X}_i, \mathbf{X}_j)$ and $\rho(\mathbf{Y}_i, \mathbf{Y}_j)$ in probability to 0, it is remarkable that the row sums of the main group on the vertical axis remain small for d as small as 5.

We study the utility of the outlier visualization plot under a normal mixture model with heterogeneous covariances. We generated a sample of $n = 100$ observations with $d = 2000$ variables from the mixture $(1 - q)\mathbb{N}_d(\mathbf{0}, \mathbf{\Sigma}_x) + q\mathbb{N}_d(\mathbf{0}, \mathbf{\Sigma}_y)$ where $q = .05$, $\mathbf{\Sigma}_x$ has unit variances and constant correlation of $\rho_x = .5$, $\mathbf{\Sigma}_y$ has variances $1, 2, \ldots, d$ and correlation of $\rho_y = .8$. The data contains 4 observations from $\mathbb{N}_d(\mathbf{0}, \mathbf{\Sigma}_y)$ component and 96 observations from $\mathbb{N}_d(\mathbf{0}, \mathbf{\Sigma}_x)$ component. The four observations are located at positions 22, 26, 46, and 96. Figure 10.2 shows the outlier detection plot where the four observations from the second component of the mixture distribution stand out for visual detection.

10.4 Interpoint Distance Outlier Test

Let $\mathbf{S} = (S_1, S_2, S_3, \ldots, S_n)'$ denote the vector of the row totals, where $S_r = \sum_{c=1}^{n} [\boldsymbol{\rho}]_{rc}$ is the sum of the dissimilarities for row r of the dissimilarity matrix $\boldsymbol{\rho}$ for $r = 1, \ldots, n$. Hence, S_r is the total dissimilarity of the r-th observation from all the other observations. If the r-th observation is an outlier, S_r will be large compared to other row totals.

We scan $\mathbf{S}$ to obtain its maximum $S_{(n)}$ and define the test statistic for testing H_0 that there is no outliers ($k = 0$) in favor of H_a that there is $k = 1$ outlier in the dataset by

$$S_{(n)} = \max(\mathbf{S}) = \max_{1 \le r \le n} \Big(\sum_{c=1}^{n} [\boldsymbol{\rho}]_{rc}\Big).$$

We reject H_0 when the d_{value} of $S_{(n)}$ is smaller than α where α is the level of significance. The distribution of $S_{(n)}$ is complicated since it relates to the maximum of n dependent and positive random variables. We use the percentile bootstrap to obtain the distribution of the test statistic. We refer to this test as the IPD outlier test or IDOT. Algorithm 1 specifies the steps for finding outliers in matrix $\mathbb{X}$. Type I error rate is controlled simultaneously using the Bonferroni bound. To check the top k rows of the dissimilarity matrix, one sets $\alpha' = \alpha/k$. Once the suspect outlier is identified we need to obtain its significance. Algorithm 1 uses the M list, supplied by Algorithm 2, which holds the sampling distribution of $S_{(n)}$ under H_0. Once the suspect outlier is found significant, it is removed from the dataset, and the method is sequentially applied to the remaining observations. The iterations continue until the current suspect outlier is no longer significant.

Algorithm 10.4.1 Algorithm 1: Outlier Detection via the data matrix $\mathbb{X}$

Input: Data matrix $\mathbb{X}$; values of n, α', and E
Output: Outliers in $\mathbb{X}$

Initialize: MoreOutliers = True, $k = 0$

while MoreOutliers **do**
    Set MoreOutliers = False
    Compute $\boldsymbol{\rho} = \text{Dist}(\mathbb{X})$
    Scan **S** to obtain its maximum $S_{(n)}$ and set $Test_{stat} = S_{(n)}$
    Use Algorithm 2 to obtain the list M
    Compute

$$P_{value} = \frac{1}{E}\sum_{i=1}^{E} \mathbb{I}(M_i > Test_{stat})$$

    if $P_{value} < \alpha'$ **then**
        $k = k + 1$
        Remove the observation with the largest row-sum vector in **S** from $\mathbb{X}$
        $n = n - 1$, MoreOutliers = True
    end if
end while

Algorithm 2 estimates the distribution of $S_{(n)}$ under H_0. We first remove the corresponding suspect observation from the data matrix $\mathbb{X}$ and then apply bootstrap resampling.

Algorithm 10.4.2 Algorithm 2: Bootstrp distribution of $S_{(n)}$ under H_0

Input: Current data matrix $\mathbb{X}$; number of bootstrap samples E; current n
Output: Sampling distribution values of $S_{(n)}$ stored in M

Remove the suspect outlier with the largest row-sum from $\mathbb{X}$
for $j = 1, \ldots, E$ **do**
    Obtain a bootstrap sample from the data matrix $\mathbb{X}$
    Compute the dissimilarity matrix $\boldsymbol{\rho}$ and the row-sum vector **S**
    Scan **S** to obtain its maximum $S_{(n)}$ and set $M[j] = S_{(n)}$
end for
Return M

The IDOT method is non-parametric, assumes k is unknown, and is applied sequentially until no outliers remains in $\mathbb{X}$. Algorithm 2 with $n-k-1$ observations obtains the sampling distribution of $S_{(n-k)}$ for $k = 0, \ldots, n/2 - 1$ under H_0 by the percentile bootstrap. IDOT is a data-driven detection method and sequentially finds those observations whose total dissimilarity from all the other observations is the largest. Type I error rate is controlled simultaneously using the Bonferroni bound.

Example 10.4.1 *We use the dissimilarity indices ρ and δ with outliers in high dimension. We generate $N = 8$ independent observations $\mathbf{z}_1, \ldots, \mathbf{z}_8 \in \mathbb{R}^{100}$, where each coordinate is*

$$z_{ij} \sim \chi^2(1), \quad i = 1, \ldots, 8, \quad j = 1, \ldots, 100.$$

We create two outliers by scaling $\tilde{\mathbf{z}}_7 = 4 \times \mathbf{z}_7$, $\tilde{\mathbf{z}}_8 = 8 \times \mathbf{z}_8$, and keep the first 6 observations unchanged $\tilde{\mathbf{z}}_i = \mathbf{z}_i$, $i = 1, \ldots, 6$. The final dataset is $\tilde{\mathbb{Z}}_8 = \{\tilde{\mathbf{z}}_1, \ldots, \tilde{\mathbf{z}}_8\}$.

We compute the 8×8 matrix $\mathbf{D}_\rho$ with entries

$$[\mathbf{D}_\rho]_{ij} = \rho(\tilde{\mathbf{z}}_i, \tilde{\mathbf{z}}_j), \quad i, j = 1, \ldots, 8.$$

Distance matrix $\mathbf{D}_\rho$:

$$\begin{bmatrix} 0 & 0.89 & 0.91 & 0.87 & 0.90 & 0.88 & 3.21 & 6.57 \\ 0.89 & 0 & 0.93 & 0.90 & 0.91 & 0.89 & 3.35 & 6.75 \\ 0.91 & 0.93 & 0 & 0.92 & 0.91 & 0.90 & 3.30 & 6.70 \\ 0.87 & 0.90 & 0.92 & 0 & 0.89 & 0.88 & 3.18 & 6.53 \\ 0.90 & 0.91 & 0.91 & 0.89 & 0 & 0.90 & 3.20 & 6.56 \\ 0.88 & 0.89 & 0.90 & 0.88 & 0.90 & 0 & 3.15 & 6.50 \\ 3.21 & 3.35 & 3.30 & 3.18 & 3.20 & 3.15 & 0 & 6.85 \\ 6.57 & 6.75 & 6.70 & 6.53 & 6.56 & 6.50 & 6.85 & 0 \end{bmatrix}$$

Row sums:

$$\textit{RowSums}(\mathbf{D}_\rho) = [12.63, 12.87, 13.06, 12.63, 12.66, 12.50, 29.59, 42.06].$$

Distance matrix $\mathbf{D}_\delta$:

$$\begin{bmatrix} 0 & 0.78 & 0.79 & 0.75 & 0.78 & 0.77 & 2.91 & 6.01 \\ 0.78 & 0 & 0.80 & 0.77 & 0.79 & 0.77 & 3.02 & 6.15 \\ 0.79 & 0.80 & 0 & 0.78 & 0.77 & 0.76 & 3.00 & 6.12 \\ 0.75 & 0.77 & 0.78 & 0 & 0.75 & 0.75 & 2.87 & 5.95 \\ 0.78 & 0.79 & 0.77 & 0.75 & 0 & 0.77 & 2.90 & 5.98 \\ 0.77 & 0.77 & 0.76 & 0.75 & 0.77 & 0 & 2.85 & 5.92 \\ 2.91 & 3.02 & 3.00 & 2.87 & 2.90 & 2.85 & 0 & 6.30 \\ 6.01 & 6.15 & 6.12 & 5.95 & 5.98 & 5.92 & 6.30 & 0 \end{bmatrix}$$

Row sums:

$$\textit{RowSums}(\mathbf{D}_\delta) = [11.19, 11.28, 11.27, 10.89, 11.17, 11.02, 22.91, 43.43].$$

The outliers (rows 7 and 8) have much larger distances to other points, reflected by the larger entries and row sums in both $\mathbf{D}_\rho$ and $\mathbf{D}_\delta$. The measures ρ and δ effectively capture the presence of outliers in high dimensions. The row sums provide a useful summary measure for ranking observations by outlyingness.

10.5 Robust Kernel-based Outlier Factor

The Robust Kernel-based Outlier Factor (RKOF), introduced by Gao et al. [114], is a non-parametric outlier detection method that compares the local kernel density at a data point $\mathbf{X}$ to a weighted kernel density over its r-neighborhood. The `OutlierDetection` package in R includes a function `dens` that implements RKOF. It outputs both an outlier score and a bootstrap-based p-value for each observation. It is defined as

$$\text{RKOF}(\mathbf{X}) = \frac{\hat{f}_r(\mathbf{X})}{\hat{f}_{\text{local}}(\mathbf{X})}, \tag{10.1}$$

where,

- $\hat{f}_{\text{local}}(\mathbf{X})$ is the local kernel density estimate at point $\mathbf{X}$,
- $\hat{f}_r(\mathbf{X})$ is the weighted average of kernel density estimates in the r-distance neighborhood of $\mathbf{X}$.

A high value of RKOF($\mathbf{X}$) indicates that the point is located in a sparse region relative to a denser surrounding neighborhood, and is thus likely to be an outlier.

Let $\mathbf{X}_1, \ldots, \mathbf{X}_n$ be observations in $\mathbb{R}^d$. For a given point $\mathbf{X}_i$.

- Compute $\hat{f}_{\text{local}}(\mathbf{X}_i)$ using a kernel function K_h and bandwidth h.
- Identify a neighborhood $N_r(\mathbf{X}_i)$ of r nearest neighbors.
- Compute the neighborhood density as a weighted average,

$$\hat{f}_r(\mathbf{X}_i) = \sum_{\mathbf{X}_j \in N_r(\mathbf{X}_i)} w_{ij} \cdot \hat{f}_{\text{local}}(\mathbf{X}_j), \tag{10.2}$$

where w_{ij} are normalized similarity-based weights (e.g., exponential of negative distance).

Example 10.5.1 *Consider a simple two-dimensional dataset with 6 points,*

Point	X_1	X_2	*Comment*
$\mathbf{X}_1$	*1.0*	*1.0*	*Regular point*
$\mathbf{X}_2$	*1.1*	*1.2*	*Regular point*
$\mathbf{X}_3$	*0.9*	*0.8*	*Regular point*
$\mathbf{X}_4$	*1.2*	*1.1*	*Regular point*
$\mathbf{X}_5$	*1.0*	*0.9*	*Regular point*
$\mathbf{X}_6$	*4.5*	*4.5*	*Potential outlier*

Using a normal kernel and setting $r = 3$,

- *For* $\mathbf{X}_1$ *to* $\mathbf{X}_5$, $\hat{f}_{local}$ *and* $\hat{f}_r$ *will be high and similar, so RKOF* ≈ 1.
- *For* $\mathbf{X}_6$, $\hat{f}_{local}$ *will be low due to sparse surroundings, while* $\hat{f}_r$ *will be relatively high if the neighborhood includes denser points with lower distances to each other. Thus, RKOF(*$\mathbf{X}_6$*)* $\gg 1$.

Gao et al. [114] define the robust kernel-based outlier factor (RKOF) as the ratio of the density estimate of the r-distance neighborhood of $\mathbf{X}$ to the local density estimate of $\mathbf{X}$. The larger $RKOF(\mathbf{X})$, the more likely that $\mathbf{X}$ is an outlier. The smaller the local kernel density of $\mathbf{X}$, and the larger the density of its neighborhood, the larger its outlier factor.

10.6 Generalized Dispersion

The generalized dispersion method (Jin et al. [161]) computes the dispersion matrix for each observation without considering the current observation. Given data $X = \{\mathbf{x}_1, \mathbf{x}_2, \ldots, \mathbf{x}_n\}$, $\mathbf{x}_i \in \mathbb{R}^d$, the dispersion of a subset $S \subset X$ is defined as

$$D(S) = \frac{1}{|S|} \sum_{\mathbf{x}_i \in S} \|\mathbf{x}_i - \bar{\mathbf{x}}_S\|^2,$$

where $|S|$ denotes the cardinality of the set S and $\bar{\mathbf{x}}_S = \frac{1}{|S|} \sum_{\mathbf{x}_i \in S} \mathbf{x}_i$.

To obtain the contribution of a point $\mathbf{x}_i \in X$, define its dispersion contribution as

$$\Delta_i = D(X) - D(X \setminus \{\mathbf{x}_i\}).$$

Large values of Δ_i indicate that removing $\mathbf{x}_i$ greatly reduces dispersion; hence, it is likely an outlier. Small or negative Δ_i values indicate that $\mathbf{x}_i$ conforms well to the data.

We rank all points by Δ_i. The top-k points are identified as outliers. P-values is computed using the vector of the proportion of times an outlier exceeds the bootstrap critical value. Since the generalized dispersion is based on the covariance matrix, it is only applicable when $n > d$.

Consider the six 4-dimensional points,

$$\mathbf{x}_1 = (1,1,1,1), \quad \mathbf{x}_2 = (2,2,2,2), \quad \mathbf{x}_3 = (2,1,2,1),$$
$$\mathbf{x}_4 = (3,3,3,3), \quad \mathbf{x}_5 = (2,2,3,2), \quad \mathbf{x}_6 = (20,20,20,20).$$

We compute

$$\bar{\mathbf{x}} = \frac{1}{6}\sum_{i=1}^{6} \mathbf{x}_i = \left(\frac{30}{6}, \frac{29}{6}, \frac{31}{6}, \frac{29}{6}\right) \approx (5, 4.83, 5.17, 4.83).$$

The dispersion is $D(X) = \frac{1}{6}\sum_{i=1}^{6} \|\mathbf{x}_i - \bar{\mathbf{x}}\|^2$. The squared distance for $\mathbf{x}_6$ is ≈ 832, dominating the dispersion. By removing $\mathbf{x}_6$,

$$\bar{\mathbf{x}}_{-6} = (2, 2, 2.2, 2), \qquad D(X \setminus \{\mathbf{x}_6\}) \approx 0.7.$$

The contribution is $\Delta_6 = D(X) - D(X \setminus \{\mathbf{x}_6\}) \gg 0$, so $\mathbf{x}_6$ is identified as the outlier.

10.7 Mahalanobis Method

Closely related to the Wilks test for outlier detection, Mahalanobis distance method (Mahalanobis [207]) computes the statistical distance of an observation $\mathbf{X}$ from the sample average: $M(\mathbf{X}) = n(\mathbf{X} - \bar{\mathbf{X}})'\mathbf{S}^{-1}(\mathbf{X} - \bar{\mathbf{X}})$, where $\mathbf{X}$ is a d-variate vector of observations, $\bar{\mathbf{X}}$ is the vector of means and $\mathbf{S}$ is the sample covariance matrix based on n observations and $n > d$. As $n \to \infty$, M has a $\chi^2(d)$ distribution with d degrees of freedom. Any value above the upper $1 - \alpha$ quantile of $\chi^2(d)$ is labeled an outlier. The Mahalanobis method is only applicable when $n > d$.

Hautamaki et al. [142] proposed two density-based methods for outlier detection in multivariate data: the Average r-th Nearest Neighbor (NN) and the r-th Nearest Neighbor (rNN). Both rely on distances among data points to identify sparse or isolated observations as outliers.

10.8 Average r-th Nearest Neighbor (NN) Method

Let $\mathbf{X}_1, \ldots, \mathbf{X}_n \in \mathbb{R}^d$ be observations. Define $d_r(\mathbf{X}_i)$ as the distance from $\mathbf{X}_i$ to its r-th nearest neighbor: $d_r(\mathbf{X}_i) = \|\mathbf{X}_i - \mathbf{X}_{i(r)}\|$, where $\mathbf{X}_{i(r)}$ is the r-th nearest neighbor of $\mathbf{X}_i$. The average distance is

$$\mathrm{Avg}_r(\mathbf{X}_i) = \frac{1}{r}\sum_{j=1}^{r} \|\mathbf{X}_i - \mathbf{X}_{i(j)}\|,$$

where $\mathbf{X}_{i(j)}$ denotes the j-th nearest neighbor of $\mathbf{X}_i$.

Points with large $\mathrm{Avg}_r(\mathbf{X}_i)$ are potential outliers. Thresholds are determined via bootstrapping to compute critical values and p-values.

10.9 r-th Nearest Neighbor (rNN) Method

Construct a directed graph $G_r = (V, E)$ where each vertex v_i corresponds to an observation $\mathbf{X}_i$. There is a directed edge from $\mathbf{X}_i$ to each of its r nearest neighbors. The edge weight is the Euclidean distance. The in-degree $\deg^-(\mathbf{X}_i)$ is the number of times $\mathbf{X}_i$ appears in the neighborhoods of other points.

A point $\mathbf{X}_i$ is considered an outlier if $\deg^-(\mathbf{X}_i) \leq T$, where T is a threshold parameter. Critical values and p-values are again estimated using bootstrapping.

Example 10.9.1 *Consider the 2D dataset in Example 10.5.1.*
NN Method:

- $\mathbf{X}_6$ *has a large average distance to its r nearest neighbors compared to the others.*
- *It will be flagged as an outlier if its score exceeds the bootstrapped critical value.*

rNN Method:

- $\mathbf{X}_6$ *appears in the neighborhoods of very few (possibly zero) other points.*
- *Its in-degree will be low. If* $\deg^-(\mathbf{X}_6) \leq T$*, it is flagged as an outlier.*

10.10 Performance Study

We compare the performance of IDOT with five other outlier detection methods in this section. In order to evaluate the level and the power of these methods and to study their distributional robustness, we generate samples from the multivariate normal distribution

$\mathbb{N}_d(\boldsymbol{\mu}, \boldsymbol{\Sigma})$. Further sampling experiments under the multivariate t distribution, a mixture of two bivariate normal distributions with heterogeneous covariances, and the skew-normal distribution appear in Modarres [232]. Critical values of $S_{(n)}$ are based on 1000 bootstrap resamples. Estimated level of significance and power values are based on 1000 replications of the experiment. Some interesting approaches to outlier detection are implemented in R package 'OutlierDetection' (Tiwari and Akanksha [339]). We will compare the performance of IDOT against the existing methods.

For the following experiment, we compared six methods of outlier detection, namely IDOT, RKOF, Generalized Dispersion (GD), Mahalanobis (M), average Nearest Neighbors (NN), 6) r-th nearest Neighbor (rNN). We tested the hypothesis that there is no outliers at level $\alpha = 0.05$ against the hypothesis that there is at least one outlier. The simulated datasets contain $NO = 0, 1, \ldots, 5$ actual number of outliers. A maximum of 5 observations are shifted by 5, 10, 15, 20, and 25 units, respectively. These values represent the outliers that each method attempts to detect.

We generated a sample of size n from $\mathbb{N}(\mathbf{0}, \boldsymbol{\Sigma})$ where $\boldsymbol{\Sigma}$ has unit variances and constant correlation $\rho = 0.5$ with $n = 100$ and $d = 5$. Table 10.1 shows the results of the experiment when there are 0, 1, and 2 outliers in the dataset. The average number of the outliers (ANO) detected for each method appears under the column *ANO*. The column NOF shows the number of outliers found. For example, when there is no outlier in the sample ($NO = 0$) the IDOT method finds zero outliers in 945 samples, 1 outlier in 52 samples, and 2 outliers in 3 samples, with an average number of outliers of 0.058. IDOT shows the best performance terms of the ANO and NOF. The worst performer is the Mahalanobis method. Due to the

TABLE 10.1: *Normal Data Sets*: The proportion of times the outlier detection methods detect 0, 1, 2, 3, 4, and 5 or more outliers when there are NO (number of outliers) in 1000 multivariate normal datasets with $n = 100$ and $d = 5$. Number of Outliers Found (NOF) and the Average number of outliers (ANO) are also reported.

Method	NO	ANO			NOF			
			0	1	2	3	4	≥ 5
IDOT	0	0.058	0.945	0.052	0.003			
RKOF	0	2.258	0.016	0.091	0.512	0.381		
GD	0	2.169	0.015	0.103	0.580	0.302		
M	0	4.419	0.006	0.036	0.085	0.190	0.210	0.482
NN	0	2.172	0.012	0.089	0.614	0.285		
rNN	0	2.159	0.010	0.106	0.599	0.285		
IDOT	1	1.056	0.003	0.945	0.054	0.001		
RKOF	1	2.254	0.010	0.102	0.512	0.376		
GD	1	2.147	0.021	0.127	0.536	0.316		
M	1	4.433	0.008	0.026	0.089	0.184	0.208	0.485
NN	1	2.130	0.008	0.125	0.596	0.271		
rNN	1	2.160	0.017	0.102	0.585	0.296		
IDOT	2	2.047			0.954	0.045	0.010	
RKOF	2	2.700	0.007	0.103	0.503	0.387		
GD	2	2.131	0.013	0.122	0.586	0.279		
M	2	4.407	0.005	0.024	0.102	0.165	0.228	0.476
NN	2	2.133	0.013	0.122	0.584	0.281		
rNN	2	2.158	0.008	0.102	0.614	0.276		

presence of outliers, the Mahalanobis method shows the effects of swamping. The other methods offer suboptimal performance compared with IDOT. For example, with 2 outliers in the dataset ($NO = 2$), IDOT and rNN have ANO values of 2.047 and 2.158, respectively. However, IDOT finds 2 outliers in 954 cases whereas rNN finds 2 outliers in 614 samples. IDOT method shows little variability compared to other outlier detection techniques.

Modarres [232] demonstrates the performance of the detection methods when $n = 100$ and $d = 200$. Since GD and the Mahalanobis methods require $n > d$ they are removed from the experiment. Among the remaining four methods, RKOF, NN and rNN find the average of the number of outliers to be $ANO = 2.2$ when no outlier is present. The same pattern is apparent for other values of NO. On the other hand, IDOT outperforms the other methods of outlier detection.

10.11 Analysis of Lumber Dataset

Johnson and Wichern [162] discuss the dataset lumber. The data were obtained by taking four measures of stiffness of each of 30 boards. The first measurement, X_1 is taken by sending a shock down the board, the second measurement, X_2, is taken while vibrating the board, and the last two measurements X_3 and X_4 are obtained from static tests. In the left panel of Figure 10.3, we see a 3-dimensional plot of the variables. After standardizing the variables, they use the Mahalanobis distance and identify observation numbers 16 and 9 as outliers. We use six methods of outlier detection, namely IDOT, RKOF, GD, M, NN, and rNN to identify outliers of the lumber dataset.

The methods RKOF, Depth, GD, and NN, do not report any outliers. The rNN method identifies observation 9, which appears in color red in the Figures, with a p-value of 0.04 as an outlier. The Mahalanobis method identifies observations 9, 16, and 21 with p-values of 0.02, 0.01, 0.04, respectively, as outliers. IDOT identifies observation 9 as the only outlier with p-value of 0. The next suspect outlier is observation number 28, which is not statistically significant with a p-value of 0.16. In the right panel of Figure 10.3, you see the row sums of the dissimilarity for the lumber dataset where observation 9 has the highest peak and

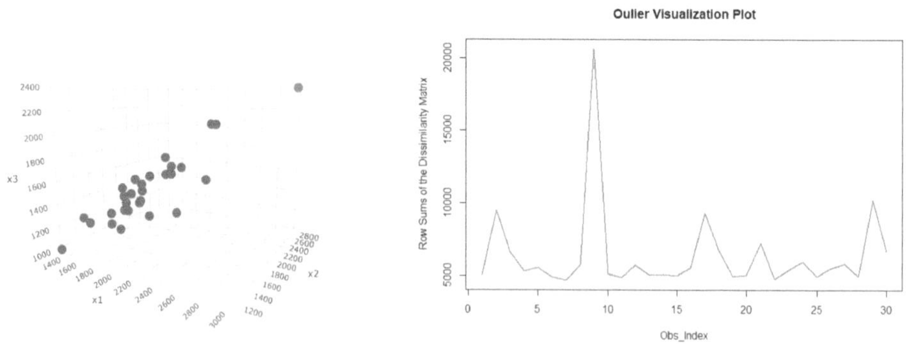

FIGURE 10.3: Left panel: Four measures of stiffness of 30 boards of lumber. Right panel: Row sums of the dissimilarities for the lumber data. The highest peak is at observation number 9.

stands out. A simultaneous plot of the empirical cumulative distribution functions (ECDFs) of the IPDs shows the ECDF of observation number 9 is far from the other ECDFs.

10.12 Summary

This chapter focuses on high-dimensional outlier detection, where traditional methods such as the Mahalanobis or Wilks approaches fail due to singular covariance matrices, masking effects, and sensitivity to extreme values when $d \gg N$. The chapter introduces robust alternatives, including minimum covariance determinant, minimum volume ellipsoid, and component-wise medians, which diminish the influence of outliers while remaining computationally feasible. Distribution-free techniques based on IPDs, nearest-neighbor statistics, and kernel-based outlier factors are developed to detect anomalous observations without relying on strict parametric assumptions. The theoretical foundations of these methods, including generalized dispersion measures and r-th nearest neighbor statistics, are described alongside approaches for assessing statistical significance using bootstrap resampling.

A central contribution is the Interpoint Distance Outlier Test (IDOT), which ranks observations by their total dissimilarity from all other points and identifies extreme values as potential outliers. The null distribution of the IDOT statistic is obtained via the percentile bootstrap, ensuring that resampling is performed under the null hypothesis by temporarily removing suspected outliers. Simulation studies under normal, t_4, skew-normal, and mixture distributions demonstrate that IDOT exhibits superior power and accuracy compared to alternative high-dimensional outlier detection methods. Visual tools, including dissimilarity plots, further enhance interpretability and facilitate the identification of anomalous observations in complex multivariate datasets.

The chapter illustrates the utility of high-dimensional outlier detection in real-world contexts such as finance, health monitoring, and manufacturing quality control. Comparisons across multiple methods highlight the robustness and computational efficiency of IDOT and related procedures. Future research directions include resampling strategies for kernel density estimators of dissimilarities and exploration of non-parametric bootstrap techniques, ensuring these approaches remain effective as both the sample size and dimensionality grow. Collectively, the methodologies presented provide a comprehensive and practical framework for robustly identifying outliers in high-dimensional data.

10.13 Exercises

Exercise 10.1 *Generate a sample of* $n = 100$ *observations from* $\mathbb{N}_d(\mathbf{0}, \mathbf{\Sigma})$ *with* $d = 5$ *variables. The covariance matrix has unit variances and constant correlation* $\rho = 0.5$. *Shift* $k = 3$ *observations at positions* 20, 40, *and* 60 *by* 5, 10, *and* 15 *units, respectively. Replicate the outlier detection plot in Figure 10.1.*

Exercise 10.2 *Examine the utility of the outlier visualization plot under a normal mixture model with heterogeneous covariances. Generate a sample of* $n = 100$ *observations with* $d = 2000$ *variables from the mixture* $(1-q)\mathbb{N}_d(\mathbf{0}, \mathbf{\Sigma}_x)+q\mathbb{N}_d(\mathbf{0}, \mathbf{\Sigma}_y)$ *where* $q = .05$, $\mathbf{\Sigma}_x$ *has unit variances and constant correlation of* $\rho_x = .5$, $\mathbf{\Sigma}_y$ *has variances* $1, 2, \ldots, d$ *and correlation of* $\rho_y = .8$. *How many observations appear from* $\mathbb{N}_d(\mathbf{0}, \mathbf{\Sigma}_y)$ *in the sample? Identify those observations and replicate the outlier detection plot in Figure 10.2.*

Exercise 10.3 *Let $d = 5$, $n = 5$, and $r = 2$. Consider the following dataset*

$$\mathbf{X}_1 = (0,0,0,0,0), \quad \mathbf{X}_2 = (1,1,1,1,1),$$
$$\mathbf{X}_3 = (0.9,1.1,1.0,1.0,1.1), \quad \mathbf{X}_4 = (10,10,10,10,10)$$
$$\mathbf{X}_5 = (0.95,0.95,1.05,1.05,1.0)$$

Compute the pairwise Euclidean distances and apply outlier detection methods,

a) Average r-th Nearest Neighbor Method,

b) r-th Nearest Neighbor (rNN) Method.

Exercise 10.4 *Outlier detection using dissimilarity measures ρ and δ. Two groups $\mathbb{X}$ and $\mathbb{Y}$, each with 3 observations in dimension $d = 6$.*

$$\mathbf{X}_1 = (1.624, -0.612, -0.528, -1.073, 0.865, -2.302)$$
$$\mathbf{X}_2 = (1.745, -0.761, 0.319, -0.249, 1.462, -2.060)$$
$$\mathbf{X}_3 = (-0.322, -0.384, 1.134, -1.100, -0.172, -0.878)$$
$$\mathbf{Y}_1 = (0.042, 0.583, -1.101, 1.145, 0.902, 0.502)$$
$$\mathbf{Y}_2 = (0.901, -0.684, -0.123, -0.936, -0.268, 0.530)$$
$$\mathbf{Y}_3 = (-0.692, -0.397, -0.687, -0.845, -0.671, -0.013)$$

Euclidean Distance Matrix is D

	$\mathbf{X}_1$	$\mathbf{X}_2$	$\mathbf{X}_3$	$\mathbf{Y}_1$	$\mathbf{Y}_2$	$\mathbf{Y}_3$
$\mathbf{X}_1$	*0.00*	*1.36*	*3.12*	*4.13*	*3.16*	*3.62*
$\mathbf{X}_2$	*1.36*	*0.00*	*3.14*	*3.94*	*3.33*	*4.02*
$\mathbf{X}_3$	*3.12*	*3.14*	*0.00*	*3.76*	*2.28*	*2.12*
$\mathbf{Y}_1$	*4.13*	*3.94*	*3.76*	*0.00*	*3.00*	*2.89*
$\mathbf{Y}_2$	*3.16*	*3.33*	*2.28*	*3.00*	*0.00*	*1.84*
$\mathbf{Y}_3$	*3.62*	*4.02*	*2.12*	*2.89*	*1.84*	*0.00*

- *Compute the distance matrix for ρ. Check for outliers.*
- *Compute the distance matrix for δ. Check for outliers.*

Exercise 10.5 *Consider the $m \times m$ distance matrix with k outliers represented as the first k rows of the distance matrix*

$$\mathbf{D}(\mathbb{X}) = \begin{array}{c} \\ X_1 \\ X_2 \\ \vdots \\ X_k \\ X_{k+1} \\ \vdots \\ X_m \end{array} \begin{array}{c} \begin{array}{cccccccc} X_1 & X_2 & \ldots & X_k & X_{k+1} & \ldots & X_m \end{array} \\ \begin{pmatrix} 0 & cd & cd & cd & cd & \ldots & cd \\ cd & 0 & cd & cd & cd & \ldots & cd \\ \vdots & \vdots & \vdots & \vdots & \vdots & \vdots & \vdots \\ cd & \ldots & cd & 0 & cd & \ldots & cd \\ cd & \ldots & cd & cd & 0 & \ldots & d \\ \vdots & \vdots & \vdots & \vdots & \vdots & \vdots & \vdots \\ cd & \ldots & cd & cd & d & \ldots & 0 \end{pmatrix} \end{array}, \tag{10.3}$$

where the remaining $m - k$ observations are equidistant. Using the IDOT method, how big should c become to detect k outliers?

Exercise 10.6 *Use the dissimilarity index δ in the IDOT method and recompute Tables 10.1.*

11

High-Dimensional Tests of Independence

11.1 Abstract

The chapter examines the problem of testing statistical independence in high-dimensional settings, where the number of variables substantially exceeds the sample size (d>>n). Such situations arise routinely in modern scientific applications, including genomics, neuroimaging, finance, machine learning, robotics, and network analysis, where understanding dependence structures is fundamental for inference, variable selection, and model construction. Classical procedures for testing independence, including Pearson's correlation, Wilks' likelihood ratio test, and related multivariate methods, rely on asymptotic theory developed under the assumption of fixed dimension and increasing sample size. Consequently, these methods often become unstable or inapplicable in high-dimensional, low-sample-size settings due to singular covariance matrices, loss of statistical power, and computational limitations.

To address these challenges, the chapter develops a unified framework for independence testing based on interpoint distances (IPDs). Rather than analyzing the original high-dimensional observations directly, the proposed methodology transforms multivariate data into collections of one-dimensional pairwise distances, thereby preserving essential dependence information while avoiding the curse of dimensionality. The central insight is that, under broad conditions, dependence between two random vectors is reflected in the dependence between their corresponding interpoint distance distributions. This reduction enables the construction of flexible, computationally efficient, and largely distribution-free procedures that remain effective even when the dimensionality far exceeds the sample size.

Building upon this framework, the chapter introduces seven novel IPD-based test statistics derived from complementary characteristics of the distance distributions. These include statistics based on empirical distribution functions, cut-point methodology, odds ratios, quadrant counts, distance correlations, and two normalizing transformations of interpoint distances. Each statistic captures different manifestations of dependence and collectively provides a comprehensive toolkit capable of detecting both linear and highly non-linear relationships. The proposed procedures are compared with several well-established competitors, including distance correlation (dCor), projection covariance (Pcov), and the Heller-Heller-Gorfine (HHG) test, all of which have been designed to identify complex dependence structures beyond simple linear association.

The chapter further develops the theoretical foundations of the proposed methodology through the introduction of the Correlation Association Test Statistics (CATS) framework. The statistics CATS_δ and CATS_ρ are formulated as V-statistics, allowing rigorous asymptotic analysis via Hoeffding decompositions. Under the null hypothesis of independence, both statistics are shown to be asymptotically normal, with limiting distributions determined by low-order components of the underlying distance kernels. These theoretical results

DOI: 10.1201/9781003789581-11

provide a solid inferential basis while avoiding many of the restrictive assumptions required by classical covariance-based methods.

Extensive simulation studies demonstrate that the proposed IPD and CATS procedures maintain accurate control of Type I error while exhibiting high statistical power across a broad range of dependence structures, including non-linear, heteroscedastic, mixture-induced, and cluster-based alternatives. In many challenging high-dimensional scenarios, the proposed methods outperform existing procedures in both sensitivity and computational efficiency. Their ability to operate effectively when d greatly exceeds n makes them particularly attractive for contemporary applications involving genomic expression data, functional brain imaging, financial markets, and other complex multivariate systems.

Overall, the chapter establishes interpoint-distance methodology as a powerful and versatile framework for modern independence testing. By combining geometric representations of multivariate data with rigorous asymptotic theory and practical computational algorithms, it provides a robust alternative to traditional covariance-based techniques. The chapter concludes by highlighting promising directions for future research, including extensions to functional and manifold-valued data, non-Euclidean spaces, and scalable algorithms for massive datasets, thereby laying the groundwork for a broad new class of high-dimensional independence tests.

11.2 Introduction

With numerous applications, independence testing is fundamental in statistics and has been extensively studied in the past several decades. However, there is still a need for methods of testing independence in a number of statistical problems where the sample size is smaller than the dimension. High-dimensional datasets are commonplace in finance, robotics, network analysis, medical imaging, machine learning, microarray and DNA analysis where independence testing is a central issue. Typical examples include microarray analysis where one assumes independence in the expression levels among different genes and functional magnetic resonance imaging research where one is interested in test the independence of voxels in different parts of the brain. Independent component analysis (ICA) is another application where one decomposes multivariate data into mutually independent components. The ICA model is subject to a constraint that at most one of these components is normal as required for model identifiability.

Classical tests of independence are based on the asymptotic theory that the sample size n approaches infinity while the dimension d is fixed. Therefore, they are not effective in settings where d approaches infinity. The main concern is testing the hypothesis of independence when the d is much larger than n. We present testing procedures that use the IPDs, which are inherently one-dimensional. These methods are effective for large d and fixed n where $d \gg nn$.

Let $\{(\mathbf{X}_i, \mathbf{Y}_i)\}_{i=1}^n$ be independent and identically distributed copies of the random vector pairs $(\mathbf{X}, \mathbf{Y})$ with joint distribution function (DF) F in $\mathbb{R}^d$ and marginal distributions G_x and G_y in $\mathbb{R}^{d_1}$ and $\mathbb{R}^{d_2}$, respectively, where $d = d_1 + d_2$. The aim in this work is to investigate whether $\mathbf{X}$ and $\mathbf{Y}$ are independent by testing the null hypothesis

$$H_0 : F(\mathbf{X}, \mathbf{Y}) = G_x(\mathbf{X})\ G_y(\mathbf{Y}) \quad \text{for all } (\mathbf{X}, \mathbf{Y}) \text{ in } \mathbb{R}^d, \tag{11.1}$$

against general alternatives

$$H_a : F(\mathbf{X}, \mathbf{Y}) \neq G_x(\mathbf{X})\ G_y(\mathbf{Y}) \quad \text{for at least one } (\mathbf{X}, \mathbf{Y}) \text{ in } \mathbb{R}^d, \tag{11.2}$$

where d is larger than n.

Consider the random vector $\mathbf{Z}$ such that $\mathbf{Z}' = [\mathbf{X}', \mathbf{Y}']$. For observation vectors that are normally distributed with covariance matrix $\mathbf{\Sigma} = \begin{bmatrix} \mathbf{\Sigma}_{xx} & \mathbf{\Sigma}_{xy} \\ \mathbf{\Sigma}_{yx} & \mathbf{\Sigma}_{yy} \end{bmatrix}$, the null hypothesis of independence is equivalent to $\mathbf{\Sigma}_{xy} = \mathbf{0}$. More generally, under any distribution where lack of covariance implies independence, testing H_0 is equivalent to testing that the covariance $\mathbf{\Sigma}_{xy} = \mathbf{0}$. Among the earliest problems in the development of multivariate techniques, the classical test of independence is based on the Pearson correlation coefficient ρ when $d_1 = d_2 = 1$. Wilks [355] shows that the likelihood ratio test of H_0 is $V = \frac{|\mathbf{R}|}{|\mathbf{R}_{xx}| \times |\mathbf{R}_{yy}|}$ where $\mathbf{R}$, $\mathbf{R_{xx}}$, and $\mathbf{R_{yy}}$ are the corresponding sample correlation matrices. Anderson [10] shows that H_0 is equivalent to the regression of one set on the other being zero. Moreover, one does not require normality of $\mathbf{X}$ to obtain the distribution of V under H_0. The likelihood ratio test is strictly unbiased and is invariant under linear transformations of $\mathbf{X}$ and $\mathbf{Y}$. The distribution of V under the null hypothesis can be written as the product of beta random variables. Assuming finite fourth moments, and under the null hypothesis $H_0 : \mathbf{X} \perp \mathbf{Y}$, the likelihood–ratio test statistic satisfies

$$-\left(n - 1 - \frac{d_1 + d_2 + 1}{2}\right) \log V \xrightarrow{d} \chi^2_{d_1 d_2}.$$

Example 11.2.1 *Let $d_1 = d_2 = 3$ and $n = 50$. Define*

$$\mathbf{Z} = (\mathbf{X}', \mathbf{Y}')' \in \mathbb{R}^6, \qquad \mathbf{Z} \sim \mathbb{N}_6(\mathbf{0}, \mathbf{I}_6).$$

Under this model, the null hypothesis $H_0 : \mathbf{X} \perp \mathbf{Y}$ is true.

Suppose the sample correlation matrix is

$$\mathbf{R} = \begin{pmatrix} \mathbf{R}_{xx} & \mathbf{R}_{xy} \\ \mathbf{R}_{yx} & \mathbf{R}_{yy} \end{pmatrix} = \begin{pmatrix} 1 & 0.1 & -0.05 & 0.02 & -0.03 & 0.04 \\ 0.1 & 1 & 0.06 & -0.01 & 0.00 & 0.02 \\ -0.05 & 0.06 & 1 & 0.05 & 0.01 & -0.02 \\ 0.02 & -0.01 & 0.05 & 1 & 0.09 & 0.00 \\ -0.03 & 0.00 & 0.01 & 0.09 & 1 & -0.07 \\ 0.04 & 0.02 & -0.02 & 0.00 & -0.07 & 1 \end{pmatrix}.$$

The determinants are approximately

$$|\mathbf{R}| \approx 0.842, \quad |\mathbf{R}_{xx}| \approx 0.951, \quad |\mathbf{R}_{yy}| \approx 0.938.$$

Wilks' Lambda is

$$V = \frac{|\mathbf{R}|}{|\mathbf{R}_{xx}||\mathbf{R}_{yy}|} \approx 0.944.$$

The likelihood ratio test statistic is

$$-\left(n - 1 - \frac{d_1 + d_2 + 1}{2}\right) \log V = -45.5 \log(0.944) \approx 2.64.$$

Under H_0,

$$-\left(n - 1 - \frac{d_1 + d_2 + 1}{2}\right) \log V \xrightarrow{d} \chi^2_{d_1 d_2} = \chi^2_9.$$

The critical value at level $\alpha = 0.05$ is

$$\chi^2_{9,0.95} \approx 16.92.$$

Since $2.64 < 16.92$, we fail to reject H_0. There is no evidence of dependence between $\mathbf{X}$ and $\mathbf{Y}$ at the 5% level.

Hoeffding [149] discusses a Cramer-von Mises type non-parametric test for bivariate independence that is robust to non–monotone dependence. Blum et al. (BKR) [38]obtained the asymptotic null distribution of an asymptotically equivalent form of Hoeffding statistic. Given n bivariate pairs (x_i, y_i), the BKR statistic is defined as

$$B_n = \frac{1}{n}\sum_{i=1}^{n}(\hat{F}(x_i, y_i) - \hat{G}_x(x_i)\hat{G}_y(y_i))^2, \tag{11.3}$$

where $\hat{F}$, $\hat{G}_x$, and $\hat{G}_y$ denote the joint and the marginal empirical distribution functions, respectively. The classical dependence measures such as Pearson, Spearman or Kendall correlations have been used for detecting linear or monotone dependence structures. Han et al. [137] study two families of distribution-free test statistics, which include Kendall's and Spearman's correlations. See also Leung et al. [187] and Mao [209].

Puri and Sen [274] introduce a non-parametric counterpart to Wilks' test, constructed through component-wise ranking. Friedman and Rafsky [110] develop graph-based methodologies for testing independence. Extending the univariate quadrant test of Blomqvist [37], Gieser and Randles [118] propose a non-parametric procedure founded on interdirection counts. Feuerverger [99] presents a consistent bivariate rank-based test of independence. Taskinen et al. ([334] [335]) extend these ideas to the multivariate setting, proposing tests based on spatial signs and ranks. Modarres [226] introduces a novel independence test derived from the likelihood of cut-points. Bagkavos and Patil [18] formulate a bivariate independence test leveraging the property that, under independence, the conditional quantiles of one variable given the other remain invariant. Fan et al. [95]) propose a test based on a Cramér–von Mises type functional, constructed from a process defined using empirical characteristic functions. Perhaps the most widely adopted multivariate test of independence is that based on distance covariance, introduced by Székely et al. [331], which employs characteristic functions to reduce high-dimensional data to one dimension. Zhu et al. [370] note potential limitations of distance covariance in high-dimensional contexts and propose projection covariance as a suitable alternative.

This chapter presents novel methods of testing H_0 based on the sample IPDs when d is larger than n and compares the methods with existing tests in the literature. We introduce seven new test statistics using the EDF of IPDs, the method of cut-point, odds ratio, correlation of IPDs, EDF in each quadrant, and two normalizing transformations of IPDs. We consider comparisons of the new tests with Dcor, HHG, and Pcov. The last section is devoted to summary and conclusions. The R code of the test statistics considered in this chapter is available from the corresponding author.

11.3 IPD Tests of Independence

Guo and Modarres [126] consider non-parametric tests of independence based on IPDs. Consider two random vectors $\mathbf{X}$ and $\mathbf{Y}$. We, $d_{(X)}$, to denote the IPD between $\mathbf{X}$s. Similarly, the random variable $d_{(Y)}$ represents the IPD of $\mathbf{Y}$s, and $d_{(XY)}$ represents cross-sample IPDs. We are interested in testing

$$H_0' : \mathbb{P}(d_{(X)} \leq x, d_{(Y)} \leq y) = \mathbb{P}(d_{(X)} \leq x)\mathbb{P}(d_{(Y)} \leq y) \text{ for all } (x, y) \in \mathbb{R}^2. \tag{11.4}$$

There are $m = n(n-1)/2$ pairs of IPDs $(d_{(x)ij}, d_{(y)ij})$. We refer to any $\mathbf{X}$ IPD with random variable $d_{(X)}$ with observed values $\{d_{(x)ij}\}$ for $1 \leq i < j \leq n$. Similarly, we refer to any **Y** IPD with random variable $d_{(Y)}$ with observed values $d_{(y)ij}$. Two **X** IPDs $d_{(x)ij}$ and $d_{(x)kh}$ are dependent if they have an index in common. With n observations, there are $\frac{n(n-1)(n-2)}{2}$ dependent pairs of distances. The following theorem shows that under multivariate normal distribution, the independence between two vectors is equivalent to the independence between the square IPDs of the two vectors.

Theorem 11.3.1 *Let* $(\mathbf{X}_1, \mathbf{Y}_1), (\mathbf{X}_2, \mathbf{Y}_2)$ *be i.i.d. copies of random vector* $(\mathbf{X}, \mathbf{Y})$ *with joint multivariate normal distribution* F *in* $\mathbb{R}^d$ *and marginal distributions* G_x *and* G_y *in* $\mathbb{R}^{d_1}$ *and* $\mathbb{R}^{d_2}$*, respectively, where* $d = d_1 + d_2$*. Suppose* $d^2_{(X)} = \|\mathbf{X}_1 - \mathbf{X}_2\|^2$*,* $d^2_{(Y)} = \|\mathbf{Y}_1 - \mathbf{Y}_2\|^2$*. The vectors* $\mathbf{X}$ *and* $\mathbf{Y}$ *are independent if and only if the IPDs* $d^2_{(X)}$ *and* $d^2_{(Y)}$ *are independent.*

The following proposition shows that when the increase in dimension involves correlated variables in two random vectors, the covariance between the IPDs diverges.

Proposition 11.3.1 *Let* $(\mathbf{X}_1, \mathbf{Y}_1), (\mathbf{X}_2, \mathbf{Y}_2)$ *be i.i.d. copies of random vector* $(\mathbf{X}, \mathbf{Y})$ *with joint multivariate normal distribution* F *in* $\mathbb{R}^d$ *and marginal distributions* G_x *and* G_y *in* $\mathbb{R}^{d_1}$ *and* $\mathbb{R}^{d_2}$*, respectively, where* $d = d_1 + d_2$*. Suppose for every variable* x_i *in* $\mathbf{X}$*, there is at least one variable (denoted as* $y_{(x_i)}$*) in* $\mathbf{Y}$ *such that the absolute value of their covariance is larger than a positive constant* ϵ*, i.e.* $|\mathbb{C}ov(x_i, y_{(x_i)})| > \epsilon > 0$*, for* $i = 1, \ldots, d_1$*. Let* $d^2_{(X)} = \|\mathbf{X}_1 - \mathbf{X}_2\|^2$*,* $d^2_{(Y)} = \|\mathbf{Y}_1 - \mathbf{Y}_2\|^2$*. When* d_1 *tends to infinity, the covariance between* $d^2_{(X)}$ *and* $d^2_{(Y)}$ *tends to infinity.*

Let $\mathbf{\Sigma}_x$ and $\mathbf{\Sigma}_y$ be the covariance matrices of **X** and **Y**, respectively. The correlation coefficient between the two random vectors (Robert et al. [282]) is defined as

$$\mathbb{C}or(\mathbf{X}, \mathbf{Y}) = \frac{\mathrm{tr}(\Sigma'_{\mathrm{xy}}\Sigma_{\mathrm{xy}})}{\sqrt{\mathrm{tr}(\Sigma^2_{\mathrm{x}})\mathrm{tr}(\Sigma^2_{\mathrm{y}})}}, \tag{11.5}$$

It can be shown that under a multivariate normal distribution, the correlation coefficient between two random vectors coincides with the correlation between their corresponding IPDs, that is,

$$\mathbb{C}or(\mathbf{X}, \mathbf{Y}) = \mathbb{C}or(d^2_{(X)}, d^2_{(Y)}).$$

However, this equivalence generally does not hold for other distributions, particularly when the dependence between **X** and **Y** is non-linear. Consider the Box–Cox power transformation (Freeman and Modarres [107]) defined for each variable X_j as

$$Y_j = \begin{cases} \dfrac{X_j^{\lambda_j} - 1}{\lambda_j}, & \lambda_j \neq 0, \\ \ln X_j, & \lambda_j = 0, \end{cases}$$

where Y_j corresponds to the transformed variable in $\mathbf{Y} = (Y_1, Y_2, \ldots, Y_d)$ and X_j belongs to $\mathbf{X} = (X_1, X_2, \ldots, X_d)$. If the transformed random vector **Y** follows a d-variate normal distribution with mean vector $\boldsymbol{\mu}$ and covariance matrix $\mathbf{\Sigma}$, then the original random vector **X**, obtained via the inverse Box–Cox transformations, is said to follow a multivariate power normal distribution. Freeman and Modarres [107] show that the p.d.f. of a multivariate power normal distribution is

$$f(\mathbf{X} \mid \boldsymbol{\lambda}, \boldsymbol{\mu}, \mathbf{\Sigma}) = \frac{1}{K} \cdot \frac{1}{(2\pi)^{d/2} \mid \mathbf{\Sigma} \mid^{1/2}} \prod_{j=1}^{d} X_j^{\lambda_j - 1} \cdot \exp\left[-\frac{1}{2}(\mathbf{Y}_j - \boldsymbol{\mu})' \Sigma^{-1} (\mathbf{Y}_j - \boldsymbol{\mu})\right],$$

where K is a normalizing constant that depends on $\boldsymbol{\lambda}, \boldsymbol{\mu}$, and $\boldsymbol{\Sigma}$. Practitioners often assume $K = 1$. Under multivariate power normal distribution, the dependence between the variables is no longer linear. The following theorem shows that such non–linear dependence is equivalent to the dependence between the IPDs.

Theorem 11.3.2 *Let $(\mathbf{W}_1, \mathbf{Z}_1), (\mathbf{W}_2, \mathbf{Z}_2)$ be i.i.d. copies of random vector $(\mathbf{W}, \mathbf{Z})$ with a joint Power-normal distribution in $\mathbb{R}^d$ and marginal distributions G_w and G_z in $\mathbb{R}^{d_1}$ and $\mathbb{R}^{d_2}$, respectively, where $d = d_1 + d_2$. Suppose $d^2_{(W)} = \|\mathbf{W}_1 - \mathbf{W}_2\|^2$, $d^2_{(Z)} = \|\mathbf{Z}_1 - \mathbf{Z}_2\|^2$. The vectors $\mathbf{W}$ and $\mathbf{Z}$ are independent if and only if the IPDs $d^2_{(W)}$ and $d^2_{(Z)}$ are independent.*

This result shows that the statistics based on IPDs can not only detect linear dependence but also find non–linear relationships that is induced on the power normal scale. Independence under other distributions requires further research. However, in the simulations, we see that the tests based on IPD also have good power for other types of dependence.

11.3.1 IPD Correlation Test ($\mathrm{Cor}_{\mathrm{ipd}}$)

The classical and most powerful test of independence under bivariate normality is based on the sample Pearson correlation coefficient r. Using a geometric argument, Fisher [100] derived the exact distribution of the correlation coefficient. Pitman [271] obtained the permutation distribution of r, demonstrated that it is symmetric with mean zero and variance $1/(n-1)$, and established that r is asymptotically normal as the sample size tends to infinity. Owing to its distinction as a uniformly most powerful unbiased and uniformly most powerful invariant test for assessing independence in bivariate normal samples, the statistic r is specifically tailored to detect linear dependence.

Let $\{\mathbf{Z}_i = [\mathbf{X}_i', \mathbf{Y}_i']'\}_{i=1}^n$ denote a random sample in $\mathbb{R}^d$. Under the regularity conditions stated in Theorem 1 of Li [192], as the dimension d diverges to infinity, the squared IPDs $\|\mathbf{Z}_i - \mathbf{Z}_j\|^2$, for $1 \le i < j \le n$, converge in distribution to a normal random variable. It follows that the random vector $(d_{(X)}, d_{(Y)})$ converges in distribution to a bivariate normal distribution as $d \to \infty$.

Denote the mean vector of $(d_{(X)}, d_{(Y)})$ with $(\mu_{d(x)}, \mu_{d(y)})$ and estimate the mean distances by $\bar{d}_{(x)} = \frac{1}{m}\sum_{i=1}^n \sum_{j=i+1}^n d_{(x)ij}$ and $\bar{d}_{(y)} = \frac{1}{m}\sum_{i=1}^n \sum_{j=i+1}^n d_{(y)ij}$, respectively. Using the $m = n(n-1)/2$ observed pairs $(d_{(x)ij}, d_{(y)ij})$, we estimate the Pearson correlation between $d_{(X)}$ and $d_{(Y)}$ with

$$\mathrm{Cor}_{\mathrm{ipd}} = \frac{1}{s_x s_y}\{\frac{1}{m}\sum_{i=1}^{n-1}\sum_{j=i+1}^{n} d_{(x)ij} d_{(y)ij} - \bar{d}_{(x)}\bar{d}_{(y)}\}, \tag{11.6}$$

where s_x and s_y are the standard deviations of $(d_{(x)ij})$ and $(d_{(x)ij})$ for $1 \le i < j \le n$, respectively. The range of the statistics $\mathrm{Cor}_{\mathrm{ipd}}$ is $-1 \le \mathrm{Cor}_{\mathrm{ipd}} \le 1$. If the alternative hypothesis is positive correlation between the IPDs, we reject the null hypothesis for large values of $\mathrm{Cor}_{\mathrm{ipd}}$. For negative correlation as alternative, the null hypothesis is rejected for a small value of $\mathrm{Cor}_{\mathrm{ipd}}$. If the alternative hypothesis is unknown, we consider the absolute value of the statistic $\mathrm{Cor}_{\mathrm{ipd}}$. We obtain the sampling distribution of $\mathrm{Cor}_{\mathrm{ipd}}$ from random permutations.

Suppose $\{\mathbf{Z}_i = [\mathbf{X}_i, \mathbf{Y}_i]\}_{i=1}^n$ is a sample of d-dimensional random vectors that are drawn from normal distribution with mean $(\boldsymbol{\mu}_x, \boldsymbol{\mu}_y)$ and covariance matrix $\boldsymbol{\Sigma} = \begin{bmatrix} \boldsymbol{\Sigma}_{xx} & \boldsymbol{\Sigma}_{xy} \\ \boldsymbol{\Sigma}_{yx} & \boldsymbol{\Sigma}_{yy} \end{bmatrix}$. The following lemma shows that the distribution of square IPD between any two random vectors within the $\mathbf{X}$ sample is a linear combination of d_1 independent central chi-square random variables with 1 degree of freedom.

Lemma 11.3.1 *Let $\mathbf{X}_1, \mathbf{X}_2$ be i.i.d. copies of random vectors $\mathbf{X}$ with multivariate normal distribution G_x with mean $\boldsymbol{\mu}_x$ and covariance $\boldsymbol{\Sigma}_{xx}$. Suppose $d^2_{(X)} = \|\mathbf{X}_1 - \mathbf{X}_2\|^2$. Then $d^2_{(X)} = \sum_{t=1}^{d_1} \lambda_t U_t^2$ where $\lambda_1, \ldots, \lambda_{d_1}$ are eigenvalues of $2\boldsymbol{\Sigma}_{xx}$ and $U_t^2 \sim \chi^2(1)$.*

Based on Lemma 11.3.1, the expected value and variance of the IPDs for $d^2_{(X)}$ are $2tr\boldsymbol{\Sigma}_{xx}$ and $8tr\boldsymbol{\Sigma}^2_{xx}$, respectively. Similarly, the expected value and variance of $d^2_{(Y)}$ are $2tr\boldsymbol{\Sigma}_{yy}$ and $8tr\boldsymbol{\Sigma}^2_{yy}$, respectively. The proof of Theorem 11.3.1 shows that $\mathbb{C}\mathrm{ov}\left(\mathrm{d}^2_{(\mathrm{X})}, \mathrm{d}^2_{(\mathrm{Y})}\right) = 8\mathrm{tr}(\Sigma'_{\mathrm{xy}}\Sigma_{\mathrm{xy}})$. Thus, the correlation $\mathbb{C}\mathrm{or}\left(\mathrm{d}^2_{(\mathrm{X})}, \mathrm{d}^2_{(\mathrm{Y})}\right)$ is given by (11.5).

Let r_{ts} be the sample correlation between the tth component of $\mathbf{X}$ and sth component of $\mathbf{Y}$ vectors. Schott [301] proposed the statistic $T = \sum_{t=1}^{d_1}\sum_{s=1}^{d_2} r_{ts}^2$ to test the independence of two normal vectors. Under multivariate normality, the asymptotic distribution of this statistic is considered by Schott [302] and Mao [208]. The non-parametric versions of this statistic using Spearman's and Kendall's correlations have also been examined in the literature. Compared with the statistic that use the correlations between the components of $\mathbf{X}$ and $\mathbf{Y}$ directly, the estimation of $\mathbb{C}\mathrm{or}\left(\mathrm{d}^2_{(\mathrm{X})}, \mathrm{d}^2_{(\mathrm{Y})}\right)$ not only considers the information from $\boldsymbol{\Sigma}_{xy}$, but also includes the correlations within the random vector.

Example 11.3.1 *Distribution of Cor_{IPD} under normality.*

Let $\mathbf{W} = (\mathbf{X}, \mathbf{Y}) \in \mathbb{R}^d$ be a multivariate random vector such that $\mathbf{X} \in \mathbb{R}^{d_1}$, $\mathbf{Y} \in \mathbb{R}^{d_2}$, and $d_1 + d_2 = d$. Suppose

$$\mathbf{W} \sim \mathbb{N}_d(\boldsymbol{\mu}, \boldsymbol{\Sigma}), \quad \text{where } \boldsymbol{\mu} = (\boldsymbol{\mu}_1, \boldsymbol{\mu}_2) \quad \boldsymbol{\Sigma} = (1-\rho)\mathbf{I}_d + \rho\mathbf{J}_d,$$

i.e., $\boldsymbol{\Sigma}$ has 1s on the diagonal and ρ in all off-diagonal positions. We have n i.i.d. samples $(\mathbf{X}_1, \mathbf{Y}_1), \ldots, (\mathbf{X}_n, \mathbf{Y}_n)$ from this distribution. Define the IPDs,

$$d_{x,ij} = \|\mathbf{X}_i - \mathbf{X}_j\|, \quad d_{y,ij} = \|\mathbf{Y}_i - \mathbf{Y}_j\|, \quad 1 \le i < j \le n.$$

Let $m = \binom{n}{2}$ denote the number of unique unordered pairs. Then define

$$Cor_{ipd} = \frac{1}{m} \sum_{1 \le i < j \le n} \frac{(d_{x,ij} - \bar{d}_x)(d_{y,ij} - \bar{d}_y)}{s_x s_y},$$

$$\bar{d}_x = \frac{1}{m}\sum_{i<j} d_{x,ij}, \quad \bar{d}_y = \frac{1}{m}\sum_{i<j} d_{y,ij}, \quad s_x^2 = \frac{1}{m}\sum_{i<j}(d_{x,ij} - \bar{d}_x)^2, \quad s_y^2 = \frac{1}{m}\sum_{i<j}(d_{y,ij} - \bar{d}_y)^2.$$

Let us consider the distribution of $\|\mathbf{X}_i - \mathbf{X}_j\|^2$ and note that $\mathbf{X}_i - \mathbf{X}_j \sim \mathbb{N}_{d_1}(0, 2\boldsymbol{\Sigma}_x)$, where $\boldsymbol{\Sigma}_x = (1-\rho)\mathbf{I}_{d_1} + \rho\mathbf{J}_{d_1}$ is the $d_1 \times d_1$.

We diagonalize $\boldsymbol{\Sigma}_x$ using its spectral decomposition. Since $\boldsymbol{\Sigma}_x$ has eigenvalue $\lambda_1 = 1 - \rho + \rho d_1$ and $\lambda_2 = 1 - \rho$ with multiplicities 1 and $d_1 - 1$. Hence, the quadratic form $\|\mathbf{X}_i - \mathbf{X}_j\|^2$ is distributed as a weighted sum of independent chi-square random variables,

$$\|\mathbf{X}_i - \mathbf{X}_j\|^2 = 2 \cdot \|Z_x\|^2 \sim 2\left[\lambda_1 Z_1^2 + \sum_{k=2}^{d_1} \lambda_2 Z_k^2\right],$$

where $Z_k \overset{iid}{\sim} \mathbb{N}(0,1)$. This is a written as

$$\|\mathbf{X}_i - \mathbf{X}_j\|^2 \sim \sum_{k=1}^{d_1} a_k \chi_1^2, \quad \text{with } a_1 = 2\lambda_1,\ a_2 = \cdots = a_{d_1} = 2\lambda_2.$$

Similarly, $\|\mathbf{Y}_i - \mathbf{Y}_j\|^2$ has a linear combination of chi-square distributions with weights derived from the eigenvalues of $\mathbf{\Sigma}_y$.

We now consider the asymptotic distribution of Cor_{ipd} based on the theory of U-statistics. Define: $U = \{d_{x,ij}\}, \quad V = \{d_{y,ij}\}, \quad 1 \le i < j \le n$. Both U and V are U-statistics of degree 2.

Let $h(\mathbf{Z}_i, \mathbf{Z}_j) = \|\mathbf{X}_i - \mathbf{X}_j\|$ and similarly for $g(\mathbf{Z}_i, \mathbf{Z}_j) = \|\mathbf{Y}_i - \mathbf{Y}_j\|$, where $\mathbf{Z}_i = (\mathbf{X}_i, \mathbf{Y}_i)$. The asymptotic distribution of the correlation of two U-statistics is given by (Serfling [303]),

$$Cor_{ipd} \approx \frac{\mathbb{E}[(d_{x,ij} - \mu_x)(d_{y,ij} - \mu_y)]}{\sqrt{\mathbb{E}[(d_{x,ij} - \mu_x)^2] \cdot \mathbb{E}[(d_{y,ij} - \mu_y)^2]}} + O_{\mathbb{P}}\left(\frac{1}{\sqrt{m}}\right),$$

where $\mu_x = \mathbb{E}[d_{x,ij}]$, $\mu_y = \mathbb{E}[d_{y,ij}]$. Thus, the asymptotic distribution is

$$Cor_{ipd} \overset{approx}{\sim} N\left(\rho_{IPD}, \frac{1}{m}\right), \quad \text{where } \rho_{ipd} = \frac{Cov(d_{x,ij}, d_{y,ij})}{\sqrt{Var(d_{x,ij}) \cdot Var(d_{y,ij})}}.$$

This follows because of the number of terms $m = \binom{n}{2}$ involved, and the correlation of two symmetric kernel U-statistics converges at rate $1/m$ to its population value. Under the null hypothesis of independence between $\mathbf{X}$ and $\mathbf{Y}$, we have

$$Cov(d_{x,ij}, d_{y,ij}) = 0 \quad \Rightarrow \quad \mathbb{E}[Cor_{ipd}] = 0.$$

Each pair $(d_{x,ij}, d_{y,ij})$ is a function of independent normal vectors. Although $d_{x,ij}$ and $d_{x,kl}$ are not independent when they share an index, the entire collection $\{d_{x,ij}\}$ satisfies the dependency structure required for the asymptotic normality of U-statistics.

Hence, by Hoeffding's theorem for U-statistics (and its extension to correlation-type statistics), the Central Limit Theorem yields

$$Cor_{ipd} \xrightarrow{d} \mathrm{N}(0, \sigma^2), \quad \text{with } \sigma^2 = \frac{1}{m} + o\left(\frac{1}{m}\right).$$

11.3.2 IPD EDF Method (EDF)

Denote the marginal DFs of $d_{(X)}$, and $d_{(Y)}$ and their joint DF by $H_x(x) = \mathbb{P}(d_{(X)} \le x)$, $H_y(y) = \mathbb{P}(d_{(Y)} \le y)$ and $H_{xy}(x, y) = \mathbb{P}(d_{(X)} \le x, d_{(Y)} \le y)$, respectively. Let $\mathbb{I}(\cdot)$ denote the indicator function and estimate the DFs by

$$\begin{aligned}
\hat{H}_x(x) &= \frac{1}{m}\sum_{s=1}^{n-1}\sum_{t=s+1}^{n} \mathbb{I}(d_{(x)st} \le x), \\
\hat{H}_y(y) &= \frac{1}{m}\sum_{s=1}^{n-1}\sum_{t=s+1}^{n} \mathbb{I}(d_{(y)st} \le y), \\
\hat{H}_{xy}(x, y) &= \frac{1}{m}\sum_{s=1}^{n-1}\sum_{t=s+1}^{n} \mathbb{I}(d_{(x)st} \le x) \times \mathbb{I}(d_{(y)st} \le y),
\end{aligned} \tag{11.7}$$

where $m = n(n-1)/2$. We obtain m estimates of the joint and marginal distributions of IPDs at $(d_{(x)ij}, d_{(y)ij})$ for $1 \le i < j \le n$. To measure the inability of bivariate DF H to factorize into a product of marginal DFs at each pair of IPDs $(d_{(x)ij}, d_{(y)ij})$ we estimate the total squared deviations from independence and define the test statistic

$$EDF = \sum_{i=1}^{n}\sum_{j=i+1}^{n} [\hat{H}_{xy}(d_{(x)ij}, d_{(y)ij}) - \hat{H}_x(d_{(x)ij})\, \hat{H}_y(d_{(y)ij})]^2 \tag{11.8}$$

The range of the statistic EDF is $0 \leq EDF \leq n(n-1)/2$. Under H_0 we expect EDF to be small, and we reject H_0 for large values of EDF.

The statistic EDF is conceptually related to the statistic B_n defined in equation 11.3, but it differs in a fundamental way. Whereas B_n is computed using the n observed i.i.d. pairs (x_i, y_i), the m pairs $(d_{(x)ij}, d_{(y)ij})$ used to construct EDF are inherently dependent. Consequently, the null distribution of EDF cannot be derived using standard asymptotic theory for independent and identically distributed observations. To address this issue, inference for EDF is instead carried out via random permutation testing, which is described below.

Let R denote the total number of permutations. For the rth permutation, $r = 1, \ldots, R$, we randomly reorder the observations $\{\mathbf{Y}_i\}_{i=1}^n$ to obtain $\{\mathbf{Y}_i^{(r)}\}_{i=1}^n$. We then form the paired sample $\{(\mathbf{X}_i, \mathbf{Y}_i^{(r)})\}_{i=1}^n$ and compute the corresponding test statistic, denoted by $EDF^{(r)}$. Under the null hypothesis of independence between $\mathbf{X}$ and $\mathbf{Y}$, the distribution of $EDF^{(r)}$ is identical to that of EDF. Consequently, the empirical null distribution of EDF can be approximated using the collection of permuted statistics $\{EDF^{(r)}\}_{r=1}^R$.

If all possible permutations are considered, the resulting permutation test yields exact critical values. However, except for very small sample sizes, exhaustive enumeration of all permutations is computationally infeasible. Instead, an approximate test is performed by generating R random permutations. Since the null hypothesis is rejected for large values of EDF, the associated p-value is computed as

$$p\text{-value}_{EDF} = \frac{1}{R}\sum_{r=1}^{R} \mathbb{I}\left(EDF^{(r)} \geq EDF\right).$$

11.3.3 IPD Cut-point Method (CP)

Modarres [226] proposes a test statistic constructed as the sum of likelihood ratio statistics for testing independence in a collection of 2×2 contingency tables defined at the n sample cut-points (x_i, y_i), for testing the null hypothesis H_0 in (11.1) when $d_1 = d_2 = 1$. The IPD cut-point method extends this approach to the problem of testing independence in high-dimensional settings by utilizing IPDs.

Consider an observed IPD pair $(d_{(x)ij}, d_{(y)ij})$ in the plane, and the four regions determined by the two lines passing through this point and parallel to the coordinate axes. Let

$$\bar{H}_{xy}(d_{(x)}, d_{(y)}) = 1 - H_x(d_{(x)}) - H_y(d_{(y)}) + H_{xy}(d_{(x)}, d_{(y)})$$

denote the bivariate survival function of the IPDs. Table 11.1 summarizes the theoretical probabilities, along with their empirical estimates, corresponding to observations falling into

TABLE 11.1: The 2×2 Table of Probabilities and Estimates at $(d_{(x)ij}, d_{(y)ij})$.

Prob.	$d_{(y)} \leq d_{(y)ij}$	$d_{(y)} > d_{(y)ij}$	
$d_{(X)} \leq d_{(x)ij}$ Estimates	$H_{xy}(d_{(x)ij}, d_{(y)ij})$ $m_{11(ij)}/m$	$H_x(d_{(x)ij}) - H_{xy}(d_{(x)ij}, d_{(y)ij})$ $m_{12(ij)}/m$	$H_x(d_{(x)ij})$ $m_{1+(ij)}/m$
$d_{(X)} > d_{(x)ij}$ Estimates	$H_y(d_{(y)ij}) - H_{xy}(d_{(x)ij}, d_{(y)ij})$ $m_{21(ij)}/m$	$\bar{H}_{xy}(d_{(x)ij}, d_{(y)ij})$ $m_{22(ij)}/m$	$1 - H_x(d_{(x)ij})$ $m_{2+(ij)}/m$
 Estimates	$H_y(d_{(y)ij})$ $m_{+1(ij)}/m$	$1 - H_y(d_{(y)ij})$ $m_{+2(ij)}/m$	1 1

each of the four regions surrounding the cut-point $(d_{(x)ij}, d_{(y)ij})$. Under random sampling, the joint distribution of the counts in these four quadrants follows a multinomial distribution for each fixed cut-point $(d_{(x)ij}, d_{(y)ij})$.

With each observation $(d_{(x)ij}, d_{(y)ij}), 1 \leq i < j \leq n$, we associate a 2×2 table with elements $m_{rk(ij)}/m$ for $r, k = 1, 2$ and calculate

$$\begin{aligned}
m_{11(ij)} &= \sum_{s=1}^{n-1} \sum_{t=s+1}^{n} \mathbb{I}(d_{(x)st} \leq d_{(x)ij}) \times \mathbb{I}(d_{(y)st} \leq d_{(y)ij}), \\
m_{12(ij)} &= \sum_{s=1}^{n-1} \sum_{t=s+1}^{n} \mathbb{I}(d_{(x)st} \leq d_{(x)ij}) \times \mathbb{I}(d_{(y)st} > d_{(y)ij}), \\
m_{21(ij)} &= \sum_{s=1}^{n-1} \sum_{t=s+1}^{n} \mathbb{I}(d_{(x)st} > d_{(x)ij}) \times \mathbb{I}(d_{(y)st} \leq d_{(y)ij}), \\
m_{22(ij)} &= \sum_{s=1}^{n-1} \sum_{t=s+1}^{n} \mathbb{I}(d_{(x)st} > d_{(x)ij}) \times \mathbb{I}(d_{(y)st} > d_{(y)ij}), \\
m_{1+(ij)} &= m_{11(ij)} + m_{12(ij)}, \\
m_{2+(ij)} &= m_{21(ij)} + m_{22(ij)}, \\
m_{+1(ij)} &= m_{11(ij)} + m_{21(ij)}, \\
m_{+2(ij)} &= m_{12(ij)} + m_{22(ij)}.
\end{aligned} \tag{11.9}$$

Recall that $m = n(n-1)/2$ is the total number of the IPD pairs. It is not difficult to show that $m_{11(ij)} + m_{12(ij)} + m_{21(ij)} + m_{22(ij)} = m$. Furthermore, let $c_{ij} = m!/(m_{11(ij)}m_{12(ij)}m_{21(ij)}m_{22(ij)})$. The likelihood at each fixed observation $(d_{(x)ij}, d_{(y)ij})$ is defined as

$$\begin{aligned}
L_{ij} = c_{ij}(H_{xy}(d_{(x)ij}, d_{(y)ij})^{m_{11(ij)}} \times (H_x(d_{(x)ij}) - H_{xy}(d_{(x)ij}))^{m_{12(ij)}} \times \\
(H_y(d_{(y)ij}) - H_{xy}(d_{(x)ij}, d_{(y)ij}))^{m_{21(ij)}} \times (\bar{H}_{xy}(d_{(x)ij}, d_{(y)ij}))^{m_{22(ij)}}.
\end{aligned} \tag{11.10}$$

The likelihood ratio test of independence at each $(d_{(x)ij}, d_{(y)ij})$ is

$$\lambda_{ij} = \prod_{r=1}^{2} \prod_{s=1}^{2} [\frac{m_{r+(ij)}m_{+s(ij)}}{m\ m_{rs(ij)}}]^{m_{rs(ij)}}. \tag{11.11}$$

The cut-point test statistic is Cd$= \sum_{i=1}^{n} \sum_{j=i+1}^{n} -2\log(\lambda_{ij})$ or

$$\text{CP} = 2\sum_{i=1}^{n} \sum_{j=i+1}^{n} \sum_{r=1}^{2} \sum_{s=1}^{2} m_{rs(ij)} \log \frac{m\ m_{rs(ij)}}{m_{r+(ij)}m_{+s(ij)}}. \tag{11.12}$$

When $n \to \infty$ and for fixed d the asymptotic distribution of $-2\log(\lambda_{ij})$ is χ^2 with one degree of freedom. Furthermore, λ_{ij}'s are dependent as they depend on the cut-points $(d_{(x)ij}, d_{(y)ij})$. The statistic CP computes the total likelihood of all m cut-points, and the m individual likelihood ratio test statistics $-2\log(\lambda_{ij})$ are available to further classify regions of positive (negative) or no dependence. In practice, one uses random permutation testing to obtain the sampling distribution of CP.

Example 11.3.2 *Let $n = 3$ with observations be $X = (2,4,7)$, $Y = (3,6,8)$. The $m = \binom{3}{2} = 3$ IPDs are*

$$(d_{(x)12}, d_{(y)12}) = (2,3), \qquad (d_{(x)13}, d_{(y)13}) = (5,5), \qquad (d_{(x)23}, d_{(y)23}) = (3,2).$$

Consider the cut-point $(2, 3)$. *The quadrant counts are*

$$(m_{11}, m_{12}, m_{21}, m_{22}) = (1, 0, 1, 1).$$

The marginal totals are $m_{1+} = 1, \quad m_{2+} = 2, \quad m_{+1} = 2, \quad m_{+2} = 1$. *The individual likelihood-ratio contribution is*

$$\begin{aligned} -2\log(\lambda) &= 2\left[\log\left(\frac{3}{2}\right) + \log\left(\frac{3}{4}\right) + \log\left(\frac{3}{2}\right)\right] \\ &= 2\log\left(\frac{27}{16}\right) \approx 1.0465. \end{aligned}$$

The same calculation is performed for the remaining cut-points, and the cut-point statistic CP *is obtained by summing all* $-2\log(\lambda_{ij})$ *values.*

11.3.4 IPD Odds Ratio (*Odds*)

Fleiss et al. [103] describe an odds ratio test for the independence of in 2×2 tables. Agresti [3] proposed energized odds ratios for ordinal data. In this section, we apply odds ratios to the IPDs. Consider the odds ratio of the 2×2 table (Table 11.1), at each point $(d_{(x)ij}, d_{(y)ij})$ we have the odds ratio as

$$\theta_{(xy)ij} = \frac{H_{xy}(d_{(x)ij}, d_{(y)ij})\bar{H}_{xy}(d_{(x)ij}, d_{(y)ij})}{[H_x(d_{(x)ij}) - H_{xy}(d_{(x)ij}, d_{(y)ij})][H_y(d_{(y)ij}) - H_{xy}(d_{(x)ij}, d_{(y)ij})]}, \tag{11.13}$$

where $0 < \theta_{(xy)ij} < \infty$. If the odds ratio $\theta_{(xy)ij}$ is larger than 1, then the IPD of $\mathbf{X}s$ and $\mathbf{Y}s$ at $(d_{(x)ij}, d_{(y)ij})$ are associated in the sense that the IPD of $\mathbf{X}s$ being $d_{(x)ij}$ raises the odds of $\mathbf{Y}$ IPD being $d_{(y)ij}$. If $\theta_{(xy)ij}$ is less than 1, then $\mathbf{X}$ and $\mathbf{Y}$ IPD are negatively correlated. Under independence, the odds ratios $\theta_{(xy)ij} = 1$ for all $(d_{(x)ij}, d_{(y)ij})$ in $\mathbb{R}^2$. The 2×2 table of probabilities and estimates at $(d_{(x)ij}, d_{(y)ij})$ appear in Table 11.1. Suppose we estimate the odds ratio at each $(d_{(x)ij}, d_{(y)ij})$ with $\hat{\theta}_{(xy)ij} = (m_{11(ij)}m_{22(ij)})/(m_{12(ij)}m_{21(ij)})$. We consider the average odds-ratio and propose the statistic

$$Odds = \frac{1}{m}\sum_{i=1}^{n}\sum_{j=i+1}^{n} \hat{\theta}_{(xy)ij}. \tag{11.14}$$

If the alternative hypothesis is positive correlation between the IPDs, the null hypothesis of independence is rejected for large values of *Odds*. If we consider the alternative as negative correlation between the IPDs, we reject the null hypothesis for small values of *Odds*. Otherwise, we consider the difference between *Odds* and 1. The distribution of *Odds* under the null is calculated based on the permutation principle.

11.3.5 IPD Quadrant Test (Quad)

While the CDF of a univariate random variable is uniquely defined, no such uniqueness holds for multivariate random vectors. In the univariate case ($d = 1$), the distribution function is uniquely determined by the relation $\mathbb{P}(X \leq x) = 1 - \mathbb{P}(X > x)$. This property no longer extends to dimensions $d \geq 2$. Indeed, for a d-dimensional random vector, there are $2^d - 1$ distinct ways to define a distribution function, corresponding to the different combinations of inequalities across components.

In the bivariate setting, the CDF of a pair (x, y) may be defined according to how probability mass is accumulated over the four regions determined by the point (x, y). Extending this idea to the multivariate context, we employ IPDs and define distribution functions such as $\mathbb{P}(d_{(X)} \leq x, d_{(Y)} \leq y)$, $\mathbb{P}(d_{(X)} \leq x, d_{(Y)} > y)$, $\mathbb{P}(d_{(X)} > x, d_{(Y)} \leq y)$, and

$\mathbb{P}(d_{(X)} > x, d_{(Y)} > y)$. In testing the independence between $\mathbf{X}$ and $\mathbf{Y}$, all four possible orderings of the IPDs are taken into account. The criteria to measure the dependence at each (x, y) are

$$\begin{aligned}
\theta_1(x,y) &= \mathbb{P}(d_{(X)} \le x, d_{(Y)} \le y) - \mathbb{P}(d_{(X)} \le x)\ \mathbb{P}(d_{(Y)} \le y),\\
\theta_2(x,y) &= \mathbb{P}(d_{(X)} \le x, d_{(Y)} > y) - \mathbb{P}(d_{(X)} \le x)\ \mathbb{P}(d_{(Y)} > y),\\
\theta_3(x,y) &= \mathbb{P}(d_{(X)} > x, d_{(Y)} \le y) - \mathbb{P}(d_{(X)} > x)\ \mathbb{P}(d_{(Y)} \le y),\\
\theta_4(x,y) &= \mathbb{P}(d_{(X)} > x, d_{(Y)} > y) - \mathbb{P}(d_{(X)} > x)\ \mathbb{P}(d_{(Y)} > y).
\end{aligned}$$

It is clear that $\sum_{k=1}^{4} \theta_k(x, y) = 0$. Under the null hypothesis of independence, we have $\theta_k(x, y) = 0$ for $k = 1, \ldots, 4$. The null hypothesis is rejected if either one of $\theta_1(x, y), \ldots, \theta_4(x, y)$ has large absolute value. We estimate θ_k with $\hat{\theta}_k$ using

$$\begin{aligned}
\hat{\theta}_1(x,y) &= \hat{H}_{xy}(x,y) - \hat{H}_x(x)\hat{H}_y(y),\\
\hat{\theta}_2(x,y) &= \hat{H}_{2xy}(x,y) - \hat{H}_x(x)(1 - \hat{H}_y(y)),\\
\hat{\theta}_3(x,y) &= \hat{H}_{3xy}(x,y) - (1 - \hat{H}_x(x))\hat{H}_y(y),\\
\hat{\theta}_4(x,y) &= \hat{H}_{4xy}(x,y) - (1 - \hat{H}_x(x))(1 - \hat{H}_y(y)),
\end{aligned}$$

where

$$\begin{aligned}
\hat{H}_{2xy}(x,y) &= \frac{1}{m}\sum_{s=1}^{n}\sum_{t=s+1}^{n} \mathbb{I}(d_{(x)st} \le x)\mathbb{I}(d_{(y)st} > y),\\
\hat{H}_{3xy}(x,y) &= \frac{1}{m}\sum_{s=1}^{n}\sum_{t=s+1}^{n} \mathbb{I}(d_{(x)st} > x)\mathbb{I}(d_{(y)st} \le y),\\
\hat{H}_{4xy}(x,y) &= \frac{1}{m}\sum_{s=1}^{n}\sum_{t=s+1}^{n} \mathbb{I}(d_{(x)st} > x)\mathbb{I}(d_{(y)st} > y).
\end{aligned}$$

To test H_0 we will use the test statistic

$$Quad = \max_{1 \le i < j \le n, 1 \le k \le 4} \hat{\theta}_k^2(d_{(x)ij}, d_{(y)ij}). \tag{11.15}$$

Ranging over the four quadrants and n data points, the statistic $Quad$ is the maximum discrepancy between the estimated c.d.f from its factorized version. The range of $Quad$ is $0 \le Quad \le 1/4$, and we reject the null hypothesis for a large value of $Quad$. We use random permutations to obtain the distribution of $Quad$.

Example 11.3.3 *Quad statistic for testing independence with* $n = 3$ *points. Suppose we have observed data,*

$$X_1 = 1, \quad X_2 = 3, \quad X_3 = 5 \quad Y_1 = 2, \quad Y_2 = 4, \quad Y_3 = 8.$$

Compute all pairwise IPDs,

$$\begin{aligned}
d_{(x)12} &= |1-3| = 2, & d_{(y)12} &= |2-4| = 2,\\
d_{(x)13} &= |1-5| = 4, & d_{(y)13} &= |2-8| = 6,\\
d_{(x)23} &= |3-5| = 2, & d_{(y)23} &= |4-8| = 4.
\end{aligned}$$

Let $m = \binom{3}{2} = 3$ *be the number of IPD pairs. The IPD pairs are* $(2, 2), \quad (4, 6), \quad (2, 4)$. *We evaluate the statistic* $Quad = \max_{i<j,k} \hat{\theta}_k^2(d_{(x)ij}, d_{(y)ij})$. *Computation at Cut-point* $(x, y) = (2, 2)$, *count how many of the 3 IPD pairs fall into each of the 4 quadrants.*

- $\hat{H}_x(2) = \frac{2}{3}$ *(since 2 values ≤ 2)*
- $\hat{H}_y(2) = \frac{1}{3}$ *(since only one value ≤ 2)*
- $\hat{H}_{xy}(2,2) = \frac{1}{3}$ *(only $(2,2)$ in bottom-left)*
- $\hat{H}_{2xy}(2,2) = \frac{1}{3}$ *($(2,4)$ has $x \le 2$, $y > 2$)*
- $\hat{H}_{3xy}(2,2) = \frac{0}{3}$ *(no point with $x > 2$, $y \le 2$)*
- $\hat{H}_{4xy}(2,2) = \frac{1}{3}$ *($(4,6)$ has $x > 2$, $y > 2$)*

Compute $\hat{\theta}_k$ at $(2,2)$,

$$\begin{aligned}
\hat{\theta}_1 &= \hat{H}_{xy} - \hat{H}_x\hat{H}_y = \frac{1}{3} - \frac{2}{3}\cdot\frac{1}{3} = \frac{1}{3} - \frac{2}{9} = \frac{1}{9},\\
\hat{\theta}_2 &= \hat{H}_{2xy} - \hat{H}_x(1-\hat{H}_y) = \frac{1}{3} - \frac{2}{3}\cdot\frac{2}{3} = \frac{1}{3} - \frac{4}{9} = -\frac{1}{9},\\
\hat{\theta}_3 &= \hat{H}_{3xy} - (1-\hat{H}_x)\hat{H}_y = 0 - \frac{1}{3}\cdot\frac{1}{3} = -\frac{1}{9},\\
\hat{\theta}_4 &= \hat{H}_{4xy} - (1-\hat{H}_x)(1-\hat{H}_y) = \frac{1}{3} - \frac{1}{3}\cdot\frac{2}{3} = \frac{1}{3} - \frac{2}{9} = \frac{1}{9}.
\end{aligned}$$

Contributions to Quad,

$$\hat{\theta}_k^2 = \left(\frac{1}{9}\right)^2 = \frac{1}{81}, \quad \left(-\frac{1}{9}\right)^2 = \frac{1}{81} \Rightarrow \max_k \hat{\theta}_k^2 = \frac{1}{81} \approx 0.0123.$$

Similar steps can be followed for cut-points $(4,6)$ and $(2,4)$. Then compute Quad as the maximum over all 12 values of $\hat{\theta}_k^2$. The test statistic Quad captures the strongest evidence of departure from independence across all quadrants and cut-points. The null distribution is computed using permutation of the $\mathbf{Y}$ values.

11.3.6 IPD Transformation Statistic

In this section, we introduce two additional test statistics constructed by transforming the marginal distributions to normality. It is well known that marginal normalization can substantially enhance the power of normal-theory tests for independence; Fisher's Z-transformation provides a classic example of this principle. Kowalski and Tarter [174] study the properties of normal-theory tests for independence when applied to observations from bivariate distributions constrained to have normal marginals, and demonstrate that such tests exhibit increased power when conducted on the transformed random vectors.

For a random variable X with a continuous DF H_x, the transformation $W_x = \Phi^{-1}(H_x(X))$, where Φ^{-1} is the standard normal quantile function, produces normally distributed variates. We will compute the empirical density function of the IPDs using (11.7) and then form

$$\begin{aligned}
Trans &= \mathrm{Cor}_{\mathrm{ipd}}\left(\Phi^{-1}(\hat{H}_x(x)), \Phi^{-1}(\hat{H}_y(y))\right),\\
&= \frac{1}{\hat{\sigma}_1\hat{\sigma}_2}\{\frac{1}{m}\sum_{i=1}^{n-1}\sum_{j=i+1}^{n} \Phi^{-1}(\hat{H}_x(d_{(x)ij})\Phi^{-1}(\hat{H}_y(d_{(y)ij}) - \hat{\mu}_1\hat{\mu}_2\},
\end{aligned} \tag{11.16}$$

where $m = n(n-1)/2$, $\hat{\mu}_1$ and $\hat{\sigma}_1$ are the mean and variance of $\Phi^{-1}(\hat{H}_x(d_{(x)ij})$, $\hat{\mu}_2$ and $\hat{\sigma}_2$ are the mean and variance of $\Phi^{-1}(\hat{H}_y(d_{(y)ij})$ for $1 \le i < j \le n$. It measures the correlation between IPDs of **X** and **Y** after normalizing transformations of the margins. The distribution of $Trans$ is obtained by permutation method.

Alternative transformations to normality have also been proposed in the literature, including the cube root transformation. Freeman and Modarres [106] investigate the efficiency and power of normal-theory tests of independence applied after a Box–Cox transformation, and derive an explicit expression relating the correlation between the transformed variables to the correlation on the normal scale. They demonstrate that conducting the test of independence based on the Pearson correlation coefficient after transformation to normality is uniformly more efficient. Wilson and Hilferty [357] study the cube root transformation of a chi-squared random variable and show that it yields an approximate normal distribution.

Lemma 11.3.1 shows that the distribution of IPDs drawn from multivariate normal vectors is a linear combination of chi-squared random variables, which is often approximated using a chi-squared distribution. The distribution of IPDs is of chi-squared type for observations from other distributions (Song and Modarres [319]). Hence, we consider a test based on the correlation of the cube root of the IPDs. Let the random vectors $d_{(x)}^{1/3}$ and $d_{(y)}^{1/3}$ denote the cube root of the sample IPDs. The cube root transformation computes

$$\mathrm{IPD}_{cube} = \mathbb{C}\mathrm{or}\left(\mathrm{d}_{(\mathrm{x})}^{1/3}, \mathrm{d}_{(\mathrm{y})}^{1/3}\right). \tag{11.17}$$

The approach to calculate the correlation estimate is similar to equation (11.16). We use random permutations to obtain the approximate sampling distribution of IPD_{cube}.

11.4 Distance Correlation

Székely et al. [331] propose distance correlation to measure the dependence between random vectors. Suppose $f_{X,Y}(t,s)$, $f_X(t)$, and $f_Y(s)$ are the joint and marginal characteristic functions of the random vectors **X** and **Y** with finite first moments. The distance covariance (dCov) between **X** and **Y** and their distance variance are defined by

$$\begin{aligned}
\mathscr{V}(\mathbf{X},\mathbf{Y}) &= \|f_{X,Y}(t,s) - f_X(t)f_Y(s)\|_{L_2},\\
\mathscr{V}(\mathbf{X}) &= \|f_{X,X}(t,s) - f_X(t)f_X(s)\|_{L_2},\\
\mathscr{V}(\mathbf{Y}) &= \|f_{Y,Y}(t,s) - f_Y(t)f_Y(s)\|_{L_2},
\end{aligned}$$

where $\|\cdot\|_{L_2}$ is a weighted L_2 norm of the difference between the densities in it. The distance correlation (Dcor) is defined as

$$\mathscr{R}(X,Y) = \begin{cases} \frac{\mathscr{V}(\mathbf{X},\mathbf{Y})}{\sqrt{\mathscr{V}(\mathbf{X})\mathscr{V}(\mathbf{Y})}}, & \mathscr{V}(\mathbf{X})\mathscr{V}(\mathbf{Y}) > 0,\\ 0, & \mathscr{V}(\mathbf{X})\mathscr{V}(\mathbf{Y}) = 0. \end{cases} \tag{11.18}$$

Székely et al. [331] prove that when Euclidean norms of **X** and **Y** all have finite expectations, the random vectors **X** and **Y** are independent if and only if $\mathscr{R}(\mathbf{X},\mathbf{Y}) = 0$. Motivated by this property, they propose the empirical distance correlation $\mathscr{R}_n(\mathbb{X},\mathbb{Y})$ for sample $\mathbb{X} = \{\mathbf{X}_i\}_{i=1}^n$ and $\mathbb{Y} = \{\mathbf{Y}_i\}_{i=1}^n$.

The method finds the distance matrices of **X** and **Y** samples and centers them so each has column and row means equal to zero. By averaging the entries of component-wise products of the two centered distance matrices, the squared distance covariance between the two variables is found. Define

$$\begin{aligned}
a_{kl} &= \|\mathbf{X}_k - \mathbf{X}_l\|, \quad \bar{a}_{k\cdot} = \frac{1}{n}\sum_{l=1}^{n} a_{kl}, \\
\bar{a}_{\cdot l} &= \frac{1}{n}\sum_{k=1}^{n} a_{kl}, \quad \bar{a}_{\cdot\cdot} = \frac{1}{n^2}\sum_{k,l=1}^{n} a_{kl}, \\
A_{kl} &= a_{kl} - \bar{a}_{k\cdot} - \bar{a}_{\cdot l} + \bar{a}_{\cdot\cdot}, \\
b_{kl} &= \|\mathbf{Y}_k - \mathbf{Y}_l\|, \quad \bar{b}_{k\cdot} = \frac{1}{n}\sum_{l=1}^{n} b_{kl}, \quad \bar{b}_{\cdot l} = \frac{1}{n}\sum_{k=1}^{n} b_{kl} \quad \bar{b}_{\cdot\cdot} = \frac{1}{n^2}\sum_{k,l=1}^{n} b_{kl}, \\
B_{kl} &= b_{kl} - \bar{b}_{k\cdot} - \bar{b}_{\cdot l} + \bar{b}_{\cdot\cdot}, \text{ for } k, l = 1, \ldots, n.
\end{aligned} \tag{11.19}$$

The empirical distance covariance and the empirical distance variance are

$$\begin{aligned}
\mathscr{V}_n(\mathbb{X}, \mathbb{Y}) &= \frac{1}{n}\Big(\sum_{k,l=1}^{n} A_{kl}B_{kl}\Big)^{1/2}, \\
\mathscr{V}_n(\mathbb{X}) &= \frac{1}{n}\Big(\sum_{k,l=1}^{n} A_{kl}^2\Big)^{1/2}, \\
\mathscr{V}_n(\mathbb{Y}) &= \frac{1}{n}\Big(\sum_{k,l=1}^{n} B_{kl}^2\Big)^{1/2}.
\end{aligned}$$

The empirical distance correlation is

$$\mathscr{R}_n(\mathbb{X}, \mathbb{Y}) = \begin{cases} \frac{\mathscr{V}_n(\mathbb{X},\mathbb{Y})}{\sqrt{\mathscr{V}_n(\mathbb{X})\mathscr{V}_n(\mathbb{Y})}}, & \mathscr{V}_n(\mathbb{X})\mathscr{V}_n(\mathbb{Y}) > 0, \\ 0, & \mathscr{V}_n(\mathbb{X})\mathscr{V}_n(\mathbb{Y}) = 0. \end{cases}$$

Székely et al. [331] prove that almost surely, $\lim_{n\to\infty} \mathscr{R}_n^2(\mathbb{X}, \mathbb{Y}) = \mathscr{R}(X, Y)$. Therefore, it is straightforward to use $\mathscr{R}_n$ to measure the independence based on sample $\mathbb{X}, \mathbb{Y}$. The R package "energy" is built for independent tests with distance correlation, and we use Dcor to denote this method.

Example 11.4.1 *Suppose*

$$\mathbb{X} = \begin{bmatrix} 1 & 0 & 0 & 0 & 0 \\ 0 & 1 & 0 & 0 & 0 \\ 0 & 0 & 1 & 0 & 0 \\ 1 & 1 & 0 & 0 & 0 \\ 0 & 1 & 1 & 0 & 0 \\ 1 & 0 & 1 & 0 & 0 \end{bmatrix}, \quad \mathbb{Y} = \begin{bmatrix} 1 & 1 & 1 & 1 & 1 \\ 1 & 1 & 1 & 1 & 0 \\ 1 & 1 & 1 & 0 & 0 \\ 0 & 0 & 0 & 0 & 1 \\ 0 & 0 & 0 & 1 & 1 \\ 0 & 0 & 0 & 1 & 0 \end{bmatrix}.$$

Compute the Euclidean distances between all pairs of rows,

$$A = [a_{kl}] = \begin{bmatrix} 0 & \sqrt{2} & \sqrt{2} & 1 & \sqrt{3} & \sqrt{2} \\ \sqrt{2} & 0 & \sqrt{2} & 1 & 1 & \sqrt{3} \\ \sqrt{2} & \sqrt{2} & 0 & \sqrt{3} & 1 & 1 \\ 1 & 1 & \sqrt{3} & 0 & \sqrt{2} & \sqrt{2} \\ \sqrt{3} & 1 & 1 & \sqrt{2} & 0 & \sqrt{2} \\ \sqrt{2} & \sqrt{3} & 1 & \sqrt{2} & \sqrt{2} & 0 \end{bmatrix},$$

$$B = [b_{kl}] = \begin{bmatrix} 0 & 1 & \sqrt{2} & \sqrt{5} & 2 & \sqrt{3} \\ 1 & 0 & 1 & 2 & \sqrt{3} & \sqrt{2} \\ \sqrt{2} & 1 & 0 & \sqrt{3} & 2 & 1 \\ \sqrt{5} & 2 & \sqrt{3} & 0 & 1 & 1 \\ 2 & \sqrt{3} & 2 & 1 & 0 & 1 \\ \sqrt{3} & \sqrt{2} & 1 & 1 & 1 & 0 \end{bmatrix}.$$

Compute row means, column means, and grand means to center A and B into A_{kl} and B_{kl}. These centered matrices are symmetric with zero diagonals.

Using the centered matrices, we compute

$$\sum_{k,l} A_{kl}^2 \approx 17.75, \qquad \sum_{k,l} B_{kl}^2 \approx 27.19, \qquad \sum_{k,l} A_{kl}B_{kl} \approx 21.85.$$

Then the empirical distance variances and covariance are

$$\mathscr{V}_n(\mathbb{X}) = \frac{\sqrt{17.75}}{6} \approx 0.7029,$$
$$\mathscr{V}_n(\mathbb{Y}) = \frac{\sqrt{27.19}}{6} \approx 0.8695,$$
$$\mathscr{V}_n(\mathbb{X}, \mathbb{Y}) = \frac{\sqrt{21.85}}{6} \approx 0.7801.$$

Finally, the empirical distance correlation is

$$\mathscr{R}_n(\mathbb{X}, \mathbb{Y}) = \frac{0.7801}{\sqrt{0.7029 \cdot 0.8695}} \approx 0.9996 \approx 1.$$

This indicates very strong dependence between $\mathbb{X}$ and $\mathbb{Y}$.

11.5 Test of Heller-Heller-Gorfin

The HHG test, proposed by Heller, Heller, and Gorfine [145], is a consistent and distribution-free test for independence between random vectors $\mathbf{X} \in \mathbb{R}^{d_x}$ and $\mathbf{Y} \in \mathbb{R}^{d_y}$. It is based on analyzing local neighborhoods around each observation, capturing both linear and complex non-linear dependence structures. Heller, Heller, and Gorfine [145] construct an R package named "HHG". The test statistic is computed in $O(n^2 \log(n))$.

The HHG test is based on the idea that if two random variables $\mathbf{X}$ and $\mathbf{Y}$ are independent, then the relative ordering of pairwise distances among observations in $\mathbf{X}$ should be unrelated to the relative ordering of distances among observations in $\mathbf{Y}$. Conversely, if $\mathbf{X}$ and $\mathbf{Y}$ are dependent, then local geometric structures in $\mathbf{X}$, such as which points are close or far from a given observation, will tend to be reflected in $\mathbf{Y}$.

Given n i.i.d. observations $(\mathbf{X}_i, \mathbf{Y}_i)$ for $i = 1, \ldots, n$, the HHG test constructs a test statistic by considering all unordered pairs (i, j) with $i \neq j$, and computing a 2×2 contingency table for each pair based on the distances between the points. For each pair (i, j), we define

- For every $k \neq i, j$:
  - $A_k = 1$ if $\|\mathbf{X}_k - \mathbf{X}_i\| < \|\mathbf{X}_k - \mathbf{X}_j\|$, else $A_k = 2$
  - $B_k = 1$ if $\|\mathbf{Y}_k - \mathbf{Y}_i\| < \|\mathbf{Y}_k - \mathbf{Y}_j\|$, else $B_k = 2$

- Let $m_{ab(ij)}$ denote the number of $k \neq i, j$ such that $A_k = a$ and $B_k = b$, for $a, b \in \{1, 2\}$

We define the marginal counts as

$$\begin{aligned} m_{1+(ij)} &= m_{11(ij)} + m_{12(ij)}, & m_{2+(ij)} &= m_{21(ij)} + m_{22(ij)}, \\ m_{+1(ij)} &= m_{11(ij)} + m_{21(ij)}, & m_{+2(ij)} &= m_{12(ij)} + m_{22(ij)}. \end{aligned}$$

Then, the HHG test statistic is

$$HHG = (n-2) \sum_{i=1}^{n} \sum_{\substack{j=1 \\ j \neq i}}^{n} \frac{\left\{ m_{12(ij)} m_{21(ij)} - m_{11(ij)} m_{22(ij)} \right\}^2}{m_{1+(ij)} m_{2+(ij)} m_{+1(ij)} m_{+2(ij)}}.$$

This test measures whether there is association between proximity in $\mathbf{X}$ and proximity in $\mathbf{Y}$. If $\mathbf{X}$ and $\mathbf{Y}$ are independent, proximity in $\mathbf{X}$ should not imply closeness in $\mathbf{Y}$, and the 2×2 tables should be close to uniform. Large deviations from uniformity indicate dependence.

One can alternatively obtain the HHG test statistic in terms of distance matrices. Let $\{(\mathbf{X}_i, \mathbf{Y}_i)\}_{i=1}^n$ be i.i.d. observations. Define the pairwise distance matrices

$$\mathbf{D}_X(i,j) = \|\mathbf{X}_i - \mathbf{X}_j\|, \qquad \mathbf{D}_Y(i,j) = \|\mathbf{Y}_i - \mathbf{Y}_j\|, \quad i, j = 1, \ldots, n.$$

For each anchor point i, define the rank matrices $\mathbf{R}_X$ and $\mathbf{R}_Y$ by ranking the distances

$$\{\mathbf{D}_X(i,j) : j \neq i\}, \qquad \{\mathbf{D}_Y(i,j) : j \neq i\}$$

in increasing order. Ties are handled using midranks.

For small samples, the HHG test statistic is constructed by counting concordant triples. For all ordered triples (i, j, k) with distinct indices, define

$$A_{ijk} = \begin{cases} 1, & \text{if } \mathbf{R}_X(i,j) = \mathbf{R}_X(i,k) \text{ and } \mathbf{R}_Y(i,j) = \mathbf{R}_Y(i,k), \\ 0, & \text{otherwise.} \end{cases}$$

The test statistic is then

$$T = \sum_{i \neq j \neq k} A_{ijk}.$$

For general sample size n, the HHG test statistic is

$$T_n = \sum_{i=1}^{n} \sum_{\substack{j=1 \\ j \neq i}}^{n} \sum_{\substack{k=1 \\ k \neq i,\, k \neq j}}^{n} \mathbb{I}\Big(\mathbf{R}_X(i,j) = \mathbf{R}_X(i,k) \text{ and } \mathbf{R}_Y(i,j) = \mathbf{R}_Y(i,k) \Big).$$

This statistic aggregates, for each anchor point i, the agreement between the local rank structure induced by $\mathbf{X}$ and that induced by $\mathbf{Y}$. Under the null hypothesis of independence, the distribution of T_n is obtained by permutation of the $\mathbf{Y}$-values relative to the $\mathbf{X}$-values. The p-value is computed as the proportion of permutations yielding a statistic at least as large as the observed one.

Although the HHG test can be presented in different forms, the rank–triple formulation and the 2×2 contingency–table formulation define the same test statistic and are equivalent.

In the rank–based formulation, one fixes an anchor point i and compares the relative distances $\|\mathbf{X}_i - \mathbf{X}_j\|$ and $\|\mathbf{X}_i - \mathbf{X}_k\|$, and similarly for $\mathbf{Y}$. A concordant triple (i, j, k) occurs when the ordering of distances from i to j and k agrees in both $\mathbf{X}$ and $\mathbf{Y}$. Thus, the statistic counts local agreements in distance rankings around each anchor point.

In the contingency–table formulation, one fixes an unordered pair (i, j) and uses it to partition all remaining points $k \neq i, j$ into two groups according to whether they are closer to i or to j. This partition is performed independently in $\mathbf{X}$ and in $\mathbf{Y}$, yielding a 2×2 table whose cells record the number of points whose relative closeness to i and j agrees or disagrees across the two spaces.

Each entry of the 2×2 table aggregates many pairwise distance comparisons of the form

$$\|\mathbf{X}_k - \mathbf{X}_i\| < \|\mathbf{X}_k - \mathbf{X}_j\| \quad \text{and} \quad \|\mathbf{Y}_k - \mathbf{X}_i\| < \|\mathbf{Y}_k - \mathbf{Y}_j\|,$$

which are equivalent to rank comparisons of distances. Consequently, the chi–square statistic computed from the table measures the degree of concordance between distance orderings in $\mathbf{X}$ and $\mathbf{Y}$ for the pair (i, j).

Summing the chi–square statistics over all pairs (i, j) is therefore equivalent to aggregating concordant and discordant local distance comparisons over all triples (i, j, k). The contingency–table formulation provides a computationally efficient aggregation of the same local rank information that is counted explicitly in the triple–based formulation. Hence, both formulations capture the same notion of local geometric agreement between $\mathbf{X}$ and $\mathbf{Y}$ and define the same HHG test of independence.

Example 11.5.1 *We simulate six observations $(\mathbf{X}_i, Y_i)$ where each $\mathbf{X}_i$ is a 5-dimensional vector and Y_i is scalar,*

i	X_{i1}	X_{i2}	X_{i3}	X_{i4}	X_{i5}	Y_i
1	1.0	2.1	1.5	0.9	1.2	3.1
2	1.3	1.8	1.6	1.0	1.1	2.9
3	2.5	2.7	2.0	1.9	2.2	4.5
4	3.0	3.2	2.8	3.1	3.0	5.0
5	0.5	0.8	0.6	0.4	0.7	1.0
6	2.1	2.0	1.9	2.2	2.3	4.0

Example Calculation for Pair $(i = 1, j = 2)$.
We consider the remaining points $k \in \{3, 4, 5, 6\}$ and compute the following for each

- *Compute $\|\mathbf{X}_k - \mathbf{X}_1\|$ and $\|\mathbf{X}_k - \mathbf{X}_2\|$*
- *Compute $\|Y_k - Y_1\|$ and $\|Y_k - Y_2\|$*
- *Determine A_k and B_k*

Suppose after these computations, the classification is

k	3	4	5	6
A_k	2	2	1	1
B_k	2	1	1	2

It follows that the counts are

	$B_k = 1$	$B_k = 2$
$A_k = 1$	$m_{11} = 1$	$m_{12} = 1$
$A_k = 2$	$m_{21} = 1$	$m_{22} = 1$

Compute marginal sums we get

$$m_{1+} = 1 + 1 = 2, \quad m_{2+} = 1 + 1 = 2$$
$$m_{+1} = 1 + 1 = 2, \quad m_{+2} = 1 + 1 = 2$$

Now compute contribution of this pair to the HHG statistic,

$$Term_{(1,2)} = \frac{(6-2)\cdot(1\cdot1-1\cdot1)^2}{2\cdot2\cdot2\cdot2} = \frac{4\cdot0^2}{16} = 0$$

Note that this calculation is repeated for each of the $n(n-1)$ ordered pairs and the final HHG statistic is the sum of all these pairwise contributions. Under the null hypothesis of independence, the statistic has a known permutation distribution.

Example 11.5.2 *Let $n=3$, and suppose we observe $\mathbf{X} = (1,2,3)'$ and $\mathbf{Y} = (4,5,6)'$.*
We test $H_0 : \mathbf{X}$ and $\mathbf{Y}$ are independent versus $H_a : \mathbf{X}$ and $\mathbf{Y}$ are dependent.
The distance and rank matrices are

$$\mathbf{D}_X = \begin{bmatrix} 0 & 1 & 2 \\ 1 & 0 & 1 \\ 2 & 1 & 0 \end{bmatrix}, \qquad \mathbf{R}_X = \begin{bmatrix} 1 & 2 & 3 \\ 2 & 1 & 2 \\ 3 & 2 & 1 \end{bmatrix},$$

$$\mathbf{D}_Y = \begin{bmatrix} 0 & 1 & 2 \\ 1 & 0 & 1 \\ 2 & 1 & 0 \end{bmatrix}, \qquad \mathbf{R}_Y = \begin{bmatrix} 1 & 2 & 3 \\ 2 & 1 & 2 \\ 3 & 2 & 1 \end{bmatrix}.$$

For $n=3$, there are 6 such ordered triples. In this example, no concordant triples occur, and hence $T=0$.

11.6 Projection Correlation

To handle high-dimensional data, Zhu et al. [370] project the data into low dimension and then apply Pearson correlation to test independence. They show that two random vectors are independent if and only if any of their projections are independent. Motivated by the equivalence, they defined the squared projection covariance between two random vectors $\mathbf{X}$ and $\mathbf{Y}$ and proposed the test statistic as

$$Pcov = n^{-3} \sum_{k,l,r=1}^{n} A_{klr} B_{klr},$$

where

$$a_{klr} = \arccos\left\{ \frac{(\mathbf{X}_k - \mathbf{X}_r)'(\mathbf{X}_l - \mathbf{X}_r)}{\|\mathbf{X}_k - \mathbf{X}_r\| \|\mathbf{X}_l - \mathbf{X}_r\|} \right\},$$

$$\bar{a}_{k\cdot r} = n^{-1}\sum_{l=1}^{n} a_{klr}, \quad \bar{a}_{\cdot lr} = n^{-1}\sum_{k=1}^{n} a_{klr}, \quad \bar{a}_{\cdot\cdot r} = n^{-2}\sum_{k=1}^{n}\sum_{l=1}^{n} a_{klr},$$

$$A_{klr} = a_{klr} - \bar{a}_{k\cdot r} - \bar{a}_{\cdot lr} - \bar{a}_{\cdot\cdot r},$$

$$b_{klr} = \arccos\left\{ \frac{(\mathbf{Y}_k - \mathbf{Y}_r)'(\mathbf{Y}_l - \mathbf{Y}_r)}{\|\mathbf{Y}_k - \mathbf{Y}_r\| \|\mathbf{Y}_l - \mathbf{Y}_r\|} \right\},$$

$$\bar{b}_{k\cdot r} = n^{-1}\sum_{l=1}^{n} b_{klr}, \quad \bar{b}_{\cdot lr} = n^{-1}\sum_{k=1}^{n} b_{klr},$$

$$\bar{b}_{\cdot\cdot r} = n^{-2}\sum_{k=1}^{n}\sum_{l=1}^{n} b_{klr},$$

$$B_{klr} = b_{klr} - \bar{b}_{k\cdot r} - \bar{b}_{\cdot lr} - \bar{b}_{\cdot\cdot r}.$$

Example 11.6.1 *Squared projection covariance (Pcov) example with $n = 3$. Let the observed vectors be*

$$\mathbf{X}_1 = (1,2)', \qquad \mathbf{X}_2 = (3,4)', \qquad \mathbf{X}_3 = (5,6)'$$
$$\mathbf{Y}_1 = (2,1)', \qquad \mathbf{Y}_2 = (4,3)', \qquad \mathbf{Y}_3 = (6,5)'$$

Compute angles a_{klr} and b_{klr}; We compute all 27 angles. For instance, for $(k,l,r) = (1,2,3)$ we get

$$\mathbf{X}_1 - \mathbf{X}_3 = (-4,-4), \quad \mathbf{X}_2 - \mathbf{X}_3 = (-2,-2),$$

Dot product $= (-4)(-2)+(-4)(-2) = 16$, *Norms* $= \|\mathbf{X}_1-\mathbf{X}_3\| = \sqrt{32}$, $\|\mathbf{X}_2-\mathbf{X}_3\| = \sqrt{8}$,

$$\cos(\theta) = \frac{16}{\sqrt{32}\cdot\sqrt{8}} = \frac{16}{\sqrt{256}} = 1 \Rightarrow a_{123} = \arccos(1) = 0$$

Another example, $(k,l,r) = (1,3,2)$ we get

$$\mathbf{X}_1 - \mathbf{X}_2 = (-2,-2), \quad \mathbf{X}_3 - \mathbf{X}_2 = (2,2),$$

$$\textit{Dot product} = (-2)(2) + (-2)(2) = -8, \quad \textit{Norms} = \sqrt{8}, \sqrt{8},$$

$$cos(\theta) = \frac{-8}{8} = -1 \Rightarrow a_{132} = \arccos(-1) = \pi.$$

The same is repeated for all k, l, r and for $\mathbf{Y}$ values to get b_{klr}. Results are

$$a_{111} = 0, \quad a_{112} = 0, \quad a_{113} = 0$$
$$a_{121} = 0, \quad a_{122} = 0, \quad a_{123} = 0$$
$$a_{131} = 0, \quad a_{132} = \pi, \quad a_{133} = 0$$
$$\ldots \quad \textit{(remaining } a_{klr} \textit{ and } b_{klr} \textit{ computed similarly)}$$

Center the angles, for each fixed r, we compute

$$\bar{a}_{k\cdot r} = \frac{1}{3}\sum_{l=1}^{3} a_{klr}, \quad \bar{a}_{\cdot lr} = \frac{1}{3}\sum_{k=1}^{3} a_{klr}, \quad \bar{a}_{\cdot\cdot r} = \frac{1}{9}\sum_{k=1}^{3}\sum_{l=1}^{3} a_{klr}$$
$$A_{klr} = a_{klr} - \bar{a}_{k\cdot r} - \bar{a}_{\cdot lr} + \bar{a}_{\cdot\cdot r}$$

The same for B_{klr} from b_{klr} values.

$$Pcov = \frac{1}{27}\sum_{k=1}^{3}\sum_{l=1}^{3}\sum_{r=1}^{3} A_{klr}B_{klr}$$

Using computed values, suppose

$$\sum A_{klr}B_{klr} \approx 5.463 \quad \Rightarrow \quad Pcov \approx \frac{5.463}{27} \approx 0.2023$$

This detailed example illustrates how to compute the projection covariance statistic using angular distances and their centered versions. This is a consistent test of independence for high-dimensional vectors based on projections.

11.7 Comparisons

A desirable method should maintain its power in testing against all form of dependence on the data. We consider some monotone dependence structures under different distributions in section 11.7.1 and some non–monotone dependence structures in section 11.7.2. Three existing methods are described in section 11.4 and will be used for comparison.

11.7.1 Sampling Experiment

In this section, we compare the proposed methods with existing approaches implemented in the R packages `energy` and `HHG`, as well as with the test statistic proposed by Zhu et al. [370]. The dependence structures considered include both linear (correlated) relationships and more unusual, non-standard patterns of dependence. All tests are conducted at a nominal significance level of $\alpha = 0.05$.

The simulation studies are carried out under the following multivariate distributions: (a) normal, (b) folded normal (Leone et al. [185]), (c) log-normal, (d) Cauchy, (e) t, and (f) double exponential. Each simulation scenario consists of 1000 Monte Carlo replications, with a sample size of $n = 20$ for each replication. We consider dimensions $d_1 = d_2 \in \{50, 100, 200\}$.

The data-generating mechanism proceeds as follows. First, we generate $d = d_1 + d_2$ dimensional observations $\{\mathbf{Z}_i\}_{i=1}^n$ from a multivariate normal distribution with mean vector $\boldsymbol{\mu}$ and covariance matrix $\boldsymbol{\Sigma}$. The first d_1 components of $\mathbf{Z}_i$ are assigned to $\mathbf{X}_i$, and the remaining d_2 components are assigned to $\mathbf{Y}_i$, so that $\mathbf{Z}_i = (\mathbf{X}_i, \mathbf{Y}_i)$, a column vector. Let

$$\boldsymbol{\Sigma} = \begin{bmatrix} \boldsymbol{\Sigma}_{xx} & \boldsymbol{\Sigma}_{xy} \\ \boldsymbol{\Sigma}_{yx} & \boldsymbol{\Sigma}_{yy} \end{bmatrix},$$

where $\boldsymbol{\Sigma}_{xy}$ is a $d_1 \times d_2$ cross-covariance matrix. By varying $\boldsymbol{\Sigma}_{xy}$, we generate paired samples $\{(\mathbf{X}_i, \mathbf{Y}_i)\}_{i=1}^n$ exhibiting different degrees of dependence.

By applying either the absolute value or the exponential transformation to the components of $\mathbf{X}_i$ and $\mathbf{Y}_i$, for $i = 1, \ldots, n$, we obtain samples from the multivariate folded normal and log-normal distributions, respectively. The covariance matrix is specified as $\boldsymbol{\Sigma} = (\sigma_{ij})$, where $\sigma_{ii} = 1+\rho$ and $\sigma_{ij} = \rho$ for $i \neq j$, $i, j = 1, \ldots, d$. The parameter ρ controls the strength of dependence and is varied over the set $\rho \in \{0, 0.1, 0.2, 0.3\}$. When $\rho = 0$, the samples are generated under the null hypothesis of independence, and the resulting simulation outcomes provide estimates of the Type I error rates.

Table 11.2 reports the estimated Type I error rates and empirical powers of the test statistics under the multivariate normal model. It is observed that all procedures adequately control the Type I error, with estimated significance levels closely matching the nominal level of 0.05. With respect to power, the distance covariance (Dcor) and projection covariance (Pcov) tests exhibit the strongest performance, followed by the Cor_{ipd} and Cor_{cube} procedures. Overall, all ten test statistics perform satisfactorily under multivariate normality, with most power estimates exceeding 0.9 even for alternatives close to independence. Moreover, the power of all methods increases as the dimensions d_1 and d_2 grow.

Guo and Modarres [126] present further simulation results for the multivariate folded normal distribution. The estimated significance levels are close to the nominal level of 0.05, indicating satisfactory control of the Type I error for all procedures. In terms of power, the Cor_{ipd}, Cor_{cube}, and *Trans* tests exhibit the strongest performance, followed by the *Odds* and CP methods. Notably, the seven newly proposed procedures uniformly outperform Dcor, HHG, and Pcov in this setting. As expected, increasing the dimensionality leads to improved power across all testing methods. Moreover, in comparison with the multivariate normal case, all tests experience a reduction in power under the folded normal distribution.

Under the multivariate log-normal distribution, all methods maintain appropriate control of the Type I error. With respect to power, the Pcov test clearly outperforms all competing approaches. Although the CP and *Trans* tests are less powerful than Pcov, they remain competitive relative to the remaining methods. As in the previous scenarios, increasing the dimension results in higher power for all testing procedures.

TABLE 11.2: The type-I error and power of different statistics under the multivariate normal distribution.

ρ	d_1	*EDF*	*Odds*	CP	Cor$_{\text{ipd}}$	Cor_{cube}	*Quad*	*Trans*	*Dcor*	*HHG*	*Pcov*
0	50	0.051	0.035	0.046	0.042	0.042	0.049	0.042	0.043	0.041	0.049
	100	0.060	0.054	0.067	0.065	0.064	0.047	0.068	0.045	0.054	0.041
	200	0.048	0.063	0.050	0.061	0.059	0.057	0.059	0.046	0.042	0.045
0.1	50	0.431	0.566	0.505	0.583	0.560	0.386	0.536	0.792	0.507	0.763
	100	0.731	0.837	0.795	0.859	0.841	0.673	0.817	0.971	0.808	0.961
	200	0.93	0.959	0.949	0.972	0.967	0.905	0.959	0.999	0.958	0.997
0.2	50	0.893	0.943	0.928	0.953	0.946	0.876	0.931	0.991	0.939	0.987
	100	0.978	0.989	0.985	0.993	0.991	0.972	0.988	1	0.992	1
	200	0.998	1	0.998	1	1	0.997	0.999	1	1	1
0.3	50	0.983	0.993	0.989	0.995	0.994	0.978	0.993	1	0.992	1
	100	0.995	0.998	0.998	0.999	0.999	0.994	0.997	1	0.999	1
	200	1	1	1	1	1	1	1	1	1	1

They further present simulation results for the multivariate Cauchy, t, and double exponential distributions, which are characterized by heavy tails. Under the Cauchy distribution, all tests maintain Type I error rates near the nominal level of 0.05. In terms of power, the EDF and CP methods perform the best, while the remaining methods exhibit comparable performance. Notably, increasing the dimensionality has little effect on the performance of the tests in this setting.

Under the t distribution, all tests continue to control the Type I error adequately. Regarding power, EDF and CP again demonstrate the strongest performance, followed by the *Odds* and *Trans* methods. Similar to the Cauchy case, the testing power changes only slightly as the dimension of the vectors increases.

For the multivariate double exponential distribution, all tests maintain Type I error rates close to the nominal level of 0.05. In terms of power, the *Trans* test exhibits the highest performance, followed by the *Odds* and CP methods. Notably, in this scenario, increasing the dimensionality results in a decrease in power, which contrasts with the patterns observed in the other distributions. This suggests that the choice of norm used to compute the IPDs may influence the performance of the tests, and further investigation is warranted to determine the optimal norm for this setting.

11.7.2 Special Dependence

In an interesting study, Newton [249] presents seven distinct patterns of dependence and conducts simulations to evaluate the performance of the energy statistic proposed by Székely et al. [331] for testing independence. Newton's simulations are reported for the case $d_1 =$

$d_2 = 1$ with sample size $n = 500$. The specific dependence patterns considered are illustrated in Figure 11.1. Notably, all of these structures have zero Pearson correlation. Newton's results demonstrate that the energy test is capable of detecting non-linear dependence beyond what the Pearson correlation can capture.

We examine the same dependence structures, but under scenarios with high dimensionality and small sample sizes. Let $\{(\mathbf{X}_i, \mathbf{Y}_i)\}_{i=1}^n$ denote the observations. Suppose (x_i, y_i) represents a point in any of the specified dependence patterns for $i = 1, \ldots, n$. The observations are generated as

$$\mathbf{X}_i = x_i \mathbf{1}_{d_1}, \quad \mathbf{Y}_i = y_i \mathbf{1}_{d_2},$$

where $\mathbf{1}_{d_1}$ denotes a $d_1 \times 1$ vector of ones. In the simulations, we set $d_1 = d_2 = 100$ and consider sample sizes $n \in \{20, 30\}$. Each scenario is replicated 1000 times.

Table 11.3 displays the power of the statistics under special dependence. One observes that distance correlation (Dcor) and projection covariance (Pcov) lose power when the sample size is small. The EDF, CP, and *Quad* methods are sensitive to the 1st, 4th, 5th, and 6th dependence structures. The *Odds* test can detect the 1st, 4th, and 5th patterns when $n = 30$. For the 4th and 5th dependence structures, nearly all methods except Dcor and Pcov perform well. However, all methods have difficulty detecting the 2nd, 3rd, and 7th dependence structures under these small-sample conditions.

FIGURE 11.1: The special patterns.

TABLE 11.3: The power of different statistics for independence test under special dependence structure.

Type	n	*EDF*	*Odds*	CP	Cor_{ipd}	Cor_{cube}	*Quad*	*Trans*	*Dcor*	*HHG*	*Pcov*
1	20	0.588	0.229	0.569	0	0	0.549	0	0	0.046	0.008
	30	0.998	0.809	1	0	0.141	0.980	0.086	0.002	0.620	0.161
2	20	0.073	0.026	0.0850	0.019	0.031	0.074	0.021	0.040	0.066	0.035
	30	0.080	0.014	0.143	0.004	0.013	0.094	0.006	0.040	0.100	0.027
3	20	0.105	0.009	0.193	0.001	0.010	0.115	0.003	0.026	0.119	0.021
	30	0.325	0.007	0.553	0	0.001	0.253	0.001	0.034	0.255	0.019
4	20	0.697	0.522	0.627	0.47	0.644	0.631	0.538	0.097	0.569	0.209
	30	0.910	0.862	0.895	0.745	0.886	0.891	0.793	0.434	0.864	0.473
5	20	0.989	0.987	0.987	0.96	0.93	0.937	0.935	0.132	0.813	0.273
	30	1	1	1	1	1	1	1	0.161	0.994	0.362
6	20	0.620	0.006	0.775	0	0	0.342	0	0	0.069	0
	30	0.997	0.003	1	0	0	0.994	0	0	0.397	0.003
7	20	0.048	0.048	0.054	0.045	0.048	0.048	0.048	0.041	0.040	0.049
	30	0.041	0.048	0.037	0.050	0.049	0.041	0.046	0.056	0.043	0.053

11.7.3 Discussion

We have compared the various testing methods across different distributions and dependence structures. No single method consistently dominates all others in every scenario. Generally, tests based on IPDs exhibit superior performance in detecting non-linear dependence.

In practical applications, if prior knowledge about the underlying distribution or the form of dependence is available, one may choose the corresponding optimal testing method. Otherwise, an ensemble approach combining multiple methods and making a decision via majority or weighted voting can help reduce the risk of relying on a single method.

The primary limitation of IPD-based methods lies in their computational cost. For a sample of size n, the time complexity of computing IPDs is $O(n^2)$, and even simple operations, such as calculating the mean of IPDs, also require $O(n^2)$ time. The time complexities of all methods are summarized in Table 11.4. Methods with complexity exceeding $O(n^2)$ may become computationally prohibitive for large datasets.

TABLE 11.4: Time Complexity.

EDF	$Odds$	CP	Cor$_{\text{ipd}}$	Cor_{cube}	$Quad$	$Trans$	$Dcor$	HHG	$Pcov$
$O(n^4)$	$O(n^4)$	$O(n^4)$	$O(n^2)$	$O(n^2)$	$O(n^4)$	$O(n^2)$	$O(n^2)$	$O(n^2\log(n))$	$O(n^3)$

11.8 Dissimilarity-based Independence Testing

Assessing independence in high-dimensional random vectors is a fundamental statistical problem, made challenging when the sample size is small relative to the dimension. Classical measures, such as distance covariance or kernel-based statistics, often become unstable or require careful tuning in such settings. We propose non-parametric independence tests based on dissimilarities that avoid kernel tuning and remain valid under heterogeneous marginal scales. Formulated as V-statistics, we derive Hoeffding decompositions and establish asymptotic normality under independence, with explicit variance expressions. The statistics are consistent against all fixed alternatives and have non-trivial power against nonlinear, heteroscedastic, and latent-structure dependence. Simulations show accurate type-I error control, stability as dimension grows, and superior power compared to existing methods. Applications to four benchmark gene expression datasets demonstrate practical effectiveness. This methodology provides a theoretically grounded, robust, and computationally practical tool for high-dimensional, low-sample-size independence testing.

Székely et al. [331] popularized distance covariance, transforming high-dimensional data via characteristic functions, though Zhu et al. [370] highlighted limitations in very high dimensions and proposed the projection covariance. In regression analysis, the conditional mean $\mathbb{E}[Y|X]$ varies with X. Classical tests of independence may fail if the relationship is non-linear or if the conditional variance also changes. Even if $\mathbb{E}[Y|X]$ is constant, the conditional covariance $\mathbb{C}ov(Y|X = x)$ may vary with x. Classical correlation is insensitive to this scenario. Yet, heteroscedasticity can lead to important structures. When a latent categorical variable S influences both X and Y, marginally X and Y become dependent even though $X \perp Y|S$. Detecting this form of dependence is crucial for e.g. latent class models, batch-effect correction, and cluster analysis. In higher dimensions, classical correlation is not directly applicable since data often exhibit non–linear, non–monotonic, or heteroscedastic behavior. Dependence measures such as distance covariance or dissimilarity–based correlation are necessary because they satisfy the property that the dependence measure equals zero if and only if $X \perp Y$.

In causal inference, many instrumental–variable methods assume instruments are independent of unobserved confounders, where violation may be exposed by dependence tests. In high-dimensional data with genomics features X may affect outcomes Y in complex ways (latent clusters, heterogeneity); detecting dependence beyond linear association is vital. In machine learning and feature screening, dependence–based screening (e.g., selecting features X that are dependent on Y) benefits from powerful independence tests beyond Pearson correlation.

In this section, we introduce the statistic CATS_δ, which measures discrepancies in interpoint distances across cross-coordinate partitions defined by the joint sample. We also present CATS_ρ, which retains the same structural framework but incorporates normalized correlations of distance-based components, improving sensitivity to certain non-linear alternatives. The contributions of this section are as follows:

- The CATS statistics for testing independence are introduced, derived from geometric properties of interpoint distances with minimal distributional assumptions.
- The asymptotic null distributions of CATS_δ and CATS_ρ are derived, proving asymptotic normality and providing explicit variance formulas.
- Consistency against all fixed alternatives is established, and power properties are derived under certain local dependencies that classical covariance tests fail to detect.
- The CATS statistics remain valid and stable when the dimension exceeds or is comparable to the sample size.
- The CATS framework is compared to distance covariance, kernel-based independence measures, and IPD-based methods, clarifying theoretical relationships and advantages.

11.8.1 Distance Covariance and Energy Statistics

Modarres [231] discusses high-dimensional measures of dissimilarity for testing equality of distributions. In the next section, we use dissimilarities for testing independence. Székely et al. [331] propose distance correlation to measure the dependence between random vectors. Let $\mathbf{X} \in \mathbb{R}^{d_1}$ and $\mathbf{Y} \in \mathbb{R}^{d_2}$ be random vectors with finite first moments. The distance covariance between $\mathbf{X}$ and $\mathbf{Y}$ is defined by

$$\mathcal{V}^2(\mathbf{X},\mathbf{Y}) = \int_{\mathbb{R}^{d_1+d_2}} \left|\varphi_{\mathbf{X},\mathbf{Y}}(t,s) - \varphi_{\mathbf{X}}(t)\varphi_{\mathbf{Y}}(s)\right|^2 \, w(t,s)\,dt\,ds, \tag{11.20}$$

where $\varphi_{\mathbf{X},\mathbf{Y}}(t,s)$ is the joint characteristic function of $(\mathbf{X},\mathbf{Y})$, and $\varphi_{\mathbf{X}}(t)$ and $\varphi_{\mathbf{Y}}(s)$ are the marginal characteristic functions and $w(t,s)$ is the weight function ensuring the distance covariance integral remains finite. The integrand compares the observed joint structure to what the joint structure would look like if $\mathbf{X}$ and $\mathbf{Y}$ were independent. If $\mathbf{X}$ and $\mathbf{Y}$ are independent, then the joint characteristic function factors as $\varphi_{\mathbf{X},\mathbf{Y}}(t,s) = \varphi_{\mathbf{X}}(t)\,\varphi_{\mathbf{Y}}(s)$, for all (t,s). In that case the integrand is zero everywhere, and $\mathcal{V}^2(\mathbf{X},\mathbf{Y}) = 0$. If there is any dependence, even non-linear or non-monotone, the integrand is not zero, and $\mathcal{V}^2(\mathbf{X},\mathbf{Y}) > 0$. This gives a direct link between the analytic definition and independence testing. Energy statistic and *dCov* are related. Indeed, the following representation of distance covariance aligns it with the energy statistic framework.

$$\mathcal{V}^2(\mathbf{X},\mathbf{Y}) = \mathbb{E}\big[\|\mathbf{X}-\mathbf{X}_1\|\,\|\mathbf{Y}-\mathbf{Y}_1\|\big] + \mathbb{E}\big[\|\mathbf{X}-\mathbf{X}_1\|\big]\,\mathbb{E}\big[\|\mathbf{Y}-\mathbf{Y}_1\|\big] - 2\,\mathbb{E}\big[\|\mathbf{X}-\mathbf{X}_1\|\,\|\mathbf{Y}-\mathbf{Y}_2\|\big],$$

where $(\mathbf{X},\mathbf{Y})$, $(\mathbf{X}_1,\mathbf{Y}_1)$, and $(\mathbf{X}_1,\mathbf{Y}_2)$ are independent copies of $(\mathbf{X},\mathbf{Y})$. The expectation $\mathbb{E}\big[\|\mathbf{X}-\mathbf{X}_1\|\,\|\mathbf{Y}-\mathbf{Y}_1\|\big]$ captures how much large differences in $\mathbf{X}$ tend to align with large differences in $\mathbf{Y}$. If changes in $\mathbf{X}$ and changes in $\mathbf{Y}$ move together, this term is large. The product $\mathbb{E}\big[\|\mathbf{X}-\mathbf{X}_1\|\big]\,\mathbb{E}\big[\|\mathbf{Y}-\mathbf{Y}_1\|\big]$ adjusts for marginal spread. It removes the effect of variation that comes from $\mathbf{X}$ alone or from $\mathbf{Y}$ alone. The last term $\mathbb{E}\big[\|\mathbf{X}-\mathbf{X}_1\|\,\|\mathbf{Y}-\mathbf{Y}_2\|\big]$ uses cross-pairs that are independent across coordinates and provides a baseline. It reflects what we would see if the $\mathbf{X}$ and the $\mathbf{Y}$ were unrelated. We will use this connection in the next section to present CATS statistics.

11.8.2 CATS Statistics

Since Euclidean distances are affected by the concentration phenomena, we propose to use a dissimilarity measure that takes advantage of the phenomena when testing independence. Let $\mathbf{Z} = (\mathbf{X}, \mathbf{Y})$ where $X \in \mathbb{R}^{d_1}$ and $Y \in \mathbb{R}^{d_2}$ where $d = d_1 + d_2$. Let $\mathbf{Z}_i = (\mathbf{X}_i, \mathbf{Y}_i) \in \mathbb{R}^d$ be n i.i.d. copies of $\mathbf{Z}$ with continuous probability distributions F, mean vectors $\boldsymbol{\mu}$, and covariance matrices $\boldsymbol{\Sigma}$. We test $H_0 : X \in \mathbb{R}^{d_1}$ and $Y \in \mathbb{R}^{d_2}$ are statistically independent.

11.8.3 Dissimilarity δ

For observation $\mathbf{x} = (x_1, \ldots, x_d)$ define

$$\bar{x}^d = \frac{1}{d}\sum_{t=1}^{d} x_t, \quad [s^d_{(x)}]^2 = \frac{1}{d}\sum_{t=1}^{d}(x_t - \bar{x}^d)^2.$$

For distinct observations $\mathbf{x}_i, \mathbf{y}_j, \mathbf{z}_k$, define modified distances

$$\|\mathbf{x}_i - \mathbf{z}_k\|_D = \sqrt{(\bar{x}^d_i - \bar{z}^d_k)^2 + (s^d_{(x)i} - s^d_{(z)k})^2}.$$

MADMD dissimilarity is given by

$$\delta(x_i, y_j) = \frac{1}{n-2}\sum_{k \neq i,j} \big|\|\mathbf{x}_i - \mathbf{z}_k\|_D - \|\mathbf{y}_j - \mathbf{z}_k\|_D\big|.$$

Since δ measures the mean absolute differences of modified distances $\|\mathbf{x}_i - \mathbf{z}_k\|_D$ and $\|\mathbf{y}_j - \mathbf{z}_k\|_D$, we refer to it as MADMD. The dissimilarity index uses the modified distances based on the lower bound of the Euclidean distances (Modarres [231]). Let $\delta(\mathbf{x}_i, \mathbf{y}_j)$ be the dissimilarity measure defined via MADMD. We define the double-centered matrices

$$A_{ij} = \delta_X(\mathbf{x}_i, \mathbf{x}_j) - \bar{\delta}_{i\cdot} - \bar{\delta}_{\cdot j} + \bar{\delta}_{\cdot\cdot}, \quad B_{ij} = \delta_Y(\mathbf{y}_i, \mathbf{y}_j) - \bar{\delta}_{i\cdot} - \bar{\delta}_{\cdot j} + \bar{\delta}_{\cdot\cdot}, \tag{11.21}$$

and define the centered dissimilarity functions

$$\tilde{\delta}_X(\mathbf{x}_i, \mathbf{x}_j) = \delta(\mathbf{x}_i, \mathbf{x}_j) - \frac{1}{n}\sum_{l=1}^{n}\delta(\mathbf{x}_i, \mathbf{x}_l) - \frac{1}{n}\sum_{l=1}^{n}\delta(\mathbf{x}_l, \mathbf{x}_j) + \frac{1}{n^2}\sum_{l,m=1}^{n}\delta(\mathbf{x}_l, \mathbf{x}_m), \tag{11.22}$$

and similarly for Y,

$$\tilde{\delta}_Y(\mathbf{y}_i, \mathbf{y}_j) = \delta(\mathbf{y}_i, \mathbf{y}_j) - \frac{1}{n}\sum_{l=1}^{n}\delta(\mathbf{y}_i, \mathbf{y}_l) - \frac{1}{n}\sum_{l=1}^{n}\delta(\mathbf{y}_l, \mathbf{y}_j) + \frac{1}{n^2}\sum_{l,m=1}^{n}\delta(\mathbf{y}_l, \mathbf{y}_m). \tag{11.23}$$

Looking at the right side of (11.21), terms ensure the matrix $\mathbf{A}$ is doubly centered (both row and column sums are zero). We now define a Cross-Association Test Statistic (CATS) using MADMD to measure independence,

$$\text{CATS}_\delta(X, Y) = \frac{1}{n^2}\sum_{i,j=1}^{n} A_{ij}B_{ij}. \tag{11.24}$$

The population version with centered $\tilde{\delta}$ is given by

$$V_{\delta}^2(\mathbf{X}, \mathbf{Y}) = \mathbb{E}[\tilde{\delta}_X(\mathbf{X}, \mathbf{X}')\tilde{\delta}_Y(\mathbf{Y}, \mathbf{Y}')],$$

where

$$\tilde{\delta}_X(\mathbf{x}, \mathbf{x}') = \delta_X(\mathbf{x}, \mathbf{x}') - \mathbb{E}[\delta_X(\mathbf{x}, \mathbf{X}')] - \mathbb{E}[\delta_X(\mathbf{X}, \mathbf{x}')] + \mathbb{E}[\delta_X(\mathbf{X}, \mathbf{X}')].$$

The corresponding sample version is

$$\hat{V}_{\delta}^2(\mathbf{X}, \mathbf{Y}) = \frac{1}{n^2} \sum_{i,j=1}^{n} \tilde{\delta}_X(\mathbf{x}_i, \mathbf{x}_j)\, \tilde{\delta}_Y(\mathbf{y}_i, \mathbf{y}_j). \tag{11.25}$$

This is the same as $\mathrm{CATS}_{\delta}(X, Y)$. We define the dissimilarity correlation by

$$R_{\delta}^2(\mathbf{X}, \mathbf{Y}) = \frac{V_{\delta}^2(\mathbf{X}, \mathbf{Y})}{V_{\delta}(\mathbf{X}, \mathbf{X})\, V_{\delta}(\mathbf{Y}, \mathbf{Y})}. \tag{11.26}$$

If $\mathbf{X}$ and $\mathbf{Y}$ are independent, then

$$\mathbb{E}\Big[\tilde{\delta}_X(\mathbf{X}, \mathbf{X}')\tilde{\delta}_Y(\mathbf{Y}, \mathbf{Y}')\Big] = \mathbb{E}\Big[\tilde{\delta}_X(\mathbf{X}, \mathbf{X}')\Big]\, \mathbb{E}\Big[\tilde{\delta}_Y(\mathbf{Y}, \mathbf{Y}')\Big] = 0,$$

hence, $V_{\delta}^2(\mathbf{X}, \mathbf{Y}) = 0$.

Theorem 11.8.1 (Asymptotic null distribution of $\hat{V}_{\delta}^2$.) *Assume that $(\mathbf{X}_i, \mathbf{Y}_i)$, $i = 1, \ldots, n$, are i.i.d. copies of $(\mathbf{X}, \mathbf{Y})$ valued in $\mathbb{R}^{d_1+d_2}$, with finite second moments, and that the dissimilarity kernels $\tilde{\delta}_X$ and $\tilde{\delta}_Y$ are square-integrable and positive definite. Then, under the null hypothesis of independence H_0,*

$$n\, \hat{V}_{\delta}^2(\mathbf{X}, \mathbf{Y}) \xrightarrow{d} \sum_{r=1}^{\infty} \lambda_r Z_r^2, \tag{11.27}$$

where $\{Z_r\}_{r\geq 1}$ are i.i.d. standard normal variables and $\{\lambda_r\}_{r\geq 1}$ are non-negative constants depending on the eigenvalues associated with the kernel $K_{\delta}((\mathbf{x}, \mathbf{y}), (\mathbf{x}', \mathbf{y}')) = \tilde{\delta}_X(\mathbf{x}, \mathbf{x}')\, \tilde{\delta}_Y(\mathbf{y}, \mathbf{y}')$.

Proof: The statistic $\hat{V}_{\delta}^2(\mathbf{X}, \mathbf{Y})$ is a symmetric V-statistic with kernel

$$h((\mathbf{X}_i, \mathbf{Y}_i), (\mathbf{X}_j, \mathbf{Y}_j)) = \tilde{\delta}_X(\mathbf{X}_i, \mathbf{X}_j)\, \tilde{\delta}_Y(\mathbf{Y}_i, \mathbf{Y}_j).$$

Under H_0, we have $\mathbb{E}[h((\mathbf{X}, \mathbf{Y}), (\mathbf{X}', \mathbf{Y}'))] = 0$ by the centering of $\tilde{\delta}_X$ and $\tilde{\delta}_Y$. Hence, the kernel h is degenerate of order one, i.e. $\mathbb{E}[\, h((\mathbf{X}, \mathbf{Y}), (\mathbf{X}', \mathbf{Y}')) \mid \mathbf{X}, \mathbf{Y}\,] = 0$. For a degenerate V-statistic of order 2 with kernel h (possessing finite second moments), the classical limit theorem for degenerate V-statistics (Serfling [303]) states that

$$n\, \hat{V}_{\delta}^2(\mathbf{X}, \mathbf{Y}) \xrightarrow{d} \sum_{r=1}^{\infty} \lambda_r Z_r^2,$$

where $\{\lambda_r\}$ are the non-zero eigenvalues of the operator T_h defined on $L_2(F(\mathbf{X}, \mathbf{Y}))$ by

$$(T_h f)(\mathbf{x}, \mathbf{y}) = \int h((\mathbf{x}, \mathbf{y}), (\mathbf{x}', \mathbf{y}')) f(\mathbf{x}', \mathbf{y}')\, dF(\mathbf{x}', \mathbf{y}').$$

Under H_0, $h((\mathbf{x},\mathbf{y}),(\mathbf{x}',\mathbf{y}')) = \tilde{\delta}_X(\mathbf{x},\mathbf{x}')\tilde{\delta}_Y(\mathbf{y},\mathbf{y}')$ is a product kernel. The spectrum of T_h is thus $\{\lambda_k^{(X)}\lambda_\ell^{(Y)}\}$ where $\lambda_k^{(X)}$ and $\lambda_\ell^{(Y)}$ are the eigenvalues of the centered kernels $\tilde{\delta}_X$ and $\tilde{\delta}_Y$, respectively. Therefore, the limiting null distribution has the form

$$n\,\hat{V}_\delta^2(\mathbf{X},\mathbf{Y}) \xrightarrow{d} \sum_{k,\ell\geq 1} \lambda_k^{(X)}\lambda_\ell^{(Y)} Z_{k\ell}^2,$$

where $Z_{k\ell} \sim \mathbb{N}(0,1)$ i.i.d., and the result follows after relabeling $\{\lambda_k^{(X)}\lambda_\ell^{(Y)}\}$ as the sequence $\{\lambda_r\}_{r\geq 1}$. □

This provides a way to approximate the null distribution of the test statistic using empirical eigenvalues. We now discuss the Gaussian case with the following result.

Theorem 11.8.2 : *Let (X,Y) be jointly multivariate normal with*

$$\begin{pmatrix}\mathbf{X}\\ \mathbf{Y}\end{pmatrix} \sim \mathbb{N}_{d_1+d_2}\left(\begin{pmatrix}\mu_X\\ \mu_Y\end{pmatrix}, \begin{pmatrix}\mathbf{\Sigma}_X & \mathbf{\Sigma}_{XY}\\ \mathbf{\Sigma}_{XY}' & \mathbf{\Sigma}_Y\end{pmatrix}\right),$$

and correlation matrix $\mathbf{\Gamma} = (\gamma)_{ij} = \mathbf{\Sigma}_X^{-1/2}\mathbf{\Sigma}_{XY}\mathbf{\Sigma}_Y^{-1/2}$. We have $V_\delta^2(\mathbf{X},\mathbf{Y}) = 0$ if and only if $\mathbf{\Gamma} = \mathbf{I}$. That is, $\gamma_{ij} = 0$ for $i \neq j$.

This theorem states that, in the Gaussian case, $V_\delta^2(\mathbf{X},\mathbf{Y})$ detects exactly the same type of structure as the usual correlation matrix. Under dependence ($\mathbf{\Gamma} \neq \mathbf{I}$), we have $V_\delta^2(\mathbf{X},\mathbf{Y}) > 0$, and it increases with $||\mathbf{\Gamma}||_F$. Under a fixed alternative H_a with $V_\delta^2(\mathbf{X},\mathbf{Y}) > 0$, the kernel is typically non-degenerate and the V-statistic has a non-zero (first-order) projection. Using the Hoeffding decomposition (Serfling [303]) gives the asymptotic normality under fixed alternative. If $V_\delta^2(\mathbf{X},\mathbf{Y}) > 0$ and the kernel has finite second moments, then

$$\sqrt{n}(\mathrm{CATS}_\delta - V_\delta^2) \xrightarrow{d} \mathbb{N}(0,\sigma^2),$$

where

$$\sigma^2 = 4\,\mathbb{V}ar\big(\mathbb{E}[\tilde{\delta}_X(\mathbf{X},\mathbf{X}')\tilde{\delta}_Y(\mathbf{Y},\mathbf{Y}')|\mathbf{X},\mathbf{Y}]\big).$$

Using the null critical value derived from the null-limit, denote $q_{n,\alpha}$ the $1-\alpha$ quantile of $n\ \mathrm{CATS}_\delta$ under H_0. Equivalently, set the test threshold for CATS_δ as $c_{n,\alpha} = q_{n,\alpha}/n$. Then the approximate (large-n) power of the level-α test is

$$\text{Power} \approx 1 - \Phi\left(\frac{c_{n,\alpha} - V_\delta^2}{\sigma/\sqrt{n}}\right) = 1 - \Phi\left(\frac{q_{n,\alpha}/n - V_\delta^2}{\sigma/\sqrt{n}}\right).$$

For large n, $q_{n,\alpha}/n \to 0$ (since the null-limit is $O_p(1)$ for $n \cdot \mathrm{CATS}_\delta$), so the simple large-sample approximation is

$$\text{Power} \approx \Phi\left(\frac{\sqrt{n}\,V_\delta^2}{\sigma}\right),$$

which indicates that power grows as $\sqrt{n}\,V_\delta^2/\sigma$ increases. One can estimate the key quantities from data as follows.

Algorithm 11.8.1 Computing CATS

1: **Input:** Matrices A and B derived from the data; sample size n
2: **Output:** Test statistic and variance estimates for CATS

3: **Step 1:** Compute the centered matrices A and B

4: **Step 2:** Estimate the leading marginal eigenvalue via SVD of A and B

5: **Step 3:** Estimate the rank-one approximation

$$\lambda_1 \approx \hat{\lambda}_1^{(A)}\hat{\lambda}_1^{(B)}$$

6: **Step 4:** Estimate σ^2, the asymptotic variance under the alternative

$$\hat{\sigma}^2 = 4 \cdot \frac{1}{n}\sum_{i=1}^{n}\left(\frac{1}{n}\sum_{j=1}^{n} h(z_i, z_j) - \hat{V}_\delta^2\right)^2$$

where

$$h(z_i, z_j) = \tilde{a}_{ij}\tilde{b}_{ij}, \quad \hat{V}_\delta^2 = \text{CATS}_\delta$$

7: **Step 5:** Use permutation or eigen-approximation for critical values; for large n, use the asymptotic normal approximation

11.8.4 Dissimilarity ρ

Pal et al. [261] and Sarkar et al. [299] use the data cloud $\mathbf{Z}$ to measure the dissimilarity between two observations $\mathbf{x}$ and $\mathbf{y}$ by

$$\rho(\mathbf{x}, \mathbf{y}) = \frac{1}{N-2}\frac{1}{\sqrt{d}}\sum_{\mathbf{z}\backslash\{\mathbf{x},\mathbf{y}\}} \big|\|\mathbf{x}-\mathbf{z}\| - \|\mathbf{y}-\mathbf{z}\|\big|. \tag{11.28}$$

The dissimilarity index is based on the mean absolute difference of distances (MADD) and uses the scaling factor $\frac{1}{\sqrt{d}}$ since IPDs usually increase with the rate of $\sqrt{d}$ (Hall et al. [133]). We define the double-centered matrices

$$A_{ij} = \rho_X(\mathbf{x}_i, \mathbf{x}_j) - \bar{\rho}_{i\cdot} - \bar{\rho}_{\cdot j} + \bar{\rho}_{\cdot\cdot}, \quad B_{ij} = \rho_Y(\mathbf{y}_i, \mathbf{y}_j) - \bar{\rho}_{i\cdot} - \bar{\rho}_{\cdot j} + \bar{\rho}_{\cdot\cdot}.$$

Let $\rho(\mathbf{x}_i, \mathbf{y}_j)$ be the dissimilarity measure defined via MADMD. Define the centered dissimilarity functions

$$\tilde{\rho}_X(\mathbf{x}_i, \mathbf{x}_j) = \rho(\mathbf{x}_i, \mathbf{x}_j) - \frac{1}{n}\sum_{l=1}^{n}\rho(\mathbf{x}_i, \mathbf{x}_l) - \frac{1}{n}\sum_{l=1}^{n}\rho(\mathbf{x}_l, \mathbf{x}_j) + \frac{1}{n^2}\sum_{l,m=1}^{n}\rho(\mathbf{x}_l, \mathbf{x}_m), \tag{11.29}$$

and

$$\tilde{\rho}_Y(\mathbf{y}_i, \mathbf{y}_j) = \rho(\mathbf{y}_i, \mathbf{y}_j) - \frac{1}{n}\sum_{l=1}^{n}\rho(\mathbf{y}_i, \mathbf{y}_l) - \frac{1}{n}\sum_{l-1}^{n}\rho(\mathbf{y}_l, \mathbf{y}_j) + \frac{1}{n^2}\sum_{l,m=1}^{n}\rho(\mathbf{y}_l, \mathbf{y}_m). \tag{11.30}$$

CATS using MADD suggest the following to measure independence,

$$\mathrm{CATS}_\rho(\mathbf{X},\mathbf{Y}) = \frac{1}{n^2}\sum_{i,j=1}^{n} A_{ij}B_{ij}. \tag{11.31}$$

The population version with centered $\tilde{\rho}$ is given by

$$V_\rho^2(\mathbf{X},\mathbf{Y}) = \mathbb{E}[\tilde{\rho}_X(\mathbf{X},\mathbf{X}')\tilde{\rho}_Y(\mathbf{Y},\mathbf{Y}')],$$

where

$$\tilde{\rho}_X(\mathbf{x},\mathbf{x}') = \rho_X(\mathbf{x},\mathbf{x}') - \mathbb{E}[\rho_X(\mathbf{x},\mathbf{X}')] - \mathbb{E}[\rho_X(\mathbf{X},\mathbf{x}')] + \mathbb{E}[\rho_X(\mathbf{X},\mathbf{X}')].$$

Many of the properties of $V_\delta^2(\mathbf{X},\mathbf{Y})$ extends to $V_\rho^2(\mathbf{X},\mathbf{Y})$. In particular, Theorems 11.8.1 and 11.8.2 hold for $V_\rho^2(\mathbf{X},\mathbf{Y})$. Under dependence ($\gamma \neq \mathbf{I}$), $V_\rho^2(\mathbf{X},\mathbf{Y}) > 0$ and increases with $\|\gamma\|_F$. Under a fixed alternative H_a with $V_\rho^2(\mathbf{X},\mathbf{Y}) > 0$, the kernel is typically non-degenerate and the V-statistic has a non-zero (first-order) projection. If $V_\rho^2(\mathbf{X},\mathbf{Y}) > 0$ and the kernel has finite second moments, then

$$\sqrt{n}(\mathrm{CATS}_\rho - V_\rho^2) \xrightarrow{d} \mathbb{N}(0,\sigma^2),$$

where

$$\sigma^2 = 4\,\mathbb{V}ar\big(\mathbb{E}[\tilde{\rho}_X(\mathbf{X},\mathbf{X}')\tilde{\rho}_Y(\mathbf{Y},\mathbf{Y}')|\mathbf{X},\mathbf{Y}]\big).$$

What are the advantages of dissimilarity–based methods (like V_δ^2 or V_ρ^2)? These measures operate in arbitrary dimensions (d_1, d_2) and detect all types of dependence (linear, non-linear, monotonic, non-monotonic, heteroscedastic). They lead to simple sample test statistics (e.g., CATS_δ or CATS_ρ) that can be implemented via permutations. Especially in heteroscedastic settings (where mean can be constant while covariance changes), dissimilarity–based tests can succeed where classical methods fail. Furthermore, both CATS_δ and CATS_ρ are designed for high dimension, where $d \gg n$. In such settings, Euclidean distance between raw vectors can lose contrast because of distance concentration. MADMD avoids this by summarizing each observation through low-dimensional summaries such as mean and variances, and by comparing points through how they relate to the rest of the sample. This can give more stable behavior when the dimension is large.

Dependence beyond mean relationships, especially conditional or variance driven dependence, has been studied. For example the measure of conditional distance correlation was developed by Zhang et al. [367], allowing testing of $\mathbf{X} \perp \mathbf{Y}|\ \mathbf{Z}$ non-parametrically. The broader theme is that independence tests must detect not just shifts in mean but changes in variance/scale or other features. The new tests based on CATS_δ and CATS_ρ detect dependence due to shifts in mean and changes in variance/scale. We will discuss these in the next section.

11.8.5 Testing Independence in Mixture Models

Variance–driven Dependence

Although fewer works explicitly focus on mixture–driven dependence, the general paradigm of latent variable induced marginal dependence is common in cluster analysis, finite mixture modeling, batch–effect correction. The use of non-parametric dependence testing in such contexts remains a growing area. The universal consistency results for dependence measures (e.g. energy/dCov) make them suitable for mixture–cases. That is, any dependence measure that is zero only under independence implies that latent mixing, which induces dependence will be captured. Permutation tests are widely used to obtain p–values for dependence measures under the H_0. However, for large n, permutations become expensive. The chi–square approximations for Dcor (Shen et al. [306]) offer one way forward.

While much of the literature covers dependence measures in general, less attention has been devoted to the special case of variance–driven dependence in the multivariate setting using dissimilarity–based tests. The CATS statistics fit nicely into this gap, offering a tool for high–dimensional, non-linear, heteroscedastic/mixture settings.

Example 11.8.1 *Conditional Variance in a Two–Component Multivariate Normal Mixture. Let $(S, \mathbf{X}, \mathbf{Y})$ be a random vector where the latent component $S \in \{1, 2\}$ has $\mathbb{P}(S = s) = \pi_s$, $s = 1, 2$, $\pi_1 + \pi_2 = 1$. Conditional on $S = s$,*

$$\begin{pmatrix} \mathbf{X} \\ \mathbf{Y} \end{pmatrix} \Big| S = s \sim \mathbb{N}_{d_1+d_2}\left(\begin{pmatrix} \boldsymbol{\mu}_X^{(s)} \\ \boldsymbol{\mu}_Y^{(s)} \end{pmatrix}, \boldsymbol{\Sigma}^{(s)} = \begin{pmatrix} \boldsymbol{\Sigma}_{XX}^{(s)} & \boldsymbol{\Sigma}_{XY}^{(s)} \\ \boldsymbol{\Sigma}_{YX}^{(s)} & \boldsymbol{\Sigma}_{YY}^{(s)} \end{pmatrix} \right).$$

Here $\mathbf{X} \in \mathbb{R}^{d_1}$, $\mathbf{Y} \in \mathbb{R}^{d_2}$. We are especially interested in the case where within each component $\mathbf{X} \perp \mathbf{Y}$, so $\boldsymbol{\Sigma}_{XY}^{(s)} = 0$ for each s, but the mixture induces marginal dependence. In this set–up, we have

$$\mathbb{E}[\mathbf{Y}|\mathbf{X} = \mathbf{x}] = \sum_{s=1}^{2} \mathbb{P}(S = s|\mathbf{X} = \mathbf{x}) \left(\boldsymbol{\mu}_Y^{(s)} + \boldsymbol{\Sigma}_{YX}^{(s)} [\boldsymbol{\Sigma}_{XX}^{(s)}]^{-1} (\mathbf{x} - \boldsymbol{\mu}_X^{(s)}) \right).$$

If $\boldsymbol{\Sigma}_{YX}^{(s)} = 0$ and $\boldsymbol{\mu}_Y^{(s)}$ are constant or equal, then $\mathbb{E}[\mathbf{Y}|\mathbf{X} = \mathbf{x}]$ may be constant. Moreover,

$$\mathbb{C}ov(Y|\mathbf{X} = \mathbf{x}) =$$

$$\sum_{s=1}^{2} \mathbb{P}(S = s|\mathbf{X} = \mathbf{x}) \left\{ \boldsymbol{\Sigma}_{YY}^{(s)} - \boldsymbol{\Sigma}_{YX}^{(s)} [\boldsymbol{\Sigma}_{XX}^{(s)}]^{-1} \boldsymbol{\Sigma}_{XY}^{(s)} \right\} - (\mathbb{E}[\mathbf{Y}|\mathbf{X} = \mathbf{x}]) (\mathbb{E}[\mathbf{Y}|\mathbf{X} = \mathbf{x}])'.$$

When $\boldsymbol{\Sigma}_{YX}^{(s)} = 0$, and $\mathbb{E}[Y|\mathbf{X} = \mathbf{x}] = 0$, we have

$$\mathbb{C}ov(\mathbf{Y}|\mathbf{X} = \mathbf{x}) = \sum_{s=1}^{2} \mathbb{P}(S = s|\mathbf{X} = \mathbf{x}) \, \boldsymbol{\Sigma}_{YY}^{(s)}.$$

Note that when $\mathbb{E}[XY] = 0$, and $\mathbb{E}[X] = \mathbb{E}[Y] = 0$ so that $\mathbb{C}ov(X, Y) = 0$, yet $\mathbb{C}ov(Y|X)$ varies with X. This is exactly the variance–driven dependence scenario.

Example 11.8.2 *Variance-Driven Dependence. Consider a two-component bivariate normal mixture $\mathbb{P}(S = 1) = \alpha$, and $\mathbb{P}(S = 2) = 1 - \alpha$, with latent variable $S \in \{1, 2\}$. Conditional on $S = s$,*

$$(X, Y) \mid S = s \sim \mathcal{N}_2\left(\begin{pmatrix} \mu_s \\ \mu_s \end{pmatrix}, \begin{pmatrix} \sigma_s^2 & \rho\sigma_s^2 \\ \rho\sigma_s^2 & \sigma_s^2 \end{pmatrix} \right),$$

with scalar variables X, Y, correlation $\rho = 0.5$, variances $\sigma_1^2 = \sigma_2^2 = 1$, and component separation $\mu_1 = -3$, $\mu_2 = 3$. We are interested in the conditional variance of Y given $X = x$.

$$\begin{aligned} \mathbb{V}ar(Y|X = x) = &\sum_{s=1}^{2} \mathbb{P}(S = s|X = x) \mathbb{V}ar(Y|X, S = s) + \\ &\sum_{s=1}^{2} \mathbb{P}(S = s|X = x) (\mu_{Y|X}^{(s)} - \mathbb{E}[Y|X = x])^2. \end{aligned} \tag{11.32}$$

The conditional variance within each component is $\mathbb{V}ar(Y|X, S = s) = \sigma_s^2(1 - \rho^2) = 1 \cdot (1 - 0.5^2) = 0.75$. *The conditional mean within each component is* $\mu_{Y|X}^{(s)} = \rho X + (1 - \rho)\mu_s$. *So,*

$$\mu_{Y|X}^{(1)} = 0.5X - 1.5, \quad \mu_{Y|X}^{(2)} = 0.5X + 1.5.$$

For the mixture conditional mean, assume the posterior probability varies with X,

$$p_1(x) = \mathbb{P}(S = 1|X = x) = 0.5e^{-x^2/18}, \quad p_2(x) = 1 - p_1(x).$$

Then the mixture conditional mean is

$$\mathbb{E}[Y|X = x] = p_1(x)\mu_{Y|X}^{(1)} + (1 - p_1(x))\mu_{Y|X}^{(2)} = 0.5X + 1.5 - 3p_1(x).$$

The conditional variance of Y *is obtained from*

$$(\mu_{Y|X}^{(1)} - \mathbb{E}[Y|X])^2 = 9 - 18p_1 + 9p_1^2, \quad (\mu_{Y|X}^{(2)} - \mathbb{E}[Y|X])^2 = 9p_1^2$$

Plug in to get the simplified variance to get

$$\begin{aligned}\mathbb{V}ar(Y|X) &= 0.75 + p_1(9 - 18p_1 + 9p_1^2) + (1 - p_1)9p_1^2 \\ &= 0.75 + 9p_1 - 9p_1^2 = 0.75 + 9\,p_1(x)\big(1 - p_1(x)\big)\end{aligned}$$

The plot of the conditional variance appears in Figure 11.2.

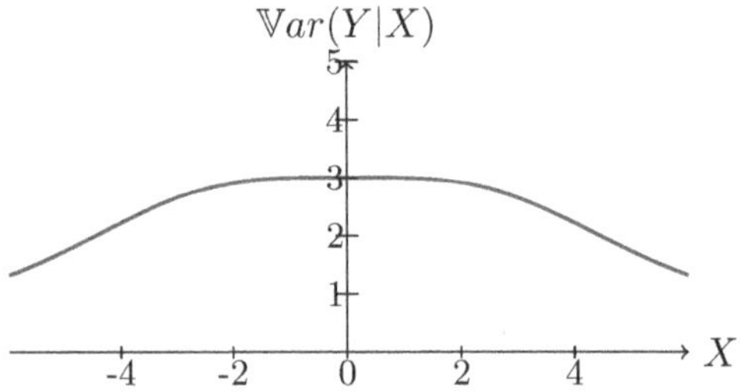

FIGURE 11.2: Conditional variance of Y given X with ($\mu_1 = -3$, $\mu_2 = 3$). The variance curve shows strong variance-driven dependence.

Mean-driven Dependence

In the mixture model, even if $X \perp Y|S$, marginally X and Y may depend via S. Hence under the true distribution, H_0 : X and Y are independent (marginally) is false in this mixture. We apply CATS to the observed sample (x_i, y_i). We compute the observed CATS_δ and obtain the null distribution by random permutations. Because the mixture induces dependence in the distribution of (X, Y), one expects $V_\delta^2(\mathbf{X}, \mathbf{Y}) > 0$ and thus CATS to be significantly positive, enabling detection of dependence even when $\mathbb{C}ov(X, Y) = 0$.

Example 11.8.3 *Consider the mixture of multivariate normal distributions*

$$(\mathbf{X}, \mathbf{Y}) \sim \sum_{s=1}^{k} \pi_s\, \mathbb{N}_{d_1+d_2}\left(\begin{pmatrix}\boldsymbol{\mu}_X^{(s)} \\ \boldsymbol{\mu}_Y^{(s)}\end{pmatrix}, \boldsymbol{\Sigma}^{(s)} = \begin{pmatrix}\boldsymbol{\Sigma}_{XX}^{(s)} & \boldsymbol{\Sigma}_{XY}^{(s)} \\ \boldsymbol{\Sigma}_{YX}^{(s)} & \boldsymbol{\Sigma}_{YY}^{(s)}\end{pmatrix}\right), \quad \sum_{s=1}^{k} \pi_s = 1, \quad \pi_s > 0.$$

The mixture marginal means are $\boldsymbol{\mu}_X = \sum_{s=1}^{k} \pi_s \boldsymbol{\mu}_X^{(s)}$, *and* $\boldsymbol{\mu}_Y = \sum_{s=1}^{k} \pi_s \boldsymbol{\mu}_Y^{(s)}$. *A standard calculation yields the decomposition*

$$\mathbb{C}ov(\mathbf{X}, \mathbf{Y}) = \sum_{s=1}^{k} \pi_s \boldsymbol{\Sigma}_{XY}^{(s)} + \sum_{s=1}^{k} \pi_s \big(\boldsymbol{\mu}_X^{(s)} - \boldsymbol{\mu}_X\big)\big(\boldsymbol{\mu}_Y^{(s)} - \boldsymbol{\mu}_Y\big)'. \tag{11.33}$$

Thus there are two distinct sources of linear dependence,

1. *Within-component covariance, coming from* $\boldsymbol{\Sigma}_{XY}^{(s)}$.

2. *Between-component mean-shift covariance, coming from* $(\boldsymbol{\mu}_X^{(s)} - \boldsymbol{\mu}_X)(\boldsymbol{\mu}_Y^{(s)} - \boldsymbol{\mu}_Y)'$.

To see the effects of mean-driven dependence, we observe that if each component has zero cross-covariance, then $\boldsymbol{\Sigma}_{XY}^{(s)} = 0$, *but the means differ across components, then only the second term in* (11.33) *contributes, and*

$$\mathbb{C}ov(\mathbf{X}, \mathbf{Y}) = \sum_{s=1}^{k} \pi_s \big(\boldsymbol{\mu}_X^{(s)} - \boldsymbol{\mu}_X\big)\big(\boldsymbol{\mu}_Y^{(s)} - \boldsymbol{\mu}_Y\big)'.$$

This is classical mean-induced dependence in mixture models.

Consider the effects of variance-driven mixtures where $\boldsymbol{\mu}_X^{(s)} = 0,\ \boldsymbol{\mu}_Y^{(s)} = 0,$ *and* $\boldsymbol{\Sigma}_{XY}^{(s)} = 0$ *for all* s. *Since the mixture means are also* 0, *both terms in* (11.33) *vanish, and* $\mathbb{C}ov(X, Y) = 0$. *Despite this, the mixture can exhibit non-linear dependence. This arises because the conditional distribution* $\mathbf{Y}|\mathbf{X} = \mathbf{x}$ *depends on* $\mathbf{x}$. *If components have different variances, then extreme values of* $\mathbf{X}$ *make it more likely that one is in a high-variance component, which changes the distribution of* $\mathbf{Y}$. *Thus, even when* $\boldsymbol{\mu}_X^{(s)} = \boldsymbol{\mu}_Y^{(s)} = 0,$ *and* $\boldsymbol{\Sigma}_{XY}^{(s)} = 0$ *the model still has dependence because the quantities* $\mathbb{V}ar(\mathbf{Y}|\mathbf{X} = \mathbf{x})$ *or more general conditional features vary with* $\mathbf{x}$. *This phenomenon is called the variance-driven dependence.*

Standard covariance-based measures fail here, since $\mathbb{C}ov(\mathbf{X}, \mathbf{Y}) = 0$. *However, statistics sensitive to conditional distributional changes detect such dependence. An example is the* CATS_δ *statistic, which is designed to capture higher-order or tail-adaptive dependence not visible through second moments.*

Example 11.8.4 *Mean-Driven Dependence in a Two-Component Mixture. We consider a two-component bivariate normal mixture,*

$$(X, Y) \sim \pi_1 \mathcal{N}_2\big(\boldsymbol{\mu}^{(1)}, \boldsymbol{\Sigma}^{(1)}\big) + \pi_2 \mathcal{N}_2\big(\boldsymbol{\mu}^{(2)}, \boldsymbol{\Sigma}^{(2)}\big),$$

with

$$\pi_1 = \pi_2 = 0.5, \quad \boldsymbol{\mu}^{(1)} = \begin{pmatrix} -3 \\ -3 \end{pmatrix}, \quad \boldsymbol{\mu}^{(2)} = \begin{pmatrix} 3 \\ 3 \end{pmatrix}, \quad \boldsymbol{\Sigma}^{(1)} = \boldsymbol{\Sigma}^{(2)} = \begin{pmatrix} 1 & 0 \\ 0 & 1 \end{pmatrix}.$$

Each component has zero cross-covariance ($\Sigma_{XY}^{(s)} = 0$*), Variances are equal (*$\sigma^2 = 1$*), Means differ (*$\mu_1 = -3, \mu_2 = 3$*) to show mean-driven dependence. Even though within each component* $X \perp Y \mid S$, *marginally* X *and* Y *depend through* S. *Mixture marginal means are*

$$\mu_X = \sum_{s=1}^{2} \pi_s \mu_X^{(s)} = 0, \quad \mu_Y = \sum_{s=1}^{2} \pi_s \mu_Y^{(s)} = 0.$$

The marginal covariance is

$$\mathbb{C}ov(X, Y) = \sum_{s=1}^{2} \pi_s \Sigma_{XY}^{(s)} + \sum_{s=1}^{2} \pi_s (\mu_X^{(s)} - \mu_X)(\mu_Y^{(s)} - \mu_Y)$$

Within-component covariances vanish, $\Sigma_{XY}^{(s)} = 0$, *and between-component mean-shift contributes,*

$$\mathbb{C}ov(X, Y) = 0.5(-3 - 0)(-3 - 0) + 0.5(3 - 0)(3 - 0) = 4.5 + 4.5 = 9$$

Large positive covariance arises purely from mean-driven dependence. The conditional mean of Y given $X = x$ in a mixture is

$$\mathbb{E}[Y|X = x] = p_1(x)\mu^{(1)}_{Y|X} + p_2(x)\mu^{(2)}_{Y|X}, \quad p_s(x) = \mathbb{P}(S = s|X = x)$$

Here, each component has unit variance, $\sigma^2 = 1$, so

$$p_1(x) = \frac{\pi_1 \exp\left(-\frac{(x-\mu_X^{(1)})^2}{2}\right)}{\pi_1 \exp\left(-\frac{(x-\mu_X^{(1)})^2}{2}\right) + \pi_2 \exp\left(-\frac{(x-\mu_X^{(2)})^2}{2}\right)} = \frac{\exp(-\frac{(x+3)^2}{2})}{\exp(-\frac{(x+3)^2}{2}) + \exp(-\frac{(x-3)^2}{2})}$$

$$p_2(x) = 1 - p_1(x)$$

Component conditional means are

$$\mu^{(1)}_{Y|X} = \mu^{(1)}_Y = -3, \quad \mu^{(2)}_{Y|X} = \mu^{(2)}_Y = 3.$$

Thus the mixture conditional mean is

$$\mathbb{E}[Y|X = x] = p_1(x) \cdot (-3) + (1 - p_1(x)) \cdot 3 = 3 - 6p_1(x)$$

The plot of the conditional mean appears in Figure 11.3. It is a smooth sigmoidal function, increasing from -3 to 3 as X increases. When $X \ll 0$, the first component dominates, $p_1(x) \approx 1$, so $\mathbb{E}[Y|X] \approx \mu^{(1)}_Y = -3$. When $X \gg 0$, the second component dominates, $p_2(x) = 1 - p_1(x) \approx 1$, so $\mathbb{E}[Y|X] \approx \mu^{(2)}_Y = 3$. For $X \approx 0$, both components contribute roughly equally, giving $\mathbb{E}[Y|X] \approx 0$.

This demonstrates mean-driven dependence: X and Y are dependent marginally, even though $X \perp Y \mid S$. The sigmoid shape arises entirely from the mixture weights $p_1(x)$ and $p_2(x)$, since the component conditional means are constant and independent of X.

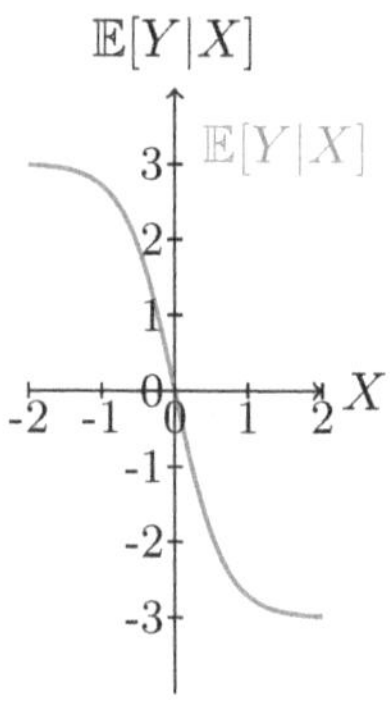

FIGURE 11.3: Conditional mean of Y given X using a safe logistic approximation for $p_1(x)$. The sigmoid curve shows dramatic mean-driven dependence: small $X \to Y \approx -3$, large $X \to Y \approx 3$.

11.9 Comparison

We compared the performance of the following test statistics for testing independence using sampling experiments with $m = n = 15$. The competing test statistics are $\text{CATS}_\delta(X, Y)$ in (11.24), $\text{CATS}_\rho(\mathbf{X}, \mathbf{Y})$ in (11.31), Dcor Method in (11.18), *HHG* in (11.5), *Pcov* in (11.6), and COR_{IPD} in (11.6).

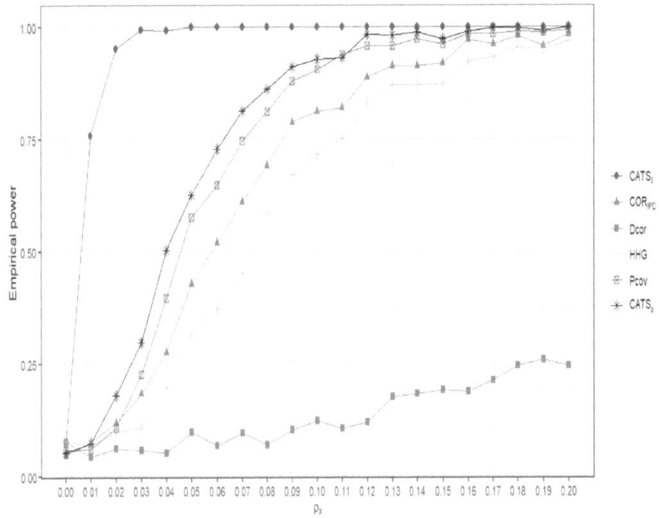

FIGURE 11.4: Empirical power for Beta marginals: $Y_j \sim \text{Beta}(5,3)$ with copula $C^{\text{G}}_{\mathbf{R}}$.

11.9.1 Experiment 1

We carried out sampling experiments with 500 replications, $d_1 = d_2 = 512$ and $m = n = 15$ to estimate the α level and the empirical power of the test statistics against various alternatives. Let $\mathbf{X} \sim \mathbb{N}_d(\boldsymbol{\mu}, \boldsymbol{\Sigma}_x)$ and $\mathbf{R} = \text{corr}(\mathbf{X})$. Define $U_j = \Phi\big((X_j - \mu_j)/\sigma_j\big) \sim \text{Unif}(0,1)$. For strictly increasing marginals F_j, set $Y_j = F_j^{-1}(U_j)$. Then the joint CDF of $\mathbf{Y}$ is $C^{\text{G}}_{\mathbf{R}}\big(F_1(y_1), \ldots, F_d(y_d)\big)$, i.e., Gaussian copula with correlation $\mathbf{R}$ and marginals F_j. Let $\boldsymbol{\mu} = 0$ and $\boldsymbol{\Sigma}_x(\rho_x)$ be a $d \times d$ covariance matrix with linearly increasing diagonal variances and a common off—diagonal covariance ρ_x. Specifically, the ith diagonal entry equals $i + \rho_x$ for $i = 1, \ldots, d$, and every off-–diagonal entry equals ρ_x.

Under a non-zero covariance alternative, to obtain the power of the test each sample is varied in $\rho_x \in \{0, 0.02, \ldots, 0.50\}$. We use the approximate permutation distribution of the test statistics and set $\alpha = 0.05$. Figures 11.4-11.8 show the power of the statistics based on samples taken from the following distributions with copula $C^{\text{G}}_{\mathbf{R}}$.

- Weibull marginals: $Y_j \sim \text{Weibull}(3,1)$.
- Exponential marginals (rate): $Y_j \sim \text{Exp}(1)$.
- Beta marginals: $Y_j \sim \text{Beta}(5,3)$.
- Student's t marginals ($df = 10$): $Y_j \sim t_{10}$.
- Uniform marginals via Normal: $Y_j = \Phi\big((X_j - \mu_j)/\sigma_j\big) \sim \text{Unif}(0,1)$.

For each of the above marginal specifications we also examined the size performance of the six competing procedures. Under the independence model ($\rho_x = 0$) we generated $m = n = 15$ observations with $d_1 = d_2 = 512$ in each replication, and used 500 Monte Carlo runs with nominal level $\alpha = 0.05$. The empirical type I error rates, computed as the proportion of rejections of $H_0 : \mathbf{X} \perp \mathbf{Y}$ over the replications, are summarized in Table 11.5.

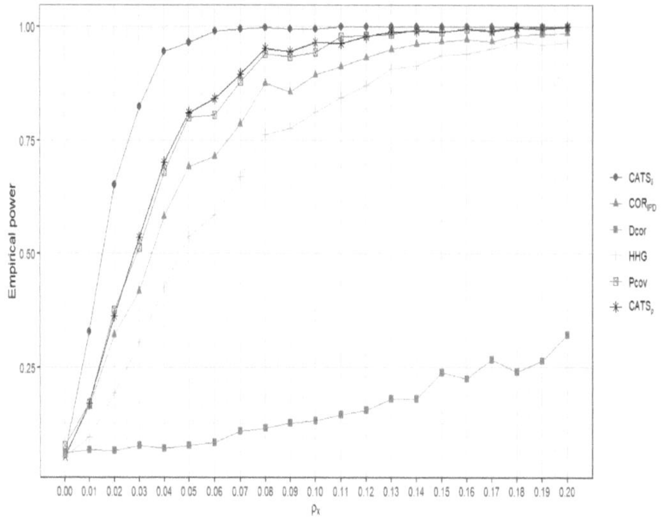

FIGURE 11.5: Empirical power for Exponential marginals with copula $C^{\mathrm{G}}_{\mathbf{R}}$.

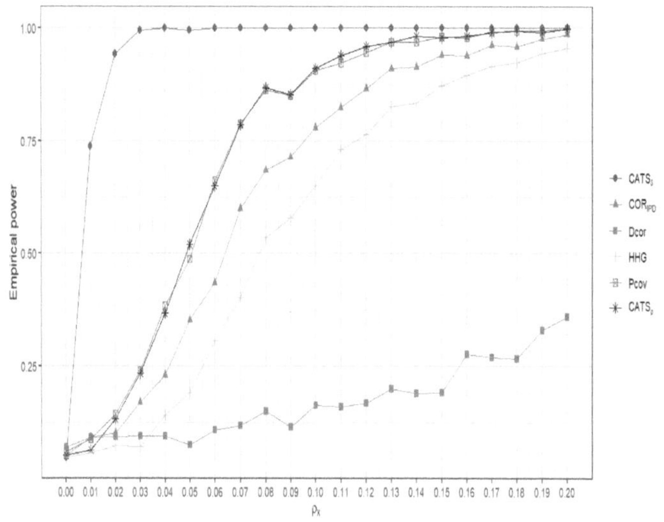

FIGURE 11.6: Empirical power for Weibull marginals $Y_j \sim$ Weibull$(3, 1)$ with copula $C^{\mathrm{G}}_{\mathbf{R}}$.

TABLE 11.5: Empirical type I error rates (nominal level $\alpha = 0.05$) with stated marginals.

Distribution	CATS_δ	CATS_ρ	Dcor Method	HHG	$Pcov$	COR_{IPD}
Exponential	0.052	0.063	0.067	0.057	0.042	0.053
Beta	0.057	0.045	0.065	0.062	0.068	0.047
Student's t_{10}	0.044	0.062	0.033	0.046	0.054	0.032
Weibull(3,1)	0.049	0.039	0.039	0.047	0.042	0.051
Uniform	0.039	0.067	0.045	0.049	0.036	0.038
χ^2_2	0.056	0.053	0.059	0.051	0.039	0.054

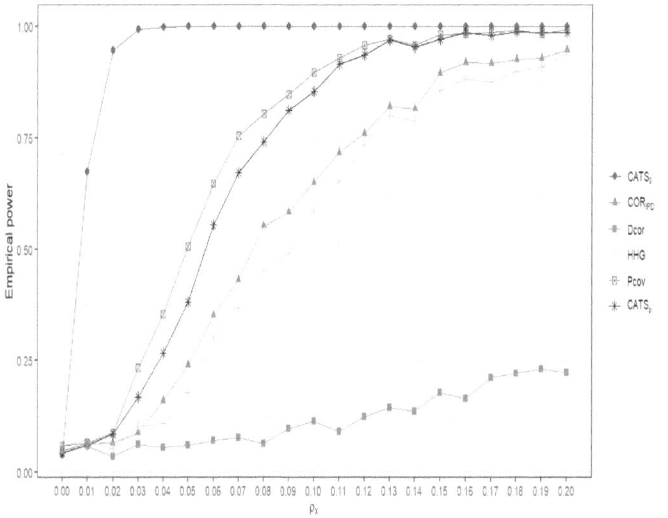

FIGURE 11.7: Empirical power for Student's t_{10} marginals with copula $C^{\mathrm{G}}_{\mathbf{R}}$.

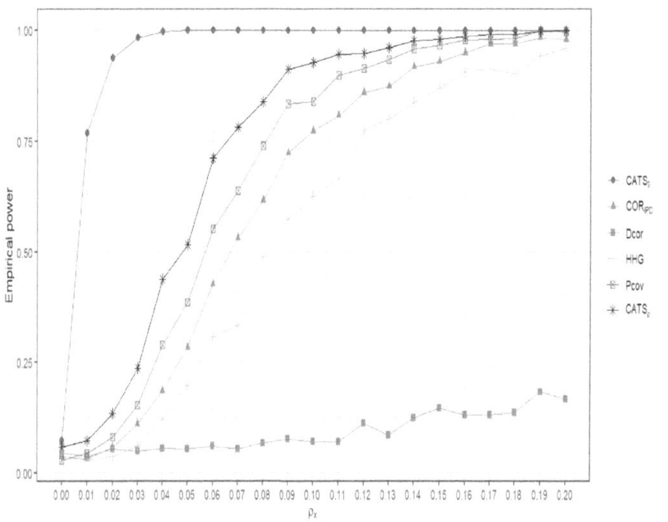

FIGURE 11.8: Empirical power for Uniform marginals Unif$(0, 1)$ with copula $C^{\mathrm{G}}_{\mathbf{R}}$.

11.9.2 Experiment 2: Mean-–driven Dependence

In the second experiment, we investigate alternatives where the dependence between X and Y is created purely through differences in component means. Throughout we keep the same simulation settings as in Experiment 1, with $d_1 = d_2 = d$ and $n = m = 15$, and we use 500 replications. The nominal level is $\alpha = 0.05$ and the permutation calibration of the test statistics is the same as in Experiment 1.

Let $Z = (\mathbf{Z}_X, \mathbf{Z}_Y)$ be a $(d_1 + d_2)$–dimensional random vector generated from a two—component Gaussian location mixture

$$\mathbf{Z} \sim \tfrac{1}{2}\,\mathbb{N}_{d_1+d_2}(\boldsymbol{\mu}^{(1)}, \mathbf{I}_{d_1+d_2}) + \tfrac{1}{2}\,\mathbb{N}_{d_1+d_2}(\boldsymbol{\mu}^{(2)}, \mathbf{I}_{d_1+d_2}),$$

where $\mathbf{Z}_X \in \mathbb{R}^{d_1}$ and $\mathbf{Z}_Y \in \mathbb{R}^{d_2}$ denote the first and the last d_1 and d_2 coordinates of $\mathbf{Z}$,

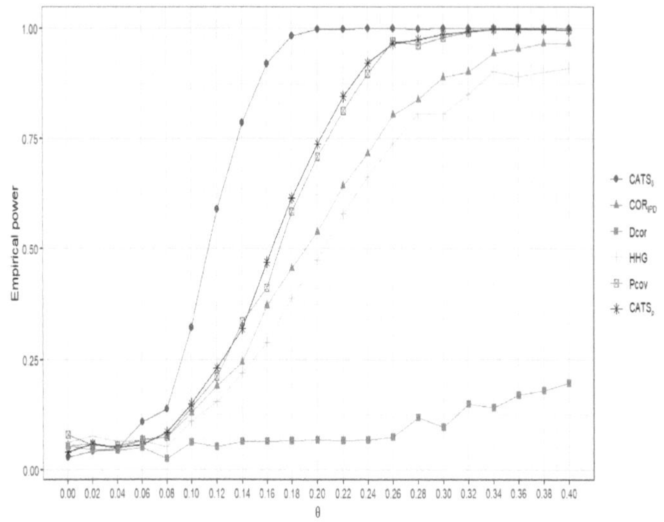

FIGURE 11.9: Empirical power of the competing tests under the mean-–driven mixture alternatives of Experiment 2 for Exponential margins.

respectively. We take

$$\boldsymbol{\mu}^{(1)} = \begin{pmatrix} \theta \mathbf{v}_X \\ \theta \mathbf{v}_Y \end{pmatrix}, \qquad \boldsymbol{\mu}^{(2)} = \begin{pmatrix} -\theta \mathbf{v}_X \\ -\theta \mathbf{v}_Y \end{pmatrix},$$

with fixed direction vectors $\mathbf{v}_X \in \mathbb{R}^{d_1}$ and $\mathbf{v}_Y \in \mathbb{R}^{d_2}$; in the simulations we use $\mathbf{v}_X = (1, \ldots, 1)'$ and $\mathbf{v}_Y = (1, \ldots, 1)'$. Conditional on the latent component label, the covariance is block-–diagonal and equal to $\mathbf{I}_{d_1+d_2}$, so that within each mixture component $\mathbf{Z}_X$ and $\mathbf{Z}_Y$ are independent and have no cross-–covariance. The dependence between $\mathbf{Z}_X$ and $\mathbf{Z}_Y$ therefore arises only through the shifts in the component means, and its strength is controlled by the scalar parameter $\theta \geq 0$. Under $\theta = 0$ the two components coincide, and $\mathbf{Z}_X$ and $\mathbf{Z}_Y$ are independent, which yields a benchmark null model for assessing the empirical size.

To obtain non-–Gaussian marginals while preserving the underlying copula, we apply the same monotone marginal transformations as in Experiment 1. Specifically, write $Z_j = (Z_j - \mu_j)/\sigma_j$ for the jth coordinate and set $U_j = \Phi(Z_j) \sim \text{Unif}(0, 1)$. For strictly increasing marginal distribution functions F_j, define $Y_j = F_j^{-1}(U_j)$. This yields a random vector (X, Y) with Gaussian copula induced by the mixture distribution of Z and with prescribed marginals. We consider the same families of marginals as in Experiment 1, including chi–square, gamma, exponential, beta, Student t, uniform via the probability integral transform, and log–normal marginals.

For each value of θ in a grid (that is, $\theta \in \{0, 0.02, \ldots, 0.38, 0.40\}$), we generate 500 independent samples from the above model and record the rejection frequencies of the competing tests. This experiment is designed to highlight the ability of the procedures to detect classical mean-–driven dependence in high—dimensional mixtures.

Figures 11.9 and 11.10 show that all procedures gain power quickly as the mean—driven signal increases. The CATS$_\delta$ statistic is most sensitive to weak dependence. Dcor method increases more slowly, and the distance correlation test is clearly less powerful for moderate values of θ. The ranking of the methods is essentially the same for exponential and Beta$(5, 3)$ marginals.

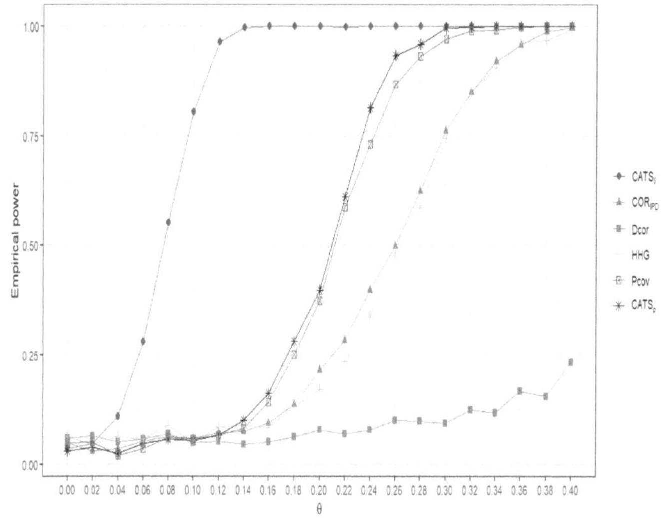

FIGURE 11.10: Empirical power of the competing tests under the mean-–driven mixture alternatives of Experiment 2 for Beta$(5, 3)$ margins.

11.9.3 Experiment 3: Variance-driven Dependence

The third experiment focuses on alternatives where the covariance between X and Y is identically zero, but the joint distribution still exhibits non-linear dependence through heterogeneous component variances. Again we use $d_1 = d_2 = d$ and $n = m = 15$ with 500 replications and nominal level $\alpha = 0.05$.

Let $\mathbf{Z} = (\mathbf{Z}_X, \mathbf{Z}_Y)$ be generated from a two-–component Gaussian scale mixture

$$\mathbf{Z} \sim \tfrac{1}{2}\,\mathbb{N}_{d_1+d_2}(\mathbf{0}, \boldsymbol{\Sigma}^{(1)}) + \tfrac{1}{2}\,\mathbb{N}_{d_1+d_2}(\mathbf{0}, \boldsymbol{\Sigma}^{(2)}),$$

where

$$\boldsymbol{\Sigma}^{(1)} = \mathbf{I}_{d_1+d_2}, \qquad \boldsymbol{\Sigma}^{(2)} = c^2\mathbf{I}_{d_1+d_2},$$

with $c > 1$. Thus both components have zero mean and block-–diagonal covariance matrices with no cross—covariance between $\mathbf{Z}_X$ and $\mathbf{Z}_Y$. Consequently, $\mathbb{C}ov(\mathbf{Z}_X, \mathbf{Z}_Y) = 0$ for all c, but the model can still exhibit dependence because the conditional distribution of $\mathbf{Z}_Y$ given $\mathbf{Z}_X = \mathbf{x}$ depends on the magnitude of $\mathbf{x}$: extreme values of $\mathbf{Z}_X$ are more likely to come from the high—variance component, which in turn changes the variability of $\mathbf{Z}_Y$. This is an example of variance-–driven dependence, where $\mathbb{V}ar(\mathbf{Z}_Y|\mathbf{Z}_X = \mathbf{x})$ varies with x even though the overall covariance vanishes. The parameter c controls the strength of this effect; in the simulations we take $c = 1+\lambda$ with λ ranging over a grid such as $\lambda \in \{0, 0.02, \dots, 0.38, 0.40\}$. When $\lambda = 0$ the two components coincide, and $\mathbf{Z}_X$ and $\mathbf{Z}_Y$ are independent.

As in the previous experiments, we obtain non—Gaussian marginals by applying monotone transformations to the Gaussian mixture. For each coordinate we set $U_j = \Phi(Z_j)$ and then define $Y_j = F_j^{-1}(U_j)$ for the same collection of marginal distribution functions F_j used in Experiment 1. This construction preserves the variance—driven dependence induced by the mixture while allowing us to examine the robustness of the tests under different marginal distributions.

For each value of λ we generate 500 samples from the above model and compute the empirical rejection probabilities of all test statistics. Since the global covariance between X and Y is zero in this experiment, tests that rely only on linear correlation are expected

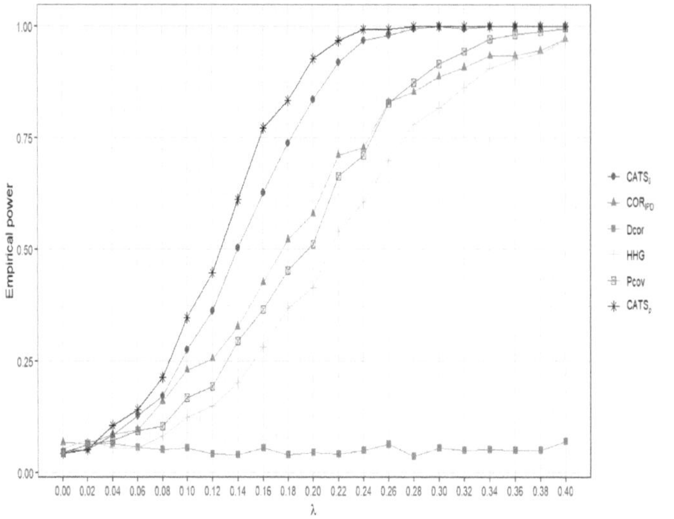

FIGURE 11.11: Empirical power of the competing tests under the variance-–driven mixture alternatives of Experiment 3 for exponential margins.

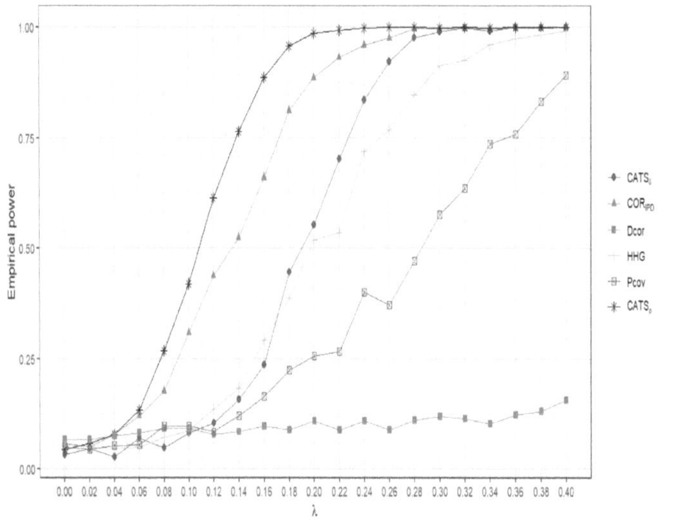

FIGURE 11.12: Empirical power of the competing tests under the variance—driven mixture alternatives of Experiment 3 for Beta(5, 3) margins.

to have little or no power, whereas procedures based on distance, angular, or IPD-–type measures should remain sensitive to this non-linear form of dependence. The horizontal axis is the dependence parameter λ, and the vertical axis is the empirical power.

Figure 11.11 and 11.12 show a clear separation between the proposed procedures and the distance correlation test under variance-–driven dependence. COR_{IPD}, CATS_{δ}, CATS_{ρ}, *Pcov* and HHG all attain high power for moderate values of λ, whereas the Rizzo—Székely distance correlation test remains close to the nominal level even under strong alternatives. The qualitative pattern is similar for exponential and Beta(5, 3) marginals, with only minor differences in the speed at which *Pcov* approaches unit power.

For n observations, the time complexity of IPDs is $O(n^2)$. The time complexities for the methods are listed in Table 11.6. The time complexity higher than $O(n^2)$ is restrictive for large datasets.

TABLE 11.6: Time Complexity.

COR_{IPD}	*Dcor*	*HHG*	*Pcov*	CATS_δ
$O(n^2)$	$O(n^2)$	$O(n^2 \log(n))$	$O(n^3)$	$O(n^2)$

11.9.4 Experiment 3: Special Dependence Revisited

We consider the special dependence of Newton [249] with seven patterns of dependence and runs simulations to compare the test statistics. The dimension of the observations are $d_1 = d_2 = 100$, and the sample size varies in $n \in \{10, 15, 20\}$. Each simulation is conducted under 1000 replications, and the results are shown in Table 11.7.

TABLE 11.7: The power for testing independence under special dependence structure.

Pattern	n	COR_{IPD}	CATS_δ	CATS_ρ	Dcor	*HHG*	*Pcov*
1	10	0.780	0.853	0.833	0.999	0.642	0.992
1	15	0.684	0.487	0.756	0.918	0.478	0.908
1	20	0.271	0.315	0.278	0.464	0.104	0.498
2	10	0.676	0.901	0.970	0.736	0.931	0.658
2	15	0.641	0.490	0.460	0.222	0.287	0.186
2	20	0.435	0.095	0.130	0.301	0.292	0.328
3	10	0.352	0.035	0.069	0.121	0.054	0.111
3	15	0.607	0.331	0.262	0.134	0.166	0.106
3	20	0.608	0.051	0.107	0.179	0.362	0.169
4	10	0.608	0.011	0.028	0.202	0.016	0.208
4	15	0.008	0.368	0.302	0.083	0.122	0.093
4	20	0.108	0.009	0.102	0.332	0.197	0.436
5	10	0.036	0.367	0.183	0.139	0.162	0.135
5	15	0.057	0.208	0.155	0.834	0.082	0.962
5	20	0.009	0.141	0.098	0.505	0.012	0.614
6	10	0.993	0.830	0.874	0.953	0.575	0.937
6	15	1.000	0.882	0.851	0.828	0.444	0.779
6	20	1.000	0.780	0.783	0.550	0.405	0.488
7	10	0.933	0.855	0.829	0.826	0.763	0.792
7	15	0.469	0.787	0.798	0.700	0.528	0.713
7	20	0.636	0.901	0.908	0.970	0.994	0.971

11.10 Benchmark Datasets

We studied the following gene expression benchmark datasets (Ramey [276]) for testing independence. For each dataset we used the first 200 features.

The Prostate Tumor gene expression dataset (Chung and Keles [69]) consists of 102 samples. There are 52 tumor and 50 normal. Each sample has 6033 gene expression measurements. Classes are encoded as 0 (Normal) and 1 (Tumor). The data were normalized, log-transformed, and standardized as described in Dettling [74] and Dettling and Bühlmann [75]. The Khan dataset (Khan et al. [167]) is a benchmark microarray dataset for pediatric small round blue cell tumors (SRBCT). It contains 53 samples with 2308 gene expression features, divided into four tumor types. We used the EWS (Ewing's Sarcoma) with 23 samples and RMS (Rhabdomyosarcoma) with 10 samples. The Chin breast cancer dataset (Chin et al. [66]) consists of 118 samples with 22215 gene expression features. Samples are labeled as Normal with 43 samples and Tumor with 75 samples. The Golub leukemia dataset (Golub et al. [121]) contains 72 samples with 7129 gene expression features, classified into ALL (Acute Lymphoblastic Leukemia) with 47 samples and AML (Acute Myeloid Leukemia) with 25 samples.

The following Table displays the p-values of the above test statistics for testing independence. Each p-value is based on 500 random permutations of the samples.

TABLE 11.8: The p-value for testing independence for benchmark datasets.

Dataset	COR_{IPD}	CATS_δ	CATS_ρ	Dcor	HHG	$Pcov$
Prostate	0.071	0.278	0.064	0.011	0.016	0.012
Khan	0.962	0.218	0.312	0.452	0.396	0.236
Chin	0.582	0.481	0.868	0.962	0.790	0.974
Golub	0.602	0.838	0.744	0.722	0.762	0.726

11.11 Summary

This chapter addresses the challenges of testing independence in high-dimensional settings, where the number of variables d may far exceed the sample size n, rendering classical methods like likelihood ratio tests and Pearson correlation ineffective. To overcome these limitations, the chapter develops a suite of tests based on IPDs, which transform multivariate observations into univariate distances while retaining dependence information. Key IPD-based methods include the empirical distribution function (EDF) test, cut-point (CP) method, odds ratio, quadrant, correlation-based, and transformation statistics. Each method utilizes the center-outward ranking and distribution of IPDs to detect dependencies between high-dimensional vectors, providing distribution-free, computationally feasible alternatives to traditional approaches.

Beyond IPD methods, the chapter explores distance correlation, projection correlation, and the Heller-Heller-Gorfine (HHG) test, which are capable of detecting non-linear or complex dependence structures in multivariate data. Comparative simulation studies evaluate the power and robustness of these tests across various multivariate distributions and specialized dependence scenarios, highlighting the advantages of IPD-based approaches in small-sample, high-dimensional contexts. Results demonstrate that the proposed tests are highly effective at identifying both linear and non-linear dependencies while maintaining computational efficiency, making them particularly suitable for genomics, neuroimaging, finance, and machine learning applications.

Practical considerations for implementing high-dimensional independence tests are discussed, including permutation-based significance assessment and visualization of dependence patterns. Collectively, the chapter provides a unified framework for robustly testing independence in modern high-dimensional data, combining theoretical rigor with practical applicability.

We also presented a new framework for testing independence using statistics derived from interpoint distance distributions. The statistics CATS_δ and CATS_ρ were formulated as V-statistics, enabling precise analytical treatment using Hoeffding decompositions and classical asymptotic tools. We established that both statistics are asymptotically normal under independence, with limiting variances depending only on low-order components of the distance kernel.

The geometric structure of the CATS statistics allows them to detect a wide range of dependence types, including non-linear and heteroscedastic forms, as well as dependence induced by latent mixture components or hidden clustering. This stands in contrast to classical correlation-based methods, which often miss such alternatives due to their reliance on linear models. Empirical results demonstrate that the proposed tests maintain accurate type-I error control and exhibit strong power across numerous dependence scenarios, outperforming several established methods when the data exhibit complex, non-Euclidean dependence. Importantly, the methods remain reliable when the dimension of the data is large relative to the sample size, making them particularly useful for contemporary high-dimensional low-sample size applications.

Future work includes extending the CATS framework to functional data, manifold-valued observations, and other non-Euclidean settings, as well as developing fast approximations for extremely large datasets. The geometric and asymptotic principles developed here offer a foundation for a broader class of independence tests that remain both theoretically rigorous and practically effective.

11.12 Proofs

Proof: Theorem 11.3.1:

The necessary part is trivial, since functions of independent random vectors are independent. To prove the sufficient part we need to show that if IPDs $d^2_{(X)} = \|\mathbf{X}_1 - \mathbf{X}_2\|^2$ and $d^2_{(Y)} = \|\mathbf{Y}_1 - \mathbf{Y}_2\|^2$ are independent, then $\mathbf{X}$ and $\mathbf{Y}$ are independent. Since the contrapositive of any true proposition is also true, we will show if $\mathbf{X}$ and $\mathbf{Y}$ are dependent, then $d^2_{(X)}$ and $d^2_{(Y)}$ are dependent.

Suppose $\mathbf{Z} = [\mathbf{X}', \mathbf{Y}']'$ is a random vector from multivariate normal distribution with mean $(\boldsymbol{\mu}_x, \boldsymbol{\mu}_y)$ and covariance matrix $\boldsymbol{\Sigma} = \begin{bmatrix} \boldsymbol{\Sigma}_{xx} & \boldsymbol{\Sigma}_{xy} \\ \boldsymbol{\Sigma}_{yx} & \boldsymbol{\Sigma}_{yy} \end{bmatrix}$. Let $\sigma_{(x)s_1s_2}, \sigma_{(y)t_1t_2}, \sigma_{(xy)st}$ be the components of $\boldsymbol{\Sigma}_{xx}, \boldsymbol{\Sigma}_{yy}, \boldsymbol{\Sigma}_{xy}$, respectively, for $s, s_1, s_2 = 1, \ldots, d_1; t, t_1, t_2 = 1, \ldots, d_2$. Suppose $\mathbf{X}$ and $\mathbf{Y}$ are dependent, we have

$$tr(\Sigma^2_{xy}) = \sum_{s=1}^{d_1} \sum_{t=1}^{d_2} \sigma^2_{(xy)st} > 0. \tag{11.34}$$

Let X_{it} be the tth variable of the random vector $\mathbf{X}_i$, and Y_{is} be the sth variable of the random vector $\mathbf{Y}_i$ for $i = 1, 2, t = 1, \ldots, d_1, s = 1, \ldots, d_2$. One can show that

$$\begin{aligned} \mathbb{C}\mathrm{ov}\left(\mathrm{d}^2_{(\mathrm{X})}, \mathrm{d}^2_{(\mathrm{Y})}\right) &= \mathbb{C}\mathrm{ov}\left(\sum_{\mathrm{t}=1}^{\mathrm{d}_1} (\mathrm{X}_{1\mathrm{t}} - \mathrm{X}_{2\mathrm{t}})^2, \sum_{\mathrm{s}=1}^{\mathrm{d}_2} (\mathrm{Y}_{1\mathrm{s}} - \mathrm{Y}_{2\mathrm{s}})^2\right) \\ &= \sum_{t=1}^{d_1} \sum_{s=1}^{d_2} 4\sigma_{(x)tt}\sigma_{(y)ss} \mathbb{C}\mathrm{ov}\left(\mathrm{W}^2_{1\mathrm{t}}, \mathrm{W}^2_{2\mathrm{s}}\right), \end{aligned}$$

where $W_{1t} = \frac{X_{1t} - X_{2t}}{\sqrt{2\sigma_{(x)tt}}} \sim \mathbb{N}(0, 1)$ and $W_{2s} = \frac{Y_{1s} - Y_{2s}}{\sqrt{2\sigma_{(y)ss}}} \sim \mathbb{N}(0, 1)$.

The covariance between W_{1t} and W_{2s} is

$$\mathbb{C}\mathrm{ov}\left(\mathrm{W}_{1\mathrm{t}}, \mathrm{W}_{2\mathrm{s}}\right) = \frac{\mathbb{C}\mathrm{ov}\left(\mathrm{X}_{1\mathrm{t}} - \mathrm{X}_{2\mathrm{t}}, \mathrm{Y}_{1\mathrm{s}} - \mathrm{Y}_{2\mathrm{s}}\right)}{4\sigma_{(x)tt}\sigma_{(y)ss}}$$

$$= \frac{\mathbb{C}\mathrm{ov}\left(\mathrm{X_{1t}}, \mathrm{Y_{1s}}\right) + \mathbb{C}\mathrm{ov}\left(\mathrm{X_{2t}}, \mathrm{Y_{2s}}\right)}{4\sigma_{(x)tt}\sigma_{(y)ss}} = \frac{1}{2}\rho_{(xy)st},$$

where $\rho_{(xy)st}$ is the correlation between the tth component of $\mathbf{X}$ and sth component of $\mathbf{Y}$ vectors. Thus, the W_{1t} and W_{2s} are jointly normal random variables with $\frac{1}{2}\rho_{(xy)st}$ correlation. Using Lemma 2.9.1, the conditional expectation and variance of W_{1t} given W_{2s} are $\mathbb{E}W_{1t}|W_{2s}) = \rho_{(xy)st}W_{2s}$, and $\mathrm{Var}(W_{1t}|W_{2s}) = 1 - \rho^2_{(xy)st}$. Thus, the covariance between W_{1t}^2 and W_{2s}^2 is $\mathbb{C}\mathrm{ov}(\mathrm{W}_{1t}^2, \mathrm{W}_{2s}^2) = \rho^2_{(\mathrm{xy})\mathrm{st}}(\mathbb{E}\mathrm{W}_{2s}^4) - 1) = 2\rho^2_{(\mathrm{xy})\mathrm{st}}$. Hence, we obtain

$$\begin{aligned}
\mathbb{C}\mathrm{ov}\left(\mathrm{d}^2_{(\mathrm{X})}, \mathrm{d}^2_{(\mathrm{Y})}\right) &= \sum_{t=1}^{d_1}\sum_{s=1}^{d_2} 4\sigma_{(x)tt}\sigma_{(y)ss}\mathbb{C}\mathrm{ov}\left(\mathrm{W}_{1t}^2, \mathrm{W}_{2s}^2\right), \\
&= \sum_{t=1}^{d_1}\sum_{s=1}^{d_2} 8\sigma_{(x)tt}\sigma_{(y)ss}\rho^2_{(xy)st} \qquad (11.35) \\
&= 8\sum_{t=1}^{d_1}\sum_{s=1}^{d_2} \mathbb{C}\mathrm{ov}^2(\mathrm{X_{1t}}, \mathrm{Y_{1s}}) = 8\mathrm{tr}(\Sigma'_{\mathrm{xy}}\Sigma_{\mathrm{xy}})
\end{aligned}$$

Under the assumption of dependence, equation (11.34) and equation (11.35), we have $\mathbb{C}\mathrm{ov}(\mathrm{d}^2_{(\mathrm{X})}, \mathrm{d}^2_{(\mathrm{Y})}) > 0$. Thus, $d^2_{(X)}$ and $d^2_{(Y)}$ are dependent. □

Proof: Proposition 11.3.1:

Let $\mathbf{\Sigma}_{xy}$ be the covariance matrix between $\mathbf{X}$ and $\mathbf{Y}$. Because every variable in $\mathbf{X}$ has a correlated variable in $\mathbf{Y}$, there are at least d_1 elements in $\mathbf{\Sigma}_{xy}$ that has larger absolute value than ϵ. Based on equation (11.35), we have $\mathbb{C}\mathrm{ov}\left(\mathrm{d}^2_{(\mathrm{X})}, \mathrm{d}^2_{(\mathrm{Y})}\right) = 2\mathrm{tr}(\Sigma'_{\mathrm{xy}}\Sigma_{\mathrm{xy}}) > 2\epsilon^2\mathrm{d}_1 \to \infty$.
□

Proof: Theorem 11.3.2:

Let ρ_x and ρ_y denote the coefficient of correlations in the bivariate power normal and bivariate normal scales, respectively. Freeman and Modarres [107] study the properties of the power normal distribution and show that

$$\rho_x = h(\rho_y) = \frac{\sum_{i=1}^{\infty} i! b_{1i} b_{2i}\ \rho_y^i}{\sqrt{(\sum_{i=1}^{\infty} b_{1i}^2 i!)(\sum_{j=1}^{\infty} b_{2j}^2 j!)}},$$

where the form of the function h depends on the mean, covariance and the transformation parameter. Here, b_{1i}, b_{2i} are defined in terms of the parameters and the Chebyshev-Hermite polynomials. It follows that $\rho_y = 0$ if and only if $\rho_x = 0$ and based on Theorem 11.3.1, the proof is completed. □

Proof: Lemma 11.3.1:

Let $\mathbf{W} = \mathbf{X}_1 - \mathbf{X}_2$, then $\mathbf{W}$ follows multivariate normal distribution with mean $\mathbf{0}$ and covariance $2\mathbf{\Sigma}_{xx}$. We have $d^2_{(X)} = \mathbf{W}'\mathbf{W}$, which is the quadratic form of $\mathbf{W}$. Let $\mathbf{Z} = (2\Sigma_{xx})^{-1/2}\mathbf{W}$. Hence, $\mathbb{C}ov(\mathbf{W}) = \mathbf{I}$. Let $\mathbf{P}$ be an orthogonal matrix that diagonalizes $2\Sigma_{xx}$ and $\mathbf{PP}' = \mathbf{I}$. That is $\mathbf{P}'2\Sigma_{xx}\mathbf{P} = diag(\lambda_1, \ldots, \lambda_{d_1})$, where $\lambda_1, \ldots, \lambda_{d_1}$ are the eigenvalues of $2\Sigma_{xx}$ and $diag(\lambda_1, \ldots, \lambda_{d_1})$ is the diagonal matrix of the eigenvalues. Let $\mathbf{U} = \mathbf{P}'\mathbf{Z}$, it follows that $\mathbf{Z} = \mathbf{PU}$. The distribution of $\mathbf{U}$ is multivariate normal distribution with mean $\mathbb{E}(\mathbf{U}) = \mathbf{0}$ and covariance $\mathbb{C}ov(\mathbf{U}) = \mathbf{I}$. Thus,

$$\begin{aligned}
d^2_{(X)} &= \mathbf{W}'\mathbf{W} = \mathbf{Z}'(2\Sigma_{xx})^{1/2}(2\Sigma_{xx})^{1/2}\mathbf{Z} \\
&= \mathbf{Z}'(2\Sigma_{xx})\mathbf{Z} = \mathbf{U}'\mathbf{P}'(2\Sigma_{xx})\mathbf{PU} \\
&= \mathbf{U}'diag(\lambda_1, \ldots, \lambda_{d_1})\mathbf{U} = \sum_{t=1}^{d_1} \lambda_t U_t^2,
\end{aligned}$$

where $\mathbf{U} = (U_1, \dots, U_{d_1})'$, and U_t^2 follows chi-square distribution for $t = 1, \dots, d_1$. □

Proof: Theorem 11.8.2:

Independence implies zero distance covariance. If $\mathbf{X} \perp \mathbf{Y}$, then $\mathbf{\Sigma}_{XY} = 0$ implies $\mathbf{\Gamma} = \mathbf{I}$. By independence, $\tilde{\delta}_X(\mathbf{X}, \mathbf{X}') \perp \tilde{\delta}_Y(\mathbf{Y}, \mathbf{Y}')$, so $V_{\tilde{\delta}}^2(\mathbf{X}, \mathbf{Y}) = \mathbb{E}[\tilde{\delta}_X(\mathbf{X}, \mathbf{X}')\tilde{\delta}_Y(\mathbf{Y}, \mathbf{Y}')] = 0$. Zero distance covariance implies independence. If $V_{\tilde{\delta}}^2(\mathbf{X}, \mathbf{Y}) = 0$, then $\tilde{\delta}_X(\mathbf{X}, \mathbf{X}')$ and $\tilde{\delta}_Y(\mathbf{Y}, \mathbf{Y}')$ are uncorrelated. For Gaussian vectors, uncorrelated implies independence. Therefore $\mathbf{X} \perp \mathbf{Y}$, which implies $\gamma_{ij} = 0$. Hence, $V_{\tilde{\delta}}^2(\mathbf{X}, \mathbf{Y}) = 0 \iff \mathbf{X} \perp \mathbf{Y} \iff \gamma_{ij} = 0$.
□

11.13 Exercises

Exercise 11.1 *Suppose* $\mathbf{V} = (X, Y, Z)$ *are distributed with correlation coefficients* $\rho_{X,Y} > 0$ *and* $\rho_{Y,Z} > 0$ *and* $\rho_{X,Z}$. *If* $\rho_{X,Y}^2 + \rho_{Y,Z}^2 > 1$, *prove that* $\rho_{X,Z} > 0$.

Exercise 11.2 *Apply the test of Heller, Heller, and Gorfine [145] with* $n = 4$, *and*

$$\mathbf{X} = (1\ 3\ 2\ 4)', \quad \mathbf{Y} = (10\ 20\ 15\ 30)'$$

Exercise 11.3 *Compute the squared projection covariance (Pcov) with the following* $n = 3$ *data points. Let the observed vectors be*

$$\mathbf{X}_1 = (2,0)', \qquad \mathbf{X}_2 = (0,2)', \qquad \mathbf{X}_3 = (1,1)'$$
$$\mathbf{Y}_1 = (1,0)', \qquad \mathbf{Y}_2 = (0,1)', \qquad \mathbf{Y}_3 = (1,1)'$$

Exercise 11.4 *Let* $n = 4$ *observations in* $\mathbb{R}^3$, *where each point is a concatenation of two subvectors:* $\mathbf{W}_i = (\mathbf{X}_i, \mathbf{Y}_i) \in \mathbb{R}^3$ *with* $\mathbf{X}_i \in \mathbb{R}^2$, $\mathbf{Y}_i \in \mathbb{R}$. *We define the following*

$$\begin{aligned} \mathbf{W}_1 &= (1,2,1), & \mathbf{W}_2 &= (3,1,2), \\ \mathbf{W}_3 &= (0,0,0), & \mathbf{W}_4 &= (2,2,3). \end{aligned}$$

We extract $\mathbf{X}_i = (w_{i1}, w_{i2})$, $\mathbf{Y}_i = w_{i3}$.

a) Compute all pairwise distances.

b) Assemble the distance vectors $\mathbf{d}_x$ *and* $\mathbf{d}_y$.

c) Compute Cor_{IPD}.

Exercise 11.5 *Székely et al. [331] propose distance correlation to measure the dependence between random vectors. What is the relationship between the distance correlation and the energy statistic?*

Exercise 11.6 *Let* $\boldsymbol{X} \in \mathbb{R}^p$ *and* $\boldsymbol{Y} \in \mathbb{R}^q$ *be centered random vectors. Write*

$$\mathbf{\Sigma}_x = \mathrm{Cov}(\boldsymbol{X}) \in \mathbb{R}^{p\times p}, \quad \mathbf{\Sigma}_y = \mathrm{Cov}(\boldsymbol{Y}) \in \mathbb{R}^{q\times q}, \quad \mathbf{\Sigma}_{xy} = \mathrm{Cov}(\boldsymbol{X}, \boldsymbol{Y}) \in \mathbb{R}^{p\times q}.$$

Robert, Cléroux and Ranger [282] define the vector correlation by

$$\mathbb{C}or(\boldsymbol{X}, \boldsymbol{Y}) = \frac{\mathrm{tr}(\Sigma_{xy}'\Sigma_{xy})}{\sqrt{\mathrm{tr}(\Sigma_x^2)\,\mathrm{tr}(\Sigma_y^2)}} = \frac{\|\Sigma_{xy}\|_{\mathrm{F}}^2}{\|\Sigma_x\|_{\mathrm{F}}\,\|\Sigma_y\|_{\mathrm{F}}}, \tag{11.36}$$

where $\|\cdot\|_{\mathrm{F}}$ *is the Frobenius norm. If* $\mathbb{E}\|\boldsymbol{X}\|^2 < \infty$ *and* $\mathbb{E}\|\boldsymbol{Y}\|^2 < \infty$, *prove that* $0 \le \mathbb{C}or(\boldsymbol{X}, \boldsymbol{Y}) \le 1$, *and* $or(\boldsymbol{X}, \boldsymbol{Y}) = or(\boldsymbol{Y}, \boldsymbol{X})$. *It is invariant to translations of* $\boldsymbol{X}, \boldsymbol{Y}$.

Exercise 11.7 *Write an R function to implement the Pcov method.*

12

High-Dimensional Change Point Detection Methods

12.1 Abstract

High-dimensional change point detection has emerged as a fundamental problem in contemporary statistical analysis, especially in settings characterized by high dimension and low sample size (HDLSS). This chapter develops a comprehensive framework for identifying both single and multiple change points in high-dimensional data streams, allowing for shifts in the mean, variance, or broader distributional properties. We propose distribution-free, proximity-based methodologies that employ novel IPD-based dissimilarity measures, and difference matrices to localize structural changes without relying on sparsity assumptions or parametric models. A range of state-of-the-art techniques is examined, including the Efficient Sparsity Adaptive Change-point Estimator (ESACE), Informative Sparse Projection (ISP), Fabiseach factorized binary search, CUSUM-type procedures, and the non-parametric E-Divisive (ecp) method. Rigorous theoretical results establish consistency and asymptotic properties of the proposed estimators, supported by simulation studies demonstrating enhanced power and robustness in both sparse and dense change regimes. Real-data applications, including an analysis of the S&P-500 index, highlight the practical effectiveness of these approaches. Overall, the chapter underscores a balance of computational efficiency, statistical soundness, and methodological flexibility, providing a unified treatment of high-dimensional change point detection with relevance to finance, genomics, and engineering.

12.2 Introduction

The objective of change point detection methods is to identify structural changes in a sequence of time-ordered multivariate observations when an underlying property of the process undergoes a change. Broadly, change point detection procedures are categorized into offline and online settings. In offline change point detection, the complete dataset is assumed to be available in advance, allowing retrospective analysis and often more elaborate statistical procedures. By contrast, online change point detection operates sequentially, updating inference in real time as new observations are collected (Guo and Modarres [128]). Although offline methodologies have been extensively studied in the literature, online approaches remain comparatively less explored and are not yet as fully developed. In the online framework, data arrive continuously and the full sample is never observed at once, giving rise to a streaming data structure. Such data streams consist of potentially unbounded sequences of transient observations, which necessitate fast, memory-efficient algorithms

DOI: 10.1201/9781003789581-12

capable of detecting changes promptly as the data evolve. The objective of the current chapter is detection of HDLSS change points in an offline setting.

We propose a distribution-free proximity–based method for high-dimensional change points that can address challenging situations when the sample size is very small compared to the dimension as in the so-called HDLSS data or when non–sparse changes may occur due to changes in many variables but with small significant magnitudes. The proposed method can detect changes in mean or variance of high-dimensional observations as well as other distributional changes. We present efficient algorithms that can detect single and multiple high-dimensional change points. We use non-parametric metrics, including a new dissimilarity measure and some new distance and difference distance matrices, to develop a procedure to estimate change point locations. We also introduce a non-parametric test to determine the significance of estimated change points. We provide theoretical guarantees for our method and demonstrate its empirical performance in comparison with some of the recent methods for high-dimensional change points. An R package called `HDDchangepoint` is developed to implement the proposed method.

Change point analysis plays a central role in a wide range of disciplines, including economics, finance, engineering, genetics, and medical research. Its primary goal is to identify meaningful structural changes in the underlying distribution governing an observed data sequence. While the change point problem has been extensively investigated in the context of low-dimensional settings, its extension to high-dimensional data remains substantially more complex, particularly as such data increasingly arise across modern application domains. In high-dimensional environments, changes may occur in either a small or large subset of variables and may vary in magnitude, yet when the sample size is limited relative to the number of variables, separating genuine distributional shifts from random fluctuations becomes a challenging statistical task.

In recent years, growing attention has been devoted to the HDLSS paradigm, in which the sample size n is extremely limited while the dimension d may be very large. The asymptotic framework associated with HDLSS data differs fundamentally from classical high-dimensional asymptotics, typically characterized by $d \gg n \to \infty$, since in the HDLSS setting the dimension $d \to \infty$ whereas the sample size n may remain fixed (e.g., Hall et al. [133]; Jung and Marron [164]; Li [192]). Change point detection becomes particularly difficult under such conditions due to the scarcity of observations. In this chapter, we demonstrate that many existing high-dimensional change point methodologies exhibit degraded performance in HDLSS scenarios, especially when the sample size is exceedingly small relative to the dimensionality.

A prevalent approach in the high-dimensional change point literature is to reduce the complexity of the problem by lowering the data dimension, thereby enabling the application of simpler procedures such as univariate change point algorithms to suitably transformed observations. Dimension reduction may be achieved through random projections (Wang and Samworth [347]; Hahn et al. [132]), geometric mappings (Grundy et al. [124]), or related techniques including principal component analysis (Xiao et al. [360]), factor models (Barigozzi et al. [20]), and regularization-based methods (Lee et al. [184]; Safikhani and Shojaie [293]). The effectiveness of this strategy typically depends on sparsity assumptions—often requiring a substantial degree of sparsity—to attain oracle-type performance. Consequently, these methods may be inadequate for non-sparse change point settings, where distributional shifts occur across many variables, each potentially with a small but collectively significant magnitude.

An alternative line of work identifies change point locations by examining dissimilarity measures computed between pairs of observations. This approach typically relies on carefully chosen distance metrics, such as IPDs or divergence measures derived from Euclidean distances (Matteson and James [221]). Related methodologies, including those proposed by

Garreau and Arlot [115] and Chu and Chen [68], make indirect use of IPDs by constructing similarity graphs and basing inference on the number of edges crossing potential change locations. The effectiveness of these procedures, particularly their statistical power, is strongly influenced by the choice of distance measures and associated test statistics. In this chapter, we seek to introduce a new, powerful framework for high-dimensional change point detection grounded in dissimilarity measures, with a particular emphasis on performance in HDLSS settings.

In recent years, a substantial body of work has emerged on change point analysis for high-dimensional data. Here, we briefly review several representative contributions with an emphasis on offline change point detection. Chen and Zhang [60], Garreau and Arlot [115], and Chu and Chen [68] proposed kernel and similarity graph—based methodologies for identifying structural changes. Wang and Samworth [347] introduced a two-stage procedure, INSPECT, which combines random projections with sparsity assumptions to estimate change points under changes in the mean, relying on normality. Enikeeva and Harchaoui [89] developed a method for detecting sparse mean shifts in high-dimensional sequences, focusing on the single change point case and assuming normal distributions. Grundy et al. [124] proposed a geometric mapping approach that projects high-dimensional observations onto two-dimensional representations based on distances and angles, also under the normality assumption. Li et al. [198] presented a data-adaptive framework that generalizes the classical CUSUM statistic through a U-statistic CUSUM matrix, assuming independence across variables and primarily addressing single change point detection. Li [189] further introduced an asymptotically distribution-free, proximity-based method for high-dimensional change point analysis. Building on the random projection framework of Wang and Samworth [347] , Hahn et al. [132] proposed a Bayesian procedure to efficiently estimate projection directions for change point detection. Yu and Chen [366] developed a bootstrap CUSUM test to enable finite-sample inference in high-dimensional settings, while Follain et al. [104] extended random projection techniques to accommodate heterogeneous missing data. Additional related methodologies are surveyed in Liu et al. [198].

Some existing methodologies and their R implementation include

12.2.1 Efficient Sparsity Adaptive Change-point Estimator (ESACE)

Consider a sequence $\{\boldsymbol{X}_t\}_{t=1}^T$. ESACE estimates (Wang and Samworth [347]) change points using the penalized least squares

$$\min_{\{\boldsymbol{\mu}_k\},\{t_k\}} \sum_{k=0}^{K} \sum_{t=t_k+1}^{t_{k+1}} \|\boldsymbol{X}_t - \boldsymbol{\mu}_k\|_2^2 + \lambda \sum_{k=1}^{K} \|\boldsymbol{\mu}_k - \boldsymbol{\mu}_{k-1}\| \tag{12.1}$$

where $\lambda > 0$ encourages sparsity in changes.

R Implementation: `HDCD` package. `https://cran.r-project.org/package=HDCD`

12.2.2 Informative Sparse Projection (ISP)

ISP detects change points in a lower-dimensional subspace. Let $\boldsymbol{v} \in \mathbb{R}^d$ be a sparse projection vector, then the projected univariate sequence is $Y_t = \boldsymbol{v}'\boldsymbol{X}_t$. CUSUM statistics are applied on Y_t to detect change points,

$$CUSUM(t) = \sqrt{\frac{(T-t)}{Tt}} \sum_{i=1}^{t} (Y_i - \bar{Y}_T) \tag{12.2}$$

R Implementation: `HDCD` package. `https://cran.r-project.org/package=HDCD`

12.2.3 Fabiseach: Factorized Binary Search

Fabiseach (Ondrus et al. [258]) uses matrix factorization $\boldsymbol{X} = \boldsymbol{W}\boldsymbol{H}$ to reduce dimensions, then applies binary search for multiple change points. For network data $\boldsymbol{X}_t \approx \boldsymbol{W}_k\boldsymbol{H}_k$, $t_{k-1} < t \leq t_k$. Change points are detected when the reconstruction error $\text{Error}(t) = \|\boldsymbol{X}_t - \boldsymbol{W}_k\boldsymbol{H}_k\|_F$ exceeds a threshold.

R Implementation: `fabisearch` package.

`https://arxiv.org/abs/2207.02986`arXiv:2207.02986

12.2.4 CUSUM-based High-Dimensional Detection

CUSUM statistic for coordinate j is

$$CUSUM_j(t) = \sqrt{\frac{(T-t)}{Tt}} \sum_{i=1}^{t} (X_{ij} - \bar{X}_{\cdot j}) \tag{12.3}$$

Aggregate statistic over coordinates is $S(t) = \max_{1 \leq j \leq d} |CUSUM_j(t)|$.

R Implementation: `InspectChangepoint` by Wang et al. [346].

12.2.5 Non-parametric E-Divisive (ecp)

The energy statistic between two segments A and B is

$$\mathcal{E}(A,B) = \frac{2}{|A||B|} \sum_{x \in A} \sum_{y \in B} \|\mathbf{x} - \mathbf{y}\| - \frac{1}{|A|^2} \sum_{x,x' \in A} \|\mathbf{x} - \mathbf{x}'\| - \frac{1}{|B|^2} \sum_{y,y' \in B} \|\mathbf{y} - \mathbf{y}'\| \tag{12.4}$$

A large $\mathcal{E}$ indicates a change point between segments (Matteson and James [221]).

R Implementation: `ecp` package. https://cran.r-project.org/web/packages/ecp/index.html

12.3 High-Dimensional Distance Change point

Drikvandi and Modarres [80] present a high-dimensional change point detection methodology based on the distance matrix. The R implementation of the `HDDchangepoint` package is at

`https://github.com/statistics11/HDDchangepoint`

Change point detection in HDLSS settings presents significant challenges and remains an underexplored area. While many existing approaches primarily address sparse change points, this method is capable of handling non-sparse, high-dimensional scenarios. Moreover, unlike conventional techniques that focus predominantly on detecting changes in the mean of high-dimensional observations, this approach can identify changes in the mean, variance, and other distributional characteristics.

In contrast to most recent high-dimensional change point detection methods, which rely on the assumption of normality (as discussed above), the framework we present is entirely distribution-free. We employ innovative non-parametric tools to accurately detect the locations of change points without imposing any parametric assumptions on the underlying variables.

Many contemporary methods either rely on bootstrap or permutation procedures or derive asymptotic distributions to assess the significance of detected change points. The methodology we present incorporates both asymptotic and permutation-based inference, providing a unified framework suitable for both small and large sample size contexts.

Conceptually, this approach identifies change point locations by analyzing an $n \times n$ distance matrix rather than the original $n \times d$ observation matrix. Since the $n \times n$ distance matrix has a dimension independent of d, this strategy not only simplifies the mathematical analysis but also enhances computational efficiency, facilitating the detection of change points in high-dimensional data.

We use the following notation throughout the chapter. For a vector $\boldsymbol{u} \in \mathbb{R}^d$, we write the L_s-norm as $\|\boldsymbol{u}\|_s = \left(\sum_{j=1}^d |\boldsymbol{u}_j|^s\right)^{1/s}$. The infinity norm is defined as $\|\boldsymbol{u}\|_\infty = \max_j |\boldsymbol{u}_j|$. We write $\surd$ as the short version of square root. For a real-value constant a, notation $|a|$ denotes the absolute value of a, while for a set C, $|C|$ denotes the cardinality of C. Also, $O(.)$ and $o(.)$ denote the usual big O and little o notation. We sometimes write $O^d(.)$ or $o^d(.)$ to emphasize them for $d \to \infty$. We write $O_{\mathbb{P}}(.)$ and $o_{\mathbb{P}}(.)$ to denote, respectively, the big O in probability and the little o in probability. We use $\xrightarrow{P}$ and $\xrightarrow{D}$ for convergence in probability and convergence in distribution, respectively. We also define some more specific notation as we develop the chapter.

Let $\boldsymbol{X}_1, \boldsymbol{X}_2, \ldots, \boldsymbol{X}_n$ denote a sequence of independent d-dimensional random vectors with corresponding, generally unknown, probability distributions $F_1, F_2, \ldots, F_n$. We consider high-dimensional settings in which the dimension d substantially exceeds the sample size n, and where the d components of each observation $\boldsymbol{X}_i = (\boldsymbol{X}_{i1}, \boldsymbol{X}_{i2}, \ldots, \boldsymbol{X}_{id})$ may exhibit dependence structures of arbitrary form. To maximize generality, our framework adopts a fully distribution-free approach, making no parametric assumptions regarding the forms of the distributions $F_1, F_2, \ldots, F_n$.

For clarity of exposition, we first present the proposed methodology for detecting a single change point in high-dimensional settings, and later generalize the framework to accommodate multiple change points in section 12.4. The single change point detection problem can be formulated as the following hypothesis testing task.

$$\begin{cases} H_0 : F_1 = F_2 = \cdots = F_n, \\ H_1^s : F_1 = \cdots = F_{\tau-1} \neq F_\tau = \cdots = F_n, \end{cases} \tag{12.5}$$

where τ represents the unknown location of the change point. To address this problem, we first obtain an estimate of τ and then evaluate the statistical significance of the identified change using a suitable testing procedure.

Let $\boldsymbol{X} = [\boldsymbol{X}_1, \boldsymbol{X}_2, \ldots, \boldsymbol{X}_n]'$ denote the full dataset as an $n \times d$ matrix. As highlighted in the introduction, detecting change points in high-dimensional settings is particularly challenging when the sample size n is small relative to the dimension d. In the proposed approach, rather than searching directly in the $n \times d$ space, we reformulate the problem in the lower-dimensional $n \times n$ space via the distance matrix, without reducing the original data dimension. This stands in contrast to dimensionality reduction techniques such as random projections (Wang and Samworth [347]), geometric mappings (Grundy et al. [124]), principal component analysis (Xiao et al. [360]), or factor models (Barigozzi et al. [20]), all of which reduce the effective dimension of the data. The method employs a proximity-based framework, using dissimilarity measures to identify change points in the original data $\boldsymbol{X}$ through its $n \times n$ distance matrix. Since the distance matrix has dimension $n \times n$, compared with $n \times d$ for $\boldsymbol{X}$, this formulation is both mathematically and computationally more tractable, particularly in HDLSS scenarios where $n \ll d$. The details of this methodology are provided in the subsequent sections.

Let $d_{ij} := d(\boldsymbol{X}_i, \boldsymbol{X}_j)$ denote a dissimilarity measure between observations $\boldsymbol{X}_i$ and $\boldsymbol{X}_j$, where $d(\cdot, \cdot)$ represents an appropriate distance function for high-dimensional vectors. Since the proposed methodology is compatible with any suitable dissimilarity metric, a detailed discussion of the choice of distance measure is deferred until after the main framework has been presented. Using the pairwise dissimilarities d_{ij}, we construct the $n \times n$ distance matrix for the observations $\boldsymbol{X}_1, \boldsymbol{X}_2, \ldots, \boldsymbol{X}_n$ as

$$\mathbf{D} = \begin{bmatrix} d_{11} & d_{12} & \ldots & d_{1n} \\ \vdots & \vdots & \ddots & \vdots \\ d_{n1} & d_{n2} & \ldots & d_{nn} \end{bmatrix},$$

where the diagonal entries satisfy $d_{11} = \cdots = d_{nn} = 0$.

To estimate the location of the change point τ, we propose a procedure that identifies $\hat{\tau}$ by locating the maximum variation between consecutive observations in the distance matrix $\mathbf{D}$. Specifically, we define the distance differences between each pair of observations as

$$\begin{aligned} \Delta_{ij} &:= |d_{ij} - d_{i,j-1}|, \quad && i = 1, \ldots, n,\ j = 2, \ldots, n, \\ \Delta_{i1} &:= 0, && i = 1, \ldots, n. \end{aligned}$$

This yields the $n \times n$ difference distance matrix

$$\boldsymbol{\Delta} = \begin{bmatrix} 0 & \Delta_{12} & \ldots & \Delta_{1,n-1} & \Delta_{1n} \\ \vdots & \vdots & \ddots & \vdots & \vdots \\ 0 & \Delta_{n2} & \ldots & \Delta_{n,n-1} & \Delta_{nn} \end{bmatrix}.$$

We then define the change point estimate $\hat{\tau}$ based on the column sums of Δ as

$$\hat{\tau} = \underset{1 \le j \le n}{\arg\max} \left\{ \frac{1}{n} \sum_{i=1}^{n} \Delta_{ij} \right\}. \tag{12.6}$$

Here, $\hat{\tau}$ corresponds to the location of the maximal slope among the column sums of the difference distance matrix $\boldsymbol{\Delta}$ (a simple illustrative example is provided in Figure 12.1). If all column averages satisfy $\frac{1}{n}\sum_{i=1}^{n} \Delta_{ij} = 0$ or are equal for $j = 1, \ldots, n$, we set $\hat{\tau} = \emptyset$, where the empty set $\emptyset$ indicates that no change point is detected.

It is well known that the standard Euclidean distance performs poorly in high-dimensional settings. To address this issue, Hall et al. [133] proposed a modified Euclidean distance, obtained by dividing the standard distance by $\sqrt{d}$, to ensure proper convergence in high dimensions. Within our framework, one may employ the modified Euclidean distance

$$d_{ij} = d(\boldsymbol{X}_i, \boldsymbol{X}_j) = d^{-1/2} \|\boldsymbol{X}_i - \boldsymbol{X}_j\|_2, \tag{12.7}$$

or other L_s-norm distances. A more general dissimilarity measure can further incorporate information from all remaining $n - 2$ observations, defined as

$$d_{ij} = d(\boldsymbol{X}_i, \boldsymbol{X}_j) = \frac{1}{n-2} \sum_{l \neq i,j} \Big| \|\boldsymbol{X}_i - \boldsymbol{X}_l\|_D - \|\boldsymbol{X}_j - \boldsymbol{X}_l\|_D \Big|, \tag{12.8}$$

where $\|\cdot - \cdot\|_D$ denotes a suitable distance function. We propose two straightforward options for this choice. A multivariate option is to use either the modified Euclidean distance, $\|\boldsymbol{X}_u - \boldsymbol{X}_l\|_D = d^{-1/2}\|\boldsymbol{X}_u - \boldsymbol{X}_l\|_2$, or the modified L_1-norm, $\|\boldsymbol{X}_u - \boldsymbol{X}_l\|_D = d^{-1}\|\boldsymbol{X}_u - \boldsymbol{X}_l\|_1$. Alternatively, a univariate option defines a distance based on differences in the sample mean and variance of each observation. Let

$$\bar{\boldsymbol{X}}_i^d = \frac{1}{d} \sum_{j=1}^{d} \boldsymbol{X}_{ij}, \qquad S_{\mathbf{X}_i}^d = \sqrt{\frac{1}{d} \sum_{j=1}^{d} (\boldsymbol{X}_{ij} - \bar{\boldsymbol{X}}_i^d)^2}$$

denote the mean and standard deviation of the i-th observation $\boldsymbol{X}_i$, $i = 1, \ldots, n$. We then define

$$\|\boldsymbol{X}_u - \boldsymbol{X}_l\|_D = \sqrt{(\bar{\boldsymbol{X}}_u^d - \bar{\boldsymbol{X}}_l^d)^2 + (S_{\boldsymbol{X}_u}^d - S_{\boldsymbol{X}_l}^d)^2}, \tag{12.9}$$

which captures the combined differences in mean and variance between each pair of observations.

The following theorem establishes that the dissimilarity measure $d(\boldsymbol{X}_i, \boldsymbol{X}_j)$ defined in (12.8) is a pseudo-metric. Specifically, it satisfies the properties i) $d(\boldsymbol{X}_i, \boldsymbol{X}_j) \geq 0$, ii) $d(\boldsymbol{X}_i, \boldsymbol{X}_i) = 0$, iii) $d(\boldsymbol{X}_i, \boldsymbol{X}_j) = d(\boldsymbol{X}_j, \boldsymbol{X}_i)$, iv) $d(\boldsymbol{X}_i, \boldsymbol{X}_j) \leq d(\boldsymbol{X}_i, \boldsymbol{X}_l) + d(\boldsymbol{X}_j, \boldsymbol{X}_l)$. The proof of this result, along with all other technical proofs, is provided in section 12.9.

Theorem 12.3.1 *The dissimilarity measure $d(\boldsymbol{X}_i, \boldsymbol{X}_j)$ in (12.8) is a pseudo metric on $\boldsymbol{X}$ for $n \geq 3$.*

The following theorem indicates that proximity in terms of the modified Euclidean distance implies proximity in terms of $d(\boldsymbol{X}_i, \boldsymbol{X}_j)$, but the converse implication does not hold.

Theorem 12.3.2 *Consider the dissimilarity measure $d(\boldsymbol{X}_i, \boldsymbol{X}_j)$ in (12.8) and the Euclidean distance $\|\boldsymbol{X}_i - \boldsymbol{X}_j\|_2$ for each pair of observations $\boldsymbol{X}_i$ and $\boldsymbol{X}_j$. We have $d(\boldsymbol{X}_i, \boldsymbol{X}_j) \leq \frac{1}{\sqrt{d}}\|\boldsymbol{X}_i - \boldsymbol{X}_j\|_2$.*

In section 12.5, we show that $d(\boldsymbol{X}_i, \boldsymbol{X}_j) \to 0$ if $\boldsymbol{X}_i$ and $\boldsymbol{X}_j$ have the same distribution. The computational cost of the dissimilarity measure $d(\boldsymbol{X}_i, \boldsymbol{X}_j)$ is of order $O(nd)$. The cost of computing $d(\boldsymbol{X}_i, \boldsymbol{X}_j)$ for all $1 \leq i < j \leq n$ is $O(n^3 d)$ with n not being fixed. The cost is $O(d)$ with fixed n as in HDLSS data.

We illustrate the proposed method with a simple example containing a single change point. We generate 5 observations from a standard multivariate normal distribution with $d = 500$, and an additional 5 observations from the same distribution but with the mean shifted by 0.5. The left panel of Figures 12.1 displays the elements of the distance matrix D computed using the dissimilarity measure in (12.8), with the univariate distance function based on differences in sample means and variances, and the right panel of Figures 12.1 shows the multivariate distance function based on the modified Euclidean distance. In both cases, observations from the first distribution exhibit greater dissimilarity with those from the second distribution, demonstrating that both distance measures capture the change point effectively. The left panel of The left panel in Figure 12.2 shows the corresponding difference distance matrix $\boldsymbol{\Delta}$, highlighting that the change point location (position 6) corresponds to the largest Δ_{ij} values. The right panel of Figure 12.2 presents the column sums of $\boldsymbol{\Delta}$, which correctly identify the change point location in agreement with the proposed estimator $\hat{\tau}$ in (12.6). In section 12.5, we formally prove that $\hat{\tau}$ is consistent for the true change point τ_0 under suitable assumptions.

Figures show an illustrative example with one change point: plots for all elements of the distance matrix D obtained using the dissimilarity measure $d(\boldsymbol{X}_i, \boldsymbol{X}_j)$ in (12.8) with both the univariate and multivariate distance functions, as well as for all column sums of the difference distance matrix $\boldsymbol{\Delta}$.

We now construct a test statistic based on the estimated change point $\hat{\tau}$ to evaluate whether it is statistically significant. The proposed statistic incorporates information from the dissimilarities d_{ij} and the difference matrix Δ_{ij} both before and after the estimated change point. To formalize this, and in the context of the hypothesis test in (12.5), we define two sets of indices corresponding to observations before and after $\hat{\tau}$,

$$\nabla_{\hat{\tau}}^- := \{j : 1 \leq j \leq \hat{\tau} - 1\}, \qquad \nabla_{\hat{\tau}}^+ := \{j : \hat{\tau} \leq j \leq n\}.$$

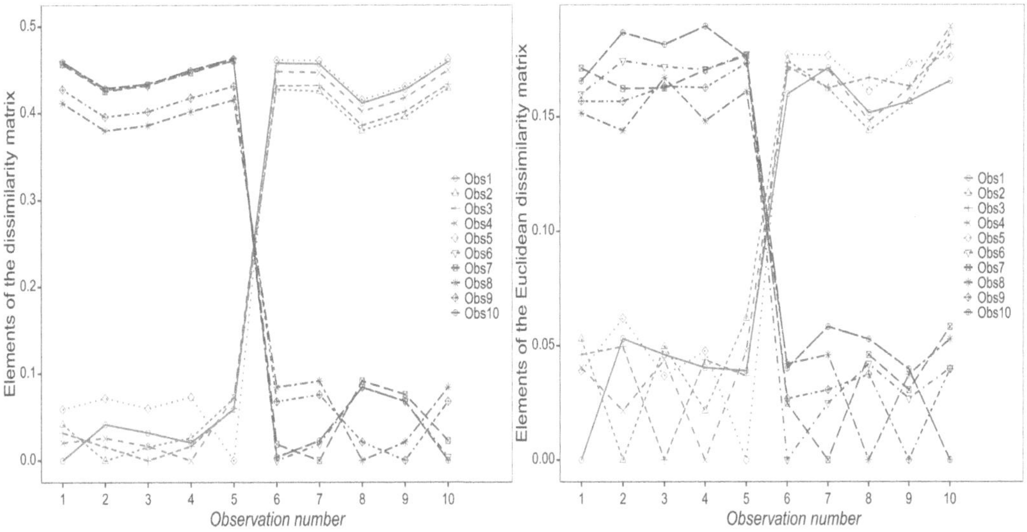

FIGURE 12.1: Left panel: For matrix D using 12.9. Right Panel: For matrix D using 12.7.

The two sets $\nabla_{\hat{\tau}}^{-}$ and $\nabla_{\hat{\tau}}^{+}$ are visualized in Figure 12.3, showing their connection with the change point candidate $\hat{\tau}$. We then propose the following test statistic for testing the significance of change point estimate $\hat{\tau}$

$$T(\hat{\tau}) = \frac{1}{n|\nabla_{\hat{\tau}}^{-}||\nabla_{\hat{\tau}}^{+}|} \sum_{i=1}^{n} \sum_{j \in \nabla_{\hat{\tau}}^{-}} \sum_{j' \in \nabla_{\hat{\tau}}^{+}} \left(d_{ij} - d_{ij'}\right)^2, \tag{12.10}$$

where $|\nabla_{\hat{\tau}}^{-}| = \hat{\tau} - 1$ and $|\nabla_{\hat{\tau}}^{+}| = n - \hat{\tau} + 1$ are the cardinalities of $\nabla_{\hat{\tau}}^{-}$ and $\nabla_{\hat{\tau}}^{+}$, respectively.

We investigate the theoretical properties of the statistic $T(\hat{\tau})$ in section 12.5. Intuitively, the test statistic $T(\hat{\tau})$ measures the differences between the distances among observations preceding the change point and those following it. Under the null hypothesis, these differences are expected to be small, whereas in the presence of a change point, $T(\hat{\tau})$ deviates from zero. Specifically, we will show that, as $d \to \infty$,

$$T(\hat{\tau})\big|_{H_0} = o_{\mathbb{P}}(1), \qquad T(\hat{\tau})\big|_{H_1^s} = \kappa_{\tau_0}^2 + o_{\mathbb{P}}(1),$$

where κ_{τ_0} is defined in Theorem 12.5.1, and τ_0 denotes the true change point. We can therefore carry out the hypothesis test for a single change point and reject the null hypothesis H_0 for large values of $T(\hat{\tau})$. For large n, we conduct the test using the asymptotic distribution of $T(\hat{\tau})$ under H_0, when $n, d \to \infty$. In Theorem 12.5.5, we will prove that the asymptotic distribution of $T(\hat{\tau})$, denoted by $G_T(\cdot)$, is a normal distribution as $n, d \to \infty$ under some assumptions, so that

$$\frac{n|\nabla_{\hat{\tau}}^{-}||\nabla_{\hat{\tau}}^{+}|\left(T(\hat{\tau}) - \frac{1}{n|\nabla_{\hat{\tau}}^{-}||\nabla_{\hat{\tau}}^{+}|}\sum_{i=1}^{n}\sum_{j\in\nabla_{\hat{\tau}}^{-}}\sum_{j'\in\nabla_{\hat{\tau}}^{+}} m_{ijj'}\right)}{\sqrt{\sum_{i=1}^{n}\sum_{j\in\nabla_{\hat{\tau}}^{-}}\sum_{j'\in\nabla_{\hat{\tau}}^{+}}\sum_{k=1}^{n}\sum_{l\in\nabla_{\hat{\tau}}^{-}}\sum_{l'\in\nabla_{\hat{\tau}}^{+}} c_{ijj'kll'}}} \xrightarrow{D} \mathbb{N}(0,1),$$

under H_0, where $m_{ijj'}$ and $c_{ijj'kll'}$ are specified in the statement of Theorem 12.5.5.

For small sample sizes n, we employ a permutation procedure based on $T(\hat{\tau})$ to perform the test. Permutation testing is particularly appropriate here because, under H_0, all observations share the same distribution and are therefore exchangeable. In each permutation step, we randomly permute the indices of observations before and after the estimated

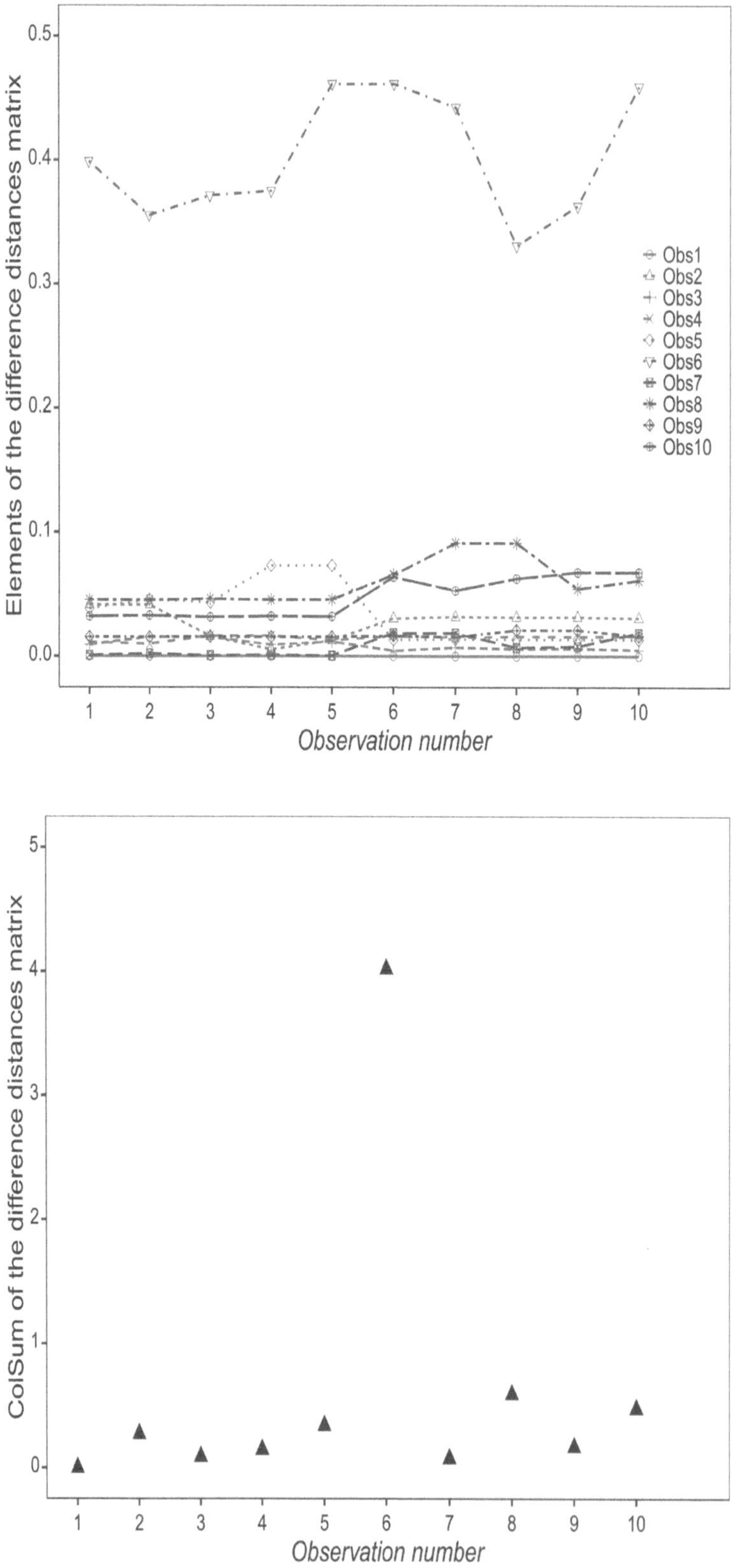

FIGURE 12.2: Left panel: Difference distance matrix for $\mathbf{\Delta}$. Right panel: Column sum of difference distance matrix $\mathbf{\Delta}$

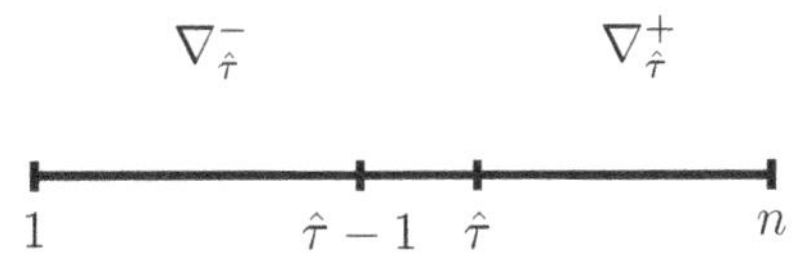

FIGURE 12.3: The sets $\nabla_{\hat{\tau}}^{-}$ and $\nabla_{\hat{\tau}}^{+}$ with the estimated change point location $\hat{\tau}$.

change point $\hat{\tau}$, while keeping the change point location fixed, to generate a permuted sample. Based on R such permutations, the approximate permutation distribution of the test statistic, denoted $G_{T_R}(t)$, is defined as

$$G_{T_R}(t) = \frac{1}{R}\sum_{r=1}^{R} \mathbb{I}(T_r(\hat{\tau}) \leq t), \tag{12.11}$$

where $\mathbb{I}(\cdot)$ is the indicator function and $T_r(\hat{\tau})$ denotes the test statistic computed for the r-th permuted sample. The p-value of the permutation test, denoted by d_{perm}, is given by

$$p_{\text{perm}} = 1 - G_{T_R}(T_{\text{obs}}(\hat{\tau})) = \frac{1}{R}\sum_{r=1}^{R} \mathbb{I}(T_r(\hat{\tau}) > T_{\text{obs}}(\hat{\tau})), \tag{12.12}$$

where $T_{\text{obs}}(\hat{\tau})$ is the test statistic for the observed sample. The theoretical results show that $\mathbb{P}_{H_0}(p_{\text{perm}} \leq \alpha) \leq \alpha$ for all $0 \leq \alpha \leq 1$. We will show the asymptotic and permutation tests are equivalent asymptotically under H_0, that is $G_{T_R}(t) \xrightarrow{D} G_T(t)$ for all t as $n, d \to \infty$. The following Algorithm summarizes the proposed method for single change point detection in high-dimensional data.

Algorithm 12.3.1 Algorithm A: Single Change Point Detection

Input: A data sequence or matrix of observations

$$\boldsymbol{X} = [\boldsymbol{X}_1, \boldsymbol{X}_2, \ldots, \boldsymbol{X}_n]^T$$

Output: Change point estimate $\hat{\tau}$, or "NA" if there is no significant change point

Step 1: Calculate the $n \times n$ distance matrix D for the $n \times d$ data matrix $\boldsymbol{X}$
Step 2: Calculate the $n \times n$ difference distance matrix $\boldsymbol{\Delta}$
Step 3: Calculate the change point estimate $\hat{\tau}$. If $\hat{\tau} = \emptyset$, stop the algorithm and return "NA" for no detected change point. Otherwise, proceed to the next step.
Step 4: Calculate the test statistic $T(\hat{\tau})$
Step 5: Apply the permutation test based on $T(\hat{\tau})$ if n is small, or the asymptotic test if n is large, to test the significance of the change point estimate $\hat{\tau}$
Step 6: If the test is significant, return the change point estimate $\hat{\tau}$. Otherwise, return "NA" for no significant change point

In addition to estimating and testing for the change point τ, one can also construct a confidence interval for its location using the change point estimate $\hat{\tau}$. Deriving the exact or asymptotic distribution of the change point estimate (12.6) is challenging due to the presence of absolute value functions in both the definition of Δ_{ij} and the dissimilarity measure d_{ij} in (12.8). We therefore adopt a permutation-based approach to construct a confidence interval for τ. Unlike the permutation procedure described previously, which is performed under H_0, here we generate permuted samples by independently permuting the observations before and after the change point τ within each respective segment. This is justified because observations preceding the change point share the same distribution, as do those following it. For each of R permutation samples, we compute the change point estimate (12.6), denoting these permutation estimates by $\hat{\tau}_1^*, \ldots, \hat{\tau}_R^*$. A $100(1-\alpha)\%$ confidence interval for the change point location τ is then given by

$$\left(2\hat{\tau} - \hat{\tau}^*_{(1-\alpha/2)},\, 2\hat{\tau} - \hat{\tau}^*_{(\alpha/2)}\right), \tag{12.13}$$

where $\hat{\tau}^*_{(\alpha/2)}$ and $\hat{\tau}^*_{(1-\alpha/2)}$ denote the $(\alpha/2)$-th and $(1-\alpha/2)$-th percentiles of the R ordered permutation estimates, respectively.

12.4 Multiple Change Point Detection

In this section, we extend the approach to detect multiple change points in high-dimensional data. The problem of detecting multiple change points in general can be formulated as

$$\begin{cases} H_0 : F_1 = F_2 = \ldots = F_n \\ H_1^m : F_1 = \ldots = F_{\tau_1 - 1} \neq F_{\tau_1} = \ldots = F_{\tau_2 - 1} \neq F_{\tau_2} = \ldots = F_{\tau_s - 1} \neq F_{\tau_s} = \ldots = F_n, \end{cases} \tag{12.14}$$

where $1 < \tau_1 < \tau_2 < \ldots < \tau_s < n$ are the unknown change point locations and s is the number of change points, which is also unknown. If the above null hypothesis is rejected, the main objective will be to find the s change point estimates $\hat{\tau}_1, \hat{\tau}_2, \ldots, \hat{\tau}_s$.

Analogous to the illustrative example in Figure 12.1 for a single change point, Figure 12.4 demonstrates how the dissimilarity measure (12.8) and the proposed approach based on the difference distance matrix $\boldsymbol{\Delta}$ can facilitate the detection of multiple change points (here using the same settings as the previous example, but with two true change points at locations 6 and 11). To address the multiple change point problem, we employ a recursive binary segmentation procedure grounded in our single change point detection method. We also illustrate its combination with the wild binary segmentation procedure of Fryzlewicz [111] at the end of this section. The proposed single change point method is applied sequentially to identify all significant change points in the data according to either the asymptotic or permutation test. Suppose that s change points are detected sequentially and denoted by $\hat{\gamma}_k$, $k = 1, \ldots, s$. We adopt the notation $\hat{\gamma}_k$ because, when multiple change points are present, the points detected via binary segmentation are not necessarily in increasing order. The sorted change points are then given by $(\hat{\tau}_1, \hat{\tau}_2, \ldots, \hat{\tau}_s) = \text{sort}(\hat{\gamma}_1, \hat{\gamma}_2, \ldots, \hat{\gamma}_s)$.

Plots of all elements of the distance matrix D, computed using the dissimilarity measure $d(\boldsymbol{X}_i, \boldsymbol{X}_j)$ in (12.8), are shown for both the univariate and multivariate distance functions in Figure 12.5. Additionally, plots of all column sums of the difference distance matrix $\boldsymbol{\Delta}$ are provided to illustrate the detection of multiple change points.

The recursive binary segmentation procedure begins by applying the single change point algorithm to the full data sequence $[\boldsymbol{X}_1, \boldsymbol{X}_2, \ldots, \boldsymbol{X}_n]$. If a change point is detected, the sequence is split into two segments sub-sequences, preceding and following the detected change point, and the algorithm is then applied separately to each sub-sequence to identify any additional change points. Let b_k and e_k denote the starting and ending indices of a given sub-sequence, $[\boldsymbol{X}_{b_k}, \boldsymbol{X}_{b_k+1}, \ldots, \boldsymbol{X}_{e_k}]$, for the purpose of detecting a potential change point $\hat{\gamma}_k$. If a change point $\hat{\gamma}_k$ is identified within this sub-sequence, the data is further partitioned

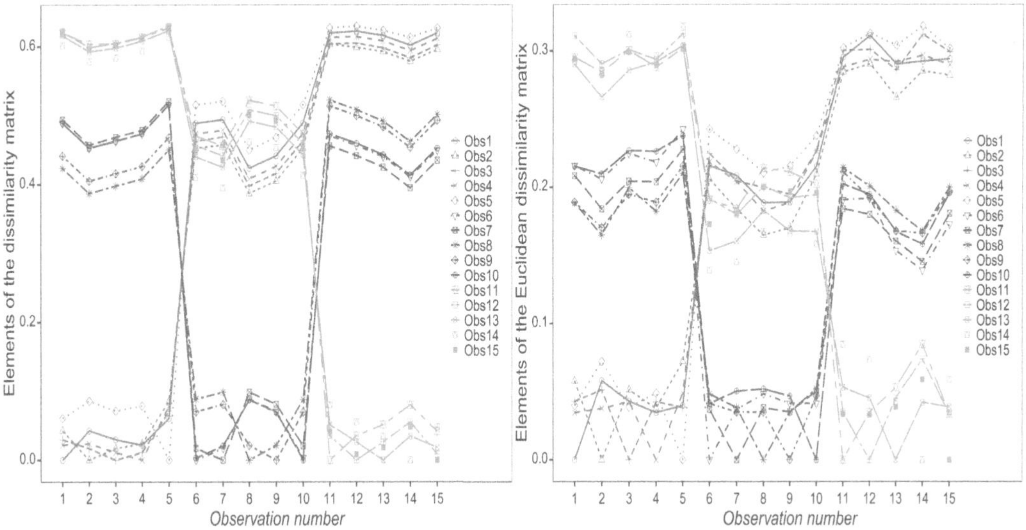

FIGURE 12.4: Left panel: For matrix D using 12.9. Right Panel: For matrix D using 12.7.

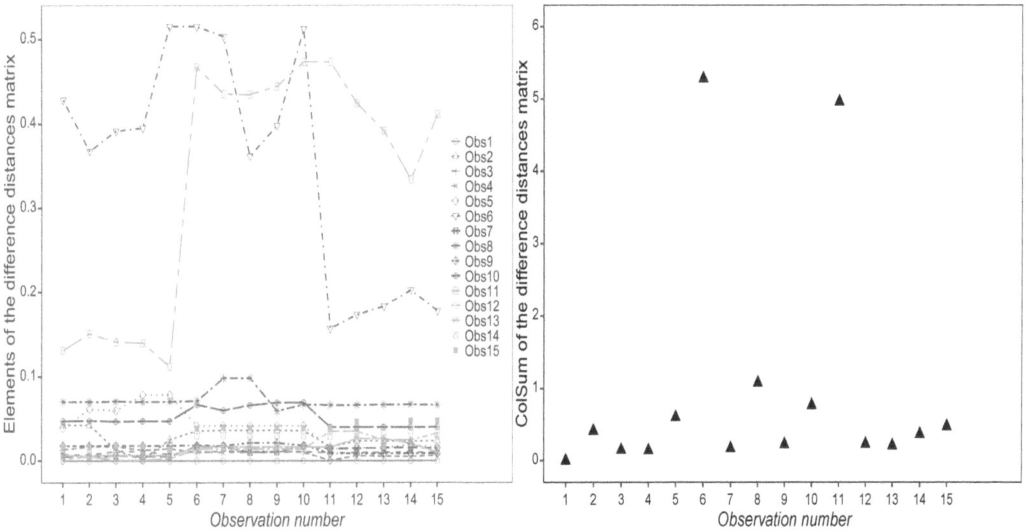

FIGURE 12.5: Left panel: Difference distance matrix for $\boldsymbol{\Delta}$. Right Panel: Column sum of difference distance matrix $\boldsymbol{\Delta}$.

into two new sub-sequences: $[\boldsymbol{X}_{b_k}, \boldsymbol{X}_{b_k+1}, \ldots, \boldsymbol{X}_{\hat{\gamma}_k-1}]$ and $[\boldsymbol{X}_{\hat{\gamma}_k}, \boldsymbol{X}_{\hat{\gamma}_k+1}, \ldots, \boldsymbol{X}_{e_k}]$. The single change point algorithm is then recursively applied to each of these sub-sequences. This process continues iteratively until no additional change points are detected.

For multiple change point detection, we extend the formula of the single change point estimate (12.6) and write the estimate of the change point location γ_k as follows

$$\hat{\gamma}_k = \underset{b_k \le j \le e_k}{\arg\max} \left\{ \frac{1}{e_k - b_k + 1} \sum_{i=b_k}^{e_k} \Delta_{ij} \right\}, \tag{12.15}$$

where we note that $b_1 = 1$ and $e_1 = n$, implying $e_1 - b_1 + 1 = n$. When checking the significance of the change point estimate $\hat{\gamma}_k$, we need to update both ∇^- and ∇^+ according to the change point estimate $\hat{\gamma}_k$. For this, the two sets of indices for observations before and after the change point estimate $\hat{\gamma}_k$ are expressed as

$$\nabla^-_{\hat{\gamma}_k} = \{j : b_k \le j \le \hat{\gamma}_k - 1\}, \quad \nabla^+_{\hat{\gamma}_k} = \{j : \hat{\gamma}_k \le j \le e_k\}.$$

To conduct the test for significance of $\hat{\gamma}_k$, the test statistic can be defined similarly as

$$T(\hat{\gamma}_k) = \frac{1}{(e_k - b_k + 1)|\nabla^-_{\hat{\gamma}_k}||\nabla^+_{\hat{\gamma}_k}|} \sum_{i=b_k}^{e_k} \sum_{j \in \nabla^-_{\hat{\gamma}_k}} \sum_{j' \in \nabla^+_{\hat{\gamma}_k}} (d_{ij} - d_{ij'})^2,$$

which simplifies to the test statistic (12.10) when $k = 1$. Algorithm B summarizes our method for multiple change point detection.

Algorithm 12.4.1 Algorithm B: Multiple Change Point Detection

1: **Input:** A data sequence or matrix of observations

$$\boldsymbol{X} = [\boldsymbol{X}_1, \boldsymbol{X}_2, \ldots, \boldsymbol{X}_n]^T$$

2: **Output:** A list of significant change point estimates $\{\hat{\tau}_1, \hat{\tau}_2, \ldots, \hat{\tau}_s\}$, or "NA" if there is no significant change point

3: **Step 1:** Apply the single change point detection algorithm (Algorithm A) to the data sequence $\boldsymbol{X}$. If there is no significant change point, terminate the procedure and return "NA". Otherwise, denote the detected change point by $\hat{\gamma}_1$ and proceed to the next step.

4: **Step 2:** Split the data sequence into two sub-sequences before and after the detected change point. Apply Algorithm A to each of the two sub-sequences separately to check for further change points.

5: **Step 3:** Repeat Step 2 until no further sub-sequences contain significant change points

6: **Step 4:** Denote all detected change points by

$$\{\hat{\gamma}_1, \hat{\gamma}_2, \ldots, \hat{\gamma}_s\}$$

Return

$$(\hat{\tau}_1, \hat{\tau}_2, \ldots, \hat{\tau}_s) = \text{sort}(\hat{\gamma}_1, \hat{\gamma}_2, \ldots, \hat{\gamma}_s)$$

as the list of significant change points

We further integrate the wild binary segmentation procedure of Fryzlewicz [111] into our framework for multiple change point detection. Following the principles of wild binary segmentation, we first randomly draw a large number of intervals (b_k^*, e_k^*) from the entire domain $\{1, \ldots, n\}$, including the full interval $(1, n)$. For each interval, we identify

$$\arg\max_{b_k^* \le j \le e_k^*} \left\{ \frac{1}{e_k^* - b_k^* + 1} \sum_{i=b_k^*}^{e_k^*} \Delta_{ij} \right\}.$$

The candidate change point is chosen as the location that maximizes $\frac{1}{e_k^* - b_k^* + 1} \sum_{i=b_k^*}^{e_k^*} \Delta_{ij}$ across all sampled intervals. The significance of this candidate is then evaluated using either the asymptotic or permutation test. If the candidate is significant, the procedure is applied recursively to the segments to the left and right of the detected change point. This process continues until no further significant change points are identified. Theoretical results for wild binary segmentation indicate that it performs at least as well as standard binary segmentation. Algorithm B can be readily adapted to incorporate the wild binary segmentation procedure.

12.5 Asymptotic Results

We investigate the asymptotic properties of our proposed methodology for detecting single and multiple high-dimensional change points. The theoretical development is presented for both the multivariate modified Euclidean distance and the univariate distance based on differences in the sample mean and variance of the observations, although analogous results can be derived for other appropriate dissimilarity measures, including general L_s-norm distances. The analysis considers both the high-dimensional regime where n and d simultaneously approach infinity, as well as the HDLSS setting in which $d \to \infty$ while n remains fixed.

To develop the asymptotic theory with $d(\boldsymbol{X}_i, \boldsymbol{X}_j)$ based on the multivariate modified Euclidean distance function, according to Hall et al. [133] we make the following three assumptions on observations $\boldsymbol{X}_i = (\boldsymbol{X}_{i1}, \boldsymbol{X}_{i2}, \ldots, \boldsymbol{X}_{id})$, $i = 1, \ldots, n$.

(A1) Assume that $\max_{1\le i\le n} \max_{1\le j\le d} \mathbb{E}(\boldsymbol{X}_{ij}^4) < \infty$.

(A2) Assume $tr(\boldsymbol{\Sigma}_i)/d \to \lambda_i$ and $d^{-1/2}\|\mathbb{E}(\boldsymbol{X}_i) - \mathbb{E}(\boldsymbol{X}_j)\|_2 \to \eta_{ij}$ for $1 \le i, j \le n$.

(A3) Assume $\sum_{j=1}^{d}\sum_{\substack{j'=1 \\ j\neq j'}}^{d} mathbbCov(\boldsymbol{X}_{ij}^2, \boldsymbol{X}_{ij'}^2) = o(d^2)$.

Alternatively, for using the univariate distance function based on differences between the sample mean and variance of observations, we make the following assumptions.

(B1) Assume that $\max_{1\le i\le n} \max_{1\le j\le d} \mathbb{E}(\boldsymbol{X}_{ij}^2) < \infty$.

(B2) Let $\boldsymbol{S}_{ij}^2 = (\boldsymbol{X}_{ij} - \bar{\boldsymbol{X}}_i^d)^2$ and define $\sigma_{ij}^2 = \mathbb{E}(\boldsymbol{S}_{ij}^2)$. Assume $\max_{1\le i\le n} \max_{1\le j\le d} \sigma_{ij}^2 < \infty$.

(B3) For all $\epsilon > 0$ and $i = 1, \ldots, n$, define

$$A_i = \sum_{j=1}^{d} E\big((\boldsymbol{X}_{ij} - \bar{\mu}_{ij})^2\big), \quad B_i = \sum_{j=1}^{d} E\big((\boldsymbol{X}_{ij} - \bar{\mu}_{ij})^2 \mathbb{I}(|\boldsymbol{X}_{ij} - \bar{\mu}_{ij}| > \epsilon A_i)\big),$$

$$A_i^* = \sum_{j=1}^{d} E\big((\boldsymbol{S}_{ij}^2 - \sigma_{ij}^2)^2\big), \quad B_i^* = \sum_{j=1}^{d} E\big((\boldsymbol{S}_{ij}^2 - \sigma_{ij}^2)^2 \mathbb{I}(|\boldsymbol{S}_{ij}^2 - \sigma_{ij}^2| > \epsilon A_i^*)\big),$$

and assume $\frac{B_i}{A_i} \to 0$ and $\frac{B_i^*}{A_i^*} \to 0$ as $d \to \infty$.

(B4) Assume $\sum_{j=1}^{d}\sum_{\substack{j'=1 \\ j\neq j'}}^{d} \mathbb{C}ov(\boldsymbol{X}_{ij}, \boldsymbol{X}_{ij'}) = o(d^2)$ and $\sum_{j=1}^{d}\sum_{\substack{j'=1 \\ j\neq j'}}^{d} \mathbb{C}ov(\boldsymbol{S}_{ij}^2, \boldsymbol{S}_{ij'}^2) = o(d^2)$ as $d \to \infty$.

(B5) Assume $\sum_{i=1}^{n}\sum_{j=1}^{n}\sum_{k=1}^{n}\sum_{\substack{l=1 \\ i\neq j\neq k\neq l}}^{n} \mathbb{C}ov\big(\|\boldsymbol{X}_i - \boldsymbol{X}_j\|_D^2, \|\boldsymbol{X}_k - \boldsymbol{X}_l\|_D^2\big) = o(n^4)$ as $n \to \infty$.

These are standard assumptions required for the convergence of the distance functions in our asymptotic analysis. Assumptions (A1)–(A3) are necessary for the convergence of the modified Euclidean distance, as used by Hall et al. [133], Modarres [229], and others. In particular, Assumption (A3) implies weak dependence among variables and is only relevant in the presence of correlated variables; it is trivially satisfied when the variables are independent. Assumptions (B1) and (B2) ensure bounded second moments, which are required for the convergence of the mean $\bar{\boldsymbol{X}}_i^d$ and standard deviation $S^d_{\boldsymbol{X}_i}$. Assumption (B3) corresponds to the standard Lindeberg condition, commonly employed to apply the central limit theorem; this is a weaker requirement than the Lyapunov condition.

Assumptions (B4)–(B5) impose weak dependence among variables and are relevant only for correlated variables. Specifically, (B4) guarantees bounded covariances for the components $\boldsymbol{X}_{ij}$, a standard condition to satisfy the central limit theorem and the weak law of large numbers under dependence. Assumption (B5) is an analogous covariance condition for $\|\boldsymbol{X}_i - \boldsymbol{X}_j\|_D^2$, which can be simplified using

$$\|\boldsymbol{X}_i - \boldsymbol{X}_j\|_D^2 = (\bar{\boldsymbol{X}}_i^d - \bar{\boldsymbol{X}}_j^d)^2 + (S^d_{\boldsymbol{X}_i} - S^d_{\boldsymbol{X}_j})^2.$$

Note that Assumptions (B1) and (B2) imply

$$\sum_{i=1}^{n}\sum_{j=1}^{n}\mathbb{E}(\|\boldsymbol{X}_i - \boldsymbol{X}_j\|_D^2) < \infty.$$

Moreover, Assumptions (A3) and (B4)–(B5) hold under variables exhibiting the ρ-mixing property as well as spatial dependence, allowing for a more general dependence structure. For instance, under spatial dependence with

$$\mathbb{C}ov(\boldsymbol{X}_{ij}, \boldsymbol{X}_{ij'}) \propto \rho^{|j-j'|}, \quad |\rho| \leq 1,$$

it follows that

$$\sum_{j\neq j'}\sum_{j=1}^{d}\sum_{j'=1}^{d}\mathbb{C}ov(\boldsymbol{X}_{ij}, \boldsymbol{X}_{ij'}) = o(d^2).$$

If observations $\boldsymbol{X}_i$ and $\boldsymbol{X}_j$ have the same distribution, we can write

$$\begin{aligned}
&\lambda_i = \lambda_j := \lambda_{\nabla_\tau^-}, \ \eta_{ij} := \eta_{\nabla_\tau^-} && \forall i, j \in \nabla_\tau^-,\\
&\lambda_i = \lambda_j := \lambda_{\nabla_\tau^+}, \ \eta_{ij} := \eta_{\nabla_\tau^+} && \forall i, j \in \nabla_\tau^+,\\
&\bar{\mu}_i = \bar{\mu}_j := \bar{\mu}_{\nabla_\tau^-}, \ \sigma_i = \sigma_j := \sigma_{\nabla_\tau^-} && \forall i, j \in \nabla_\tau^-,\\
&\bar{\mu}_i = \bar{\mu}_j := \bar{\mu}_{\nabla_\tau^+}, \ \sigma_i = \sigma_j := \sigma_{\nabla_\tau^+} && \forall i, j \in \nabla_\tau^+,
\end{aligned}$$

in which $\bar{\mu}_i = \mathbb{E}(\bar{\boldsymbol{X}}_i^d)$ and $\sigma_i = \sqrt{\mathbb{E}((S^d_{\boldsymbol{X}_i})^2)}$ for $i = 1, \ldots, n$. Note that $\bar{\mu}_i = d^{-1}\sum_{j=1}^{d}\mu_{ij}$ and $\sigma_i^2 = d^{-1}\sum_{j=1}^{d}\sigma_{ij}^2$. The following theorem concerns the asymptotic behaviour of the dissimilarity measure $d(\boldsymbol{X}_i, \boldsymbol{X}_j)$ for high-dimensional observations.

Theorem 12.5.1 *Consider the data sequence* $\boldsymbol{X}_1, \boldsymbol{X}_2, \ldots, \boldsymbol{X}_n$ *and dissimilarity measure* $d(\boldsymbol{X}_i, \boldsymbol{X}_j)$ *in* (12.8). *Suppose that* τ *is a change point so that* $F_1 = \ldots = F_{\tau-1} \neq F_\tau = \ldots = F_n$. *As* $d \to \infty$, *we have*

$$\begin{aligned}
&d(\boldsymbol{X}_i, \boldsymbol{X}_j) = o_{\mathbb{P}}(1) && \forall\, i \in \nabla_\tau^-, j \in \nabla_\tau^- \ \textit{or} \ \forall\, i \in \nabla_\tau^+, j \in \nabla_\tau^+,\\
&d(\boldsymbol{X}_i, \boldsymbol{X}_j) = \kappa_\tau + o_{\mathbb{P}}(1) && \forall\, i \in \nabla_\tau^-, j \in \nabla_\tau^+ \ \textit{or} \ \forall\, i \in \nabla_\tau^+, j \in \nabla_\tau^-,
\end{aligned}$$

where for the modified Euclidean distance function, under Assumptions (A1)-(A3),

$$\begin{aligned}
\kappa_\tau = {} & \frac{1}{N-2}\\
& \times \Big\{(|\nabla_\tau^-| - 1)\big|\sqrt{\lambda^2_{\nabla_\tau^-} + \lambda^2_{\nabla_\tau^+} + \eta^2_{\nabla_\tau^-\nabla_\tau^+}} - \sqrt{2\lambda^2_{\nabla_\tau^-}}\big| + (|\nabla_\tau^+| - 1)\big|\sqrt{\lambda^2_{\nabla_\tau^-} + \lambda^2_{\nabla_\tau^+} + \eta^2_{\nabla_\tau^-\nabla_\tau^+}} - \sqrt{2\lambda^2_{\nabla_\tau^+}}\big|\Big\},
\end{aligned}$$

while for the univariate distance function based on differences between the sample mean and variance, under Assumptions (B1)-(B4),

$$\kappa_\tau = \left|\sqrt{(\bar{\mu}_{\nabla_\tau^-} - \bar{\mu}_{\nabla_\tau^+})^2 + (\sigma_{\nabla_\tau^-} - \sigma_{\nabla_\tau^+})^2}\right|.$$

The following two theorems provide theoretical guarantees for the consistency of the proposed single change point estimator (12.6) in high-dimensional settings, both when $d \to \infty$ with n fixed and when $d \to \infty$ while n also diverges.

Theorem 12.5.2 *Suppose that there is a true single change point τ_0, $2 \leq \tau_0 \leq n$, so that $F_1 = \ldots = F_{\tau_0 - 1} \neq F_{\tau_0} = \ldots = F_n$. Under the assumptions of Theorem 12.5.1, (a) we have as $d \to \infty$, with any fixed n,*

$$\begin{cases} \frac{1}{n}\sum_{i=1}^{n} \Delta_{ij} = o_{\mathbb{P}}(1) & \text{if } j \neq \tau_0, \\ \frac{1}{n}\sum_{i=1}^{n} \Delta_{ij} = \kappa_{\tau_0} + o_{\mathbb{P}}(1) & \text{if } j = \tau_0, \end{cases}$$

where $\kappa_{\tau_0} = O(1)$ under Assumptions (A1)-(A2) or (B1)-(B2).
(b) it follows $\hat{\tau} - \tau_0 = o_{\mathbb{P}}(1)$, hence the change point estimate $\hat{\tau} = \arg\max_{1 \leq j \leq n}\left\{\frac{1}{n}\sum_{i=1}^{n} \Delta_{ij}\right\}$ is consistent for τ_0 as $d \to \infty$.

From this theorem, the consistency of the proposed change point estimate holds when $d \to \infty$ and n is fixed as in HDLSS data. In the following result, we show that the consistency also holds when $n \to \infty$ as well, as one expects.

Theorem 12.5.3 *Under the conditions in Theorem 12.5.2, we have as $n \to \infty$ and $d \to \infty$*

$$\max_{1 \leq j \leq n}\left\{\frac{1}{n}\sum_{i=1}^{n} \Delta_{ij} - \mathbb{I}(j = \tau_0)\kappa_{\tau_0}\right\} = o_{\mathbb{P}}(1).$$

The next theorem studies the asymptotic limit of the proposed test statistic (12.10) under the null hypothesis H_0 as well as under the alternative hypothesis H_1^s.

Theorem 12.5.4 *Suppose that there is a true single change point τ_0, $2 \leq \tau_0 \leq n$, and consider the statistic $T(\hat{\tau})$ in (12.10) based on the change point estimate $\hat{\tau} = \arg\max_{1 \leq j \leq n}\left\{\frac{1}{n}\sum_{i=1}^{n} \Delta_{ij}\right\}$. Under the above assumptions, we have as $d \to \infty$*

$$T(\hat{\tau})\big|_{H_0} = o_{\mathbb{P}}(1), \qquad T(\hat{\tau})\big|_{H_1^s} = \kappa_{\tau_0}^2 + o_{\mathbb{P}}(1).$$

The following theorem proves that the asymptotic distribution of the test statistic (12.10) is a normal distribution under the above conditions when both n and d go to infinity.

Theorem 12.5.5 *Consider the test statistic $T(\hat{\tau})$ in (12.10) for testing a significant change point. Under the null hypothesis H_0 and the above assumptions, we have as $n \to \infty$ and $d \to \infty$*

$$T_{sd}(\hat{\tau}) := \frac{n|\nabla_{\hat{\tau}}^-||\nabla_{\hat{\tau}}^+|\Big(T(\hat{\tau}) - \frac{1}{n|\nabla_{\hat{\tau}}^-||\nabla_{\hat{\tau}}^+|}\sum_{i=1}^{n}\sum_{j\in\nabla_{\hat{\tau}}^-}\sum_{j'\in\nabla_{\hat{\tau}}^+} m_{ijj'}\Big)}{\sqrt{\sum_{i=1}^{n}\sum_{j\in\nabla_{\hat{\tau}}^-}\sum_{j'\in\nabla_{\hat{\tau}}^+}\sum_{k=1}^{n}\sum_{l\in\nabla_{\hat{\tau}}^-}\sum_{l'\in\nabla_{\hat{\tau}}^+} c_{ijj'kll'}}} \xrightarrow{D} \mathbb{N}(0,1),$$

where $T_{sd}(\hat{\tau})$ is defined as the standardized test statistic. For the univariate distance function we have $m_{ijj'} = v_{ij} + v_{ij}^* + v_{ij'} + v_{ij'}^* + o^d(1)$ and

$$\begin{aligned} c_{ijj'kll'} &= \mathbb{C}ov\big(\|\boldsymbol{X}_i - \boldsymbol{X}_j\|_D^2 + \|\boldsymbol{X}_i - \boldsymbol{X}_{j'}\|_D^2, \|\boldsymbol{X}_k - \boldsymbol{X}_l\|_D^2 + \|\boldsymbol{X}_k - \boldsymbol{X}_{l'}\|_D^2\big) \\ &= \mathbb{C}ov\big(\|\boldsymbol{X}_i - \boldsymbol{X}_j\|_D^2, \|\boldsymbol{X}_k - \boldsymbol{X}_l\|_D^2\big) + \mathbb{C}ov\big(\|\boldsymbol{X}_i - \boldsymbol{X}_j\|_D^2, \|\boldsymbol{X}_k - \boldsymbol{X}_{l'}\|_D^2\big) \\ &\quad + \mathbb{C}ov\big(\|\boldsymbol{X}_i - \boldsymbol{X}_{j'}\|_D^2, \|\boldsymbol{X}_k - \boldsymbol{X}_l\|_D^2\big) + \mathbb{C}ov\big(\|\boldsymbol{X}_i - \boldsymbol{X}_{j'}\|_D^2, \|\boldsymbol{X}_k - \boldsymbol{X}_{l'}\|_D^2\big), \end{aligned}$$

where

$$\mathbb{C}ov\big(\|\boldsymbol{X}_i - \boldsymbol{X}_j\|_D^2, \|\boldsymbol{X}_k - \boldsymbol{X}_l\|_D^2\big) = \begin{cases} 2(v_{ij}^2 + v_{ij}^{*2}) + o^d(1) & \text{if } i \neq j, k \neq l, i = k, j = l, \\ (v_{ii}^{*2} + v_{ii}^{*2})/2 + o^d(1) & \text{if } i \neq j, k \neq l, i = k, j \neq l, \\ (v_{jj}^{*2} + v_{jj}^{*2})/2 + o^d(1) & \text{if } i \neq j, k \neq l, i \neq k, j = l, \\ o^d(1) & \text{otherwise,} \end{cases}$$

and $v_{ij} = \mathbb{V}ar(\bar{\boldsymbol{X}}_i^d) + \mathbb{V}ar(\bar{\boldsymbol{X}}_j^d)$ *and* $v_{ij}^* = \mathbb{V}ar(S_{\boldsymbol{X}_i}^d) + \mathbb{V}ar(S_{\boldsymbol{X}_j}^d)$. *For the multivariate modified Euclidean distance function we have* $m_{ijj'} = \mathbb{E}(d^{-1}\|\boldsymbol{X}_i - \boldsymbol{X}_j\|_2^2) + \mathbb{E}(d^{-1}\|\boldsymbol{X}_i - \boldsymbol{X}_{j'}\|_2^2)$ *and*

$$\begin{aligned} c_{ijj'kll'} &= \mathbb{C}ov\big(d^{-1}\|\boldsymbol{X}_i - \boldsymbol{X}_j\|_2^2 + d^{-1}\|\boldsymbol{X}_i - \boldsymbol{X}_{j'}\|_2^2, d^{-1}\|\boldsymbol{X}_k - \boldsymbol{X}_l\|_2^2 + d^{-1}\|\boldsymbol{X}_k - \boldsymbol{X}_{l'}\|_2^2\big) \\ &= \mathbb{C}ov\big(d^{-1}\|\boldsymbol{X}_i - \boldsymbol{X}_j\|_2^2, d^{-1}\|\boldsymbol{X}_k - \boldsymbol{X}_l\|_2^2\big) + \mathbb{C}ov\big(d^{-1}\|\boldsymbol{X}_i - \boldsymbol{X}_j\|_2^2, d^{-1}\|\boldsymbol{X}_k - \boldsymbol{X}_{l'}\|_2^2\big) \\ &\quad + \mathbb{C}ov\big(d^{-1}\|\boldsymbol{X}_i - \boldsymbol{X}_{j'}\|_2^2, d^{-1}\|\boldsymbol{X}_k - \boldsymbol{X}_l\|_2^2\big) + \mathbb{C}ov\big(d^{-1}\|\boldsymbol{X}_i - \boldsymbol{X}_{j'}\|_2^2, \|\boldsymbol{X}_k - \boldsymbol{X}_{l'}\|_2^2\big). \end{aligned}$$

Remark.

We employ estimates of the constants v_{ij} and v_{ij}^* to compute $m_{ijj'}$ and $c_{ijj'kll'}$. By Slutsky's theorem, the result of Theorem 12.5.5 continues to hold when consistent estimates of v_{ij} and v_{ij}^* are substituted. In the univariate distance case, a consistent estimate of $\mathbb{V}ar(\bar{\boldsymbol{X}}_i^d)$ in v_{ij} is obtained as follows (under Assumption (B4)):

$$\mathbb{V}ar(\bar{\boldsymbol{X}}_i^d) = \frac{1}{d^2}\sum_{l=1}^{d}\sum_{l'=1}^{d}\mathbb{C}ov(\boldsymbol{X}_{il}, \boldsymbol{X}_{il'}) = \frac{1}{d^2}\sum_{l=1}^{d}\mathbb{V}ar(\boldsymbol{X}_{il}) + \frac{o(d^2)}{d^2},$$

and

$$d^{-1}\sum_{l=1}^{d}(\boldsymbol{X}_{il} - \bar{\boldsymbol{X}}_i^d)^2 - d^{-1}\sum_{l=1}^{d}\mathbb{V}ar(\boldsymbol{X}_{il}) \xrightarrow{d} 0.$$

Similarly, a consistent estimate of $\mathbb{V}ar(S_{\boldsymbol{X}_i}^d)$ in v_{ij}^* can be obtained by first applying (under Assumption (B4)) the analogous covariance decomposition for the sample variance.

$$\begin{aligned} \mathbb{V}ar\big((S_{\boldsymbol{X}_i}^d)^2\big) &= \frac{1}{d^2}\sum_{l=1}^{d}\sum_{l'=1}^{d}\mathbb{C}ov\big((\boldsymbol{X}_{il} - \bar{\boldsymbol{X}}_i^d)^2, (\boldsymbol{X}_{il'} - \bar{\boldsymbol{X}}_i^d)^2\big) \\ &= \frac{1}{d^2}\sum_{l=1}^{d}\mathbb{V}ar\big((\boldsymbol{X}_{il} - \bar{\boldsymbol{X}}_i^d)^2\big) + \frac{o(d^2)}{d^2} \\ &= \frac{1}{d^2}\sum_{l=1}^{d}\mathrm{E}\big((\boldsymbol{X}_{il} - \bar{\boldsymbol{X}}_i^d)^4\big) - \frac{1}{d^2}\sum_{l=1}^{d}\big(\mathbb{V}ar(\boldsymbol{X}_{il})\big)^2 + o(1) \\ &= \frac{1}{d^2}\sum_{l=1}^{d}\mathrm{E}\big((\boldsymbol{X}_{il} - \bar{\boldsymbol{X}}_i^d)^4\big) - d\big(\mathbb{V}ar(\bar{\boldsymbol{X}}_i^d)\big)^2 + o(1) \end{aligned} \tag{12.16}$$

and $d^{-1}\sum_{l=1}^{d}(\boldsymbol{X}_{il} - \bar{\boldsymbol{X}}_i^d)^4 - d^{-1}\sum_{l=1}^{d}\mathrm{E}\big((\boldsymbol{X}_{il} - \bar{\boldsymbol{X}}_i^d)^4\big) \xrightarrow{P} 0$, and then applying the delta method to obtain the required estimate as $\widehat{\mathbb{V}ar}(S_{\boldsymbol{X}_i}^d) = \frac{1}{4(S_{\boldsymbol{X}_i}^d)^2}\widehat{\mathbb{V}ar}\big((S_{\boldsymbol{X}_i}^d)^2\big)$. Note that the last equality in (12.16) is obtained by applying the Taylor expansion $\big(\mathbb{V}ar(\boldsymbol{X}_{il})\big)^2 = \big(d\mathbb{V}ar(\bar{\boldsymbol{X}}_i^d)\big)^2 + 2d\mathbb{V}ar(\bar{\boldsymbol{X}}_i^d)\big(\mathbb{V}ar(\boldsymbol{X}_{il}) - d\mathbb{V}ar(\bar{\boldsymbol{X}}_i^d)\big) + o(d)$. Also for the modified Euclidean distance case, under Assumption (A3), we use the consistent estimates of $\mathbb{E}(\|\boldsymbol{X}_i - \boldsymbol{X}_j\|_2^2)$ and $\mathbb{C}ov\big(\|\boldsymbol{X}_i - \boldsymbol{X}_j\|_2^2, \|\boldsymbol{X}_k - \boldsymbol{X}_l\|_2^2\big)$.

We now study the optimality and detection power of the proposed method and obtain conditions for optimal detection rates and full power.

Theorem 12.5.6 *Consider the conditions in Theorem 12.5.1 and Theorem 12.5.5. We have, as $d \to \infty$,*

$$\mathbb{P}_{H_0}(|T_{sd}(\hat{\tau})| > Z_{\alpha/2}) \to \alpha,$$
$$\mathbb{P}_{H_1^s}(|T_{sd}(\hat{\tau})| > Z_{\alpha/2}) = 1 - \Phi\big(Z_{\alpha/2}\frac{c}{c^*} - \frac{n^2\kappa_{\tau_0}^2}{c^*}\big) + \Phi\big(-Z_{\alpha/2}\frac{c}{c^*} - \frac{n^2\kappa_{\tau_0}^2}{c^*}\big) + o(1),$$

where Φ is the CDF of the standard normal distribution, $Z_{\alpha/2}$ is the corresponding critical point of the standard normal distribution, $c = \sqrt{\sum_{i=1}^n \sum_{j\in\nabla_{\hat{\tau}}^-} \sum_{j'\in\nabla_{\hat{\tau}}^+} \sum_{k=1}^n \sum_{l\in\nabla_{\hat{\tau}}^-} \sum_{l'\in\nabla_{\hat{\tau}}^+}}$ $c_{ijj'kll'}$ with $c_{ijj'kll'}$ given in Theorem 12.5.5, and

$$c^* = \sqrt{\sum_{i=1}^n \sum_{j\in\nabla_{\hat{\tau}}^-} \sum_{j'\in\nabla_{\hat{\tau}}^+} \sum_{k=1}^n \sum_{l\in\nabla_{\hat{\tau}}^-} \sum_{l'\in\nabla_{\hat{\tau}}^+}} c^*_{ijj'kll'}$$

with $c^*_{ijj'kll'} = c_{ijj'kll'} + b_{ijkl} + b_{ijkl'} + b_{ij'kl} + b_{ij'kl'}$ where b_{ijkl} is specified in the proof of theorem.

Under Assumptions (B4)-(B5) or (A3) the covariances are bounded asymptotically, so c and c^* are finite. So, with $\kappa_{\tau_0}^2 > 0$, the test is consistent and has full power as $n \to \infty$, that is,

$$\mathbb{P}_{H_1^s}(|T_{\mathrm{sd}}(\hat{\tau})| > Z_{\alpha/2}) \to 1, \quad \text{as } n \to \infty \text{ and } d \to \infty.$$

Also, with fixed n as in HDLSS data, the test is consistent if $\kappa_{\tau_0}^2$ diverges. The following results demonstrate this for two cases when there is a change point in mean and when there is a change in variance of observations, considering the sparsity level and the change magnitude.

Corollary 12.5.1 *Consider a change point in mean of high-dimensional observations, while variance remains unchanged, with the mean shift $\delta_k := \mu_{ik} - \mu_{jk}$, $k = 1, \ldots, d$, for all $i \in \nabla_\tau^-, j \in \nabla_\tau^+$. Let $S_0 = \{k : \delta_k \neq 0\}$ be the set of variables having a change point with $s_0 := |S_0|$ being its cardinality or the sparsity level. Then, $\kappa_{\tau_0}^2 = \frac{1}{d^2}(\sum_{k\in S_0}\delta_k)^2$ and*

$$\mathbb{P}_{H_1^s}(|T_{sd}(\hat{\tau})| > Z_{\alpha/2}) = 1 - \Phi\big(Z_{\alpha/2}\frac{c}{c^*} - \frac{n^2(\sum_{k\in S_0}\delta_k)^2/d^2}{c^*}\big) + \Phi\big(-Z_{\alpha/2}\frac{c}{c^*} - \frac{n^2(\sum_{k\in S_0}\delta_k)^2/d^2}{c^*}\big) + o(1).$$

Hence, with fixed n, the test is consistent if $(\sum_{k\in S_0}\delta_k)^2 > d^2/n^2$. In particular, when $\delta_k = \delta_{\min}$ for all $k \in S_0$ with $\delta_{\min} = \min_{k\in S_0}\delta_k$, the optimality is achieved if

$$s_0 > \frac{d}{n|\delta_{\min}|} \quad \text{or} \quad |\delta_{\min}| > \frac{d}{ns_0}.$$

Corollary 12.5.2 *Consider a change point in variance of high-dimensional observations, while mean remains unchanged, with the variance shift $\omega_k := \sigma_{ik}^2 - \sigma_{jk}^2$, $k = 1, \ldots, d$, for all $i \in \nabla_\tau^-, j \in \nabla_\tau^+$. Let $S_0 = \{k : \omega_k \neq 0\}$ be the set of variables having a change point in this case with $s_0 := |S_0|$ being the sparsity level. Then, $\kappa_{\tau_0}^2 = \frac{(\sum_{k\in S_0}\omega_k)^2}{d^2(\sigma_{\nabla_{\tau_0}^-} + \sigma_{\nabla_{\tau_0}^+})^2}$. Hence, similar to Corollary 12.5.1, with fixed n, the test is consistent if $(\sum_{k\in S_0}\omega_k)^2 > d^2/n^2$, where the assumption of finite variance implies $\sigma_{\nabla_{\tau_0}^-} + \sigma_{\nabla_{\tau_0}^+} < \infty$. In particular, when $\omega_k = \omega_{\min}$ for all $k \in S_0$ with $\omega_{\min} = \min_{k\in S_0}\omega_k$, the optimality is achieved if*

$$s_0 > \frac{d}{n|\omega_{\min}|} \quad \text{or} \quad |\omega_{\min}| > \frac{d}{ns_0}.$$

We now demonstrate the consistency of multiple change point estimates from Algorithm B for multiple high-dimensional change point detection.

Theorem 12.5.7 *Suppose there are s true change points $\tau_1^0, \tau_2^0, \ldots, \tau_s^0$, $2 \leq \tau_1^0 < \tau_2^0 < \ldots < \tau_s^0 \leq n$, so that $F_1 = \ldots = F_{\tau_1^0 - 1} \neq F_{\tau_1^0} = \ldots = F_{\tau_2^0 - 1} \neq F_{\tau_2^0} = \ldots = F_{\tau_{s-1}^0} \neq F_{\tau_s^0} = \ldots = F_n$. Assume the minimum spacing between change points satisfies $\min_{1 \leq i \leq s-1} |\tau_{i+1}^0 - \tau_i^0| \geq Mn^{\varepsilon}$ for some $M > 0$ and $\varepsilon \leq 1$. Under the above assumptions, Algorithm B returns the change point estimates $(\hat{\tau}_1, \hat{\tau}_2, \ldots, \hat{\tau}_s) = sort(\hat{\gamma}_1, \hat{\gamma}_2, \ldots, \hat{\gamma}_s)$ that satisfy $\|(\hat{\tau}_1, \hat{\tau}_2, \ldots, \hat{\tau}_s) - (\tau_1^0, \tau_2^0, \ldots, \tau_s^0)\|_\infty = o_{\mathbb{P}}(1)$.*

The condition $\min_{1 \leq i \leq s-1} |\tau_{i+1}^0 - \tau_i^0| \geq Mn^{\varepsilon}$ ensures that there are not too many change points to detect, because it implies that $s \leq \left(\min_{1 \leq i \leq s-1} |\tau_{i+1}^0 - \tau_i^0|/M\right)^{1/\varepsilon}$ since $0 \leq s < n$. The following result shows that the asymptotic test and the permutation test are equivalent asymptotically and that the permutation test is also unbiased when $n \to \infty$ and $d \to \infty$.

Theorem 12.5.8 *Consider the asymptotic test with the distribution $G_T(t)$ obtained in Theorem 12.5.5 and the permutation test with the approximate permutation distribution $G_{T_R}(t)$ in (12.11) and with the permutation-based p-value p_{perm} in (12.12). Under the above assumptions and the null hypothesis H_0 of no change points, we have as $n \to \infty$ and $d \to \infty$*

(a) $G_{T_R}(t) \xrightarrow{D} G_T(t) \quad \forall t,$

(b) $\mathbb{P}_{H_0}(\mathbf{Z} \leq \alpha) \leq \alpha \quad \forall 0 \leq \alpha \leq 1.$

12.6 Numerical Results

Drikvandi and Modarres [80] assess the performance of the proposed methods across various simulation scenarios and compare them with several recent approaches for high-dimensional change point detection. Specifically, they benchmark the methods against the non-parametric E-divisive procedure of Matteson and James [221], the random projection-based Inspect method of Wang and Samworth [347], and the non-parametric HDcpdetect method of Li et al. [190]. For all competing methods, tuning parameters are set according to the recommended defaults provided in their respective R packages.

Drikvandi and Modarres [80] evaluate performance for both single and multiple change point detection under a variety of scenarios, with particular emphasis on challenging HDLSS settings in which n is small relative to d. Both the asymptotic and permutation tests of the method are investigated using the univariate and multivariate distance functions. In the simulations, they observe that the results for the univariate and multivariate distances are generally similar; therefore, to avoid redundancy, only one set of results is presented for comparison with other methods.

12.6.1 Single Change in Mean

We only consider the scenario of a single change point in the mean of high-dimensional observations. We generate n random observations $\boldsymbol{X}_1, \boldsymbol{X}_2, \ldots, \boldsymbol{X}_n$ from a d-variate normal distribution, where the first $3n/5$ observations are drawn from $\mathbb{N}(\boldsymbol{\mu}_1, \boldsymbol{\Sigma})$ and the remaining $2n/5$ observations from $\mathbb{N}(\boldsymbol{\mu}_2, \boldsymbol{\Sigma})$. We examine various high-dimensional settings with $n \in \{45, 90\}$ and $d \in \{500, 1500\}$, and consider mean vectors $\boldsymbol{\mu}_1 = \mathbf{0}_d$ and $\boldsymbol{\mu}_2 \in \{\mathbf{0}_d, (0.1 \times$

TABLE 12.1: Frequency and average number of detected change points over 250 replications when there is one true change point, including no change point.

μ_2	n	d	Number of true change points	Method	Frequency of the detected change points		Average number of detected change points
					0	1	
	45	500	0	Permutation	0.98	0.02	0.02
	45	500	0	Asymptotic	0.98	0.02	0.02
$\mathbf{0}_d$	45	500	0	E-divisive	0.97	0.03	0.03
	45	500	0	Inspect	0.97	0.03	0.03
	45	500	0	HDcpdetect	0.98	0.02	0.02
	45	1500	0	Permutation	0.95	0.05	0.05
	45	1500	0	Asymptotic	0.95	0.05	0.05
$\mathbf{0}_d$	45	1500	0	E-divisive	0.96	0.04	0.04
	45	1500	0	Inspect	0.94	0.06	0.06
	45	1500	0	HDcpdetect	0.96	0.04	0.04
	45	500	1	Permutation	0.80	0.20	0.20
	45	500	1	Asymptotic	0.73	0.27	0.27
$(0.1 \times \mathbf{1}_{3d/4}, 0 \times \mathbf{1}_{d/4})$	45	500	1	E-divisive	0.83	0.17	0.17
	45	500	1	Inspect	0.91	0.09	0.09
	45	500	1	HDcpdetect	0.87	0.13	0.13
	45	1500	1	Permutation	0.16	0.84	0.84
	45	1500	1	Asymptotic	0.13	0.87	0.87
$(0.1 \times \mathbf{1}_{3d/4}, 0 \times \mathbf{1}_{d/4})$	45	1500	1	E-divisive	0.38	0.62	0.62
	45	1500	1	Inspect	0.93	0.07	0.07
	45	1500	1	HDcpdetect	0.49	0.51	0.51
	45	500	1	Permutation	0.12	0.88	0.88
	45	500	1	Asymptotic	0.10	0.90	0.90
$(0.2 \times \mathbf{1}_{3d/4}, 0 \times \mathbf{1}_{d/4})$	45	500	1	E-divisive	0.14	0.86	0.86
	45	500	1	Inspect	0.77	0.23	0.23
	45	500	1	HDcpdetect	0.15	0.85	0.85
	45	1500	1	Permutation	0.00	1.00	1.00
	45	1500	1	Asymptotic	0.00	1.00	1.00
$(0.2 \times \mathbf{1}_{3d/4}, 0 \times \mathbf{1}_{d/4})$	45	1500	1	E-divisive	0.02	0.98	0.98
	45	1500	1	Inspect	0.63	0.37	0.37
	45	1500	1	HDcpdetect	0.02	0.98	0.98

$\mathbf{1}_{3d/4}, 0 \times \mathbf{1}_{d/4}), (0.2 \times \mathbf{1}_{3d/4}, 0 \times \mathbf{1}_{d/4})\}$, where $\mathbf{0}_d$ and $\mathbf{1}_d$ denote d-dimensional vectors of zeros and ones, respectively.

For the covariance structure, we set $\boldsymbol{\Sigma} = \sigma_d^2 \boldsymbol{V}_d$ with $\sigma_d^2 \in \{0.5, 1\}$, and consider two forms of $\boldsymbol{V}_d$: the uncorrelated structure $\boldsymbol{V}_d = \boldsymbol{I}_d$, where $\boldsymbol{I}_d$ is the d-dimensional identity matrix, and the correlated autoregressive structure $\boldsymbol{V}_d = [V_{ij}]_{i,j=1}^d = [0.5^{|i-j|}]_{i,j=1}^d$. The true change point location is $\tau_1 = 3n/5 + 1$ in all scenarios except.

The results for the frequency and average number of detected change points are summarized in Table 12.1. It is evident that all methods perform well under the null scenario with no change point. However, in cases with a true change point, our proposed method using both the asymptotic and permutation tests outperforms the alternatives, namely E-divisive, Inspect, and HDcpdetect. Across the 250 replications, the proposed method identifies the change point with higher frequency under all scenarios considered.

We note that the Inspect method of Wang and Samworth [347] is less competitive in non-sparse high-dimensional settings, as it relies on sparsity and performs better with substantially larger sample sizes n when d is very large (Hahn et al. [132]; Follain et al. [104]). Drikvandi and Modarres [80] presents box plots of the estimated change point for each method over the 250 replications, illustrating that our proposed approach yields more accurate estimates. The performance of our method under the correlated autoregressive covariance structure is similar.

12.6.2 Multiple Changes in Mean

We next consider the case of multiple change points in the mean of high-dimensional observations, using the same simulation settings as before but now with three true change points in the data. Specifically, we generate n random observations $\boldsymbol{X}_1, \boldsymbol{X}_2, \ldots, \boldsymbol{X}_n$ from a d-variate normal distribution, where the first $3n/10$ observations are drawn from $\mathbb{N}(\boldsymbol{\mu}_1 = \boldsymbol{0}_d, \boldsymbol{\Sigma})$, the next $n/5$ observations from $\mathbb{N}(\boldsymbol{\mu}_2, \boldsymbol{\Sigma})$, the subsequent $3n/10$ observations from $\mathbb{N}(2\boldsymbol{\mu}_2, \boldsymbol{\Sigma})$, and the final $n/5$ observations from $\mathbb{N}(3\boldsymbol{\mu}_2, \boldsymbol{\Sigma})$. The three true change point locations are therefore $\tau_1 = 3n/10 + 1$, $\tau_2 = n/2 + 1$, and $\tau_3 = 4n/5 + 1$.

As in the single change point simulations, we perform 250 replications for each scenario and apply the proposed method alongside the alternative methods. We set the minimum segment length to 5. For each method, we compute both the frequency and the average number of correctly detected change points across the 250 replications, as well as the total number of detected change points, including any false positives.

Using multiple change points, Drikvandi and Modarres [80], present the frequency and average number of correctly detected change points for each method. From the results, it is evident that all methods perform well in scenarios without change points. In the presence of three true change points, our proposed method, implemented with both the asymptotic and permutation tests, consistently outperforms the competing methods across all scenarios considered. The reported frequencies indicate that the proposed approach identifies the true change points with higher accuracy compared to E-divisive (Matteson and James [221]), Inspect (Wang and Samworth [347]), and HDcpdetect (Li et al. [190]). Notably, the performance of our method, as well as E-divisive, improves with increasing dimension d, whereas Inspect is more sensitive to sample size n due to its reliance on sparsity (Hahn et al. [132]). The average number of correctly detected change points over the 250 replications is also substantially closer to the true number of change points for our method, demonstrating its robustness in multiple change point scenarios. Drikvandi and Modarres [80] also report the average total number of detected change points, including false positives, and indicate that our method neither over-detects nor under-detects change points. Based on these simulation findings, we recommend using the permutation test for relatively small datasets (approximately $n < 50$ or $d < 500$), while the asymptotic test is preferable for larger datasets due to its computational efficiency.

Drikvandi and Modarres [80] further investigate the performance of the methods for detecting a change in variance of the observations while the mean remains unchanged. They consider two scenarios in which the variance of the observations increases by 0.1 and 0.2, that is, $\boldsymbol{\Sigma}_1 = 0.5\boldsymbol{V}_d$ and $\boldsymbol{\Sigma}_2 \in \{0.6\boldsymbol{V}_d, 0.7\boldsymbol{V}_d\}$, while the mean remains constant, $\boldsymbol{\mu}_1 = \boldsymbol{\mu}_2$. From their results, it is clear that our proposed method, using both the asymptotic and permutation tests, performs well in detecting changes in variance, whereas the other methods show little to no power for detecting change points arising solely from variance shifts. Drikvandi and Modarres [80] further investigate change in distribution for fixed mean and variance.

Drikvandi and Modarres [80] further evaluate the performance of our methods in detecting high-dimensional change points when the distribution of observations is not normal. For this, we generate data from the asymmetric chi-squared distribution with one degree of freedom and scale the simulated observations to have the same mean and variance as in the previous cases with normal distribution. They show that the new method, based on both the asymptotic and permutation tests, performs better than the other methods. Unlike our method as well as E-divisive and HDcpdetect, which are distribution-free, the performance of Inspect is more affected compared to the previous results, as this method relies on the normality of observations. In the simulation experiments, a similar performance is also observed when data are generated from log-normal distribution and Student's t-distribution.

Drikvandi and Modarres [80] show the computation time of their method for multiple change point detection and evaluate the use of wild binary segmentation in their approach, as described at the end of section 12.4. We are interested in how the wild binary segmentation may improve the results. The results show that the wild binary segmentation slightly improves the results compared to the recursive binary segmentation. We note that our single change point algorithm is consistent, so it tends to produce a true change point, and thus the two binary segmentation procedures perform similarly well. As the results indicate, the use of wild binary segmentation requires considerably more computation time.

12.7 Analysis of S&P-500 Data Revisited

We apply the proposed method to a real data application using US stock return data. The dataset is available at https://www.finance.yahoo.com and can be obtained using the R package `BatchGetSymbols` for different time periods. The data considered here contains the daily closing prices of stocks from the S&P 500 index during the first half of the COVID-19 pandemic, from 1st January 2020 to 30th June 2020, resulting in $n = 125$ time points and $d = 496$ stocks. This specific period is chosen because, according to expert analysis reported in SP500-index (Statistical Research Department, 2022), the S&P 500 index exhibited high volatility, dropping by about 20% in early March 2020 and entering a bear market. While this drop represented the steepest one-day fall since 1987, the index began to recover at the start of April 2020. Stock markets fell amid the COVID-19 pandemic as investors feared a severe impact on economic growth. Figure 12.6 provides a visual summary of the standardized stock price changes over this period. A very steep drop can be observed around early March 2020, affecting the majority of stocks to varying degrees, indicating a non-sparse high-dimensional change point problem.

Change point plots for the S&P 500 data over the time period 1/1/2020 and 30/6/2020 during the COVID-19 pandemic. In plot 12.7 the two trading days, 13th and 16th March 2020, show the largest total dissimilarity from all the other days and are marked by vertical bars (in green).

Figure 12.7 displays the dissimilarity visualisation of the S&P 500 data using the dissimilarity measure $d(\boldsymbol{X}_i, \boldsymbol{X}_j)$ with both the modified Euclidean distance function and the univariate distance function, which exhibit a similar trend. The column sums of the two dissimilarity indices are shown for the trading days in the first half of 2020. The figure indicates that the trading days 13th and 16th March 2020 correspond to the largest total dissimilarity from all other observations, and these days are marked by vertical bars.

We implement our distribution-free method using our R package `HDDchangepoint` to detect significant change points in this high-dimensional dataset. Using the asymptotic test with both distance functions, our method identifies six change point locations: 51, 57, 66, 95, 106, 112. The permutation test returns four change point locations: 51, 57, 66, 95. The four common change points detected by both tests indicate consistent findings, while the permutation test appears slightly more conservative given the high dimension of the data, in line with our numerical results. The asymptotic p-values for these four significant change points are 5.64e−09, 5.06e−04, 1.27e−05, and 2.50e−07, respectively. Using a minimum segment length of 10 or 15 for binary segmentation yields the same four change points.

For comparison, the E-divisive method of Matteson and James [221] with minimum segment length 15 detects seven significant change points at locations 15, 37, 48, 68, 80, 95, 106. The random projection method of Wang and Samworth [347] finds nine change points: 18, 35, 44, 50, 65, 78, 99, 111, 124. The HDcpdetect method of Li et al. [190] identifies only two change points: 48 and 80. As observed in our simulations, HDcpdetect tends to detect

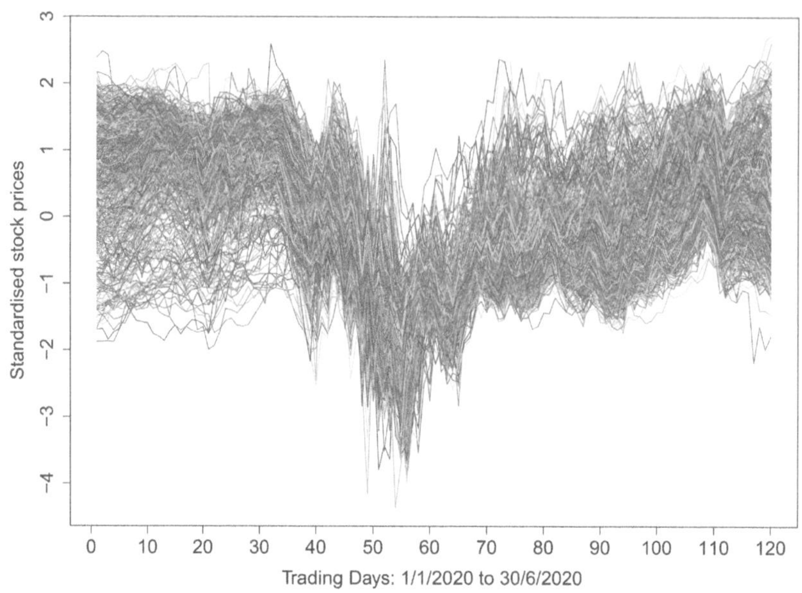

FIGURE 12.6: The prices for all the 496 stocks in the S&P 500 data over the time period 1/1/2020 and 30/6/2020. Note that the prices are standardized for visualisation.

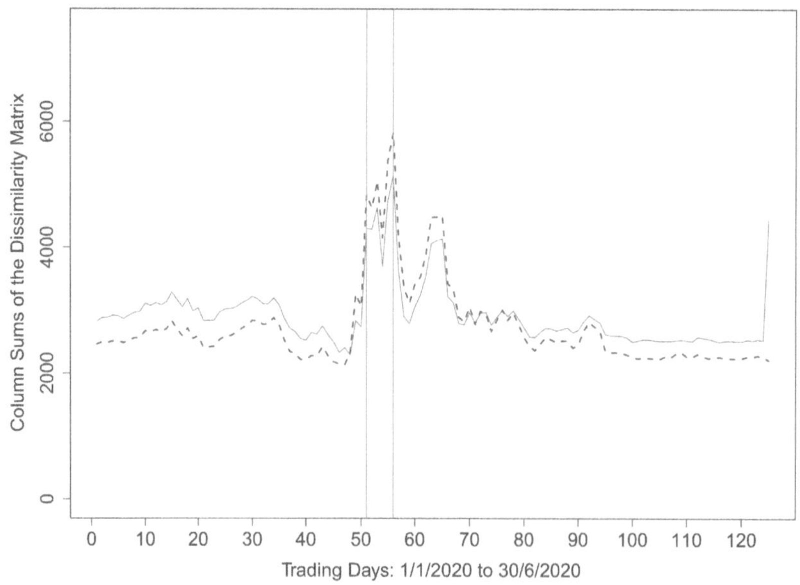

FIGURE 12.7: Dissimilarity visualisation of S&P 500 data using $d(\boldsymbol{X}_i, \boldsymbol{X}_j)$ with the modified Euclidean distance function (red solid line) and univariate distance function (blue dashed line).

TABLE 12.2: Frequency and average number of correctly detected change points over 250 replications when there are three true change points, including no change points.

μ_2	n	d	Number of true change points	Method	Frequency of the detected change points				Average number of detected change points
					0	1	2	3	
	90	500	0	Permutation	0.97	0.03	0.00	0.00	0.03
	90	500	0	Asymptotic	0.97	0.03	0.00	0.00	0.03
$\mathbf{0}_d$	90	500	0	E-divisive	0.96	0.04	0.00	0.00	0.04
	90	500	0	Inspect	0.95	0.05	0.00	0.00	0.05
	90	500	0	HDcpdetect	0.96	0.04	0.00	0.00	0.04
	90	1500	0	Permutation	0.96	0.04	0.00	0.00	0.04
	90	1500	0	Asymptotic	0.96	0.04	0.00	0.00	0.04
$\mathbf{0}_d$	90	1500	0	E-divisive	0.96	0.04	0.00	0.00	0.04
	90	1500	0	Inspect	0.94	0.06	0.00	0.00	0.06
	90	1500	0	HDcpdetect	0.94	0.06	0.00	0.00	0.06
	90	500	3	Permutation	0.32	0.46	0.20	0.02	0.92
	90	500	3	Asymptotic	0.24	0.48	0.26	0.02	1.06
$(0.1 \times \mathbf{1}_{3d/4}, 0 \times \mathbf{1}_{d/4})$	90	500	3	E-divisive	0.30	0.56	0.12	0.02	0.86
	90	500	3	Inspect	0.56	0.38	0.06	0.00	0.50
	90	500	3	HDcpdetect	0.32	0.62	0.06	0.00	0.74
	90	1500	3	Permutation	0.00	0.05	0.19	0.76	2.71
	90	1500	3	Asymptotic	0.00	0.05	0.15	0.80	2.75
$(0.1 \times \mathbf{1}_{3d/4}, 0 \times \mathbf{1}_{d/4})$	90	1500	3	E-divisive	0.05	0.25	0.46	0.24	1.89
	90	1500	3	Inspect	0.22	0.58	0.19	0.01	0.99
	90	1500	3	HDcpdetect	0.08	0.44	0.38	0.10	1.50
	90	500	3	Permutation	0.00	0.02	0.18	0.80	2.78
	90	500	3	Asymptotic	0.00	0.02	0.16	0.82	2.80
$(0.2 \times \mathbf{1}_{3d/4}, 0 \times \mathbf{1}_{d/4})$	90	500	3	E-divisive	0.00	0.02	0.24	0.74	2.72
	90	500	3	Inspect	0.06	0.30	0.36	0.28	1.86
	90	500	3	HDcpdetect	0.01	0.45	0.14	0.40	1.93
	90	1500	3	Permutation	0.00	0.00	0.00	1.00	3.00
	90	1500	3	Asymptotic	0.00	0.00	0.00	1.00	3.00
$(0.2 \times \mathbf{1}_{3d/4}, 0 \times \mathbf{1}_{d/4})$	90	1500	3	E-divisive	0.00	0.00	0.02	0.98	2.98
	90	1500	3	Inspect	0.00	0.08	0.35	0.57	2.49
	90	1500	3	HDcpdetect	0.00	0.85	0.04	0.11	1.26

fewer change points when multiple changes occur, and the random projection method shows lower accuracy in high-dimensional settings with small sample sizes (see Table 12.2). Based on these simulation results, we conclude that our method provides more accurate estimates of change point locations, particularly the change points at 51 and 57, which correspond to the steep drop in stock prices in early March 2020 due to the COVID-19 pandemic.

12.8 Summary

Chapter 12 addresses high-dimensional change point detection in HDLSS settings, as commonly encountered in finance, genomics, and engineering. The chapter develops distribution-free, proximity-based methods using IPDs, novel dissimilarity measures, and difference matrices to detect both single and multiple change points. These methods do not rely on sparsity assumptions or parametric models, making them applicable to non-sparse high-dimensional data. Key techniques include the Efficient Sparsity Adaptive Change-point Estimator (ESACE), Informative Sparse Projection (ISP), Fabiseach factorized binary search, CUSUM-type procedures, and the non-parametric E-Divisive (ecp) method, providing a unified framework for detecting shifts in mean, variance, or other distributional properties of high-dimensional data streams.

The chapter establishes the consistency and asymptotic properties of the proposed estimators. It introduces novel test statistics to formally assess the significance of estimated change points, deriving asymptotic distributions under standard moment conditions and using permutation distributions for small-sample corrections. Simulation studies demonstrate that the methods achieve high power and robustness in both sparse and dense change regimes, outperforming existing alternatives for detecting small-magnitude changes across many variables. Recursive binary segmentation and wild binary segmentation strategies are incorporated to efficiently handle multiple change points.

The R package `HDDchangepoint` implements the proposed method, including both recursive binary segmentation and wild binary segmentation, as well as the real data application. The package is available at "`https://github.com/statistics11/HDDchangepoint`". It returns significant change point estimates and their corresponding p-values, and can also be applied with any user-specified dissimilarity measure. Applications to real-world data, including the S&P-500 index, illustrate the practical effectiveness of the methods in detecting structural breaks. The chapter emphasizes a balance of statistical rigor, computational feasibility, and methodological flexibility, providing a comprehensive toolkit for high-dimensional change point detection in HDLSS contexts.

12.9 Proofs

Proof of Theorem 12.3.1

Considering the definition of $d(\boldsymbol{X}_i, \boldsymbol{X}_j)$, we can easily observe $d(\boldsymbol{X}_i, \boldsymbol{X}_j) \geq 0$, $d(\boldsymbol{X}_i, \boldsymbol{X}_i) = 0$ and $d(\boldsymbol{X}_i, \boldsymbol{X}_j) = d(\boldsymbol{X}_j, \boldsymbol{X}_i)$. To prove the triangle inequality $d(\boldsymbol{X}_i, \boldsymbol{X}_j) \leq d(\boldsymbol{X}_i, \boldsymbol{X}_l) + d(\boldsymbol{X}_j, \boldsymbol{X}_l)$, we first note that if $n = 3$,

$$\begin{aligned}
d(\boldsymbol{X}_i, \boldsymbol{X}_j) &= \big| \|\boldsymbol{X}_i - \boldsymbol{X}_l\|_D - \|\boldsymbol{X}_j - \boldsymbol{X}_l\|_D \big| \\
&= \big| \|\boldsymbol{X}_i - \boldsymbol{X}_l\|_D - \|\boldsymbol{X}_j - \boldsymbol{X}_l\|_D + \|\boldsymbol{X}_i - \boldsymbol{X}_j\|_D - \|\boldsymbol{X}_i - \boldsymbol{X}_j\|_D \big| \\
&= \big| (\|\boldsymbol{X}_i - \boldsymbol{X}_j\|_D - \|\boldsymbol{X}_j - \boldsymbol{X}_l\|_D) - (\|\boldsymbol{X}_i - \boldsymbol{X}_j\|_D - \|\boldsymbol{X}_i - \boldsymbol{X}_l\|_D) \big| \\
&= \big| (\|\boldsymbol{X}_i - \boldsymbol{X}_j\|_D - \|\boldsymbol{X}_l - \boldsymbol{X}_j\|_D) - (\|\boldsymbol{X}_j - \boldsymbol{X}_i\|_D - \|\boldsymbol{X}_l - \boldsymbol{X}_i\|_D) \big| \\
&\leq \big| \|\boldsymbol{X}_i - \boldsymbol{X}_j\|_D - \|\boldsymbol{X}_l - \boldsymbol{X}_j\|_D \big| + \big| \|\boldsymbol{X}_j - \boldsymbol{X}_i\|_D - \|\boldsymbol{X}_l - \boldsymbol{X}_i\|_D \big| \\
&\leq d(\boldsymbol{X}_i, \boldsymbol{X}_l) + d(\boldsymbol{X}_j, \boldsymbol{X}_l).
\end{aligned} \tag{12.17}$$

So $d(\boldsymbol{X}_i, \boldsymbol{X}_j) \leq d(\boldsymbol{X}_i, \boldsymbol{X}_l) + d(\boldsymbol{X}_j, \boldsymbol{X}_l)$ holds for $n = 3$. For $n \geq 4$ and any $\boldsymbol{X}_k$ with $k \geq 4$,

$$\begin{aligned}
\big| \|\boldsymbol{X}_i - \boldsymbol{X}_k\|_D - \|\boldsymbol{X}_j - \boldsymbol{X}_k\|_D \big| &= \big| \|\boldsymbol{X}_i - \boldsymbol{X}_k\|_D + \|\boldsymbol{X}_k - \boldsymbol{X}_l\|_D - \|\boldsymbol{X}_k - \boldsymbol{X}_l\|_D - \|\boldsymbol{X}_j - \boldsymbol{X}_k\|_D \big| \\
&\leq \big| \|\boldsymbol{X}_i - \boldsymbol{X}_k\|_D - \|\boldsymbol{X}_l - \boldsymbol{X}_k\|_D \big| + \big| \|\boldsymbol{X}_j - \boldsymbol{X}_k\|_D - \|\boldsymbol{X}_l - \boldsymbol{X}_k\|_D \big|.
\end{aligned} \tag{12.18}$$

Using 12.17 and 12.18, we obtain

$$\begin{aligned}
\sum_{k \neq i,j} \big| \|\boldsymbol{X}_i - \boldsymbol{X}_k\|_D - \|\boldsymbol{X}_j - \boldsymbol{X}_k\|_D \big| \leq &\sum_{k \neq i,l} \big| \|\boldsymbol{X}_i - \boldsymbol{X}_k\|_D - \|\boldsymbol{X}_l - \boldsymbol{X}_k\|_D \big| \\
&+ \sum_{k \neq j,l} \big| \|\boldsymbol{X}_j - \boldsymbol{X}_k\|_D - \|\boldsymbol{X}_l - \boldsymbol{X}_k\|_D \big|.
\end{aligned}$$

Hence, $d(\boldsymbol{X}_i, \boldsymbol{X}_j) \leq d(\boldsymbol{X}_i, \boldsymbol{X}_l) + d(\boldsymbol{X}_j, \boldsymbol{X}_l)$ for $n \geq 3$.

Proof of Theorem 12.3.2

Using the triangle inequality for $d(\boldsymbol{X}_i, \boldsymbol{X}_j)$ proved in Theorem 12.3.1, we find

$$\begin{aligned}
d(\boldsymbol{X}_i, \boldsymbol{X}_j) &= \frac{1}{n-2} \sum_{l \neq i,j} \big| \|\boldsymbol{X}_i - \boldsymbol{X}_l\|_D - \|\boldsymbol{X}_j - \boldsymbol{X}_l\|_D \big| \\
&\leq \frac{1}{n-2} \sum_{l \neq i,j} \big| \|\boldsymbol{X}_i - \boldsymbol{X}_j\|_D + \|\boldsymbol{X}_j - \boldsymbol{X}_l\|_D - \|\boldsymbol{X}_j - \boldsymbol{X}_l\|_D \big| \\
&= \frac{1}{n-2} \sum_{l \neq i,j} \|\boldsymbol{X}_i - \boldsymbol{X}_j\|_D \\
&= \|\boldsymbol{X}_i - \boldsymbol{X}_j\|_D.
\end{aligned}$$

Hence, the theorem immediately follows if the multivariate modified Euclidean distance function is used for distance function $\| \cdot - \cdot \|_D$. If the univariate distance function based on differences between the sample mean and variance of the observations is used, we can write

$$\begin{aligned}
\|\boldsymbol{X}_i - \boldsymbol{X}_j\|_2^2 &= \sum_{k=1}^{d} \left(\boldsymbol{X}_{ik} - \boldsymbol{X}_{jk}\right)^2 \\
&= \sum_{k=1}^{d} \left(\boldsymbol{X}_{ik} - \bar{\boldsymbol{X}}_i^d + \bar{\boldsymbol{X}}_i^d - \boldsymbol{X}_{jk} - \bar{\boldsymbol{X}}_j^d + \bar{\boldsymbol{X}}_j^d\right)^2 \\
&= \sum_{k=1}^{d} \left((\boldsymbol{X}_{ik} - \bar{\boldsymbol{X}}_i^d) - (\boldsymbol{X}_{jk} - \bar{\boldsymbol{X}}_j^d) + (\bar{\boldsymbol{X}}_i^d - \bar{\boldsymbol{X}}_j^d)\right)^2 \\
&= d(S_{\boldsymbol{X}_i}^d)^2 + d(S_{\boldsymbol{X}_j}^d)^2 + d(\bar{\boldsymbol{X}}_i^d - \bar{\boldsymbol{X}}_j^d)^2 - 2\sum_{k=1}^{d} (\boldsymbol{X}_{ik} - \bar{\boldsymbol{X}}_i^d)(\boldsymbol{X}_{jk} - \bar{\boldsymbol{X}}_j^d) \\
&= d(S_{\boldsymbol{X}_i}^d - S_{\boldsymbol{X}_j}^d)^2 + d(\bar{\boldsymbol{X}}_i^d - \bar{\boldsymbol{X}}_j^d)^2 - 2\sum_{k=1}^{d} (\boldsymbol{X}_{ik} - \bar{\boldsymbol{X}}_i^d)(\boldsymbol{X}_{jk} - \bar{\boldsymbol{X}}_j^d) + 2pS_{\boldsymbol{X}_i}^d S_{\boldsymbol{X}_j}^d \\
&\geq d(S_{\boldsymbol{X}_i}^d - S_{\boldsymbol{X}_j}^d)^2 + d(\bar{\boldsymbol{X}}_i^d - \bar{\boldsymbol{X}}_j^d)^2 \\
&= d\|\boldsymbol{X}_i - \boldsymbol{X}_j\|_D^2,
\end{aligned}$$

where the inequality is obtained using $\sum_{k=1}^{d} (\boldsymbol{X}_{ik} - \bar{\boldsymbol{X}}_i^d)(\boldsymbol{X}_{jk} - \bar{\boldsymbol{X}}_j^d) \leq dS_{\boldsymbol{X}_i}^d S_{\boldsymbol{X}_j}^d$ which holds by the Cauchy—Schwarz inequality. Hence, $d(\boldsymbol{X}_i, \boldsymbol{X}_j) \leq \frac{1}{\sqrt{d}} \|\boldsymbol{X}_i - \boldsymbol{X}_j\|_2$.

Proof of Theorem 12.5.1

First, we give the proof for the multivariate distance function based on the modified Euclidean distance. For this, using the weak law of large numbers and under Assumptions (A1)-(A3), it is straightforward to show (See Hall et al. [133])

$$\begin{aligned}
&d^{-1/2}\|\boldsymbol{X}_i - \boldsymbol{X}_j\|_2 \xrightarrow{P} \sqrt{2\lambda_{\nabla_\tau^-}^2} && \forall i \in \nabla_\tau^-, j \in \nabla_\tau^- \\
&d^{-1/2}\|\boldsymbol{X}_i - \boldsymbol{X}_j\|_2 \xrightarrow{P} \sqrt{2\lambda_{\nabla_\tau^+}^2} && \forall i \in \nabla_\tau^+, j \in \nabla_\tau^+ \\
&d^{-1/2}\|\boldsymbol{X}_i - \boldsymbol{X}_j\|_2 \xrightarrow{P} \sqrt{\lambda_{\nabla_\tau^-}^2 + \lambda_{\nabla_\tau^+}^2 + \eta_{\nabla_\tau^- \nabla_\tau^+}^2} && \forall i \in \nabla_\tau^-, j \in \nabla_\tau^+ \\
&d^{-1/2}\|\boldsymbol{X}_i - \boldsymbol{X}_j\|_2 \xrightarrow{P} \sqrt{\lambda_{\nabla_\tau^-}^2 + \lambda_{\nabla_\tau^+}^2 + \eta_{\nabla_\tau^- \nabla_\tau^+}^2} && \forall i \in \nabla_\tau^+, j \in \nabla_\tau^-.
\end{aligned} \tag{12.19}$$

Recall $\|\boldsymbol{X}_i - \boldsymbol{X}_j\|_D = d^{-1/2}\|\boldsymbol{X}_i - \boldsymbol{X}_j\|_2$ in this case. We then write

$$\begin{aligned} d(\boldsymbol{X}_i, \boldsymbol{X}_j) &= \frac{1}{n-2}\sum_{l\neq i,j} \big|\|\boldsymbol{X}_i - \boldsymbol{X}_l\|_D - \|\boldsymbol{X}_j - \boldsymbol{X}_l\|_D\big| \\ &= \frac{1}{n-2}\Bigg\{ \sum_{l\in\nabla_\tau^- - \{i,j\}} \big|\|\boldsymbol{X}_i - \boldsymbol{X}_l\|_D - \|\boldsymbol{X}_j - \boldsymbol{X}_l\|_D\big| \\ &\qquad + \sum_{l\in\nabla_\tau^+ - \{i,j\}} \big|\|\boldsymbol{X}_i - \boldsymbol{X}_l\|_D - \|\boldsymbol{X}_j - \boldsymbol{X}_l\|_D\big| \Bigg\}. \end{aligned} \tag{12.20}$$

Inserting (12.19) into (12.20) leads to the result stated in the theorem for $d(\boldsymbol{X}_i, \boldsymbol{X}_j)$ for each i and j belonging to ∇_τ^- or ∇_τ^+ accordingly.

Next, we provide the proof for the univariate distance function based on differences between the sample mean and variance of the observations. For this, using the weak law of large numbers and under Assumptions (B1)-(B4), we have $\bar{\boldsymbol{X}}_i^d - \bar{\mu}_{\nabla_\tau^-} \xrightarrow{P} 0 \; \forall i \in \nabla_\tau^-$ and $\bar{\boldsymbol{X}}_i^d - \bar{\mu}_{\nabla_\tau^+} \xrightarrow{P} 0 \; \forall i \in \nabla_\tau^+$ as $d \to \infty$. Similarly, $S_{\boldsymbol{X}_i}^d - \sigma_{\nabla_\tau^-} \xrightarrow{P} 0 \; \forall i \in \nabla_\tau^-$ and $S_{\boldsymbol{X}_i}^d - \sigma_{\nabla_\tau^+} \xrightarrow{P} 0 \; \forall i \in \nabla_\tau^+$ as $d \to \infty$. Hence, recalling $\|\boldsymbol{X}_i - \boldsymbol{X}_j\|_D = \sqrt{(\bar{\boldsymbol{X}}_i^d - \bar{\boldsymbol{X}}_j^d)^2 + (S_{\boldsymbol{X}_i}^d - S_{\boldsymbol{X}_j}^d)^2}$ in this case, we get

$$\begin{aligned} &\|\boldsymbol{X}_i - \boldsymbol{X}_j\|_D \xrightarrow{P} 0 && \forall i \in \nabla_\tau^-, j \in \nabla_\tau^- \\ &\|\boldsymbol{X}_i - \boldsymbol{X}_j\|_D \xrightarrow{P} 0 && \forall i \in \nabla_\tau^+, j \in \nabla_\tau^+ \\ &\|\boldsymbol{X}_i - \boldsymbol{X}_j\|_D - \big|\sqrt{(\bar{\mu}_{\nabla_\tau^-} - \bar{\mu}_{\nabla_\tau^+})^2 + (\sigma_{\nabla_\tau^-} - \sigma_{\nabla_\tau^+})^2}\big| \xrightarrow{P} 0 && \forall i \in \nabla_\tau^-, j \in \nabla_\tau^+ \\ &\|\boldsymbol{X}_i - \boldsymbol{X}_j\|_D - \big|\sqrt{(\bar{\mu}_{\nabla_\tau^-} - \bar{\mu}_{\nabla_\tau^+})^2 + (\sigma_{\nabla_\tau^-} - \sigma_{\nabla_\tau^+})^2}\big| \xrightarrow{P} 0 && \forall i \in \nabla_\tau^+, j \in \nabla_\tau^-. \end{aligned}$$

Again inserting the above into (12.20) will give the result stated in the theorem for $d(\boldsymbol{X}_i, \boldsymbol{X}_j)$ for each i and j belonging to ∇_τ^- or ∇_τ^+ accordingly.

Proof of Theorem 12.5.2

(a) Since τ_0 is the true change point, using the result of Theorem 12.5.1 it is straightforward to find that as $d \to \infty$

$$\begin{aligned} \Delta_{ij} &= o_{\mathbb{P}}(1) && \forall i = 1, \ldots, n, \; j \neq \tau_0 \\ \Delta_{ij} &= \kappa_{\tau_0} + o_{\mathbb{P}}(1) && \forall i = 1, \ldots, n, \; j = \tau_0. \end{aligned}$$

Therefore, if $j \neq \tau_0$, we have $\frac{1}{n}\sum_{i=1}^n \Delta_{ij} = o_{\mathbb{P}}(1)$. If $j = \tau_0$, we have as $d \to \infty$

$$\begin{aligned} \frac{1}{n}\sum_{i=1}^n \Delta_{ij} &= \frac{1}{n}\sum_{i=1}^n \Delta_{i,\tau_0} \\ &= \frac{1}{n}\sum_{i\in\nabla_{\tau_0}^-} \Delta_{i,\tau_0} + \frac{1}{n}\sum_{i\in\nabla_{\tau_0}^+} \Delta_{i,\tau_0} \\ &= \frac{1}{n}|\nabla_{\tau_0}^-|\kappa_{\tau_0} + o_{\mathbb{P}}(1) + \frac{1}{n}|\nabla_{\tau_0}^+|\kappa_{\tau_0} + o_{\mathbb{P}}(1) \\ &= \kappa_{\tau_0} + o_{\mathbb{P}}(1), \end{aligned}$$

where in the last step we used the fact that $|\nabla_{\tau_0}^-| + |\nabla_{\tau_0}^+| = n$.

(b) Using the result of part (a) and the fact that arg max is a continuous function, we find that as $d \to \infty$

$$\begin{aligned}
\hat{\tau} &= \arg\max_{1\le j\le n}\left\{\frac{1}{n}\sum_{i=1}^{n}\Delta_{ij}\right\}\\
&= \arg\max_{1\le j\le n}\left\{\frac{1}{n}\sum_{i=1}^{n}\Delta_{ij}\mathbb{I}(j\neq\tau_0)+\frac{1}{n}\sum_{i=1}^{n}\Delta_{ij}\mathbb{I}(j=\tau_0)\right\}\\
&= \arg\max_{1\le j\le n}\left\{\frac{1}{n}\sum_{i=1}^{n}o_{\mathbb{P}}(1)\mathbb{I}(j\neq\tau_0)+\frac{1}{n}\sum_{i=1}^{n}(\kappa_{\tau_0}+o_{\mathbb{P}}(1))\mathbb{I}(j=\tau_0)\right\}\\
&\xrightarrow{P} \arg\max_{1\le j\le n}\left\{\frac{1}{n}\sum_{i=1}^{n}\kappa_{\tau_0}\mathbb{I}(j=\tau_0)\right\}\\
&= \arg\max_{1\le j\le n}\left\{\kappa_{\tau_0}\mathbb{I}(j=\tau_0)\right\}\\
&= \tau_0,
\end{aligned}$$

and hence $\hat{\tau}-\tau_0 = o_{\mathbb{P}}(1)$, where $o_{\mathbb{P}}(1)$ follows from the third equality above. This completes the proof of theorem.

Proof of Theorem 12.5.3

First, using the result of Theorem 12.5.2, we have for any fixed n

$$\frac{1}{n}\sum_{i=1}^{n}\Delta_{ij}-\mathbb{I}(j=\tau_0)\kappa_{\tau_0}=o_{\mathbb{P}}(1)\quad \text{as } d\to\infty.$$

Furthermore, when $d\to\infty$ and with any fixed n, Δ_{ij} has a degenerate distribution at κ_{τ_0} for all $i=1,\ldots,n$ as

$$\mathbb{P}(\Delta_{ij}=\kappa_{\tau_0})\to\begin{cases}1 & \text{if } j=\tau_0\\ 0 & \text{if } j\neq\tau_0.\end{cases}$$

Thus, $\mathbb{E}(\Delta_{ij})\to\mathbb{I}(j=\tau_0)\kappa_{\tau_0}<\infty$ and $\mathbb{E}(\Delta_{ij}^2)\to\mathbb{I}(j=\tau_0)\kappa_{\tau_0}^2<\infty$ as $d\to\infty$. Then, using the weak law of large numbers when $n\to\infty$, we have

$$\frac{1}{n}\sum_{i=1}^{n}\Delta_{ij}-\frac{1}{n}\sum_{i=1}^{n}\mathbb{E}(\Delta_{ij})\xrightarrow{P}0\quad \text{as } n\to\infty. \tag{12.21}$$

Also, for any fixed n, we have as $d\to\infty$

$$\frac{1}{n}\sum_{i=1}^{n}\mathbb{E}(\Delta_{ij})\to\frac{1}{n}\sum_{i=1}^{n}\mathbb{I}(j=\tau_0)\kappa_{\tau_0}=\mathbb{I}(j=\tau_0)\kappa_{\tau_0}. \tag{12.22}$$

From (12.21) and (12.22), we get

$$\frac{1}{n}\sum_{i=1}^{n}\Delta_{ij}-\mathbb{I}(j=\tau_0)\kappa_{\tau_0}=o_{\mathbb{P}}(1)\quad \text{as } n,d\to\infty.$$

Hence, we find as $n\to\infty$ and $d\to\infty$

$$\max_{1\le j\le n}\left\{\frac{1}{n}\sum_{i=1}^{n}\Delta_{ij}-\mathbb{I}(j=\tau_0)\kappa_{\tau_0}\right\}=\max_{1\le j\le n}\left\{\{\mathbb{I}(j=\tau_0)\kappa_{\tau_0}+o_{\mathbb{P}}(1)\}-\mathbb{I}(j=\tau_0)\kappa_{\tau_0}\right\}=o_{\mathbb{P}}(1),$$

which completes the proof.

Proof of Theorem 12.5.4

Using Theorem 12.5.1, under the null hypothesis of no change points, we have as $d \to \infty$

$$d_{ij} = d(\boldsymbol{X}_i, \boldsymbol{X}_j) = o_{\mathbb{P}}(1) \qquad \forall i, j = 1, \ldots, n.$$

This implies $T(\hat{\tau}) = o_{\mathbb{P}}(1)$ under the null hypothesis H_0.
Using Theorem 12.5.1, under the alternative hypothesis H_1^s, we get as $d \to \infty$

$$\begin{aligned}
d_{ij} &= o_{\mathbb{P}}(1) && \forall i \in \nabla_{\hat{\tau}}^-, j \in \nabla_{\hat{\tau}}^- \\
d_{ij} &= o_{\mathbb{P}}(1) && \forall i \in \nabla_{\hat{\tau}}^+, j \in \nabla_{\hat{\tau}}^+ \\
d_{ij} &= \kappa_{\tau_0} + o_{\mathbb{P}}(1) && \forall i \in \nabla_{\hat{\tau}}^-, j \in \nabla_{\hat{\tau}}^+ \\
d_{ij} &= \kappa_{\tau_0} + o_{\mathbb{P}}(1) && \forall i \in \nabla_{\hat{\tau}}^+, j \in \nabla_{\hat{\tau}}^-.
\end{aligned}$$

We note that this result initially holds for the two sets $\nabla_{\tau_0}^-$ and $\nabla_{\tau_0}^+$, and consequently for $\nabla_{\hat{\tau}}^-$ and $\nabla_{\hat{\tau}}^+$ since $\hat{\tau}$ is a consistent estimate of τ_0 as shown in Theorem 12.5.2.
Using the above result, we find as $d \to \infty$

$$\begin{aligned}
T(\hat{\tau}) &= \frac{1}{n|\nabla_{\hat{\tau}}^-||\nabla_{\hat{\tau}}^+|} \sum_{i=1}^{n} \sum_{j \in \nabla_{\hat{\tau}}^-} \sum_{j' \in \nabla_{\hat{\tau}}^+} (d_{ij} - d_{ij'})^2 \\
&= \frac{1}{n|\nabla_{\hat{\tau}}^-||\nabla_{\hat{\tau}}^+|} \sum_{i=1}^{n} \sum_{j \in \nabla_{\hat{\tau}}^-} \sum_{j' \in \nabla_{\hat{\tau}}^+} d_{ij}^2 + \frac{1}{n|\nabla_{\hat{\tau}}^-||\nabla_{\hat{\tau}}^+|} \sum_{i=1}^{n} \sum_{j \in \nabla_{\hat{\tau}}^-} \sum_{j' \in \nabla_{\hat{\tau}}^+} d_{ij'}^2 \\
&\quad - \frac{2}{n|\nabla_{\hat{\tau}}^-||\nabla_{\hat{\tau}}^+|} \sum_{i=1}^{n} \sum_{j \in \nabla_{\hat{\tau}}^-} \sum_{j' \in \nabla_{\hat{\tau}}^+} d_{ij} d_{ij'} \\
&= \frac{1}{n|\nabla_{\hat{\tau}}^-|} \sum_{i=1}^{n} \sum_{j \in \nabla_{\hat{\tau}}^-} d_{ij}^2 + \frac{1}{n|\nabla_{\hat{\tau}}^+|} \sum_{i=1}^{n} \sum_{j' \in \nabla_{\hat{\tau}}^+} d_{ij'}^2 - \frac{2}{n|\nabla_{\hat{\tau}}^-||\nabla_{\hat{\tau}}^+|} \sum_{i=1}^{n} \sum_{j \in \nabla_{\hat{\tau}}^-} \sum_{j' \in \nabla_{\hat{\tau}}^+} d_{ij} d_{ij'} \\
&= \frac{1}{n|\nabla_{\hat{\tau}}^-|} \sum_{i \in \nabla_{\hat{\tau}}^-} \sum_{j \in \nabla_{\hat{\tau}}^-} d_{ij}^2 !+! \frac{1}{n|\nabla_{\hat{\tau}}^-|} \sum_{i \in \nabla_{\hat{\tau}}^+} \sum_{j \in \nabla_{\hat{\tau}}^-} d_{ij}^2 !+! \frac{1}{n|\nabla_{\hat{\tau}}^+|} \sum_{i \in \nabla_{\hat{\tau}}^-} \sum_{j' \in \nabla_{\hat{\tau}}^+} d_{ij'}^2 \\
&\quad + \frac{1}{n|\nabla_{\hat{\tau}}^+|} \sum_{i \in \nabla_{\hat{\tau}}^+} \sum_{j' \in \nabla_{\hat{\tau}}^+} d_{ij'}^2 - \frac{2}{n|\nabla_{\hat{\tau}}^-||\nabla_{\hat{\tau}}^+|} \sum_{i=1}^{n} \sum_{j \in \nabla_{\hat{\tau}}^-} \sum_{j' \in \nabla_{\hat{\tau}}^+} d_{ij} d_{ij'} \\
&= o_{\mathbb{P}}(1) + \left\{ \frac{|\nabla_{\tau_0}^+|}{n} \kappa_{\tau_0}^2 + o_{\mathbb{P}}(1) \right\} + \left\{ \frac{|\nabla_{\tau_0}^-|}{n} \kappa_{\tau_0}^2 + o_{\mathbb{P}}(1) \right\} + o_{\mathbb{P}}(1) - o_{\mathbb{P}}(1) \\
&= \frac{|\nabla_{\tau_0}^-| + |\nabla_{\tau_0}^+|}{n} \kappa_{\tau_0}^2 + o_{\mathbb{P}}(1) \\
&= \kappa_{\tau_0}^2 + o_{\mathbb{P}}(1),
\end{aligned}$$

which holds under the alternative hypothesis H_1^s.

Proof of Theorem 12.5.5

We first prove the asymptotic distribution when using the univariate distance function $\|\boldsymbol{X}_i - \boldsymbol{X}_j\|_D = \sqrt{(\bar{\boldsymbol{X}}_i^d - \bar{\boldsymbol{X}}_j^d)^2 + (S_{\boldsymbol{X}_i}^d - S_{\boldsymbol{X}_j}^d)^2}$ in (12.8). To obtain the asymptotic distribution of the test statistic $T(\hat{\tau})$ in this case, we know from the result of Theorem 12.5.4 that, under H_1^s,

$$T(\hat{\tau}) = \kappa_{\tau_0}^2 + o_{\mathbb{P}}(1),$$

and furthermore, under H_0, $T(\hat{\tau}) = o_{\mathbb{P}}(1)$ as $d \to \infty$. Let us define

$$W(\hat{\tau}) := \frac{1}{n|\nabla_{\hat{\tau}}^-||\nabla_{\hat{\tau}}^+|} \sum_{i=1}^{n} \sum_{j \in \nabla_{\hat{\tau}}^-} \sum_{j' \in \nabla_{\hat{\tau}}^+} \{\|\boldsymbol{X}_i - \boldsymbol{X}_j\|_D^2 + \|\boldsymbol{X}_i - \boldsymbol{X}_{j'}\|_D^2\},$$

which can be rewritten as

$$\begin{aligned} W(\hat{\tau}) =& \frac{1}{n|\nabla_{\hat{\tau}}^-||\nabla_{\hat{\tau}}^+|} \sum_{i \in \nabla_{\hat{\tau}}^-} \sum_{j \in \nabla_{\hat{\tau}}^-} \sum_{j' \in \nabla_{\hat{\tau}}^+} \{\|\boldsymbol{X}_i - \boldsymbol{X}_j\|_D^2 + \|\boldsymbol{X}_i - \boldsymbol{X}_{j'}\|_D^2\} \\ &+ \frac{1}{n|\nabla_{\hat{\tau}}^-||\nabla_{\hat{\tau}}^+|} \sum_{i \in \nabla_{\hat{\tau}}^+} \sum_{j \in \nabla_{\hat{\tau}}^-} \sum_{j' \in \nabla_{\hat{\tau}}^+} \{\|\boldsymbol{X}_i - \boldsymbol{X}_j\|_D^2 + \|\boldsymbol{X}_i - \boldsymbol{X}_{j'}\|_D^2\}. \end{aligned}$$

Using the asymptotic results for $\|\boldsymbol{X}_i - \boldsymbol{X}_j\|_D$ given in the proof of Theorem 12.5.1, one can find that as $d \to \infty$

$$W(\hat{\tau}) = \begin{cases} \kappa_{\tau_0}^2 + o_{\mathbb{P}}(1) & \text{under } H_1^s \\ o_{\mathbb{P}}(1) & \text{under } H_0, \end{cases}$$

and furthermore

$$\|\boldsymbol{X}_i - \boldsymbol{X}_j\|_D^2 + \|\boldsymbol{X}_i - \boldsymbol{X}_{j'}\|_D^2 = \begin{cases} \kappa_{\tau_0}^2 + o_{\mathbb{P}}(1) & i \in \nabla_{\hat{\tau}}^-, j \in \nabla_{\hat{\tau}}^-, j' \in \nabla_{\hat{\tau}}^+ \\ \kappa_{\tau_0}^2 + o_{\mathbb{P}}(1) & i \in \nabla_{\hat{\tau}}^+, j \in \nabla_{\hat{\tau}}^-, j' \in \nabla_{\hat{\tau}}^+, \end{cases}$$

and

$$(d_{ij} - d_{ij'})^2 = \begin{cases} \kappa_{\tau_0}^2 + o_{\mathbb{P}}(1) & i \in \nabla_{\hat{\tau}}^-, j \in \nabla_{\hat{\tau}}^-, j' \in \nabla_{\hat{\tau}}^+ \\ \kappa_{\tau_0}^2 + o_{\mathbb{P}}(1) & i \in \nabla_{\hat{\tau}}^+, j \in \nabla_{\hat{\tau}}^-, j' \in \nabla_{\hat{\tau}}^+. \end{cases}$$

Therefore, the asymptotic distribution of $T(\hat{\tau})$ is the same as the asymptotic distribution of $W(\hat{\tau})$ when both d and n go to infinity. It thus suffices to derive the asymptotic distribution of $W(\hat{\tau})$ as it is easier to calculate. For this, we first note that using CLT as $d \to \infty$ and under Assumptions (B1)-(B4), we have

$$\begin{aligned} v_{ij}^{-1/2}\big((\bar{\boldsymbol{X}}_i^d - \bar{\boldsymbol{X}}_j^d) - (\bar{\mu}_i - \bar{\mu}_j)\big) &\xrightarrow{D} \mathbb{N}(0,1), \\ v_{ij}^{*-1/2}\big((S_{\boldsymbol{X}_i}^d - S_{\boldsymbol{X}_j}^d) - (\sigma_i - \sigma_j)\big) &\xrightarrow{D} \mathbb{N}(0,1), \end{aligned} \tag{12.23}$$

in which

$$\begin{aligned} v_{ij} &= \mathbb{V}ar(\bar{\boldsymbol{X}}_i^d) + \mathbb{V}ar(\bar{\boldsymbol{X}}_j^d) = \frac{1}{p^2}\sum_{l=1}^{d}\sum_{l'=1}^{d} \mathbb{C}ov(\boldsymbol{X}_{il}, \boldsymbol{X}_{il'}) + \frac{1}{d^2}\sum_{l=1}^{d}\sum_{l'=1}^{d} \mathbb{C}ov(\boldsymbol{X}_{jl}, \boldsymbol{X}_{jl'}), \\ v_{ij}^* &= \mathbb{V}ar(S_{\boldsymbol{X}_i}^d) + \mathbb{V}ar(S_{\boldsymbol{X}_j}^d) = \frac{1}{4\sigma_i^2 d^2}\sum_{l=1}^{d}\sum_{l'=1}^{d} \mathbb{C}ov\big((\boldsymbol{X}_{il} - \bar{\boldsymbol{X}}_i^d)^2, (\boldsymbol{X}_{il'} - \bar{\boldsymbol{X}}_i^d)^2\big) \\ &\quad + \frac{1}{4\sigma_j^2 d^2}\sum_{l=1}^{d}\sum_{l'=1}^{d} \mathbb{C}ov\big((\boldsymbol{X}_{jl} - \bar{\boldsymbol{X}}_j^d)^2, (\boldsymbol{X}_{jl'} - \bar{\boldsymbol{X}}_j^d)^2\big). \end{aligned}$$

Note that we used the delta method to obtain the second equation in (12.23) following the asymptotic normality of $(S_{\boldsymbol{X}_i}^d)^2$ and $(S_{\boldsymbol{X}_j}^d)^2$. From (12.23), we get as $d \to \infty$

$$\begin{aligned} v_{ij}^{-1}(\bar{\boldsymbol{X}}_i^d - \bar{\boldsymbol{X}}_j^d)^2 - \chi_1^2(v_{ij}^{-1}(\bar{\mu}_i - \bar{\mu}_j)^2) &\xrightarrow{D} 0, \\ v_{ij}^{*-1}(S_{\boldsymbol{X}_i}^d - S_{\boldsymbol{X}_j}^d)^2 - \chi_1^2(v_{ij}^{*-1}(\sigma_i - \sigma_j)^2) &\xrightarrow{D} 0, \end{aligned} \tag{12.24}$$

where $\chi_a^2(b)$ denotes a non-central chi-squared variable whose asymptotic distribution is non-central chi-squared with the degrees of freedom a and the non-centrality parameter b. Since in this case $\|\boldsymbol{X}_i - \boldsymbol{X}_j\|_D^2 = (\bar{\boldsymbol{X}}_i^d - \bar{\boldsymbol{X}}_j^d)^2 + (S_{\mathbf{X}_i}^d - S_{\mathbf{X}_j}^d)^2$, using (12.24) we obtain the expectation and variance of $\|\boldsymbol{X}_i - \boldsymbol{X}_j\|_D^2$ when $d \to \infty$ as follows

$$\begin{aligned} E\big(\|\boldsymbol{X}_i - \boldsymbol{X}_j\|_D^2\big) &= \big[(v_{ij} + v_{ij}^*) + (\bar{\mu}_i - \bar{\mu}_j)^2 + (\sigma_i - \sigma_j)^2\big] + o^d(1) \\ \mathbb{V}ar\big(\|\boldsymbol{X}_i - \boldsymbol{X}_j\|_D^2\big) &= \big[2(v_{ij}^2 + v_{ij}^{*2}) + 4v_{ij}(\bar{\mu}_i - \bar{\mu}_j)^2 + 4v_{ij}^*(\sigma_i - \sigma_j)^2\big] + o^d(1). \end{aligned} \tag{12.25}$$

Furthermore, it is straightforward to see that

$$\mathbb{C}ov\big(\|\boldsymbol{X}_i - \boldsymbol{X}_j\|_D^2, \|\boldsymbol{X}_k - \boldsymbol{X}_l\|_D^2\big) =$$
$$\begin{cases} \mathbb{V}ar\big(\|\boldsymbol{X}_i - \boldsymbol{X}_j\|_D^2\big) & \text{if } i \neq j, k \neq l, i = k, j = l, \\ \mathbb{C}ov\big((\bar{\boldsymbol{X}}_i^d - \bar{\boldsymbol{X}}_j^d)^2, (\bar{\boldsymbol{X}}_i^d - \bar{\boldsymbol{X}}_l^d)^2\big) + \mathbb{C}ov\big((S_{\mathbf{X}_i}^d - S_{\mathbf{X}_j}^d)^2, (S_{\mathbf{X}_i}^d - S_{\mathbf{X}_l}^d)^2\big) & \text{if } i \neq j, k \neq l, i = k, j \neq l, \\ \mathbb{C}ov\big((\bar{\boldsymbol{X}}_i^d - \bar{\boldsymbol{X}}_j^d)^2, (\bar{\boldsymbol{X}}_k^d - \bar{\boldsymbol{X}}_j^d)^2\big) + \mathbb{C}ov\big((S_{\mathbf{X}_i}^d - S_{\mathbf{X}_j}^d)^2, (S_{\mathbf{X}_k}^d - S_{\mathbf{X}_j}^d)^2\big) & \text{if } i \neq j, k \neq l, i \neq k, j = l, \\ \mathbb{V}ar\big(\|\boldsymbol{X}_i - \boldsymbol{X}_j\|_D^2\big) & \text{if } i \neq j, k \neq l, i = l, j = k, \\ \mathbb{C}ov\big((\bar{\boldsymbol{X}}_i^d - \bar{\boldsymbol{X}}_j^d)^2, (\bar{\boldsymbol{X}}_k^d - \bar{\boldsymbol{X}}_i^d)^2\big) + \mathbb{C}ov\big((S_{\mathbf{X}_i}^d - S_{\mathbf{X}_j}^d)^2, (S_{\mathbf{X}_k}^d - S_{\mathbf{X}_i}^d)^2\big) & \text{if } i \neq j, k \neq l, i = l, j \neq k, \\ \mathbb{C}ov\big((\bar{\boldsymbol{X}}_i^d - \bar{\boldsymbol{X}}_j^d)^2, (\bar{\boldsymbol{X}}_j^d - \bar{\boldsymbol{X}}_l^d)^2\big) + \mathbb{C}ov\big((S_{\mathbf{X}_i}^d - S_{\mathbf{X}_j}^d)^2, (S_{\mathbf{X}_j}^d - S_{\mathbf{X}_l}^d)^2\big) & \text{if } i \neq j, k \neq l, i \neq l, j = k, \\ 0 & \text{otherwise,} \end{cases} \tag{12.26}$$

To calculate the covariances in (12.26), we know from the theory of normal distribution that if $(Z_1, Z_2) \sim \mathbb{N}(\mu_{Z_1}, \mu_{Z_2}, \sigma_{Z_1}^2, \sigma_{Z_2}^2, \rho_{Z_1,Z_2})$ then $\mathbb{C}ov(Z_1^2, Z_2^2) = 4\mu_{Z_1}\mu_{Z_2}\mathbb{C}ov(Z_1, Z_2) + 2\mathbb{C}ov^2(Z_1, Z_2)$, so we get

$$\begin{aligned} \mathbb{C}ov\big((\bar{\boldsymbol{X}}_i^d - \bar{\boldsymbol{X}}_j^d)^2, (\bar{\boldsymbol{X}}_i^d - \bar{\boldsymbol{X}}_l^d)^2\big) - \big[2(\bar{\mu}_i - \bar{\mu}_j)(\bar{\mu}_i - \bar{\mu}_l)v_{ii} + v_{ii}^2/2\big] \to 0 \\ \mathbb{C}ov\big((S_{\mathbf{X}_i}^d - S_{\mathbf{X}_j}^d)^2, (S_{\mathbf{X}_i}^d - S_{\mathbf{X}_l}^d)^2\big) - \big[2(\sigma_i - \sigma_j)(\sigma_i - \sigma_l)v_{ii}^* + v_{ii}^{*2}/2\big] \to 0, \end{aligned}$$

where we used the fact that $\mathbb{C}ov\big((\bar{\boldsymbol{X}}_i^d - \bar{\boldsymbol{X}}_j^d), (\bar{\boldsymbol{X}}_i^d - \bar{\boldsymbol{X}}_l^d)\big) = \mathbb{V}ar(\bar{\boldsymbol{X}}_i^d) = v_{ii}/2$ and $\mathbb{C}ov\big((S_{\mathbf{X}_i}^d - S_{\mathbf{X}_j}^d), (S_{\mathbf{X}_i}^d - S_{\mathbf{X}_l}^d)\big) = \mathbb{V}ar(S_{\mathbf{X}_i}^d) = v_{ii}^*/2$. We calculate the other covariance terms in (12.26) similarly. Now, when $n \to \infty$ we find that the asymptotic distribution of $W(\tau_0)$ converges to a normal distribution using CLT and under Assumptions (B1)-(B5), as follows

$$\frac{n|\nabla_{\tau_0}^-||\nabla_{\tau_0}^+|\Big(W(\tau_0) - \frac{1}{n|\nabla_{\tau_0}^-||\nabla_{\tau_0}^+|}\sum_{i=1}^n \sum_{j \in \nabla_{\tau_0}^-} \sum_{j' \in \nabla_{\tau_0}^+} m_{ijj'}\Big)}{\sqrt{\sum_{i=1}^n \sum_{j \in \nabla_{\tau_0}^-} \sum_{j' \in \nabla_{\tau_0}^+} \sum_{k=1}^n \sum_{l \in \nabla_{\tau_0}^-} \sum_{l' \in \nabla_{\tau_0}^+} c_{ijj'kll'}}} \xrightarrow{D} \mathbb{N}(0, 1), \tag{12.27}$$

where

$$\begin{aligned} m_{ijj'} &= E\big(\|\boldsymbol{X}_i - \boldsymbol{X}_j\|_D^2\big) + E\big(\|\boldsymbol{X}_i - \boldsymbol{X}_{j'}\|_D^2\big), \\ c_{ijj'kll'} &= \mathbb{C}ov\big(\|\boldsymbol{X}_i - \boldsymbol{X}_j\|_D^2 + \|\boldsymbol{X}_i - \boldsymbol{X}_{j'}\|_D^2, \|\boldsymbol{X}_k - \boldsymbol{X}_l\|_D^2 + \|\boldsymbol{X}_k - \boldsymbol{X}_{l'}\|_D^2\big) \\ &= \mathbb{C}ov\big(\|\boldsymbol{X}_i - \boldsymbol{X}_j\|_D^2, \|\boldsymbol{X}_k - \boldsymbol{X}_l\|_D^2\big) + \mathbb{C}ov\big(\|\boldsymbol{X}_i - \boldsymbol{X}_j\|_D^2, \|\boldsymbol{X}_k - \boldsymbol{X}_{l'}\|_D^2\big) \\ &\quad + \mathbb{C}ov\big(\|\boldsymbol{X}_i - \boldsymbol{X}_{j'}\|_D^2, \|\boldsymbol{X}_k - \boldsymbol{X}_l\|_D^2\big) + \mathbb{C}ov\big(\|\boldsymbol{X}_i - \boldsymbol{X}_{j'}\|_D^2, \|\boldsymbol{X}_k - \boldsymbol{X}_{l'}\|_D^2\big). \end{aligned}$$

For this result, in addition to Assumption (A5), we require $E\big(\|\boldsymbol{X}_i - \boldsymbol{X}_j\|_D^2\big) < \infty$ which follows from Assumptions (A1) and (A2). Letting $d \to \infty$, we can calculate both $m_{ijj'}$ and $c_{ijj'kll'}$ using the results in (12.25) and (12.26) obtained as $d \to \infty$. In particular,

$$\begin{aligned} m_{ijj'} &= E\big(\|\boldsymbol{X}_i - \boldsymbol{X}_j\|_D^2\big) + E\big(\|\boldsymbol{X}_i - \boldsymbol{X}_{j'}\|_D^2\big) \\ &= (v_{ij} + v_{ij}^* + v_{ij'} + v_{ij'}^*) + (\bar{\mu}_i - \bar{\mu}_j)^2 + (\sigma_i - \sigma_j)^2 + (\bar{\mu}_i - \bar{\mu}_{j'})^2 + (\sigma_i - \sigma_{j'})^2 + o^d(1). \end{aligned}$$

The asymptotic normality in (12.27) also holds if we replace τ_0 with $\hat{\tau}$ as it is consistent for τ_0. Therefore, we find as $n \to \infty$ and $d \to \infty$

$$\frac{n|\nabla_{\hat{\tau}}^-||\nabla_{\hat{\tau}}^+|\Big(W(\hat{\tau}) - \frac{1}{n|\nabla_{\hat{\tau}}^-||\nabla_{\hat{\tau}}^+|}\sum_{i=1}^n \sum_{j\in\nabla_{\hat{\tau}}^-}\sum_{j'\in\nabla_{\hat{\tau}}^+} m_{ijj'}\Big)}{\sqrt{\sum_{i=1}^n \sum_{j\in\nabla_{\hat{\tau}}^-}\sum_{j'\in\nabla_{\hat{\tau}}^+}\sum_{k=1}^n \sum_{l\in\nabla_{\hat{\tau}}^-}\sum_{l'\in\nabla_{\hat{\tau}}^+} c_{ijj'kll'}}} \xrightarrow{D} \mathbb{N}(0,1).$$

Under the null hypothesis H_0, the asymptotic distribution of $W(\hat{\tau})$ when $n \to \infty$ and $d \to \infty$ simplifies since $\bar{\mu}_i = \bar{\mu}_j$ and $\sigma_i = \sigma_j$ for all i and j under H_0, resulting in the following simpler formulae

$$m_{ijj'} = v_{ij} + v_{ij}^* + v_{ij'} + v_{ij'}^* + o^d(1),$$

$$\mathbb{C}ov\big(\|\boldsymbol{X}_i - \boldsymbol{X}_j\|_D^2, \|\boldsymbol{X}_k - \boldsymbol{X}_l\|_D^2\big) = \begin{cases} 2(v_{ij}^2 + v_{ij}^{*2}) + o^d(1) & \text{if } i \neq j, k \neq l, i = k, j = l, \\ (v_{ii}^{*2} + v_{ii}^{*2})/2 + o^d(1) & \text{if } i \neq j, k \neq l, i = k, j \neq l, \\ (v_{jj}^{*2} + v_{jj}^{*2})/2 + o^d(1) & \text{if } i \neq j, k \neq l, i \neq k, j = l, \\ 2(v_{ij}^2 + v_{ij}^{*2}) + o^d(1) & \text{if } i \neq j, k \neq l, i = l, j = k, \\ (v_{ii}^{*2} + v_{ii}^{*2})/2 + o^d(1) & \text{if } i \neq j, k \neq l, i = l, j \neq k, \\ (v_{jj}^{*2} + v_{jj}^{*2})/2 + o^d(1) & \text{if } i \neq j, k \neq l, i \neq l, j = k, \\ o^d(1) & \text{otherwise.} \end{cases}$$

For the multivariate modified Euclidean distance function $\|\boldsymbol{X}_i - \boldsymbol{X}_j\|_D = d^{-1/2}\|\boldsymbol{X}_i - \boldsymbol{X}_j\|_2$, we similarly define

$$W^*(\hat{\tau}) := \frac{1}{n|\nabla_{\hat{\tau}}^-||\nabla_{\hat{\tau}}^+|}\sum_{i=1}^n \sum_{j\in\nabla_{\hat{\tau}}^-}\sum_{j'\in\nabla_{\hat{\tau}}^+} \{d^{-1}\|\boldsymbol{X}_i - \boldsymbol{X}_j\|_2^2 + d^{-1}\|\boldsymbol{X}_i - \boldsymbol{X}_{j'}\|_2^2\},$$

which has the same asymptotic limit as $T(\hat{\tau})$ when both $d, n \to \infty$ under H_0. Again using CLT as above, we obtain in this case

$$\begin{aligned} m_{ijj'} &= \mathbb{E}(d^{-1}\|\boldsymbol{X}_i - \boldsymbol{X}_j\|_2^2) + \mathbb{E}(d^{-1}\|\boldsymbol{X}_i - \boldsymbol{X}_{j'}\|_2^2), \\ c_{ijj'kll'} &= \mathbb{C}ov\big(d^{-1}\|\boldsymbol{X}_i - \boldsymbol{X}_j\|_2^2 + d^{-1}\|\boldsymbol{X}_i - \boldsymbol{X}_{j'}\|_2^2, d^{-1}\|\boldsymbol{X}_k - \boldsymbol{X}_l\|_2^2 + d^{-1}\|\boldsymbol{X}_k - \boldsymbol{X}_{l'}\|_2^2\big) \\ &= \mathbb{C}ov\big(d^{-1}\|\boldsymbol{X}_i - \boldsymbol{X}_j\|_2^2, d^{-1}\|\boldsymbol{X}_k - \boldsymbol{X}_l\|_2^2\big) + \mathbb{C}ov\big(d^{-1}\|\boldsymbol{X}_i - \boldsymbol{X}_j\|_2^2, d^{-1}\|\boldsymbol{X}_k - \boldsymbol{X}_{l'}\|_2^2\big) \\ &\quad + \mathbb{C}ov\big(d^{-1}\|\boldsymbol{X}_i - \boldsymbol{X}_{j'}\|_2^2, d^{-1}\|\boldsymbol{X}_k - \boldsymbol{X}_l\|_2^2\big) + \mathbb{C}ov\big(d^{-1}\|\boldsymbol{X}_i - \boldsymbol{X}_{j'}\|_2^2, \|\boldsymbol{X}_k - \boldsymbol{X}_{l'}\|_2^2\big). \end{aligned}$$

Proof of Theorem 12.5.6

We showed in Theorem 12.5.5 that $T(\hat{\tau})$ is asymptotically normal under H_0 or under H_1^s. So, the proof is straightforward for $\mathbb{P}_{H_0}(|T_{\text{sd}}(\hat{\tau})| > Z_{\alpha/2}) \to \alpha$ using the standard normal distribution of $T_{\text{sd}}(\hat{\tau})$ under H_0. We use the notation

$$\begin{aligned} c &= \sqrt{\sum_{i=1}^n \sum_{j\in\nabla_{\hat{\tau}}^-}\sum_{j'\in\nabla_{\hat{\tau}}^+}\sum_{k=1}^n \sum_{l\in\nabla_{\hat{\tau}}^-}\sum_{l'\in\nabla_{\hat{\tau}}^+} c_{ijj'kll'}}, \\ c^* &= \sqrt{\sum_{i=1}^n \sum_{j\in\nabla_{\hat{\tau}}^-}\sum_{j'\in\nabla_{\hat{\tau}}^+}\sum_{k=1}^n \sum_{l\in\nabla_{\hat{\tau}}^-}\sum_{l'\in\nabla_{\hat{\tau}}^+} c_{ijj'kll'}^*}, \\ a_{ijj'} &= (\bar{\mu}_i - \bar{\mu}_j)^2 + (\sigma_i - \sigma_j)^2 + (\bar{\mu}_i - \bar{\mu}_{j'})^2 + (\sigma_i - \sigma_{j'})^2, \\ c_{ijj'kll'}^* &= c_{ijj'kll'} + b_{ijkl} + b_{ijkl'} + b_{ij'kl} + b_{ij'kl'}, \\ e &= \frac{1}{n|\nabla_{\hat{\tau}}^-||\nabla_{\hat{\tau}}^+|}\sum_{i=1}^n \sum_{j\in\nabla_{\hat{\tau}}^-}\sum_{j'\in\nabla_{\hat{\tau}}^+} (m_{ijj'} + a_{ijj'}). \end{aligned} \tag{12.28}$$

To prove the second result, we work with the distribution of $T_{\text{sd}}(\hat{\tau})$ under H_1^s by writing

$$\mathbb{P}_{H_1^s}(T_{\text{sd}}(\hat{\tau}) > Z_{\alpha/2}) = \mathbb{P}_{H_1^s}\Big(\frac{n|\nabla_{\hat{\tau}}^-||\nabla_{\hat{\tau}}^+|\Big(T(\hat{\tau}) - \frac{1}{n|\nabla_{\hat{\tau}}^-||\nabla_{\hat{\tau}}^+|}\sum_{i=1}^n \sum_{j\in\nabla_{\hat{\tau}}^-}\sum_{j'\in\nabla_{\hat{\tau}}^+} m_{ijj'}\Big)}{c} > Z_{\alpha/2}\Big)$$

$$= \mathbb{P}_{H_1^s}\Big(\frac{n|\nabla_{\hat{\tau}}^-||\nabla_{\hat{\tau}}^+|\Big(T(\hat{\tau}) - e\Big)}{c} > Z_{\alpha/2} - \frac{\sum_{i=1}^n \sum_{j\in\nabla_{\hat{\tau}}^-}\sum_{j'\in\nabla_{\hat{\tau}}^+} a_{ijj'}}{c}\Big)$$

$$= \mathbb{P}_{H_1^s}\Big(\frac{n|\nabla_{\hat{\tau}}^-||\nabla_{\hat{\tau}}^+|\Big(T(\hat{\tau}) - e\Big)}{c^*} > \frac{Z_{\alpha/2}c - \sum_{i=1}^n \sum_{j\in\nabla_{\hat{\tau}}^-}\sum_{j'\in\nabla_{\hat{\tau}}^+} a_{ijj'}}{c^*}\Big)$$

$$= \mathbb{P}_{H_1^s}\Big(Z > \frac{Z_{\alpha/2}c - \sum_{i=1}^n \sum_{j\in\nabla_{\hat{\tau}}^-}\sum_{j'\in\nabla_{\hat{\tau}}^+} a_{ijj'}}{c^*}\Big) + o(1),$$

where

$$b_{ijkl} = \begin{cases} 4v_{ij}(\bar{\mu}_i - \bar{\mu}_j)^2 + 4v_{ij}^*(\sigma_i - \sigma_j)^2 & \text{if } i \neq j, k \neq l, i = k, j = l, \\ 2(\bar{\mu}_i - \bar{\mu}_j)(\bar{\mu}_i - \bar{\mu}_l)v_{ii} + 2(\sigma_i - \sigma_j)(\sigma_i - \sigma_l)v_{ii}^* & \text{if } i \neq j, k \neq l, i = k, j \neq l, \\ 2(\bar{\mu}_i - \bar{\mu}_j)(\bar{\mu}_k - \bar{\mu}_j)v_{jj} + 2(\sigma_i - \sigma_j)(\sigma_k - \sigma_j)v_{jj}^* & \text{if } i \neq j, k \neq l, i \neq k, j = l, \\ 4v_{ij}(\bar{\mu}_i - \bar{\mu}_j)^2 + 4v_{ij}^*(\sigma_i - \sigma_j)^2 & \text{if } i \neq j, k \neq l, i = l, j = k, \\ 2(\bar{\mu}_i - \bar{\mu}_j)(\bar{\mu}_k - \bar{\mu}_i)v_{ii} + 2(\sigma_i - \sigma_j)(\sigma_k - \sigma_i)v_{ii}^* & \text{if } i \neq j, k \neq l, i = l, j \neq k, \\ 2(\bar{\mu}_i - \bar{\mu}_j)(\bar{\mu}_j - \bar{\mu}_l)v_{jj} + 2(\sigma_i - \sigma_j)(\sigma_j - \sigma_l)v_{jj}^* & \text{if } i \neq j, k \neq l, i \neq l, j = k, \\ 0 & \text{otherwise.} \end{cases}$$

Then, since

$$\begin{aligned}
\sum_{i=1}^n \sum_{j\in\nabla_{\hat{\tau}}^-}\sum_{j'\in\nabla_{\hat{\tau}}^+} a_{ijj'} &= \sum_{i=1}^n \sum_{j\in\nabla_{\hat{\tau}}^-}\sum_{j'\in\nabla_{\hat{\tau}}^+} (\bar{\mu}_i - \bar{\mu}_j)^2 + (\sigma_i - \sigma_j)^2 + (\bar{\mu}_i - \bar{\mu}_{j'})^2 + (\sigma_i - \sigma_{j'})^2 \\
&= \sum_{i\in\nabla_{\hat{\tau}}^-} \sum_{j\in\nabla_{\hat{\tau}}^-}\sum_{j'\in\nabla_{\hat{\tau}}^+} (\bar{\mu}_i - \bar{\mu}_j)^2 + (\sigma_i - \sigma_j)^2 + (\bar{\mu}_i - \bar{\mu}_{j'})^2 + (\sigma_i - \sigma_{j'})^2 \\
&\quad + \sum_{i\in\nabla_{\hat{\tau}}^+} \sum_{j\in\nabla_{\hat{\tau}}^-}\sum_{j'\in\nabla_{\hat{\tau}}^+} (\bar{\mu}_i - \bar{\mu}_j)^2 + (\sigma_i - \sigma_j)^2 + (\bar{\mu}_i - \bar{\mu}_{j'})^2 + (\sigma_i - \sigma_{j'})^2 \\
&= \sum_{i\in\nabla_{\hat{\tau}}^-} \sum_{j\in\nabla_{\hat{\tau}}^-}\sum_{j'\in\nabla_{\hat{\tau}}^+} (\bar{\mu}_i - \bar{\mu}_{j'})^2 + (\sigma_i - \sigma_{j'})^2 \\
&\quad + \sum_{i\in\nabla_{\hat{\tau}}^+} \sum_{j\in\nabla_{\hat{\tau}}^-}\sum_{j'\in\nabla_{\hat{\tau}}^+} (\bar{\mu}_i - \bar{\mu}_j)^2 + (\sigma_i - \sigma_j)^2 \\
&= |\nabla_{\hat{\tau}}^-| \sum_{i\in\nabla_{\hat{\tau}}^-}\sum_{j'\in\nabla_{\hat{\tau}}^+} (\bar{\mu}_i - \bar{\mu}_{j'})^2 + (\sigma_i - \sigma_{j'})^2 \\
&\quad + |\nabla_{\hat{\tau}}^+| \sum_{i\in\nabla_{\hat{\tau}}^-}\sum_{j\in\nabla_{\hat{\tau}}^-} (\bar{\mu}_i - \bar{\mu}_j)^2 + (\sigma_i - \sigma_j)^2 \\
&= |\nabla_{\hat{\tau}}^-||\nabla_{\hat{\tau}}^-||\nabla_{\hat{\tau}}^+|((\bar{\mu}_{\nabla_{\tau}^-} - \bar{\mu}_{\nabla_{\tau}^+})^2 + (\sigma_{\nabla_{\tau}^-} - \sigma_{\nabla_{\tau}^+})^2) \\
&\quad + |\nabla_{\hat{\tau}}^+||\nabla_{\hat{\tau}}^-||\nabla_{\hat{\tau}}^+|((\bar{\mu}_{\nabla_{\tau}^-} - \bar{\mu}_{\nabla_{\tau}^+})^2 + (\sigma_{\nabla_{\tau}^-} - \sigma_{\nabla_{\tau}^+})^2) \\
&= (|\nabla_{\hat{\tau}}^-| + |\nabla_{\hat{\tau}}^+|)|\nabla_{\hat{\tau}}^-||\nabla_{\hat{\tau}}^+|\kappa_{\tau_0}^2 \\
&= n|\nabla_{\hat{\tau}}^-||\nabla_{\hat{\tau}}^+|\kappa_{\tau_0}^2
\end{aligned}$$

and $|\nabla_{\hat{\tau}}^-||\nabla_{\hat{\tau}}^+| = O(n)$, we find

$$\mathbb{P}_{H_1^s}(T_{\text{sd}}(\hat{\tau}) > Z_{\alpha/2}) = \mathbb{P}_{H_1^s}\big(Z > Z_{\alpha/2}\frac{c}{c^*} - \frac{n^2\kappa_{\tau_0}^2}{c^*}\big) + o(1) = 1 - \Phi\big(Z_{\alpha/2}\frac{c}{c^*} - \frac{n^2\kappa_{\tau_0}^2}{c^*}\big) + o(1). \tag{12.29}$$

It is also clear to obtain similarly

$$\mathbb{P}_{H_1^s}(T_{\rm sd}(\hat{\tau}) < -Z_{\alpha/2}) = \mathbb{P}_{H_1^s}\big(Z < -Z_{\alpha/2}\frac{c}{c^*} - \frac{n^2\kappa_{\tau_0}^2}{c^*}\big) + o(1) = \Phi\big(-Z_{\alpha/2}\frac{c}{c^*} - \frac{n^2\kappa_{\tau_0}^2}{c^*}\big) + o(1). \tag{12.30}$$

From (12.29) and (12.30), we get

$$\mathbb{P}_{H_1^s}(|T_{\rm sd}(\hat{\tau})| > Z_{\alpha/2}) = 1 - \Phi\big(Z_{\alpha/2}\frac{c}{c^*} - \frac{n^2\kappa_{\tau_0}^2}{c^*}\big) + \Phi\big(-Z_{\alpha/2}\frac{c}{c^*} - \frac{n^2\kappa_{\tau_0}^2}{c^*}\big) + o(1).$$

Proof of Corollary 12.5.1

The proof of this corollary is straightforward using the result of Theorem 12.5.6. See also the proof of the next corollary.

Proof of Corollary 12.5.2

Using the result of Theorem 12.5.6, the proof of this corollary is also straightforward, and we just prove $\kappa_{\tau_0}^2 = \frac{(\sum_{k\in S_0}\omega_k)^2}{d^2(\sigma_{\nabla_{\tau_0}^-}+\sigma_{\nabla_{\tau_0}^+})^2}$ in this case. For this, first recall $\sigma_i^2 = d^{-1}\sum_{k=1}^d \sigma_{ik}^2$, $i = 1,\ldots,n$.

Since $\sigma_i^2 - \sigma_j^2 = (\sigma_i - \sigma_j)(\sigma_i + \sigma_j)$, we can write for $i \in \nabla_\tau^-, j \in \nabla_\tau^+$

$$(\sigma_i - \sigma_j)^2 = \frac{(\sigma_i^2 - \sigma_j^2)^2}{(\sigma_i + \sigma_j)^2} = \frac{\big(d^{-1}\sum_{k=1}^d \sigma_{ik}^2 - d^{-1}\sum_{k=1}^d \sigma_{jk}^2\big)^2}{(\sigma_i + \sigma_j)^2} = \frac{d^{-2}(\sum_{k=1}^d \omega_k)^2}{(\sigma_i + \sigma_j)^2}. \tag{12.31}$$

Considering that the mean of observations is unchanged in this case, we get from (12.31) that

$$\kappa_{\tau_0}^2 = \frac{d^{-2}(\sum_{k=1}^d \omega_k)^2}{(\sigma_{\nabla_{\tau_0}^-} + \sigma_{\nabla_{\tau_0}^+})^2} = \frac{(\sum_{k\in S_0}\omega_k)^2}{d^2(\sigma_{\nabla_{\tau_0}^-} + \sigma_{\nabla_{\tau_0}^+})^2}.$$

Proof of Theorem 12.5.7

Let us define the notation

$$\bar{\Delta}_{j,b_k,e_k} := \frac{1}{e_k - b_k + 1}\sum_{i=b_k}^{e_k} \Delta_{ij}, \qquad j = 1,\ldots,n.$$

We can then rewrite the change point estimate (12.15) as

$$\hat{\gamma}_k = \mathop{\arg\max}_{b_k \le j \le e_k} \big\{\bar{\Delta}_{j,b_k,e_k}\big\}.$$

For clarity of the proof, first consider the case when there are two true change points τ_1^0, τ_2^0, $2 \le \tau_1^0 < \tau_2^0 \le n$. Similar to the proof of Theorem 12.5.2, it is straightforward to show as $d \to \infty$ that

$$\begin{aligned} \bar{\Delta}_{j,b_k,e_k} &= o_{\mathbb{P}}(1) && j \ne \tau_1^0, \tau_2^0 \\ \bar{\Delta}_{j,b_k,e_k} &= \kappa_{\tau_1^0} + o_{\mathbb{P}}(1) && j = \tau_1^0 \\ \bar{\Delta}_{j,b_k,e_k} &= \kappa_{\tau_2^0} + o_{\mathbb{P}}(1) && j = \tau_2^0, \end{aligned} \tag{12.32}$$

where $\kappa_{\tau_i^0}$, $i = 1, 2$, are obtained by plugging in τ_i^0 in the formula of κ_τ given in Theorem 12.5.1. Note that the above result holds with any b_k and e_k. The binary segmentation starts with $k = 1$, $b_k = 1$ and $e_k = n$, so

$$\hat{\gamma}_1 = \mathop{\arg\max}_{1 \le j \le n} \big\{\bar{\Delta}_{j,1,n}\big\},$$

and using (12.32) we find as $d \to \infty$

$$\begin{aligned}
\hat{\gamma}_1 &= \arg\max_{1\le j\le n}\{\bar{\Delta}_{j,1,n}\} \\
&= \arg\max_{1\le j\le n}\{\bar{\Delta}_{j,1,n}\mathbb{I}(j\neq \tau_1^0,\tau_2^0)+\bar{\Delta}_{j,1,n}\mathbb{I}(j=\tau_1^0)+\bar{\Delta}_{j,1,n}\mathbb{I}(j=\tau_2^0)\} \\
&= \arg\max_{1\le j\le n}\{o_{\mathbb{P}}(1)\mathbb{I}(j\neq \tau_1^0,\tau_2^0)+(\kappa_{\tau_1^0}+o_{\mathbb{P}}(1))\mathbb{I}(j=\tau_1^0)+(\kappa_{\tau_2^0}+o_{\mathbb{P}}(1))\mathbb{I}(j=\tau_2^0)\} \\
&\xrightarrow{P} \arg\max_{1\le j\le n}\{\kappa_{\tau_1^0}\mathbb{I}(j=\tau_1^0)+\kappa_{\tau_2^0}\mathbb{I}(j=\tau_2^0)\}.
\end{aligned}$$

Hence, we get either $\hat{\gamma}_1 \xrightarrow{P} \tau_1^0$ or $\hat{\gamma}_1 \xrightarrow{P} \tau_2^0$, depending on which of $\kappa_{\tau_1^0}$ and $\kappa_{\tau_2^0}$ is larger. First, if $\kappa_{\tau_1^0} > \kappa_{\tau_2^0}$, we get $\hat{\gamma}_1 \xrightarrow{P} \tau_1^0$ and then split the data sequence to two sub-sequences before and after $\hat{\gamma}_1$: one with $b_2 = 1$ and $e_2 = \hat{\gamma}_1 - 1$, and one with $b_2 = \hat{\gamma}_1$ and $e_2 = n$. Then, since $\hat{\gamma}_1 \xrightarrow{P} \tau_1^0 < \tau_2^0$, we have either

$$\hat{\gamma}_2 = \arg\max_{1\le j\le \hat{\gamma}_1-1}\{\bar{\Delta}_{j,1,\hat{\gamma}_1-1}\} = \arg\max_{1\le j\le \hat{\gamma}_1-1}\{o_{\mathbb{P}}(1)\} \xrightarrow{P} \arg\max_{1\le j\le \hat{\gamma}_1-1}\{0\} = \emptyset \tag{12.33}$$

where the empty set $\emptyset$ means no change point returned, or

$$\begin{aligned}
\hat{\gamma}_2 &= \arg\max_{\hat{\gamma}_1\le j\le \hat{n}}\{\bar{\Delta}_{j,\hat{\gamma}_1,n}\} \\
&= \arg\max_{\hat{\gamma}_1\le j\le \hat{n}}\{\bar{\Delta}_{j,\hat{\gamma}_1,n}\mathbb{I}(j\neq \tau_2^0)+\bar{\Delta}_{j,\hat{\gamma}_1,n}\mathbb{I}(j=\tau_2^0)\} \\
&= \arg\max_{\hat{\gamma}_1\le j\le \hat{n}}\{o_{\mathbb{P}}(1)\mathbb{I}(j\neq \tau_2^0)+(\kappa_{\tau_2^0}+o_{\mathbb{P}}(1))\mathbb{I}(j=\tau_2^0)\} \\
&\xrightarrow{P} \arg\max_{\hat{\gamma}_1\le j\le n}\{\kappa_{\tau_2^0}\mathbb{I}(j=\tau_2^0)\} \\
&= \tau_2^0.
\end{aligned}$$

Thus, we must have $\hat{\gamma}_2 \xrightarrow{P} \tau_2^0$. Second, if $\kappa_{\tau_1^0} < \kappa_{\tau_2^0}$, then we get $\hat{\gamma}_1 \xrightarrow{P} \tau_2^0$ and we can similarly as above show that $\hat{\gamma}_2 \xrightarrow{P} \tau_1^0$. If we further split the data sub-sequences before and after the second change point estimate $\hat{\gamma}_2$, no more $\hat{\gamma}_k$ will converge by considering (12.32) and a similar calculation as in (12.33). So, denoting $(\hat{\tau}_1, \hat{\tau}_2) = \text{sort}(\hat{\gamma}_1, \hat{\gamma}_2)$, we have $\|(\hat{\tau}_1, \hat{\tau}_2) - (\tau_1^0, \tau_2^0)\|_\infty = o_{\mathbb{P}}(1)$ for this case of two true change points.

Now suppose there are s true change points $\tau_1^0, \tau_2^0, \ldots, \tau_s^0$, $2 \le \tau_1^0 < \tau_2^0 < \ldots < \tau_s^0 \le n$. Similar to the above, we obtain as $d \to \infty$

$$\begin{aligned}
\bar{\Delta}_{j,b_k,e_k} &= o_{\mathbb{P}}(1) && j\neq \tau_1^0,\tau_2^0,\ldots,\tau_s^0 \\
\bar{\Delta}_{j,b_k,e_k} &= \kappa_{\tau_1^0}+o_{\mathbb{P}}(1) && j=\tau_1^0 \\
\bar{\Delta}_{j,b_k,e_k} &= \kappa_{\tau_2^0}+o_{\mathbb{P}}(1) && j=\tau_2^0 \\
&\ \ \vdots \\
\bar{\Delta}_{j,b_k,e_k} &= \kappa_{\tau_s^0}+o_{\mathbb{P}}(1) && j=\tau_s^0.
\end{aligned} \tag{12.34}$$

The binary segmentation starts with $k = 1$, $b_k = 1$ and $e_k = n$, so

$$\hat{\gamma}_1 = \arg\max_{1\le j\le n}\{\bar{\Delta}_{j,1,n}\},$$

and using (12.34) we find as $d \to \infty$

$$\begin{aligned}
\hat{\gamma}_1 &= \arg\max_{1\le j\le n}\left\{\bar{\Delta}_{j,1,n}\right\} \\
&= \arg\max_{1\le j\le n}\left\{\bar{\Delta}_{j,1,n}\mathbb{I}(j\neq \tau_1^0,\tau_2^0,\ldots,\tau_s^0)+\sum_{i=1}^{s}\bar{\Delta}_{j,1,n}\mathbb{I}(j=\tau_i^0)\right\} \\
&= \arg\max_{1\le j\le n}\left\{o_{\mathbb{P}}(1)\mathbb{I}(j\neq \tau_1^0,\tau_2^0,\ldots,\tau_s^0)+\sum_{i=1}^{s}(\kappa_{\tau_i^0}+o_{\mathbb{P}}(1))\mathbb{I}(j=\tau_i^0)\right\} \\
&\xrightarrow{P} \arg\max_{1\le j\le n}\left\{\sum_{i=1}^{s}\kappa_{\tau_i^0}\mathbb{I}(j=\tau_i^0)\right\}.
\end{aligned}$$

Hence, we have $\hat{\gamma}_1 \xrightarrow{P} \tau_{k_1}^0$ for a $\tau_{k_1}^0$, $k_1\in\{1,\ldots,s\}$, for which $\kappa_{\tau_{k_1}^0}=\max\{\kappa_{\tau_1^0},\kappa_{\tau_2^0},\ldots,\kappa_{\tau_s^0}\}$. Then, we split the data sequence to two sub-sequences before and after $\hat{\gamma}_1$: one with $b_2=1$ and $e_2=\hat{\gamma}_1-1$, and one with $b_2=\hat{\gamma}_1$ and $e_2=n$. Applying the same process as above for the two true change points case to each of these two sub-sequences, we find for the next change point estimates $\hat{\gamma}_2$ and $\hat{\gamma}_3$ that $\hat{\gamma}_2 \xrightarrow{P} \tau_{k_2}^0$ and $\hat{\gamma}_3 \xrightarrow{P} \tau_{k_3}^0$ where we have either $\tau_{k_2}^0<\tau_{k_1}^0$ and $\tau_{k_3}^0>\tau_{k_1}^0$ or $\tau_{k_2}^0>\tau_{k_1}^0$ and $\tau_{k_3}^0<\tau_{k_1}^0$. Continuing this process, we find $\|(\hat{\tau}_1,\hat{\tau}_2,\ldots,\hat{\tau}_s)-(\tau_1^0,\tau_2^0,\ldots,\tau_s^0)\|_\infty = o_{\mathbb{P}}(1)$, as $d\to\infty$. We note that for the last two sub-sequences no additional change points will converge using the same argument as above for the case with two change points according to a similar calculation as in (12.33).

Proof of Theorem 12.5.8

(a) First, because under the null hypothesis H_0 all the variables $\boldsymbol{X}_1,\boldsymbol{X}_2,\ldots,\boldsymbol{X}_n$ have the same distribution, $T_r(\hat{\tau})$ must have the same distribution as $T(\hat{\tau})$ under H_0 for all random permutations, and hence

$$\mathbb{E}(G_{T_R}(t)) = \frac{1}{R}\sum_{r=1}^{R}\mathbb{P}\big(T_r(\hat{\tau})\le t\big)=\mathbb{P}\big(T(\hat{\tau})\le t\big).$$

This then implies $\mathbb{E}(G_{T_R}(t))\to G_T(t)$ for all t as $n\to\infty$ and $d\to\infty$ under H_0, since the asymptotic distribution of $T(\hat{\tau})$ is $G_T(t)$ using Theorem 12.5.5. To show that $G_{T_R}(t)\xrightarrow{D} G_T(t)$, it suffices to prove that $\mathbb{V}ar(G_{T_R}(t))\to 0$ as $n\to\infty$ and $d\to\infty$. This holds because

$$\mathbb{E}(G_{T_r}^2(t)) = \frac{1}{R^2}\sum_{r=1}^{R}\sum_{r'=1}^{R}\mathbb{P}\big(T_r(\hat{\tau})\le t, T_{r'}(\hat{\tau})\le t\big)=\mathbb{P}\big(T(\hat{\tau})\le t, T(\hat{\tau})\le t\big)\to G_T^2(t).$$

It should be noted that this holds only under H_0 because the variables would not have the same distribution under the alternative hypothesis.

(b) Using part (a) above, as $n\to\infty$ and $d\to\infty$, we can write for $0\le\alpha\le 1$

$$\begin{aligned}
\mathbb{P}_{H_0}(p_{\text{perm}}\le\alpha) &= \mathbb{P}_{H_0}(1-G_{T_R}(T_{\text{obs}}(\hat{\tau}))\le\alpha)\to\mathbb{P}_{H_0}(1-G_T(T_{\text{obs}}(\hat{\tau}))\le\alpha) \\
&= \mathbb{P}_{H_0}(G_T(T_{\text{obs}}(\hat{\tau}))\ge 1-\alpha)=\mathbb{P}(U\ge 1-\alpha)=\alpha,
\end{aligned}$$

where U denotes a random variable having uniform distribution on the interval (0,1).

12.10 Exercises

Exercise 12.1 *Change Point Estimation in a Sequence of Normal Random Vectors: Let* $\{\mathbf{X}_i\}_{i=1}^n \subset \mathbb{R}^d$ *be a sequence of independent but not identically distributed random vectors. Suppose that there is a single change point at index* $\eta \in \{1, \ldots, n-1\}$*, such that*

$$\mathbf{X}_i \sim \begin{cases} \mathbb{N}(\boldsymbol{\mu}_1, \mathbf{I}_d) & \textit{for } i = 1, \ldots, \eta \\ \mathbb{N}(\boldsymbol{\mu}_2, \mathbf{I}_d) & \textit{for } i = \eta + 1, \ldots, n \end{cases}$$

with $\boldsymbol{\mu}_1 \neq \boldsymbol{\mu}_2$*, and identity covariance* $\mathbf{I}_d$*. Estimate the unknown change point* η *based on observed data* $\{\mathbf{X}_i\}_{i=1}^n$*.*

Exercise 12.2 *Change point estimation in a sequence of Wishart random matrices: Let* $\{\mathbf{X}_i\}_{i=1}^n \subset \mathbb{R}^{d\times d}$ *be a sequence of independent random matrices such that*

$$\mathbf{X}_i \sim \begin{cases} \mathbb{W}_d(\Sigma_1, \nu) & \textit{for } i = 1, \ldots, \eta \\ \mathbb{W}_d(\Sigma_2, \nu) & \textit{for } i = \eta + 1, \ldots, n \end{cases}$$

Here, $\mathbb{W}_d(\Sigma, \nu)$ *denotes the d-dimensional Wishart distribution with scale matrix* $\boldsymbol{\Sigma}$ *and degrees of freedom* $\nu \geq d$*. We assume* $\boldsymbol{\Sigma}_1 \neq \boldsymbol{\Sigma}_2$*, but* ν *is known and constant. Design an algorithm to find the change point using likelihood ratio test.*

Exercise 12.3 *Change point detection in the mean using Hotelling's* T^2 *statistic. Let* $\{\mathbf{X}_i\}_{i=1}^n \subset \mathbb{R}^d$ *be independent random vectors such that*

$$\mathbf{X}_i \sim \begin{cases} \mathbb{N}_d(\boldsymbol{\mu}_1, \boldsymbol{\Sigma}) & \textit{for } i = 1, \ldots, \eta \\ \mathbb{N}_d(\boldsymbol{\mu}_2, \boldsymbol{\Sigma}) & \textit{for } i = \eta + 1, \ldots, n \end{cases}$$

Assume $\boldsymbol{\mu}_1 \neq \boldsymbol{\mu}_2$*, but the covariance matrix* $\boldsymbol{\Sigma}$ *is constant and unknown. Estimate the unknown change point* $\eta \in \{1, \ldots, n-1\}$ *using the Hotelling two-sample statistic.*

Exercise 12.4 *Change point detection using the energy statistic. Let* $\{\mathbf{X}_i\}_{i=1}^n \subset \mathbb{R}^d$ *be independent, with a single change point* $\eta \in \{1, \ldots, n-1\}$ *such that*

$$\mathbf{X}_i \sim \begin{cases} F(\boldsymbol{\mu}_1, \mathbf{I}_d), & i = 1, \ldots, \eta \\ G(\boldsymbol{\mu}_2, \mathbf{I}_d), & i = \eta + 1, \ldots, n \end{cases}$$

Assume $\boldsymbol{\mu}_1 \neq \boldsymbol{\mu}_2$ *and use the energy distance statistic to estimate* η*.*

Exercise 12.5 *Write R code to generate five observations from* $F_1 = \mathbb{N}(0, \mathbf{I})$ *with* $d = 1000$*, five observations from* $F_2 = \mathbb{N}(\mathbf{1}, \mathbf{I})$ *and five observations from* $F_3 = \mathbb{N}(\mathbf{2}, \mathbf{I})$*.*

- *Find the limiting values of* $\tilde{\delta}_{F_i,F_j}$ *for* $1 \leq i < j \leq 3$*.*
- *Obtain a display of the rows of the dissimilarity matrix* $\mathbf{D}_\delta$ *against the observation labels. Identify the change points and clusters that are formed. What is the value of the Rand index* $\mathcal{R}(\psi)$*?*

Exercise 12.6 *Change Point Estimation in a Sequence of Normal Random Vectors. Let* $\{\mathbf{X}_i\}_{i=1}^n \subset \mathbb{R}^d$ *be a sequence of independent but not identically distributed random vectors. Suppose that there is a single change point at index* $\eta \in \{1, \ldots, n-1\}$*, such that*

$$\mathbf{X}_i \sim \begin{cases} \mathbb{N}(0, \sigma_1^2 \mathbf{I}_d) & \textit{for } i = 1, \ldots, \eta, \\ \mathbb{N}(0, \sigma_2^2 \mathbf{I}_d) & \textit{for } i = \eta + 1, \ldots, n, \end{cases}$$

where $\sigma_1^2 \neq \sigma_2^2$, *and* $\mathbf{I}_d$ *is the* $d \times d$ *identity matrix.*

Estimate the unknown change point η *based on the observed data* $\{\mathbf{X}_i\}_{i=1}^n$.

Exercise 12.7 *See Exercise 3.13.*

Let $\mathbf{X} = (X_1, X_2, \ldots, X_d)' \in \mathbb{R}^d$ *be a random vector with the following properties. Mean zero:* $\mathbb{E}[X_i] = 0$ *and spatially dependent Toeplitz covariance:*

$$\mathrm{Cov}(X_i, X_j) = \sigma^2 \rho^{|i-j|}, \quad \textit{with } |\rho| \leq 1$$

Define the empirical average, $\bar{x} = \frac{1}{d}\sum_{i=1}^d X_i$. *Analyze whether* $\bar{x} \to 0$ *as* $d \to \infty$, *under the Weak and Strong Laws of Large Numbers.*

13

High-Dimensional Clustering Methods

13.1 Abstract

High-dimensional clustering presents unique challenges in modern data analysis, particularly in HDLSS settings. This chapter provides a comprehensive treatment of clustering methodologies for high-dimensional data, encompassing hierarchical, partitioning, density-based, and change point-based approaches. We introduce a novel proximity-based clustering method, the Change Point Clustering (CPC) algorithm, which leverages dissimilarity measures and distance concentration phenomena to construct clusters with theoretical guarantees of consistency as the dimension grows. The CPC method utilizes the distance difference matrix and an agreement matrix based on the Rand index to determine optimal cluster boundaries, effectively reducing the $N \times d$ data matrix to an $N \times N$ matrix for computational efficiency. Comparative studies against $k\text{M}_\rho$, HDBSCAN, and MST demonstrate superior performance in various high-dimensional distributions. The chapter also discusses strategies for estimating the number of clusters, the influence of distance metrics, and the limitations posed by correlated variables on distance concentration. Applications to real datasets, including Lymphoma data, illustrate practical utility, providing guidance for high-dimensional clustering in genomics, finance, and other complex domains.

13.2 Introduction

Clustering, with deep roots in anthropology (Driver and Kroeber [81]), numerical taxonomy (Sokal and Sneath [318]), and statistics (Lerman [186]), is one of the most fundamental techniques in data analysis, concerned with organizing observations into coherent groups. As an unsupervised learning method, cluster analysis partitions unlabeled observations into clusters such that observations within the same cluster are highly similar, while those in different clusters are markedly dissimilar. Its appeal as an exploratory tool lies in its flexibility, as it does not require assumptions about cluster sizes or distributions.

Given the lack of a universally accepted definition of a cluster, numerous definitions and associated algorithms exist in the literature. For instance, a set of 52 playing cards may be clustered according to suits, colors, or face values, among other criteria. Popular partitioning methods assign observations to non-overlapping, collectively exhaustive clusters. Clusters can be defined based on the proximity or similarity of observations to a cluster center relative to other centers. In contrast, density-based clustering identifies clusters as regions of high density, separated by areas of lower density, a method that generally requires larger sample sizes to be effective.

DOI: 10.1201/9781003789581-13

We present a novel clustering method tailored for HDLSS data. The proposed method is proximity-based and uses both the distance concentration phenomenon and the limiting behavior of dissimilarity indices to construct clusters. We describe an algorithm that sequentially orders each row of the dissimilarity matrix to detect change points, which are subsequently used to define cluster boundaries. An agreement matrix of the Rand indices for the row-wise clusters is then constructed, and the minimum of the row sums of this matrix is used to identify the optimal clustering solution.

The literature on clustering methods is extensive (Everitt et al. [92]) and encompasses approaches such as partitioning, hierarchical, density-based, grid-based, model-based, and constraint-based techniques. Popular algorithms, including k-means and k-medoids along with their numerous variants, remain widely employed. Nevertheless, relatively few clustering methods are specifically designed for the HDLSS context, where the number of observations is considerably smaller than the number of features. Methods for estimating the number of clusters prior to clustering are often necessary, as the true number of clusters K is typically unknown. Clustering is frequently formulated as a discrete optimization problem, where the objective is to determine the partition of observations that optimizes a given criterion; however, in most instances, the computational complexity of this approach is substantial.

Clustering based on distance or proximity measures constitutes one of the foundational approaches in unsupervised learning. Distance-based clustering methods operate by quantifying the similarity or dissimilarity between observations. Let $X = \{x_1, x_2, \ldots, x_n\}$ represent a set of n objects in a metric space, and let $d(x_i, x_j)$ denote a distance (or proximity) function that measures the dissimilarity between objects x_i and x_j. Commonly employed distance measures include

- Euclidean distance: $d(x_i, x_j) = \sqrt{\sum_{k=1}^{d}(x_{ik} - x_{jk})^2}$
- Manhattan distance: $d(x_i, x_j) = \sum_{k=1}^{d} |x_{ik} - x_{jk}|$
- Minkowski distance: $d(x_i, x_j) = \left(\sum_{k=1}^{d} |x_{ik} - x_{jk}|^q\right)^{1/q}$
- Cosine distance: $d(x_i, x_j) = 1 - \frac{x_i \cdot x_j}{\|x_i\| \|x_j\|}$

The choice of distance measure strongly influences the clustering result.
Hierarchical clustering constructs a tree of nested clusters using a distance matrix.

13.2.1 Agglomerative Methods

Start with n clusters (one per object) and iteratively merge clusters based on a linkage criterion. Let $d(A, B)$ denote the distance between clusters A and B.

- Single Linkage (Nearest Neighbor): $d_{\text{single}}(A, B) = \min_{x_i \in A, x_j \in B} d(x_i, x_j)$.
- Complete Linkage (Furthest Neighbor): $d_{\text{complete}}(A, B) = \max_{x_i \in A, x_j \in B} d(x_i, x_j)$
- Average Linkage: $d_{\text{average}}(A, B) = \frac{1}{|A||B|} \sum_{x_i \in A} \sum_{x_j \in B} d(x_i, x_j)$
- Centroid Method: $d_{\text{centroid}}(A, B) = \|\bar{x}_A - \bar{x}_B\|$
- Ward's Method: Minimize within-cluster variance, $\Delta(A, B) = \frac{|A||B|}{|A|+|B|} \|\bar{x}_A - \bar{x}_B\|^2$

A dendrogram is a tree diagram that records the sequences of merges or splits. Let k denote the number of clusters desired. To obtain k clusters from an agglomerative dendrogram, one must perform exactly $k-1$ merges from the n initial singletons. Conversely, cutting the dendrogram at a particular height can produce a desired number of clusters.

A key Observation is that cutting a dendrogram at k levels does *not necessarily* produce k clusters directly. Formally, if a dendrogram has n leaves,

- Cutting the dendrogram at a height that intercepts $k-1$ internal links produces k clusters.
- Each cut increases the number of clusters by one; hence, k cuts can produce $k+1$ clusters.

Example 13.2.1 (Comparison of linkage methods) *We are given the following six points in* $\mathbb{R}^5$

- $A_1, A_2, A_3 \sim \mathbb{N}_5(\mathbf{0}, I_5)$
- $\mathbf{B}_1, \mathbf{B}_2, \mathbf{B}_3 \sim \mathbb{N}_5(\mathbf{1}, I_5)$

Point	x_1	x_2	x_3	x_4	x_5
$\mathbf{A}_1$	0.497	−0.138	0.647	1.523	−0.234
$\mathbf{A}_2$	−0.234	−0.469	0.242	−0.463	0.242
$\mathbf{A}_3$	−0.908	−1.412	1.466	−0.225	0.068
$\mathbf{B}_1$	0.969	0.708	2.055	0.534	1.531
$\mathbf{B}_2$	0.532	1.543	1.536	1.534	0.578
$\mathbf{B}_3$	2.579	0.767	0.530	1.542	2.483

We construct hierarchical clustering with three linkage methods: Single linkage (nearest neighbor), Complete linkage (farthest neighbor) and Average linkage. The dendrograms appear in Figure 13.1.

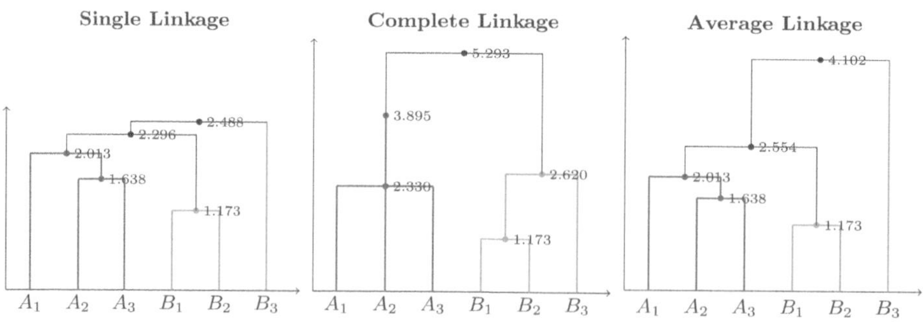

FIGURE 13.1: Comparison of single, complete, and average linkage dendrograms. Blue branches correspond to A points; red branches to B points. Dots mark merge points with annotated heights.

The height at which clusters merge is the distance defining that merge. Single linkage merges points early due to nearest-neighbor chaining. Complete linkage maintains compact clusters; A/B groups only merge at the highest distance. Average linkage balances chaining and compactness, producing intermediate merge heights.

Single linkage (nearest neighbor) uses the minimum IPD to define inter-cluster dissimilarity, which tends to form chains. In some applications such as detection of hot spots (Patil and Modarres [263]) this can be an advantage as the method can detect non-elliptical clusters. In this example, although the A and B groups originate from different distributions, the algorithm merges them before attaching the outlying point B_3.

The class of agglomerative clustering methods encompasses some of the most widely used techniques in practice. Lance and Williams [181] introduced a general formulation for a family of agglomerative algorithms by specifying the distance between a newly formed cluster $Y_{(ij)}$ *and an existing cluster* $Y_{(k)}$ *as*

$$d_{(ij)k} = \alpha_i d_{ik} + \alpha_j d_{jk} + \beta d_{ij} - \gamma |d_{ik} - d_{jk}|,$$

where d_{ij} *denotes the IPD between the* i*th and* j*th observations, and the parameter vector* $\theta = (\alpha_i, \alpha_j, \gamma)$ *uniquely determines a particular member of this family. For instance, the single-linkage (nearest-neighbor) method corresponds to* $\theta = (1/2, 0, -1/2)$*, the complete-linkage (furthest-neighbor) method to* $\theta = (1/2, 0, 1/2)$*, and the weighted average linkage method to* $\theta = (1/2, 0, 0)$.

13.2.2 Divisive Methods

Start with all points in one cluster and iteratively split clusters using distance-based criteria such as maximizing between-cluster distances. Divisive methods are less common in practice but are conceptually important.

13.3 Partitioning Methods

Partitioning methods aim to assign each object to exactly one cluster by optimizing an objective function based on distances.

13.3.1 k-Means Clustering

The objective is to minimize total within-cluster sum of squares (WCSS),

$$\min \sum_{i=1}^{k} \sum_{x \in C_i} \|x - \mu_i\|^2$$

where μ_i is the centroid of cluster C_i. It uses Euclidean distance. It is efficient, works well for compact, spherical clusters. See section 13.7

Example 13.3.1 (k-means example) *Consider the six points of Example 13.2.1.*

$$\mathbf{A}_1, \mathbf{A}_2, \mathbf{A}_3 \sim \mathbb{N}_5(0, I), \quad \mathbf{B}_1, \mathbf{B}_2, \mathbf{B}_3 \sim \mathbb{N}_5(1, I)$$

We apply k-means clustering with $k = 2$. Cluster assignments are

Point	$\mathbf{A}_1$	$\mathbf{A}_2$	$\mathbf{A}_3$	$\mathbf{B}_1$	$\mathbf{B}_2$	$\mathbf{B}_3$
Cluster	*0*	*0*	*0*	*1*	*1*	*1*

Cluster centroids are

Cluster	c_1	c_2	c_3	c_4	c_5
0	−0.215	−0.673	0.785	0.278	0.025
1	1.360	1.006	1.374	1.203	1.531

We project the six points onto the first two principal components. Figure 13.2 shows the result.

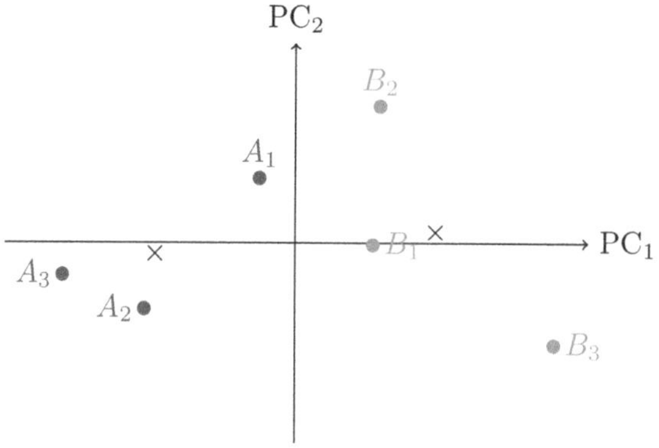

FIGURE 13.2: K-means clustering with $k = 2$. PCA projection to two dimensions. Blue points: cluster 0; red points: cluster 1. Black × symbols indicate projected centroids.

The centroids are the exact averages of the assigned points. The two clusters cleanly separate the $\mathbb{N}(\mathbf{0}, I)$ *and* $\mathbb{N}(\mathbf{1}, I)$ *samples.*

13.3.2 k-Medoid

Partitioning around medoids (PAM) uses actual data points as cluster centers (medoids) instead of centroids. We minimizes, total distance between points and cluster medoids,

$$\min \sum_{i=1}^{k} \sum_{x \in C_i} d(x, m_i).$$

The method is robust to outliers and allows arbitrary distance metrics.

13.4 Density-Based Clustering

Distance or proximity measures are also used in density-based clustering.

13.4.1 DBSCAN

Density-based spatial clustering of applications with noise (DBSCAN) defines clusters as dense regions separated by sparse regions. DBSCAN requires a) ε-neighborhood distance and b) Minimum points threshold (MinPts). Two points x_i and x_j are in the same cluster if $d(x_i, x_j) \leq \varepsilon$ and the density criterion is satisfied. DBSCAN can detects arbitrarily shaped clusters and noise points.

13.4.2 HDBSCAN

HDBSCAN (Hierarchical Density-Based Spatial Clustering of Applications with Noise) is an extension of DBSCAN that constructs a hierarchy of clusters and selects the most stable ones. It avoids the need to choose a global distance threshold and can detect clusters of varying densities. HDBSCAN uses the following key concepts in constructing clusters.

a) For a point $\mathbf{x}_i$, the core distance is the distance from $\mathbf{x}_i$ to its minPts-th closest neighbor,

$$\text{core}_k(\mathbf{x}_i) = \text{distance to the } k\text{-th nearest neighbor of } \mathbf{x}_i.$$

For example, if minPts $= 4$, we sort all other points by how far they are from $\mathbf{x}_i$, then the core distance is the distance to the 4-th closest point.

b) For points $\mathbf{x}_i$ and $\mathbf{x}_j$, the mutual reachability distance is defined as

$$d_{\text{mreach},k}(\mathbf{x}_i, \mathbf{x}_j) = \max\left\{\text{core}_k(\mathbf{x}_i), \text{core}_k(\mathbf{x}_j), \|\mathbf{x}_i - \mathbf{x}_j\|\right\}.$$

c) An MST is constructed using the mutual reachability distances as edge weights, connecting all points with the smallest total edge weight.

d) By progressively cutting the MST at increasing distances, a cluster tree (dendrogram) is created.

e) The stability of a cluster C is computed as

$$\text{Stability}(C) = \sum_{\mathbf{x} \in C} (\lambda_{\text{end}}(\mathbf{x}) - \lambda_{\text{birth}}(\mathbf{x})),$$

where $\lambda = 1/d_{\text{mreach}}$. The difference $\lambda_{\text{end}}(x) - \lambda_{\text{birth}}(\mathbf{x})$ represents the lifetime of point $\mathbf{x}$ in cluster C. In practice, the absolute value is taken for ranking.

f) The hierarchy is pruned, and the clusters with the highest stability are retained.

Example 13.4.1 *Consider 5 Points in $\mathbb{R}^5$. Let the data points in $\mathbb{R}^5$ be*

$$\mathbf{x}_1 = (0,0,0,0,0), \qquad \mathbf{x}_2 = (0,1,0,0,0), \quad \mathbf{x}_3 = (0,2,0,0,0),$$
$$\mathbf{x}_4 = (10,10,10,10,10), \qquad \mathbf{x}_5 = (11,10,10,10,10).$$

Set minPts $= 2$. Compute Core Distances, the distance to the 2nd nearest neighbor are

$$core_2(\mathbf{x}_1) = 1, \quad core_2(\mathbf{x}_2) = 1, \ldots, core_2(\mathbf{x}_5) = 1$$

Mutual Reachability Distances are computed with

$$d_{mreach}(\mathbf{x}_i, \mathbf{x}_j) = \max\{core_2(\mathbf{x}_i), core_2(\mathbf{x}_j), \|\mathbf{x}_i - \mathbf{x}_j\|\}.$$

Selected examples are

$$d_{mreach}(\mathbf{x}_1, \mathbf{x}_2) = \max\{1, 1, 1\} = 1,$$
$$d_{mreach}(\mathbf{x}_1, \mathbf{x}_3) = \max\{1, 1, 2\} = 2,$$
$$d_{mreach}(\mathbf{x}_1, \mathbf{x}_4) = \max\{1, 1, \sqrt{500}\} \approx 22.36,$$
$$d_{mreach}(\mathbf{x}_4, \mathbf{x}_5) = \max\{1, 1, 1\} = 1.$$

Using mutual reachability distances, the MST includes edges

- $\mathbf{x}_1 \leftrightarrow \mathbf{x}_2$ *with weight 1*
- $\mathbf{x}_2 \leftrightarrow \mathbf{x}_3$ *with weight 1*
- $\mathbf{x}_4 \leftrightarrow \mathbf{x}_5$ *with weight 1*
- $\mathbf{x}_3 \leftrightarrow \mathbf{x}_4$ *with weight* ≈ 20.36

At a threshold distance of 1, the MST breaks into two clusters,

$$C_1 = \{\mathbf{x}_1, \mathbf{x}_2, \mathbf{x}_3\}, \quad C_2 = \{\mathbf{x}_4, \mathbf{x}_5\}.$$

As the threshold increases, these clusters merge. For C_1 *we have*

$$\lambda_{start} = 1, \quad \lambda_{end} \approx \frac{1}{20.36} \approx 0.049.$$

$$Stability(C_1) \approx \sum_{\mathbf{x} \in C_1} |\lambda_{end} - \lambda_{start}| = 3 \cdot |0.049 - 1| \approx 2.85.$$

Similarly, compute for C_2*. The hierarchy is pruned based on these stability values. HDBSCAN successfully separates the two spatially distinct groups* C_1 *and* C_2*.*

13.5 Distance Matrix and the Rand Index

The choice of distance or proximity metric for clustering is crucial and depends on the data type. For continuous variables one uses the Euclidean distance. For Bernoulli observations, the Hamming distance (Hamming [136]) is utilized. Modarres [228] discusses the distribution of IPDs of Bernoulli vectors. Gower distance (Gower [122]) is used for mixed data types, and for high-dimensional sparse data, we discuss the MADMD dissimilarity. See section 13.6.

The input to most clustering algorithms is an $N \times d$ data matrix, where each observation is represented by a d-dimensional feature vector. Prior to clustering, the data matrix is often standardized to ensure scale homogeneity across features. Distance-based clustering methods frequently rely solely on the pairwise distances or dissimilarities between observations as input. In such cases, rather than using the original $N \times d$ data matrix, one can construct an $N \times N$ distance matrix. This transformation can provide substantial data reduction, particularly in the HDLSS setting where $d \gg N$.

It is important to note that the Euclidean distance between two vectors depends only on those vectors and does not account for the underlying probability mass distribution. In high-dimensional spaces, violations of neighborhood structure are well-documented for Euclidean and fractional distances (Aggarwal et al. [47]; Beyer et al. [29]; Francois et al.

[105]7), which can adversely affect the performance of machine learning methods that rely on such distances, including the nearest neighbor (NN) classifier. To illustrate this phenomenon, Hall et al. [133] consider two independent samples from d-dimensional normal distributions F_1 and F_2. They show that, depending on the parameter values, all observations from F_1 may have their nearest neighbors within F_1 with high probability, while observations from F_2 may have their nearest neighbors in F_1 with probability approaching 1. This effect arises from the use of Euclidean distance and the distance concentration phenomenon, and it demonstrates that neighborhood structure is not preserved as $d \to \infty$.

The distance concentration phenomenon refers to the tendency of inter-point distances (IPDs) to become nearly indistinguishable in high-dimensional spaces. Specifically, the coefficient of variation of these distances approaches zero as $d \to \infty$. As a result, the relative difference between the nearest and farthest observations diminishes, rendering the concept of a nearest neighbor (NN) increasingly meaningless. Chan and Hall [56] highlighted the challenges faced by the NN method in such settings and proposed scale-adjustment strategies.

Several clustering techniques have been developed for high-dimensional data (Ahn et al. [4]; Witten and Tibshirani [358]; Sarkar and Ghosh [299]). Hall et al. [133] demonstrate that in high-dimensional spaces, observations tend to lie near the vertices of a regular simplex, with apparent randomness arising from the random rotation of the simplex. While originally proposed for supervised learning, the maximal data piling (MDP) distance (Ahn and Marron [5]) can also be applied to cluster high-dimensional observations. However, although the MDP distance can effectively distinguish between cluster means, it fails to differentiate cluster variances when $d \to \infty$ with fixed sample size.

The Rand index [277] measures the performance of a clustering method ψ with

$$\mathcal{R}(\psi) = \binom{n}{2}^{-1} \sum_{i<j} \mathbb{I}[\mathbb{I}\{\psi(\mathbf{X}_i) = \psi(\mathbf{X}_j)\} + \mathbb{I}\{\mathcal{C}(\mathbf{X}_i) = \mathcal{C}(\mathbf{X}_j)\} = 1], \tag{13.1}$$

where $\mathbb{I}\{\cdot\}$ is the indicator function, $\mathcal{C}(\mathbf{X}_i)$ is the true cluster label of $\mathbf{X}_i$ and $\psi(\mathbf{X}_i)$ is the clustered label assigned to $\mathbf{X}_i$. Using all pairwise observations, the Rand index measures whether clustering results are consistent with the true cluster labels. One can show that $0 \le \mathcal{R}(\psi) \le 1$, with 1 indicating that the two data clusters do not concur on any pair of observations and 0 indicating that the data clusters are identical. The closer the Rand index is to 0, the better the clustering performance. A clustering method is called perfect if its Rand index is zero. Rand [277] defines two indices $\mathcal{R}(\psi)$ and $1 - \mathcal{R}(\psi)$. We use the latter definition.

The Rand Index measures agreement between two clusterings of the same set of n objects. Let $U = \{U_1, \ldots, U_r\}$ and $V = \{V_1, \ldots, V_s\}$ be two partitions. Over all $\binom{n}{2}$ unordered pairs we have

a) pairs in the same cluster in U and the same cluster in V,

b) pairs in different clusters in U and different clusters in V,

c) pairs in the same cluster in U but different clusters in V,

d) pairs in different clusters in U but the same cluster in V.

$$\mathrm{RI} = \frac{c+d}{a+b+c+d} = \frac{c+d}{\binom{n}{2}} \in [0,1].$$

Example 13.5.1 *Let $n = 5$ and suppose the cluster labels are $U = (1,1,2,2,3)$, and $V = (1,1,3,2,3)$. These labels define the clusters*

$$U_1 = \{1,2\}, \quad U_2 = \{3,4\}, \quad U_3 = \{5\},$$

$$V_1 = \{1,2\}, \quad V_2 = \{4\}, \quad V_3 = \{3,5\}.$$

A convenient way to compare the two clusterings is to construct a contingency table counting how many objects belong to each pair of clusters.

$U\backslash V$	V_1	V_2	V_3	*Row sum*
U_1	*2*	*0*	*0*	*2*
U_2	*0*	*1*	*1*	*2*
U_3	*0*	*0*	*1*	*1*
Column sum	*2*	*1*	*2*	*5*

Each cell n_{ij} counts the number of objects that belong to cluster U_i and cluster V_j simultaneously.

Pairs that are in the same cluster in both partitions correspond to pairs within the same cell of the table. Thus

$$a = \sum_{i,j} \binom{n_{ij}}{2} = \binom{2}{2} = 1.$$

Pairs in the same cluster in U but not in V are obtained from the row sums,

$$c = \sum_{i} \binom{n_{i\cdot}}{2} - a = \left[\binom{2}{2} + \binom{2}{2} + \binom{1}{2}\right] - 1 = 1.$$

Pairs in the same cluster in V but not in U are obtained from the column sums,

$$d = \sum_{j} \binom{n_{\cdot j}}{2} - a = \left[\binom{2}{2} + \binom{1}{2} + \binom{2}{2}\right] - 1 = 1.$$

Finally,

$$b = \binom{5}{2} - a - c - d = 10 - 1 - 1 - 1 = 7.$$

Hence, $(a,b,c,d) = (1,7,1,1)$. The number of disagreements is $c + d = 2$, and therefore $\mathrm{RI} = \frac{2}{10} = 0.2$.

In general, if n_{ij} denotes the contingency table entries, $n_{i\cdot}$ the row sums, and $n_{\cdot j}$ the column sums, then

$$a = \sum_{i,j} \binom{n_{ij}}{2}, \qquad c = \sum_{i} \binom{n_{i\cdot}}{2} - a,$$

$$d = \sum_{j} \binom{n_{\cdot j}}{2} - a, \qquad b = \binom{n}{2} - a - c - d.$$

These formulas allow the Rand index to be computed efficiently without enumerating all $\binom{n}{2}$ pairs.

13.6 Change Point Clustering Method

Following Wang and Modarres [348] let K be the number of clusters, n_k be the sample size of the k-th cluster for $k = 1, \ldots, K$ and $N = \sum_{k=1}^{K} n_k$. The symmetric, zero-diagonal,

$N \times N$ dissimilarity matrix between all pairs of the observations $\boldsymbol{X}_1, \boldsymbol{X}_2, \ldots, \boldsymbol{X}_N$ using dissimilarity measure d_{ij} is given by

$$\mathbf{D}(\mathbb{X}) = \begin{bmatrix} 0 & d_{12} & \ldots & d_{1N} \\ \vdots & \vdots & \ldots & \vdots \\ d_{N1} & d_{N2} & \ldots & 0 \end{bmatrix}.$$

In this section, we employ the dissimilarity matrices $\mathbf{D}(\mathbb{X})_\delta$ and $\mathbf{D}(\mathbb{X})_\rho$ to detect change points and construct clusters. The proposed clustering approach utilizes the property that $\delta(\boldsymbol{X}_i, \boldsymbol{X}_j)$ converges to 0 when $\boldsymbol{X}_i$ and $\boldsymbol{X}_j$ belong to the same group, whereas for observations from different groups, $\delta(\boldsymbol{X}_i, \boldsymbol{X}_j) \to \tilde{\delta} > 0$ as $d \to \infty$.

The strategy involves computing the distance difference matrix to identify $k-1$ change points in each row of the row-sorted distance matrix. Change point detection is based on identifying sharp changes in the slope of the line segments connecting consecutive distances. The unknown change point locations are denoted by τ_{it} for $i = 1, \ldots, N$ and $t = 1, \ldots, k-1$.

For each row of the distance matrix with entries d_{ij}, we construct the sorted row $\mathbf{R}_i = [d^s_{ij}]$ for $i, j = 1, \ldots, N$, where d^s_{ij} represents the ordered distances. The distance differences between consecutive distances and the corresponding $N \times N$ distance difference matrix are defined as

$$\begin{aligned} \Delta_{ij} &:= |d^s_{ij} - d^s_{i,j-1}|, & i = 1, \ldots, N, \ j = 2, \ldots, N, \\ \Delta_{i1} &:= 0, & i = 1, \ldots, N. \end{aligned}$$

$$\Delta = \begin{bmatrix} 0 & \Delta_{12} & \ldots & \Delta_{1N-1} & \Delta_{1N} \\ \vdots & \vdots & \ldots & \vdots & \vdots \\ 0 & \Delta_{N2} & \ldots & \Delta_{NN-1} & \Delta_{NN} \end{bmatrix}.$$

There are $k-1$ change points corresponding to k clusters. Using the i-th row of the distance difference matrix $\boldsymbol{\Delta}$, for $i = 1, \ldots, N$, our estimates of the change point locations are given by the positions of the top $k-1$ maximum values of Δ_{ij} for $j = 1, \ldots, N$. Specifically, the change points are located as

$$\hat{\tau}_{it} = \underset{1 \le j \le n-t+1}{\arg\max} \left\{ \Delta_{ij} \right\}, \tag{13.2}$$

where, after identifying each maximum, its value is swapped with the entry at the bottom of the row at location $n-t+1$ to exclude it from subsequent iterations. This procedure is repeated $k-1$ times.

The estimated change points $\hat{\tau}_{it}$ correspond to the locations of the largest slopes in the i-th row of the distance difference matrix and partition that row into k clusters. Let `Label` denote the vector of cluster labels for the observations. We denote the clustering methods as CPC_ρ and CPC_δ, respectively. Using the $k-1$ change points, each row is partitioned into k clusters, and the resulting $N \times N$ clustering matrix $\mathbf{C}$ contains the cluster labels for all rows of the distance matrix. An agreement matrix $\mathbf{A}$ is constructed based on the Rand index computed from the cluster labels. Each row of $\mathbf{A}$ indicates the consistency of the clustering obtained from that row with the clusterings obtained from all other $N-1$ rows. The final clusters correspond to the row that achieves the minimum total Rand index with all other rows.

Algorithm CPC gives detailed implementation of the change point clustering method. The computational complexity of the CPC algorithm depends on the total sample size $N = \sum_{i=1}^{k} n_i$. In Step 1, the algorithm computes the distance matrix, which requires $O(N^2)$

operations. The main loop in Step 2 is executed N times, with Step 2a involving the sorting of each row of the distance matrix, an operation of complexity $O(N \log N)$. The most computationally intensive component is the construction of the agreement matrix in Step 3, which incurs a time complexity of $O(N^3)$. The following result demonstrates that CPC$_\delta$ achieves perfect clustering in the limit as $d \to \infty$.

Algorithm 13.6.1 Algorithm CPC: High-Dimensional Change Point Clustering

1: **Step 1:** Compute the distance matrix $\mathbf{D}_\delta(\mathbb{X})$

2: **Step 2:** For each row R_i of $\mathbf{D}_\delta(\mathbb{X})$:

1. **Step 2a:** Sort R_i with the Label vector as a concomitant
2. **Step 2b:** Compute the i-th row of the distance matrix $\Delta_{ij} := |d^s_{ij} - d^s_{i,j-1}|$
3. **Step 2c:** Estimate $k-1$ change points: $\hat{\tau}_{it} = \arg\max_{1 \le j \le n} \Delta_{ij}$, for $t = 1, \ldots, k-1$
4. **Step 2d:** Construct k clusters from the change points

3: **Step 3:** Compute the agreement matrix $\mathbf{A} = [a_{ij}]$, where $a_{ij} = \text{Rand}(C_i, C_j)$

1. **Step 3a:** Compute the row sum of $\mathbf{A}$: $s_k = \sum_{j=1}^{N} a_{kj}$
2. **Step 3b:** The best clustering is $\mathcal{C}_m$, where $m = \arg\min_{1 \le k \le N} s_k$.

Theorem 13.6.1 *Suppose we use CPC$_\delta$ to cluster independent high-dimensional observations from k_0 populations $F_1, F_2, \ldots, F_{k_0}$. Under Assumptions B1-B3, if for every pair (F_i, F_j), $\bar{\mu}^d_i \neq \bar{\mu}^d_j$ or $\sigma^d_i \neq \sigma^d_j$ holds for $i \neq j$, then the Rand index of clusters obtained CPC$_\delta$ with the true clusters converges to 0 in probability as $d \to \infty$.*

The following examples show the step-by-step execution of the CPC algorithm and explore the possibility of ties among the limiting values of the distance matrices.

Example 13.6.1 *Let $\mathbf{X}_{ij}$ be i.i.d. from F_j for $i = 1, \ldots, n_j$ and $j = 1, \ldots 3$, where $N = \sum_{j=1}^{3} n_j$. We denote the three groups with X, Y and Z, where $0 < \tilde{\delta}_{xy} = \tilde{\delta}_{yz} < \tilde{\delta}_{xz}$. As $d \to \infty$, the $N \times N$ symmetric matrix $\mathbf{D}_\delta$ approaches the limiting form*

$$\tilde{\mathbf{D}}_\delta(\mathbb{X}) = \begin{bmatrix} [\mathbf{0}]_{n_1 \times n_1} & [\tilde{\delta}_{xy}]_{n_1 \times n_2} & [\tilde{\delta}_{xz}]_{n_1 \times n_3} \\ [\tilde{\delta}_{xy}]_{n_2 \times n_1} & [\mathbf{0}]_{n_2 \times n_2} & [\tilde{\delta}_{yz}]_{n_2 \times n_3} \\ [\tilde{\delta}_{xz}]_{n_3 \times n_1} & [\tilde{\delta}_{yz}]_{n_3 \times n_2} & [\mathbf{0}]_{n_3 \times n_3} \end{bmatrix}. \tag{13.3}$$

The $N \times N$ asymptotic difference dissimilarity matrix as $d \to \infty$ is

$$\tilde{\mathbf{\Delta}}_\delta = \begin{bmatrix} 0\ldots0 & \tilde{\delta}_{xy} & 0\ldots0 & |\tilde{\delta}_{xz} - \tilde{\delta}_{xy}| & 0\ldots0 \\ \vdots & \vdots & \vdots & \vdots & \vdots \\ 0\ldots0 & \tilde{\delta}_{xy} & 0\ldots0 & 0 & 0\ldots0 \\ 0\ldots0 & \tilde{\delta}_{xz} & 0\ldots0 & |\tilde{\delta}_{yz} - \tilde{\delta}_{xz}| & 0\ldots0 \\ \vdots & \vdots & \vdots & \vdots & \vdots \\ 0\ldots0 & \tilde{\delta}_{xz} & 0\ldots0 & |\tilde{\delta}_{yz} - \tilde{\delta}_{xz}| & 0\ldots0 \end{bmatrix}.$$

For the first n_1 rows, the change points occur at $\hat{\tau}_1 = n_1 + 1$ and $\hat{\tau}_2 = n_1 + n_2 + 1$, yielding the clusters

$$C_1 = \{\mathbf{X}_1, \ldots, \mathbf{X}_{n_1}\}, \quad C_2 = \{\mathbf{X}_{n_1+1}, \ldots, \mathbf{X}_{n_1+n_2}\}, \quad C_3 = \{\mathbf{X}_{n_1+n_2+1}, \ldots, \mathbf{X}_N\}.$$

For rows $n_1 + 1$ to $n_1 + n_2$, the first change point occurs at position n_1, correctly identifying cluster $C_1 = \{\mathbf{X}_1, \ldots, \mathbf{X}_{n_1}\}$. However, the second change point corresponds to the largest non-zero value, resulting in the second and third clusters being misidentified.

Similarly, for rows $n_1 + n_2 + 1$ to N, the change points occur at positions $n_1 + 1$ and $n_1 + n_2 + 1$, reproducing the same clustering as obtained for the first n_1 rows. To reconcile these discrepancies, we compute the Rand index between clusters in different rows of the clustering matrix $\mathbf{C}$. The resulting agreement matrix $\mathbf{A} = [Rand(\mathcal{C}_i, \mathcal{C}_j)]$ and its minimum row sum s_k identify the correct clustering, which corresponds to the clustering obtained from the first n_1 rows of $\tilde{\mathbf{\Delta}}$. The Rand index of these clusters with respect to the true labels is 0, indicating perfect agreement.

Example 13.6.2 *We generate $\mathbf{X}_1, \ldots, \mathbf{X}_8$ from $K = 3$ distributions in $d = 1000$,*

$$\begin{aligned} \mathbf{X}_1, \mathbf{X}_5, \mathbf{X}_7 &\sim F_1 = \mathbb{N}(\mathbf{1}_d, \mathbf{I}), \\ \mathbf{X}_2, \mathbf{X}_3, \mathbf{X}_6 &\sim F_2 = \mathbb{N}(1.5\mathbf{1}_d, \mathbf{I}), \\ \mathbf{X}_4, \mathbf{X}_8 &\sim F_3 = \mathbb{N}(2\mathbf{1}_d, \mathbf{I}). \end{aligned}$$

The limiting values of δ are $\tilde{\delta}_{F_1,F_2} = \tilde{\delta}_{F_2,F_3} = \frac{1}{2}$, $\tilde{\delta}_{F_1,F_3} = 1$. The observed dissimilarity matrix $\mathbf{D}_\delta(\mathbb{X})$ is

$$\mathbf{D}_\delta(\mathbb{X}) = \begin{array}{c} \\ X_1 \\ X_2 \\ X_3 \\ X_4 \\ X_5 \\ X_6 \\ X_7 \\ X_8 \end{array} \begin{array}{c} \begin{array}{cccccccc} X_1 & X_2 & X_3 & X_4 & X_5 & X_6 & X_7 & X_8 \end{array} \\ \begin{pmatrix} 0 & 0.4454 & 0.4098 & 0.5068 & 0.0204 & 0.4252 & 0.0199 & 0.5116 \\ 0.4454 & 0 & 0.0730 & 0.4582 & 0.4396 & 0.0530 & 0.4396 & 0.4678 \\ 0.4098 & 0.0730 & 0 & 0.5033 & 0.4040 & 0.0255 & 0.4039 & 0.5127 \\ 0.5068 & 0.4582 & 0.5033 & 0 & 0.5132 & 0.4919 & 0.5129 & 0.0153 \\ 0.0204 & 0.4396 & 0.4040 & 0.5132 & 0 & 0.4191 & 0.001664 & 0.5181 \\ 0.4252 & 0.0530 & 0.0255 & 0.4919 & 0.4191 & 0 & 0.4190 & 0.5011 \\ 0.0199 & 0.4396 & 0.4039 & 0.5129 & 0.001664 & 0.4190 & 0 & 0.5178 \\ 0.5116 & 0.4678 & 0.5127 & 0.0153 & 0.5181 & 0.5011 & 0.5178 & 0 \end{pmatrix} \end{array}.$$

We now construct, for each row, the sorted distance matrix $\mathbf{R}_i$ and corresponding differences $\mathbf{\Delta}_i$. For brevity we only list the first two and last two rows.

$$\mathbf{R}_1 = X_1 \begin{array}{c} \begin{array}{cccccccc} X_1 & X_7 & X_5 & X_3 & X_6 & X_2 & X_4 & X_8 \end{array} \\ \begin{pmatrix} 0 & 0.0199 & 0.0204 & 0.4098 & 0.4252 & 0.4454 & 0.5068 & 0.5116 \end{pmatrix} \end{array},$$

$$\mathbf{\Delta}_1 = X_1 \begin{array}{c} \begin{array}{cccccccc} X_1 & X_7 & X_5 & X_3 & X_6 & X_2 & X_4 & X_8 \end{array} \\ \begin{pmatrix} 0 & 0.0199 & 0.0015 & 0.3894 & 0.0154 & 0.0202 & 0.0614 & 0.0048 \end{pmatrix} \end{array},$$

$$\hat{\tau}_1 = (4, 7), \quad \mathcal{C}_1 = (X_1, X_7, X_5),\ \mathcal{C}_2 = (X_3, X_6, X_2),\ \mathcal{C}_3 = (X_4, X_8)$$

$$\mathbf{R}_2 = X_2 \begin{array}{c} \begin{array}{cccccccc} X_2 & X_6 & X_3 & X_7 & X_5 & X_1 & X_4 & X_8 \end{array} \\ \begin{pmatrix} 0 & 0.0530 & 0.0730 & 0.4396 & 0.4396 & 0.4454 & 0.4582 & 0.4678 \end{pmatrix} \end{array},$$

$$\mathbf{\Delta}_2 = X_2 \begin{array}{c} \begin{array}{cccccccc} X_2 & X_6 & X_3 & X_7 & X_5 & X_1 & X_4 & X_8 \end{array} \\ \begin{pmatrix} 0 & 0.0530 & 0.0200 & 0.3666 & 0.0012 & 0.0070 & 0.0140 & 0.0108 \end{pmatrix} \end{array},$$

$$\hat{\tau}_2 = (2, 4), \quad \mathcal{C}_1 = (X_2),\ \mathcal{C}_2 = (X_6, X_3),\ \mathcal{C}_3 = (X_7, X_5, X_1, X_4, X_8)$$

$$\mathbf{R}_7 = X_7 \begin{pmatrix} X_7 & X_5 & X_1 & X_3 & X_6 & X_2 & X_4 & X_8 \\ 0 & 0.001664 & 0.0199 & 0.4039 & 0.4190 & 0.4396 & 0.5129 & 0.5178 \end{pmatrix},$$

$$\mathbf{\Delta}_7 = X_7 \begin{pmatrix} X_7 & X_5 & X_1 & X_3 & X_6 & X_2 & X_4 & X_8 \\ 0 & 0.001664 & 0.0194 & 0.3840 & 0.0153 & 0.0206 & 0.0733 & 0.0049 \end{pmatrix},$$

$$\hat{\tau}_7 = (4,7), \quad \mathcal{C}_1 = (X_7, X_5, X_1), \ \mathcal{C}_2 = (X_3, X_6, X_2), \ \mathcal{C}_3 = (X_4, X_8)$$

$$\mathbf{R}_8 = X_8 \begin{pmatrix} X_8 & X_4 & X_2 & X_6 & X_1 & X_3 & X_7 & X_5 \\ 0 & 0.0153 & 0.4678 & 0.5011 & 0.5116 & 0.5127 & 0.5178 & 0.5181 \end{pmatrix},$$

$$\mathbf{\Delta}_8 = X_8 \begin{pmatrix} X_8 & X_4 & X_2 & X_6 & X_1 & X_3 & X_7 & X_5 \\ 0 & 0.0153 & 0.4525 & 0.0333 & 0.0105 & 0.0016 & 0.0051 & 0.0003 \end{pmatrix},$$

$$\hat{\tau}_8 = (3,4), \quad \mathcal{C}_1 = (X_4, X_8), \ \mathcal{C}_2 = (X_2), \ \mathcal{C}_3 = (X_1, X_3, X_7, X_6, X_5)$$

There are 8 cluster sets corresponding to each row of the matrix. To choose the final clustering, the agreement matrix $\mathbf{A}$ *of Rand indices of the cluster labels is formed. We select the clusters of the row that obtains the minimum total Rand index.*

The method is invariant with respect to the order in which observations are presented, since the observation labels are concomitant when each row of the distance matrix is ordered.

Example 13.6.3 *We generated five observations from* $F_1 = \mathbb{N}(0, \mathbf{I})$ *with* $d = 1000$*, five observations from* $F_2 = \mathbb{N}(\mathbf{1}, \mathbf{I})$*, and five observations from* $F_3 = \mathbb{N}(\mathbf{2}, \mathbf{I})$*. The limiting values of the dissimilarity indices as* $d \to \infty$ *are* $\tilde{\delta}_{F_1,F_2} = \tilde{\delta}_{F_2,F_3} = 1$*, and* $\tilde{\delta}_{F_1,F_3} = 2$*. Figure 13.3 displays the rows of the dissimilarity matrix* $\mathbf{D}_\delta$ *plotted against the observation labels. Three clusters are identified using the difference dissimilarity matrix: the first five observations correspond to cluster 1, the middle five observations form cluster 2, and the remaining five observations constitute cluster 3. The slope of the line segments connecting*

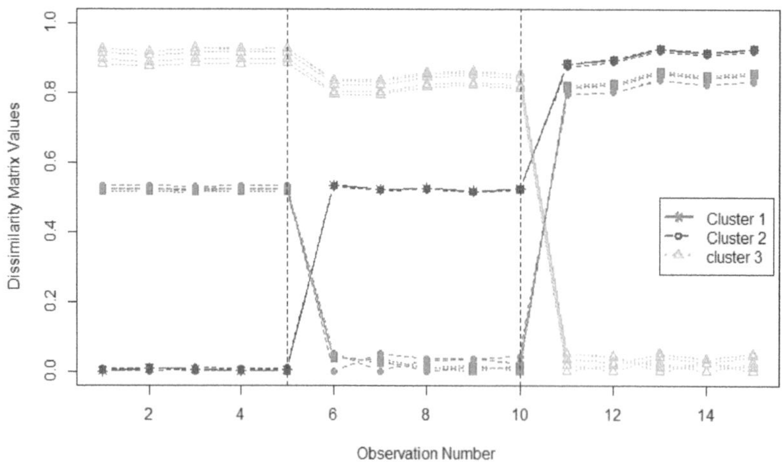

FIGURE 13.3: Display of the rows of the dissimilarity matrix $\mathbf{D}$ against the observation numbers reveals 3 clusters with $\mathbf{X}_i$ in $\mathcal{C}_1 = \{1 \le i \le 5\}$, $\mathcal{C}_2 = \{6 \le i \le 10\}$ and $\mathcal{C}_3 = \{11 \le i \le 15\}$.

consecutive clusters is maximal in the distance difference matrix, and the corresponding induced labels determine the cluster membership. The Rand index $\mathcal{R}(\psi) = 0$, indicating perfect agreement between the estimated clusters and the true labels.

13.7 k-means Clustering

The k-means clustering algorithm is described in detail by Hartigan [139]. For completeness, the procedure for high-dimensional k-means clustering is outlined below.

Algorithm 13.7.1 High-Dimensional k-Means Clustering

Input: Data points $\{x_1, x_2, \ldots, x_n\} \subset \mathbb{R}^p$, number of clusters k
Output: Cluster centers $\mathbf{C} = (c_1, \ldots, c_k)$ and cluster assignments

Step 1: Initialization
Select k initial cluster centers and denote them by $\mathbf{C} = (c_1, c_2, \ldots, c_k)$.

Step 2: Assignment
for each observation x_i **do**
    Compute distances $d(x_i, c_j)$ for $j = 1, \ldots, k$
    Assign x_i to cluster

$$\arg\min_{1 \le j \le k} d(x_i, c_j)$$

end for

Step 3: Update
for each cluster $j = 1, \ldots, k$ **do**
    Update center

$$c_j \leftarrow \frac{1}{|C_j|} \sum_{x_i \in C_j} x_i$$

end for

Step 4: Repeat
Repeat Steps 2–3 until convergence or a maximum number of iterations is reached.

The k-means clustering algorithm seeks k clusters $\mathbf{C} = (C_1, C_2, \ldots, C_k)$ with centers $\mathbf{C} = (c_1, c_2, \ldots, c_k)$ such that the function

$$\Phi(C_1, C_2, \ldots, C_k) = \sum_{r=1}^{k} \frac{1}{2|C_r|} \sum_{\mathbf{z}, \mathbf{w} \in C_r} \|\mathbf{z} - \mathbf{w}\|^2 \tag{13.4}$$

is minimized, where $|C_r|$ is the cardinality of set $\mathcal{C}_r$. For cluster C_t with center c_t and cluster size $|C_t|$, one can show that

$$\sum_{\mathbf{x}_i \in C_t} \sum_{\mathbf{x}_j \in C_t} (\mathbf{x}_i - \mathbf{x}_j)^2 = \sum_{\mathbf{x}_i \in C_t} \sum_{\mathbf{x}_j \in C_t} [(\mathbf{x}_i - c_t)^2 + (c_t - \mathbf{x}_j)^2 + 2(\mathbf{x}_i - c_t)(c_t - \mathbf{x}_j)]$$

$$= |C_t| \sum_{\mathbf{x}_i \in C_t} (\mathbf{x}_i - c_t)^2 + |C_t| \sum_{\mathbf{x}_j \in C_t} (c_t - \mathbf{x}_j)^2 + 0 = 2|C_t| \sum_{\mathbf{x}_i \in C_t} (\mathbf{x}_i - c_t)^2.$$

Therefore, $\Phi(C_1, C_2, \ldots, C_k)$ can be expressed as Equation (13.4).

Sarkar and Ghosh [299] propose a clustering method by optimizing the function

$$\Phi_\rho = \sum_{r=1}^{k} \frac{1}{2|C_r|} \sum_{\mathbf{z},\mathbf{w}\in C_r} \rho^2(\mathbf{z}, \mathbf{w}),$$

where ρ is the dissimilarity index. In step 2, for observation $\mathbf{X}$ in the dataset, we calculate its distance to cluster C_r by $\rho(\mathbf{X}, C_r) = \frac{1}{|C_r|} \sum_{\mathbf{w}\in C_r} \rho^2(\mathbf{X}, \mathbf{w})$. The observation $\mathbf{X}$ is assigned to cluster which minimizes $\rho(\mathbf{X}, C_r)$. The process is repeated until no more clusters are modified or the maximum iterations limit is reached. The k-means clustering method based on the dissimilarity index ρ will be referred as kM_ρ. Suppose that $\mathbb{X}$ consists of independent observations from k populations $F_1, F_2, \ldots, F_k$ satisfying assumptions (A1)-(A3). If for every pair of populations (F_i, F_j), $v_{ij} > 0$ or $\sigma_i^2 \neq \sigma_j^2$ holds for $i \neq j$, Sarkar and Ghosh [299] show that kM_ρ has a Rand index, which converges to 0 in probability as $d \to \infty$.

Consider optimizing the function

$$\Phi_\delta = \sum_{r=1}^{k} (2|C_r|)^{-1} \sum_{\mathbf{z},\mathbf{w}\in C_r} \delta^2(\mathbf{z}, \mathbf{w}),$$

where δ is the dissimilarity index. One begins with k clusters and iteratively updates them. In step 2, the dissimilarity between observation $\mathbf{X}$ and cluster C_r is given by

$$\delta(\mathbf{X}, C_r) = |C_r|^{-1} \sum_{\mathbf{w}\in C_r} \delta^2(\mathbf{X}, \mathbf{w}).$$

The observation $\mathbf{X}$ will be assigned to the cluster which minimize $\delta(\mathbf{X}, C_r)$, and the procedure is repeated until no more groups are modified or the maximum number of iterations is reached. The k-means clustering method based on δ will be referred as kM_δ. One can show that clustering based on kM_δ is also perfect.

Theorem 13.7.1 *Suppose kM_δ is used to cluster independent observations from $F_1, F_2, \ldots, F_k$. Under Assumptions B1-B3, if for every pair (F_i, F_j), $\bar{\mu}_i \neq \bar{\mu}_j$ or $\sigma_i^2 \neq \sigma_j^2$ holds for $i \neq j$, then the Rand index converges to 0 in probability as $d \to \infty$.*

13.8 Estimating the Number of Clusters

In practice, the number of clusters must typically be estimated from the data. Several methods have been proposed for this purpose, including those of Calinski and Harabasz [48], Hartigan [139], Krzanowski and Lai [178], Rousseeuw [287], Tibshirani et al. [337], Sugar and James [326], Wang [345], and Huang et al. [114].

A common approach begins by applying a clustering algorithm to partition the data into a fixed number k of clusters. These clusters are then used to compute a cluster stability index $L(k)$. Performing a grid search over different values of k to optimize $L(k)$ yields an estimate of the number of clusters. Here, we employ the KL statistic (Krzanowski and Lai [178]), which is defined as follows. Let the base clustering algorithm produce clusters $C_1, \ldots, C_k$. The KL statistic for k clusters is given by

$$KL(k) = \frac{\text{Diff}(k)}{\text{Diff}(k+1)}, \quad \text{where} \quad \text{Diff}(k) = (k-1)^{2/d} W_{k-1} - k^{2/d} W_k,$$

and W_k denotes the within-group sum of squares, defined as

$$W_k = \sum_{j=1}^{k} \frac{1}{2|C_j|} \sum_{Z,W \in C_j} \|Z - W\|^2.$$

The statistic is evaluated for $k = 2, \ldots, K$, and the estimated number of clusters is

$$\hat{k}_{KL} = \arg\max_{2 \le k \le K} KL(k).$$

13.9 Perfect and Order Preserving Property

Following Sarkar and Ghosh [299], we define the perfect and order preserving (POP) property as follows. For any fixed k, let

$$C_{\Phi}^{(k)} = (C_1^{(k)}, \ldots, C_k^{(k)})$$

denote the k clusters estimated by a clustering algorithm Φ applied to the sample space $\mathcal{X}$, which consists of observations drawn from k_0 populations. Let $C_{\mathcal{X}}$ represent the true clustering of the sample. We call this clustering perfect and order preserving (POP) at k_0 if the following conditions hold.

(i) The clustering algorithm Φ is perfect, i.e., for $k = k_0$, $Rand(C_{\Phi}^{(k)}, C_{\mathcal{X}}) = 0$.

(ii) For any $k < k_0$ and for all $i = 1, \ldots, k_0$, there exists $1 \le j \le k$ such that $C_i^{(k_0)} \subseteq C_j^{(k)}$.

(iii) For any $k > k_0$ and for all $i = 1, \ldots, k$, there exists $1 \le j \le k_0$ such that $C_i^{(k)} \subseteq C_j^{(k_0)}$.

Suppose independent observations $\mathbf{U}$ and $\mathbf{V}$ follow distributions F_i and F_j, respectively. Following Hardy's notation, Sarkar and Ghosh [299] denote the asymptotic order of $\rho(\mathbf{U}, \mathbf{V})$ as $\phi_{ij}(d)$. That is, $\rho(\mathbf{U}, \mathbf{V}) \overset{\mathrm{P}}{\asymp} \phi_{ij}(d)$, i.e., as $d \to \infty$, $\Pr\{0 < \rho(\mathbf{U}, \mathbf{V})/\phi_{ij}(d) < \infty\} \to 0$. Sarkar and Ghosh [299] assume that

Assumption(A4): $\phi_{ii}(d) \overset{\mathrm{P}}{\asymp} \phi_{-}(d)$ for all $i = 1, \ldots, k_0$ and $\phi_{ij}(d) \overset{\mathrm{P}}{\asymp} \phi_{+}(d)$ for $i \ne j$, where $\phi_{-}(d) = \mathbf{o}(\phi_{+}(d))$.

Similarly, we denote the asymptotic order of $\delta(\mathbf{U}, \mathbf{V})$ as $\psi_{ij}(d)$. Let $\delta(\mathbf{U}, \mathbf{V}) \overset{\mathrm{P}}{\asymp} \psi_{ij}(d)$, i.e., as $d \to \infty$, $\Pr\{0 < \delta(\mathbf{U}, \mathbf{V})/\psi_{ij}(d) < \infty\} \to 0$. We assume

Assumption(B4): $\psi_{ii}(d) \overset{\mathrm{P}}{\asymp} \psi_{-}(d)$ for all $i = 1, \ldots, k_0$ and $\psi_{ij}(d) \overset{\mathrm{P}}{\asymp} \psi_{+}(d)$ for $i \ne j$, where $\psi_{-}(d) = \mathbf{o}(\psi_{+}(d))$.

In fact, assumptions (A4) and (B4) indicate that if $\mathbf{U}$ and $\mathbf{V}$ come from the same population, we have $\rho(\mathbf{U}, \mathbf{V}) = \delta(\mathbf{U}, \mathbf{V}) = \mathbf{o}_{\mathbb{P}}(1)$. Hence, as $d \to \infty$, $\phi_{-}(d)$ and $\psi_{-}(d)$ converge to 0. If $\mathbf{U}$ and $\mathbf{V}$ come from different populations, the limiting value $\phi_{+}(d)$ and $\psi_{+}(d)$ remains bounded away from 0 as $d \to \infty$. Suppose that the base clustering algorithm is POP at k_0. Sarkar and Ghosh [299] prove that if we have observations from $k_0 \ge 2$ populations, which satisfy assumption (B4), then as $d \to \infty$, number of clusters estimated by KL statistic with dissimilarity index ρ converges to k_0 in probability. The next result shows that the estimated number of clusters by KL statistic with dissimilarity index δ converges to k_0 in probability.

Theorem 13.9.1 *Suppose that we have observations from $k_0 \geq 2$ populations which satisfy assumption (A4), and the base clustering algorithm is POP at k_0. As $d \to \infty$, number of clusters estimated by KL statistic with dissimilarity index δ converges to k_0 in probability.*

Sarkar and Ghosh [299] have demonstrated that the k-means clustering algorithm, when modified with the dissimilarity index ρ, satisfies the POP property. Consequently, it can recover the true number of clusters k_0 using the KL statistic with probability 1 as $d \to \infty$. Similarly, it can be shown that the k-means algorithm modified with the dissimilarity index δ also satisfies the POP property, and therefore can recover k_0 by the KL statistic with probability 1 as $d \to \infty$.

The following lemma establishes that, under the CPC algorithm framework, the number of clusters estimated by the KL statistic converges in probability to the true number k_0.

Lemma 13.9.1 *For independent observations from k_0 populations $F_1, \dots, F_{k_0}$, if there exists i, such that, $\tilde{\delta}(F_i, F_j)$ are not equal for all $j \neq i$, then the CPC_δ is POP, and therefore the KL algorithms with CPC_δ provides the true number of clusters k_0 as $d \to \infty$.*

13.10 Comparisons

In this section we compare the performance of four other clustering methods.

- The R implementation of HDBSCAN (Campello et al. [49]) computes the hierarchy of all DBSCAN clusters.
- We also considered clustering methods based on the MST. The observations form a complete graph with N vertices $\mathbf{X}_i$ edges with weights d_{ij}. A spanning tree is a connected subgraph of an edge-weighted graph with no cycles. The MST is the tree with the minimum total edge weight. By cutting off $k-1$ edges of the MST we obtain k clusters. One can use the dissimilarity indices ρ and δ to obtain clustering methods MST_E, MST_ρ, and MST_δ.
- CPC$_\rho$, and CPC$_\delta$, High-Dimensional Distance Difference method with ρ and δ.
- kM_ρ, kM_δ, and k-means clustering using the index ρ and δ, and the Euclidean distance.

We use $d = 2^i$ features for $i = 4, \dots, 11$ in each experiment, with 500 replications, to cluster the observations. The Rand index between the clustering labels obtained by each method and the true labels was computed for each replication and averaged over all 500 replications. For each experiment, the number of clusters was first estimated using the KL statistic (Krzanowski and Lai [178]).

When F_i has mean $\mu_i \mathbf{1}$ and covariance $\mathbf{I}$, with $\mu_i \in \boldsymbol{\mu}_x = \{1, 1.5, 2, 2.5, 3\}$, the limiting values of the dissimilarity indices as $d \to \infty$ are

$$\tilde{\delta}_{F_1,F_2} = \tilde{\delta}_{F_2,F_3} = \tilde{\delta}_{F_3,F_4} = \tilde{\delta}_{F_4,F_5} = \frac{1}{2}, \quad \tilde{\delta}_{F_1,F_3} = \tilde{\delta}_{F_2,F_4} = \tilde{\delta}_{F_3,F_5} = 1, \quad \tilde{\delta}_{F_1,F_4} = \tilde{\delta}_{F_2,F_5} = \frac{3}{2}.$$

We generated $n_i = 8$ observations $\mathbf{X}_i$ from $\mathbb{N}(\mu_i \mathbf{1}, \mathbf{I})$ where $\mu_i \in \boldsymbol{\mu}_x$ for $i = 1, \dots, K = 5$. The Rand indices of the nine clustering methods appear in Table 13.1. shows that the CPC, kM, and MST methods produce smaller Rand index values when using the dissimilarity

TABLE 13.1: **Normal and Absolute Normal Samples**. Table entries show the Rand indices with the correct labels for different clustering methods.

Dist.	d	CPC_ρ	CPC_δ	kM_ρ	kM_δ	k-means	$HDBS$	MST_E	MST_ρ	MST_δ
Normal	16	0.329	0.244	0.230	0.184	0.216	0.760	0.718	0.390	0.256
	32	0.250	0.190	0.175	0.146	0.178	0.761	0.695	0.372	0.206
	64	0.160	0.106	0.126	0.098	0.146	0.670	0.557	0.228	0.129
	128	0.100	0.042	0.078	0.039	0.104	0.689	0.539	0.071	0.013
	512	0.014	0.002	0.002	0.002	0.058	0.389	0.361	0	0
	1024	0.018	0	0.001	0.004	0.054	0.222	0.289	0	0
	2048	0.007	0	0	0	0.065	0.036	0.224	0	0
Abs(Normal)	16	0.326	0.243	0.216	0.200	0.219	0.744	0.703	0.381	0.276
	32	0.280	0.196	0.173	0.153	0.190	0.727	0.599	0.357	0.223
	64	0.185	0.117	0.128	0.104	0.159	0.782	0.554	0.257	0.138
	128	0.081	0.042	0.097	0.050	0.116	0.782	0.516	0.101	0.030
	512	0.006	0.005	0.007	0.002	0.061	0.647	0.444	0	0
	1024	0.001	0	0.001	0.001	0.054	0.429	0.389	0	0
	2048	0.001	0	0.001	0	0.062	0.242	0.299	0	0

measure δ compared to ρ. It is observed that $k\text{M}_\delta$ performs slightly better than CPC_δ; however, both methods achieve perfect clustering as the dimension d becomes large. The excellent performance of MST_δ is also noteworthy. In contrast, the clustering methods k-means, HDBSCAN, and MST_E perform poorly.

Wang and Modarres [348] carry out further experiments under various distributions. These simulation studies highlight the superiority of using the dissimilarity measures ρ and δ in high-dimensional settings. The results indicate that CPC can outperform kM, particularly under heavy-tailed distributions.

Table 13.2 summarizes the performance of all clustering algorithms under various distributions. All relevant code for the algorithms and simulation studies has been uploaded to a GitHub repository to facilitate reproduction and further research. The repository contains the simulation program and a program to create IPD plots. The GitHub account is available at: `https://github.com/rezamodarres/High-Dimensional-Classification`.

TABLE 13.2: Classification of clustering algorithms as Excellent, Marginal, and Poor under different distributions for HDLSSS data.

Dist.	CPC_ρ	CPC_δ	kM_ρ	kM_δ	k-means	$HDBS$	MST_E	MST_ρ	MST_δ
Normal	E	E	E	E	M	M	P	E	E
Abs(Normal)	E	E	E	E	M	P	P	E	E
Bernoulli	E	E	E	E	P	P	P	E	E
MSN	E	E	E	E	P	M	P	E	E
t_3	M	E	M	M	P	P	P	M	M
χ^2_3	P	E	P	E	P	P	P	P	E
Scaled Normal	E	E	E	E	P	P	P	E	E

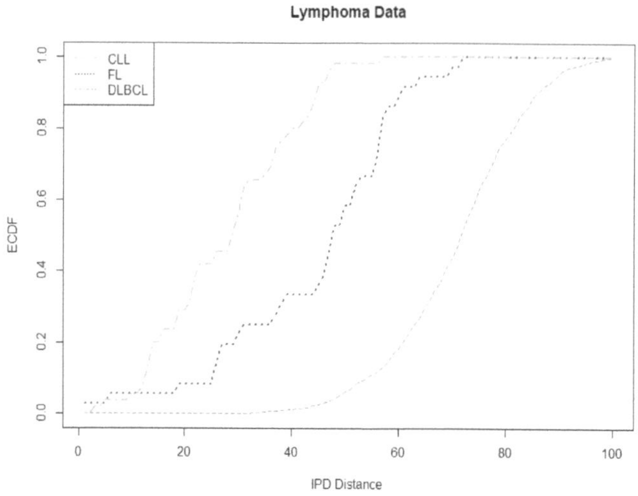

FIGURE 13.4: ECDFs for average IPDs of CLL (top), FL (middle), and DLBCL groups (bottom).

13.11 Analysis of Lymphoma Dataset

We analyzed the `lymphoma` dataset to evaluate the clustering methods on a real dataset. The dataset is available in the R package `Splus`. The `lymphoma` dataset was originally analyzed by Alizadeh et al. to identify distinct types of lymphoma. It contains expression levels of $d = 4026$ genes for 42 diffuse large B-cell lymphoma (DLBCL), 9 follicular lymphoma (FL), and 11 chronic lymphocytic leukemia (CLL) samples.

The average IPDs are 81.25 for DLBCL, 63.26 for FL, and 51.92 for CLL. Modarres [229] introduced a method for visualizing multivariate data using a simultaneous plot of the empirical cumulative distribution functions (ECDFs) of the IPDs. Figure 13.4 shows the ECDFs of the average IPDs for the three groups. The ECDFs are clearly distinct, suggesting that the groups arise from different distributions. The plot of relative variance (RV) in Equation (3.3) as a function of d is shown in Figure 13.5. The plot indicates that the RV tends to zero as d increases, confirming the behavior expected in high-dimensional settings.

The Rand indices for the clustering methods compared to the true labels as a function of d are reported in Table 13.3. For the lymphoma dataset, the best-performing clustering methods are CPC_ρ and MST_ρ, achieving Rand indices of 0.052 and estimated numbers of clusters 2 and 3, respectively. Other clustering methods, including kM_ρ, kM_δ, and methods based on δ, perform poorly.

13.12 Summary

Chapter 13 focuses on high-dimensional clustering in HDLSS settings, where the number of features d far exceeds the number of observations N. It provides a comprehensive treatment

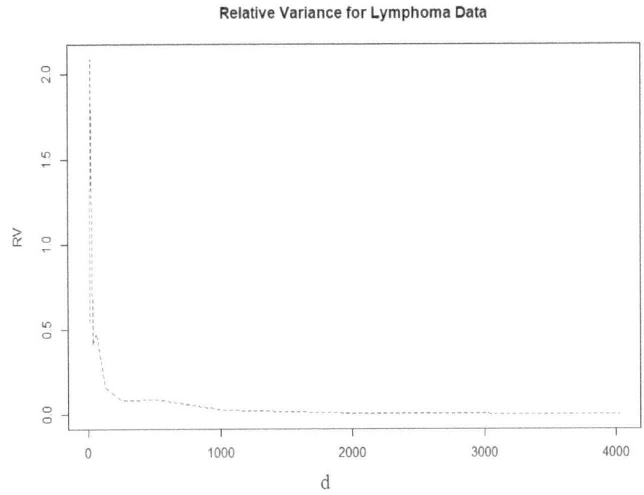

FIGURE 13.5: Plot of the Relative Variance for the Lymphoma Data.

TABLE 13.3: **Lymphoma Data Set**. Table entries show the Rand indices with the correct labels for different clustering methods.

d	CPC_ρ	CPC_δ	kM_ρ	kM_δ	k-means	$HDBS$	MST_E	MST_ρ	MST_δ
100	0.428	0.549	0.410	0.487	0.375	0.258	0.493	0.419	0.427
500	0.428	0.439	0.337	0.474	0.403	0.345	0.546	0.395	0.508
1000	0.044	0.093	0.233	0.356	0.393	0.222	0.203	0.200	0.093
2000	0.066	0.218	0.238	0.403	0.360	0.166	0.126	0.200	0.085
4026	0.052	0.521	0.274	0.513	0.307	0.115	0.106	0.052	0.521

of clustering methodologies, including hierarchical (agglomerative and divisive), partitioning (k-means, k-medoids), density-based (DBSCAN, HDBSCAN), and change point-based approaches. A major contribution is the Change Point Clustering (CPC) algorithm, which exploits dissimilarity measures and the distance concentration phenomenon to detect cluster boundaries consistently as the dimension grows. CPC reduces the original $N \times d$ data matrix to an $N \times N$ dissimilarity matrix, improving computational efficiency, and uses the distance difference matrix together with an agreement matrix based on the Rand index to determine the final clusters. Comparative studies show that CPC outperforms other high-dimensional clustering methods, including kM_ρ, HDBSCAN, and MST, under a variety of distributional settings.

CPC identifies clusters by detecting change points in the slopes of consecutive distances in the distance difference matrix. For k clusters, the $k - 1$ largest change points in each row define preliminary clusters, which are then aggregated using the Rand index agreement matrix to obtain the final clustering solution. The chapter emphasizes dissimilarity indices ρ (Sarkar and Ghosh [299]) and δ (Modarres [231]), which leverage distance concentration so that intra-cluster dissimilarities approach zero and inter-cluster dissimilarities converge to positive constants as $d \to \infty$. This ensures that CPC achieves perfect clustering asymptotically, and simulations confirm its superior performance across high-dimensional distributions.

The chapter also addresses practical considerations such as estimating the number of clusters, the impact of correlated variables on distance concentration, and diagnostics like relative variance plots to assess the presence of distance concentration. Weak dependence among variables, for example via ρ-mixing processes, satisfies the theoretical assumptions of CPC, while i.i.d. samples naturally induce distance concentration. Overall, chapter 13 presents a robust, computationally feasible, and theoretically sound framework for high-dimensional clustering, with applications in genomics, finance, and other complex domains.

13.13 Proofs

Proof of Theorem 13.6.1: The number of i.i.d. observations from population F_i is denoted as n_i. Let $\tilde{\delta} ij$ denote the asymptotic value of the dissimilarity between F_i and F_j. It follows from Assumptions B1-B3 that $\tilde{\delta_{ij}} > 0$ holds for all $i \neq j$ as $d \to \infty$, since $\bar{\mu}_i^d \neq \bar{\mu}_j^d$ or $\sigma_i^d \neq \sigma_j^d$. Consider the i-th row of the ordered distance matrix R_i = order $(\delta(X_i, X_1), \ldots, \delta(X_i, X_n))$. Note that the i-th observation will be placed at the first location since $\delta(X_i, X_i) = 0$ and the i-th row of distance difference matrix Δ_i, starts with a zero element since $\min R_i = \delta(X_i, X_i) = 0$. Next, we order $(\tilde{\delta}_{i1}, \ldots, \tilde{\delta}_{ik_0})$ and denoted the ordered sequence and the corresponding populations as $(\tilde{\delta}_{i1^*}, \ldots, \tilde{\delta}^*_{ik_0})$ and $(i_1, i_2, \ldots, i_k)$, respectively. It follows that the asymptotic form of Δ_i is $\tilde{\Delta}_i = (\mathbf{0}_{n_{i_1}}, |\tilde{\delta}^*_{i1} - 0|, \mathbf{0}_{n_{i_2}-1}, |\tilde{\delta}^*_{i2} - \tilde{\delta}^*_{i1}|, \mathbf{0}_{n_{i_3}-1}, \ldots, |\tilde{\delta}^*_{ik_0} - \tilde{\delta}^*_{i(k_0-1)}|, \mathbf{0}_{n_{i_{k_0}}-1})$. The form of $\tilde{\Delta}_i$ indicates the following:

1) If $\tilde{\delta}^*_{im} \neq \tilde{\delta}^*_{in}$ holds for all $n \neq m$, then the i-th row of the ordered distance matrix can provide us with the correct $k_0 - 1$ of change points and k_0 clusters with probability 1 as $d \to \infty$.

2) If there exist some m_i, such that $\tilde{\delta}^*_{im_i} = \tilde{\delta}^*_{i(m_i-1)}$, then the i-th row of the ordered distance matrix can only provide with the correct $k_0 - 2$ of change points and $k_0 - 1$ clusters with probability 1 as $d \to \infty$. In fact, even though there are only $k_0 - 2$ change points are true, the change point involving the i-th observation is always selected since $|\tilde{\delta}^*_{i1} - 0| > 0$ holds. Hence, the clusters obtained from observation i tends to provide a partially correct result, that is, the cluster including observation i is correct with probability 1 as $d \to \infty$. The agreement matrix that contains Rand indices of pairwise clusters aims to capture all such cases.

Consider the clusters obtained from i-th and j-th rows of the difference matrix as $\mathcal{C}_i$ and $\mathcal{C}_j$, respectively. The agreement on the i-th observation is measured with Rand($\mathcal{C}_i$,$\mathcal{C}_j$) for all $i \neq j$. If $\mathcal{C}_j$ agrees with the $\mathcal{C}_i$ on observation i; i.e. observations i and j locate the same change point, and the result is a zero-term in Rand($\mathcal{C}_i$,$\mathcal{C}_j$), then the cluster in $\mathcal{C}_i$ that includes observation i is correct with probability 1 as $d \to \infty$. Otherwise, the disagreement between $\mathcal{C}_j$ and $\mathcal{C}_i$ in terms of observation i would result in a positive term in Rand($\mathcal{C}_i$,$\mathcal{C}_j$). Hence, the row sums of the agreement matrix provide a consensus for all observations. The minimum value of row sums indicates the best clustering with highest agreement. That is, this cluster is generated by all the correct change points. Hence, the Rand index converges to 0 as $d \to \infty$. □

Proof of Theorem 13.7.1: Under assumptions 3.6.3, if $\bar{\mu}_i^d \neq \bar{\mu}_j^d$ or $\sigma_i^d \neq \sigma_j^d$ holds for all $i \neq j$, then $\delta(i,j) > 0$ for every $i \neq j$. For any cluster $\mathcal{C}_k$, as $d \to \infty$, and an observation $\mathbf{z}$, it follows that $\delta(\mathbf{z}, \mathcal{C}_k) = |\mathcal{C}_k|^{-1} \sum_{\mathbf{w} \in \mathcal{C}_k} \delta^2(\mathbf{z}, \mathbf{w}) \to 0 \; as \; d \to \infty$, if and only if $\mathbf{z}$ and all observations in $\mathcal{C}_k$ are from the same population. If $\mathbf{z}$ and all observations in $\mathcal{C}_k$ are from the same population, for all $\mathbf{w} \in C_k$, as $d \to \infty$, then $\delta^2(\mathbf{z}, \mathbf{w})$ converges to 0 in probability. Hence, $\delta(\mathbf{z}, C_k) \to 0$. On the other hand, if $\delta(\mathbf{z}, C_k) \to 0$ as $d \to \infty$, then $\delta^2(\mathbf{z}, \mathbf{w}) \to 0$ must hold for all $\mathbf{z}, \mathbf{w} \in C_k$, that is, $\mathbf{z}$ and all observations in $\mathcal{C}_k$ are from the same population. For the optimized clustering $\Phi_\delta^*(C_1, \ldots, C_{k_0}) = \min\{\Phi_\delta\}$, if each $\mathcal{C}_k$ contains observations from the same population, then $\Phi_\delta^*(C_1, \ldots, C_{k_0})$ converges to 0 as $d \to \infty$. Otherwise, it converges in probability to a strictly positive quantity. Therefore, when the true number of the clusters k_0 is known, for the minimization of $\Phi_\delta^*(C_1, \ldots, C_{k_0})$, each cluster $\mathcal{C}_k$ must contain all observations from a single population with probability converging to 1 as $d \to \infty$.

Proof of Theorem 13.9.1: The KL statistic can be written as $\mathrm{KL}(k) = |(\Lambda_{k-1} - \Lambda_k)/(\Lambda_k - \Lambda_{k+1})|$, where $\Lambda_k = k^{2/d} W_k = k^{2/d} \sum_{j=1}^{k} \frac{1}{2|C_j|} \sum_{\mathbf{Z}, \mathbf{W} \in C_j} \delta_0^2(\mathbf{Z}, \mathbf{W})$. We consider three different cases as follows. For any $k < k_0$, since the base clustering algorithm is POP at k_0, there exists at least one estimated cluster which contains observations from two different populations, and no two clusters contain observations from the same population. Denote $\delta(C_i) = (|C_i||C_i - 1|)^{-1} \sum_{\mathbf{U}, \mathbf{V} \in C_i} \delta(\mathbf{U}, \mathbf{V})$ and $\Delta(C_i, C_j) = (|C_i||C_j|)^{-1} \sum_{\mathbf{U} \in C_i, \mathbf{V} \in C_j} \delta(\mathbf{U}, \mathbf{V})$. So, under (B4), for some i, we have $\Delta(C_i) \asymp \psi_+(d)$ and $\Delta(C_i, C_j) \asymp \psi_+(d)$ for all $i \neq j$. Thus, $B_k^\circ = \min_{1 \leq i < j \leq k} \Delta(C_i, C_j) \overset{P}{\asymp} \psi_+(d)$ and $W_k^\circ = \max_{1 \leq i < j \leq k} \Delta(C_i) \overset{P}{\asymp} \psi_+(d)$.

For $k > k_0$, no cluster contains observations from two different populations, while there exists at least two clusters that contain observations from the same population given by the second property of POP algorithm. Hence, $\Delta(C_i) \overset{P}{\asymp} \psi_-(d)$ for every i and $\Delta(C_i, C_j) \overset{P}{\asymp} \psi_-(d)$ for some $i \neq j$. Thus, we have $B_k^\circ \overset{P}{\asymp} \psi_-(d)$ and $W_k^\circ \overset{P}{\asymp} \psi_-(d)$, which implies $D(k) \overset{P}{\asymp} 1$.

For $k = k_0$, no cluster contains observations from two different populations. Hence, $\Delta(C_i) \overset{P}{\asymp} \psi_-(d)$ for every i and $\Delta(C_i, C_j) \overset{P}{\asymp} \psi_+(d)$ for all $i \neq j$. Thus, we have $B_k^\circ \overset{P}{\asymp} \psi_+(d)$ and $W_k^\circ \overset{P}{\asymp} \psi_-(d)$, which implies $D(k) \overset{P}{\asymp} (\psi_+(d)/\psi_-(d)))$. Now combining these three cases and considering $\psi_-(d) = \mathbf{o}(\psi_+(d))$, one can show that $\Pr(D(k_0) > D(k)$, for $\forall k \neq k_0) \to 1$ as $d \to \infty$. Further, for any fixed k,we have $k^{2/d} \to 1$ as $d \to \infty$. That is, for any $k \geq 1$, $\Lambda_k \overset{P}{\asymp} W_k$. One can show that $\mathrm{KL}(k) \overset{P}{\asymp} (\psi_+(d)/\psi_-(d)))$ for $k = k_0$, and $\mathrm{KL}(k) \overset{P}{\asymp} 1$ for all $k \neq k_0$. Therefore, $\Pr(\mathrm{KL}(k_0) > \mathrm{KL}(k)$, for $\forall k \neq k_0) \to 1$ as $d \to \infty$. Hence, the number of clusters estimated by KL statistic with dissimilarity index δ converges to k_0 in probability, that is $k_{KL}^* \to k_0$ as $d \to \infty$. Using the POP property of the CPC algorithm, as $d \to \infty$, the estimated number of the clusters by CPC and KL statistics converges to k_0 in probability. □

Proof of Lemma 13.9.1: The first property is supported by Theorem 13.6.1. The CPC_δ clustering is perfect when $k = k_0$. For any $k < k_0$, the algorithm provides k clusters from $k - 1$ change points. The remaining $(k_0 - k)$ of the change points and the resulting clusters are determined by the agreement matrix. With respect to the ordered distance matrix, all the observations are well-organized in order as $d \to \infty$. For all $i = 1, \ldots, k_0$, there exists $1 \leq j \leq k$ such that $C_i^{(k_0)} \subseteq C_j^{(k)}$. For any $k > k_0$, the algorithm is a two-step process, with the first step, it successfully provides k_0 clusters. For the second step, the algorithm seeks $k - k_0$ change points that do not provide correct positions. Regardless, the clustering with k clusters will be a finer partition of true clustering. That is, for $i = 1, \ldots, k$, there exists $1 \leq j \leq k_0$ such that $C_i^{(k)} \subseteq C_j^{(k_0)}$. Therefore, CPC_δ is POP, the KL algorithms with CPC_δ can provide the true number of the clusters k_0. □

13.14 Exercises

Exercise 13.1 *Consider $K = 3$ with $n_1 = n_2 = n_3 = n$ and suppose we obtain independent samples from the d-variate normal distribution,* $\mathbf{X}_i \sim F_1 = \mathbb{N}_d(\mathbf{0}, \mathbf{I})$, $\mathbf{Y}_i \sim F_2 = \mathbb{N}_d(\mathbf{1}_d, \mathbf{I})$, *and* $\mathbf{Z}_i \sim F_3 = \mathbb{N}_d(-\mathbf{1}_d, \mathbf{I})$, *for* $i = 1, \ldots n$.

a) *Show that when $n_1 = n_2$, $\tilde{\rho}(F_1, F_2)$ is independent of sample sizes and equals*

$$\tilde{\rho}(F_1, F_2) = |\sqrt{2}\sigma_F - \sqrt{\sigma_F^2 + \sigma_{F_2}^2 + \nu_{xy}^2}|,$$

where $\nu_{xy}^2 = \frac{1}{d}\|\boldsymbol{\mu}_x - \boldsymbol{\mu}_y\|^2$.

b) *Show that* $\nu_{xy}^2 = \nu_{xz}^2 = 1$, $\nu_{yz}^2 = 4$, *and* $\sigma_F^2 = tr(\boldsymbol{\Sigma}_x)/d = 1 = \sigma_{F_2}^2 = \sigma_{F_3}^2$. *Hence,* $\tilde{\rho}(F_1, F_2) = \tilde{\rho}(F_1, F_3) = |\sqrt{2} - \sqrt{3}|$, *and* $\tilde{\rho}(F_2, F_3) = |\sqrt{2} - \sqrt{6}|$. *The limiting value of δ index, is independent of the sample sizes.*

c) *Show that* $\bar{\mu}_x^d = \frac{1}{d}\mathbf{1}_d'\boldsymbol{\mu}_x = 0$, $\bar{\mu}_y^d = 1$ *and* $\bar{\mu}_z^d = -1$. *Moreover,* $\nu_x^2 = \nu_y^2 = \nu_z^2 = 0$ *and using* $[\sigma_x^d]^2 = \sigma_F^2 + \nu_x^2$ *obtain* $\sigma_x^d = \sigma_y^d = \sigma_z^d = 1$. *Hence,* $\tilde{\delta}(F_1, F_2) = \tilde{\delta}(F_1, F_3) = 1$ *and* $\tilde{\delta}(F_2, F_3) = 2$.

Exercise 13.2 *Suppose $\mathbf{X}$ and $\mathbf{Y}$ contain d independent Bernoulli variables, probability vectors* $\mathbf{R} = r\mathbf{1}_d$ *and* $\mathbf{S} = s\mathbf{1}_d$. *Let* $\mathbb{X} = \{\mathbf{X}_i\}_{i=1}^{n_1}$ *and* $\mathbb{Y} = \{\mathbf{Y}_j\}_{j=1}^{n_2}$ *be i.i.d. copies of $\mathbf{X}$ and $\mathbf{Y}$, respectively.*

a) *Show that* $\frac{1}{\sqrt{d}}\|\mathbf{X}_1 - \mathbf{X}_2\|$ *converges to* $\sqrt{2r(1-r)}$, *and* $\frac{1}{\sqrt{d}}\|\mathbf{Y}_1 - \mathbf{Y}_2\|$ *converges to* $\sqrt{2s(1-s)}$, *whereas* $\frac{1}{\sqrt{d}}\|\mathbf{X}_1 - \mathbf{Y}_1\|$ *converges to* $\sqrt{r(1-r) + s(1-s) + (r-s)^2}$ *in probability as $d \to \infty$.*

b) *When $n_1 = n_2$, show that* $\tilde{\rho}(F_1, F_2) = |\sqrt{2r(r-1)} - \sqrt{r(r-1) + s(s-1) + (r-s)^2}|$. *It follows that* $\bar{\mu}_x^d = r$, $\bar{\mu}_y^d = s$. *Moreover,* $\nu_x^2 = \nu_y^2 = 0$ *and* $\sigma_x^d = r(r-1)$ *and* $\sigma_y^d = s(1-s)$. *Hence,* $\tilde{\delta}(F_1, F_2) = |\sqrt{(r-s)^2 + [\sqrt{r(1-r)} - \sqrt{s(s-1)}]^2}|$.

Exercise 13.3 *See Sarkar and Ghosh [299] on perfect clustering. Suppose $U_d(a, b)$ denote the uniform distribution over the region* $\{\mathbf{X} \in \mathbb{R}^d : a\sqrt{d} \leq \|\mathbf{X}\| \leq d\sqrt{d}\}$. *Prove that if* $\mathbf{U} \sim U_d(a_1, b_1)$ *and* $\mathbf{V} \sim U_d(a_2, b_21)$ *are independent random vectors then* $\frac{1}{d}\|\mathbf{U} - \mathbf{V}\|^2 \to b_1^2 + b_2^2$ *in probability as $d \to \infty$.*

Exercise 13.4 *Given the true cluster labels* $\mathcal{C} = (1, 1, 1, 2, 2, 2)$ *and predicted cluster labels* $\psi = (1, 1, 2, 2, 3, 3)$, *compute the Rand index.*

Exercise 13.5 *For $n = 5$ with true labels* $\mathcal{C} = (1, 1, 2, 2, 2)$ *and predicted labels* $\psi = (1, 2, 2, 1, 2)$, *compute the Rand index.*

Exercise 13.6 *Show that the Rand index always lies between 0 and 1.*

Exercise 13.7 *Suppose all data points are placed in a single cluster by ψ. What is the Rand index in terms of the number of true clusters?*

Exercise 13.8 *Prove that the Rand index is symmetric, that is, interchanging labels of ψ or $\mathcal{C}$ does not affect its value.*

Exercise 13.9 *Given two arrays labels1 and labels2, write an R function to compute the Rand index.*

14

Analysis of Distance Matrices

14.1 Abstract

Distance matrices and their eigen-structures constitute a fundamental framework for high-dimensional data analysis, particularly when the original observations are unavailable or inherently relational. This chapter systematically investigates constant, two-sample, and HDLSS distance matrices, with a focus on outlier detection, imputation of missing distances, change point analysis, and the evaluation of distributional equality. We introduce and analyze two dissimilarity measures, δ and ρ, alongside the L_1-norm Euclidean distance, deriving their asymptotic eigenvalues for scenarios with one to three groups of observations. Using principal coordinate analysis and Frobenius norm-based statistics, we establish connections between eigenvalues, energy statistics, and tests for shift, scale, and joint shift-scale alternatives. The chapter elucidates the impact of outliers on distance matrices, highlighting potential distortions of eigenvalues and the limitations of existing methods. Moreover, we demonstrate that, in high-dimensional settings, distances within a group converge to a constant structure, enabling rigorous statistical tests for the equality of K distributions. Applications span bioinformatics, machine learning, cognitive science, and pattern recognition, providing robust tools for dimensionality reduction, clustering, and hypothesis testing. The theoretical developments are complemented with practical guidance for interpreting eigenvalue-based measures and for handling HDLSS data.

14.2 Introduction

Distance (or dissimilarity) matrices appear in data analysis when the original observations are not directly available or meaningful, only pairwise relationships (similarities or dissimilarities) are known, the data arises from subjective evaluations (e.g., human similarity ratings), and when the objects being compared are complex such as graphs, strings, sequences, images, etc. and when the data are naturally non–Euclidean. Examples include, edit distances between strings or DNA sequences, perceptual similarity between colors or sounds, and shortest-path distances in a graph or network, among others. There are several reasons why we might only have a distance matrix and not the original data, including, Nature of the data: Some datasets are inherently relational, Dimensionality concerns: Original vectors may lie in very high-dimensional or unknown spaces, Data collection method: Humans often rate similarity without providing additional data, Privacy or abstraction: Only pairwise comparisons are shared due to privacy, proprietary models, or simplification, and Data transformation: Original features may have been transformed or lost in preprocessing.

DOI: 10.1201/9781003789581-14

Data can manifest in various forms, including distance or dissimilarity matrices. In the realm of large-scale data, where both the number of observations and variables are vast, raw data may be challenging to store, process, or interpret directly. For relational data structures, such as social networks or transportation systems, distance matrices serve as fundamental representations of pairwise distances. These matrices are indispensable across numerous scientific domains, offering a structured approach to quantifying dissimilarity between elements within a dataset.

Distance matrices play a prominent role in a wide range of statistical learning and data analysis methodologies, including clustering techniques such as k-means and hierarchical clustering. They are also fundamental to dimensionality reduction methods, most notably multidimensional scaling, whose objective is to embed high-dimensional data into a lower-dimensional space while approximately preserving pairwise distances for visualization and subsequent analysis. In document clustering, distance matrices enable the organization and classification of textual data by grouping documents with similar content. In computer vision, they are used to compare feature vectors extracted from images, thereby supporting applications such as facial recognition, in which identity matching is determined by distances between facial features. In bioinformatics, distance matrices provide a natural framework for quantifying pairwise dissimilarities among DNA, RNA, or protein sequences. They also arise in optimization problems, including shortest-path and routing problems. From a graph-theoretic perspective, a distance matrix implicitly encodes adjacency relationships among observations, with observations represented as vertices of a graph. The eigenvalues of the distance matrix yield important information about the underlying connectivity structure: in particular, the number of distinct eigenvalues is closely related to the number of connected components in the graph representation of the data.

A distance matrix is an $N \times N$ symmetric matrix that contains the pairwise distances. A distance matrix allows one to transform the $N \times d$ data matrix to an $N \times N$ matrix of distances. This transformation provides a large data reduction when $d \gg N$ as is the case for HDLSS data. A distance matrix has time complexity of $O(dN^2)$ as it requires computing $N(N-1)/2$ IPDs. A dissimilarity matrix may not always satisfy the triangle inequality and is hence a semi-metric. See a discussion on metrics, semi-metric and dissimilarity functions see section 3.3.

Let $\mathbb{X} = \{\mathbf{X}_{i=1}^N\}$ be a set of N i.i.d. vectors from distribution F. The IPDs between $\mathbf{X}_i$ and $\mathbf{X}_j$ is given by $d_{ij} = \|\mathbf{X}_i - \mathbf{X}_j\|$. A distance matrix $\mathbf{D}(\mathbb{X})$ has zeroes on the main diagonal and d_{ij} on the off-diagonal, $i \neq j = 1, \ldots, N$. Menger [223] defines the squared volume of the N-simplex in terms of the Euclidean distance matrix $\mathbf{D}(\mathbb{X})$. The squared volume of the N-simplex is always non–negative and zero only if the observations are affinely dependent. Menger [223] proves that any symmetric matrix of non–negative real numbers with zeros on the diagonal is a Euclidean matrix. Glick [120] studied separation measures whose lower and upper bounds are functions of all pairwise IPDs. Using distance matrices, Modarres [229] proposes methods for visualization and comparison of high-dimensional observations and Modarres [232] discusses high-dimensional outlier detection methods.

Let $\mathbf{D}^{(1)}$ and $\mathbf{D}^{(2)}$ be distance matrices on the same set of points. We examine operations that produce another distance matrix $\mathbf{D}'$. Operations that preserve distance matrix properties include,

1. Scalar multiplication: $\mathbf{D}' = c\mathbf{D}, \quad c > 0$. It preserves all metric properties.

2. Power transformation: $\mathbf{D}'_{ij} = \mathbf{D}_{ij}^{\alpha}, \quad 0 < \alpha \leq 1$. Triangle inequality is preserved when $\alpha \in (0, 1]$, since the power function is concave and increasing.

3. Addition of distance matrices: $\mathbf{D}' = \mathbf{D}^{(1)} + \mathbf{D}^{(2)}$ is a valid distance matrix.

4. Weighted sum: $\mathbf{D}' = \sum_{k=1}^{K} w_k \mathbf{D}^{(k)}, \quad w_k \geq 0$ remains a valid distance matrix if each $\mathbf{D}^{(k)}$ is one.

5. Entry-wise function: $\mathbf{D}'_{ij} = f(\mathbf{D}_{ij}), f : [0, \infty) \to [0, \infty)$ must be concave, monotone increasing, and $f(0) = 0$. Examples are $f(x) = \sqrt{x}, \log(1 + x), x^{\alpha}$ with $\alpha \in (0, 1]$.

The following transformations are conditionally valid or invalid.

1. Element-wise square: $\mathbf{D}'_{ij} = \mathbf{D}^2_{ij}$ may violate the triangle inequality, and thus not a metric.

2. Entry-wise Max/Min: $\mathbf{D}'_{ij} = \max(\mathbf{D}_{ij}, \epsilon)$ or $\min(\mathbf{D}_{ij}, \tau)$. These can violate metric properties unless carefully constrained (e.g., $\epsilon = 0$).

3. Hadamard product $\mathbf{D}'_{ij} = \mathbf{D}_{ij} \cdot M_{ij}$ may not ensure symmetry or triangle inequality.

4. Function of matrix elements: $\mathbf{D}'_{ij} = e^{-\mathbf{D}_{ij}}$, typically produces a similarity matrix, not a distance matrix.

Can we recover the original observations from a distance matrix? In general, a distance matrix encodes pairwise distances but not absolute positions. Hence, it loses information about rotation, translation, and reflection. The resulting observations may correspond to infinitely many point configurations (due to symmetry). We can often embed the distance matrix into a Euclidean space to approximate the original configuration. Common methods include, Multidimensional Scaling (MDS), which finds a Euclidean embedding that best preserves distances, Isomap, which handles non-linear manifolds using geodesic distances, and t-SNE, and UMAP, which perform non-linear dimensionality reduction techniques preserving local structures. The limitations are that the embedding is not unique, some distance matrices are not Euclidean (i.e., no valid embedding exists in any $\mathbb{R}^n$), and numerical approximations may distort distances.

The relationships between eigenvalues of the distance matrices and outliers are largely unexplored. We show a discrepancy in the sizes of eigenvalues of the distance matrix that is contaminated with outliers. We present test statistics to determine when a distance matrix has a constant structure. We relate the eigenvalues of the two-sample distance matrix to the tests of equality of distributions, study dissimilarity matrices and their eigenvalues in high dimensional and low-sample-size setting.

The aim of this chapter is to study distance matrices and their eigenvalue structures for proximity–based outlier and change point detection methods. Outliers are those observations that exhibit different characteristics than the majority of the observations in the dataset and adversely affect model selection and parameter estimation. Outlier detection methods find applications in early warning systems, public health, fraud detection, finance, geographic information systems, and data mining. Distance matrices exploit the spatial proximity of the observations to detect outliers. Given m independent vectors $\mathbf{X}_1, \mathbf{X}_2, \ldots, \mathbf{X}_m$ in $\mathbb{R}^d$ from an unknown distribution F with mean $\boldsymbol{\mu}_x$ and covariance matrix $\boldsymbol{\Sigma}_x$, investigators are often interested in testing $H_0 : \mathbf{X}_i \sim F(\boldsymbol{\mu}_x, \boldsymbol{\Sigma}_x), i = 1, \ldots, m$, against $H_a : \mathbf{X}_i \sim F(\boldsymbol{\mu}_x, \boldsymbol{\Sigma}_x)$, for some i, and $\mathbf{X}_j \sim F(c_j\boldsymbol{\mu}_x + \mathbf{e}_j, c_j^2\boldsymbol{\Sigma}_x)$, for all $j \neq i$. The slippage parameters $\mathbf{e}_j$, the mean vector $\boldsymbol{\mu}_x$, the scale parameters c_j, and the covariance $\boldsymbol{\Sigma}_x$ matrix are unknown, so are the indices j of the outliers.

Change point analysis concerns detection of changes in the distribution of the observations that are ordered by time or location (Chen and Gupta [61]). Let $\mathbf{X}_1, \mathbf{X}_2, \ldots, \mathbf{X}_m$ be a sequence of independent random vectors with probability distributions $F_1, F_2, \ldots, F_m$, respectively. Consider the null hypothesis $K_0 : F_1 = F_2 = \ldots = F_m$ and the alternative hypotheses K_a that there are k change points where $F_{\tau_i} \neq F_{\tau_i+1}$ for $i = 1, \ldots, k$ and

$1 < \tau_1 < \ldots < \tau_k < m$ are the unknown change points. Both the outlier and change point detection problems share a common goal of detecting the observations that create a disruption in the behavior of data under the null hypothesis. Most methods of high-dimensional change point and outlier detection are based on the distance matrix.

The origin of the distance matrix dates back to Cayley [53]. Interpoint distances (IPDs) are extensively used in clustering methods and shape analysis where shapes are the finite vectors of coordinates characterizing the shapes. Distance matrices are the building blocks of many popular machine learning algorithms, including the k-nearest neighbors for supervised learning and k-means clustering for unsupervised learning. Glick [120] studies separation measures whose lower and upper bounds are functions of all pairwise IPDs that are contained in the distance matrix. Guo and Modarres [125] use IPDs for high-dimensional classification, Modarres [229] proposes methods for visualization and comparison of high-dimensional observations, and Modarres [232] discusses high-dimensional outlier detection methods. Atkinson et al. [17] use IPDs to compare an observed sample with a high-dimensional simulated sample. IPDs-based methods have been increasingly common in the literature of hypothesis testing, in particular for two-sample comparison problems and case-control studies, see Jureckova and Kalina [165] and Marozzi [214]. Interpoint distances allow one to transform the $m \times d$ data matrix to an $m \times m$ matrix of distances. This transformation provides a large data reduction when $d \gg m$.

The distance concentration phenomenon (Francois et al. [105]) describes the effect of high dimension d on IPDS. In the HDLSS setting, the sparsity of the observations has unfavorable effects on the performance of most outlier detection methods. Angiulli [11] shows that the interpoint distances concentrate around a single value and become almost indistinguishable in high-dimensional spaces. The coefficient of variation of the IPDs tends to zero as $d \to \infty$. Hall, Marron and Neeman [133] show that high-dimensional data have a tendency to lie deterministically at the vertices of a regular simplex and the randomness in the data are the result of a random rotation of this simplex.

Suppose $\mathbf{X}_i$ are i.i.d. F with mean $\boldsymbol{\mu}$ and covariance $\boldsymbol{\Sigma}$, for $i = 1, \ldots, m$. One can show that $d_{(x)ij} = \frac{1}{\sqrt{d}}\|\mathbf{X}_i - \mathbf{X}_j\| = \sqrt{2tr(\boldsymbol{\Sigma})} + O_{\mathbb{P}}(1)$ and $\frac{1}{\sqrt{d}}\|\mathbf{X}_i - \mathbf{X}_j\|$ converges to $\frac{1}{\sqrt{d}}\mathbb{E}\|\mathbf{X}_i - \mathbf{X}_j\| = \sqrt{2}\sigma_F$ where $tr(\boldsymbol{\Sigma})/d \to \sigma_F$ as for $i \neq j = 1, \ldots, m$ as $d \to \infty$. Hence, all pairwise distances are nearly the same when $d \to \infty$. It is important to state the assumptions that underpin the theory of IPDs. We consider the situations when m and d approach infinity, as well as when one of them tends to infinity and the other remains finite. For example, we assume $m \to \infty$ for test S in section 14.3. A permutation test approach is advocated for the test statistic S in section 14.3. In general, when m is relatively small, a permutation-based approach for IPDs is feasible, especially when using random permutations. The asymptotic approach for HDLSS setting is favored for larger m when $d \to \infty$. The assumptions and results for HDLSS settings when $d \gg m$ are stated in section 14.7.

We assume that $\mathbf{X} = \{\mathbf{X}_i\}_{i=1}^m$ and $\mathbf{Y} = \{\mathbf{Y}_j\}_{j=1}^n$ are independent samples of d-dimensional random vectors that are drawn from distribution functions F_1 and F_2, with means and covariances $\boldsymbol{\mu}_x$, $\boldsymbol{\Sigma}_x = (\sigma)_{ij}$, $\boldsymbol{\mu}_y$ and $\boldsymbol{\Sigma}_y$, respectively. Let $N = m + n$. We determine whether a distance matrix has constant structure and offer test statistics based on the eigenvalues of the distance matrix and the asymptotic normality of sample average of the IPDS. We study the eigenvalue decomposition of constant distance matrices. We examine the distance matrix when there is an outlying observation due to shift in the mean and change in the scales of the observations. We determine the eigenvalues of this matrix and relate them to the shift and scale parameters. We consider the distance matrix when there are two independent samples and examine the eigenvalues of the two-sample distance matrix. We also investigate the dissimilarity matrices and their eigenvalues for outlier and change point detection methods in HDLSS setting. We identify the outliers and change points based on the row sums of the distance matrix. We present a dissimilarity visualization plot for outlier and change point detection and use it for visual inspection.

14.3 Constant Distance Matrix

The use of distance matrices has become increasingly prominent in the hypothesis testing literature, particularly for two-sample comparison problems. In the context of testing distributional equality, Friedman and Rafsky [109] introduced a runs test based on the minimum spanning tree, and Henze and Penrose [147] subsequently established that this test is both asymptotically distribution-free and consistent. Further developments include the work of Bickel and Breiman [32] and Schilling [300], who proposed nearest-neighbor-based tests that also enjoy consistency and asymptotic distribution-free properties. In the high-dimensional setting, Chen and Qin [64] developed tests specifically designed to address challenges arising from increasing dimensionality. Permutation-based procedures employing IPDs have been proposed by Székely and Rizzo ([333], [330]), Baringhaus and Franz [21], and Biswas and Ghosh [35]. Rosenbaum [284] introduced an exact, distribution-free test grounded in adjacency structures. In addition, Modarres and Song [239] provide a comprehensive treatment of IPDs, including their theoretical properties, practical applications, and visualization techniques.

Consider i.i.d. vectors $\mathbf{X}_i$ and $\mathbf{X}_j$ from F_1. The IPD between $\mathbf{X}_i$ and $\mathbf{X}_j$ is given by $d_{ij} = \|\mathbf{X}_i - \mathbf{X}_j\|$. The distance $\mathbf{D}(\mathbb{X})$ is an $m \times m$ symmetric matrix with zeroes on the main diagonal and d_{ij} on the off-diagonal, $i \neq j = 1, \ldots, m$ (Modarres [237]). In a sample of m independent $\mathbf{X}_i$, there are $m(m-1)/2$ IPDs, of which any two IPDs d_{ij} and $d_{i'j'}$ are said to be independent if and only if $(i, j) \neq (i', j')$. IPDs are asymptotically pairwise independent as m becomes large (Shurygin [309]). Consider the $m \times m$ symmetric matrix of distances $\mathbf{D}(\mathbb{X})$ with elements d_{ij} when $i \neq j$ and zero on the main diagonal,

$$\mathbf{D}(\mathbb{X}) = \begin{array}{c} \\ X_1 \\ X_2 \\ X_3 \\ \\ X_m \end{array} \begin{array}{c} \begin{array}{ccccc} X_1 & X_2 & X_3 & \ldots & X_m \end{array} \\ \begin{pmatrix} 0 & d_{12} & d_{13} & \ldots & d_{1m} \\ d_{21} & 0 & d_{23} & \ldots & d_{2m} \\ d_{31} & d_{32} & 0 & \ldots & d_{3m} \\ \ldots & \ldots & \ldots & \ldots & \ldots \\ d_{m1} & d_{m2} & d_{m3} & \ldots & 0 \end{pmatrix} \end{array}. \tag{14.1}$$

The expected distance matrix $\mathbb{E}(\mathbf{D}(\mathbb{X}))$ has zero diagonal entries and finite off-diagonal elements $\mathbb{E}(d_{ij}) = \mathbb{E}\|\mathbf{X}_i - \mathbf{X}_j\|$. We assume that the fourth moments of the components of $\mathbf{X} = (X_1, \ldots, X_d)'$ are uniformly bounded. We estimate $\mathbf{\Sigma}_x$ with S_x, the sample covariance matrix. Under elliptical distributions (Muirhead [244]), one can show that

$$\sqrt{m-1}\,(diag(S_x) - diag(\mathbf{\Sigma}_x)) \to \mathbb{N}_d(\mathbf{0}, \mathbf{\Gamma} = (\gamma)_{ii,kk}) \tag{14.2}$$

where $\gamma_{ii,ii} = \mathbb{V}ar(s_{ii}) = 3(\kappa+1)\sigma_{ii}^2$ and covariances

$$\gamma_{ii,kk} = \mathbb{C}ov(s_{ii}, s_{kk}) = (\kappa+1)(\sigma_{ii}\sigma_{kk} + 2\sigma_{ik}^2), \tag{14.3}$$

as $m \to \infty$. The kurtosis parameter κ relates

$$\sigma_{ijkl} = \mathbb{E}[(\mathbf{X} - \mu)_i(\mathbf{X} - \mu)_j(\mathbf{X} - \mu)_k(\mathbf{X} - \mu)_l \tag{14.4}$$

to $\sigma_{ij} = \mathbb{E}[(\mathbf{X} - \mu)_i(\mathbf{X} - \mu)_j]$ by

$$\sigma_{ijkl} = (\kappa+1)(\sigma_{ij}\sigma_{kl} + \sigma_{ik}\sigma_{jl} + \sigma_{il}\sigma_{jk}). \tag{14.5}$$

Since $\sigma_{iiii} = 3(\kappa+1)\sigma_{ii}^2$, elliptical distributions have the same marginal kurtosis

$$\kappa = \frac{1}{3}\Big(\frac{\sigma_{iiii}}{\sigma_{ii}^2} - 3\Big).$$

Let $h(\mathbf{S}_x)$ be continuous and scalar-valued function of $\mathbf{S}_x$ with continuous first and second order derivatives in a neighborhood of $\mathbf{S}_x = \boldsymbol{\Sigma}_x$. It follows from the delta method that $\sqrt{m}(h(\mathbf{S}_x) - h(\boldsymbol{\Sigma}_x))$ are asymptotically normal as $m \to \infty$, so that

$$\sqrt{m}\left(\bar{d}_x^{\,2} - 2tr(\boldsymbol{\Sigma}_x)\right) \to N\left(0, 8(1+\kappa)tr\left(\boldsymbol{\Sigma}_x^2\right) + 4\kappa\left(tr\boldsymbol{\Sigma}_x\right)^2\right). \tag{14.6}$$

Under the normal model, the excess kurtosis satisfies $\kappa = 0$, and the variance of the squared mean IPD is given by $\mathbb{V}\mathrm{ar}(\bar{d}_x^2) = \frac{8}{m}\,\mathrm{tr}(\boldsymbol{\Sigma}_x^2)$. This result facilitates hypothesis testing and the construction of confidence intervals for $\mathbb{E}(\bar{d}_x^2)$ in the asymptotic regime $m \to \infty$.

The Euclidean distance geometry problem concerns the reconstruction of a data configuration matrix from information on pairwise IPDs. Gower [123] introduced principal coordinate analysis, whose objective is to recover the original observation matrix through an eigenvalue decomposition of the distance matrix; see also Mardia, Kent and Bibby [211]. The Gower distance is widely used in practice and has become particularly popular in quality monitoring applications(Adegoke et al. [2]).

This approach is invariant under translations and rotations of the data configuration. Let $\mathbf{I}_m$ denote the identity matrix of order m, and let $\mathbf{J}_m = \mathbf{1}_m\mathbf{1}_m'$ denote the $m \times m$ matrix of ones. To obtain a singular value decomposition of the distance matrix for the purpose of reconstructing the original observations, the following steps are carried out.

Consider the $m \times m$ symmetric distance matrix $\mathbf{D}(\mathbb{X})$ with zero on the main diagonal and d everywhere else. It is not difficult to show the eigenvalues of $\mathbf{D}(\mathbb{X})$ are $\lambda_i = \frac{1}{2}d^2$ for $i = 1, \ldots, m-1$ and $\lambda_m = 0$. The ratio $\lambda_1/\lambda_2 = 1$ for constant distance matrices. Large deviations from this ratio indicates that the distance matrix is not constant.

Example 14.3.1 *Suppose $\mathbf{X}_i \sim F(\boldsymbol{\mu}, \sigma_i^2\mathbf{I}_d)$ for $i = 1, \ldots, m$. One can show that*

$$\mathbb{E}(\mathbf{D}_{ij}(\mathbf{X})) = \mathbb{E}(\|\mathbf{X}_i - \mathbf{X}_j\|) = \sqrt{d(\sigma_i^2 + \sigma_j^2)} \text{ for } i \neq j = 1, \ldots, m,$$

with zero for $i = j$. Hence, the observations are i.i.d. when $\sigma_i = \sigma$ for $i = 1, \ldots, m$.

We are interested in testing the hypothesis that observations in $\mathbf{X}$ are equidistant,

$$H_0 : \mathbb{E}(\mathbf{D}(\mathbb{X})) = \mathbb{E}(d)(\mathbf{J}_m - \mathbf{I}_m) \text{ against } H_a : \mathbb{E}(\mathbf{D}(\mathbb{X})) \neq \mathbb{E}(d)(\mathbf{J}_m - \mathbf{I}_m). \tag{14.7}$$

The hypothesis H_0 states that the non–zero eigenvalues of the expected distance matrix are all equal. A natural statistic for testing H_0 is based on (14.6). Since $\mathbb{E}(\bar{d}_x^{\,2}) = 2tr(\boldsymbol{\Sigma}_x)$ it follows that

$$T = m\left(\bar{d}_x^{\,2} - 2tr(\boldsymbol{\Sigma}_x)\right)^2 / 8tr\left(\boldsymbol{\Sigma}_x^2\right) \sim \chi^2(1). \tag{14.8}$$

One may construct a test based on the internal structure of the distance matrix as follows. The average of the off-diagonal elements in the kth row of $\mathbf{D}$ and the overall average of the off-diagonal elements of $\mathbf{D}$ are denoted with

$$\bar{d}_k = \frac{1}{m-1}\sum_{i \neq k = 1}^{m} d_{ij}, \qquad \bar{d} = \frac{2}{m(m-1)}\sum_{i=1}^{m}\sum_{j=i+1}^{m} d_{ij}, \tag{14.9}$$

respectively. We reject H_0 in favor of H_a for large values of

$$S = \sum_{i=1}^{m}\sum_{j=i+1}^{m-1}(d_{ij} - \bar{d})^2 - \sum_{k=1}^{m}(\bar{d}_k - \bar{d})^2. \tag{14.10}$$

One may also reject H_0 in favor of H_a for large values of the ratio of the first two eigenvalues of the distance matrix,

$$R = \frac{\hat{\lambda}_1}{\hat{\lambda}_2}. \tag{14.11}$$

The sampling distribution of S and R can be obtained by the non-parametric bootstrap of the distance matrix as follows. Let the main diagonal of $\mathbf{D}$ be set to zero. We resample columns $\mathbf{X}_i = (X_{1i}, \ldots, X_{mi})$ and $\mathbf{X}_j = (X_{1j}, \ldots, X_{mj})$ of the data matrix, to obtain the bootstrap estimate of d_{ij} for $i = 1, \ldots, m$ and $j = i + 1, \ldots, m - 1$.

14.4 Distance Matrix with Outliers

Consider the distance $m \times m$ matrix of interest

$$\mathbf{D}(\mathbb{X}) = \begin{bmatrix} 0 & cd+e & cd+e & \ldots & cd+e \\ cd+e & 0 & d & \ldots & d \\ cd+e & d & 0 & \ldots & d \\ \vdots & \vdots & \vdots & \vdots & \vdots \\ cd+e & d & d & \ldots & 0 \end{bmatrix}. \tag{14.12}$$

To study an alternative to the constant distance matrix, suppose the distance between the first and the remaining observations scaled by c and shifted by $e > 0$ with $c \neq 0$ while the rest of the observations are equidistant. One can show that the eigenvalues of (14.12) are

$$\lambda_1 = \frac{m-1}{m}(cd+e)^2 - \frac{m-2}{2m}d^2, \quad \lambda_2 = \ldots = \lambda_{m-1} = \frac{1}{2}d^2 \text{ and } \lambda_m = 0. \tag{14.13}$$

The ratio

$$\lambda_1/\lambda_2 = \frac{2(m-1)}{m}\frac{(cd+e)^2}{d^2} - \frac{m-2}{m} \geq 1$$

with equality holding when $c = 1$ and $e = 0$. When $e = 0$ the ratio $\lambda_1/\lambda_2 = \frac{2(m-1)}{m}c^2 - \frac{m-2}{m} > 1$ unless $|c| \leq 1$. When $c = 0$ the ratio $\lambda_1/\lambda_2 > 1$ when $e > d$. It is noteworthy that when the distance matrix is not constant, there is a discrepancy in the sizes of eigenvalues and $\lambda_1/\lambda_2 \gg 1$.

Example 14.4.1 *Let $\mathbf{X}_1, \ldots, \mathbf{X}_m$ be a random sample from F with mean $\mathbf{0}$ and covariance $\sigma^2\mathbf{I}_d$. Suppose $\mathbf{X}_1$ is multiplied by a constant $c > 0$. One can show that $\mathbb{E}(\mathbf{D}_{ij}(\mathbf{X})) = \sigma\sqrt{d(c^2+1)}$ for $i = 1$ and $j = 2, \ldots, m$ and $i = 2, \ldots, m$ and $j = 1$ whereas $\mathbb{E}(\mathbf{D}_{ij}(\mathbf{X})) = \sqrt{2d}\sigma$ for $i \neq j = 2, \ldots, m$. Under H_0, we have $c = 1$ and under H_a we have $c > 1$.*

Example 14.4.2 *Let $\mathbf{X}_1, \ldots, \mathbf{X}_m$ be a random sample from F with zero mean and covariance $\mathbf{I}_d$. Suppose $\mathbf{X}_1$ is multiplied by a constant scalar $\sigma_1 > 0$ and shifted by $\boldsymbol{\mu}$ so that $\mathbf{X}_1$ has mean $\boldsymbol{\mu}$ and covariance $\sigma_1^2\mathbf{I}_d$. It is not difficult to show that $\mathbb{E}(\mathbf{D}^2_{ij}(\mathbf{X})) = d(\sigma_1^2 + 1) + \|\boldsymbol{\mu}\|^2$ for $i = 1$ and $j = 2, \ldots, m$ and $i = 2, \ldots, m$ and $j = 1$ whereas $\mathbb{E}(\mathbf{D}^2_{ij}(\mathbf{X})) = 2d$ for $i \neq j = 2, \ldots, m$. The expected distance matrix is equidistant when $\sigma_1 = 1$ and $\boldsymbol{\mu} = \mathbf{0}$.*

Example 14.4.3 *To test the null hypothesis H_0 that the distance matrix has a constant structure, we used the test statistics T, S, and R in the following sampling experiment. We generated m observations from $\mathbb{N}_d(\mathbf{0}, \mathbf{I}_d)$. We transformed one observation by shift e and* *scale c. We obtained the sampling distribution of S and R under H_0 using non-parametric bootstrap of the distance matrix with* $e = 0$ *and* $c = 1$. *We used* $\bar{d} = \frac{2}{m(m-1)} \sum_{i=1}^{m} \sum_{j=i+1}^{m} d_{ij}$ *and set* $\mathbf{A} = (a_{ij}) = -\frac{1}{2}\bar{d}^2$. *We used* $m = (15, 25)$ *and* $d = (100, 400)$.

Table 14.1 shows the proportion of times the test statistics S, T, and R reject the hypothesis of constant distance matrix in 1000 multivariate normal datasets against the alternative of a single outlier. It also displays the ratio $\bar{\lambda}_1/\bar{\lambda}_2$ *where* $\bar{\lambda}_i$ *is the average of the ith eigenvalues for* $i = 1, 2$. *Under* H_0 *where* $c = 1$ *and* $e = 0$, *this ratio is close to unity whereas under a single outlier model, where* $c > 1$ *and* $e > 0$, *the ratio becomes inflated. The test statistics maintain their* $\alpha = 0.05$ *level. The test statistic R is more powerful than T under scale changes. The test statistic S appears to be more powerful than T and R for all the alternatives considered. The proposed method is robust with respect to distributional assumptions. Similar results to Table 14.1 were obtained for non–normal data.*

TABLE 14.1: *Outlier Detection*: The proportion of times the test statistics S, T, and R reject the hypothesis of constant distance matrix in 1000 multivariate normal datasets.

m	d	c	e	P_S	P_T	P_R	$\frac{\bar{\lambda}_1}{\bar{\lambda}_2}$	m	d	c	e	P_S	P_T	P_R	$\frac{\bar{\lambda}_1}{\bar{\lambda}_2}$
15	100	1.0	0.0	0.052	0.054	0.052	1.15	25	100	1.0	0.0	0.054	0.040	0.044	1.12
		1.0	0.4	0.038	0.054	0.032	1.15			1.0	0.4	0.066	0.048	0.048	1.13
		1.2	0.0	0.330	0.122	0.092	1.18			1.2	0.0	0.232	0.122	0.056	1.14
		1.2	0.4	0.286	0.132	0.070	1.17			1.2	0.4	0.202	0.082	0.046	1.14
		1.6	0.0	0.972	0.670	0.786	1.53			1.6	0.0	0.970	0.522	0.672	1.38
		1.6	0.4	0.976	0.578	0.800	1.53			1.6	0.4	0.968	0.420	0.722	1.38
	400	1.0	0.0	0.070	0.038	0.060	1.08		400	1.0	0.0	0.062	0.034	0.054	1.07
		1.0	0.4	0.068	0.060	0.052	1.08			1.0	0.4	0.040	0.042	0.062	1.07
		1.2	0.0	0.826	0.346	0.464	1.15			1.2	0.0	0.730	0.194	0.274	1.10
		1.2	0.4	0.846	0.312	0.450	1.16			1.2	0.4	0.746	0.196	0.298	1.11
		1.6	0.0	1.000	0.998	1.000	1.83			1.6	0.0	1.000	0.940	1.000	1.68
		1.6	0.4	1.000	0.976	1.000	1.82			1.6	0.4	1.000	0.952	1.000	1.67

The discrepancy in the sizes of the first two eigenvalues of the distance matrix as measured with $R = \hat{\lambda}_1/\hat{\lambda}_2 \gg 1$ indicates that the distance matrix is not constant. However, it does not identify the source of discrepancy. Modarres [232] proposes a proximity–based, non-parametric outlier detection method based on the row totals of the distance matrix. See chapter 10. We study the relationships between the eigenvalues of distance matrices and outliers in (14.13). We show a discrepancy in the sizes of eigenvalues of the distance matrices that are contaminated with outliers. Modarres [235] proposes an approach in which one observation is sequentially removed from a dataset of m observations, the corresponding reduced distance matrix is computed, and its leading eigenvalue is extracted. This procedure yields a collection of m leading eigenvalues, each associated with the deletion of a single observation. The resulting list may be used to identify the observation whose removal produces the smallest leading eigenvalue. The potential outlier is then identified as the observation whose deletion leads to the largest reduction in the first eigenvalue. This methodology can be extended to the detection of multiple outliers by iteratively removing the identified outlying observation and reapplying the procedure to the reduced distance matrix.

14.5 Imputing Distance and Dissimilarity Matrices

Distance and dissimilarity matrices are central objects in statistics, machine learning, bioinformatics (Troyanskaya, et al. [341]), facial recognition, multidimensional scaling (MDS), kernel methods, and network analysis. In real applications, however, some distances may be missing due to incomplete data collection, partially observed data, computational constraints, or privacy restrictions. When distances are missing, many standard algorithms either fail or implicitly ignore incomplete observations, leading to biased results. As a consequence, principled methods for estimating missing distances are essential. We present two methods for imputing missing distances, Row/Column median imputation and k-nearest neighbor (kNN) imputation. A random sample $\mathbf{X}_1, \ldots, \mathbf{X}_n$ gives rise to the distance matrix $\mathbf{D} = (d)_{ij}$ for $i \neq j = 1, \ldots, n$. If $\mathbf{X}_i$ or $\mathbf{X}_j$ has missing components, then $(d)_{ij}$ is missing. Here, a missing component is denoted with NA.

14.5.1 Row/Column Median Imputation

Let d_{ij} be missing. Define

$$\begin{aligned} \tilde{d}_{i\cdot} &= \text{median}\{d_{ik} : k \neq i,\ d_{ik} \text{ observed}\}, \\ \tilde{d}_{\cdot j} &= \text{median}\{d_{kj} : k \neq j,\ d_{kj} \text{ observed}, \end{aligned}$$

where $\tilde{d}_{i\cdot}$ is the median distance when $\mathbf{X}_i$ is fixed and $\tilde{d}_{\cdot j}$ is the median distance when $\mathbf{X}_j$ is fixed. By fixing $\mathbf{X}_i$ the distances within that row are independent. Similarly, by fixing $\mathbf{X}_j$ the distances within that column are independent. The imputed value is

$$\hat{d}_{ij}^{\text{median}} = \frac{\tilde{d}_{i\cdot} + \tilde{d}_{\cdot j}}{2}.$$

Diagonal zeros are excluded from all median computations.

14.5.2 k-NN Imputation

For a missing entry d_{ij}:

1. Identify the k rows whose observed distance profiles are closest to row i.

2. Retain those neighbors with observed distances to object j.

3. Impute $\hat{d_{ij}^{\text{NN}}} = \frac{1}{k} \sum_{r \in \text{NN}_k(i)} d_{rj}$.

In Step 1, to obtain the k nearest neighbors to row i, one can order the list of $\|\mathbf{X}_{i.} - \mathbf{X}_{j.}\|$ and take the smallest k values in the list to obtain the rows. In Step 2, we retain k distances d_{rj} where r is the selected row number.

This method preserves some local structure of the distance matrix and is more robust than mean imputation to heterogeneity. However, its performance is sensitive to the choice of k, and if many distances are missing, the method can become unstable. Computational cost also grows with the size of the matrix.

For missing index set M, the mean squared error is computed as

$$\text{MSE} = \frac{1}{|M|} \sum_{(i,j) \in M} (d_{ij}^{\text{true}} - \hat{d}_{ij})^2.$$

Example 14.5.1 *Suppose we have the* 3×3 *Distance Matrix,*

$$\mathbf{D} = \begin{pmatrix} 0 & 2 & NA \\ 2 & 0 & 4 \\ NA & 4 & 0 \end{pmatrix}$$

For median-based imputation we see that the median of row 1 is 2 and median of row 3 is 4. Thus,

$$\hat{d}_{13}^{median} = (2+4)/2 = 3.$$

For nearest neighbor imputation with $k = 1$, *we observe that the nearest neighbor to row 1 is row 2. Since* $d_{23} = 4$ *we get* $\hat{d}_{13}^{NN} = 4$. *If the true missing value is* $d_{13}^{true} = 3.5$, *then*

$$MSE_{median} = (3.5 - 3)^2 = 0.25, \quad MSE_{NN} = (3.5 - 4)^2 = 0.25.$$

Example 14.5.2 *Consider the* 5×5 *distance matrix*

$$\mathbf{D} = \begin{pmatrix} 0 & 2 & 3 & 4 & 6 \\ 2 & 0 & 5 & NA & 7 \\ 3 & 5 & 0 & 6 & NA \\ 4 & NA & 6 & 0 & 8 \\ 6 & 7 & NA & 8 & 0 \end{pmatrix}$$

For median-based imputation we compute

$$\hat{d}_{24}^{median} = (5+6)/2 = 5.5, \quad \hat{d}_{35}^{median} = (5+7)/2 = 6.$$

For nearest neighbor imputation with $k = 2$ *we observe that*

$$\hat{d}_{24}^{NN} = (4+6)/2 = 5, \quad \hat{d}_{35}^{NN} = (6+7)/2 = 6.5.$$

If we assume $d_{24}^{true} = 6$, $d_{35}^{true} = 5.5$, *then* $MSE_{median} = 0.25$ *and* $MSE_{NN} = 1$.

Notice that when n is even, excluding the 0 distance there remains $n - 1$ distances per row. With n-1 distances the median is at position $(n-1-1)/2$. Up to $(n-2)/2$ distances can be missing per row, but the matrix is symmetric so with n rows the total number of missing values in the entire distance matrix can be up to $n/2(n-2)/2 = n(n-2)/4$. Since there are $\binom{n}{2} = n(n-1)/2$ total IPDs, the ratio of the number of missing values to total IPDs is $\frac{n-2}{2(n-1)}$. This ratio rapidly approaches $\frac{1}{2}$. Hence, the Median imputation can impute distance matrices where up to half of its entries are missing. Median-based imputation provides a simple and robust method.

14.5.3 Mean Imputation

Mean imputation is the simplest approach to handle missing distances. Each missing entry d_{ij} is replaced by the mean of the observed distances in the matrix. This method is straightforward, computationally inexpensive, and ensures symmetry if applied carefully to both d_{ij} and d_{ji}. However, it ignores the underlying geometry of the points, and tends to shrink variability, which can bias downstream analyses, such as clustering or MDS.

Despite its simplicity, mean imputation is often used as a baseline method. It is moderately effective when only a small fraction of distances are missing, and the matrix is relatively homogeneous. Its main limitations are high sensitivity to outliers and a high-breakdown point, meaning a single extreme observed distance can disproportionately affect all imputed values.

14.5.4 Matrix Completion via Nuclear Norm Minimization

Matrix completion aims to recover missing entries assuming the matrix has an approximate low-rank structure. For a partially observed distance matrix $\mathbf{D}$, the nuclear norm minimization problem is

$$\min_{\hat{\mathbf{D}}} \|\hat{\mathbf{D}}\|_* \quad \text{subject to observed } d_{ij},$$

where $\|\hat{\mathbf{D}}\|_*$, the nuclear norm, is the sum of the singular values of $\hat{\mathbf{D}}$. The nuclear norm acts as a convex surrogate for matrix rank, promoting low-rank solutions [50].

This method is effective when distances can be represented in a low-dimensional latent space. It generally produces low MSE and is relatively robust to moderate amounts of missing data. Its main limitations are computational cost and the assumption that the distance matrix is approximately low-rank, which may not hold for highly irregular or noisy data.

14.5.5 Multidimensional Scaling (MDS) Iterative Imputation

MDS iterative imputation uses classical or non-metric MDS to estimate the coordinates of points in a low-dimensional space, using observed distances. Missing distances are initially replaced by reasonable guesses (e.g., the mean), and then the algorithm alternates between estimating coordinates and recomputing distances until convergence.

This method preserves global geometric structure, making it particularly suitable for visualization or downstream geometric analyses. Its performance degrades if too many distances are missing or if the missing entries are not randomly distributed. Convergence can also be slow in large matrices.

14.5.6 Regression-Based Imputation

Regression imputation models a missing distance d_{ij} as a function of other observed distances in the matrix. For instance, one might predict d_{ij} based on the distances from points i and j to other points using linear regression or other parametric models [196]. The predicted value replaces the missing entry.

Regression imputation can capture dependencies between distances and often yields lower MSE than naive mean imputation. Its main limitations are sensitivity to model misspecification and lower robustness to outliers. Non-linear relationships may require more flexible models, such as generalized additive models or machine learning approaches.

14.5.7 Random Forest Imputation

Random forest imputation uses an ensemble of regression trees to predict missing distances from observed entries. For each missing d_{ij}, the algorithm constructs multiple regression trees using other distances as predictors and averages the results.

This method can capture non-linear relationships, is robust to outliers, and often produces low MSE. It is computationally intensive and requires careful handling to maintain the symmetry of the distance matrix. It generally performs well even with moderate to high fractions of missing distances.

14.6 Properties of the Methods

Table 14.2 summarizes the main properties of the six methods in terms of mean squared error (MSE), robustness to outliers, and breakdown behavior.

TABLE 14.2: Summary of properties for six missing distance estimation methods.

Method	MSE	Robustness	Breakdown
Row/Column Median	Low	High	High
Mean Imputation	Moderate	Low	High sensitivity to outliers
KNN Imputation	Low—Moderate	Moderate	Sensitive to choice of k
Matrix Completion	Low	Moderate	Stable if low-rank assumption
holds			
MDS Iterative	Low	Moderate	Can fail if too many missing
entries			
Regression Imputation	Low	Low—Moderate	Sensitive to model
misspecification			
Random Forest	Low	High	Generally stable, robust to
non-linearity			

14.6.1 Comparison

A random sample of size n is generated from $\mathbb{N}(\mathbf{0}, \mathbf{I})$ and $\chi^2_{(1)}$ distributions. The number of missing distances is $t = \{2, 3, \ldots, 10\}$, the sample size n is $\{15, 20, 30\}$ and $d = 100$. Euclidean distances are computed, and t values of the off-diagonal entries are removed, and imputation is performed. For NN we used the nearest 5 neighbors. The experiment is replicated $R = 1000$ times. The mean squared error is computed per missing entry and averaged over R experiments. The MSEs for the normal distribution appear in Table 14.3. It is clear that Median method produces smaller MSE than kNN and is more accurate imputation method for missing distances. Median-based imputation presents a simple and robust technique, while kNN imputation adapts to local structure and may incur higher variance. Choice of method depends on data structure, number of missing entries, and downstream analysis needs.

TABLE 14.3: Comparison of the MSE of Median and kNN methods for different values of t and n

	Median			kNN		
t	$n = 15$	$n = 20$	$n = 30$	$n = 15$	$n = 20$	$n = 30$
2	0.741	0.696	0.656	1.519	1.811	2.276
3	0.684	0.634	0.651	1.384	1.695	2.226
4	0.681	0.653	0.682	1.462	1.787	2.321
5	0.686	0.654	0.662	1.386	1.745	2.336
6	0.695	0.671	0.632	1.379	1.761	2.203
7	0.668	0.640	0.640	1.382	1.729	2.216
8	0.678	0.648	0.637	1.366	1.743	2.215
9	0.687	0.660	0.647	1.360	1.657	2.226
10	0.683	0.658	0.664	1.329	1.693	2.233

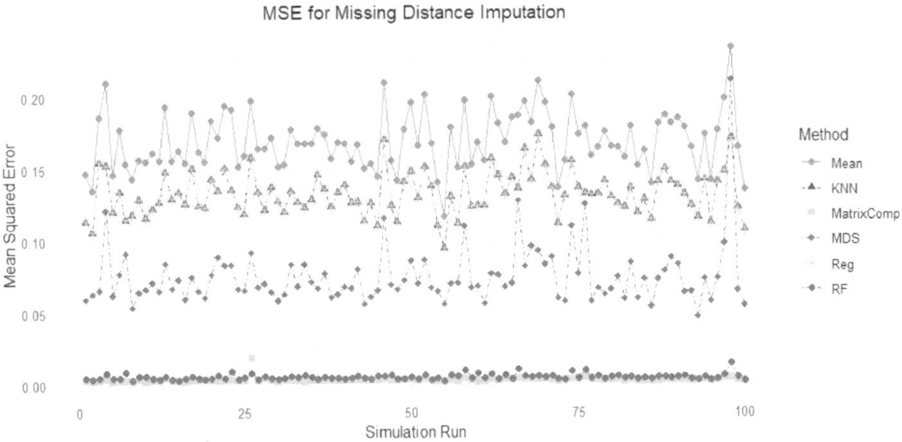

FIGURE 14.1: MSEs of the imputation methods with 20% missing distances.

Figure 14.1 shows the MSE of 6 methods of estimating the missing components of a $d\times$ distance matrix with 20 percent of its values set to missing at random. We generated $m = 50$ samples with $d = 100$ and repeated the experiment 100 times. The methods are named Mean, KNN, MatrixComp, MDS ,Reg , and RF. The plotted MSEs clearly show that the random forest and matrix completion methods are the best performing imputations in this case. The order in terms of MSE is Mean>KNN=Reg>MDS>RF>MatrixComp.

In the following sections, we identify the eigenvalues of the distance matrix for two samples and relate them to test for equality of distributions. We present statistics to test whether a distance matrix has constant structure. We have shown that outliers inflate the eigenvalues of the distance matrix. We examine the dissimilarity matrices and their eigenvalues for outlier and change point detection methods in HDLSS setting and present a dissimilarity visualization plot.

We establish that, as $d \to \infty$, the asymptotic form of the pairwise distances within a given group of observations converges to a constant distance matrix. We analyze the Euclidean distance matrix together with two dissimilarity matrices, derive their respective asymptotic representations, and investigate their spectral decompositions. An alternative formulation of these matrices is proposed, leading to substantial improvements in computational efficiency. We further clarify the connection between the eigenvalues of these distance and dissimilarity matrices and classical statistical procedures for testing the equality of high-dimensional distributions. In particular, we show that the equality of K distributions is equivalent to the associated distance matrix possessing a constant structure. For the cases of two and three observation groups, we explicitly derive the eigenvalues of the corresponding asymptotic dissimilarity matrices. Building on these limiting forms and their eigen-structures, we introduce new test statistics for assessing the equality of K distributions. These statistics are motivated by IPDs but avoid their direct computation, thereby offering both theoretical insight and computational advantages.

14.7 Two-sample Distance Matrix

Suppose $\mathbf{X}_i$ are i.i.d. F_1 for $i = 1, \ldots, m$ independently of $\mathbf{Y}_j$ i.i.d. F_2 for $j = 1, \ldots, n$. Let $N = m + n$ and consider the $N \times N$ symmetric matrix of within and between sample distances

$$\mathbf{D}_E(\mathbb{X}, \mathbb{Y}) = \begin{bmatrix} \mathbf{R}_{m\times m} & \mathbf{H}_{m\times n} \\ \mathbf{H}_{n\times m} & \mathbf{C}_{n\times n} \end{bmatrix} = \begin{pmatrix} r\,(\mathbf{1}_{m\times m} - \mathbf{I}_m) & h\,\mathbf{1}_{m\times n} \\ h\,\mathbf{1}_{n\times m} & c\,(\mathbf{1}_{n\times n} - \mathbf{I}_n) \end{pmatrix}. \tag{14.14}$$

The elements of $\mathbf{D}_E(\mathbb{X}, \mathbb{Y})$ are $d_{(x)ij} = \|\mathbf{X}_i - \mathbf{X}_j\|$ in the $\mathbf{R}$ matrix, $d_{(y)ij} = \|\mathbf{Y}_i - \mathbf{Y}_j\|$ in the $\mathbf{C}$ matrix and $d_{(xy)ij} = \|\mathbf{X}_i - \mathbf{Y}_j\|$ in the $\mathbf{H}$ matrix. The eigenvalues of (14.14) are given by

$$\lambda_1 = \frac{1}{2N}[2mnh^2 - n(m-1)r^2 - m(n-1)c^2], \tag{14.15}$$

$$\lambda_2 = \ldots = \lambda_m = \frac{1}{2}r^2, \quad \lambda_{m+1} = \ldots = \lambda_{N-1} = \frac{1}{2}c^2 \text{ and } \lambda_N = 0.$$

Recall that $tr\,(\mathbf{\Sigma}_1)\,/d \to \sigma^2_{F_1}$ and $tr\,(\mathbf{\Sigma}_2)\,/d \to \sigma^2_{F_2}$ by Assumption (A1) of 3.6.3. As $d \to \infty$, $\mathbb{E}(d_x) \to r_1 = \sqrt{2}\sigma_{F_1}$, $\mathbb{E}(d_y) \to r_2 = \sqrt{2}\sigma_{F_2}$ and $\mathbb{E}(d_{xy}) \to h$. Maa, Pearl, and Bartoskzyński [206] prove that $h \overset{d}{=} r_1 \overset{d}{=} r_2$ if and only if $F_1 = F_2$ where d stands for equality in distributions. Under the null hypothesis $K_0 : F_1 = F_2$, we have $h = r_1 = r_2$ so that $\lambda_i = \frac{1}{2}r^2$ for $i = 1, \ldots, N-1$ and $\lambda_N = 0$, resulting in a constant distance matrix. Under shift alternatives, IPDs are invariant so we can establish $\mathbb{E}(\hat{h}^2) > \mathbb{E}(\hat{r}_1) = \mathbb{E}(\hat{r}_2)]$. In fact, Modarres [229] establishes that the distribution of $\hat{h}$ is stochastically larger than that of $\hat{r}_1$ and $\hat{r}_2$.

Suppose $\mathbf{X}_i \sim F_1 = \mathbb{N}_d(\mathbf{0}, \mathbf{I}_d)$ and $\mathbf{Y}_j \sim F_2 = \mathbb{N}_d(\boldsymbol{\mu}, \mathbf{\Sigma}_y)$ for $i = 1, \ldots m$ and $j = 1, \ldots n$. One can show that $r_1^2 = d^2_{(x)ij} = \|\mathbf{X}_i - \mathbf{X}_j\|^2$ is distributed as $2\chi^2(d)$ so that $\mathbb{E}(r_1^2) = 2d$. Similarly, one can show that $r_2^2 = d^2_{(y)ij} = \|\mathbf{Y}_i - \mathbf{Y}_j\|^2$ is distributed as a linear combination of independent $\chi^2(1)$ variables; i.e. $\sum_{i=1}^d \nu_i\chi^2(1)$ where ν_i are the eigenvalues of $2\mathbf{\Sigma}_y$ with $\mathbb{E}(r_2^2) = 2tr(\mathbf{\Sigma}_y)$. Moreover, $h^2 = d^2_{(xy)ij} = \|\mathbf{X}_i - \mathbf{Y}_j\|^2$ is distributed as a linear combination of independent non–central $\chi^2(1)$ variables; i.e. $\sum_{i=1}^d \tau_i\chi^2_{\delta_i}(1)$ where τ_i are the eigenvalues of $(\mathbf{I}_d + \mathbf{\Sigma}_y)$, δ_i are non–centrality parameters and $\mathbb{E}(h^2) = tr(\mathbf{I}_d + \mathbf{\Sigma}_y) + \|\boldsymbol{\mu}\|^2$. Hence, $F_1 = F_2$ if and only if $\boldsymbol{\mu} = \mathbf{0}$ and $\mathbf{\Sigma}_y = \mathbf{I}_d$, providing a constant distance matrix. When $\boldsymbol{\mu} > \mathbf{0}$ and $\mathbf{\Sigma}_y = \mathbf{I}_d$, we obtain $\mathbb{E}(h^2) > [\mathbb{E}(r_1^2) = \mathbb{E}(r_2^2)]$.

14.7.1 HDLSS Distance Matrix

Mardia, Kent and Bibby [211] show that any symmetric distance (dissimilarity) matrix can be used to obtain the decomposition. For example one may use the L_1 or L_2 norms as the dissimilarity index to compute the distances. It is clear that the eigenvalues and vectors of $\mathbf{D}$ are a function of the distance measure that is utilized; hence, we use $\mathbf{D}_E$, $\mathbf{D}_\rho$ and $\mathbf{D}_\delta$ to denote the distance matrix based on the Euclidean distance, dissimilarities ρ and δ, respectively. We discuss dissimilarity matrices that are used in HDLSS comparison of two independent samples and their eigenvalues when $d \to \infty$.

As $d \to \infty$, the distance matrix of ρ dissimilarities becomes

$$\mathbf{D}_\rho(\mathbb{X}, \mathbb{Y}) = \begin{bmatrix} 0_{m\times m} & \mathbf{H}_{m\times n} \\ \mathbf{H}_{n\times m} & 0_{n\times n} \end{bmatrix}, \tag{14.16}$$

where $\mathbf{H}_{ij} = h = \tilde{\rho}$. Hence, the eigenvalues of $\mathbb{E}(\mathbf{D}_\rho(\mathbb{X}))$, as $d \to \infty$ are $\lambda_1 = \frac{mn}{N}\tilde{\rho}^2$, and 0 with multiplicity $N-1$. As $d \to \infty$, the distance matrix of δ dissimilarities is in the form of matrix in (14.16) where $h = \tilde{\delta}$. Hence, the eigenvalues of $\mathbb{E}(\mathbf{D}_\delta(\mathbb{X}))$, as $d \to \infty$ are $\lambda_1 = \frac{mn}{N}((\mu_x - \mu_y)^2 + (\sigma_x - \sigma_y)^2)$, and 0 with multiplicity $N-1$.

14.8 Dissimilarity Matrices

Several authors have demonstrated that the Euclidean distance is an inadequate measure of separation in high-dimensional settings due to the distance concentration phenomenon. To alleviate this issue, Hall et al. [133] proposed scaling Euclidean distances by a factor of $1/\sqrt{d}$ in order to ensure convergence as $d \to \infty$. Let $\mathbf{X}$ and $\mathbf{Y}$ be independent d-variate random vectors drawn from distributions F_1 and F_2, with respective mean vectors $\boldsymbol{\mu}_1$ and $\boldsymbol{\mu}_2$ and covariance matrices $\mathbf{\Sigma}_1$ and $\mathbf{\Sigma}_2$. Let $\mathbb{X} = \{\mathbf{X}_i\}_{i=1}^{n_1}$ and $\mathbb{Y} = \{\mathbf{Y}_j\}_{j=1}^{n_2}$ denote independent and identically distributed samples from $\mathbf{X}$ and $\mathbf{Y}$, respectively, and let $\mathbb{Z}$ represent the pooled sample of size $N = n_1 + n_2$. Whereas the Euclidean distance captures information from only a single pair of observations, we introduce an index that additionally incorporates information from the remaining $N - 2$ observations in the combined sample,

$$d_{ij} = d(\boldsymbol{Z}_i, \boldsymbol{Z}_j) = \frac{1}{N-2} \sum_{l \neq i,j} \left| \|\boldsymbol{Z}_i - \boldsymbol{Z}_l\|_D - \|\boldsymbol{Z}_j - \boldsymbol{Z}_l\|_D \right|, \tag{14.17}$$

where $\|\cdot\|_D$ is any appropriate distance function such as $\|\boldsymbol{Z}_i - \boldsymbol{Z}_l\|_D = d^{-1/2}\|\boldsymbol{Z}_u - \boldsymbol{Z}_l\|_2$ or L_1-norm $\|\boldsymbol{Z}_i - \boldsymbol{Z}_l\|_D = d^{-1}\|\boldsymbol{Z}_u - \boldsymbol{Z}_l\|_1$. To obtain a more efficient formulation for d_{ij}, define the ordered set of distances: $d_i^{(l)} = \|\boldsymbol{Z}_i - \boldsymbol{Z}_l\|_D$ for each $l \neq i, j$. The values $d_i^{(l)}$ can be arranged in increasing order to obtain $d_i^{(1)} \leq d_i^{(2)} \leq \ldots \leq d_i^{(N-2)}$. Similarly, we order $d_j^{(l)}$ to obtain $d_j^{(1)} \leq d_j^{(2)} \leq \cdots \leq d_j^{(N-2)}$. Since d_{ij} is the mean of absolute differences, we have $d_{ij} = \frac{1}{N-2}\sum_{l=1}^{N-2} |d_i^{(l)} - d_j^{(l)}|$. This reduces the number of operations from $O(dN^2)$ that uses brute-force pairwise comparisons to $O(dNlogN)$, which is required for sorting N values. In the following discussion, we denote three asymptotic matrices as $d \to \infty$: L_1-norm distance matrix, $\mathbf{D}_E$, and the dissimilarity matrices $\mathbf{D}_\rho$ and $\mathbf{D}_\delta$.

14.8.1 Dissimilarity Matrix ρ

Sarkar and Ghosh [298] use the $\|\boldsymbol{Z}_i - \boldsymbol{Z}_l\|_D = d^{-1/2}\|\boldsymbol{Z}_i - \boldsymbol{Z}_l\|_2$ to define the dissimilarity measure

$$\rho(\boldsymbol{Z}_i, \boldsymbol{Z}_j) = \frac{1}{\sqrt{d}(N-2)} \sum_{l \neq i,j} \left| \|\boldsymbol{Z}_i - \boldsymbol{Z}_l\|_2 - \|\boldsymbol{Z}_j - \boldsymbol{Z}_l\|_2 \right|. \tag{14.18}$$

Let $h_{12} = \sqrt{\sigma_{F_1}^2 + \sigma_{F_2}^2 + \nu_{12}^2}$. Under Assumptions 3.6.3, Sarkar and Ghosh [298] show that as $d \to \infty$, $\rho(\mathbf{X}_1, \mathbf{X}_2)$ and $\rho(\mathbf{Y}_1, \mathbf{Y}_2)$ converge in probability to 0, while $\rho(\mathbf{X}_1, \mathbf{Y}_1)$ converges in probability to

$$\tilde{\rho}(F_1, F_2) = \frac{1}{N-2}[(n_1 - 1)|\sqrt{2}\sigma_{F_1} - h_{12}| + (n_2 - 1)|h_{12} - \sqrt{2}\sigma_{F_2}|] \geq 0, \tag{14.19}$$

which is 0 if and only if $\sigma_{F_1} = \sigma_{F_2}$ and $\nu_{12}^2 = 0$. As $d \to \infty$, $\mathbb{E}[\mathbf{D}_\rho(X_i, X_j)] \to 0$, $\mathbb{E}[\mathbf{D}_\rho(Y_i, Y_j)] \to 0$, and $\mathbb{E}[\mathbf{D}_\rho(X_i, Y_j)] \to \tilde{\rho}(F_1, F_2)$, in probability. The expected dissimilarity matrix based on ρ is

$$\mathbb{E}[\mathbf{D}_\rho(F_1, F_2)] = \begin{pmatrix} 0 & \tilde{\rho}(F_1, F_2) \\ \tilde{\rho}(F_1, F_2) & 0 \end{pmatrix}. \tag{14.20}$$

The only non–zero eigenvalue of the distance matrix $\mathbb{E}[\mathbf{D}_\rho]$ is $\lambda = \tilde{\rho}^2(F_1, F_2)$.

14.8.2 Dissimilarity Matrix δ

Consider the $N \times d$ combined data matrix of $\mathbb{Z}$. We define the mean and standard deviations of each row as follows,

$$\mathbb{Z} = \begin{bmatrix} Z_{11} & Z_{12} & \dots & Z_{1d} \\ \vdots & \vdots & \vdots & \vdots \\ Z_{N1} & Z_{N2} & \dots & Z_{Nd} \end{bmatrix} \Longrightarrow \begin{bmatrix} \bar{Z}_1^d & S_{\mathbf{Z}_1}^d \\ \vdots & \vdots \\ \bar{Z}_N^d & S_{\mathbf{Z}_N}^d \end{bmatrix},$$

where, the mean and standard deviation of the i-th observation $\boldsymbol{Z}_i, i = 1, \dots, N$ are

$$\bar{Z}_i^d = d^{-1} \sum_{j=1}^{d} \mathbf{Z}_{ij} \text{ and } S_{\mathbf{Z}_i}^d = \sqrt{\sum_{j=1}^{d} \left(\mathbf{Z}_{ij} - \bar{Z}_i^d\right)^2 / d}. \tag{14.21}$$

We similarly define $\bar{X}_i^d$ and $S_{\mathbf{X}_i}^d$, and $\bar{Y}_i^d$ and $S_{\mathbf{Y}_i}^d$ Note that $\bar{\mu}_1^d = \mathbb{E}(\bar{X}_i^d)$ and $(\sigma_1^d)^2 = \mathbb{E}(S_{\mathbf{X_i}}^d)^2$. Similarly, $\bar{\mu}_2^d = \mathbb{E}(\bar{Y}_i^d)$ and $(\sigma_2^d)^2 = \mathbb{E}(S_{\mathbf{Y_i}}^d)^2$. Modarres [231] defines the following modified distances based on the lower bound of the Euclidean distance and proposes the dissimilarity measure δ, where

$$\|\boldsymbol{Z}_i - \boldsymbol{Z}_l\|_D = \sqrt{(\bar{Z}_i^d - \bar{Z}_l^d)^2 + (S_{\boldsymbol{Z}_i}^d - S_{\boldsymbol{Z}_l}^d)^2}$$

$$\|\boldsymbol{Z}_j - \boldsymbol{Z}_l\|_D = \sqrt{(\bar{Z}_j^d - \bar{Z}_l^d)^2 + (S_{\boldsymbol{Z}_j}^d - S_{\boldsymbol{Z}_l}^d)^2},$$

$$\delta(\boldsymbol{Z}_i, \boldsymbol{Z}_j) = \frac{1}{(N-2)} \sum_{l \neq i,j} \big| \|\boldsymbol{Z}_i - \boldsymbol{Z}_l\|_D - \|\boldsymbol{Z}_j - \boldsymbol{Z}_l\|_D \big|. \tag{14.22}$$

Modarres [231] shows that as $d \to \infty$, under Assumptions 3.6.3, $\delta(\mathbf{X}_1, \mathbf{X}_2)$ and $\delta(\mathbf{Y}_1, \mathbf{Y}_2)$ converge in probability to 0, while $\delta(\mathbf{X}_1, \mathbf{Y}_1)$ converges in probability to

$$\tilde{\delta}(F_1, F_2) = |\sqrt{(\bar{\mu}_1^d - \bar{\mu}_2^d)^2 + (\sigma_1^d - \sigma_2^d)^2}| \geq 0, \tag{14.23}$$

which equals 0 if and only if $\bar{\mu}_1^d = \bar{\mu}_2^d$, and $\sigma_1^d = \sigma_2^d$.

Assumptions 3.6.3 and 3.6.3 are needed to establish convergence in distribution of the dissimilarity indices, as $d \to \infty$. When the variables are highly correlated C of (A3) is order $o(d^2)$. Hence, (A3) and (B3) place weak dependence among the variables. Assumption (A3) holds for stochastic processes with the ρ-mixing property and the process need not be stationary. When the observations are i.i.d., then (A3)-(B3) hold trivially.

Consider the $N \times N$ dissimilarity matrix of $\mathbb{Z}$ samples using δ,

$$\mathbf{D}_\delta(Z_i, Z_j) = [(\bar{Z}_i^d - \bar{Z}_j^d)^2 + (S_{\boldsymbol{Z}_i}^d - S_{\boldsymbol{Z}_j}^d)^2]^{1/2}, \quad i, j = 1, \dots, N. \tag{14.24}$$

As $d \to \infty$, $\mathbb{E}[\mathbf{D}_\delta(X_i, X_j)] \to 0$, $\mathbb{E}[\mathbf{D}_\delta(Y_i, Y_j)] \to 0$, and $\mathbb{E}[\mathbf{D}_\delta(X_i, Y_j)] \to \tilde{\delta}(F_1, F_2)$ in probability. The expected dissimilarity matrix based on δ is

$$\mathbb{E}[\mathbf{D}_\delta(F_1, F_2)] = \begin{pmatrix} 0 & \tilde{\delta}(F_1, F_2) \\ \tilde{\delta}(F_1, F_2) & 0 \end{pmatrix}. \tag{14.25}$$

The only non–zero eigenvalue of the distance matrix $\mathbb{E}[\mathbf{D}_\delta]$ is $\lambda = \tilde{\delta}^2(F_1, F_2)$. Wang and Modarres [348] present a clustering method for HDLSS data. Using the dissimilarity matrix they describe an algorithm that orders each row of the dissimilarity matrix to estimate the change points, which define cluster boundaries. They prove that the method achieves perfect clustering as $d \to \infty$ for a fixed sample size.

14.8.3 L_1-norm Euclidean Distance

Consider the expected L_1-norm distance matrix $\mathbb{E}[\mathbf{D}_E]$ as $d \to \infty$. Recall that $tr\,(\mathbf{\Sigma}_1)\,/d \to \sigma^2_{F_1}$ and $tr\,(\mathbf{\Sigma}_2)\,/d \to \sigma^2_{F_2}$ by Assumption (A1). As $d \to \infty$, $\mathbb{E}(d_x) \to r_1 = \sqrt{2}\sigma_{F_1}$, $\mathbb{E}(d_y) \to r_2 = \sqrt{2}\sigma_{F_2}$ and $\mathbb{E}(d_{xy}) \to h_{12}$ as $d \to \infty$. The asymptotic expected dissimilarity matrix based on L_1-norm is

$$\mathbb{E}[\mathbf{D}_E(\mathbb{X}, \mathbb{Y})] = \begin{pmatrix} \sqrt{2}\sigma_{F_1}(\mathbf{1}_{n_1 \times n_1} - \mathbf{I}_{n_1}) & h_{12}\,\mathbf{1}_{n_1 \times n_2} \\ h_{12}\,\mathbf{1}_{n_2 \times n_1} & \sqrt{2}\sigma_{F_2}(\mathbf{1}_{n_2 \times n_2} - \mathbf{I}_{n_2}) \end{pmatrix}. \tag{14.26}$$

As $d \to \infty$, $EN_E \to 2h_{12} - r_1 - r_2$ and $BG_E \to (h_{12} - r_1)^2 + (h_{12} - r_2)^2$.

14.9 Decomposition of Distance Matrices

A distance matrix possesses a rank that reflects the intrinsic dimensionality of the observations in $\mathbb{R}^N$. Although a distance matrix always has N eigenvalues, the number of distinct eigenvalues can provide valuable information regarding the structure, separation, or connectivity among groups of observations. Gower [123] introduced principal coordinate analysis to recover the original observation matrix via an eigenvalue decomposition of the distance matrix (see also Mardia, Kent, and Bibby [211]). This method is invariant under translations and rotations of the data. Let $\mathbf{I}_N$ denote the identity matrix of order N, and let $\mathbf{J}_N = \mathbf{1}_N\mathbf{1}'_N = \mathbf{1}_{N \times N}$ denote the $N \times N$ matrix of ones. To perform a singular value decomposition (SVD) of a distance matrix $\mathbf{D}$ for reconstructing the original observations, the following procedure is employed. Define the centering matrix

$$\mathbf{P} = \mathbf{I}_N - \frac{1}{N}\mathbf{J}_N,$$

which is both symmetric and idempotent. It then follows that the matrices $\mathbf{A}$ and $\mathbf{B}$ (defined in the usual PCA framework) share the same set of eigenvalues, facilitating the recovery of the original configuration.

Algorithm 14.9.1 Algorithm A: Decomposition of Distance Matrices

1: **Step 1:** Let

$$\mathbf{D} = (d_{ij}) \quad \text{and} \quad \mathbf{A} = (a_{ij}), \quad a_{ij} = -\frac{1}{2}d^2_{ij}, \quad i, j = 1, \ldots, N$$

2: **Step 2:** Let $\mathbf{B} = \mathbf{P}\,\mathbf{A}\,\mathbf{P}$, with eigenvalues $\mathbf{\Lambda} = \text{diag}(\lambda_1, \ldots, \lambda_N)$ and eigenvectors

$$\mathbf{\Gamma} = (\gamma_1, \ldots, \gamma_N)$$

Assuming $\lambda_1 \geq \lambda_2 \geq \ldots \geq \lambda_N$, we have $\mathbf{B} = \mathbf{\Gamma}\,\mathbf{\Lambda}\,\mathbf{\Gamma}'$.

3: **Step 3:** Given the $N \times d$ data matrix $\mathbf{X}$, we have

$$\mathbf{X}\mathbf{X}' = \mathbf{B} = \mathbf{\Gamma}\,\mathbf{\Lambda}\,\mathbf{\Gamma}'$$

Matrix $\mathbf{B}$ is symmetric and double-centered with row and column sums of zero. The rank of $\mathbf{B}$ is $N - 1$ since double-centering removes the mean component. Hence, it has $N - 1$ non–zero and one zero eigenvalues.

14.9.1 One Group Distances

The $n_1 \times n_1$ expected distance matrix is

$$\mathbb{E}(\mathbf{D}_E(\mathbb{X})) = r_1(\mathbf{1}_{n_1 \times n_1} - \mathbf{I}_{n_1}), \tag{14.27}$$

where $r_1 = \sqrt{2}\sigma_{F_1}$. This matrix has 0 on the main diagonal and r_1 on the off-diagonal. The eigenvalues of the constant distance matrix are $\lambda_i = \frac{1}{2}r_1^2$ for $i = 1, \ldots, n_1 - 1$ and $\lambda_{n_1} = 0$. The gaps or the difference between two successive eigenvalues are zero for constant distance matrices. The eigenvalues of the distance matrix and the gaps they generate can indicate multimodality. Modarres [233] considers the discrepancy in the eigenvalues of a distance matrix and offers a test for constancy of the distance matrix.

Example 14.9.1 *Suppose $\mathbf{X}_i$ is drawn independently from F_1 with mean $\boldsymbol{\mu}$ and covariance $\boldsymbol{\Sigma}_i$ for $i = 1, \ldots, N$. One can verify that $\mathbb{E}\|\mathbf{X}_i - \mathbf{X}_j\|^2 = \sum_{t=1}^d \lambda_{tij}$ where λ_{tij} is the t-th eigenvalue of $\boldsymbol{\Sigma}_i + \boldsymbol{\Sigma}_j$ and the observations are equidistant when $\boldsymbol{\Sigma}_i = \boldsymbol{\Sigma}_j = \boldsymbol{\Sigma}$ for $i \neq j = 1, \ldots, N$. Hence, the common distance is $\mathbb{E}\|\mathbf{X}_i - \mathbf{X}_j\|^2 = 2tr(\boldsymbol{\Sigma})$ and $\frac{1}{\sqrt{d}}\mathbb{E}\|\mathbf{X}_i - \mathbf{X}_j\|$ converges to $\sqrt{2}\sigma_F$ as $d \to \infty$, where F is the common DF.*

14.9.2 Two–Group Distances

Suppose that $\mathbf{X} = \{\mathbf{X}_i\}_{i=1}^{n_1}$ and $\mathbf{Y} = \{\mathbf{Y}_j\}_{j=1}^{n_2}$ are independent samples of d-dimension from DFs F_1 and F_2, with means and covariances $\boldsymbol{\mu}_1$, $\boldsymbol{\Sigma}_1$, $\boldsymbol{\mu}_2$ and $\boldsymbol{\Sigma}_2$, respectively. Consider the $N \times N$ matrix of within and between sample distances

$$\mathbb{E}[\mathbf{D}_E(\mathbb{X}, \mathbb{Y})] = \begin{pmatrix} r_1(\mathbf{1}_{n_1 \times n_1} - \mathbf{I}_{n_1}) & h_{12}\ \mathbf{1}_{n_1 \times n_2} \\ h_{12}\ \mathbf{1}_{n_2 \times n_1} & r_2(\mathbf{1}_{n_2 \times n_2} - \mathbf{I}_{n_2}) \end{pmatrix}. \tag{14.28}$$

The matrices $h_{12}\mathbf{1}_{n_1 \times n_2}$ and $h_{12}\mathbf{1}_{n_1 \times n_2}$ are of rank 1 because their entries are identical. Since the distance matrix is symmetric the eigenvalues will be real. The following theorem investigates the eigenvalues of (14.28). First, we show that with two groups of observations there are 3 main eigenvalues associated with the asymptotic distance matrix.

Theorem 14.9.1 *The eigenvalues of the distance matrix $\mathbf{B} = \mathbf{PAP}$, where $a_{ij} = -\frac{1}{2}d_{ij}^2$ in (14.28), are*

$$\lambda_1 = 0, \quad \lambda_2 = \frac{1}{2}r_1^2, \quad \lambda_3 = \frac{1}{2}r_2^2,$$

with multiplicities $n_1 - 1$ and $n_2 - 1$, respectively, and

$$\lambda_4 = \frac{1}{2N}[2n_1n_2h_{12}^2 - n_2(n_1 - 1)r_1^2 - n_1(n_2 - 1)r_2^2). \tag{14.29}$$

To keep $\lambda_4 > 0$ one must ensure $h_{12}^2 > \frac{n_1-1}{2n_1}r_1^2 + \frac{n_2-1}{2n_2}r_2^2$.

Notice when $r_1 = r_2 = h_{12}$ the distance matrix has a constant structure with only one positive eigenvalue given by λ_2. When $\lambda_4 < 0$ there are only two positive eigenvalues and the distance matrix can be summarized in two dimensions, and when $\lambda_4 > 0$ one can summarize the distance matrix in three dimensions. The number of distinct eigenvalues is often associated with the number of independent directions of variation in the data cloud. Note that the eigenvalues of $\mathbb{E}[\mathbf{D}_E(\mathbb{X}, \mathbb{Y})]$ as $d \to \infty$ are

$$\lambda_1 = 0, \quad \lambda_2 = \frac{1}{2}r_1^2 = \sigma_{F_1}^2, \quad \lambda_3 = \frac{1}{2}r_2^2 = \sigma_{F_2}^2$$

and

$$\lambda_4 = \frac{1}{N}(n_2\sigma_{F_1}^2 + n_1\sigma_{F_2}^2 + \frac{n_1 n_2}{d}\|\boldsymbol{\mu}_1 - \boldsymbol{\mu}_2\|^2),$$

as $d \to \infty$. The distance matrix has a block structure with four eigenvalues. The zero eigenvalue and its eigenvector correspond to a direction along which the matrix has no variation. Two eigenvalues λ_1 and λ_2 and their eigenvectors reveal the internal variation within each group and directions of those variations. The last eigenvalue reflects variation between the groups and its eigenvector provides the direction of that variation.

Theorem 14.9.2 *The expected L_1-norm distance matrix in* (14.28) *has a constant structure if and only if $F_1 = F_2$.*

When these is a shift in the observations of F_1 or F_2 the within sample IPDs remain invariant whereas h_{12} increases. Modarres [229] proves that distribution of h_{12} is stochastically larger than that of r_1 and r_2 under shift alternatives. As $d \to \infty$, $EN_E \to 2h_{12} - r_1 - r_2$ and $BG_E \to (h_{12} - r_1)^2 + (h_{12} - r_2)^2$. In each case, $F_1 = F_2$ holds if and only if $h_{12} = r_1 = r_2$. That is, when the expected L_1-norm distance matrix has a constant structure.

Example 14.9.2 *Suppose the distance matrix $\mathbb{E}(\mathbf{D}_E(\mathbb{X}))$ is affected by $k < N/2$ outliers. Suppose the first k observations are multiplied by some $c > 0$. The first k observations are regarded as outliers for large values of c. The distance matrix will k outliers is of the form*

$$\mathbb{E}[\mathbf{D}_E(\mathbb{X})] = \begin{pmatrix} c\, r_1(\mathbf{1}_{k\times k} - \mathbf{I}_k) & c\, h_{12}\, \mathbf{1}_{k\times N-k} \\ c\, h_{12}\, \mathbf{1}_{N-k\times k} & r_2(\mathbf{1}_{N-k\times N-k} - \mathbf{I}_{N-k}) \end{pmatrix}. \quad (14.30)$$

The matrix $\mathbb{E}(\mathbf{D}_E(\mathbb{X}))$ has the form of the matrix in (14.28) *with $n_1 = k$, $n_2 = N - k$, and $c\, r_1$ and $c\, h_{12}$ in place of r_1 and h_{12}, respectively. Hence, the eigenvalues of $\mathbb{E}(\mathbf{D}_E(\mathbb{X}))$ are*

$$\lambda_1 = 0 \quad \lambda_2 = \frac{1}{2}c^2 r_1^2, \quad \lambda_3 = \frac{1}{2}r_2^2,$$

with multiplicities $k - 1$ and $N - k - 1$, respectively, and

$$\lambda_4 = \frac{1}{2N}[2k(N-k)c^2 h_{12}^2 - (N-k)(k-1)c^2 r_1^2 - k(N-k-1)r_2^2]. \quad (14.31)$$

There is a discrepancy in the sizes of eigenvalues of a distance matrix that is contaminated with outliers. The energy and BG test statistics with samples of sizes k and $N - k$ become $EN_E = 2c\, h_{12} - c\, r_1 - r_2$ and $(c\, h_{12} - c\, r_1)^2 + (c\, h_{12} - r_2)^2$. When $c \neq 1$, both statistics are affected by outliers.

14.9.3 K Group Distances

Suppose $\mathbb{X} = \{\mathbf{X}_1, \ldots, \mathbf{X}_{n_1}\}$, $\mathbb{Y} = \{\mathbf{Y}_1, \ldots, \mathbf{Y}_{n_2}\}$, and $\mathbb{Z} = \{\mathbf{Z}_1, \ldots, \mathbf{Z}_{n_3}\}$ be i.i.d. samples from F_i with means $\boldsymbol{\mu}_i$ and covariance $\boldsymbol{\Sigma}_i$ for $i = 1, 2, 3$. Let $N = n_1 + n_2 + n_3$. The three groups are independent. To obtain the asymptotic behavior of the dissimilarity matrix δ we extend assumptions (B1-B3) to obtain

- (B1′) There exist constants $\sigma_{F_i}^2 > 0$, such that $tr\,(\boldsymbol{\Sigma}_i)\,/d \to \sigma_{F_i}^2$, and $\frac{1}{d}\|\boldsymbol{\mu}_i - \bar{\mu}_i \mathbf{1}_d)\|^2 \to \nu_i^2$, as $d \to \infty$, for $i = 1, 2, 3$.
- (B2′) The fourth moments of the components of $\mathbf{X} = (X_1, \ldots, X_d)'$ and $\mathbf{Y} = (Y_1, \ldots, Y_d)'$ and $\mathbf{Z} = (Z_1, \ldots, Z_d)'$ are uniformly bounded.
- (B3′) $T_i = \sum_{r=1}^{d} \sum_{t=r+1}^{d} \sigma_{irt}$ are of the order $o(d^2)$ for $i = 1, 2, 3$.

To obtain the asymptotic behavior of the dissimilarity matrix ρ we extend assumptions (A1-A3) to obtain

- (A1′) There exist constants $\sigma^2_{F_i}$ such that $tr(\mathbf{\Sigma}_i)/d \to \sigma^2_{F_i}$ as $d \to \infty$ for $i \neq j = 1, 2, 3$, and $\frac{1}{d}\|\boldsymbol{\mu}_i - \boldsymbol{\mu}_j\|^2 \to \nu^2_{ij}$.
- (A2′) The fourth moments of the components of $\mathbf{X} = (X_1, \ldots, X_d)'$ and $\mathbf{Y} = (Y_1, \ldots, Y_d)'$ and $\mathbf{Z} = (Z_1, \ldots, Z_d)'$ are uniformly bounded.
- (A3′) For $\mathbf{W} = \mathbf{X}_1 - \mathbf{X}_2$, $\mathbf{Y}_1 - \mathbf{Y}_2$ and $\mathbf{X}_1 - \mathbf{Y}_2$, let $C_{12} = \sum_{1 \le s \neq t \le d} \mathbb{C}or\left(\mathbf{W}_s^2, \mathbf{W}_t^2\right)$. For $\mathbf{W} = \mathbf{X}_1 - \mathbf{X}_2$, $\mathbf{Z}_1 - \mathbf{Z}_2$ and $\mathbf{X}_1 - \mathbf{Z}_2$, let $C_{13} = \sum_{1 \le s \neq t \le d} \mathbb{C}or\left(\mathbf{W}_s^2, \mathbf{W}_t^2\right)$. For $\mathbf{W} = \mathbf{Y}_1 - \mathbf{Y}_2$, $\mathbf{Z}_1 - \mathbf{Z}_2$ and $\mathbf{Y}_1 - \mathbf{Z}_2$, let $C_{23} = \sum_{1 \le s \neq t \le d} \mathbb{C}or\left(\mathbf{W}_s^2, \mathbf{W}_t^2\right)$. We assume C_{12}, C_{13} and C_{23} are of the order $o(d^2)$.

Recall that $\bar{\mu}_1^d = \mathbb{E}(\bar{X}_i^d)$ and $(\sigma_1^d)^2 = \mathbb{E}(S^d_{\mathbf{X}_i})^2$. Similarly, $\bar{\mu}_2^d = \mathbb{E}(\bar{Y}_i^d)$ and $(\sigma_2^d)^2 = \mathbb{E}(S^d_{\mathbf{Y}_i})^2$ and $\bar{\mu}_2^d = \mathbb{E}(\bar{Z}_i^d)$ and $(\sigma_3^d)^2 = \mathbb{E}(S^d_{\mathbf{Z}_i})^2$. As $d \to \infty$, $\delta(\mathbf{X}_1, \mathbf{X}_2)$, $\delta(\mathbf{Y}_1, \mathbf{Y}_2)$ and $\delta(\mathbf{Z}_1, \mathbf{Z}_2)$ converge in probability to 0, while the between group distances converge in probability to

$$\tilde{\delta}(F_i, F_j) = |\sqrt{(\bar{\mu}_i^d - \bar{\mu}_j^d)^2 + (\sigma_i^d - \sigma_j^d)^2}| \geq 0, \tag{14.32}$$

which equal 0 if and only if $\bar{\mu}_1^d = \bar{\mu}_2^d = \bar{\mu}_3^d$, and $\sigma_1^d = \sigma_2^d = \sigma_3^d$. Denoting $\tilde{\delta}^2(F_i, F_j)$ with $\tilde{\delta}^2_{ij}$, the expected dissimilarity matrix based on δ as $d \to \infty$ for three groups is

$$\mathbb{E}[\mathbf{D}_\delta(F_1, F_2, F_3)] = \begin{pmatrix} 0 & \tilde{\delta}_{12} & \tilde{\delta}_{13} \\ \tilde{\delta}_{12} & 0 & \tilde{\delta}_{23} \\ \tilde{\delta}_{13} & \tilde{\delta}_{23} & 0 \end{pmatrix}. \tag{14.33}$$

For the dissimilarity index ρ, as $d \to \infty$, $\rho(\mathbf{X}_1, \mathbf{X}_2)$, $\rho(\mathbf{Y}_1, \mathbf{Y}_2)$ and $\rho(\mathbf{Z}_1, \mathbf{Z}_2)$ converge in probability to 0. Let $h_{ij} = \sqrt{\sigma^2_{F_i} + \sigma^2_{F_j} + \nu^2_{ij}}$ for $(i,j) = \{(1,2), (1,3), (2,3)\}$. As $d \to \infty$ we have

$$\tilde{\rho}(F_i, F_j) = \frac{1}{N-2}[(n_i - 1)|\sqrt{2}\sigma_{F_i} - h_{ij}| + (n_j - 1)|h_{ij} - \sqrt{2}\sigma_{F_j}|] \geq 0, \tag{14.34}$$

which are 0 if and only if $\sigma_{F_1} = \sigma_{F_2} = \sigma_{F_3}$ and $\nu^2_{12} = \nu^2_{13} = \nu^2_{23} = 0$. Denoting $\tilde{\rho}^2(F_i, F_j)$ with $\tilde{\rho}^2_{ij}$, the expected dissimilarity matrix based on ρ as $d \to \infty$ for three groups is

$$\mathbb{E}[\mathbf{D}_\rho(F_1, F_2, F_3)] = \begin{pmatrix} 0 & \tilde{\rho}_{12} & \tilde{\rho}_{13} \\ \tilde{\rho}_{12} & 0 & \tilde{\rho}_{23} \\ \tilde{\rho}_{13} & \tilde{\rho}_{23} & 0 \end{pmatrix}. \tag{14.35}$$

The expected L_1-norm distance matrix for 3 groups is

$$\mathbb{E}[\mathbf{D}_E(\mathbb{X}, \mathbb{Y}, \mathbb{Z})] = \begin{pmatrix} r_1(\mathbf{1}_{n_1 \times n_1} - \mathbf{I}_{n_1}) & h_{12}\,\mathbf{1}_{n_1 \times n_2} & h_{13}\,\mathbf{1}_{n_1 \times n_3} \\ h_{12}\,\mathbf{1}_{n_2 \times n_1} & r_2(\mathbf{1}_{n_2 \times n_2} - \mathbf{I}_{n_2}) & h_{23}\,\mathbf{1}_{n_2 \times n_3} \\ h_{13}\,\mathbf{1}_{n_3 \times n_1} & h_{23}\,\mathbf{1}_{n_3 \times n_2} & r_3(\mathbf{1}_{n_3 \times n_3} - \mathbf{I}_{n_3}), \end{pmatrix} \tag{14.36}$$

where $r_i = \sqrt{2}\sigma_{F_i}$, for $i = 1, 2, 3$ as $d \to \infty$.

Theorem 14.9.3 *The eigenvalues of the distance matrix* $\mathbf{B} = \mathbf{PAP}$, *where* $a_{ij} = -\frac{1}{2}d^2_{ij}$ *in* (14.36), *are zero,* $\lambda_i = \frac{1}{2}r_i^2$, *with multiplicities* $n_i - 1$, *for* $i = 1, 2, 3$. *The other two roots are the solution of* $N\lambda^2 + b\lambda + c = 0$, *where*

$$b = \frac{1}{2}\sum_{i=1}^{3}(n_i - 1)(N - n_i)r_i^2 - n_1 n_2 h_{12}^2 - n_1 n_3 h_{13}^2 - n_2 n_3 h_{23}^2, \tag{14.37}$$

$$\begin{aligned}c =& \frac{1}{4}\left[n_3(n_1-1)(n_2-1)r_1^2r_2^2 + n_2(n_1-1)(n_3-1)r_1^2r_3^2 + n_1(n_2-1)(n_3-1)r_2^2r_3^2\right] \\ &- \frac{1}{2}\left[n_2n_3(n_1-1)r_1^2h_{23}^2 + n_1n_3(n_2-1)r_2^2h_{13}^2 + n_1n_2(n_3-1)r_3^2h_{12}^2\right] \\ &- \frac{1}{2}n_1n_2n_3\left[h_{12}^2h_{13}^2 + h_{12}^2h_{23}^2 + h_{13}^2h_{23}^2 + \frac{1}{2}(h_{12}^4 + h_{13}^4 + h_{23}^4)\right].\end{aligned} \tag{14.38}$$

14.10 Eigenvalues of Euclidean Matrices

We study the eigenvalues of group-structured squared distance matrix $\mathbf{D}_E$ and its centered counterparts, motivated by principal coordinates analysis. The main questions are

1. How do within-group distances determine eigenvalues?
2. How does between-group interaction reduce to a smaller eigenvalue problem?
3. Can we compute explicit patterns for any number of groups K using trace, determinant, and principal minors?

Distance Matrix $\mathbf{D}_E$

Let $\mathbf{D} = (d_{ij}) \in \mathbb{R}^{N\times N}$ denote the distance matrix of N points partitioned into K groups of sizes $n_1, \ldots, n_K$:

$$d_{ij} = \begin{cases} 0, & i = j, \\ r_\ell, & i \neq j \text{ in group } \ell, \\ h_{\ell m}, & i \in \ell, j \in m, \ell \neq m \end{cases}$$

Let $\mathbf{1} \in \mathbb{R}^N$ be the vector of ones and define

$$\mathbf{A} = -\frac{1}{2}\mathbf{D}^{(2)}, \qquad \mathbf{B} = \mathbf{PAP}, \qquad \mathbf{P} = \mathbf{I} - \frac{1}{N}\mathbf{11}'.$$

Within-Group Eigenvalues

For a group i, vectors $\mathbf{x}$ supported on group i with $\mathbf{1}'\mathbf{x} = 0$ must satisfy

$$\mathbf{A}_i = -\frac{1}{2}r_i^2(\mathbf{11}' - \mathbf{I}) \implies \mathbf{A}_i\mathbf{x} = \frac{1}{2}r_i^2\mathbf{x}.$$

Centering does not change $\mathbf{x}$, so that $\mathbf{Bx} = \frac{1}{2}r_i^2\mathbf{x}$, with multiplicity $n_i - 1$. Hence,

$$\lambda_{\text{within}} = \frac{1}{2}r_i^2, \ n_i - 1 \text{ times per group } i$$

Between-Group Reduction Matrix

Consider the distance matrix $\mathbf{D}_E$ in (14.42). Its reduced $K \times K$ matrix is

$$\mathbf{M}_{ij}^{(A)} = \begin{cases} -\frac{1}{2}r_i^2(n_i - 1), & i = j, \\ -\frac{1}{2}h_{ij}^2 n_j, & i \neq j \end{cases}$$

with weights $w_i = n_i/N$, $\mathbf{w} = [w_1, \ldots, w_K]'$, and centered matrices

$$\widetilde{\mathbf{M}} = \mathbf{P}_K \mathbf{M}^{(A)} \mathbf{P}_K, \quad \mathbf{P}_K = \mathbf{I} - \mathbf{1}\mathbf{w}'$$

The eigenvalues of $\widetilde{\mathbf{M}}$ give the between-group eigenvalues.

Trace, Determinant, and Principal Minors

For small K, we can explicitly write the characteristic polynomial in terms of trace, determinant, and sums of principal minors, $\det(\widetilde{\mathbf{M}} - \lambda\mathbf{I}) = 0$.

Example 14.10.1 *Consider* $K = 3$ *and let* $\widetilde{\mathbf{M}} = [a_{ij}]$ *where* $a_{ii} = -\frac{1}{2}r_i^2(n_i - 1)$ *and* $a_{ij} = -\frac{1}{2}h_{ij}^2 n_j, i \neq j$. *Characteristic polynomial is*

$$\det(\widetilde{\mathbf{M}} - \lambda\mathbf{I}) = 0 \Rightarrow \lambda^3 - (tr)\lambda^2 + (\textit{sum of } 2 \times 2 \textit{ principal minors})\lambda - \det = 0$$

The trace is $tr(\widetilde{\mathbf{M}}) = a_{11} + a_{22} + a_{33}$. *To find the sum of* 2×2 *principal minors, we choose each pair of rows and columns and find*

$$M_{12} = a_{11}a_{22} - a_{12}a_{21}, \quad M_{13} = a_{11}a_{33} - a_{13}a_{31}, \quad M_{23} = a_{22}a_{33} - a_{23}a_{32} \tag{14.39}$$

So that the sum of 2×2 *principal minors is* $M_{12} + M_{13} + M_{23}$. *The determinant is*

$$\det(\widetilde{\mathbf{M}}) = a_{11}a_{22}a_{33} + a_{12}a_{23}a_{31} + a_{13}a_{21}a_{32} - a_{11}a_{23}a_{32} - a_{12}a_{21}a_{33} - a_{13}a_{22}a_{31}.$$

We solve the characteristic polynomial $\lambda^3 - (tr)\lambda^2 + (\textit{sum of } 2 \times 2 \textit{ minors})\lambda - \det = 0$. *One zero eigenvalue arises from centering. The remaining two eigenvalues are obtained by factoring out* λ.

14.10.1 General K Case

Let $\widetilde{\mathbf{M}}$ be $K \times K$ centered reduced matrix. The characteristic polynomial:

$$\det(\widetilde{\mathbf{M}} - \lambda\mathbf{I}) = \lambda^K - s_1\lambda^{K-1} + s_2\lambda^{K-2} - s_3\lambda^{K-3} + \cdots + (-1)^K s_K = 0$$

where

- $s_1 = \text{tr}(\widetilde{\mathbf{M}})$ (sum of eigenvalues)
- $s_2 = \sum_{\text{all } 2 \times 2 \text{ principal minors}}$ (sum of products of eigenvalues in pairs)
- $s_3 = \sum_{\text{all } 3 \times 3 \text{ principal minors}}$ (sum of triple products)
- $s_K = \det(\widetilde{\mathbf{M}})$ (product of all eigenvalues)

Centering implies one eigenvalue is zero. The remaining $K-1$ eigenvalues are determined by $s_1, \ldots, s_{K-1}$ (traces of powers, sums of minors, and determinant).

Example 14.10.2 *Let us consider* $K = 4$ *groups with sizes* n_1, n_2, n_3, n_4 *and distances* r_i *within groups,* h_{ij} *between groups. The centered reduced matrix is:*

$$\widetilde{\mathbf{M}} = [a_{ij}] \quad a_{ii} = -\frac{1}{2}r_i^2(n_i - 1), \quad a_{ij} = -\frac{1}{2}h_{ij}^2 n_j \ (i \neq j)$$

The characteristic polynomial is $\lambda^4 - s_1\lambda^3 + s_2\lambda^2 - s_3\lambda + s_4 = 0$, *where*

- $s_1 = tr(\widetilde{\mathbf{M}}) = a_{11} + a_{22} + a_{33} + a_{44}$
- $s_2 = \sum_{2 \times 2 \; principal \; minors}$
- $s_3 = \sum_{3 \times 3 \; principal \; minors}$
- $s_4 = \det(\widetilde{\mathbf{M}})$

Let $s_2 = M_{12} + M_{13} + M_{14} + M_{23} + M_{24} + M_{34}$ *where the* 2×2 *principal minors, using all pairs of rows/columns, are*

$$\begin{aligned} M_{12} &= a_{11}a_{22} - a_{12}a_{21}, \quad & M_{13} &= a_{11}a_{33} - a_{13}a_{31}, \quad & M_{14} &= a_{11}a_{44} - a_{14}a_{41}, \\ M_{23} &= a_{22}a_{33} - a_{23}a_{32}, \quad & M_{24} &= a_{22}a_{44} - a_{24}a_{42}, \quad & M_{34} &= a_{33}a_{44} - a_{34}a_{43}, \end{aligned}$$

The 3×3 *principal minors are* M_{123}*,* M_{124}*,* M_{134}*,* M_{234} *and*

$$s_3 = M_{123} + M_{124} + M_{134} + M_{234}$$

Let S_4 *is the set of all permutations of* $\{1, 2, 3, 4\}$*. The determinant is*

$$s_4 = \det(\widetilde{\mathbf{M}}) = \sum_{\sigma \in S_4} sign(\sigma) \prod_{i=1}^{4} a_{i,\sigma(i)}.$$

We solve the characteristic polynomial $\lambda^4 - s_1\lambda^3 + s_2\lambda^2 - s_3\lambda + s_4 = 0$*. Centering implies one eigenvalue is zero. Factoring* λ*, we solve the cubic for the remaining three between-group eigenvalues.*

This procedure can be extended to any K using the same symmetric polynomial approach. For symbolic or numeric computation, all minors can be computed easily. The eigenvalues correspond to variances along coordinates representing between-group variation. Within-group eigenvalues are $\frac{1}{2}r_i^2$, multiplicity $n_i - 1$. Zero eigenvalue corresponds to the overall mean. Between-group eigenvalues are computed from symmetric polynomials of $\widetilde{\mathbf{M}}$. All coefficients are explicitly in terms of roots of characteristic polynomial of $\widetilde{\mathbf{M}}$, expressible via trace, sums of principal minors, determinant.

Example 14.10.3 *Consider the symmetric case with equal group sizes and distances,* $n_i = n$*,* $r_i = r$*,* $h_{ij} = h$ *for* $i \neq j = 1, \ldots K$*. The* $K \times K$ *reduced matrix* $\mathbf{M}^{(A)}$ *has* $-\frac{1}{2}(n-1)r^2$ *on the main-diagonal and off-diagonal is* $-\frac{1}{2}nh^2$*. Centering uses weights* $w_i = n/N = 1/K$*. The centered reduced matrix is*

$$\widetilde{\mathbf{M}} = (\mathbf{I} - \mathbf{1}w')\mathbf{M}^{(A)}(\mathbf{I} - w\mathbf{1}')$$

The eigenvalues of the centered matrix are obtained as follows. We have an eigenvector along $\mathbf{1}$ *(uniform across groups) with* $\lambda_0 = 0$*. The eigenvectors orthogonal to* $\mathbf{1}$ *sum to zero across groups). Let* $\mathbf{v}$ *satisfy* $\sum_i v_i = 0$*. Action of* $\mathbf{M}^{(A)}$ *on such* $\mathbf{v}$ *is*

$$\begin{aligned} (\mathbf{M}^{(A)}\mathbf{v})_i &= -\frac{1}{2}(n-1)r^2 v_i - \frac{1}{2}nh^2 \sum_{j \neq i} v_j \\ &= -\frac{1}{2}(n-1)r^2 v_i - \frac{1}{2}nh^2 \Big(- v_i \Big) \quad (since \sum_j v_j = 0) \\ &= -\frac{1}{2}(n-1)r^2 v_i + \frac{1}{2}nh^2 v_i = \frac{1}{2}(nh^2 - (n-1)r^2)v_i \end{aligned}$$

The multiplicity is $K-1$ (all orthogonal directions to uniform vector). The within-group eigenvalues are $\lambda_{within} = \frac{1}{2}r^2$, with multiplicity $n-1$ per block. The total number is $K(n-1)$. Hence,

$$\textit{Eigenvalues of } \mathbf{B} = \begin{cases} \frac{1}{2}r^2, & \textit{multiplicity } K(n-1) \text{ (within-block)} \\ 0, & \textit{multiplicity } 1 \text{ (uniform vector)} \\ \frac{1}{2}(nh^2-(n-1)r^2), & \textit{multiplicity } K-1 \text{ (between-block)} \end{cases}$$

The pattern is extremely simple due to symmetry. Within-group variation depends only on r^2. Between-group variation depends on the difference between scaled h^2 and r^2. This formula generalizes for any number of groups K with equal sizes and distances.

Example 14.10.4 *Consider three groups with sizes n_1, n_2, n_3, within-group distances r_1, r_2, r_3, and between-group distances h_{12}, h_{13}, h_{23}. The centered reduced matrix is:*

$$\widetilde{\mathbf{M}} = [a_{ij}], \quad a_{ii} = -\frac{1}{2}r_i^2(n_i-1), \quad a_{ij} = -\frac{1}{2}h_{ij}^2 n_j$$

The cubic characteristic polynomial is $\det(\widetilde{\mathbf{M}} - \lambda\mathbf{I}) = \lambda^3 - s_1\lambda^2 + s_2\lambda - s_3 = 0$ *where*

$$\begin{aligned} s_1 &= tr(\widetilde{\mathbf{M}}) = a_{11} + a_{22} + a_{33} \\ s_2 &= \textit{sum of all } 2\times 2 \textit{ principal minors} = M_{12} + M_{13} + M_{23} \\ s_3 &= \det(\widetilde{\mathbf{M}}) \end{aligned}$$

with 2×2 minors,

$$M_{12} = a_{11}a_{22} - a_{12}a_{21}, \quad M_{13} = a_{11}a_{33} - a_{13}a_{31}, \quad M_{23} = a_{22}a_{33} - a_{23}a_{32}. \tag{14.40}$$

Centering implies one eigenvalue is zero ($\lambda_0 = 0$). Factor λ out. Now we only need to solve the quadratic $\lambda^2 - s_1\lambda + s_2 = 0$. Quadratic formula gives $\lambda = \frac{s_1 \pm \sqrt{s_1^2 - 4s_2}}{2}$. These are the two non-zero between-group eigenvalues. All coefficients s_1, s_2 are expressed explicitly in terms of r_i, h_{ij}, n_i,

$$\begin{aligned} s_1 &= -\frac{1}{2}\left[(n_1-1)r_1^2 + (n_2-1)r_2^2 + (n_3-1)r_3^2\right] \\ s_2 &= \sum_{i<j}(a_{ii}a_{jj} - a_{ij}a_{ji}) = M_{12} + M_{13} + M_{23} \\ M_{12} &= \frac{1}{4}\left[r_1^2(n_1-1)r_2^2(n_2-1) - n_1n_2h_{12}^4\right], \\ M_{13} &= \frac{1}{4}\left[r_1^2(n_1-1)r_3^2(n_3-1) - n_1n_3h_{13}^4\right], \\ M_{23} &= \frac{1}{4}\left[r_2^2(n_2-1)r_3^2(n_3-1) - n_2n_3h_{23}^4\right]. \end{aligned}$$

Then, $s_2 = M_{12} + M_{13} + M_{23}$. Hence,

$$\textit{Eigenvalues of } \mathbf{B} = \begin{cases} \lambda_{within} = \frac{1}{2}r_i^2, & \textit{multiplicity } n_i - 1 \textit{ per group} \\ \lambda_0 = 0, & \textit{uniform vector} \\ \lambda_{1,2} = \frac{s_1 \pm \sqrt{s_1^2 - 4s_2}}{2}, & \textit{between-group directions} \end{cases}$$

This fully expresses all eigenvalues in terms of r_i, h_{ij}, n_i for $K = 3$.

Example 14.10.5 *Consider three groups with sizes n_1, n_2, n_3, within-group distances r_1, r_2, r_3, and between-group distances h_{12}, h_{13}, h_{23}. The centered reduced matrix is*

$$\widetilde{\mathbf{M}} = [a_{ij}], \quad a_{ii} = -\frac{1}{2}r_i^2(n_i - 1), \quad a_{ij} = -\frac{1}{2}h_{ij}^2 n_j$$

From previous derivations, vectors supported on a single group and summing to zero have eigenvalues,

$$\lambda_i = \frac{1}{2}r_i^2, \quad \textit{with multiplicity } n_i - 1, \quad i = 1, 2, 3$$

These correspond to within-group variation. Centering ensures one eigenvalue is exactly zero (uniform vector across groups), $\lambda_0 = 0$. The remaining two eigenvalues arise from group interactions. They are the roots of a quadratic: $N\lambda^2 + b\lambda + c = 0$ where b and c are given in Theorem 14.9.3. Here $N = n_1 + n_2 + n_3$. This fully accounts for all weights and interactions in the centered matrix. The two between-group eigenvalues are

$$\lambda_{4,5} = \frac{-b \pm \sqrt{b^2 - 4Nc}}{2N}$$

These are exact, including effects of n_i and h_{ij}. Hence,

$$\textit{Eigenvalues of } \mathbf{B} = \begin{cases} \frac{1}{2}r_1^2, & \textit{multiplicity } n_1 - 1 \\ \frac{1}{2}r_2^2, & \textit{multiplicity } n_2 - 1 \\ \frac{1}{2}r_3^2, & \textit{multiplicity } n_3 - 1 \\ 0, & \textit{uniform vector} \\ \lambda_{4,5} = \textit{roots of } N\lambda^2 + b\lambda + c & \textit{(between-group)} \end{cases}$$

This formula exactly matches the results in (14.37)–(14.38). Centering reduces the cubic to a quadratic, simplifying computation. The multiplicities clearly separate within-group, zero, and between-group directions. This method generalizes: for $K > 3$, the remaining eigenvalues are roots of a degree-$(K-1)$ polynomial after factoring out $\lambda = 0$.

14.10.2 Eigenvectors for General K

We now describe the full set of eigenvectors of the matrix $\mathbf{B} = \mathbf{PAP}$ for an arbitrary number of groups K. The structure mirrors the eigenvalue decomposition and splits into three distinct families, within-group eigenvectors, the global (zero) eigenvector, and between-group eigenvectors.

For within-block eigenvectors, fix a group i of size n_i. Consider vectors supported only on block i, $\mathbf{x}' = (0\ldots, \mathbf{x}_i \ldots 0)$, with $\mathbf{1}'_{n_i}\mathbf{x}_i = 0$. That is, $\mathbf{x}_i \in \mathbb{R}^{n_i}$ satisfies $\sum_{j \in \text{group } i} x_j = 0$. From previous derivations $\mathbf{Bx} = \frac{1}{2}r_i^2\mathbf{x}$. Thus, within-group eigenvectors $\mathbf{x}$ supported on block i, $\mathbf{1}'\mathbf{x} = 0$, $\lambda = \frac{1}{2}r_i^2$. Dimension of $n_i - 1$ per block , so that the total is $\sum_{i=1}^{K}(n_i - 1)$. The global eigenvector (zero eigenvalue) corresponds to the vector of all ones, $\mathbf{1}' = (\mathbf{1}_{n_1} \cdots \mathbf{1}_{n_2} \cdots \mathbf{1}_{n_K})$ and satisfies $\mathbf{P1} = 0 \quad \Rightarrow \quad \mathbf{B1} = 0$. Thus, $\lambda = 0$ with eigenvector of $\mathbf{1}$. This corresponds to the centering constraint (removal of the mean).

The between-block eigenvectors are the most important non-trivial directions. We consider vectors that are constant within each block, $\mathbf{x}' = (c_1\mathbf{1}_{n_1} \cdots c_2\mathbf{1}_{n_2} \cdots c_K\mathbf{1}_{n_K})$. Define the reduced vector, $\mathbf{c} = (c_1, \ldots, c_K)'$. To reduce to $K \times K$ problem, applying $\mathbf{A}$ gives

$$\mathbf{Ax} \longleftrightarrow \mathbf{M}^{(A)}\mathbf{c}$$

where

$$\mathbf{M}_{ij}^{(A)} = \begin{cases} -\frac{1}{2}r_i^2(n_i - 1), & i = j, \\ -\frac{1}{2}h_{ij}^2 n_j, & i \neq j. \end{cases}$$

Because of $\mathbf{P}$, admissible vectors must satisfy

$$\mathbf{1}'\mathbf{x} = 0 \iff \sum_{i=1}^{K} n_i c_i = 0.$$

Equivalently, $\mathbf{w}'\mathbf{c} = 0, \quad w_i = \frac{n_i}{N}$. The between-group eigenvectors are obtained by solving $\widetilde{\mathbf{M}}\mathbf{c} = \lambda\mathbf{c}, \quad$ with $\mathbf{w}'\mathbf{c} = 0$, where

$$\widetilde{\mathbf{M}} = (\mathbf{I} - \mathbf{1}\mathbf{w}')\mathbf{M}^{(A)}(\mathbf{I} - \mathbf{w}\mathbf{1}').$$

Lifting back to $\mathbb{R}^N$, each eigenvector $\mathbf{c}$ produces a full eigenvector, $\mathbf{x}' = (c_1\mathbf{1}_{n_1}, \ldots, c_K\mathbf{1}_{n_K})$ The total dimension is

$$\sum_{i=1}^{K}(n_i - 1) + 1 + (K - 1) = \left(\sum_{i=1}^{K} n_i\right) = N.$$

Thus all eigenvectors are accounted for.

Assume $n_i = n$, $i = r$, and $h_{ij} = h$. Then, all weights are equal, $w_i = \frac{1}{K}$. The constraint becomes $\sum_{i=1}^{K} c_i = 0$.

A convenient orthogonal basis for this subspace is given by contrast vectors

$$\begin{aligned} \mathbf{c}^{(1)} &= (1, -1, 0, \ldots, 0), \\ \mathbf{c}^{(2)} &= (1, 1, -2, 0, \ldots, 0), \\ \mathbf{c}^{(3)} &= (1, 1, 1, -3, 0, \ldots, 0), \\ \mathbf{c}^{(K-1)} &= (1, 1, \ldots, 1, -(K - 1)). \end{aligned}$$

Each satisfies $\sum_{i=1}^{K} c_i = 0$.

Lifting to full eigenvectors, each contrast vector gives $\mathbf{x}^{(\ell)} = (c_1^{(\ell)}\mathbf{1}_n, c_2^{(\ell)}\mathbf{1}_n, \ldots, c_K^{(\ell)}\mathbf{1}_n)'$. All these vectors share the same eigenvalue $\lambda = \frac{1}{2}\left(nh^2 - (n-1)r^2\right)$ with multiplicity $K - 1$.

The within-group eigenvectors capture variation inside clusters. The between-group eigenvectors capture differences between cluster means. In the symmetric case, these reduce to simple contrasts, analogous to ANOVA contrasts. The zero eigenvector corresponds to translation invariance removed by centering.

14.11 Testing the Equality of $K \geq 2$ Distributions

Let $\mathbb{X}^{(k)} = \{\mathbf{X}_i^{(k)}\}_{i=1}^{n_i}$ denote the i.i.d. observations of the k-th group for $k = 1, \ldots, K$ and $N = \sum_{i=1}^{K} n_i$. We are interested in testing

$$H_0 : F_1 = F_2 = \ldots = F_K, \tag{14.41}$$

against $H_a : F_i \neq F_j$ for at least one $i \neq j = 1, \ldots, K$.

14.11.1 L_1-norm Distance Matrix

The expected L_1-norm matrix $\mathbf{D}_E = \mathbb{E}[\mathbf{D}_E(\mathbb{X}^{(1)}, \ldots, \mathbb{X}^{(K)})]$ is given by

$$\mathbf{D}_E = \begin{pmatrix} r_1(\mathbf{1}_{n_1\times n_1} - \mathbf{I}_{n_1}) & h_{12}\ \mathbf{1}_{n_1\times n_2} & h_{13}\ \mathbf{1}_{n_1\times n_3} & \cdots & h_{1K}\ \mathbf{1}_{n_1\times n_K} \\ h_{12}\ \mathbf{1}_{n_2\times n_1} & r_2(\mathbf{1}_{n_2\times n_2} - \mathbf{I}_{n_2}) & h_{23}\ \mathbf{1}_{n_2\times n_3} & \cdots & h_{2K}\ \mathbf{1}_{n_2\times n_K} \\ \vdots & \vdots & \cdots & \vdots & \vdots \\ h_{1K}\ \mathbf{1}_{n_K\times n_1} & h_{2K}\ \mathbf{1}_{n_K\times n_2} & h_{3K}\ \mathbf{1}_{n_K\times n_3} & \cdots & r_K(\mathbf{1}_{n_K\times n_K} - \mathbf{I}_{n_K}) \end{pmatrix} \tag{14.42}$$

where $r_i = \sqrt{2}\sigma_{F_i}$, for $i = 1, 2, \ldots, K$ and $h_{ij} = \sqrt{\sigma^2_{F_i} + \sigma^2_{F_j} + \nu^2_{ij}}$ as $d \to \infty$.

Theorem 14.11.1 *The expected L_1-norm distance matrix $\mathbf{D}_E$ in* (14.42) *has a constant structure if and only if $F_1 = F_2 = \ldots = F_K$.*

We are interested in testing whether within and between distances are the same so that the distance matrix has a constant structure with common value d. Hence, we test the hypothesis

$$K_0 : \mathbf{D}_E = \mathbb{E}(d)(\mathbf{J}_N - \mathbf{I}_N) \text{ against } K_a : \mathbf{D}_E \neq \mathbb{E}(d)(\mathbf{J}_N - \mathbf{I}_N). \tag{14.43}$$

We propose three test statistics for testing K_0. The first test is based on the internal structure of the distance matrix as follows. The average of the off-diagonal elements in the kth row of $\mathbf{D}_E$ and the overall average of the off-diagonal elements of $\mathbf{D}_E$ are denoted with

$$\bar{d}_k = \frac{1}{N-1}\sum_{i\neq k=1}^{N} d_{ij}, \text{ and } \quad \bar{d} = \frac{2}{N(N-1)}\sum_{i=1}^{N}\sum_{j=i+1}^{N} d_{ij}, \tag{14.44}$$

respectively. We reject K_0 in favor of H_a for large values of

$$S = \sum_{i=1}^{N}\sum_{j=i+1}^{N-1}(d_{ij} - \bar{d})^2 - \sum_{k=1}^{N}(\bar{d}_k - \bar{d})^2. \tag{14.45}$$

Since the $N \times N$ expected L_1-norm distance matrix with K groups has 0 eigenvalue with multiplicity 1, $n_i - 1$ eigenvalues $\lambda_i = \frac{1}{2}r_i^2$ for $i = 1, \ldots, K$, there are $K - 1$ other roots of the characteristic equation $|\mathbf{B} - \lambda\mathbf{I_N}| = 0$.

When $K = 2$ or $K = 3$ another test statistic compares the plug-in estimate of the adjacent theoretical eigenvalues of $\mathbb{E}[\mathbf{D}_E(\mathbb{X}, \mathbb{Y})]$ given in Theorems 14.9.1 and 14.9.3. Suppose $\lambda_{(1)} \leq \lambda_{(2)} \leq \lambda_{(3)} \leq \lambda_{(4)} \leq \lambda_{(5)}$. The test is rejected for large values of

$$E = \sum_{i=2}^{T}(\lambda_{(i+1)} - \lambda_{(i)})^2, \tag{14.46}$$

where $T = 3$ for $K = 2$ and when $T = 4$ for $K = 3$.

The next test is based on the observation that in a constant distance matrix the gaps between successive eigenvalues is zero. We know that the eigenvalues of the constant distance matrix in (14.27) are $\lambda_1 = 0$ and $\lambda_i = \frac{1}{2}r_1^2$ for $i = 2, \ldots, N$ with multiplicity N-1. We construct

$$\Delta = \sum_{i=3}^{N}(\hat{\lambda}_{(i)} - \hat{\lambda}_{(i-1)})^2, \tag{14.47}$$

where $\hat{\lambda}_{(1)} = 0 < \hat{\lambda}_{(2)} \leq \hat{\lambda}_{(3)} \leq \ldots \leq \hat{\lambda}_{(N)}$ are the ordered eigenvalues of the data matrix. We reject K_0 for large values of Δ. When $K = 2$, Δ and E are the same.

Example 14.11.1 *Let there be k groups, each of size n, total $N = kn$. Assume within group scale $\sigma_F = 1$, so that $r = \sqrt{2}$, and between-group separation $h = \sqrt{2+\nu^2}$. The k distributions have equal scale and equidistant means. The expected L_1-distance matrix is the $N \times N$ matrix*

$$\boldsymbol{D}_E = \begin{pmatrix} r(\mathbf{1}_n\mathbf{1}_n' - \boldsymbol{I}_n) & h\,\mathbf{1}_{n\times n} & \dots & h\,\mathbf{1}_{n\times n} \\ h\,\mathbf{1}_{n\times n} & r(\mathbf{1}_n\mathbf{1}_n' - \boldsymbol{I}_n) & \dots & h\,\mathbf{1}_{n\times n} \\ \vdots & \vdots & \dots & \vdots \\ h\,\mathbf{1}_{n\times n} & h\,\mathbf{1}_{n\times n} & \dots & r(\mathbf{1}_n\mathbf{1}_n' - \boldsymbol{I}_n) \end{pmatrix}.$$

We form the classical MDS kernel

$$\boldsymbol{B} = -\frac{1}{2}\boldsymbol{P}(\boldsymbol{D}_E \cdot \boldsymbol{D}_E)\boldsymbol{P}, \qquad \boldsymbol{P} = \boldsymbol{I}_N - \frac{1}{N}\mathbf{1}_N\mathbf{1}_N'.$$

The eigenvalues of $\boldsymbol{B}$ are $\frac{1}{2}r^2$ for within-group distances with multiplicity $k(n-1)$, $\frac{1}{k}\big(n\,h^2 - (n-1)r^2\big)$ for between-group contrasts with multiplicity $k-1$, and zero for centering with multiplicity 1.

14.11.2 Tests Based on Dissimilarities δ and ρ

Consider the $K \times K$ asymptotic dissimilarity matrix based on δ is,

$$\mathbb{E}(\mathbf{D}_\delta) = \begin{pmatrix} 0 & \tilde{\delta}_{12} & \tilde{\delta}_{13} & \dots & \tilde{\delta}_{1K} \\ \tilde{\delta}_{12} & 0 & \tilde{\delta}_{23} & \dots & \tilde{\delta}_{2K} \\ \tilde{\delta}_{13} & \tilde{\delta}_{23} & 0 & \dots & \tilde{\delta}_{3K} \\ \vdots & \vdots & \vdots & \dots & \vdots \\ \tilde{\delta}_{1K} & \tilde{\delta}_{2K} & \tilde{\delta}_{3K} & \dots & 0 \end{pmatrix}. \tag{14.48}$$

One obtains the asymptotic dissimilarity for $\mathbb{E}(\mathbf{D}_\rho)$ by replacing $\tilde{\delta}$ with $\tilde{\rho}$. The method of finding the eigenvalues of matrices $\mathbb{E}(\mathbf{D}_\rho)$ and $\mathbb{E}(\mathbf{D}_\rho)$ are similar since they share the same structure. We present the method for $\mathbb{E}(\mathbf{D}_\delta)$, with characteristic polynomial

$$\mathbb{P}(\lambda) = |\mathbb{E}(\mathbf{D}_\delta) - \lambda\mathbf{I}| = (-1)^K\lambda^K + c_1\lambda^{K-1} + c_2\lambda^{K-2} + \ldots + c_{K-1}\lambda + c_K \tag{14.49}$$

It can be verified that for distance matrices

$$c_1 = -tr(\mathbb{E}(\mathbf{D}_\delta) = 0 \text{ and } c_K = (-1)^K \det(\mathbb{E}(\mathbf{D}_\delta) = 0.$$

Let $t_k = tr(\mathbb{E}(\mathbf{D}_\delta^k)$. For upper-diagonal matrices, Krishnapriyan [175] finds c_r by solving the following recursive relationship

$$c_r = -\frac{1}{r}\sum_{i=1}^{r} c_{r-i}t_i$$

for $1 \le r \le K$ where $c_0 = 0$.

Example 14.11.2 *Consider the null hypothesis that distributions are equally distant. That is, $\tilde{\delta}_{ij} = d$ for $i \le j = 1, \ldots, K$, where $d > 0$ is the common separation distance. It is not difficult to show that $\mathbb{P}(\lambda) = |d(\mathbf{J}_K - \mathbf{I}_K) - \lambda\mathbf{I}_K| = (\lambda - d)^{K-1}(\lambda - d + Kd)$. The eigenvalues are $\lambda_1 = d$ with multiplicity $K-1$ and $\lambda_2 = d(1-K) < 0$. A similar result holds for equidistant distributions using $\mathbb{E}(\mathbf{D}_\rho)$.*

When $K = 2$ the eigenvalues of $\mathbb{E}(\mathbf{D}_\delta)$ and $\mathbb{E}(\mathbf{D}_\rho)$ are $\tilde{\delta}_{12}^2$ and $\tilde{\rho}_{12}^2$, respectively. Hence, test statistics based on these indices are $A_\delta = \tilde{\delta}_{12}^2$, and $A_\rho = \tilde{\rho}_{12}^2$. Suppose $K = 3$ let $\alpha = \tilde{\delta}_{12}^2 + \tilde{\delta}_{13}^2 + \tilde{\delta}_{23}^2$. The characteristic equation for the asymptotic distance matrix based on index δ given in (14.35) is $-\lambda^3 - \alpha\lambda = 0$. The eigenvalues are $\lambda_1 = 0$ and $\lambda_2 = \lambda_3 = \pm\sqrt{\alpha}$. The hypothesis

$$H_0 : F_1 = F_2 = F_3 \Longleftrightarrow L_0 : \tilde{\delta}_{12}^2 = \tilde{\delta}_{13}^2 = \tilde{\delta}_{23}^2 = 0.$$

The same holds for $\mathbb{E}(\mathbf{D}_\rho)$. Hence,

$$H_0 : F_1 = F_2 = F_3 \Longleftrightarrow M_0 : \tilde{\rho}_{12}^2 = \tilde{\rho}_{13}^2 = \tilde{\rho}_{23}^2 = 0.$$

The test statistics are

$$\begin{aligned} A_\delta &= \hat{\tilde{\delta}}_{12}^2 + \hat{\tilde{\delta}}_{13}^2 + \hat{\tilde{\delta}}_{23}^2, \\ A_\rho &= \hat{\tilde{\rho}}_{12}^2 + \hat{\tilde{\rho}}_{13}^2 + \hat{\tilde{\rho}}_{23}^2. \end{aligned}$$

We reject L_0 for large values of A_δ since $\tilde{\delta}_{ij} = 0$ if and only if H_0 holds. The same argument applies to A_ρ. We use A_δ as a test statistic for testing L_0 when $K = 3$.

When $K = 4$ the characteristic equation is $\mathbb{P}(\lambda) = \lambda^4 + c_2\lambda^2 + c_3\lambda = 0$, where

$$\begin{aligned} c_2 &= -\frac{1}{2}t_2 = -\left(\delta_{12}^2 + \delta_{13}^2 + \delta_{14}^2 + \delta_{23}^2 + \delta_{24}^2 + \delta_{34}^2\right), \\ c_3 &= -\frac{1}{3}t_3 = -\frac{2}{3}(\delta_{12}\delta_{23}\delta_{13} + \delta_{12}\delta_{24}\delta_{14} + \delta_{13}\delta_{34}\delta_{14} + \delta_{23}\delta_{34}\delta_{24}). \end{aligned} \tag{14.50}$$

One root is $\lambda = 0$ and the other roots are the solution of $\lambda^3 + c_2\lambda + c_3 = 0$. Notice that $c_2 = c_3 = 0$ if and only if $\delta_{ij}^2 = 0$. Hence, when $K = 4$ we use the statistics

$$\begin{aligned} A_\delta &= \hat{\tilde{\delta}}_{12}^2 + \hat{\tilde{\delta}}_{13}^2 + \hat{\tilde{\delta}}_{14}^2 + \hat{\tilde{\delta}}_{23}^2 + \hat{\tilde{\delta}}_{24}^2 + \hat{\tilde{\delta}}_{34}^2, \\ A_\rho &= \hat{\tilde{\rho}}_{12}^2 + \hat{\tilde{\rho}}_{13}^2 + \hat{\tilde{\rho}}_{14}^2 + \hat{\tilde{\rho}}_{23}^2 + \hat{\tilde{\rho}}_{24}^2 + \hat{\tilde{\rho}}_{34}^2. \end{aligned} \tag{14.51}$$

To develop a K-sample test, suppose the mean and standard deviation of the groups are denoted by $\bar{X}_i^{d(k)}$ and $S_{\mathbf{X}_i}^{d(k)}$ for $k = 1, \ldots, K$. We denote the grand mean and the grand standard deviation of the k-th group with

$$\hat{\mu}_k = \frac{1}{n_k}\sum_{i=1}^{n_i} \bar{X}_i^{d(k)} \text{ and } \hat{\sigma}_k = \frac{1}{n_k}\sum_{i=1}^{n_i} S_{\mathbf{X}_i}^{d(k)}. \tag{14.52}$$

The null hypothesis $H_0 : F_1 = F_2 = \ldots = F_K$ is equivalent to $L_0 : \tilde{\delta}_{ij} = 0$ and $M_0 : \tilde{\rho}_{ij} = 0$ for $1 \leq i < j = 1, \ldots K$. The test statistic for comparing $K \geq 2$ groups is the sum of $K(K-1)/2$ two-way comparison and is given by

$$\begin{aligned} A_\delta &= \frac{1}{2}tr(\hat{\mathbb{E}}(\mathbf{D}_\delta)\hat{\mathbb{E}}(\mathbf{D}_\delta)') = \sum_{t=1}^{K}\sum_{s>t=1}^{K} \hat{\tilde{\delta}}_{ts}^2, \\ A_\rho &= \frac{1}{2}tr(\hat{\mathbb{E}}(\mathbf{D}_\rho)\hat{\mathbb{E}}(\mathbf{D}_\rho)') = \sum_{t=1}^{K}\sum_{s>t=1}^{K} \hat{\tilde{\rho}}_{ts}. \end{aligned} \tag{14.53}$$

The sampling distribution of A_δ and A_ρ can be computed using permutations of the observations. Large $\hat{\tilde{\delta}}_{ts}$ or $\hat{\tilde{\rho}}_{ts}$ indicates discrepancies between groups t and s can be responsible for rejection of H_0.

The energy statistic and the test of Biswas and Ghosh [35] can be extended to the $K \geq 2$ samples case by sum of the pairwise comparisons. Comparing groups (F_i, F_j), let $\bar{d}_{ij}$ be the average of IPDS between them and denote the average of groups i and j with $\bar{d}_i$ and $\bar{d}_j$. The test statistics are given by,

$$
\begin{aligned}
EN_E &= \sum_{1\le i<j\le K} 2\bar{d}_{ij} - \bar{d}_i - \bar{d}_j, \\
BG_E &= \sum_{1\le i<j\le K} (\bar{d}_{(ij)} - \bar{d}_{(i)})^2 + (\bar{d}_{(ij)} - \bar{d}_{(j)})^2.
\end{aligned}
\tag{14.54}
$$

The null hypothesis is rejected for large values of the test statistics. The computational complexity of K sample testing based on the energy or BG statistics is $O(dK^2N^2)$. The computational complexity of A_δ and A_ρ is $O(dK^2N)$.

14.12 Christensen Dataset Revisited

We will illustrate the methods using the microarray dataset of Christensen et al. [67]. We analyze a high-dimensional dataset consisting of blood and placenta samples using Euclidean distance matrices. The goal is to determine whether the two groups arise from distinct distributions. The analysis is based on the eigen-structure of the double-centered distance matrix $\mathbf{B}$. By classical multidimensional scaling, $\mathbf{B} = \mathbf{XX}'$, and its eigenvalues represent variances along orthogonal directions.

The top five eigenvalues for both groups are 1228.1, 199, 67.1, 25, 19.9. The first eigenvalue is substantially larger than the others, indicating a dominant direction of variation. The top five eigenvalues of blood group are 16.7, 16.3, 13.5, 12.3, 9.9. The top five eigenvalues placenta group are 20.7, 18.4, 14.1, 12.3, 11.3. We estimate $r_1 = 4.838$, $r_2 = 5.046$, and $h_{12} = 10.13$. Note that $h_{12} > r_1 \approxeq r_2$. The similarity $r_1 \approx r_2$ indicates that both groups have comparable internal variability. Thus, differences between groups are not due to scale differences but due to location differences. The between-group distance is approximately twice the within-group distance. This indicates strong separation between the two groups.

The scree plot in Figure 14.2 shows a sharp drop after the first eigenvalue. This confirms that a single dominant direction captures most of the variation, corresponding to between-group separation. Figure 14.3 shows that the first eigenvector clearly separates the two

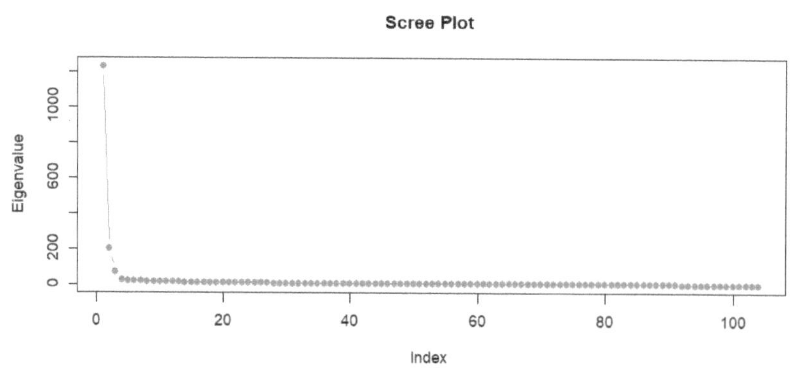

FIGURE 14.2: Scree plot of eigenvalues.

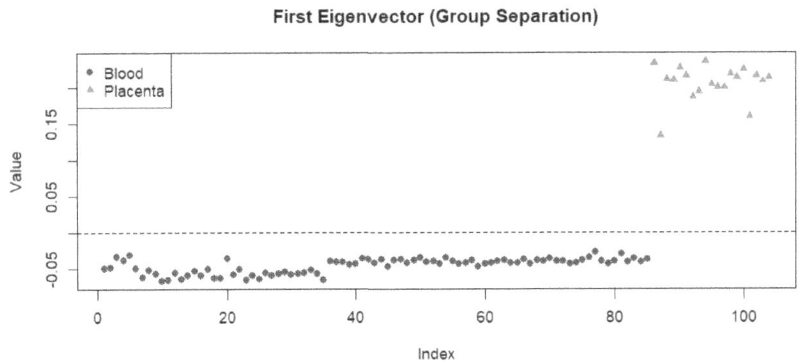

FIGURE 14.3: First eigenvector values by group.

groups, blood samples cluster on one side, while placenta samples cluster on the other. This provides direct evidence of separation.

The multidimensional scaling plot in Figure 14.4 is a plot of $\frac{1}{\lambda_i}\mathbf{e}_i$ for $i = 1, 2$, where $\lambda_1 > \lambda_2$ are the first two largest eigenvalues of $\mathbf{B}$ and $\mathbf{e}_i$ is the corresponding first two entries of the eigenvectors. The plot shows separation primarily along the first dimension, with little additional separation along the second dimension. This confirms that the data is effectively one-dimensional in terms of group structure. The between-group distances are shifted to the right relative to within-group distances. This demonstrates that observations from different groups are systematically farther apart.

Figure 14.5 shows the ECDFs of the average IPDs for the two groups. The average distances in blood group is 3.29, for placenta group is 4.85, and for between-group is 9.72. The three ECDFs are clearly different, indicating that they arise from different distributions. The blood and placenta samples exhibit strong statistical separation. The eigenvalue structure of the distance matrix, combined with graphical and asymptotic analysis, demonstrates that the two groups arise from distinct distributions.

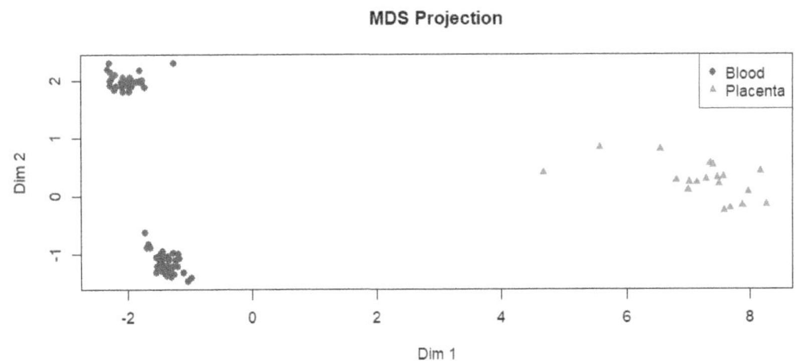

FIGURE 14.4: Two-dimensional MDS projection.

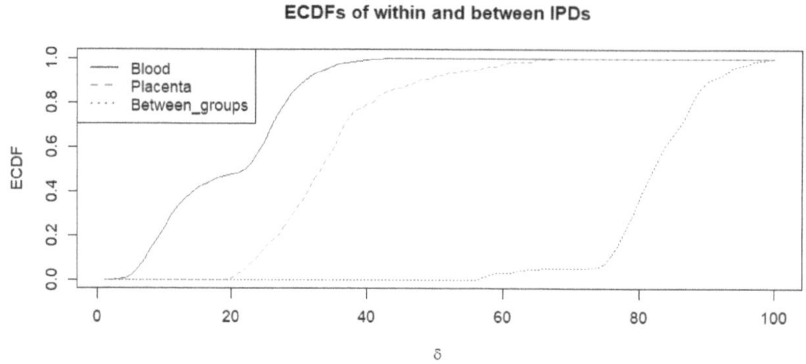

FIGURE 14.5: CDF of of within and between distances.

14.13 Other Asymptotic Regimes

While the focus of this work has been on high-dimensional low-sample-size (HDLSS) data, where the sample size is fixed and $d \to \infty$, it is important to study the asymptotic behavior of the dissimilarity measures δ and ρ under other asymptotic regimes. The matrix $\mathbb{E}(\mathbf{D}_E)$ is made up of the components of ρ. The analysis is extended to the energy and BG_E statistics because as $d \to \infty$, $EN_E \to 2h_{12} - r_1 - r_2$ and $BG_E \to (h_{12} - r_1)^2 + (h_{12} - r_2)^2$.

14.13.0.1 Asymptotic Analysis for δ

The assumptions (B1)–(B3) ensure that the row-level statistics $(\bar{Z}_i^d, S_{\mathbf{Z}_i}^d)$ converge to population constants as $d \to \infty$. Modarres [231] shows that as $d \to \infty$, $\delta(\mathbf{X}_1, \mathbf{X}_2)$, and $\delta(\mathbf{Y}_1, \mathbf{Y}_2)$, tend to zero in probability while $\delta(\mathbf{X}_1, \mathbf{Y}_1) \xrightarrow{p} \tilde{\delta}(F_1, F_2)$. This limit equals zero only when $\bar{\mu}_1^d = \bar{\mu}_2^d$ and $\sigma_1^d = \sigma_2^d$.
Case 1: $n_1/d \to c_1$ and $n_2/d \to c_2$ with $c_1, c_2 \in (0, 1)$

Now consider the intermediate regime where both sample sizes and dimension grow at comparable rates such that $\frac{n_1}{d} \to c_1 \in (0, 1)$, and $\frac{n_2}{d} \to c_2 \in (0, 1)$. Here, $N/d = c_1 + c_2 = c < \infty$. In this regime, the averaging factor $1/(N-2)$ in δ remains of order $O(1/d)$. Thus, $\delta(\mathbf{Z}_i, \mathbf{Z}_j) = O_{\mathbb{P}}(1)$, and δ converges to a finite, non-degenerate limit. Within-group dissimilarities behave as before. That is, $\delta(\mathbf{X}_i, \mathbf{X}_j) \xrightarrow{p} 0$, and $\delta(\mathbf{Y}_i, \mathbf{Y}_j) \xrightarrow{p} 0$, while cross-group dissimilarities retain the positive limit, $\delta(\mathbf{X}_i, \mathbf{Y}_j) \xrightarrow{p} \tilde{\delta}(F_1, F_2) = \left|\sqrt{(\bar{\mu}_1^d - \bar{\mu}_2^d)^2 + (\sigma_1^d - \sigma_2^d)^2}\right|$. Thus, when n_1/d and n_2/d converge to constants, the δ measure remains informative and preserves separability between the two populations.
Case 2: $n_1/d \to 0$ and $n_2/d \to 0$ (Both Samples Very Small)

If both n_1 and n_2 are negligible relative to the dimension d, $\frac{n_1}{d} \to 0$, $\frac{n_2}{d} \to 0$, then the number of terms in the average δ is very small compared to d. Consequently, both the numerator and the denominator in $\delta(\mathbf{Z}_i, \mathbf{Z}_j)$ are $o(d)$, yielding $\delta(\mathbf{Z}_i, \mathbf{Z}_j) \xrightarrow{p} 0$ for all pairs i, j. When both sample sizes are extremely small compared to the dimension, there is insufficient averaging in the double sum, so the dissimilarity measure δ collapses to zero. Cross-group differences are no longer detectable in this limit.

14.13.0.2 Asymptotic Analysis for ρ

Let us consider $d \to \infty$ and different scaling regimes for n_1 and n_2.
Case 1: Assume that $n_1 \sim c_1 d, \quad n_2 \sim c_2 d, \quad c_1, c_2 \in (0,1), \quad N \sim (c_1 + c_2)d$.
It follows that

$$\tilde{\rho}(F_1, F_2) \approx \frac{c_1|\sqrt{2}\sigma_{F_1} - h_{12}| + c_2|h_{12} - \sqrt{2}\sigma_{F_2}|}{c_1 + c_2}.$$

Both classes contribute proportionally and ρ converges to a finite positive constant (unless $\sigma_{F_1} = \sigma_{F_2}$ and $\nu_{12} = 0$).
Case 2: Suppose $n_1/d \to 0$, $n_2/d \to c_2 \in (0,1)$.
Here n_1 is negligible relative to d, n_2 grows linearly with d,

$$(n_1 - 1)|\sqrt{2}\sigma_{F_1} - h_{12}| = o(d), \quad (n_2 - 1)|h_{12} - \sqrt{2}\sigma_{F_2}| \sim O(d), \quad N - 2 \sim n_2.$$

Hence, we have $\tilde{\rho}(F_1, F_2) \to |h_{12} - \sqrt{2}\sigma_{F_2}|$.
Case 3: Assume that $n_1/d \to c_1 \in (0,1)$, $n_2/d \to 0$.
Now n_2 is negligible,

$$(n_1 - 1)|\sqrt{2}\sigma_{F_1} - h_{12}| \sim O(d), \quad (n_2 - 1)|h_{12} - \sqrt{2}\sigma_{F_2}| = o(d), \quad N - 2 \sim n_1.$$

Hence, $\tilde{\rho}(F_1, F_2) \to |\sqrt{2}\sigma_{F_1} - h_{12}|$.
Case 4: Assume that $n_1/d \to 0$, $n_2/d \to 0$.
If both n_1 and n_2 are negligible relative to d, then

$$(n_1 - 1)|\sqrt{2}\sigma_{F_1} - h_{12}| = o(d), \quad (n_2 - 1)|h_{12} - \sqrt{2}\sigma_{F_2}| = o(d), \quad N - 2 = o(d).$$

Hence, the numerator and denominator are both $o(d)$, so $\tilde{\rho}(F_1, F_2) \to 0$. If both classes are very small compared to d, the contribution of the double sum vanishes, and ρ converges to zero.

14.13.0.3 Analysis of EN

Let $\mathbf{X}_1, \ldots, \mathbf{X}_{n_1} \sim F_1$ and $\mathbf{Y}_1, \ldots, \mathbf{Y}_{n_2} \sim F_2$ be independent samples in $\mathbb{R}^d$. Denote population means and covariances by $\mu_X = \mathbb{E}(\mathbf{X}_i)$, $\mu_Y = \mathbb{E}(\mathbf{Y}_j)$, $\Sigma_X = \mathbb{C}ov(\mathbf{X}_i)$, $\Sigma_Y = \mathbb{C}ov(\mathbf{Y}_j)$. Define $\delta_\mu = \mu_X - \mu_Y$, $\Delta_\Sigma = \Sigma_X - \Sigma_Y$. Assume $\mathbf{X}_i = \mu_X + \Sigma_X^{1/2} Z_i$, $\mathbf{Y}_j = \mu_Y + \Sigma_Y^{1/2} W_j$, where the coordinates of Z_i, W_j are i.i.d., mean zero, variance 1, and finite fourth moments.

Under the above model and moment conditions, as $d, n_1, n_2 \to \infty$, Chakraborty and Zhang [54] show that

$$EN_E = \frac{1}{\sqrt{2d}}\Big(\|\delta_\mu\|^2 + \frac{1}{2}\|\Sigma_X^{1/2} - \Sigma_Y^{1/2}\|_F^2\Big) + O\mathbb{P}\Big(\frac{1}{\sqrt{dn_1}} + \frac{1}{\sqrt{dn_2}}\Big).$$

When $n_i/d \to 0$, the stochastic term $O_{\mathbb{P}}((dn_i)^{-1/2})$ vanishes faster than the deterministic leading term. Thus, under $H_0 : F_1 = F_2$, $E_n = O_{\mathbb{P}}(d^{-1/2})$, and the centered and scaled statistic degenerates. Hence, we have

$$\frac{EN_E - \mathbb{E}EN_E}{\sqrt{\mathbb{V}ar(EN_E)}} \xrightarrow{P} 0.$$

Hence, in extremely high dimensions, EN_E detects primarily mean differences $\|\delta_\mu\|^2$.

14.13.0.4 Analysis of BG_E

It is not difficult to show that $2BG_E = EN_E^2 + (\bar{d}_{(x)} - \bar{d}_{(y)})^2$.
Case 1: When $n_1/d \to c$ and $n_2/d \to c$ for some $c \in (0,1)$.

The stochastic terms $O\mathbb{P}\left(\frac{1}{\sqrt{dn_i}}\right) = O\mathbb{P}(d^{-1})$ are of smaller order than the deterministic leading term, but not negligible. Both deterministic and stochastic contributions scale as $O(d^{-1/2})$, and hence the statistic retains a non-trivial variance. Therefore, the BG statistic will converge to a non-degenerate limit under H_0, maintaining sensitivity to both mean and covariance differences. In particular, BG_E remains consistent for fixed mean or scale differences as $d, n_1, n_2 \to \infty$ with $n_i/d \to c \in (0, 1)$.
Case 2: When $n_1/d \to 0$ and $n_2/d \to 0$.

In this regime, the stochastic term $O_{\mathbb{P}}((dn_i)^{-1/2})$ vanishes much faster than the deterministic term. Thus, under $H_0 : F_1 = F_2$, we have $E_n = O_{\mathbb{P}}(d^{-1/2})$, and EN_E degenerates. Consequently, the Biswas—Ghosh statistic BG_E also degenerates under H_0 in this regime, losing power to detect general alternatives.

14.14 Comparisons

In this section we describe sampling scenarios where we compare the following statistics in terms of observed significance level and power, i) energy statistic EN in (14.54), ii) BG statistic given in (14.54), iii) Statistic S given in (14.45), iv) Statistic Δ given in (14.47), v) Statistic E given in (14.46), and vi) Statistics A_δ and A_ρ given in (14.53). We used random permutations to obtain the sampling distributions of the statistics under the null and the alternative hypotheses. We used $\alpha = 0.05$. Under H_0 the observations are exchangeable. To obtain its permutation distribution, we generate $R = 1000$ random permutation. We describe the method for statistic Δ. The other statistics are similarly treated. The approximate permutation distribution of the test statistic, denoted by H_Δ, is

$$H_\Delta(t) = \frac{1}{R}\sum_{r=1}^{R} \mathbb{I}(\Delta_r \leq t),$$

where $\mathbb{I}(\cdot)$ is the indicator function and Δ_r is the test statistic calculated for the r-th permutation sample. The p-value of the permutation test, denoted by d_{perm}, is given by

$$d_{\text{perm}} = 1 - H_\Delta(\text{obs}) = \frac{1}{R}\sum_{r=1}^{R} \mathbb{I}(\Delta_r > \Delta(\text{obs})),$$

where $\Delta(\text{obs})$ is the test statistic for the observed sample. The results are based on 1000 simulations.

14.14.1 Testing $H_0 : F_1 = F_2$

We test $H_0 : F_1 = F_2$ by generating samples of size $n_1 = 20$ from $F_1 \sim \mathbb{N}_d(\mathbf{0}, \mathbf{I})$ and $n_2 = 20$ from $F_2 \sim \mathbb{N}_d(\mu\mathbf{1}, \sigma^2{}_i\mathbf{I})$ where $d = 100$, $\mu_i = 0$, and $\sigma_i = 1$ under H_0. First we consider a shift alternative where $\mu \in \{0, 0.01, 0.02, \ldots, 0.4\}$ and $\sigma_i = 1$. Figure 14.6 displays the power of the above statistics as a function of μ. The range of the observed significance levels is $(0.043, 0.056)$. The power ordering is $A_\delta > \Delta = E \approxeq EN > BG > A_\delta > S$. Statistic A_δ dominates the other tests in terms of power. The constancy of distances S is the least powerful compared to the other four statistics. Similar performance is noted for Δ, EN and E statistics.

Modarres [231] considers the power of the above statistics as a function of $\sigma^2 \in \{1, 1.01, \ldots, 0.25\}$ for scale alternatives. He finds that observed significance levels is in the range $(0.041, 0.050)$ and the power ordering $BG > A_\delta > S > A_\rho > \Delta = E \approxeq EN$. The energy statistics and E show little power under scale alternatives. Modarres [237] also investigates the power of the statistics for testing $H_0 : F_1 = F_2$ against both shift and scale alternatives. Generated samples under the alternative have $\mu \in \{0, 0.01, \ldots, 0.25\}$ and $\sigma^2 \in \{1, 1.01, 1.02, \ldots, 0.25\}$. He reports the power ordering $A_\delta > BG > S \approxeq \Delta = E \approxeq EN > A_\rho$. It is interesting that A_δ performs better than BG when shift changes are added.

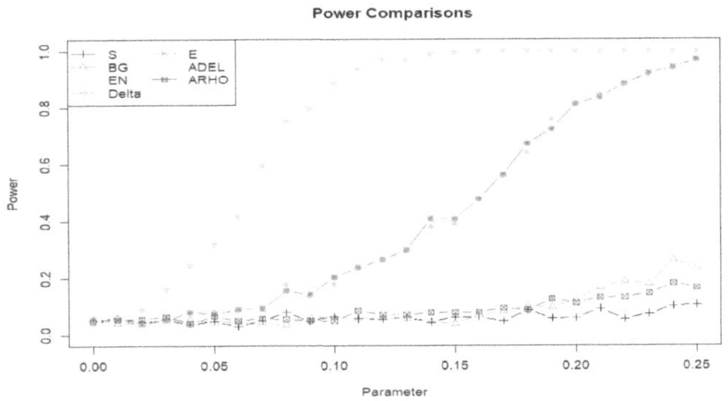

FIGURE 14.6: Power for $H_0 : F_1 = F_2$ under shift alternatives.

14.14.2 Test for the Equality of Several Groups

Modarres [231] further found the power of the above statistics as a function of μ for testing $H_0 : F_1 = F_2 = F_3$. Generated samples are of size $n_i = 15$ from $F_i \sim \mathbb{N}_d(\mu_i \mathbf{1}, \sigma_i^2 \mathbf{I})$. Under H_0, $\mu_i = 0$, and $\sigma_i = 1$. Modarres [231] considers a shift alternative where $\mu \in \{0, 0.01, 0.02, \ldots, 0.5\}$ and $\sigma_i = 1$. The range of the observed significance levels is $(0.043, 0.056)$. He reports the power ordering $A_\delta \gg EN > E > BG > S > A_\rho > \Delta$ from the simulation results. Statistic A_δ dominates the other tests in terms of power. The statistic Δ is the least powerful.

Modarres [231] also found the power of the test statistics for testing $H_0 : F_1 = F_2 = F_3$ under scale alternatives where $\sigma_i \in \{1, 1.01, \ldots, 1.5\}$. The range of the observed significance levels is $(0.041, 0.059)$. From simulation results he reports the power ordering $E > BG > A_\delta > A_\rho = S > \Delta > EN$. Statistic E performs better than other statistics for scale alternatives. The energy statistic is the least powerful.

Modarres [231] investigated the power of the test statistics for testing $H_0 : F_1 = F_2 = F_3$ under shift and scale alternatives where $\mu \in \{0, 0.01, 0.02, \ldots, 0.25\}$ and $\sigma_i \in \{1, 1.01, \ldots, 1.5\}$. The range of the observed significance levels is $(0.035, 0.053)$. He reports the power ordering $A_\delta > E > BG > S > S > A_\rho > \Delta$ from the simulation results. Statistic A_δ performs well for shift and scale alternatives. The statistic Δ is the least powerful statistics.

Assume samples of size $n_i = 15$ are generated from $F_i \sim \mathbb{N}_d(\mu_i \mathbf{1}, \sigma_i^2 \mathbf{I})$ where $d = 100$, $\mu_i = 0$, and $\sigma_i = 1$ under $H_0 : F_1 = F_2 = F_3 = F_4$. Modarres [231] compares A_δ, BG, EN, S, A_ρ with $K = 4$ groups of observations. Under shift alternatives, the range of the observed levels is $(0.048, 0.055)$, which is close to $\alpha = 0.05$. He reports the power ordering $A_\delta \gg EN \gg BG > S > A_\rho$ from the simulation results. Statistic A_δ dominates the other statistics. The A_ρ is the least powerful compared. Under scale alternatives, the range of the observed significance levels is $(0.041, 0.056)$. We report the power ordering

$BG > A_\delta > A_\rho > S > EN$. Statistic BG performs well for scale alternatives. The statistic EN is the least powerful.

Under shift and scale alternatives we found the power of testing $H_0 : F_1 = F_2 = F_3 = F_4$ where $\mu \in \{0, 0.01, 0.02, \ldots, 0.25\}$ and $\sigma_i \in \{1, 1.01, \ldots, 1.5\}$. The range of the observed significance levels is $(0.039, 0.057)$. From the simulation results we report the power ordering is $A_\delta \gg BG > S > A_\rho > EN$. Statistic A_δ dominates the other tests in terms of power. The energy statistic is the least powerful.

14.15 Summary

Chapter 14 provides a comprehensive analysis of distance matrices in high-dimensional settings, emphasizing their eigen-structures and applications in statistical inference. The chapter examines constant distance matrices, distance matrix imputation, two-sample and HDLSS distance matrices, and explores the impact of outliers on these structures. Two dissimilarity measures, δ and ρ, alongside the L_1-norm Euclidean distance, are studied in detail. The asymptotic eigenvalues of these matrices are derived for scenarios with one to three groups of observations, and their relationships to classical statistical measures, including the energy statistic (Rizzo and Székely [333]; Baringhaus and Franz [21]) and the Biswas-Ghosh [35] test statistic, are established. These results form the foundation for rigorous hypothesis testing, including assessments of equality across K distributions and sensitivity to location shifts, scale transformations, and combined shift-scale changes.

The chapter highlights that, as the dimension $d \to \infty$, pairwise distances within a group converge to a constant structure, providing a basis for formal statistical tests. Principal coordinate analysis is used to decompose distance and dissimilarity matrices, with eigenvalues elucidating both within-group constancy and between-group differences. The analysis demonstrates that pure scale transformations primarily influence the roots $\lambda = r_i^2/2$, whereas components involving h_{ij} capture joint shift-scale effects. Furthermore, the presence of outliers can significantly distort the eigenvalues, illustrating the limited robustness of conventional statistics, including the energy and BG statistics, under such contamination.

The eigenvalues of $\mathbb{E}(\mathbf{D}_\delta)$ and $\mathbb{E}(\mathbf{D}_\rho)$ are intrinsically linked to the trace terms $\mathrm{tr}\big(\mathbb{E}(\mathbf{D}_\delta)^r\big)$ and $\mathrm{tr}\big(\mathbb{E}(\mathbf{D}_\rho)^r\big)$, respectively, for $r = 1, \ldots, K$. For $K = 2, 3$, we derive these eigenvalues explicitly. The test statistics A_δ and A_ρ are defined as the sum of squared pairwise comparisons, equivalently representing one-half of the squared Frobenius norm of the matrices $\mathbb{E}(\mathbf{D}_\delta)$ and $\mathbb{E}(\mathbf{D}_\rho)$, respectively. Since the squared Frobenius norm equals the sum of the squared eigenvalues, these statistics quantify the total variability captured by the distance matrices. In particular, A_δ, which incorporates both group means and variances, exhibits strong power against shift and shift-scale alternatives. An intuitive yet fundamental observation is that, for constant distance matrices, all differences between successive eigenvalues vanish. This property motivates the introduction of the statistic Δ, which explicitly measures deviations from this constant structure.

Overall, chapter 14 integrates theoretical derivations, practical diagnostics, and asymptotic results to provide a robust framework for high-dimensional distance matrix analysis. It offers tools for outlier detection, hypothesis testing for equality of distributions, and dimensionality reduction. Applications span bioinformatics, machine learning, cognitive science, and pattern recognition, emphasizing both methodological rigor and practical relevance for interpreting complex, high-dimensional relational data.

14.16 Proofs

In the following discussion we will need the next two results. The determinant of an a $N \times N$ block matrix $\mathbf{B} = (\mathbf{B}_{ij})$, where $\mathbf{B}_{ij}$ is $n_i \times n_j$ matrix for $i = j = 1, 2$ is given by

$$\begin{aligned} |\mathbf{B}| &= |\mathbf{B}_{22}|\,|\mathbf{B}_{11} - \mathbf{B}_{21}\mathbf{B}_{22}^{-1}\mathbf{B}_{12}| \quad \text{for } |B_{22} \neq 0 \\ &= |\mathbf{B}_{11}|\,|\mathbf{B}_{22} - \mathbf{B}_{21}\mathbf{B}_{11}^{-1}\mathbf{B}_{12}| \quad \text{for } |\mathbf{B}_{11}| \neq 0, \end{aligned} \tag{14.55}$$

where $\mathbf{B}_{11} - \mathbf{B}_{21}\mathbf{B}_{22}^{-1}\mathbf{B}_{12}$ is the Schur complement. Suppose $\mathbf{M} = (a-t)\mathbf{I}_N + t\mathbf{1}_{N\times N}$ is an $N \times N$ matrix with a on the diagonal and t on the off-diagonal. The eigenvalues of $\mathbf{M}$ are $\lambda_1 = a - t + Nt$ and $\lambda_2 = a - t$, with multiplicity $N-1$. Its determinant is $|\mathbf{M}| = (a-t)^{N-1}(a - t + Nt)$, and its inverse is

$$\mathbf{M}^{-1} = \frac{1}{a-t}(\mathbf{I}_N - \frac{t}{a-t+Nt}\mathbf{1}_{N\times N}). \tag{14.56}$$

Proof: 14.9.1

Let $N = n_1 + n_2$ and $\mathbf{1}_{n_1\times n_2} = \mathbf{1}_{n_1}\mathbf{1}'_{n_2}$ be an $n_1 \times n_2$ matrix of 1's. We use Algorithm A to find the eigenvalues. Consider $a_{ij} = -\frac{1}{2}\mathbb{E}(d_{ij}^2)$ where

$$\mathbf{A} = -\frac{1}{2}\begin{pmatrix} r_1(\mathbf{1}_{n_1\times n_1} - \mathbf{I}_{n_1}) & h_{12}\mathbf{1}_{n_1\times n_2} \\ h_{12}\mathbf{1}_{n_2\times n_1} & r_2(\mathbf{1}_{n_2\times n_2} - \mathbf{I}_{n_2}). \end{pmatrix} \tag{14.57}$$

Recall that $\mathbf{J}_N = \mathbf{1}_{N\times N}$ is the $N \times N$ matrix of 1's. One can verify that

$$\mathbf{B} = (\mathbf{I}_N - \mathbf{J}_N)\,\mathbf{A}\,(\mathbf{I}_N - \mathbf{J}_N) = \mathbf{A} - \frac{1}{N}\mathbf{J}_N\mathbf{A} - \frac{1}{N}\mathbf{A}\mathbf{J}_N + \frac{1}{N^2}\mathbf{J}_N\mathbf{A}\mathbf{J}_N \tag{14.58}$$

$$\mathbf{B}(i,j) = \mathbf{A}(i,j) - \frac{1}{N}\sum_{k=1}^{N}\mathbf{A}(i,k) - \frac{1}{N}\sum_{k=1}^{N}\mathbf{A}(k,j) + \frac{1}{N^2}\sum_{k=1}^{N}\sum_{t=1}^{N}\mathbf{A}(k,t) \tag{14.59}$$

The second term is the mean of the elements in the i-th row of $\mathbf{A}$. The third term is the mean of the elements in the j-th column of$\mathbf{A}$., and the last term is the grand mean of all elements in $\mathbf{A}$. These terms ensure the matrix is doubly centered (both row and column sums are zero). Furthermore, one can show that

$$\mathbf{J}_N\mathbf{A} = -\frac{1}{2}\left(((n_1-1)r_1^2 + n_1h_{12}^2)\mathbf{1}_{N\times n_1} \quad (n_2h_{12}^2 + (n_2-1)r_2^2)\mathbf{1}_{N\times n_2}\right)_{N\times N} \tag{14.60}$$

$$\mathbf{J}_N\mathbf{A}\mathbf{J}_N = -\frac{1}{2}\left(n_1(n_1-1)r_1^2 + 2n_1n_2h_{12}^2 + n_2(n_2-1)r_2^2\right)\mathbf{J}_N. \tag{14.61}$$

Suppose we partition the $N \times N$ matrix $\mathbf{B}$ as

$$\mathbf{B} = -\frac{1}{2}\begin{pmatrix} (b_{11} - b_{11}^*)\mathbf{I}_{n_1} + b_{11}^*\mathbf{1}_{n_1\times n_1} & b_{12}\mathbf{1}_{n_1\times n_2} \\ b_{12}\mathbf{1}_{n_2\times n_1} & (b_{22} - b_{22}^*)\mathbf{I}_{n_2} + b_{22}^*\mathbf{1}_{n_2\times n_2} \end{pmatrix} \tag{14.62}$$

Matrix $\mathbf{B}$ is symmetric double-centered with row and column sum of zero. The rank of $\mathbf{B}$ is $N-1$ since double-centering removes the mean component. Hence, it has $N-1$ non-zero and one zero eigenvalue. Using (14.59), the entries of $\mathbf{B}$ matrix are

$$
\begin{aligned}
-\frac{1}{2}b_{11} &= \frac{1}{N}\left((n_1-1)r_1^2+n_2h_{12}^2\right)-T\\
-\frac{1}{2}b_{22} &= \frac{1}{N}\left(n_1h_{12}^2+(n_2-1)r_2^2\right)-T\\
-\frac{1}{2}b_{11}^* &= -\frac{1}{2}r_1^2+\frac{1}{N}\left((n_1-1)r_1^2+n_2h_{12}^2\right)-T\\
-\frac{1}{2}b_{22}^* &= -\frac{1}{2}r_2^2+\frac{1}{N}\left(n_1h_{12}^2+(n_2-1)r_2^2\right)-T\\
-\frac{1}{2}b_{12} &= \frac{1}{2N}\left((n_1-1)r_1^2+(n_2-1)r_2^2\right)-T,
\end{aligned}
\tag{14.63}
$$

where $T=\frac{1}{2N^2}\left(n_1(n_1-1)r_1^2+2n_1n_2h_{12}^2+n_2(n_2-1)r_2^2\right)$.

Pre and post multiplying the $\mathbf{B}$ matrix by the centering matrix $\mathbf{P}$ has the effect of subtracting the rows means and the column means and adding the grand mean of $\mathbf{A}$ to each a_{ij}. Hence, $\mathbf{B}$ has zero row and column sums. It follows that

$$
\begin{cases}
b_{11}+(n_1-1)b_{11}^*+n_2b_{12}=0,\\
n_1b_{12}+(n_2-1)b_{22}^*+b_{22}=0.
\end{cases}
\tag{14.64}
$$

The eigenvalues of $-\mathbf{2B}$ are obtained from

$$
|-2\mathbf{B}-\lambda\mathbf{I}_N|=0=-\lambda(b_{11}-b_{11}^*-\lambda)^{n_1-1}(b_{22}-b_{22}^*-\lambda)^{n_2-1}(-Nb_{12}-\lambda)=0
\tag{14.65}
$$

The eigenvalues of $\mathbf{B}$ after multiplying the roots by $-\frac{1}{2}$ are

$$
\begin{aligned}
\lambda &= 0,\\
\lambda_1 &= \frac{b_{11}^*-b_{11}}{2}=\frac{1}{2}r_1^2 \quad \text{with multiplicity } n_1-1\\
\lambda_2 &= \frac{b_{22}^*-b_{22}}{2}=\frac{1}{2}r_2^2 \quad \text{with multiplicity } n_2-1\\
\lambda_3 &= -\frac{N}{2}b_{12}=\frac{1}{2N}[2n_1n_2h_{12}^2-n_2(n_1-1)r_1^2-n_1(n_2-1)r_2^2]
\end{aligned}
\tag{14.66}
$$

□ **Proof:** 14.9.2

Recall that $tr\left(\boldsymbol{\Sigma}_1\right)/d\to\sigma_{F_1}^2$ and $tr\left(\boldsymbol{\Sigma}_2\right)/d\to\sigma_{F_2}^2$ by Assumption (A1). As $d\to\infty$ one can verify that $\mathbb{E}(\bar{d}_x)\to r_1=\sqrt{2}\sigma_{F_1}$, $\mathbb{E}(\bar{d}_y)\to r_2=\sqrt{2}\sigma_{F_2}$ and $\mathbb{E}(\bar{d}_{xy})\to h_{12}$ as $d\to\infty$. Hence, if $F_1=F_2$, then it trivially follows that $h_{12}=r_1=r_2$. If $h_{12}=r_1=r_2$, then, we $2h_{12}-r_1-r_2=0$. Baringhaus and Franz [21] prove that for F_1 and F_2 with finite expected norms, $2h_{12}-r_1-r_2\geq 0$ where equality holds if and only if F_1 and F_2 are identical. So, $h_{12}=r_1=r_2$ implies $F_1=F_2$. □

Proof: 14.9.3

Consider $a_{ij}=-\frac{1}{2}\mathbb{E}(d_{ij}^2)$, where

$$
\mathbf{A}=-\frac{1}{2}\begin{pmatrix}
r_1(\mathbf{1}_{n_1\times n_1}-\mathbf{I}_{n_1}) & h_{12}\mathbf{1}_{n_1\times n_2} & h_{13}\mathbf{1}_{n_1\times n_3}\\
h_{12}\mathbf{1}_{n_2\times n_1} & r_2(\mathbf{1}_{n_2n_2}-\mathbf{I}_{n_2}) & h_{23}\mathbf{1}_{n_2\times n_3}\\
h_{13}\mathbf{1}_{n_3\times n_1} & h_{23}\mathbf{1}_{n_3\times n_2} & r_3(\mathbf{1}_{n_3\times n_3}-\mathbf{I}_{n_3})
\end{pmatrix}_{N\times N}
\tag{14.67}
$$

To obtain $\mathbf{B}=\mathbf{A}-\frac{1}{N}\mathbf{J}_N\mathbf{A}-\frac{1}{N}\mathbf{A}\mathbf{J}_N+\frac{1}{N^2}\mathbf{J}_N\mathbf{A}\mathbf{J}_N$, we can show that

$$
\mathbf{J}_N\mathbf{A}=-\frac{1}{2}\begin{pmatrix}c_1\,\mathbf{1}_{N\times n_1} & c_2\,\mathbf{1}_{N\times n_2} & c_3\,\mathbf{1}_{N\times n_3}\end{pmatrix}
\tag{14.68}
$$

$$\begin{aligned} c_1 &= (n_1-1)r_1^2 + n_2 h_{12}^2 + n_3 h_{13}^2, \\ c_2 &= n_1 h_{12}^2 + (n_2-1)r_2^2 + n_3 h_{23}^2, \\ c_3 &= n_1 h_{13}^2 + n_2 h_{23}^2 + (n_3-1)r_3^2. \end{aligned} \tag{14.69}$$

Thus, $\mathbf{J}_N\mathbf{A}$ has n_1 columns of c_1, n_2 columns of c_2 and n_3 columns of c_3. One obtains

$$\mathbf{A}\mathbf{J}_N = -\frac{1}{2}\begin{pmatrix} c_1\mathbf{1}_{n_1\times N} \\ c_2\mathbf{1}_{n_2\times N} \\ c_3\mathbf{1}_{n_3\times N} \end{pmatrix} \tag{14.70}$$

Thus, the matrix $\mathbf{A}\mathbf{J}_N$ has n_1 rows of c_1, n_2 rows of c_2 and n_3 rows of c_3. It follows that $\mathbf{J}_N\ \mathbf{A}\ \mathbf{J}_N = -\frac{1}{2}\ T\ \mathbf{J}_N$ where $T = n_1c_1 + n_2c_2 + n_3c_3$. We partition $\mathbf{B}$ as

$$\mathbf{B} = -\frac{1}{2}\begin{pmatrix} (b_{11}-b_{11}^*)\mathbf{I}_{n_1} + b_{11}^*\mathbf{1}_{n_1\times n_1} & b_{12}\mathbf{1}_{n_1\times n_2} & b_{13}\mathbf{1}_{n_1\times n_3} \\ b_{12}\mathbf{1}_{n_2\times n_1} & (b_{22}-b_{22}^*)\mathbf{I}_{n_2} + b_{11}^*\mathbf{1}_{n_2\times n_2}) & b_{23}\mathbf{1}_{n_2\times n_3} \\ b_{13}\mathbf{1}_{n_3\times n_1} & b_{23}\mathbf{1}_{n_3\times n_2} & (b_{33}-b_{33}^*)\mathbf{I}_{n_3} + b_{33}^*\mathbf{1}_{n_3\times n_3}) \end{pmatrix} \tag{14.71}$$

Using $\mathbf{B} = \mathbf{PAP}$, the entries of $\mathbf{B}$ matrix are

$$\begin{aligned} -\frac{1}{2}b_{ii} &= \frac{1}{N}c_i - \frac{1}{2N^2}T, \text{ for } i = 1,2,3, \\ -\frac{1}{2}b_{ii}^* &= -\frac{1}{2}r_i^2 + \frac{1}{N}c_i - \frac{1}{2N^2}T, \text{ for } i = 1,2,3, \\ -\frac{1}{2}b_{ij} &= -\frac{1}{2}h_{ij}^2 + \frac{1}{2N}c_i + \frac{1}{2N}c_j - \frac{1}{2N^2}T, \text{ for } (i,j) = (1,2),(1,3),(2,3). \end{aligned} \tag{14.72}$$

Since $\mathbf{B}$ has has zero row and column sums it follows that

$$\begin{cases} b_{11} + (n_1-1)b_{11}^* + n_2b_{12} + n_3b_{13} = 0, \\ n_1b_{12} + b_{22} + (n_2-1)b_{22}^* + n_3b_{23} = 0, \\ n_1b_{13} + n_2b_{23} + b_{33} + (n_3-1)b_{33}^* = 0. \end{cases} \tag{14.73}$$

The characteristic polynomial is

$$|-2\mathbf{B} - \lambda\mathbf{I}_N| = \left|\begin{pmatrix} \mathbf{d}_{11} & b_{12}\mathbf{1}_{n_1n_2} & b_{13}\mathbf{1}_{n_1n_3} \\ b_{12}\mathbf{1}_{n_2n_1} & \mathbf{d}_{22} & b_{23}\mathbf{1}_{n_2n_3} \\ b_{13}\mathbf{1}_{n_3n_1} & b_{23}\mathbf{1}_{n_3n_2} & \mathbf{d}_{33} \end{pmatrix}\right| \tag{14.74}$$

where for $i = 1,2,3$, $\mathbf{d}_{ii} = (b_{ii} - b_{ii}^* - \lambda)\mathbf{I}_{n_i} + b_{ii}^*\mathbf{1}_{n_i\times n_1}$. Hence, $|-2\mathbf{B} - \lambda\mathbf{I}_N|$ equals

$$|\mathbf{d}_{11}|\left|\begin{pmatrix} \mathbf{d}_{22} & b_{23}\mathbf{1}_{n_2\times n_3} \\ b_{23}\mathbf{1}_{n_3\times n_2} & \mathbf{d}_{33} \end{pmatrix} - \begin{pmatrix} b_{12}\mathbf{1}_{n_2\times n_1} \\ \\ b_{13}\mathbf{1}_{n_3\times n_1} \end{pmatrix} \mathbf{d}_{11}^{-1}\left(b_{12}\mathbf{1}_{n_1\times n_2} \quad b_{13}\mathbf{1}_{n_1\times n_3}.\right)\right| \tag{14.75}$$

Suppose that $b_{12} = b_{13} = b_{23} = 0$. It is easy to see that the eigenvalues of $\mathbf{d}_{ii}$ are for $i = 1,2,3$,

$$\lambda_i = -\frac{1}{2}(b_{ii} - b_{ii}^*) = \frac{1}{2}r_i^2 \text{ with multiplicity } n_i - 1. \tag{14.76}$$

Since the rank of $\mathbf{A}$ is $N-1$ there is a zero eigenvalue, and the remaining two eigenvalues are the roots of $N\lambda^2 + b\lambda + c = 0$, where b and c are defined in (14.37) and (14.38), respectively. The quadratic equation is determined by dividing $|-2\mathbf{B} - \lambda\mathbf{I}_N|$ by $\prod_{i=1}^3(\lambda_i - r_i^2/2)$. □

Proof: 14.11.1

By Theorem 14.9.2 we know $H_0 : F_1 = F_2$ is equivalent to testing $K_0 : r_1 = r_2 = h_{12}$. Following Theorem 14.9.2 and comparing two distributions at a time, it is not difficult to show that testing H_0 is equivalent to testing K_0. If H_0 is true then K_0 follows trivially since $F_i = F_j$ ensures $r_i = r_j = h_{ij}$ for $i \neq j = 1, \ldots, K$. If K_0 is true, then $2h_{ij} - r_i - r_j = 0$. We know from Baringhaus and Franz [21] that for F_i and F_j with finite expected norm, $2h_{ij} - r_i - r_j \geq 0$, where equality holds if $F_i = F_j$. Hence, testing H_0 is equivalent to testing K_0. □

14.17 Exercises

Exercise 14.1 *Let* $\mathbf{X}_1$, $\mathbf{X}_2$, *and* $\mathbf{X}_3$ *be three points in* $\mathbb{R}^d$ *with the constant* 3×3 *distance matrix* $\mathbf{D}$,

$$\mathbf{D} = \begin{bmatrix} 0 & d & d \\ d & 0 & d \\ d & d & 0 \end{bmatrix}$$

Define the double centering matrix $B = -\frac{1}{2}\mathbf{PAP}$, *where* $\mathbf{A}$ *is the matrix of squared distances,* $a_{ij} = -\frac{1}{2}d_{ij}^2$, *and* $\mathbf{P} = \mathbf{I} - \frac{1}{3}\mathbf{1}\mathbf{1}'$ *is the centering matrix. Compute the eigenvalues and eigenvectors of* B *assuming* $d^2 = 6$. *How are the distances represented in* $\mathbb{R}^2$? *Is there any loss of information by representing the distances in* $\mathbb{R}^2$?

Exercise 14.2 *Consider the* $m \times m$ *distance matrix with* k *outliers represented as the first* k *rows of the distance matrix*

$$\mathbf{D}(\mathbb{X}) = \begin{matrix} & X_1 & X_2 & \ldots & X_k & X_{k+1} & \ldots & X_m \\ X_1 & 0 & cd & cd & cd & cd & \ldots & cd \\ X_2 & cd & 0 & cd & cd & cd & \ldots & cd \\ \vdots & \vdots & \vdots & \vdots & \vdots & \vdots & \vdots & \vdots \\ X_k & cd & \ldots & cd & 0 & cd & \ldots & cd \\ X_{k+1} & cd & \ldots & cd & cd & 0 & \ldots & d \\ \vdots & \vdots & \vdots & \vdots & \vdots & \vdots & \vdots & \\ X_m & cd & \ldots & cd & cd & d & \ldots & 0 \end{matrix}, \tag{14.77}$$

where the remaining $m-k$ *observations are equidistant. Find the eigenvalues of* $\mathbf{D}(\mathbb{X})$. *What condition ensures that* $\lambda_1 > \lambda_2$?

Exercise 14.3 *Suppose* $\mathbf{X}_i$ *are iid* $\mathrm{N}_d(\boldsymbol{\mu}_1, \boldsymbol{\Sigma}_1)$ *under the hypothesis of no outlier, for* $i = 1, \ldots, m$. *Suppose there is one outlier in the sample at observation* $\mathbf{X}_1$ *so that* $\mathbf{X}_i$ *are iid* $\mathrm{N}_d(\boldsymbol{\mu}_1, \boldsymbol{\Sigma}_1)$, *for* $i = 2, \ldots, m$ *and while the outlier arises from a 2-component normal mixture* $\mathbf{X}_1 \sim \alpha\, \mathrm{N}_d(\boldsymbol{\mu}_1, \boldsymbol{\Sigma}_1) + (1-\alpha)\, \mathrm{N}_d(\boldsymbol{\mu}_2, \boldsymbol{\Sigma}_2)$. *Let* d_{1j}^2 *be the IPD between* $\mathbf{X}_2, \ldots, \mathbf{X}_{m-1}$ *and* $\mathbf{X}_1$. *To study a simple case of scale contamination consider* $\boldsymbol{\mu}_1 = \boldsymbol{\mu}_2$ *with* $\boldsymbol{\Sigma}_1 = \boldsymbol{I}_d$ *and* $\boldsymbol{\Sigma}_2 = \sigma_2^2\boldsymbol{I}_d$. *Suppose* $\mathbb{E}_{H_0}(\mathbf{D}(\mathbb{X}))$ *is a constant distance matrix with components* $d_{ij}^2 = 2p$ *for* $i \neq j = 1, \ldots, m$ *and* $\mathbb{E}_{H_a}(\mathbf{D}(\mathbb{X}))$ *is a constant distance matrix with components* $\mathbb{E}(d_{1j}^2)$ *for* $i \neq j = 1, \ldots, m$.

a) Find the distribution of d_{ij}^2 *for* $i \neq j = 1, \ldots, m$.

b) Find the distribution of d_{1j}^2 *for* $j = 2, \ldots, m$.

c) Simplify the distribution of d_{1j}^2 *for* $j = 2, \ldots, m$ *and find* $\mathbb{E}(d_{1j}^2)$ *for the scale contamination model.*

d) What are the largest two eigenvalues of $\mathbb{E}_{H_0}(\mathbf{D}(\mathbb{X}))$ *and* $\mathbb{E}_{H_a}(\mathbf{D}(\mathbb{X}))$.

Exercise 14.4 *Expected Dissimilarity Matrices when* $d \to \infty$.

Suppose $\mathbb{X} = \{\mathbf{X}_1, \ldots, \mathbf{X}_{n_1}\}$, $\mathbb{Y} = \{\mathbf{Y}_1, \ldots, \mathbf{Y}_{n_2}\}$, *and* $\mathbb{Z} = \{\mathbf{Z}_1, \ldots, \mathbf{Z}_{n_3}\}$ *be i.i.d. samples from* F_i *with means* $\boldsymbol{\mu}_i$ *and covariance* $\boldsymbol{\Sigma}_i = \sigma^2 \mathbf{I}_d$ *for* $i = 1, 2, 3$. *Let* $N = n_1 + n_2 + n_3$. *We are interested in testing* $H_0 : F_1 = F_2 = F_3$, *against* $H_a : F_i \neq F_j$ *for at least one* $i \neq j = 1, \ldots, 3$. *Assume* $n_1 = n_2 = n_3 = n$, $\boldsymbol{\mu}_1 = \mathbf{0}$, *and* $\boldsymbol{\mu}_2 = \boldsymbol{\mu}_3 = \boldsymbol{\mu}$. *Construct the expected* L_1*-norm matrix* $\mathbf{D}_E$ *in* (14.42), $\mathbb{E}(\mathbf{D}_\delta)$ *in* (14.36), *and* $\mathbb{E}(\mathbf{D}_\delta)$ *in* (14.35).

Exercise 14.5 *Find the Eigenvalues and eigenvectors of the expected distance matrices in Exercise 14.4.*

Exercise 14.6 *Expected Dissimilarity Matrices for Uniform Distributions as* $d \to \infty$.

Let $\mathbf{X}_i = (X_{i1}, \ldots, X_{id})$ *be a d-dimensional random vector with i.i.d. components from* $U(a_i, b_i)$, *for* $i = 1, 2, 3, 4$. *Define*

$$\mu_i = \mathbb{E}[X_{ij}] = \frac{a_i + b_i}{2}, \quad \sigma_i^2 = \mathbb{V}ar(X_{ij}) = \frac{(b_i - a_i)^2}{12}$$

Let n *samples be drawn from each group* $\mathbb{X}_i$. *Define the normalized Euclidean distance*

$$\mathbf{D}(\mathbf{X}, \mathbf{Y}) = \frac{\|\mathbf{X} - \mathbf{Y}\|_2}{\sqrt{d}}.$$

Find the expected distance matrices $\mathbf{D}_E$, $\mathbf{D}_\rho$, *and* $\mathbf{D}_\delta$ *under the limit* $d \to \infty$.

Exercise 14.7 *Eigenvalue Analysis of Dissimilarity Matrices from Uniform Distributions Consider four populations*

$$\mathbb{X}_i = (X_{i1}, \ldots, X_{id})' \sim Uniform(a_i, b_i)^d$$

with, $a_1 = a_2 = a_3 = 0, \quad b_1 = b_2 = b_3 = 1, \quad a_4 = 2, \quad b_4 = 3.$

Hence, $\mu_{1,2,3} = 0.5, \quad \mu_4 = 2.5, \quad \sigma^2_{1,2,3,4} = \frac{1}{12}$ *and*

$$\nu_{i4}^2 = d(\mu_i - \mu_4)^2 = 4d \quad for\ i = 1, 2, 3.$$

Find the eigenvalues and eigenvectors of the expected distance matrices $\mathbf{D}_E$, $\mathbf{D}_\rho$, *and* $\mathbf{D}_\delta$ *under the limit* $d \to \infty$.

Exercise 14.8 *Expected Dissimilarity Matrices for Chi-Squared Populations*

Let $\mathbb{X}^{(i)} = \{\mathbf{X}_1^{(i)}, \ldots, \mathbf{X}_n^{(i)}\} \subset \mathbb{R}^d$, *where each group* $i \in \{1, 2, 3, 4\}$ *consists of* n *i.i.d. vectors with independent components from* $\chi^2(d_i)$, *and the distance between two vectors* $\mathbf{X}, \mathbf{Y} \in \mathbb{R}^d$ *is defined as*

$$D(\mathbf{X}, \mathbf{Y}) = \frac{\|\mathbf{X} - \mathbf{Y}\|_2}{\sqrt{d}}.$$

Find the expected distance matrices $\mathbf{D}_E$, $\mathbf{D}_\rho$, *and* $\mathbf{D}_\delta$ *under the limit* $d \to \infty$.

Exercise 14.9 *Eigen analysis of Dissimilarity Matrices for Chi-Squared Distributions.*

Let $X_i \sim \chi^2(1)$ *for* $i = 1, 2, 3$, *and* $X_4 \sim \chi^2(d)$, *with* $d \to \infty$.

We draw n *i.i.d. observations from each distribution. The distance between samples is defined as*

$$D(x, y) = \frac{\|x - y\|_2}{\sqrt{d}}.$$

We consider three dissimilarity matrices,

- $\mathbf{D}_E$: *Expected Euclidean distances.*
- $\mathbf{D}_\rho$: *Based on a definition similar to energy distances.*
- $\mathbf{D}_\delta$: *Based on modified distances using summary statistics.*

Find the eigenvalues and vectors of the expected distance matrices $\mathbf{D}_E$, $\mathbf{D}_\rho$, *and* $\mathbf{D}_\delta$ *under the limit* $d \to \infty$.

Exercise 14.10 *Expected Distance and Dissimilarity Matrices for k Groups with Mixture Contamination.*

Let $\mathbb{X}_i = \{\mathbf{X}_{i1}, \mathbf{X}_{i2}, \ldots, \mathbf{X}_{in}\}$ *denote the sample from group* i, *for* $i = 1, 2, \ldots, k$. *All observations lie in* $\mathbb{R}^d$, *and the total sample size is* $N = nk$.

- *For groups* $i = 2, 3, \ldots, k$, *the* n *observations are i.i.d. from the multivariate normal distribution,*
$$\mathbf{X}_{ij} \overset{i.i.d.}{\sim} \mathrm{N}_d(\mu_2 \mathbf{1}_d, \sigma_2^2 \mathbf{I}_d), \quad j = 1, \ldots, n.$$
- *The first group is contaminated with a mixture,*
$$\mathbf{X}_{1j} \sim \alpha\, \mathrm{N}_d(\mu_1 \mathbf{1}_d, \sigma_1^2 \mathbf{I}_d) + (1 - \alpha)\, \mathrm{N}_d(\mu_2 \mathbf{1}_d, \sigma_2^2 \mathbf{I}_d), \quad j = 1, \ldots, n.$$

We consider the normalized Euclidean distance

$$D(\mathbf{X}, \mathbf{Y}) = \frac{\|\mathbf{X} - \mathbf{Y}\|_2}{\sqrt{d}}.$$

We consider three dissimilarity matrices $\mathbf{D}_E$, $\mathbf{D}_\rho$, *and* $\mathbf{D}_\delta$. *Find the expected distance matrices* $\mathbf{D}_E$, $\mathbf{D}_\rho$, *and* $\mathbf{D}_\delta$ *under the limit* $d \to \infty$.

Exercise 14.11 *Let* $\mathbf{X} \sim \mathrm{N}(\boldsymbol{\mu}_1, \mathbf{I}_d)$, $\mathbf{Y} \sim \mathrm{N}(\boldsymbol{\mu}_2, \mathbf{I}_d)$, *and* $\mathbf{Z} \sim \mathrm{N}(\boldsymbol{\mu}_3, \mathbf{I}_d)$ *be three independent multivariate normal random vectors in* $\mathbb{R}^d$. *Under what conditions do they have equal expected distances? Let*

$$\boldsymbol{\mu}_1 = \mathbf{0}, \quad \boldsymbol{\mu}_2 = a\mathbf{e}_1, \quad \boldsymbol{\mu}_3 = \frac{a}{2}\mathbf{e}_1 + \frac{a\sqrt{3}}{2}\mathbf{e}_2,$$

where $a > 0$ *and* $\mathbf{e}_i$ *is the* i*-th standard basis vector. Verify that* $\mathbf{X}, \mathbf{Y}$ *and* Z *have equal expected distances.*

Exercise 14.12 *Let* $\mathbf{X} \sim \mathit{Unif}(a_1, b_1)^d$, $\mathbf{Y} \sim \mathit{Unif}(a_2, b_2)^d$, *and* $\mathbf{Z} \sim \mathit{Unif}(a_3, b_3)^d$, *where each random vector has independent components uniformly distributed over a hyper rectangle.*

Assume all intervals have equal length ℓ, *i.e.,* $b_i - a_i = \ell$ *for* $i = 1, 2, 3$. *Then, each marginal has*

$$\mathbb{E}[X_i] = \frac{a_i + b_i}{2}, \quad \mathbb{V}ar(X_i) = \frac{\ell^2}{12}.$$

Let $\ell = 2$ *and define*

$$\begin{aligned}
(a_1, b_1) &= (-1, 1)^d \quad \Rightarrow \boldsymbol{\mu}_1 = \mathbf{0}, \\
(a_2, b_2) &= (1, 3)^d \quad \Rightarrow \boldsymbol{\mu}_2 = 2\mathbf{1}, \\
(a_3, b_3) &= (-3, -1)^d \quad \Rightarrow \boldsymbol{\mu}_3 = -2\mathbf{1}.
\end{aligned}$$

Show that the means form an equilateral triangle.

Exercise 14.13 *Consider the expected L_1-distance matrix $\mathbf{D}_E$ for $K = 3$ groups of equal size n,*

$$\mathbf{D}_E = \begin{pmatrix} 2(\mathbf{1}_n - \mathbf{I}_n) & h\,\mathbf{1}_{n\times n} & h, \mathbf{1}_{n\times n} \\ h\,\mathbf{1}_{n\times n} & 2(\mathbf{1}_n - \mathbf{I}_n) & h\,\mathbf{1}_{n\times n} \\ h\,\mathbf{1}_{n\times n} & h\,\mathbf{1}_{n\times n} & 2(\mathbf{1}_n - \mathbf{I}_n) \end{pmatrix}.$$

Assume $r_i = \sqrt{2}\,\sigma_{F_i} = 2$ (so $\sigma^2_{F_i} = 2$ for all i) and $\nu_{ij} = 1$, hence

$$h = \sqrt{\sigma^2_{F_i} + \sigma^2_{F_j} + \nu^2_{ij}} = \sqrt{2+2+1} = \sqrt{5} \qquad (i \neq j).$$

What are the eigenvalues of $\mathbf{D}_E$?

Exercise 14.14 *Prove* (14.2), *i.e.*

$$\sqrt{m}\left(\bar{d}_x^{\,2} - 2tr(\mathbf{\Sigma}_x)\right) \rightarrow N\left(0, 8(1+\kappa)tr\left(\mathbf{\Sigma}_x^2\right) + 4\kappa\left(tr\mathbf{\Sigma}_x\right)^2\right). \qquad (14.78)$$

.

Exercise 14.15 *Let $\mathbf{D}_1 \in \mathbb{R}^{m\times m}$ and $\mathbf{D}_2 \in \mathbb{R}^{n\times n}$ be symmetric distance matrices with zero diagonal entries, i.e., $\mathbf{D}_1 = \mathbf{D}_1'$, $\mathbf{D}_2 = \mathbf{D}_2'$, $(\mathbf{D}_1)_{ii} = 0$, $(\mathbf{D}_2)_{jj} = 0$. Define $\mathbf{D} = \mathbf{D}_1 \otimes \mathbf{D}_2 \in \mathbb{R}^{mn\times mn}$ as the Kronecker product of $\mathbf{D}_1$ and $\mathbf{D}_2$. Prove that $\mathbf{D}$ is symmetric and has zeros on the diagonal, i.e., $\mathbf{D} = \mathbf{D}'$, and $\mathbf{D}_{ii} = 0 \quad \forall i = 1, \ldots, mn$.*

Exercise 14.16 *Let $\mathbf{D}_1 \in \mathbb{R}^{N_1\times N_1}$ and $\mathbf{D}_2 \in \mathbb{R}^{N_2\times N_2}$ be symmetric expected distance matrices based on two-group models. Assume $\mathbf{D}_1$ has n_1 samples from group 1, and n_2 samples from group 2, with $N_1 = n_1 + n_2$. Similarly, $\mathbf{D}_2$ corresponds to m_1, m_2 with $N_2 = m_1 + m_2$. We have*

$$\mathbf{D}_1 = \begin{pmatrix} r_1(\mathbf{1}_{n_1}\mathbf{1}_{n_1}' - \mathbf{I}_{n_1}) & h_{12}\mathbf{1}_{n_1\times n_2} \\ h_{12}\mathbf{1}_{n_2\times n_1} & r_2(\mathbf{1}_{n_2}\mathbf{1}_{n_2}' - \mathbf{I}_{n_2}) \end{pmatrix},$$

$$\mathbf{D}_2 = \begin{pmatrix} s_1(\mathbf{1}_{m_1}\mathbf{1}_{m_1}' - \mathbf{I}_{m_1}) & g_{12}\mathbf{1}_{m_1\times m_2} \\ g_{12}\mathbf{1}_{m_2\times m_1} & s_2(\mathbf{1}_{m_2}\mathbf{1}_{m_2}' - \mathbf{I}_{m_2}) \end{pmatrix}.$$

The matrix $M \in \mathbb{R}^{(N_1N_2)\times(N_1N_2)}$ is given by $\mathbf{M} = \mathbf{D}_1 \otimes \mathbf{D}_2$. Prove that $\mathbf{M}$ is a distance matrix and find its eigenvalues.

Exercise 14.17 *Write an R function to replicate Tables 14.3.*

15

Testing Multivariate Symmetry

15.1 Abstract

This chapter develops modern statistical methodology for testing multivariate symmetry, with primary emphasis on diagonal and angular symmetry in moderate and high dimensions. Multivariate symmetry underpins a wide range of inferential procedures, including robust location testing, regression diagnostics, and distributional modeling within elliptical and related families. After reviewing major notions of symmetry, spherical, elliptical, diagonal, angular, and depth-based, we focus on testing the fundamental hypothesis $H_0 : \mathbf{X} \stackrel{D}{=} -\mathbf{X}$, which guarantees invariance under coordinate-wise sign changes and forms the basis of many non-parametric procedures. We introduce two novel test statistics designed for high-dimensional settings: a modified energy test constructed using a new dissimilarity index δ that exploits IPD concentration, and a complementary test based on observation means. The proposed methods are rotation invariant, permutation calibrated, and shown to be consistent as the dimension diverges with fixed sample size. Extensive simulation studies demonstrate substantial power gains over existing symmetry tests, particularly under heavy-tailed, skewed, and HDLSS regimes. Applications to gene expression datasets illustrate the practical relevance of the methods and reveal clear departures from diagonal symmetry. Collectively, the chapter provides a unified theoretical and applied framework for multivariate symmetry testing in contemporary high dimensional data analysis.

15.2 Introduction

Fang et al. [96] eloquently remark, "Symmetry goes, of course, to the heart of the theory and philosophy of the laws of nature and mathematics." While symmetry admits a clear definition in the univariate setting, multiple notions of multivariate symmetry have been introduced, including spherical, elliptical, diagonal, and angular symmetry, listed here in decreasing order of stringency.

Multivariate symmetry underpins a wide array of statistical methodologies, encompassing distributional modeling, shape analysis, classification, hypothesis testing, and robust estimation. Let $\mathbf{X} \in \mathbb{R}^d$ denote a random vector with distribution function F. Imposing symmetry assumptions on F can streamline inferential procedures, enhance robustness, and, in some cases, characterize entire families of distributions. Various formalizations of symmetry in $\mathbb{R}^d$ have been proposed to capture specific structural or geometric invariances inherent in F. Such symmetries correspond to invariance under transformations including reflections, rotations, or permutations, and they are central to both the theoretical foundations and practical applications of multivariate analysis. Suppose the random vector $\mathbf{X} \in \mathbb{R}^d$ has density function $f(\mathbf{X})$, distribution function $F(\mathbf{X})$, and is centered at the origin. A

DOI: 10.1201/9781003789581-15

distribution is said to be diagonally symmetric about the origin if

$$F(\mathbf{X}) = F(-\mathbf{X}).$$

Equivalently, $\mathbf{X}$ is diagonally symmetric if $H_0 : \mathbf{X} \stackrel{D}{=} -\mathbf{X}$ holds, where D denotes equality in distribution. Consequently, the density satisfies $f(\mathbf{X}) = f(-\mathbf{X})$. Moreover, $\mathbf{X}$ is diagonally symmetric about the origin if and only if all linear projections of $\mathbf{X}$ onto lines through the origin have symmetric univariate distributions.

Because the null hypothesis $H_0 : \mathbf{X} \stackrel{D}{=} -\mathbf{X}$ implies invariance to coordinate–wise sign changes, a natural calibration is randomization by sign–flip permutations. Neuhaus and Zhu [248] formalize this idea for diagonal symmetry and develop exact finite-sample procedures. Testing multivariate diagonal symmetry is foundational because it underpins common model assumptions (e.g., elliptical families, multivariate normal models, and regression error structures), simplifies analysis by forcing odd moments to vanish (in particular, a zero mean when finite), and supports robustness for sign tests and rank–based procedures.

Fang, Kotz, and Ng [96] present a comprehensive monograph on multivariate symmetric distributions, highlighting their significance in non-parametric multivariate inference. Serfling [304] discusses several prominent notions of multivariate symmetry, including spherical, elliptical, central, and angular symmetry. Diagonal symmetry has been referred to by various terms in the literature, such as equal symmetry (Hollander [152]; Randles and Wolfe [278], radial symmetry (Nelsen [247]; Modarres [225]), and axial symmetry (Maritz [212]).

In his exposition of several forms of bivariate symmetry, Nelsen [247] emphasizes that radial symmetry is independent of the functional form of the symmetric marginal distributions. A notable subclass of diagonally symmetric distributions, the elliptically contoured family, encompasses several well-known distributions, including the multivariate normal, symmetric Kotz-type (Kotz [173]), Pearson types II, IV, and VII, as well as the Laplace, t, and Cauchy distributions.

Why Symmetry Matters

Testing whether a multivariate distribution exhibits symmetry is motivated by several considerations:

- Model assumptions: Many statistical models, such as elliptical distributions, multivariate normal models, and regression error structures, rely on the assumption of diagonal symmetry. Verifying this assumption is essential to ensure the validity of subsequent inference.
- Simplification: Symmetric data imply the vanishing of all odd moments; in particular, the mean is zero whenever it exists. This property reduces the complexity of modeling and estimation.
- Robustness: Symmetry underlies the theoretical justification of numerous robust procedures, including sign tests and rank-based multivariate methods.

Practical contexts in which diagonal symmetry is particularly important include

1. Finance and econometrics: Asset returns are often modeled as symmetric around zero. Detecting asymmetry can reveal skewness, risk imbalances, or structural breaks in financial series.

2. Multivariate statistical modeling: Elliptical families, such as the multivariate normal, t, and Cauchy distributions, depend on symmetry. Testing for symmetry helps assess the suitability of these models.

3. Regression diagnostics: Many linear and generalized linear models assume that residual vectors are symmetric. Deviations from symmetry may indicate model misspecification or unmodeled effects.

4. Signal processing and communications: Noise distributions are often assumed to be symmetric with mean zero. Testing this assumption is critical for the design and validation of practical systems.

5. Image analysis and computer vision: Feature vectors or wavelet coefficients are frequently expected to be symmetric. Assessing symmetry is useful in texture analysis, pattern recognition, and anomaly detection.

6. Biostatistics and genetics: In certain studies, effect sizes or gene expression contrasts are assumed symmetric under the null hypothesis of no directional effect, guiding hypothesis testing procedures.

7. Directional and spatial statistics: Symmetry is a natural property in isotropic or centrally balanced spatial processes, facilitating modeling and inference.

The assessment of diagonal symmetry is fundamental for confirming the validity of assumptions underlying many standard statistical models, maintaining robustness in inference, and guiding the selection of suitable analytical techniques in applied settings. This property is particularly central to one-sample multivariate location tests, including the multivariate sign and Wilcoxon signed-rank procedures.

Beyond rank-based methods, permutation-based approaches provide a rigorous framework for multivariate hypothesis testing, delivering exact or asymptotically justified results under the assumption of exchangeability. Pesarin and Salmaso [269] establish the finite-sample reliability of combination-based permutation tests and propose a general strategy for synthesizing information across dependent variables or projected components. More recently, Bonnini, Assegie, and Trzcinska [42] extend these methodologies to high-dimensional contexts, encompassing multivariate two-sample scenarios, and offer practical guidance on function aggregation and control of multiple comparisons. These methodological advancements form the foundation for the projection-aggregation calibration strategy employed in this work.

For symmetric distributions, it is natural to seek a measure of central tendency that coincides with the point of symmetry. The assumption of diagonal symmetry has been widely employed in the construction of multivariate location tests (Puri and Sen [274]; Hettmansperger [148]; Maritz [212]). In particular, it plays a critical role in robust location inference under heavy-tailed distributions.

Depth functions and other center-outward ranking methods have been developed with diagonal symmetry as a guiding principle. Székely and Sen [332] note that diagonal symmetry imposes weaker restrictions than alternative notions of multivariate symmetry, such as total, spherical, or elliptical symmetry. For instance, a mixture of elliptically symmetric distributions does not necessarily retain elliptical symmetry, yet it remains diagonally symmetric.

Let $\|\mathbf{X}\| = (\mathbf{X}'\mathbf{X})^{1/2}$ denote the Euclidean norm of $\mathbf{X}$. Following Wang and Modarres [349], consider a random sample $\mathbb{X} = \{\mathbf{X}_i\}_{i=1}^N$ drawn from F. The hypothesis of diagonal symmetry can be formulated as

$$H_0 : \mathbf{X} \overset{D}{=} -\mathbf{X}, \tag{15.1}$$

with the alternative

$$H_a : \mathbf{X} \overset{D}{\neq} -\mathbf{X}.$$

We are also concerned with assessing angular symmetry in high-dimensional settings (Liu [199]). A random vector $\mathbf{X}$ is said to be angularly symmetric about the origin if

$$\frac{\mathbf{X}}{\|\mathbf{X}\|} \overset{D}{=} -\frac{\mathbf{X}}{\|\mathbf{X}\|}.$$

Equivalently, $\mathbf{X}$ exhibits angular symmetry if the normalized vector $\frac{\mathbf{X}}{\|\mathbf{X}\|}$ is diagonally symmetric. Compared to diagonal symmetry, angular symmetry constitutes a weaker form of multivariate symmetry, imposing fewer constraints on the underlying distribution.

In the context of multivariate location inference, defining the location functional with respect to the center of angular symmetry can be advantageous. Moreover, because of its close relationship with half-space symmetry, angular symmetry underlies the construction of the half-space depth function and its associated methodologies.

15.3 Types of Symmetry

While the notion of symmetry is unique when $d = 1$, there are different types of symmetry when $d > 1$. We briefly outline common symmetry types and their key applications before turning our attention to tests of spherical, angular and radial symmetry:

15.3.1 Elliptical Symmetry

$\mathbf{X}$ has an elliptically symmetric distribution about $\boldsymbol{\mu} \in \mathbb{R}^d$ if

$$\mathbf{X} \overset{d}{=} \boldsymbol{\mu} + R\mathbf{A}\mathbf{U},$$

where $R > 0$ is a scalar random variable, $\mathbf{A}$ is a non–singular $d \times d$ matrix, and $\mathbf{U} \sim \text{Unif}(\mathbb{S}^{d-1})$. This type of symmetry underlies elliptical models like the multivariate normal or t-distributions. It is important in robust estimation, multivariate tests, and PCA. The multivariate t-distribution is elliptically symmetric about $\boldsymbol{\mu}$.

15.3.2 Permutation Symmetry

$\mathbf{X} = (X_1, \ldots, X_d)'$ is permutation symmetric if for any permutation π,

$$(X_1, \ldots, X_d) \overset{d}{=} (X_{\pi(1)}, \ldots, X_{\pi(d)}).$$

This type of symmetry is fundamental to Bayesian modeling, permutation tests, and non-parametric rank-based methods. A vector of i.i.d. standard normal variables is permutation symmetric.

15.3.3 Affine Symmetry

F is affinely symmetric about $\boldsymbol{\mu}$ if

$$\mathbf{X} \overset{d}{=} \boldsymbol{\mu} + \mathbf{A}(\mathbf{X} - \boldsymbol{\mu}),$$

for some invertible matrix $\mathbf{A}$ satisfying $\mathbf{A} = \mathbf{A}^{-1}$. This type of symmetry is used in Hotelling's T^2 test, affine-invariant estimators, and multivariate outlier detection. Multivariate normal distributions with symmetric covariance matrices are affinely symmetric.

15.3.4 Projective Symmetry

F is projectively symmetric about $\boldsymbol{\theta}$ if for all unit vectors $\mathbf{u} \in \mathbb{S}^{d-1}$,

$$\mathbf{u}'(\mathbf{X} - \boldsymbol{\theta}) \stackrel{d}{=} -\mathbf{u}'(\mathbf{X} - \boldsymbol{\theta}).$$

Projective Symmetry is used in projection pursuit and tests based on one-dimensional projections. Spherically symmetric distributions are projectively symmetric.

15.3.5 Depth-Based Symmetry

Given a statistical depth function $D(\cdot; F)$, F is depth-symmetric about $\boldsymbol{\theta}$ if

$$D(\boldsymbol{\theta} + \mathbf{X}; F) = D(\boldsymbol{\theta} - \mathbf{X}; F) \quad \text{for all } \mathbf{X} \in \mathbb{R}^d.$$

This type of symmetry provides robust, non-parametric inference using data depth, multivariate medians, and ordering. The Tukey halfspace and lens depth functions are symmetric about the median for depth-symmetric distributions.

15.3.6 Spherical Symmetry

F is spherically symmetric about $\boldsymbol{\mu}$ if

$$\mathbf{X} - \boldsymbol{\mu} \stackrel{d}{=} \mathbf{O}(\mathbf{X} - \boldsymbol{\mu})$$

for every orthogonal matrix $\mathbf{O} \in \mathbb{R}^{d \times d}$. This type of symmetry is is found in directional data, and rotationally invariant models. The multivariate standard normal distribution is spherically symmetric about $\mathbf{0}$.

15.3.7 Radial Symmetry

F is radially symmetric about $\boldsymbol{\mu}$ if the distribution of $\|\mathbf{X} - \boldsymbol{\mu}\|$ determines F, i.e.,

$$\mathbf{X} - \boldsymbol{\mu} \stackrel{d}{=} -(\mathbf{X} - \boldsymbol{\mu}).$$

This type of symmetry is important in depth functions.

15.3.8 Angular Symmetry

A unit vector $\mathbf{V} = (\mathbf{X} - \boldsymbol{\mu})/\|\mathbf{X} - \boldsymbol{\mu}\|$ is angularly symmetric if $\mathbf{V} \stackrel{d}{=} -\mathbf{V}$. This type of symmetry is relevant in geophysics, shape theory, and dispersion analysis. Uniform distributions on the sphere have angular symmetry.

15.3.9 Half-Space Symmetry

F is half-space symmetric about $\boldsymbol{\theta}$ if for every direction $\mathbf{u} \in \mathbb{S}^{d-1}$,

$$\mathbb{P}(\mathbf{u}'\mathbf{X} \leq \mathbf{u}'\boldsymbol{\theta}) = \frac{1}{2}.$$

This type of symmetry is is the basis for Tukey depth, convex hull analysis, and symmetry tests. A centrally symmetric distribution has half-space symmetry.

15.4 Tests of diagonal symmetry

A substantial body of work has focused on testing the hypothesis

$$H_0 : \mathbf{X} \stackrel{D}{=} -\mathbf{X}.$$

Aki [7] proposed a rotation-invariant procedure for assessing diagonal symmetry based on the empirical distribution function. Ghosh and Ruymgaart [117] utilized empirical characteristic functions to construct a corresponding test. Székely and Sen [332] characterized diagonal symmetry through a degree-2 kernel and developed a test for the case of unknown location, employing related U-processes, while also investigating its asymptotic behavior as $N \to \infty$. Yang and Modarres [349] further discuss methodologies for testing diagonal symmetry.

Székely and Rizzo [333] introduced the concept of the energy distance, and in a subsequent work, Székely and Rizzo [329] demonstrated that tests for diagonal symmetry can be viewed as a special case of the two-sample energy test. Dyckerhoff et al. [85] proposed a bivariate diagonal symmetry test based on data depth. Einmahl and Gan [87] constructed two rotation-invariant tests utilizing empirical measures of opposite regions. Sang and Dang [295] developed a jackknife empirical likelihood approach for testing diagonal symmetry, while Chen, Meintanis, and Zhu [58] proposed a test grounded in the distance between characteristic functions. Dehghan and Faridrohani ([71], [72], [73]) extended this approach using center-outward depth rankings.

Suppose $\mathbf{X}$, $\mathbf{X}'$ are independent and identically distributed (i.i.d.) vectors taken from distribution function (DF) F_1 and $\mathbf{Y}$, $\mathbf{Y}'$ be i.i.d. vectors taken from DF F_2. To measure the separation between F_1 and F_2 Baringhaus and Franz [21] and Székely and Rizzo [333] introduce the energy distance

$$\mathbb{D}(\mathbf{X}, \mathbf{Y}) = 2\mathbb{E}\|\mathbf{X} - \mathbf{Y}\| - \mathbb{E}\|\mathbf{X} - \mathbf{X}'\| - \mathbb{E}\|\mathbf{Y} - \mathbf{Y}'\| \geq 0 \tag{15.2}$$

with equality holding when $L_0 : F_1 = F_2$ is true. Buja et al. [46] and Székely and Móri [328] prove that $\mathbb{E}(\|\mathbf{X} + \mathbf{X}'\| - \mathbb{E}\|\mathbf{X} - \mathbf{X}'\|) \geq 0$ with equality holding if and only if the distribution of $\mathbf{X}$ and $-\mathbf{X}$ are equal. Since $\mathbb{D}(\mathbf{X}, -\mathbf{X}') = 2(\mathbb{E}\|\mathbf{X} + \mathbf{X}'\| - \mathbb{E}\|\mathbf{X} - \mathbf{X}'\|)$ it follows that H_0 is equivalent to testing

$$K_0 : \mathbb{E}(\|\mathbf{X} + \mathbf{X}'\| - \|\mathbf{X} - \mathbf{X}'\|) = 0. \tag{15.3}$$

15.4.1 Test of Székely and Móri

Székely and Móri [328] suggest a consistent test against general alternatives based on the statistic

$$T_n = \frac{\sum\limits_{1 \leq i < j \leq n} (\|\mathbf{X}_i + \mathbf{X}_j\| - \|\mathbf{X}_i - \mathbf{X}_j\|)}{\sum\limits_{1 \leq i \leq n} \|\mathbf{X}_i\|}. \tag{15.4}$$

One can apply the chi-squared test criterion in Székely and Bakirov [327] and reject H_0 at significance level α if $T_n \geq (\Phi^{-1}(1 - \alpha))^2$, where Φ is the standard normal DF.

15.4.2 Test of Sang

Sang [294] uses the energy distance between the distributions of $\mathbf{X}$ and $-\mathbf{X}'$ as $2(\mathbb{E}\|\mathbf{X}+\mathbf{X}'\|-\mathbb{E}\|\mathbf{X}-\mathbf{X}'\|)$ and proposes a statistic for testing high dimensional diagonal symmetry under K_0. Let

$$U_n = \binom{n}{2}^{-1} \sum_{1\le i<j\le n} (\|\mathbf{X}_i+\mathbf{X}_j\| - \|\mathbf{X}_i - \mathbf{X}_j\|). \tag{15.5}$$

An unbiased variance of U_n is

$$\hat{\sigma}_0^2 = \binom{n}{2}^{-2} \sum_{1\le i<j\le n} (\|\mathbf{X}_i+\mathbf{X}_j\| - \|\mathbf{X}_i - \mathbf{X}_j\|)^2. \tag{15.6}$$

Sang [294] proves that under some regularity conditions as d and n both diverge,

$$Z = \frac{U_n}{\hat{\sigma}_0} \tag{15.7}$$

converges to a standard normal distribution.

We offer two test statistics next. The first test is based on the energy test using the dissimilarity δ instead of the Euclidean distance. The second test is based on the normalized difference of the grand observation means, which does not require IPD computations.

15.4.3 Energy Test using δ Dissimilarity

Consider a random partition of the sample $\mathbb{X} = \{\mathbf{X}_i\}_{i=1}^N$ into two approximately equal-sized subsets of sizes m and n, so that $N = m+n$. Let $\mathbb{X}_1 = \{\mathbf{X}_i\}_{i=1}^m$ be viewed as a sample from the original distribution F_1, and let $\mathbb{Y} = -\{\mathbf{X}_i\}_{i=m+1}^N$ represent the corresponding sample of mirror images, drawn from the distribution F_2. Under this construction, testing for diagonal symmetry is equivalent to testing the equality of distributions $L_0 : F_1 = F_2$. This equivalence allows one to utilize high-dimensional two-sample procedures to assess diagonal symmetry. In particular, we introduce a novel high-dimensional dissimilarity measure δ and combine it with the energy statistic to develop new tests for diagonal symmetry, particularly sensitive to shifts in the center.

It is important to note that these procedures are not invariant under permutations of the sample vectors $\mathbf{X}_1, \ldots, \mathbf{X}_N$ due to the sample-splitting step. Sample-splitting is, however, a widely adopted strategy in contemporary statistical methodology. Techniques such as dimension reduction, model selection, and cross-validation are frequently applied to randomly chosen subsamples, resulting in statistics that are inherently non-deterministic: different analysts can compute different values from the same dataset depending on the split. Theoretical foundations for hypothesis testing under sample-splitting have been developed in Wasserman et al. [351] and DiCiccio et al. [76].

In high-dimensional settings, the conventional Euclidean distance becomes less informative due to the concentration of measure phenomenon. To address this, a scaled Euclidean distance, normalized by the factor $1/\sqrt{d}$, has been proposed to ensure meaningful convergence as the dimension d increases (Hall et al. [133]). However, both the standard and scaled Euclidean distances rely solely on pairwise comparisons between vectors, potentially ignoring higher-order structural information. In this work, we adopt the dissimilarity index δ to capture more comprehensive relationships among observations.

Suppose $\mathbb{X} = \{\mathbf{X}_i\}_{i=1}^m$ and $\mathbb{Y} = \{\mathbf{Y}_i\}_{i=1}^n$ represent i.i.d. vectors in $\mathbb{R}^d$ from distribution functions (DFs) F_1 and F_2, respectively. We are interested in testing the null hypothesis $L_0 : F_1 = F_2$ when $d \gg N$. The IPDs using Euclidean distances are defined in section 2.5.

Suppose $\delta_{(x)kl}$, $\delta_{(y)kl}$, and $\delta_{(xy)kl}$ are the IPDs between the $\mathbb{X}$ sample, the $\mathbb{Y}$ sample, and across samples using the dissimilarity index δ defined by

$$\bar{\delta}_{(x)} = \frac{2}{m(m-1)} \sum_{1 \leq k < l \leq m} \delta_{(x)kl}, \tag{15.8}$$

$$\bar{\delta}_{(y)} = \frac{2}{n(n-1)} \sum_{1 \leq k < l \leq n} \delta_{(y)kl}, \tag{15.9}$$

$$\bar{\delta}_{(xy)} = \frac{1}{mn} \sum_{k=1}^{m} \sum_{l=1}^{n} \delta_{(xy)kl}. \tag{15.10}$$

Using the Euclidean distance, the statistics $R_{(Euc)}$ for testing diagonal symmetry is

$$R_{(Euc)} = 2\bar{d}_{(xy)} - \bar{d}_{(x)} - \bar{d}_{(y)}. \tag{15.11}$$

The hypothesis L_0 is rejected for large values of $R_{(Euc)}$, and we obtain its sampling distribution by permutation testing.

To provide a theoretical foundation for the calibration employed in our procedures, we formalize the validity of the permutation approach under the assumption of diagonal symmetry. Under the null hypothesis H_0, the sample is invariant to diagonal symmetry transformations, which implies that the associated permuted statistics are conditionally exchangeable. This invariance guarantees that the permutation distribution serves as a valid reference for inference. The following result makes these properties precise and underpins the validity of all subsequent p-values and critical regions.

Theorem 15.4.1 *Consider the permutation test with the approximate permutation distribution $\hat{G}_{R_\delta}(t)$ and the permutation-based p-value p_{perm}. Assume assumptions A1–A3 hold. Under the null hypothesis H_0 of diagonal symmetry, we have as $n \to \infty$ and $d \to \infty$*

(a) $\hat{G}_{R_\delta}(t) \xrightarrow{D} G_{R_\delta}(t) \quad \forall t,$

(b) $\mathbb{P}_{H_0}(p_{perm} \leq \alpha) \leq \alpha \quad \forall\ 0 \leq \alpha \leq 1.$

Based on R permutations, the approximate permutation distribution of the test statistic, denoted by $\hat{G}_{R_\delta}(t)$, is defined as

$$\hat{G}_{R_\delta}(t) = \frac{1}{R} \sum_{r=1}^{R} \mathbb{I}\big(R_\delta^{(r)} \leq t\big), \tag{15.12}$$

where $\mathbb{I}(\cdot)$ is the indicator function and $R_\delta^{(r)}$ is the test statistic calculated for the r–th permutation sample. The p–value of the permutation test, denoted by p_{perm}, is given by

$$p_{\text{perm}} = 1 - \hat{G}_{R_\delta}(R_\delta(\text{obs})) = \frac{1}{R} \sum_{r=1}^{R} \mathbb{I}\big(R_\delta^{(r)} > R_\delta(\text{obs})\big), \tag{15.13}$$

where $R_\delta(\text{obs})$ is the test statistic for the observed sample. The theoretical results show that $\mathbb{P}_{H_0}(p_{\text{perm}} \leq \alpha) \leq \alpha$ for all $0 \leq \alpha \leq 1$. Theorem 15.4.1 establishes the validity of permutation-based calibration under diagonal symmetry. Specifically, under H_0, the collection of permuted statistics is conditionally exchangeable, which ensures that the exact permutation distribution produces a level-α test with correct size. Furthermore, as the sample size increases, the empirical permutation cumulative distribution function (CDF) converges to the exact permutation CDF, implying that Monte Carlo approximations of the permutation test become increasingly accurate. Importantly, this result holds for any test statistic that is invariant under diagonal symmetry transformations.

The dissimilarity index δ, when combined with the energy test framework, yields a powerful procedure for assessing high-dimensional symmetry. Accordingly, a consistent test against general alternatives can be constructed using the statistic

$$R_{(\delta)} = 2\bar{\delta}_{(xy)} - \bar{\delta}_{(x)} - \bar{\delta}_{(y)}. \tag{15.14}$$

A related approach is discussed by Sarkar and Ghosh [298], who extend the energy statistic to high-dimensional two-sample tests using the scaled Euclidean distance $\|\mathbf{X}_i - \mathbf{X}_j\|_D = d^{-1/2}\|\mathbf{X}_i - \mathbf{X}_j\|$ and IPDs.

Under the null hypothesis L_0, large values of $R_{(\delta)}$ indicate departures from symmetry, and its sampling distribution can be approximated using permutation methods. In the following, we establish the high-dimensional consistency of this modified energy test based on the dissimilarity index δ.

Theorem 15.4.2 *Suppose that $\mathbb{X} = \{\mathbf{X}_1, \ldots, \mathbf{X}_m\} \sim F_1$ and $\mathbb{Y} = \{\mathbf{Y}_1, \ldots, \mathbf{Y}_n\} \sim F_2$ are independent random vectors, where F_1 and F_2 satisfy Assumptions A1–A3 with $\bar{\mu}_x^d \neq \bar{\mu}_y^d$ or $\sigma_x^d \neq \sigma_y^d$. Assuming that $\binom{N}{m} > \frac{2}{\alpha}$, where α is the level of the test, the power of $R_{(\delta)}$ converges to 1 as $d \to \infty$.*

Theorem 15.4.2 shows that the energy test based on the dissimilarity index δ for high-dimensional symmetry is consistent. The statistics $R_{(\delta)}$ converges in probability to $2\tilde{\delta}(F_1, F_2) = 2|\bar{\mu}_x^d - \bar{\mu}_y^d|$ as $d \to \infty$. For small N, we use the permutation distribution of $R_{(\delta)}$ to construct the test. Under L_0 all observations are exchangeable. To obtain its permutation distribution we generate R random permutation. The approximate permutation distribution of the test statistic, denoted by $H_{R_{(\delta)}}$, is a function of the test statistic

$$H_{R_{(\delta)}}(t) = \frac{1}{R}\sum_{r=1}^{R} \mathbb{I}\big(R^r_{(\delta)} \leq t\big), \tag{15.15}$$

where $\mathbb{I}(\cdot)$ is the indicator function and $R^r_{(\delta)}$ is the test statistic calculated for the r-th permutation sample. The p-value of the permutation test, denoted by d_{perm}, is given by

$$p_{\text{perm}} = 1 - H_{R_{(\delta)}}(\text{obs}) = \frac{1}{R}\sum_{r=1}^{R} \mathbb{I}\big(R^r_{(\delta)} > R_{(\delta)}(\text{obs})\big), \tag{15.16}$$

where $R_{(\delta)}(\text{obs})$ is the test statistic for the observed sample.

15.4.4 Test Based on the Observation Means

A major computational bottleneck in many methods that rely on IPDs is the calculation of these distances. For a dataset with N observations, computing all IPDs requires $O(dN^2)$ operations. Approaches such as those of Székely and Móri [328] and Sang [294] fall into this category. When N is in the hundreds, these computations remain manageable even for large d; however, when both N and d are large, the computational burden becomes prohibitive.

To address this, we propose a new test statistic for $L_0 : F_1 = F_2$ with computational complexity $O(dN)$, based on the squared asymptotic value of the dissimilarity index δ as $d \to \infty$. Modarres [233] shows that the eigenvalues of the expected two-sample dissimilarity matrices based on δ and ρ converge to $\frac{mn}{N}\tilde{\delta}^2$ and $\frac{mn}{N}\tilde{\rho}^2$, respectively, in the high-dimensional limit. Sarkar and Gosh [298] use $\tilde{\rho}^2$, while Wang and Modarres [349] employ $\tilde{\delta}^2$ for clustering high-dimensional observations.

The following Lemma presents the limiting value of the squared dissimilarity index as $d \to \infty$. Notably, this limit depends solely on the mean vectors of the populations and is independent of the observation variances.

Lemma 15.4.1 *Suppose* $\mathbf{X} \sim F_1$ *with mean* $\boldsymbol{\mu}$ *and covariance matrix* $\boldsymbol{\Sigma}$ *and let* $\mathbf{Y} = -\mathbf{X} \sim F_2$. *Let* $\bar{\mu} = \frac{1}{d}\sum_{t=1}^{d} \boldsymbol{\mu}_t$ *be the mean of* $\boldsymbol{\mu}$. *As* $d \to \infty$, *the limiting value of the dissimilarity index is* $\tilde{\delta}^2(F_1, F_2) = 4\bar{\mu}^2 \geq 0$, *with equality holding when* $\mathbf{X}$ *diagonally symmetric.*

Following Sang [294] we consider two mixture distributions.

Example 15.4.1 *Suppose* $\mathbf{Z}_1 = (Z_{11}, Z_{12}, \ldots, Z_{1d})$ *where* Z_{1j} *is i.i.d. from a uniform* $(-1, 1)$ *distribution. Suppose* $\mathbf{Z}_2 = (Z_{21}, Z_{22}, \ldots, Z_{2d})$ *where* Z_{2j} *is i.i.d. from a uniform* $(-1, 1 + a)$ *distribution where* $a \geq 0$ *measures the shift. Let* $\mathbf{Z} = \theta\mathbf{Z}_1 + (1 - \theta)\mathbf{Z}_2$. *It is not difficult to show that* $\bar{\mu}_x^d = \frac{1}{2}a(1 - \theta)$ *and* $\bar{\mu}_y^d = -\frac{1}{2}a(1 - \theta)$. *It follows that* $\tilde{\delta}^2(F_1, F_2) = 4a^2(1 - \theta)^2 \geq 0$. *One can verify that when* $\theta = 1$, *the distribution is diagonally symmetric and when* $\theta = 0$ *the distribution is symmetric only if* $a = 0$. *For values of* θ *in the range* $0 < \theta < 1$, $\mathbf{Z}$ *is not diagonally symmetric and* $\tilde{\delta}^2(F_1, F_2) > 0$.

Example 15.4.2 *Suppose* $\mathbf{E} = (E_1, E_2, \ldots, E_d)$ *where for* $i = 1, \ldots, d$, E_i*'s are i.i.d. from* $Exp(1)$. *Consider* N *samples from the mixture* $\mathbf{Z} = (1 - \theta)\mathbb{N}(\mathbf{0}, \mathbf{I}_d) + \theta\mathbf{E}$ *where the mixture proportion* $\theta \in [0, 1]$. *Suppose* $m = n = N/2$ *and let* $\mathbf{X}_i = \mathbf{Z}_i$ *for* $i = 1, \ldots, m$ *and* $\mathbf{Y}_j = -\mathbf{Z}_{m+j}$ *for* $j = 1, \ldots, n$. *Hence, the sample* $\mathbf{Y}_j$ *contains the mirror images. It is not difficult to show that* $\bar{\mu}_x^d = \theta$ *and* $\bar{\mu}_y^d = -\theta$. *We can also obtain* $\mathbb{V}ar(\bar{X}^d) = \frac{1}{d}[(1-\theta)^2 + 2\theta^2] = \mathbb{V}ar(\bar{Y}^d)$. *It follows that* $\tilde{\delta}^2(F_1, F_2) = 4\theta^2 \geq 0$. *If* $\theta = 0$, *then all the observations come from* $\mathbb{N}(\mathbf{0}, \mathbf{I}_d)$, *which is diagonally symmetric. For values of* θ *in the range* $0 < \theta \leq 1$, $\mathbf{Z}$ *is not diagonally symmetric and* $\tilde{\delta}^2(F_1, F_2) > 0$.

Theorem 15.4.3 *Let* $\{\mathbf{X}_i\}_{i=1}^m$ *be* m *independent copies of* $\mathbf{X}$ *with mean* $\boldsymbol{\mu}_x$ *and covariance matrix* $\boldsymbol{\Sigma}_x$. *Define the grand mean across replicates and coordinates*

$$\bar{X}^d = \frac{1}{md}\sum_{i=1}^{m}\sum_{t=1}^{d} X_{it}.$$

Then, as $m \to \infty$ *and* $d \to \infty$,

$$\sqrt{\frac{md^2}{tr(\boldsymbol{\Sigma}_x)}}\left(\bar{X}^d - \frac{1}{d}\mathbf{1}_d'\boldsymbol{\mu}_x\right) \xrightarrow{d} \mathbb{N}(0,\, 1).$$

Suppose $\{\mathbf{Y}_j\}_{j=1}^n$ are n independent copies of $\mathbf{Y}$, which has mean $\boldsymbol{\mu}_y$ and covariance matrix $\boldsymbol{\Sigma}_y$. Similarly, $\bar{Y}^d$ is asymptotically normal as $m \to \infty$ and $d \to \infty$ with mean $\mathbb{E}(\bar{Y}^d) = \frac{1}{d}\mathbf{1}_d'\boldsymbol{\mu}_y$ and $\mathbb{V}ar(\bar{Y}^d)$.

Suppose we estimate $tr(\boldsymbol{\Sigma}_x)$ and $tr(\boldsymbol{\Sigma}_y)$ with $tr(\mathbf{S}_x)$ and $tr(\mathbf{S}_y)$, respectively. It follows that $\mathbb{E}(\bar{X}^d - \bar{Y}^d) = \frac{1}{d}\mathbf{1}_d'(\boldsymbol{\mu}_x - \boldsymbol{\mu}_y) = 0$ under the null hypothesis. Since $\bar{X}^d$ and $\bar{Y}^d$ are independent $W = \frac{\bar{X}^d - \bar{Y}^d}{\hat{\sigma}}$ follows standard normal distribution as $m \to \infty$ and $d \to \infty$ where

$$\hat{\sigma}^2 = \frac{1}{d^2}(tr(\mathbf{S}_x)/m + tr(\mathbf{S}_y)/n) \tag{15.17}$$

is the pooled variance for σ^2. We propose the following statistic for testing diagonal symmetry,

$$A_D = \left(\frac{\bar{X}^d - \bar{Y}^d}{\hat{\sigma}}\right)^2, \tag{15.18}$$

which, as $\min(n, m) \to \infty$ and $d \to \infty$, follows a χ^2 distribution with 1 degree of freedom.

The null hypothesis H_0 is rejected for large values of the statistic A_D. For small sample sizes N, the permutation distribution of A_D can be used to obtain exact critical values. Importantly, the computation of A_D does not require the calculation of IPDs, and its computational complexity is $O(Nd)$. This makes the statistic particularly advantageous when both N and d are large.

According to Theorem 15.4.3, we reject L_0 if

$$A_D > \chi_1^2(1 - \alpha),$$

where $\chi_1^2(1 - \alpha)$ denotes the $(1 - \alpha)100\%$ percentile of a chi-square distribution with 1 degree of freedom. This result circumvents the computational burden of permutation tests.

The power function of A_D is defined as

$$\mathbb{P}_n(\alpha) = \mathbb{P}(A_D > \chi_1^2(1 - \alpha) \mid L_1),$$

where L_1 represents the alternative hypothesis. In the following, we establish the asymptotic distribution of A_D under L_1 and demonstrate its consistency.

Theorem 15.4.4 *Consider the sample $\mathbb{X} = \{\mathbf{X}_i\}_{i=1}^m$ arising from distribution F_1 with mean $\boldsymbol{\mu}_x$ and the sample $\mathbb{Y} = -\{\mathbf{X}_i\}_{i=m+1}^N$ with mean $\boldsymbol{\mu}_y$ arising from the population of mirror images F_2 in $\mathbb{R}^d$. Under the alternative hypothesis L_1, as $\min(n, m) \to \infty$ and $d \to \infty$, A_D has a non–central χ^2 distribution with 1 degree of freedom and non–centrality parameter η^2/σ^2, where $\eta = \frac{1}{d}\mathbf{1}_d'(\boldsymbol{\mu}_x - \boldsymbol{\mu}_y)$ and σ^2 is the common variance defined following* (15.17)*. Furthermore,*

$$\mathbb{P}_n(\alpha) = \mathbb{P}(A_D > \chi_1^2(1 - \alpha)|L_1) \to 1. \tag{15.19}$$

To observe the finite sample properties of A_D we performed the Shapiro-Wilk test of normality (Royston [292]) on W for $N \in \{20, 30, 40, 50, 60\}$ and seven distributions discussed. We simulated 500 values of the W and report the p-values in Table 15.1. Note that $m = n = N/2$ and $d = 50$ in all cases. It is evident from the p-values that W is normally distributed for small sample sizes. Hence, Table 15.1 verifies that A_D has a chi_1^2 distribution for finite sample sizes.

TABLE 15.1: P-values for W using the Shaprio-Wilk test of normality.

N	20	30	40	50	60
Normal	0.578	0.191	0.190	0.937	0.410
Uniform	0.273	0.599	0.406	0.737	0.396
t_6	0.450	0.833	0.782	0.990	0.942
Beta	0.461	0.838	0.908	0.827	0.439
Logistic	0.656	0.248	0.277	0.378	0.781
Skew Normal	0.061	0.282	0.233	0.846	0.718
Corr. Normal	0.944	0.418	0.995	0.233	0.472

15.5 Comparisons

We compared the performance of the following test statistics for testing diagonal symmetry using sampling experiments with $m = n = 20$. The test statistics are

- T_n in (15.4), $Z = \frac{U_n}{\hat{\sigma}_0}$ in (15.7), $R_{(Euc)}$ in (15.11), $R_{(\delta)}$ in (15.14) and A_D in (15.18).

15.5.1 Experiment 1

We conducted simulation experiments with 500 replications, setting the dimension $d = 512$ and the sample size $N = 40$, to evaluate the empirical α-level and power of the proposed test statistics under various alternatives. To assess power under a location shift, each generated sample was displaced by $\mu = \{0, 0.01, 0.02, \ldots, 0.1\}$. Although the test statistics are asymptotically normal as both n and d grow, we employed the approximate permutation distribution to determine critical values and set $\alpha = 0.05$. The observed significance levels closely matched the nominal level of 0.05. Figures 15.1 present the empirical power curves for the test statistics based on independent samples drawn from $\mathbb{N}(\mu\ \mathbf{1}_d, \mathbf{I})$.

Wang and Modarres [349] further investigated the empirical power curves based on independent samples drawn from Uniform t_6, Beta, Logistic, Skew-Normal, Normal-Exponential mixture described in Example 15.4.2, among others. Their findings are summarized below. In all cases they found the power ordering $A_D > R_{(\delta)} > R_{(\text{Euc})}$, with the remaining statistics exhibiting substantially lower power. Although the energy test based on the Euclidean distance, $R_{(\text{Euc})}$, performed reasonably well, its power was consistently lower than that of A_D and $R_{(\delta)}$, which uses the dissimilarity measure δ. The statistics Z and T_n demonstrate satisfactory power under uniform and beta distributions, but in all scenarios, the newly proposed statistics A_D and $R_{(\delta)}$ clearly outperform the others in detecting departures from diagonal symmetry. Overall, the power of Z and T_n is considerably lower than that of A_D and $R_{(\delta)}$.

The permutation group consists of all diagonal symmetry transformations of fixed size R. When 2^N is computationally feasible, we recommend using the exact permutation distribution, which provides an exact level-α test under H_0 by Theorem 15.4.1. In Experiment 1, we set $m = n = 20$ and $d = 512$, so that $N = 40$. For $N = 40$, we employ Monte Carlo permutation p-values based on R randomly selected diagonal symmetry transformations. By Theorem 15.4.1, these p-values are uniform for any finite R, and are therefore conservative. We choose R large enough that the variability at level α is negligible. Specifically, with $R = 1000$, the binomial standard error at $\alpha = 0.05$ is $\sqrt{\frac{\alpha(1-\alpha)}{R}} \approx 0.00689$, which is sufficiently small to ensure that conclusions remain stable. Exact enumeration is applied only when 2^N is computationally manageable; otherwise, the Monte Carlo approximation accurately reproduces the permutation distribution for our statistics under the settings of Experiments 1 and 2.

To further verify that our permutation-calibrated procedures maintain the nominal significance level, we conducted a focused size assessment under the null hypothesis. We found that all the estimated Type I errors lie within these bands, indicating satisfactory control of the Type I error.

15.5.2 Experiment 2

We fix the total sample size at $N = 40$ and vary the dimension $d = 2^i$ for $i \in \{4, \ldots, 10\}$, applying a uniform shift $\mu = 0.05$ across the nine distributions considered in Experiment 1. Under independence, the empirical power of A_D and $R_{(\delta)}$ increases with d and approaches one rapidly. In contrast, $R_{(\text{Euc})}$, T_n, and Z improve more gradually and, in several models, plateau well below power of 1, reflecting the loss of contrast for raw Euclidean IPDs due to the concentration of distances in high dimensions.

We found that tail behavior influences the rate at which power rises but does not alter the qualitative ranking. For light and moderately heavy-tailed distributions (Normal, Uniform, Beta, Logistic, t_6), A_D dominates, $R_{(\delta)}$ is a close second, and both substantially outperform $R_{(\text{Euc})}$, T_n, and Z. Skewed and mixture alternatives exhibit the same ordering, with A_D and $R_{(\delta)}$ remaining highly sensitive to departures from diagonal symmetry.

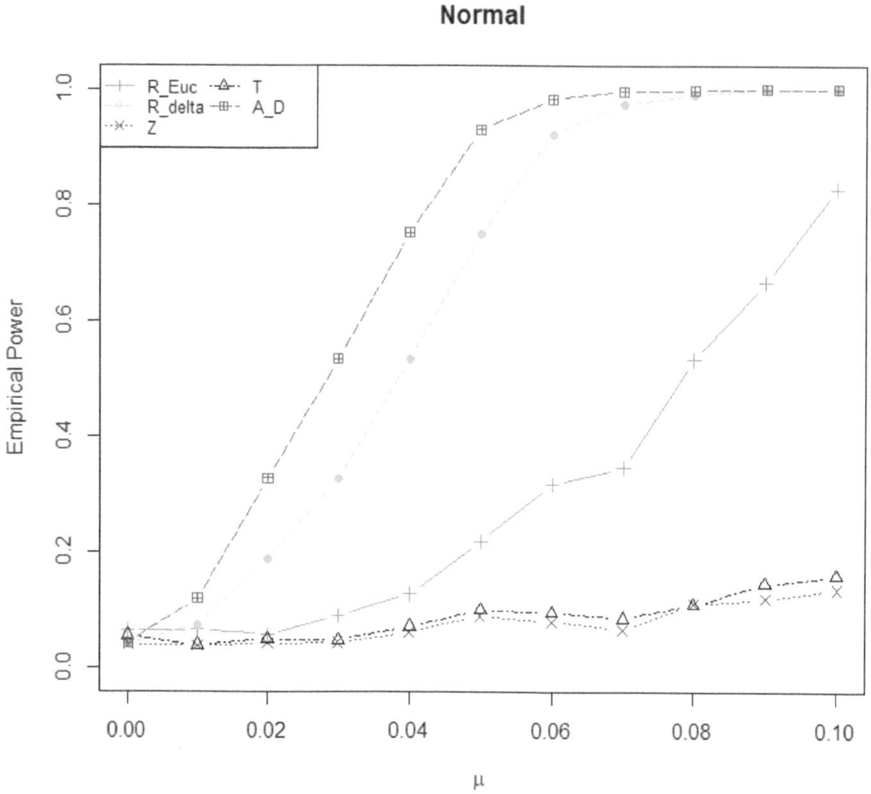

FIGURE 15.1: $\mathbb{N}(\mathbf{0}, \mathbf{I})$.

Dependence modifies the relative performance in predictable ways. Under spatial dependence, A_D and $R_{(\delta)}$ retain monotone power gains and continue to approach one as d increases, while the other statistics lag behind. In the Normal–Exponential design $R_{(\delta)}$ surpasses A_D, indicating an advantage of the dissimilarity aggregation when a strong common component exists across coordinates. Overall, across all distributions and dependence structures, the power ranking is stable: $A_D \gtrsim R_{(\delta)} \gg R_{(\text{Euc})} \gtrsim T_n \gtrsim Z$, with the notable exception of the Normal–Exponential scenario where $R_{(\delta)}$ becomes superior.

15.6 Summary

Chapter 15 develops modern statistical methods for testing multivariate symmetry, with a focus on diagonal and angular symmetry in moderate and high-dimensional settings. Multivariate symmetry is fundamental to numerous inferential tasks, including robust location testing, regression diagnostics, and modeling within elliptical or related distributional families. After reviewing major notions of symmetry, spherical, elliptical, half-space, diagonal, and angular, the chapter concentrates on testing the hypothesis $H_0 : \mathbf{X} \stackrel{D}{=} -\mathbf{X}$, which ensures invariance under coordinate-wise sign changes and underpins many non-parametric procedures.

Two novel test statistics are introduced for high-dimensional applications. The first is a modified energy test employing a new dissimilarity index δ that exploits the concentration of IPDs in high dimensions, enhancing the classical energy statistic of Székely and Móri [328]. The second test is based on the means of observations, providing a complementary approach. Both tests are rotation invariant, permutation calibrated, and provably consistent as the dimension increases while the sample size remains fixed. Simulations show that these methods achieve substantial power improvements over existing symmetry tests, particularly in heavy-tailed, skewed, and high-dimensional low-sample-size (HDLSS) regimes.

Applications to gene expression and other high-dimensional datasets illustrate the practical utility of the proposed methods, revealing clear departures from diagonal symmetry and highlighting the sensitivity of the dissimilarity-enhanced energy statistic. Unreported simulations further suggest that high-dimensional tests for angular symmetry, based on δ and the A_D statistic, also achieve very high power. Overall, the chapter provides a unified theoretical and applied framework for multivariate symmetry testing, combining methodological rigor, computational efficiency, and practical relevance for contemporary high dimensional data analysis.

15.7 Proofs

Proof of Theorem 15.4.1

Since under the null hypothesis H_0 all the vectors $\boldsymbol{X}_1, \boldsymbol{X}_2, \ldots, \boldsymbol{X}_n$ have the same distribution, $R_\delta^{(r)}$ must have the same distribution as R_δ under H_0 for all random permutations, and hence

$$\mathbb{E}(\hat{G}_{R_\delta}(t)) = \frac{1}{R}\sum_{r=1}^{R} \mathbb{P}(R_\delta^{(r)} \leq t) = \mathbb{P}(R_\delta \leq t).$$

This then implies $\mathbb{E}(\hat{G}_{R_\delta}(t)) \to G_{R_\delta}(t)$ for all t as $n \to \infty$ and $d \to \infty$ under H_0, where the distribution of R_δ is $G_{R_\delta}(t)$. We prove that $\text{Var}(\hat{G}_{R_\delta}(t)) \to 0$ as $n \to \infty$ and $d \to \infty$. This holds because

$$\mathbb{E}(\hat{G}^2_{R_\delta^{(r)}}(t)) = \frac{1}{R^2}\sum_{r=1}^{R}\sum_{r'=1}^{R} \mathbb{P}(R_\delta^{(r)} \leq t, {R_\delta}^{(r')} \leq t) = \mathbb{P}(R_\delta \leq t, R_\delta \leq t) \to G^2_{R_\delta}(t).$$

(b) Using part (a) above, as $n \to \infty$ and $d \to \infty$, we can write for $0 \leq \alpha \leq 1$

$$\begin{aligned}\mathbb{P}_{H_0}(p_{\text{perm}} \leq \alpha) &= \mathbb{P}_{H_0}(1 - \hat{G}_{R_\delta}(R_\delta(\text{obs}) \leq \alpha) \to \mathbb{P}_{H_0}(1 - G_{R_\delta}(R_\delta(\text{obs}) \leq \alpha) \\ &= \mathbb{P}_{H_0}(G_{R_\delta}(R_\delta(\text{obs}) \geq 1 - \alpha)) = \mathbb{P})U \geq 1 - \alpha) = \alpha,\end{aligned}$$

where U denotes a random variable having uniform distribution on the interval (0,1).

Proof of Theorem 15.4.2

The dissimilarity between $\mathbf{X}_i$ and $\mathbf{X}_j$ in $\mathbb{X}$, approaches 0 in probability as $d \to \infty$. Similarly, the dissimilarity between $\mathbf{Y}_i$ and $\mathbf{Y}_j$ in $\mathbb{Y}$, approaches 0 in probability as $d \to \infty$. It follows that $\bar{\delta}_{(x)}$ and $\bar{\delta}_{(y)}$ tend to 0 in probability as d diverges. Hence, the statistics $R_{(\delta)}$ converges in probability to $2\tilde{\delta}(F_1, F_2)$ as $d \to \infty$. Let $Z_N = (\mathbf{x}_1, \ldots \mathbf{x}_m, \mathbf{y}_1, \ldots, \mathbf{y}_n)$ be the set of all observations. One can show

$$\mathbb{P}\{R_{(\delta)} \geq 2\tilde{\delta}(F_1, F_2) | Z_N\} < \alpha$$

for almost every Z_N as d diverges. Without loss of generality, we assume that $m \leq n$. Consider a permutation of Z_N where r of the $\mathbf{X}$'s are labeled F_1. Hence, $(m-r)$ $\mathbf{X}$'s are labeled F_2, $(m-r)$ $\mathbf{Y}$'s are labeled F_1 and $(n-m+r)$ $\mathbf{Y}$'s are labeled F_2. Recall that the statistics is $R_{(\delta)} = 2\bar{\delta}_{(xy)} - \bar{\delta}_{(x)} - \bar{\delta}_{(y)}$. Denote the test statistics of the permutation test as $R^{(r)}_{(\delta)} = 2\bar{\delta}^{(r)}_{(xy)} - \bar{\delta}^{(r)}_{(x)} - \bar{\delta}^{(r)}_{(y)}$. It follows that the permutation distribution of $R_{(\delta)}$ as $d \to \infty$ converges in probability to

$$R^{(r)}_{(\delta)} = [\frac{2\{r(n-m+r)+(m-r)^2\}}{mn} - \frac{2r(m-r)}{m(m-1)} - \frac{2(m-r)(n-m+r)}{n(n-1)}]\tilde{\delta}(F_1, F_2). \tag{15.20}$$

One can show that $R^{(r)}_{(\delta)} \leq 2\tilde{\delta}(F_1, F_2)$ holds for all $r \leq m$, as it is as an increasing function of r that takes its maximum value of $R^{(m)}_{(\delta)} = 2\tilde{\delta}(F_1, F_2)$ at $r = m$. Furthermore, when $m = n$, and $r = 0$, we obtain $R^{(m)}_{(\delta)} = 2\tilde{\delta}(F_1, F_2)$. Combining these two cases, one obtains

$$\mathbb{P}\{R_{(\delta)} \geq 2\tilde{\delta}(F_1, F_2)|Z_N\} \leq 2m!n!/N! < \alpha.$$

On the other hand, as $d \to \infty$, under the permutation distribution, the test statistics takes the value $2\tilde{\delta}(F_1, F_2)$ or higher with probability tending to $2/\binom{N}{m}$. Hence, for all N satisfying $2/\binom{N}{m} < \alpha$, the proposed test would reject the null hypothesis with probability tending to 1 as $d \to \infty$.

Proof of Lemma 15.4.1

It is not difficult to show that the reflected sample has mean $-\boldsymbol{\mu}$ and covariance $\boldsymbol{\Sigma}$. In fact, all the even marginal moments of $\mathbf{X}$ equal those of the reflected images $\mathbf{Y}$. It follows that $\sigma^2_{F_1} = \sigma^2_{F_2}$ and $\nu^2_x = \nu^2_y$ so that $\sigma^d_x = \sigma^d_y$ as $d \to \infty$. It is trivial to show $\bar{\mu}^d_x = \frac{1}{d}\mathbf{1}'_d\boldsymbol{\mu} = \bar{\mu}$, and $\bar{\mu}^d_y = \frac{-1}{d}\mathbf{1}'_d\boldsymbol{\mu} = -\bar{\mu}$. When diagonal symmetry holds, one can show that the limiting value of the dissimilarity index is 0 as $d \to \infty$. Hence, $\tilde{\delta}^2(F_1, F_2) = 4\bar{\mu}^2 \geq 0$. When $\mathbf{X}$ is diagonally symmetric, $\boldsymbol{\mu} = \mathbf{0}$ and $\tilde{\delta}^2(F_1, F_2) = 0$.

Proof of Theorem 15.4.3

We define the sample mean over the m copies and d components as

$$\bar{X}^d = \frac{1}{md}\sum_{i=1}^{m}\sum_{t=1}^{d} X_{it}.$$

This can be rewritten as

$$\bar{X}^d = \frac{1}{d}\sum_{t=1}^{d} \bar{X}_{t\cdot},$$

where $\bar{X}_{t\cdot} = \frac{1}{m}\sum_{i=1}^{m} X_{it}$ is the sample mean of the t–th component of $\mathbf{X}$. Since the m observations on the t–th component, $\{X_{it}\}_{i=1}^{m}$, are i.i.d., by the Central Limit Theorem, we know that as $m \to \infty$,

$$\bar{X}_{t\cdot} \xrightarrow{d} \mathrm{N}(\mu_t, \frac{\sigma_t^2}{m}),$$

where σ_t^2 is the variance of the t–th component of $\mathbf{X}$. Therefore, as $m \to \infty$, the sample mean $\bar{X}_{t\cdot}$ for each t is asymptotically normal with mean μ_t and variance $\frac{\sigma_t^2}{m}$. Now, the grand mean $\bar{X}^d$ is the average of d correlated asymptotically normal variables. Therefore, by the properties of the multivariate Central Limit Theorem, $\bar{X}^d$ is asymptotically normal as $m \to \infty$. The expectation of $\bar{X}^d$ is computed as follows

$$\mathbb{E}(\bar{X}^d) = \frac{1}{md}\sum_{i=1}^{m}\sum_{t=1}^{d}\mathbb{E}(X_{it}) = \frac{1}{d}\sum_{t=1}^{d}\mu_t = \frac{1}{d}\mathbf{1}'_d\boldsymbol{\mu}_x.$$

The variance of $\bar{X}^d$ is given by

$$\mathbb{V}ar(\bar{X}^d) = \mathbb{V}ar\left(\frac{1}{md}\sum_{i=1}^{m}\sum_{t=1}^{d}X_{it}\right) = \frac{1}{m^2d^2}\sum_{i=1}^{m}\sum_{t=1}^{d}\sum_{r=1}^{d}\mathbb{C}ov(X_{it}, X_{ir}).$$

We break this into the sum of variances and covariances,

$$\mathbb{V}ar(\bar{X}^d) = \frac{1}{m^2d^2}\sum_{i=1}^{m}\left(\sum_{t=1}^{d}\text{Var}(X_{it}) + 2\sum_{1\le t<r\le d}\mathbb{C}ov(X_{it}, X_{ir})\right).$$

Using the fact that $\text{Var}(X_{it}) = \sigma_t^2$ and $\mathbb{C}ov(X_{it}, X_{ir}) = \sigma_{tr}$ (the (t,r)–th element of the covariance matrix $\boldsymbol{\Sigma}_x$), we can simplify this to

$$\mathbb{V}ar(\bar{X}^d) = \frac{1}{m^2d^2}\sum_{i=1}^{m}\left(\sum_{t=1}^{d}\sigma_t^2 + 2\sum_{1\le t<r\le d}\sigma_{tr}\right).$$

This sum can be split into two parts

$$\mathbb{V}ar(\bar{X}^d) = \frac{1}{md^2}\sum_{t=1}^{d}\sigma_t^2 + \frac{1}{md^2}\sum_{1\le t<r\le d}2\sigma_{tr}.$$

The first term is

$$\frac{1}{md^2}\sum_{t=1}^{d}\sigma_t^2 = \frac{1}{md^2}\text{tr}(\boldsymbol{\Sigma}_x).$$

The second term is $O(d^2)$ as $d \to \infty$, and hence we have

$$\mathbb{V}ar(\bar{X}^d) = \frac{1}{md^2}\text{tr}(\boldsymbol{\Sigma}_x) + O\left(\frac{1}{md^2}\right).$$

Thus, as $m, d \to \infty$, the variance converges to

$$\mathbb{V}ar(\bar{X}^d) \to \frac{1}{md^2}\text{tr}(\boldsymbol{\Sigma}_x).$$

We have shown that as $m \to \infty$ and $d \to \infty$, the random variable $\bar{X}^d$ is asymptotically normal with mean $\frac{1}{d}\mathbf{1}'_d\boldsymbol{\mu}_x$ and variance,

$$\mathbb{V}ar(\bar{X}^d) = \frac{1}{md^2}\text{tr}(\boldsymbol{\Sigma}_x).$$

It follows that $\mathbb{V}ar(\bar{X}^d)$ converge to 0 as $m \to \infty$ and $d \to \infty$.

Proof of Theorem 15.4.4

Under L_1, the images have mean vector $\boldsymbol{\mu}_x$ and the mirror images have the mean vector $\boldsymbol{\mu}_y \neq -\boldsymbol{\mu}_x$. By Theorem 15.4.3, $\mathbb{E}(\bar{X}^d) = \frac{1}{d}\mathbf{1}'_d\boldsymbol{\mu}_x$ whereas $\mathbb{E}(\bar{Y}^d) = \frac{1}{d}\mathbf{1}'_d\boldsymbol{\mu}_y$. Furthermore, $\bar{X}^d - \bar{Y}^d$ remains asymptotically normal with mean $\eta = \mathbb{E}(\bar{X}^d - \bar{Y}^d) = \frac{1}{d}\mathbf{1}'_d(\boldsymbol{\mu}_x - \boldsymbol{\mu}_y)$ and variance σ^2 as $\min(n, m) \to \infty$ and $d \to \infty$. Consequently, A_D has a non–central χ^2 distribution with one degree of freedom and non–centrality parameter η^2/σ^2 under L_1. As the non–centrality parameter becomes larger under L_1, the power of A_D tends to 1.

15.8 Exercises

Exercise 15.1 *Conditional Distributions under Spherical Symmetry*

Let $\mathbf{X} = (X_1, \ldots, X_d)' \in \mathbb{R}^d$ *be a continuous random vector with a spherically symmetric distribution about the origin. Define the squared radius* $R^2 = \|\mathbf{X}\|^2 = \sum_{j=1}^d X_j^2$. *Then define recursively for* $j = 1, \ldots, d-1$,

$$B_j = \frac{X_j^2}{R^2 - \sum_{i=1}^{j-1} X_i^2},$$

with the convention that the empty sum is zero for $j = 1$.

For each $j = 1, \ldots, d-1$, *prove that the conditional random variable* $B_j \mid X_1, \ldots, X_{j-1}$ *follows a Beta distribution,*

$$B_j \sim \text{Beta}\left(\frac{1}{2}, \frac{d-j}{2}\right),$$

and the B_j*'s are mutually independent.*

Exercise 15.2 *Let* $X, X' \in \mathbb{R}^d$ *be independent and identically distributed random vectors. Define the quantity*

$$\mathcal{E}(X) := \mathbb{E}\|X + X'\| - \mathbb{E}\|X - X'\|.$$

Prove that $\mathcal{E}(X) \geq 0$, *with equality if and only if* $X \stackrel{d}{=} -X$ *(i.e., X is diagonally symmetric about the origin).*

Exercise 15.3 *The distance-skewness coefficient, introduced by Székely and Móri [328], is defined as*

$$\text{dSkew}(\mathbf{X}) = \begin{cases} 1 - \dfrac{\mathbb{E}\|\mathbf{X} - \mathbf{X}'\|}{\mathbb{E}\|\mathbf{X} + \mathbf{X}'\|}, & \mathbb{E}\|\mathbf{X} + \mathbf{X}'\| > 0, \\ 1, & \mathbb{E}\|\mathbf{X} + \mathbf{X}'\| = 0, \end{cases}$$

where $\mathbf{X}, \mathbf{X}' \in \mathbb{R}^d$ *are independent copies. This statistic equals zero if and only if* $\mathbf{X} \stackrel{d}{=} -\mathbf{X}$, *i.e.,* $\mathbf{X}$ *is diagonally symmetric about the origin.*

a) Prove $0 \leq \text{dSkew}(\mathbf{X}) \leq 1$, *with equality to 0 if and only if* $\mathbf{X} \stackrel{d}{=} -\mathbf{X}$.

b) Let $\mathbf{X} \sim \alpha \text{N}(0, \mathbf{I}_d) + (1-\alpha)\text{N}(0, \sigma^2 \mathbf{I}_d)$, *for* $0 < \alpha < 1$, $\sigma^2 \neq 1$. *Is* $\mathbf{X}$ *diagonally symmetric? Compute* $\text{dSkew}(\mathbf{X})$.

c) Find the distribution of $X_i \pm X_i'$?

Exercise 15.4 *Marginal Symmetry vs. Diagonal Symmetry.*

Consider a random vector $\mathbf{X} = (X_1, X_2, \ldots, X_d) \in \mathbb{R}^d$ *where,* $X_1 \sim \text{N}(0,1)$,*and* $X_2, \ldots, X_d \sim$ *Uniform*$(-1/2, 1/2)$. *Assume all components are independent.*

a) Are the marginals diagonally symmetric?

b) Does $\mathbf{X}$ *possess diagonal symmetry? Explain.*

Exercise 15.5 *Let* $\mathbf{X} = (x_1, x_2, \ldots, x_d)' \in \mathbb{R}^d$ *be a random vector such that each component* $x_i \sim \text{Beta}(\alpha_i, \beta_i)$ *independently.*

Define the centered vector,

$$\mathbf{Y} = \mathbf{X} - \mathbb{E}[\mathbf{X}] \quad \textit{where } \mu_i = \mathbb{E}[x_i] = \frac{\alpha_i}{\alpha_i + \beta_i}.$$

a) Is the centered vector $\mathbf{Y}$ *diagonally symmetric about the origin?*

b) Is the distribution of $\mathbf{Y}$ *diagonally symmetric?*

c) State conditions under which $\mathbf{Y}$ *is diagonally symmetric.*

Exercise 15.6 *Show that the multivariate normal distribution satisfies the following types of symmetry: a) Elliptical, b) Spherical, c) Affine, and d) Projective.*

Exercise 15.7 *Show that the Multivariate* t *Distribution satisfies elliptical and projective symmetry.*

Exercise 15.8 *Under what conditions does the multinomial distribution satisfy permutation symmetry?*

Exercise 15.9 *Under what conditions is the Wishart distribution affinely symmetric?*

Exercise 15.10 *Show that the Dirichlet distribution is permutation and half-Space symmetric under equal parameters.*

Bibliography

[1] L. Addario-Berry, N. Broutin, and B. Reed. The diameter of the minimum spanning tree of a complete graph. In P. Chassaing, editor, Proceedings of the Fourth Colloquium on Mathematics and Computer Science, Discrete Mathematics and Theoretical Computer Science, pages 237–248, 2006.

[2] N. A. Adegoke, J. O. Ajadi, A. Mukherjee, and S. A. Abbasi. Nonparametric multivariate covariance chart for monitoring individual observations. Computers & Industrial Engineering, 167:108025, 2022.

[3] A. Agresti. Generalized odds ratios for ordinal data. Biometrics, pages 59–67, 1980.

[4] J. Ahn, M. H. Lee, and Y. J. Yoon. Clustering high dimension, low sample size data using the maximal data piling distance. Statistica Sinica, 22:443–464, 2012.

[5] J. Ahn and J. S. Marron. The maximal data piling direction for discrimination. Biometrika, 97:254–259, 2010.

[6] J. Ahn, J. S. Marron, K. M. Muller, and Y. Chi. The high-dimension, low-sample-size geometric representation holds under mild conditions. Biometrika, 94:760–766, 2007.

[7] S. Aki. Some test statistics based on the martingale term of the empirical distribution function. Annals of the Institute of Statistical Mathematics, 38:1–21, 1986.

[8] Greg Aloupis. Geometric measures of data depth. In DIMACS Series in Discrete Mathematics and Theoretical Computer Science, volume 72, pages 147–158. American Mathematical Society, 2006.

[9] C. Amado, A. M. Bianco, G. Boente, and A. M. Pires. Robust bootstrap: An alternative to bootstrapping robust estimators. REVSTAT, 12(2):169–197, 2014.

[10] T. W. Anderson. An Introduction to Multivariate Statistical Analysis. Wiley-Interscience, New Jersey, 2003.

[11] F. Angiulli. On the behavior of intrinsically high-dimensional spaces: Distances, direct and reverse nearest neighbors, and hubness. Journal of Machine Learning Research, 18:1–60, 2018.

[12] Miguel Arcones and Evarist Giné. Limit theorems for u-processes. The Annals of Probability, 21, 07 1993.

[13] S. Aryal, K. M. Ting, T. Washio, and G. Haffari. Data-dependent dissimilarity measure: an effective alternative to geometric distance measures. Knowledge and Information Systems, 53:479–506, 2017.

[14] B. Aslan and G. Zeoh. New test for the multivariate two-sample problem based on the concept of minimum energy. Journal of Statistical Computation and Simulation, 75(2):109–119, 2005.

[15] R. Assuncão, A. Costa, A. Tavares, and S. Ferreira. Fast detection of arbitrarily shaped disease clusters. Statistics in Medicine, 25:723–742, 2006.

[16] A. Asuncion and D. J. Newman. Uci machine learning repository. University of California, Irvine, School of Information and Computer Science, 2007. Available at http://www.ics.uci.edu/ mlearn/MLRepository.html.

[17] E. N. Atkinson, B. W. Brown, and J. R. Thompson. Parallel algorithms for fixed seed simulation based parameter estimation. In Computer Science and Statistics: Proceedings of the 21st Symposium on the Interface, pages 256–261, 1989.

[18] D. Bagkavos and P. N. Patil. A new test of independence for bivariate observations. Journal of Multivariate Analysis, 160:117–133, 2017.

[19] Z. Bai and H. Saranadasa. Effect of high dimension: by an example of a two sample problem. Statistica Sinica, 6:311–329, 1996.

[20] M. Barigozzi, H. Cho, and P. Fryzlewicz. Simultaneous multiple change-point and factor analysis for high-dimensional time series. Journal of Econometrics, 206:187–225, 2018.

[21] L. Baringhaus and C. Franz. On a new multivariate two-sample test. Journal of Multivariate Analysis, 88(1):190–206, 2004.

[22] V. Barnett. The ordering of multivariate data. Journal of the Royal Statistical Society, Series A, 139(3):318–355, 1976.

[23] V. Barnett and T. Lewis. Outliers in Statistical Data. John Wiley & Sons, London, 1994.

[24] M. Barni, V. Cappellini, and A. Mecocci. Fast vector median filter based on euclidean norm approximation. IEEE Signal Processing Letters, 1(6):92–94, 1994.

[25] R. Bartoszyński, D. K. Pearl, and J. Lawrence. A multidimensional goodness of fit test based on interpoint distances. Journal of the American Statistical Association, 97:577–586, 1997.

[26] K. J. Beath. A finite mixture method for outlier detection and robustness in meta-analysis. Research Synthesis Methods, 5(4):285–293, 2014.

[27] D. Bernoulli. Reprinted in biometrika (1961). Biometrika, 48:1–18, 1961. Original work published in 1777.

[28] J. R. Berrendero, A. Cuevas, and B. Pateiro-López. Shape classification based on interpoint distance distributions. Journal of Multivariate Analysis, 146:237–247, 2016.

[29] K. Beyer, J. Goldstein, R. Ramakrishnan, and U. Shaft. When is nearest neighbor meaningful? In Database Theory — ICDT, volume 1540 of Lecture Notes in Computer Science, pages 217–235, Berlin, Germany, 1999. Springer.

[30] M. Bhandary. On robustness of the test for detection of multivariate outliers. Statistics, 41(5):377–384, 2007.

[31] G. Biau and D. M. Mason. High-dimensional p-norms, 2013. arXiv:1311.0587v1 [math.ST].

[32] P. J. Bickel and L. Breiman. Sums of functions of nearest neighbor distances, moment bounds, limit theorems and a goodness of fit test. The Annals of Probability, 11:185–214, 1983.

[33] Peter J Bickel and Elizaveta Levina. Covariance regularization by thresholding. The Annals of Statistics, 36(6):2577–2604, 2008.

[34] P. Billingsley. Probability and measure. Wiley, New York, 3rd edition, 1995.

[35] M. Biswas and A. K. Ghosh. A nonparametric two-sample test applicable to high dimensional data. Journal of Multivariate Analysis, 123:160–171, 2014.

[36] M. Biswas, M. Mukhopadhyay, and A. K. Ghosh. A distribution-free two-sample run test applicable to high-dimensional data. Biometrika, 101(4):913–926, 2014.

[37] N. Blomqvist. On a measure of dependence between two random variables. The Annals of Mathematical Statistics, 21(4):593–600, 1950.

[38] J. R. Blum, J. Kiefer, and M. Rosenblatt. Distribution free tests of independence based on the sample distribution function. Annals of Mathematical Statistics, 32:485–498, 1961.

[39] D. Böhning. Computer-Assisted Analysis of Mixtures and Applications: Meta-Analysis, Disease Mapping and Others. Chapman & Hall/CRC, Boca Raton, 2000.

[40] M. Bonetti, K. L. Olson, K. D. Mandl, and M. Pagano. Parametric models for interpoint distances and their use in biosurveillance. In Proceedings of the American Statistical Association, 2003.

[41] M. Bonetti and M. Pagano. The interpoint distance distribution as a descriptor of point patterns, with an application to spatial disease clustering. Statistics in Medicine, 24(5):753–773, 2005.

[42] S. Bonnini, G. M. Assegie, and K. Trzcinska. Review about the permutation approach in hypothesis testing. Mathematics, 12(17):2617, 2024.

[43] O. Borúvka. O jistem problemu minimalnim. Moravske Prirodovedecke Spolecnosti, 3:37–58, 1926.

[44] P. Borysov, J. Hannig, and J. S. Marron. Asymptotics of hierarchical clustering for growing dimension. Journal of Multivariate Analysis, 124:456–479, 2014.

[45] T. Brijs, D. Karlis, G. Swinnen, K. Vanhoof, G. Wets, and P. Manchanda. A multivariate poisson mixture model for marketing applications. Statistica Neerlandica, 58(3):322–348, 2004.

[46] A. Buja, B. F. Logan, J. A. Reeds, and L. A. Shepp. Inequalities and positive definite functions arising from a problem in multidimensional scaling. The Annals of Statistics, 22:406–438, 1994.

[47] Aggarwal C. C., A. Hinneburg, and D. A. Keim. On the surprising behavior of distance metrics in high dimensional spaces. In International Conference on Database Theory, 2001.

[48] R. B. Calinski and J. Harabasz. A dendrite method of cluster analysis. Communications in Statistics, 3:1–27, 1974.

[49] R. Campello, D. Moulavi, and J. Sander. Density-based clustering based on hierarchical density estimates. In Proceedings of the 17th Pacific-Asia Conference on Knowledge Discovery in Databases (PAKDD 2013), volume 7819 of Lecture Notes in Computer Science, page 160, 2013.

[50] E. J. Candès and B. Recht. Exact matrix completion via convex optimization. Foundations of Computational Mathematics, 9(6):717–772, 2009.

[51] C. Caroni and P. Prescott. On rohlf's method for the detection of outliers in multivariate data. Journal of Multivariate Analysis, 52:295–307, 1995.

[52] C. Caroni and P. Prescott. Inapplicability of asymptotic results on the minimal spanning tree in statistical testing. Journal of Multivariate Analysis, 83:487–492, 2002.

[53] A. Cayley. On a theorem in the geometry of positions. Cambridge Mathematical Journal, 2:267–271, 1841.

[54] S. Chakraborty and X. Zhang. A new framework for distance and kernel-based metrics in high dimensions. Electronic Journal of Statistics, 15:5455–5522, 2021.

[55] J. M. Chambers, W. S. Cleveland, B. Kleiner, and P. A. Tukey. Graphical Methods for Data Analysis. Wadsworth, Belmont, 1983.

[56] Y.-B. Chan and P. Hall. Scale adjustments for classifiers in high-dimensional, low sample size settings. Biometrika, 96:469–478, 2009.

[57] D. Chatterjee. The concentration phenomenon: an elementary introduction, 2022. Lecture Notes, Systems & Control Engineering, IIT Bombay. Available at https://www.sc.iitb.ac.in/ chatterjee.

[58] F. Chen, S. G. Meintanis, and L. Zhu. On some characterizations and multidimensional criteria for testing homogeneity, symmetry and independence. Journal of Multivariate Analysis, 173:125–144, 2019.

[59] H. Chen, J. Chen, and J. D. Kalbfleisch. A modified likelihood ratio test for homogeneity in finite mixture models. Journal of the Royal Statistical Society: Series B, 63(1):19–29, 2001.

[60] H. Chen and N. Zhang. Graph-based change-point detection. The Annals of Statistics, 43(1):139–176, 2015.

[61] J. Chen and A. K. Gupta. Parametric statistical change point analysis: with applications to genetics, medicine, and finance. Springer, 2012.

[62] J. Chen and P. Li. Hypothesis test for normal mixture models: The em approach. The Annals of Statistics, 37(5A):2523–2542, 2009.

[63] L. S. Chen, D. Paul, R. L. Prentice, and P. Wang. A regularized hotelling's t^2 test for pathway analysis in proteomic studies. Journal of the American Statistical Association, 106(496):1345–1360, 2011.

[64] S. X. Chen and Y. L. Qin. A two-sample test for high-dimensional data with applications to gene-set testing. The Annals of Statistics, 38(2):808–835, 2010.

[65] A. Childs and N. Balakrishnan. Some approximations to the multivariate hypergeometric distribution with applications to hypothesis testing. Computational Statistics and Data Analysis, 35(2):137–154, 2000.

[66] K. Chin, et al. Genomic and transcriptional aberrations linked to breast cancer pathophysiologies. Cancer Cell. 10(6):529–541, 2006.

[67] B. C. Christensen, E. A. Houseman, C. J. Marsit, et al. Aging and environmental exposures alter tissue-specific dna methylation dependent upon cpg island context. PLoS Genetics, 2009. doi:10.1371/journal.pgen.1000602.

[68] L. Chu, H. Chen. Asymptotic distribution-free change-point detection for multivariate and non-Euclidean data Ann. Statist. 47(1):382–414, 2019. DOI: 10.1214/18-AOS1691.

[69] D. Chung and S. Keles. Sparse partial least squares classification for high dimensional data. Statistical Applications in Genetics and Molecular Biology, 9(1):Article 17, 2010.

[70] C. Cortes and V. Vapnik. Support-vector networks. Machine Learning, 20(3):273–297, 1995.

[71] S. Dehghan and M. R. Faridrohani. Affine invariant depth-based tests for the multivariate one-sample location problem. Test, 28(3):671–693, 2019.

[72] S. Dehghan and M. R. Faridrohani. Depth-based signed-rank tests for bivariate central symmetry. REVSTAT-Statistical Journal, 18:109–129, 2020.

[73] S. Dehghan, M. R. Faridrohani, and Z. Barzegar. Testing for diagonal symmetry based on center-outward ranking. Statistical Papers, 64:255–283, 2022.

[74] M. Dettling. Bagboosting for tumor classification with gene expression data. Bioinformatics, 20:3583–3593, 2004.

[75] Markus Dettling and Peter Bühlmann. Supervised clustering of genes. Genome Biology, 3:research0069.1–research0069.15, 2002.

[76] C. J. DiCiccio, T. J. DiCiccio, and J. P. Romano. Exact tests via multiple data splitting. Statistics and Probability Letters, 166:108865, 2020.

[77] D. P. Doane. Aesthetic frequency classification. American Statistician, 30:181–183, 1976.

[78] D. L. Donoho and M. Gasko. Breakdown properties of location estimates based on half-space depth and projected outlyingness. The Annals of Statistics, 20:1803–1827, 1992.

[79] David L Donoho. Breakdown properties of multivariate location estimators. Qualifying paper, Harvard University, Boston, MA, 1982.

[80] R. Drikvandi and R. Modarres. A distribution-free method for change point detection in non-sparse high dimensional data. Journal of Computational and Graphical Statistics, pages 1–30, 2024.

[81] H. E. Driver and A. L. Kroeber. Quantitative expression of cultural relationships, volume 31. University of California Publications in Archaeology and Ethnology, 1932.

[82] R. J. Durrant and A. Kabán. When is 'nearest neighbor' meaningful: A converse theorem and implications. Journal of Complexity, 25(4):385–397, 2009.

[83] S. Dutta and A. K. Ghosh. On some transformations of high dimension, low sample size data for nearest neighbor classification. Machine Learning, 102(1):57–83, 2016.

[84] S. Dutta and S. Ghosh. On a class of transformations for multivariate data visualization. Computational Statistics & Data Analysis, 50(10):2667–2684, 2006.

[85] R. Dyckerhoff, C. Ley, and D. Paindaveine. Depth-based runs tests for bivariate central symmetry. Annals of the Institute of Statistical Mathematics, 67:917–941, 2015.

[86] B. Efron and R. J. Tibshirani. An Introduction to the Bootstrap. Chapman & Hall/CRC, Washington, D.C., 1993.

[87] J. H. Einmahl and Z. Gan. Testing for central symmetry. Journal of Statistical Planning and Inference, 169:27–33, 2016.

[88] R. T. Elmore, T. P. Hettmansperger, and F. Xuan. Spherical data depth and a multivariate median. In R. Y. Liu, R. Serfling, and D. L. Souvaine, editors, Proceedings of Data Depth: Robust Multivariate Analysis, Computational Geometry and Applications, pages 87–101, Rhode Island: American Mathematical Society, 2006.

[89] F. Enikeeva and Z. Harchaoui. High-dimensional change-point detection under sparse alternatives. The Annals of Statistics, 47:2051–2079, 2019.

[90] Z. Erlich, R. Gelbard, and I. Spiegler. Data mining by means of binary representation: a model for similarity and clustering. Information System Frontiers, 2:187–197, 2002.

[91] B. S. Everitt. A monte carlo investigation of the robustness of hotelling's one- and two-sample t^2 tests. Journal of the American Statistical Association, 74(365):48–51, 1979.

[92] B. S. Everitt, S. Landau, M. Leese, and D. Stahl. Cluster Analysis. Wiley, 5th edition, 2011.

[93] J. Fan. Test of significance based on wavelet thresholding and neyman's truncation. Journal of the American Statistical Association, 91:674–688, 1996.

[94] J. Fan, R. Samworth, and Y. Wu. Ultrahigh dimensional feature selection: Beyond the linear model. Journal of Machine Learning Research, 10:2013–2038, 2009.

[95] Y. Fan, P. Lafaye de Micheaux, S. Penev, and D. Salopeka. Multivariate nonparametric test of independence. Journal of Multivariate Analysis, 153:189–210, 2017.

[96] K. Fang, S. Kotz, and K. Ng. Symmetric Multivariate and Related Distributions. Chapman & Hall, New York, NY, 1990.

[97] S. E. Feinberg. The use of chi-squared statistics for categorical data problems. Journal of the Royal Statistical Society. Series B, 41:54–64, 1979.

[98] M. Fernández-Delgado, E. Cernadas, S. Barro, and D. Amorim. Do we need hundreds of classifiers to solve real world classification problems? Journal of Machine Learning Research, 15(1):3133–3181, 2014.

[99] A. Feuerverger. A consistent test for bivariate independence. International Statistical Review, 63(3):419–433, 1993.

[100] R. A. Fisher. Frequency distribution of the values of the correlation coefficient in samples from an infinitely large population. Biometrika, 10:507–521, 1915.

[101] R. A. Fisher. The use of multiple measurements in taxonomic problems. Annals of Eugenics, 7(2):179–188, 1936.

[102] E. Fix and J. L. J. Hodges. Discriminatory analysis: nonparametric discrimination: consistency properties. International Statistical Review, 57:238–247, 1989.

[103] J. L. Fleiss, B. Levin, and M. C. Paik. Statistical methods for rates and proportions. John Wiley & Sons, 2013.

[104] B. Follain, T. Wang, and R. J. Samworth. High-dimensional changepoint estimation with heterogeneous missingness. Journal of the Royal Statistical Society Series B, 84:1023–1055, 2022.

[105] D. Francois, V. Wertz, and M. Verleysen. The concentration of fractional distances. IEEE Transactions on Knowledge and Data Engineering, 19:873–886, 2007.

[106] J. Freeman and R. Modarres. Efficiency of test for independence after box-cox transformation. Journal of Multivariate Analysis, 95:107–118, 2005.

[107] J. Freeman and R. Modarres. Inverse box-cox: the power normal distribution. Statistics and Probability Letters, 76(8):764–772, 2006.

[108] E. W. Frees. Estimating densities of functions of observations. Journal of the American Statistical Association, 89(426):517–525, 1994.

[109] J. H. Friedman and L. C. Rafsky. Multivariate generalizations of the wald-wolfowitz and smirnov two-sample tests. The Annals of Statistics, 7(4):697–717, 1979.

[110] J. H. Friedman and L. C. Rafsky. Graph-theoretic measures of multivariate association and prediction. Annals of Statistics, 11(2):377–391, 1983.

[111] P. Fryzlewicz. Wild binary segmentation for multiple change-point detection. The Annals of Statistics, 42:2243–2281, 2014.

[112] K. R. Gabriel. The biplot graphic display of matrices with application to principal component analysis. Biometrika, 58(3):453–467, 1971.

[113] R. K. Gabriel and R. R. Sokal. A new statistical approach to geographic variation analysis. Systematic Biology, 18:259–278, 1969.

[114] J. Gao, W. Hu, Z. Zhang, X. Zhang, and O. Wu. Rkof: Robust kernel-based local outlier detection. In J. Z. Huang, L. Cao, and J. Srivastava, editors, Advances in Knowledge Discovery and Data Mining, PAKDD 2011, Lecture Notes in Computer Science, Volume 6635. Springer, Berlin, 2011.

[115] D. Garreau and S. Arlot. Consistent change-point detection with kernels. Electronic Journal of Statistics, 12:4440–4486, 2018.

[116] A. K. Ghosh and P. Chaudhuri. On maximum depth and related classifiers. Scandinavian Journal of Statistics, 32(2):327–350, 2005.

[117] S. Ghosh and F. H. Ruymgaart. Applications of empirical characteristic functions in some multivariate problems. Canadian Journal of Statistics, 20:429–440, 1992.

[118] P. W. Gieser and R. H. Randles. A nonparametric test of independence between two vectors. Journal of the American Statistical Association, 92:561–567, 1997.

[119] A. Giovagnoli and H. P. Wynn. Multivariate dispersion orderings. Statistics & Probability Letters, 22:325–332, 1995.

[120] N. Glick. Measurements of separation among probability densities or random variables. The Canadian Journal of Statistics, 3(2):267–276, 1975.

[121] Todd R. Golub, Donna K. Slonim, Pablo Tamayo, et al. Molecular classification of cancer: Class discovery and class prediction by gene expression monitoring. Science, 286(5439):531–537, 1999.

[122] J. C. Gower. A general coefficient of similarity and some of its properties. Biometrics, 27(4):857–871, 1971.

[123] John C Gower. Some distance properties of latent root and vector methods used in multivariate analysis. Biometrika, 53(3-4):325–338, 1966.

[124] T. Grundy, R. Killick, and G. Mihaylov. High-dimensional changepoint detection via a geometrically inspired mapping. Statistics and Computing, 30:1155–1166, 2020.

[125] L. Guo and R. Modarres. Interpoint distance classification of high dimensional discrete observations. International Statistical Review, 2018.

[126] L. Guo and R. Modarres. Nonparametric tests of independence based on interpoint distances. Journal of Nonparametric Statistics, 32(1):225–245, 2020.

[127] L. Guo and R. Modarres. Testing the equality of matrix distributions. Statistical Methods & Applications, 29:289–307, 2020.

[128] L. Guo and R. Modarres. Two multivariate online change detection models. Journal of Applied Statistics, 49(2):427–448, 2022.

[129] A. K. Gupta and D. K. Nagar. Matrix variate distributions, volume 104. CRC Press, 1999.

[130] A. S. Hadi. Identifying multiple outliers in multivariate data. Journal of the Royal Statistical Society, Series B, 54(3):761–771, 1992.

[131] A. S. Hadi, A. H. M. R. Imon, and M. Werner. Detection of outliers. The Wiley Interdisciplinary Reviews: Computational Statistics, 1:57–70, 2009.

[132] G. Hahn, P. Fearnhead, and I. A. Eckley. Bayesproject: Fast computation of a projection direction for multivariate change point detection. Statistics and Computing, 30:1691–1705, 2020.

[133] P. Hall, J. S. Marron, and A. Neeman. Geometric representation of high dimension, low sample size data. Journal of the Royal Statistical Society, Series B, 67:427–444, 2005.

[134] P. Hall and N. Tajvidi. Permutation tests for equality of distributions in high-dimensional settings. Biometrika, 89(2):359–374, 2002.

[135] P. Hall, D. M. Titterington, and J. H. Xue. Median-based classifiers for high-dimensional data. Journal of the American Statistical Association, 104(488):1597–1608, 2009.

[136] R. W. Hamming. Error detecting and error correcting codes. Bell System Technical Journal, 26(2):147–160, 1950.

[137] F. Han, Z. Chen, and H. Liu. Distribution-free tests of independence in high dimensions. Biometrika, 104(4):813–828, 2017.

[138] D. J. Hand and K. Yu. Idiot's bayes not so stupid after all? International Statistical Review, 69(3):385–398, 2001.

[139] J. A. Hartigan. Cluster Algorithms. Wiley, New York, NY, USA, 1975.

[140] J. A. Hartigan and S. Mohanty. The runt test for multimodality. Journal of Classification, 9:63–70, 1992.

[141] T. Hastie, R. Tibshirani, and J. Friedman. The Elements of Statistical Learning: Data Mining, Inference, and Prediction, Second Edition. Springer, 2009.

[142] V. Hautamaki, I. Karkkainen, and P. Franti. Outlier detection using k-nearest neighbour graph. In IAPR International Conference on Pattern Recognition (ICPR'04), Volume 3, pages 430–433, Cambridge, UK, 2004.

[143] D. Hawkins. Identification of Outliers. Chapman & Hall, London, 1980.

[144] Y. He, T. Jiang, J. Wen, and G. Xu. Likelihood ratio test in multivariate linear regression: from low to high dimension. Statistica Sinica, 31(3):1215–1238, 2021.

[145] R. Heller, Y. Heller, and M. Gorfine. A consistent test of association based on the ranks of distances. Biometrika, 100(2):503–510, 2013.

[146] N. Henze. A multivariate two-sample test based on the number of nearest neighbor type coincidences. The Annals of Statistics, 16(2):772–783, 1988.

[147] N. Henze and M. D. Penrose. On the multivariate runs test. The Annals of Statistics, 27(1):290–298, 1999.

[148] T. P. Hettmansperger. Statistical Inference Based on Ranks. John Wiley & Sons, NY, 1984.

[149] W. Hoeffding. A non-parametric test of independence. Annals of Mathematical Statistics, 19:546–557, 1948.

[150] Wassily Hoeffding. A class of statistics with asymptotically normal distribution. The Annals of Mathematical Statistics, 19(3):293–325, 1948.

[151] P. Holgate. Estimation for the bivariate poisson distribution. Biometrika, 51(1-2):241–287, 1964.

[152] M. Hollander. Certain uncorrelated nonparametric test statistics. Journal of the American Statistical Association, 68:707–714, 1968.

[153] D. H. Hong and S. H. Lee. A note on the weak law of large numbers for exchangeable random variables. Communications of the Korean Mathematical Society, 13(2):385–391, 1998.

[154] H. Hotelling. The generalization of student's ratio. Annals of Mathematical Statistics, 2(3):360–378, 1931.

[155] H. Hotelling. The Generalization of Student's Ratio. The Annals of Mathematical Statistics, 2(3):360 – 378, 1931.

[156] J. Hu and Z. D. Bai. A review of 20 years of naive tests of significance for high-dimensional mean vectors and covariance matrices. Science China Mathematics, 59(1):2281–2300, 2016.

[157] S. Huang, T. Tong, and H. Zhao. Bias-corrected diagonal discriminant rules for high-dimensional classification. Biometrics, 66(4):1096–1106, 2010.

[158] Y. Huang, C. Li, R. Li, and S. Yang. An overview of tests on high-dimensional means. Journal of Multivariate Analysis, 188:1–14, 2022.

[159] A. Inselberg. The plane with parallel coordinates. The Visual Computer, Special Issue on Computational Geometry, 1:69–91, 1985.

[160] T. Iwashita. Asymptotic null and nonnull distribution of hotelling's t2-statistic under the elliptical distribution. Journal of Statistical Planning and Inference, 61(30):85–104, 1997.

[161] W. Jin, A. Tung, and J. Li. The generalized dispersion method for outlier detection. In International Conference on Database Theory, 2001.

[162] R. A. Johnson and D. W. Wichern. Applied Multivariate Statistical Analysis. Prentice Hall, New Jersey, 2007.

[163] S. W. Joshi and G. P. Patil. Certain structural properties of the sum-symmetric power series distributions. The Indian Journal of Statistics, Series A, 33(2):175–184, 1971.

[164] S. Jung and J. S. Marron. Pca consistency in high dimension, low sample size context. Journal of Multivariate Analysis, 37:4104–4130, 2009.

[165] J. Jurečková and J. Kalina. Nonparametric multivariate rank tests and their unbiasedness. Bernoulli, 18(1):229–251, 2012.

[166] D. Karlis and L. Meligkotsidou. Finite mixtures of multivariate poisson distributions with application. Journal of Statistical Planning and Inference, 137:1942–1960, 2007.

[167] J. Khan, J. S. Wei, M. Ringnér et al. Classification and diagnostic prediction of cancers using gene expression profiling and artificial neural networks. Nature Medicine, 7(6):673–679, 2001.

[168] D. G. Kirkpatrick and J. D. Radke. A framework for computational morphology. In G. T. Toussaint, editor, Computational Geometry, pages 217–248. North-Holland, Amsterdam, 1985.

[169] Matthäus Kleindessner and Ulrike von Luxburg. Lens depth function and k-relative neighborhood graph: Versatile tools for ordinal data analysis. Journal of Machine Learning Research, 18:1–52, 2017.

[170] V. F. Kolchin and B. A. Sevastyanov. Random Allocations. VNU Science Press, 1978.

[171] A. D. Kolesnik. The explicit probability distribution of the sum of two telegraph processes. Stochastics and Dynamics, 15(2), 2014. arXiv:1402.6866.

[172] G. Koshevoy and K. Mosler. Zonoid trimming for multivariate distributions. The Annals of Statistics, 25(5):1998–2017, 1997.

[173] S. Kotz. Multivariate distributions at a cross road. In G. P. Patil, S. Kotz, and J. K. Ord, editors, Statistical Distributions in Scientific Work, volume 1. D. Reidel Publishing Company, 1975.

[174] C. J. Kowalski and M. E. Tarter. Coordinate transformations to normality and the power of normal tests for independence. Biometrika, 56(1):139–148, 1969.

[175] H. K. Krishnapriyan. On evaluating the characteristic polynomial through symmetric functions. Journal of Chemical Information and Modeling, 35:196–198, 1995.

[176] J. B. Kruskal. On the shortest spanning subtree of a graph and a traveling salesman problem. Proceedings of the American Mathematical Society, 7:48–50, 1956.

[177] W. J. Krzanowski. Principles of Multivariate Analysis. Oxford University Press, Oxford, 1988.

[178] W. J. Krzanowski and Y. T. Lai. A criterion for determining the number of groups in a data set using sum-of-squares clustering. Biometrics, 44(1):23–34, 1988.

[179] L. A. Kurgan, K. J. Cios, R. Tadeusiewicz, M. Ogiela, and L. S. Goodenday. Knowledge discovery approach to automated cardiac spect diagnosis. Artificial Intelligence in Medicine, 23(2):149–169, 2001.

[180] P. A. Lachenbruch and M. R. Mickey. Estimation of error rates in discriminant analysis. Technometrics, 10(1):1–11, 1968.

[181] G. N. Lance and W. T. Williams. A general theory of classificatory sorting strategies: 1. hierarchical systems. The Computer Journal, 9(4):373–380, 1967.

[182] T. Lange, K. Mosler, and P. Mozharovskyi. Fast nonparametric classification based on data depth. Statistical Papers, 55:49–69, 2014.

[183] O. Ledoit and M. Wolf. A well-conditioned estimator for large-dimensional covariance matrices. Journal of Multivariate Analysis, 88:265–411, 2004.

[184] S. Lee, M. H. Seo, and Y. Shin. The lasso for high dimensional regression with a possible change point. Journal of the Royal Statistical Society: Series B, 78:193–210, 2016.

[185] F. C. Leone, L. S. Nelson, and R. B. Nottingham. The folded normal distribution. Technometrics, 3(4):543–550, 1961.

[186] I. C. Lerman. Les bases de la classification automatique. Gauthier-Villars, Paris, 1970.

[187] D. Leung and M. Drton. Testing independence in high dimensions with sums of squares of rank correlations. The Annals of Statistics, 46(1):280–307, 2018.

[188] H. Li, A. Aue, and D. Paul. High-dimensional general linear hypothesis tests via non-linear spectral shrinkage. Bernoulli, 26(4):2541–2571, 2020.

[189] J. Li. Asymptotic distribution-free change-point detection based on interpoint distances for high-dimensional data. Journal of Nonparametric Statistics, 32:157–184, 2020.

[190] J. Li, M. Xu, P. S. Zhong, and L. Li. Change point detection in the mean of high-dimensional time series data under dependence. arXiv:1903.07006, 2019.

[191] J. Li and H. Zha. Two-way poisson mixture models for simultaneous document classification and word clustering. Computational Statistics and Data Analysis, 50(1):163–180, 2006.

[192] J. Li. Asymptotic normality of interpoint distances for high dimensional data with applications to the two sample problem. Biometrika, 105(3):529–546, 2018.

[193] P. Li, J. Chen, and P. Marriott. Non-finite fisher information and homogeneity: an em approach. Biometrika, 96(2):411–426, 2009.

[194] S. M. Liao and M. Akritas. Test-based classification: A linkage between classification and statistical testing. Statistics and Probability Letters, 77(12):1269–1281, 2007.

[195] B. G. Lindsay. Mixture models: theory geometry and applications. NSF-CBMS Regional Conference Series in Probability and Statistics, Vol.5. Inst. Math. Stat., Hayward, 1995.

[196] R. J. A. Little and D. B. Rubin. Statistical Analysis with Missing Data. NSF-CBMS Regional Conference Series in Probability and Statistics, Vol.5. Wiley, 2002.

[197] B. Liu, X. Zhang, and Y. Liu. High dimensional change point inference: Recent developments and extensions. Journal of Multivariate Analysis, pages 1–19, 2022.

[198] Bin Liu, Cheng Zhou, Xinsheng Zhang, and Yufeng Liu. A unified data-adaptive framework for high dimensional change point detection. Journal of the Royal Statistical Society: Series B, 82:933–963, 2020.

[199] R. Y. Liu. On a notion of data depth based on random simplices. The Annals of Statistics, 18:405–414, 1990.

[200] R. Y. Liu, J. M. Parelius, and K. Singh. Multivariate analysis by data depth: descriptive statistics, graphics and inference (with discussion). The Annals of Statistics, 27:783–858, 1999.

[201] Z. Liu and R. Modarres. Lens data depth and median. Journal of Nonparametric Statistics, 23:1063–1074, 2011.

[202] W. S. Lok and S. M. S. Lee. A new statistical depth function with applications to multimodal data. Journal of Nonparametric Statistics, 23(3):617–631, 2011.

[203] M. Lopes, L. Jacob, and M. J. Wainwright. A more powerful two-sample test in high dimensions using random projection. In J. Shawe-Taylor, R. Zemel, P. Bartlett, F. Pereira, and K. Q. Weinberger, editors, Advances in Neural Information Processing Systems, 2011.

[204] P. J. Lopuhaa, H. P. amd Rousseeuw. Breakdown Points of Affine Equivariant Estimators of Multivariate Location and Covariance Matrices. The Annals of Statistics, 19(1):229–248, 1991.

[205] M. W. Lukens. Examination of statistical outlier structure in high dimension using interpoint distance densities and multivariate rankings. PhD thesis, George Mason University, 2004. Unpublished Ph.D. dissertation.

[206] J. F. Maa, D. K. Pearl, and R. Bartoszyński. Reducing multidimensional two-sample data to one-dimensional interpoint comparisons. The Annals of Statistics, 24(3):1069–1074, 1996.

[207] P. C. Mahalanobis. On the generalised distance in statistics. Proceedings of the National Institute of Sciences of India, 2:49–55, 1936.

[208] G. Mao. A new test of independence for high-dimensional data. Statist. Prob. Lett., 93:14–18, 2014.

[209] G. Mao. Robust test for independence in high dimensions. Commun. Statist. Theory Methods, 46:10036–10050, 2017.

[210] D. J. Marchette. Random Graphs For Statistical Pattern Recognition. Wiley-Interscience, New Jersey, 2004.

[211] K. V. Mardia, J. T. Kent, and J. M. Bibby. Multivariate Analysis. Academic Press, London, 1979.

[212] J. S. Maritz. Distribution-free Statistical Methods. Chapman & Hall, NY, 2 edition, 1995.

[213] M. Marozzi. Multivariate tests based on interpoint distances with application to magnetic resonance imaging. Statistical Methods in Medical Research, 25, 2014.

[214] M. Marozzi. Multivariate multidistance tests for high-dimensional low sample size case-control studies. Statistical Medicine, 34:1511–1526, 2015.

[215] M. Marozzi. Multivariate tests based on interpoint distances with application to magnetic resonance imaging. Statistical Methods in Medical Research, 25(6):2593–2610, 2016.

[216] M. Marozzi, M. Mukherjee, and J. Kalina. Interpoint distance tests for high-dimensional comparison studies. Journal of Applied Statistics, 47(4):653–665, 2020.

[217] A. W. Marshall and I. Olkin. A family of bivariate distribution generated by the bivariate bernoulli distribution. Journal of the American Statistical Association, 80(390):332–338, 1995.

[218] M. A. Martin and S. Roberts. An evaluation of bootstrap methods for outlier detection in least squares regression. Journal of Applied Statistics, 33(7):703–720, 2006.

[219] H. Martini and K. Swanepoel. Low-degree minimal spanning tree in normed spaces. Applied Mathematics Letters, 19:122–125, 2006.

[220] A. M. Mathai and S. B. Provost. Quadratic forms in Random Variables. Marcel Dekker, New York, 1992.

[221] D. S. Matteson and N. A. James. A nonparametric approach for multiple change point analysis of multivariate data. Journal of the American Statistical Association, 109:334–345, 2014.

[222] G. McLachlan and D. Peel. Finite Mixture Models. Wiley, New York, 2000.

[223] K. Menger. New foundation of euclidean geometry. American Journal of Mathematics, 53:721–745, 1931.

[224] D. Meyer and F. T. Wien. Support vector machines: The interface to libsvm in package e1071, 2001.

[225] R. Modarres. Estimation of a bivariate symmetric distribution function. Statistics and Probability Letters, 63:25–34, 2003.

[226] R. Modarres. A test of independence based on the likelihood of cut-points. Communications in Statistics: Simulation and Computation, 36:817–825, 2007.

[227] R. Modarres. High-dimensional generation of bernoulli random vectors. Statistics and Probability Letters, 81(8):1136–1142, 2011.

[228] R. Modarres. On the interpoint distances of bernoulli vectors. Statistics and Probability Letters, 84:215–222, 2014.

[229] R. Modarres. Graphical comparison of high dimensional distributions. International Statistical Review, 88(3):698–714, 2020.

[230] R. Modarres. On the blocks of interpoint distances. Journal of the Iranian Statistical Society, 20(1):197–218, 2021.

[231] R. Modarres. A high dimensional measure of dissimilarity. Computational Statistics and Data Analysis, 2022.

[232] R. Modarres. Nonparametric tests for detection of high dimensional outliers. Journal of Nonparametric Statistics, 34(1):206–227, 2022.

[233] R. Modarres. Analysis of distance matrices. Statistics and Probability Letters, 193:109720, 2023.

[234] R. Modarres. Nonparametric classification of high dimensional observations. Statistical Papers, 64(6):1833–1859, 2023.

[235] R. Modarres. Using eigenvalues of distance matrices for outlier detection. Intelligent Data Analysis, 28(4):871–889, 2023.

[236] R. Modarres. Hotelling t^2 test in high dimensions with application to wilks outlier method. Statistical Papers, 65(8):5203–5218, 2024.

[237] R. Modarres. Statistical analysis of distance matrices. Statistics and Computing, 36(43). https://doi.org/10.1007/s11222-025-10795-4.

[238] R. Modarres. Testing the equality of high dimensional distributions. Computational Statistics and Data Analysis, 2025. 212(C) https://doi.org/10.1016/j.csda.2025.108245

[239] R. Modarres and Y. Song. Interpoint distances: Applications, properties, and visualization. Applied Stochastic Models in Business and Industry, 36(6):1147–1168, 2020.

[240] R. Modarres and Y. Song. Multivariate power series interpoint distances. Statistical Methods & Applications, 29(4):955–982, 2020.

[241] P. K. Mondal, M. Biswas, and A. K. Ghosh. On high dimensional two-sample tests based on nearest neighbors. Journal of Multivariate Analysis, 141:168–178, 2015.

[242] J. G. Morel and N. K. Nagaraj. A finite mixture distribution for modeling multinomial extra variation. Biometrika, 80(2):363–371, 1993.

[243] K. Mosler. Multivariate Dispersion, Central Regions and Depth. Lecture Notes in Statistics. Springer, Berlin, 2002.

[244] R. J. Muirhead. Aspects of Multivariate Statistical Theory. John Wiley & Sons, New York, 1982.

[245] M. Nagode. Multivariate normal mixture modeling, clustering and classification with the rebmix package. arXiv:1801.08788, 2018.

[246] J. A. Nelder and R. Mead. A simplex method for function minimization. The Computer Journal, 7(4):308–313, 1965.

[247] R. B. Nelsen. Some concepts of bivariate symmetry. Journal of Nonparametric Statistics, 3:95–101, 1993.

[248] G. Neuhaus and L. X. Zhu. Permutation tests for reflected symmetry. Journal of Multivariate Analysis, 67(2):129–153, 1998.

[249] M. A. Newton. Introducing the discussion paper by székely and rizzo. arXiv:1010.3575, 2010.

[250] A. Nijenhuis and H. S. Wilf. Combinatorial Algorithms. Academic Press, New York, 2 edition, 1978.

[251] X. Niu, P. Li, and P. Zhang. Testing homogeneity in a multivariate mixture model. The Canadian Journal of Statistics, 39(2):218–238, 2011.

[252] A. Noack. A class of random variables with discrete distributions. Annals of Mathematical Statistics, 21(1):127–132, 1950.

[253] J. Nocedal and J.S. Wright. Numerical Optimization. Springer-Verlag, New York, 1999.

[254] H. Oja. Descriptive statistics for multivariate distributions. Statistics and Probability Letters, 1:327–333, 1983.

[255] H. Oja. Multivariate Nonparametric Methods with R: An Approach Based on Spatial Signs and Ranks. Springer, New York, 2010.

[256] H. Oja and R. H. Randles. Multivariate nonparametric tests. Statistical Science, 19:598–605, 2004.

[257] I. Olkin and J. L. Tomsky. A new class of multivariate tests based on the union-intersection principle. The Annals of Statistics, 9(4):792–802, 1981.

[258] M. Ondrus, E. Olds, and I. Cribben. Factorized binary search: change point detection in the network structure of multivariate high-dimensional time series. arXiv:2103.06347, 2013.

[259] B. O'Neill. Some useful moment results in sampling problems. The American Statistician, 68(4):282–296, 2014.

[260] R. Osada, T. Funkhouser, B. Chazelle, and D. Dobkin. Shape distributions. ACM Transactions on Graphics, 21(4):807–832, 2002.

[261] A. K. Pal, P. K. Mondal, and A. K. Ghosh. High dimensional nearest neighbor classification based on differences of inter-point distances. Pattern Recognition Letters, 74:1–8, 2016.

[262] G. P. Patil. On sampling with replacement from populations with multiple characters. The Indian Journal of Statistics, Series B, 30(3/4):355–366, 1968.

[263] G. P. Patil and R. Modarres. Hotspot detection with bivariate data. Journal of Statistical Planning and Inference, 137(11):3643–3654, 2007.

[264] G. P. Patil, R. Modarres, W. L. Myers, and P. Patankar. Spatially constrained clustering and upper level set scan hotspot detection in surveillance geoinformatics. Environmental and Ecological Statistics, 13:365–377, 2006.

[265] G. P. Patil and C. Taillie. Upper level set statistic for detecting arbitrarily shaped hotspots. Environmental and Ecological Statistics, 11:183–197, 2004.

[266] D. Peña and F. Prieto. Multivariate outlier detection and robust covariance matrix estimation. Technometrics, 43(3):286–310, 2001.

[267] K. Pearson. Contributions to the theory of mathematical evolution. Philosophical Transactions of the Royal Society of London, A(185):71–110, 1894.

[268] K. I. Penny and I. T. Jolliffe. A comparison of multivariate outlier detection methods for clinical laboratory safety data. Journal of the Royal Statistical Society: Series D (The Statistician), 50(4):295–307, 2001.

[269] F. Pesarin and L. Salmaso. Permutation Tests for Complex Data. Wiley, Chichester, 2010.

[270] A. Petrie. Graph-theoretic multisample tests of equality in distribution for high dimensional data. Computational Statistics and Data Analysis, 96:145–158, 2016.

[271] E. J. G. Pitman. Significance tests which may be applied to samples from any populations. ii. the correlation coefficient test. Supplement to the Journal of the Royal Statistical Society, 4:225–232, 1937.

[272] C. E. Priebe. Adaptive mixtures. Journal of the American Statistical Association, 89(427):796–806, 1994.

[273] R. C. Prim. Shortest connection networks and some generalizations. Bell System Technology Journal, 36:1389–1401, 1957.

[274] M. L. Puri and P. K. Sen. Nonparametric Methods in Multivariate Analysis. John Wiley and Sons, 1971.

[275] M. Radovanovic, A. Nanopoulos, and M. Ivanovic. Hubs in space: Popular nearest neighbors in high-dimensional data. Journal of Machine Learning Research, 11:2487–2531, 2010.

[276] J. A. Ramey. Datamicroarray: A collection of benchmark microarray gene expression datasets. GitHub repository, 2015.

[277] W. M. Rand. Objective criteria for the evaluation of clustering methods. Journal of the American Statistical Association, 66(336):846–850, 1971.

[278] R. H. Randles and D. A. Wolf. Introduction to the Theory of Nonparametric Statistics. John Wiley & Sons, NY, 1979.

[279] B. D. Ripley. Modeling spatial patterns. Journal of the Royal Statistical Society, Series B, 39:172–192, 1976.

[280] B. D. Ripley. The second-order analysis of stationary point processes. Journal of Applied Probability, 13(2):255–266, 1976.

[281] K. Ro, C. Zou, and Z. Wang. Outlier detection for high-dimensional data. Biometrika, 102(3):589—-599, 2015.

[282] P. Robert, R. Cléroux, and N. Ranger. Some results on vector correlation. Computational Statistics and Data Analysis, 3:25–32, 1985.

[283] F. Rohlf. Generalization of the gap test for the detection of multivariate outliers. Biometrics, 31:93–101, 1975.

[284] P. R. Rosenbaum. An exact distribution-free test comparing two multivariate distributions based on adjacency. Journal of the Royal Statistical Society, Series B, 67:515–530, 2005.

[285] M. Rosenblatt. Remarks on a multivariate transformation. Annals of Mathematical Statistics, 23(3):470–472, 1952.

[286] P. J. Rousseeuw. Multivariate estimation with high breakdown point. In W. Grossmann, G. Pflug, I. Vincze, and W. Wertz, editors, Mathematical Statistics and Applications, Volume B, pages 283–297. Reidel Publishing Company, Dordrecht, 1985.

[287] P. J. Rousseeuw. Silhouettes: A graphical aid to the interpretation and validation of cluster analysis. Journal of Computational and Applied Mathematics, 20:53–65, 1987.

[288] S. Roy, J. R. Choudhury, and S. Dutta. On some fast and robust classifiers for high dimension, low sample size data. In Proceedings of The 25th International Conference on Artificial Intelligence and Statistics, Vol. 151, PMLR, pages 9943–9968, 2022.

[289] S. Roy, S. Sarkar, S. Dutta, and A. K. Ghosh. On generalizations of some distance-based classifiers for hdlss data. Journal of Machine Learning Research, 23:1–41, 2022.

[290] S. N. Roy. Some Aspects of Multivariate Analysis. Wiley, New York, 1959.

[291] J. A. Royle and W. A. Link. A general class of multinomial mixture models for anuran calling survey data. Ecology, 86(9):2505–2512, 2005.

[292] P. Royston. Algorithm as 181: The test for normality. Applied Statistics, 31:176–180, 1982.

[293] A. Safikhani and A. Shojaie. Joint structural break detection and parameter estimation in high-dimensional nonstationary var models. Journal of Econometrics, 117:251–264, 2022.

[294] Y. Sang. Test for diagonal symmetry in high dimension. Statistics and Probability Letters, 2024. https://api.semanticscholar.org/CorpusID:264818421.

[295] Y. Sang and X. Dang. Empirical likelihood test for diagonal symmetry. Statistics and Probability Letters, 156:108595, 2020.

[296] T. Sapatinas. Identifiability of mixtures of power-series distributions and related characterizations. Annals of the Institute of Statistical Mathematics, 47(3):447–459, 1995.

[297] S. Sarkar, R. Biswas, and A. K. Ghosh. High dimensional two-sample tests based on a new class of dissimilarity indices. Technical report, Technical Report, 2020.

[298] S. Sarkar and A. K. Ghosh. On some high-dimensional two-sample tests based on averages of inter-point distances. Stat, 7(1):e187, 2018.

[299] S. Sarkar and A. K. Ghosh. On perfect clustering of high dimension, low sample size data. IEEE Transactions on Pattern Analysis and Machine Intelligence, 42(9):2257–2272, 2020.

[300] M. F. Schilling. Multivariate two-sample tests based on nearest neighbors. Journal of the American Statistical Association, 81:799–806, 1986.

[301] J. R. Schott. Testing for complete independence in high dimensions. Biometrika, 92:951–956, 2005.

[302] J. R. Schott. A test for independence of two sets of variables when the number of variables is large relative to the sample size. Statistics and Probability Letters, 78:3096–3102, 2008.

[303] R. Serfling. Approximation Theorems of Mathematical Statistics. John Wiley & Sons, Inc., 1980.

[304] R. Serfling. Multivariate Symmetry and Asymmetry. John Wiley & Sons, Inc., 2006. https://doi.org/10.1002/0471667196.ess5011.

[305] R. Serfling. Equivariance and invariance properties of multivariate quantile and related functions, and the role of standardisation. Journal of Nonparametric Statistics, 22(7):915–936, 2010.

[306] C. Shen, S. Panda, and J. T. Vogelstein. The chi-square test of distance correlation. Journal of Computational and Graphical Statistics, 31(1):254–262, 2021.

[307] Y. Shen, Z. Lin, and J. Zhu. Shrinkage-based regularization tests for high-dimensional data with application to gene set analysis. Computational Statistics & Data Analysis, 55(7):2221–2233, 2011.

[308] W. Shin. A multinomial change-point theory in the context of diagnosis code searching. PhD thesis, Texas Tech University, 1987. Unpublished Ph.D dissertation.

[309] A. M. Shurygin. Using interpoint distances for pattern recognition. Pattern Recognition and Image Analysis, 16(4):726–729, 2006.

[310] M. Sibuya, I. Yoshimura, and R. Shimizu. Negative multinomial distribution. Annals of the Institute of Statistical Mathematics, 16(1):409–426, 1964.

[311] B. Silverman and T. Brown. Short distances, flat triangles and poisson limits. Journal of Applied Probability, 15(4):816–826, 1978.

[312] J. W. Silverstein. The smallest eigenvalue of a large dimensional wishart matrix. Annals of Probability, 13:1364–1368, 1985.

[313] J. B. Simaika. An optimum property of two statistical tests. Biometrika, 32:70–80, 1941.

[314] K. Singh. A notion of majority depth. Technical report, Department of Statistics, Rutgers University, 1991.

[315] K. Singh and M. Xie. Bootlier-plot: Bootstrap based outlier detection plot. Sankhya: The Indian Journal of Statistics, 65(3):532–559, 2003.

[316] B. K. Sinha and Wieand. Union-intersection test for the mean vector when the covariance matrix is totally reducible. Journal of the American Statistical Association, 74(366a):340–343, 1979.

[317] C. G. Small. Multidimensional medians arising from geodesics on graphs. The Annals of Statistics, 25:478–494, 1997.

[318] R. R. Sokal and P. H. Sneath. Principles of Numerical Taxonomy. Freeman, San Francisco, London, 1963.

[319] Y. Song and R. Modarres. Interpoint distance test of homogeneity for multivariate mixture models. International Statistical Review, pages 1–26, 2019.

[320] Yu Song. Interpoint distance distributions and their applications. PhD thesis, George Washington University, 2018. Unpublished Ph.D dissertation.

[321] M. S. Srivastava. A test for the mean vector with fewer observations than the dimension under non-normality. Journal of Multivariate Analysis, 100:518–532, 2009.

[322] M. S. Srivastava and M. Du. A test for the mean vector with fewer observations than the dimension. Journal of Multivariate Analysis, 99(3):386–402, 2008.

[323] W. A Stahel. Breakdown of covariance estimators. Technical Report Research Report 31, Fachgruppe für Statistik, ETH Zürich, Zürich, 1981.

[324] W. Stuetzle. Estimating the cluster tree of a density by analyzing the minimal spanning tree of a sample. Journal of Classification, 20:25–47, 2003.

[325] H. A. Sturges. The choice of a class interval. Journal of the American Statistical Association, pages 65–66, 1926.

[326] C. A. Sugar and G. M. James. Finding the number of clusters in a dataset. Journal of the American Statistical Association, 98(463):750–763, 2003.

[327] G. J. Székely and N. K. Bakirov. Extremal probabilities for gaussian quadratic forms. Probability Theory and Related Fields, 126:184–202, 2003.

[328] G. J. Székely and T. F. Móri. A characteristic measure of asymmetry and its application for testing diagonal symmetry. Communications in Statistics–Theory and Methods, 30:1633–1639, 2001.

[329] G. J. Székely and M. L. Rizzo. Energy statistics: a class of statistics based on distances. Journal of Statistical Planning and Inference, 143(8):1249–1272, 2013.

[330] G. J. Székely and M. L. Rizzo. The energy of data. Annual Review of Statistics and Its Application, 4:447–479, 2017.

[331] G. J. Székely, M. L. Rizzo, and N. K. Bakirov. Measuring and testing dependence by correlation of distances. Annals of Statistics, 35(6):2769–2794, 2007.

[332] G. J. Székely and P. K. Sen. Characterization of diagonal symmetry: location unknown, and a test based on allied u-processes. Journal of Statistical Planning and Inference, 102:349–358, 2002.

[333] G. L. Székely and M. L. Rizzo. Testing for equal distributions in high dimensions. InterStat, 2004. Issue 5.

[334] S. Taskinen, H. Oja, and R. H. Randles. Sign test of independence between two random vectors. Statistics and Probability Letters, 62(1):9–21, 2003.

[335] S. Taskinen, H. Oja, and R. H. Randles. Multivariate nonparametric tests of independence. Journal of the American Statistical Association, 100(471):916–925, 2005.

[336] M. Thulin. A high-dimensional two-sample test for the mean using random subspaces. Computational Statistics & Data Analysis, 74:26–38, 2014.

[337] R. Tibshirani, G. Walther, and T. Hastie. Estimating the number of clusters in a data set via the gap statistic. Journal of the Royal Statistical Society, Series B, 63(2):411–423, 2001.

[338] K. M. Ting, Y. Zhu, M. Carman, and Z.-H. Zhou. Overcoming key weaknesses of distance-based neighborhood methods using a data dependent dissimilarity measure. In Proceedings of the 22nd ACM SIGKDD International Conference on Knowledge Discovery and Data Mining, pages 1205–1214, 2016.

[339] V. Tiwari and A. Kashikar. R package outlierdetection. `https://cran.r-project.org/web/packages/OutlierDetection/OutlierDetection.pdf`, 2019.

[340] G. T. Toussaint. The relative neighborhood graph of a finite planar set. Pattern Recognition, 12:261–268, 1980.

[341] O. Troyanskaya and et al. Missing value estimation methods for dna microarrays. Bioinformatics, 17(6):520–525, 2001.

[342] J. W. Tukey. Mathematics and the picturing of data. In R. D. James, editor, Proceedings of the International Congress of Mathematicians, volume 2, pages 523–531, Montreal, Canadian Mathematical Congress, 1975.

[343] Y. Vardi and C. H. Zhang. The multivariate l1-median and associated data depth. The Annals of Statistics, 28(2):461–482, 2000.

[344] L. A. Waller and D. Zelterman. Log-linear modeling with the negative multinomial distribution. Biometrics, 53(3):971–982, 1997.

[345] J. Wang. Consistent selection of the number of clusters via cross-validation. Biometrika, 97(4):893–904, 2010.

[346] T. Wang, B. Follain, and R. Samworth. High-dimensional changepoint estimation via sparse projection. R Package 'InspectChangepoint', 2025.

[347] T. Wang and R. Samworth. High dimensional change point estimation via sparse projection. Journal of the Royal Statistical Society, Series B, 80:57–83, 2018.

[348] Y. Wang and R. Modarres. Clustering of high-dimensional observations. Journal of Nonparametric Statistics, pages 319–343, 2024.

[349] Y. Wang and R. Modarres. Testing high dimensional diagonal symmetry. Statistical Papers, 2025.

[350] Jr. Ward, J. H. Hierarchical grouping to optimize an objective function. Journal of the American Statistical Association, 58:236–244, 1963.

[351] L. Wasserman, A. Ramdas, and S. Balakrishnan. Universal inference. Proceedings of the National Academy of Sciences, 117:16880–16890, 2020.

[352] E. Wegman. Hyperdimensional data analysis using parallel coordinates. Journal of the American Statistical Association, 85:664–675, 1990.

[353] M. West, C. Blanchette, H. Fressman, E. Huang, S. Ishida, R. Spang, H. Zuan, J. R. Marks, and J. R. Nevins. Predicting the clinical status of human breast cancer using gene expression profiles. Proceedings of the National Academy of Sciences, 98:11462–11467, 2001.

[354] J. D. Wilber, J. K. Ghosh, Nakatsu, S. M. Brouder, and R. W. Doerge. Variable selection in high-dimensional multivariate binary data with application to the analysis of microbial community dna fingerprints. Biometrics, 58(2):378–386, 2002.

[355] S. S. Wilks. On the independence of k sets of normally distributed statistical variables. Econometrica, 3:309–326, 1935.

[356] S. S. Wilks. Multivariate statistical outliers. Sankhya, 25:407–426, 1963.

[357] E. B. Wilson and M. M. Hilferty. The distribution of chi-squared. Proceedings of the National Academy of Sciences, USA, 17:684, 1931.

[358] D. M. Witten and R. Tibshirani. A framework for feature selection in clustering. Journal of the American Statistical Association, 105:713–726, 2010.

[359] Y. Wu, M. G. Genton, and L. A. Stefanski. A multivariate two-sample mean test for small sample size and missing data. Biometrics, 62(3):877–885, 2006.

[360] W. Xiao, X. Huang, F. He, J. Silva, S. Emrani, and A. Chaudhuri. Online robust principal component analysis with change point detection. IEEE Transactions on Multimedia, 22:59–68, 2019.

[361] P. Xu, G. N. Brock, and R. S. Parrish. Modified linear discriminant analysis approaches for classification of high-dimensional microarray data. Computational Statistics & Data Analysis, 53(5):1674–1687, 2009.

[362] M. Yang and R. Modarres. Multivariate tests of uniformity. Statistical Papers, 58:627–639, 2017.

[363] M. Yang and R. Modarres. β-skeleton depth functions and medians. Communications in Statistics - Theory and Methods, 47(20):5127–5143, 2018.

[364] M. Yang, R. Modarres, and L. Guo. Depth functions and multi-dimensional medians on minimal spanning trees. Journal of Applied Statistics, 47(2):323–336, 2020.

[365] K. Yata and M. Aoshima. Geometric consistency of principal component scores for high-dimensional mixture models and its application. Scandinavian Journal of Statistics, 47:899–921, 2020.

[366] M. Yu and X. Chen. Finite sample change point inference and identification for high-dimensional mean vectors. Journal of the Royal Statistical Society, Series B, 83:247–270, 2021.

[367] Hui Zhang, Haibing Yan, Xiu Cheng, et al. Conditional distance correlation: A novel dependence measure and test of conditional independence. Journal of the American Statistical Association, 110(512):1726–1734, 2016.

[368] P. Zhang, X. Wang, and P. X. K. Song. Clustering categorical data based on distance vectors. Journal of the American Statistical Association, 101(473):355–367, 2006.

[369] H. Zhou. Distance, dissimilarity index, and network community structure. Physical Review E, Statistical, Nonlinear, and Soft Matter Physics, 67(6), 2003.

[370] L. Zhu, K. Xu, R. Li, and W. Zhong. Projection correlation between two random vectors. Biometrika, 104(4):829–843, 2017.

[371] V. A. Zorich. Mathematical Analysis, Vol. II. Universitext. Springer, Heidelberg, 2nd edition, 2016.

[372] Y. Zuo and R. Serfling. General notions of statistical depth function. The Annals of Statistics, 28:461–482, 2000.

Index

For Product Safety Concerns and Information please contact our
EU representative GPSR@taylorandfrancis.com Taylor & Francis
Verlag GmbH, Kaufingerstraße 24, 80331 München, Germany